Theory of Plates and Shells

Christian Mittelstedt

Theory of Plates and Shells

Springer Vieweg

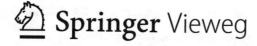

Christian Mittelstedt
Fachbereich Maschinenbau, Technical
University of Darmstadt
Darmstadt, Germany

ISBN 978-3-662-66807-8 ISBN 978-3-662-66805-4 (eBook)
https://doi.org/10.1007/978-3-662-66805-4

This Springer Vieweg imprint is published by the registered company Springer-Verlag GmbH, DE, part of Springer Nature.
The registered company address is: Heidelberger Platz 3, 14197 Berlin, Germany

In memory of my father

Gerd Mittelstedt

Preface

This book is the result of lectures and training sessions I have given over the years, both at various universities and in an industrial context. It provides an overview of the statics of disk, plate and shell structures, but also includes layered structures, i.e. the so-called laminates. This book is to be understood as a further contribution to a book series dealing with structural mechanics in lightweight construction and design, which started with the book "Structural Mechanics in Lightweight Engineering", also published by Springer in 2021.

This book is divided into four parts. Part I contains a presentation of the fundamentals of elasticity theory and the energy methods of structural mechanics, which are absolutely necessary for the understanding of this book. Part II addresses disk structures, i.e., plane thin structures that are loaded in their plane. In addition to a presentation of disk theory in Cartesian and polar coordinates, approximation methods for disk structures and anisotropic disks are also discussed. Part III is devoted to plates, i.e., plane thin structures whose loading leads to bending. In addition to formulations in Cartesian and polar coordinates for the shear-rigid plate, classical approximation methods are discussed. The considerations are extended later to higher-order theories as well as plate buckling and geometrically nonlinear analysis and a short introduction to the theory of laminated plates. The book concludes with Part IV, which deals with shells of revolution. Here, the classical division into membrane theory and bending theory is essentially followed. The focus in all explanations is on the description of the theoretical basics and the provision of analytical solutions, since in my experience it is the latter that provide a deep understanding of the presented material. A very general basic knowledge of mechanics and mathematics is required, as taught in the corresponding courses of study in the first years of engineering courses.

This book is intended for students at universities in the disciplines of mechanical engineering, civil engineering, aerospace engineering, lightweight construction and design and in all other disciplines in which disk, plate and shell structures play a role. Furthermore, I hope that the book will be useful to engineers in research and in practice and will establish itself as a reference work over the years.

Finally, I hope you will enjoy reading the contents of this book. Of course, I am happy to receive questions, suggestions and any other feedback!

Christian Mittelstedt

Darmstadt
Germany
summer 2022

Contents

Part I Fundamentals

1 Basics of Elasticity Theory 3
1.1 Introduction 3
1.2 Stress State 4
1.2.1 Stress Vector and Stress Tensor 4
1.2.2 Transformation Rules 8
1.2.3 Principal Stresses, Invariants, Mohr's Circles 10
1.2.4 Equilibrium Conditions 12
1.3 Deformations and Strains 14
1.3.1 Introduction 14
1.3.2 Green-Lagrangian Strain Tensor 17
1.3.3 Von-Kármán Strains 18
1.3.4 Infinitesimal Strain Tensor 19
1.3.5 Compatibility Equations 21
1.4 Constitutive Law 22
1.4.1 Introduction 22
1.4.2 The Generalized Hooke's Law 22
1.4.3 Strain Energy 25
1.5 Boundary Value Problems 27
1.6 Material Symmetries 29
1.6.1 Full Anisotropy 29
1.6.2 Monotropic Material 30
1.6.3 Orthogonal Anisotropy/Orthotropy 32
1.6.4 Transversal Isotropy 34
1.6.5 Isotropy 35
1.6.6 Representation in Engineering Constants 35
1.7 Transformation Rules 38
1.8 Representation of the Basic Equations in Cylindrical Coordinates 40
1.9 Plane Problems 44
1.9.1 Plane Strain State 44

1.9.2 Plane Stress State ... 46
1.9.3 Stress Transformation ... 48
1.9.4 Formulation for Orthotropic Materials ... 51
1.9.5 Formulation in Polar Coordinates ... 55
References ... 57

2 Energy Methods of Elastostatics ... 59
2.1 Work and Energy ... 59
2.1.1 Introduction ... 59
2.1.2 Inner and Outer Work ... 61
2.1.3 Principle of Work and Energy and the Law of Conservation of Energy ... 62
2.1.4 Strain Energy and Complementary Strain Energy ... 63
2.1.5 General Principle of Work and Energy of Elastostatics ... 72
2.2 The Principle of Virtual Displacements ... 74
2.2.1 Virtual Displacements and Virtual Work ... 74
2.2.2 The Principle of Virtual Displacements ... 77
2.2.3 Analysis Rules for the Variational Operator δ ... 77
2.2.4 Formulation for the Continuum ... 79
2.2.5 Application to the Rod ... 80
2.2.6 Application to the Euler-Bernoulli Beam ... 82
2.3 Principle of the Stationary Value of the Total Elastic Potential ... 83
2.3.1 Introduction ... 83
2.3.2 Application to the Rod ... 84
2.3.3 Application to the Euler-Bernoulli Beam ... 87
2.4 Approximation Methods of Elastostatics ... 89
2.4.1 The Ritz Method ... 90
2.4.2 The Galerkin Method ... 93
References ... 95

Part II Disks

3 Isotropic Disks in Cartesian Coordinates ... 99
3.1 Introduction ... 99
3.2 Fundamentals ... 101
3.2.1 Basic Equations ... 101
3.2.2 The Displacement Method ... 103
3.2.3 The Force Method ... 104
3.2.4 Boundary Conditions ... 108
3.3 Energetic Consideration ... 109
3.3.1 Strain Energy ... 109
3.3.2 Energetic Derivation of the Basic Equations ... 110

3.3.3 Disks with Arbitrary Boundaries ... 114
3.4 Elementary Solutions ... 117
3.4.1 Solutions of the Disk Equation ... 117
3.4.2 Elementary Cases ... 120
3.5 Beam-type Disks ... 123
3.6 St. Venant's Principle ... 132
3.7 The Isotropic Half-Plane ... 134
3.7.1 Decay Behaviour of Boundary Perturbations ... 134
3.7.2 The Half-Plane Under Periodic Boundary Load ... 139
3.7.3 The Half-Plane Under Non-periodic Load ... 145
3.8 The Effective Width ... 151
3.8.1 Effective Width of Flanges of Beams Under Bending ... 151
3.8.2 Effective Width for Load Introductions ... 157
References ... 161

4 Isotropic Disks in Polar Coordinates ... 163
4.1 Fundamentals ... 163
4.1.1 Basic Equations ... 163
4.1.2 The Displacement Method ... 166
4.1.3 The Force Method ... 167
4.2 Energetic Consideration ... 170
4.2.1 Strain Energy ... 170
4.2.2 Energetic Derivation of the Basic Equations ... 172
4.3 Elementary Cases ... 178
4.4 Rotationally Symmetric Disks ... 179
4.5 Non-rotationally Symmetric Circular Disks ... 186
4.6 Wedge-shaped Disks ... 190
4.7 Disks with Circular Holes ... 192
References ... 198

5 Approximation Methods for Isotropic Disks ... 201
5.1 The Displacement-Based Ritz Method ... 201
5.2 The Force-Based Ritz Method ... 211
5.3 Finite Elements for Disks ... 214
References ... 218

6 Anisotropic Disks ... 219
6.1 Basic Equations ... 219
6.1.1 Cartesian Coordinates ... 219
6.1.2 Polar Coordinates ... 223
6.2 Elementary Cases ... 226
6.3 Beam-type Disks ... 228
6.4 Decay Behaviour of Edge Perturbations ... 232
6.5 Orthotropic Circular Ring Disks ... 235
6.6 Orthotropic Circular Arc Disks ... 236

6.7 Layered Circular Ring Disks 243
6.8 Layered Circular Arc Disks 245
References 249

Part III Plates

7 Kirchhoff Plate Theory in Cartesian Coordinates 253
7.1 Introduction 253
7.2 The Kirchhoff Plate Theory 254
7.2.1 Assumptions, Kinematics and Displacement Field 254
7.2.2 Strain and Stress Field 257
7.2.3 Force and Moment Flows, Constitutive Law 260
7.2.4 Transformation Rules 264
7.3 Effective Stiffnesses for Selected Plate Structures 265
7.3.1 Homogeneous Plate of Orthotropic Material 266
7.3.2 Homogeneous Plate of Isotropic Material 266
7.3.3 Reinforced Concrete Plate 266
7.3.4 Isotropic Plate Reinforced by Equidistant Stiffeners 267
7.3.5 Isotropic Plate Reinforced by Equidistant Ribs 267
7.3.6 Corrugated Metal Sheet 268
7.3.7 Symmetrical Cross-Ply Composite Laminate 268
7.4 Basic Equations of Plate Bending in Cartesian Coordinates 270
7.4.1 Displacement Differential Equation 270
7.4.2 Equivalent Transverse Shear Forces 272
7.4.3 Boundary Conditions 275
7.5 Elementary Solutions of the Plate Equation 278
7.6 Bending of Plate Strips 280
7.7 Navier Solution for Static Plate Bending Problems 283
7.7.1 Determination of the Plate Deflection 283
7.7.2 Moments, Forces and Stresses of the Plate 286
7.7.3 Special Load Cases 288
7.8 Lévy-type solutions for static plate bending problems 291
7.8.1 Introduction 291
7.8.2 Orthotropic Plates 291
7.8.3 Isotropic Plates 295
7.9 Energetic Consideration of Plate Bending 298
7.9.1 Principle of the Minimum of the Total Elastic Potential 298
7.9.2 Principle of Virtual Displacements 301
7.9.3 Plate with Arbitrary Boundary 303
7.10 Plate on Elastic Foundation 307
7.11 The Membrane 310
References 311

8 Approximation Methods for the Kirchhoff Plate ... 313
8.1 The Ritz Method ... 313
8.2 The Galerkin Method ... 329
8.3 The Finite Element Method ... 332
References ... 332

9 Kirchhoff Plate Theory in Polar Coordinates ... 333
9.1 Transition to Polar Coordinates ... 333
9.2 Basic Equations ... 335
9.3 Rotationally Symmetric Bending of Circular Plates ... 340
9.3.1 Basic Equations ... 340
9.3.2 Plates Under Constant Surface Load ... 341
9.3.3 Plates Under Centric Point Force ... 344
9.3.4 Plate Under Edge Moments ... 347
9.3.5 Plate Under Partial Load ... 348
9.3.6 Circular Ring Plates ... 350
9.4 Asymmetric Bending of Circular Plates ... 353
9.5 Strain Energy ... 356
References ... 357

10 Higher-order Plate Theories ... 359
10.1 First-Order Shear Deformation Theory ... 360
10.1.1 Kinematics and Constitutive Equations ... 360
10.1.2 Determination of the Shear Correction Factor K ... 364
10.1.3 Equilibrium and Boundary Conditions ... 365
10.1.4 Strain Energy ... 368
10.1.5 Bending of Plate Strips ... 368
10.1.6 Navier Solution ... 371
10.1.7 Lévy-type solutions ... 372
10.1.8 The Ritz Method ... 372
10.2 Third-Order Shear Deformation Theory According to Reddy ... 376
10.2.1 Kinematics ... 376
10.2.2 Strains and Constitutive Equations ... 378
10.2.3 Equilibrium Conditions ... 382
10.2.4 Navier Solution ... 385
10.2.5 The Ritz Method ... 387
References ... 390

11 Plate Buckling ... 391
11.1 Basic Equations ... 391
11.2 Navier Solution ... 395
11.2.1 Biaxial Load ... 398
11.3 Energy Methods for the Solution of Plate Buckling Problems ... 401

11.3.1 Introduction ... 401
11.3.2 The Rayleigh Quotient ... 403
11.3.3 The Ritz Method ... 408
References ... 413

12 Geometrically Nonlinear Analysis ... 415
12.1 Kirchhoff Plate Theory ... 415
12.1.1 Energetic Consideration ... 415
12.1.2 Th. V. Kármán equations ... 421
12.1.3 Discussion of the Boundary Terms ... 424
12.1.4 Inner and External Potential ... 426
12.1.5 Special Cases ... 430
12.2 Bending of Plates with Large Deflections ... 432
12.2.1 Solution by Series Expansion ... 432
12.2.2 The Galerkin Method ... 433
12.2.3 The Ritz Method ... 434
12.3 First-Order Shear Deformation Theory ... 440
References ... 444

13 Laminated Plates ... 445
13.1 Introduction ... 445
13.2 Classical Laminated Plate Theory ... 446
13.2.1 Introduction ... 446
13.2.2 Assumptions and Kinematics ... 447
13.2.3 Strains and Stresses ... 450
13.3 Constitutive Law ... 453
13.4 Coupling Effects ... 455
13.4.1 Shear Coupling ... 456
13.4.2 Bending-Twisting Coupling ... 457
13.4.3 Bending-extension Coupling ... 457
13.5 Special Laminates ... 458
13.5.1 Isotropic Single Layer ... 458
13.5.2 Orthotropic Single Layer ... 459
13.5.3 Anisotropic Single Layer/Off-axis Layer ... 460
13.5.4 Symmetric Laminates ... 460
13.5.5 Cross-ply Laminates ... 460
13.5.6 Angle-ply Laminates ... 461
13.5.7 Quasi-isotropic Laminates ... 462
13.6 Basic Equations and Boundary Conditions ... 463
13.6.1 Equilibrium Conditions ... 463
13.6.2 Displacement Differential Equations ... 465
13.6.3 Boundary Conditions ... 467
13.7 Navier Solutions ... 468
13.7.1 Bending of a Symmetric Cross-Ply Laminate ... 468

13.7.2 Bending of an Unsymmetric Cross-Ply Laminate $[(0°/90°)_N]$ 469
13.7.3 Bending of an Unsymmetric Angle-ply Laminate $[(\pm\theta)_N]$ 472
References 474

Part IV Shells

14 Introduction to Shell Structures 477
14.1 Introduction 477
14.2 Shells of Revolution 478
14.3 Load Cases 481
14.4 Classical Shell Theory 483
14.4.1 Assumptions 483
14.4.2 Stresses; Force and Moment Quantities 484
14.4.3 Strains and Displacements 487
References 488

15 Membrane Theory of Shells of Revolution 491
15.1 Assumptions 491
15.2 Equilibrium Conditions for Shells of Revolution 494
15.2.1 Equilibrium Conditions 494
15.2.2 Rotational Symmetric Load 496
15.3 Selected Solutions for Shells of Revolution 499
15.3.1 Circular Cylindrical Shells 499
15.3.2 Spherical Shells 502
15.3.3 Conical Shells 507
15.4 Kinematics of Shells of Revolution 508
15.5 Constitutive Equations 513
15.6 Displacement Solutions for Rotationally Symmetric Loads 514
15.7 Energetic Derivation of the Basic Equations 518
References 521

16 Bending Theory of Shells of Revolution 523
16.1 Basic Equations 523
16.1.1 Equilibrium Conditions 523
16.1.2 Kinematic Equations 527
16.1.3 Constitutive Equations 531
16.1.4 Displacement Differential Equations for the Circular Cylindrical Shell 532
16.1.5 Boundary Conditions under Rotationally Symmetric Load 532
16.2 Container Theory of the Circular Cylindrical Shell 533
16.2.1 Basic Equations 533
16.2.2 The Container Equation 537
16.2.3 Solutions for the Container Equation 538

16.3 The Force Method 544
16.4 Edge Perturbations of the Spherical Shell 547
16.5 Edge Perturbations of Arbitrary Shells of Revolution 553
16.6 Circular Cylindrical Shell under Arbitrary Load 556
16.6.1 Basic Equations 556
16.6.2 Approximation According to Donnell 562
16.6.3 Solution of the Basic Equations 563
16.6.4 Boundary Conditions 565
16.7 Laminated Shells 566
16.7.1 Basic Equations 566
16.7.2 Cross-ply Laminated Cylindrical Shells under Rotationally Symmetric Load 570
References 571

Index 573

Part I
Fundamentals

Basics of Elasticity Theory

1

In this chapter all necessary basics of linear elasticity theory of three-dimensional anisotropic bodies are discussed, which are necessary for the understanding of the contents of this book. As a special case, at the end of this chapter we also discuss the simplifications as they are assumed for plane problems. In all explanations of this chapter we restrict ourselves to the absolutely necessary contents to make this book self-contained. Detailed descriptions of the material treated here can be found e.g. in Altenbach et al. (1996), Ambartsumyan (1970), Ashton and Whitney (1970), Becker and Gross (2002), Eschenauer and Schnell (1986), Frick and Klamser (1990), Göldner et al. (1979), (1985), Hahn (1985), Jones (1975), Lekhnitskii (1968), Mittelstedt and Becker (2016), Mittelstedt (2021), or Tsai and Hahn (1980).

1.1 Introduction

A three-dimensional solid (a so-called continuum) under load is considered. Due to the applied load, this structure will develop forces in its interior, namely the so-called stresses or the so-called stress state. Such a structure will also deform due to the applied load, i.e. displacements of the body points will occur. Associated with this is the concept of strains or the strain state. In order to describe all mentioned quantitues, the following set of equations is generally required. First, we consider the kinematic equations of the problem, which establish a relationship between the displacements and the strains. Furthermore, the constitutive equations are necessary, which establish a relationship between the strains on one side and the stresses on the other side. In the case of the so-called linear elasticity, this is the generalized Hooke's law. Finally, equilibrium must also be ensured at each body point, which is described by the local equilibrium conditions.

C. Mittelstedt, *Theory of Plates and Shells*,
https://doi.org/10.1007/978-3-662-66805-4_1

1.2 Stress State

1.2.1 Stress Vector and Stress Tensor

We consider the solid body of Fig. 1.1, which is subjected to a certain load and given boundary conditions. To describe the occurring state quantities in the considered solid, a spatial Cartesian reference system x, y, z with associated displacements u, v and w is introduced, where the displacements can be functions of all three spatial directions.

$$u = u(x, y, z), \quad v = v(x, y, z), \quad w = w(x, y, z). \tag{1.1}$$

The given loads can be of various types. First of all, there are the so-called volume forces, i.e. all types of loads which are distributed spatially in the body. An example of a volume force is the dead weight of a solid. Accordingly, volume forces have the unit of a force per unit volume, e.g. [N/m^3]. The spatial components are denoted as f_x, f_y, f_z and are combined in the vector $\underline{f}$:

$$\underline{f} = \begin{pmatrix} f_x \\ f_y \\ f_z \end{pmatrix}. \tag{1.2}$$

Loads can also occur in the form of surface loads, which are correspondingly in the unit of a force per unit of surface, e.g. [N/m^2]. Surface loads can occur on the entire surface of the solid under consideration, or only on a part of it. We denote the components of a surface load as t_x, t_y, t_z and they are expressed in the vector

$$\underline{t} = \begin{pmatrix} t_x \\ t_y \\ t_z \end{pmatrix}. \tag{1.3}$$

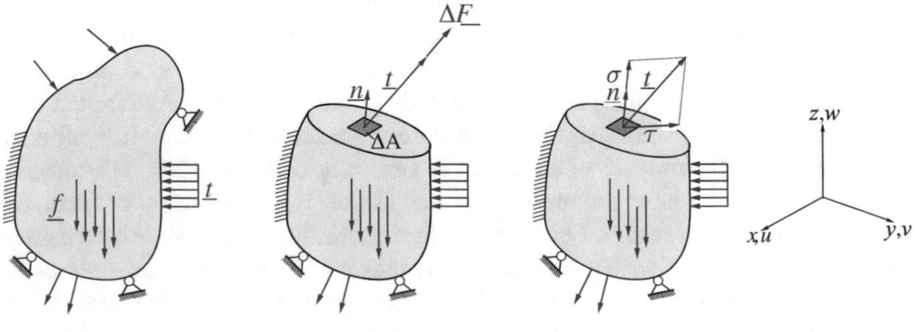

Fig. 1.1 Three-dimensional solid under load (left), section through an arbitrary body point (middle), decomposition of the stress vector into normal and shear stress (right)

Other types of loads are line loads concentrated on a line (with the unit force per unit length, e.g. [N/m]), or point loads concentrated on a point in the unit [N].

A given load on a solid causes internal forces which are related to an imaginary sectional surface. These related forces are referred to as stresses. Stresses are given in a corresponding unit, e.g. [N/m^2]. We investigate a cut at an arbitrary solid point and consider an infinitesimal cut surface ΔA (Fig. 1.1, middle). The normal vector $\underline{n}$ has the components

$$\underline{n} = \begin{pmatrix} n_x \\ n_y \\ n_z \end{pmatrix} . \tag{1.4}$$

The force $\Delta \underline{F}$ is now released on the cut surface. The stress vector is then defined as:

$$\underline{t} = \lim_{\Delta A \to 0} \frac{\Delta \underline{F}}{\Delta A} . \tag{1.5}$$

Obviously, the stress vector depends on the orientation under which the cut is made and which surface is examined. Thus, the stress vector $\underline{t}$ is a function of the normal vector $\underline{n}$:

$$\underline{t} = \underline{t}\left(\underline{n}\right) . \tag{1.6}$$

The stress vector is decomposed into two components, namely a component parallel to the normal vector and a component tangential to the section surface under consideration. These are called normal stress σ and shear stress τ, they are shown in Fig. 1.1, right. The stress components are obtained as:

$$\sigma = \underline{t} \cdot \underline{n}, \quad \tau = \sqrt{\underline{t} \cdot \underline{t} - \sigma^2} . \tag{1.7}$$

The stress state in any solid point is uniquely determined if the stress vector in three independent sections through the point is known. The normal vectors of these three sections must be linearly independent of each other. It is convenient to place these sections through the considered body point so that they are perpendicular to the three coordinate axes x, y, z. In Fig. 1.2 an infinitesimally small cube is shown, which was cut out of the considered solid in this manner. If the three stress vectors are given, then the stress vectors for any other cut orientation can be determined, as will be shown later.

Let these three intersecting surfaces be defined such that their normal vectors coincide with the coordinate axes x, y, z. The three stress vectors $\underline{t}_x$, $\underline{t}_y$, $\underline{t}_z$ can then be given as follows:

$$\underline{t}_x = \begin{pmatrix} \sigma_{xx} \\ \tau_{xy} \\ \tau_{xz} \end{pmatrix}, \quad \underline{t}_y = \begin{pmatrix} \tau_{yx} \\ \sigma_{yy} \\ \tau_{yz} \end{pmatrix}, \quad \underline{t}_z = \begin{pmatrix} \tau_{zx} \\ \tau_{zy} \\ \sigma_{zz} \end{pmatrix} . \tag{1.8}$$

The indexing of stresses is usually used in such a way that the first index indicates the direction of the surface normal of the considered section, whereas the second index indicates the direction of action of the considered stress component. Thus, stresses

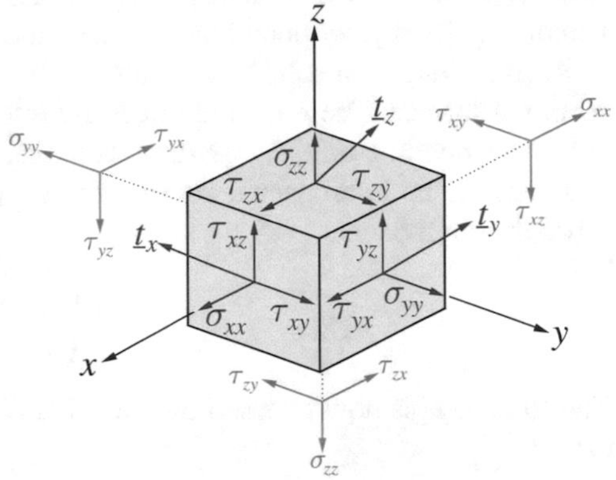

Fig. 1.2 Infinitesimal cube and associated stress components

with identical indices are normal stresses, which are perpendicular to the considered section. Stresses with two different indices are shear stresses, which are tangential to the considered section surface.

For a complete description of the stress state in a point of a solid, obviously nine stress components must be known. These are the three normal stresses σ_{xx}, σ_{yy}, σ_{zz} and the six shear stresses τ_{xy}, τ_{yx}, τ_{xz}, τ_{zx}, τ_{yz} and τ_{zy}. A stress is assumed to be positive if it points in the positive coordinate direction at a positive cut surface. A cut surface is assumed to be positive if the outward pointing normal vector points in positive coordinate direction.

In some cases, an alternative notation proves to be adequate, namely such that the reference axes are designated as x_1, x_2, x_3 and the indexing of the stress components is done accordingly. In such a notation, all stress components are often designated

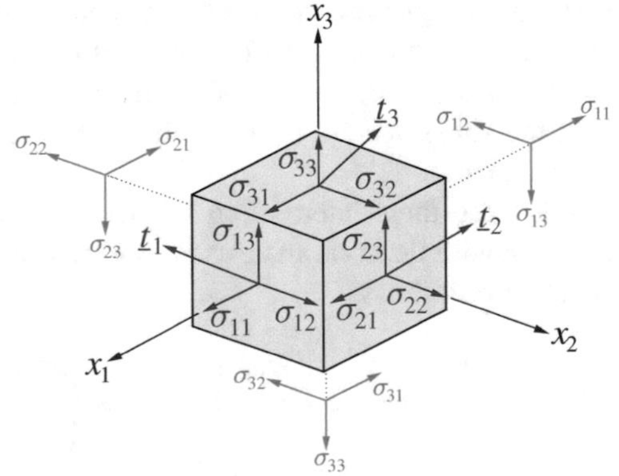

Fig. 1.3 Use of an alternative notation for reference axes and stresses on an infinitesimal cube

with σ. The three normal stresses σ_{11}, σ_{22}, σ_{33} and the six shear stresses σ_{12}, σ_{21}, σ_{13}, σ_{31}, σ_{23} and σ_{32} are then present (Fig. 1.3). The displacements are denoted as u_1, u_2, u_3, and the stress vectors are labeled $\underline{t}_1$, $\underline{t}_2$, $\underline{t}_3$:

$$\underline{t}_1 = \begin{pmatrix} \sigma_{11} \\ \sigma_{12} \\ \sigma_{13} \end{pmatrix}, \quad \underline{t}_2 = \begin{pmatrix} \sigma_{21} \\ \sigma_{22} \\ \sigma_{23} \end{pmatrix}, \quad \underline{t}_3 = \begin{pmatrix} \sigma_{31} \\ \sigma_{32} \\ \sigma_{33} \end{pmatrix}. \tag{1.9}$$

Looking at Figs. 1.2 and 1.3 respectively, it is easy to see that the moment equilibrium around the center of gravity of the infinitesimal element about all three reference axes immediately leads to the equality of associated shear stresses (i.e., those with identical but interchanged indices), so that we have

$$\tau_{xy} = \tau_{yx}, \quad \tau_{xz} = \tau_{zx}, \quad \tau_{yz} = \tau_{zy}, \tag{1.10}$$

and

$$\sigma_{12} = \sigma_{21}, \quad \sigma_{13} = \sigma_{31}, \quad \sigma_{23} = \sigma_{32}. \tag{1.11}$$

This result is applied in all that follows. As a result, the number of independent stress components is reduced to six, namely the three normal stresses σ_{xx}, σ_{yy}, σ_{zz} and σ_{11}, σ_{22}, σ_{33}, and the three shear stresses $\tau_{xy} = \tau_{yx}$, $\tau_{xz} = \tau_{zx}$, $\tau_{yz} = \tau_{zy}$ respectively $\tau_{12} = \tau_{21}$, $\tau_{13} = \tau_{31}$, $\tau_{23} = \tau_{32}$.

The stress components are combined in a symmetric matrix $\underline{\underline{\sigma}}$, the so-called Cauchy's[1] stress tensor by arranging the stress vectors $\underline{t}_x$, $\underline{t}_y$, $\underline{t}_z$ and $\underline{t}_1$, $\underline{t}_2$, $\underline{t}_3$, respectively, in columns:

$$\underline{\underline{\sigma}} = \begin{bmatrix} \sigma_{xx} & \tau_{xy} & \tau_{xz} \\ \tau_{xy} & \sigma_{yy} & \tau_{yz} \\ \tau_{xz} & \tau_{yz} & \sigma_{zz} \end{bmatrix} = \begin{bmatrix} \sigma_{11} & \sigma_{12} & \sigma_{13} \\ \sigma_{12} & \sigma_{22} & \sigma_{23} \\ \sigma_{13} & \sigma_{23} & \sigma_{33} \end{bmatrix}. \tag{1.12}$$

The Cauchy stress tensor $\underline{\underline{\sigma}}$ is a symmetric second-order tensor.

If the Cauchy stress tensor $\underline{\underline{\sigma}}$ is given at a specific body point, one can determine the stress state for any other section through the point under consideration. For the derivation of the corresponding law we consider a tetrahedron cut out of an infinitesimal cube (see Fig. 1.4). The tetrahedron is defined by the three cut areas $\mathrm{d}A_x$, $\mathrm{d}A_y$, $\mathrm{d}A_z$ and the arbitrarily oriented tetrahedral surface $\mathrm{d}A$. The normal vector of the tetrahedral surface is oriented with respect to the reference frame x, y, z under the angles φ_x, φ_y, φ_z. Let $\underline{t} = \left(t_x, t_y, t_z\right)^T$ be the stress vector with respect to the tetrahedral surface $\mathrm{d}A$. The normal vector is then given with:

$$\underline{n} = \left(n_x, n_y, n_z\right)^T = \left(\cos\varphi_x, \cos\varphi_y, \cos\varphi_z\right)^T. \tag{1.13}$$

[1] Augustin-Louis Cauchy, 1789–1857, French mathematician.

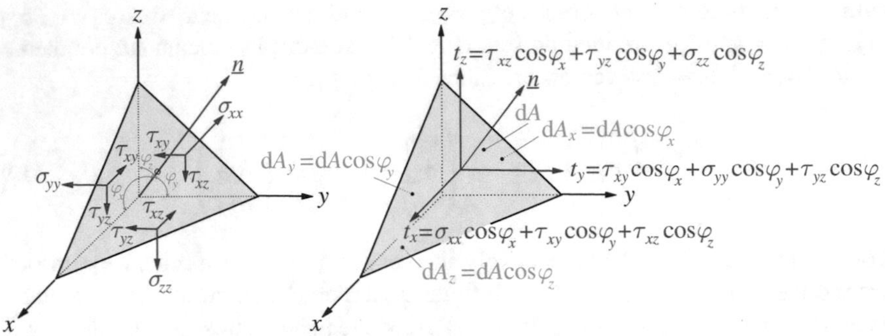

Fig. 1.4 Forces at an infinitesimal tetrahedron

The equilibrium of forces with respect to the $x-$, $y-$ and $z-$ directions then reads:

$$
\begin{aligned}
t_x \mathrm{d}A &= \sigma_{xx}\mathrm{d}A_x + \tau_{xy}\mathrm{d}A_y + \tau_{xz}\mathrm{d}A_z,\\
t_y \mathrm{d}A &= \tau_{xy}\mathrm{d}A_x + \sigma_{yy}\mathrm{d}A_y + \tau_{yz}\mathrm{d}A_z,\\
t_z \mathrm{d}A &= \tau_{xz}\mathrm{d}A_x + \tau_{yz}\mathrm{d}A_y + \sigma_{zz}\mathrm{d}A_z.
\end{aligned}
\tag{1.14}
$$

Using $\mathrm{d}A_x = \mathrm{d}An_x$, $\mathrm{d}A_y = \mathrm{d}An_y$, $\mathrm{d}A_z = \mathrm{d}An_z$, we obtain:

$$
\begin{aligned}
t_x &= \sigma_{xx}n_x + \tau_{xy}n_y + \tau_{xz}n_z,\\
t_y &= \tau_{xy}n_x + \sigma_{yy}n_y + \tau_{yz}n_z,\\
t_z &= \tau_{xz}n_x + \tau_{yz}n_y + \sigma_{zz}n_z,
\end{aligned}
\tag{1.15}
$$

respectively in a vector-matrix notation:

$$
\underline{t} = \underline{\underline{\sigma}}\,\underline{n}. \tag{1.16}
$$

If one uses the index notation (i.e. summation over equal indices is implied, with $i = 1, 2, 3$) with the reference system x_1, x_2, x_3, one obtains:

$$
t_i = \sigma_{ij} n_j. \tag{1.17}
$$

This is the so-called Cauchy's theorem. Cauchy's theorem allows the conclusion that if the stress tensor $\underline{\underline{\sigma}}$ is given at a body point, the stress tensor for any other section through that body point can be determined.

1.2.2 Transformation Rules

So far we have assumed that all stress components are given with respect to the spatial Cartesian axes x, y, z or their basis vectors $\underline{e}_x$, $\underline{e}_y$, $\underline{e}_z$. We now investigate which transformation rules apply when the orthonormal basis $\underline{e}_x, \underline{e}_y, \underline{e}_z$ is transformed

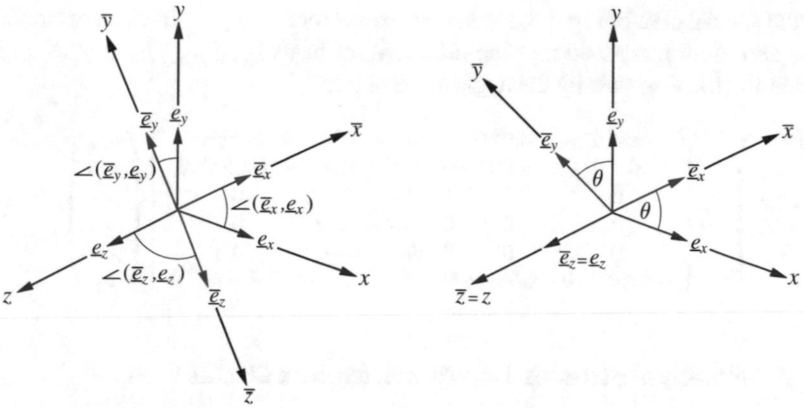

Fig. 1.5 Rotation of an orthonormal basis $\underline{e}_x, \underline{e}_y, \underline{e}_z$ into an arbitrary basis $\underline{\bar{e}}_x, \underline{\bar{e}}_y, \underline{\bar{e}}_z$ (left), rotation of an orthonormal basis $\underline{e}_x, \underline{e}_y, \underline{e}_z$ into an orthonormal basis $\underline{\bar{e}}_x, \underline{\bar{e}}_y, \underline{\bar{e}}_z$ with fixed $z-$ axis (right)

by a vector rotation into an arbitrary basis $\underline{\bar{e}}_x, \underline{\bar{e}}_y, \underline{\bar{e}}_z$ respectively the reference frame $\bar{x}, \bar{y}, \bar{z}$ (Fig. 1.5, left). The orthonormal basis $\underline{e}_x, \underline{e}_y, \underline{e}_z$ and the basis $\underline{\bar{e}}_x, \underline{\bar{e}}_y, \underline{\bar{e}}_z$ share the same origin. The basis vectors $\underline{e}_x$ and $\underline{\bar{e}}_y$ enclose the direction angle $\angle\left(\underline{\bar{e}}_y, \underline{e}_x\right)$ with the direction cosine R_{yx} (analogous for all other R_{ij}). The first index denotes the base vector from the transformed coordinate system $\bar{x}, \bar{y}, \bar{z}$, whereas the second index denotes the one from the original coordinate system x, y, z ($i, j = x, y, z$):

$$R_{ij} = \cos\left(\angle \underline{\bar{e}}_i, \underline{e}_j\right). \tag{1.18}$$

The direction cosine R_{ij} results from the scalar product of the two considered basis vectors:

$$R_{ij} = \frac{\underline{\bar{e}}_i \cdot \underline{e}_j}{\left|\underline{\bar{e}}_i\right| \left|\underline{e}_j\right|}. \tag{1.19}$$

If we now arrange the stresses in the column vector $\underline{\sigma} = (\sigma_{xx}, \sigma_{yy}, \sigma_{zz}, \tau_{yz}, \tau_{xz}, \tau_{xy})^T$, then the stress transformation from $\underline{e}_x, \underline{e}_y, \underline{e}_z$ to $\underline{\bar{e}}_x, \underline{\bar{e}}_y, \underline{\bar{e}}_z$ results from the equilibrium at the tetrahedron:

$$\underline{\bar{\sigma}} = \underline{\underline{T}}_{\sigma}\, \underline{\sigma}. \tag{1.20}$$

The transformation matrix $\underline{\underline{T}}_{\sigma}$ can be given as follows:

$$\underline{\underline{T}}_{\sigma} = \begin{bmatrix} R_{11}^2 & R_{12}^2 & R_{13}^2 & 2R_{12}R_{13} & 2R_{11}R_{13} & 2R_{11}R_{12} \\ R_{21}^2 & R_{22}^2 & R_{23}^2 & 2R_{22}R_{23} & 2R_{21}R_{23} & 2R_{21}R_{22} \\ R_{31}^2 & R_{32}^2 & R_{33}^2 & 2R_{32}R_{33} & 2R_{31}R_{33} & 2R_{31}R_{32} \\ R_{21}R_{31} & R_{22}R_{32} & R_{23}R_{33} & R_{22}R_{33}+R_{23}R_{32} & R_{21}R_{33}+R_{23}R_{31} & R_{21}R_{32}+R_{22}R_{31} \\ R_{11}R_{31} & R_{12}R_{32} & R_{13}R_{33} & R_{12}R_{33}+R_{13}R_{32} & R_{11}R_{33}+R_{13}R_{31} & R_{11}R_{32}+R_{12}R_{31} \\ R_{11}R_{21} & R_{12}R_{22} & R_{13}R_{23} & R_{12}R_{23}+R_{13}R_{22} & R_{11}R_{23}+R_{13}R_{21} & R_{11}R_{22}+R_{12}R_{21} \end{bmatrix}. \tag{1.21}$$

For the special case of Fig. 1.5, right, that the vectors $\underline{e}_x, \underline{e}_y, \underline{e}_z$ form an orthonormal basis and are transformed into the orthonormal basis $\underline{\bar{e}}_x, \underline{\bar{e}}_y, \underline{\bar{e}}_z$ by a pure rotation around the fixed $z-$axis by the angle θ we obtain:

$$\begin{pmatrix} \bar{\sigma}_{xx} \\ \bar{\sigma}_{yy} \\ \bar{\sigma}_{zz} \\ \bar{\tau}_{yz} \\ \bar{\tau}_{xz} \\ \bar{\tau}_{xy} \end{pmatrix} = \begin{bmatrix} \cos^2\theta & \sin^2\theta & 0 & 0 & 0 & 2\cos\theta\sin\theta \\ \sin^2\theta & \cos^2\theta & 0 & 0 & 0 & -2\cos\theta\sin\theta \\ 0 & 0 & 1 & 0 & 0 & 0 \\ 0 & 0 & 0 & \cos\theta & -\sin\theta & 0 \\ 0 & 0 & 0 & \sin\theta & \cos\theta & 0 \\ -\cos\theta\sin\theta & \cos\theta\sin\theta & 0 & 0 & 0 & \cos^2\theta - \sin^2\theta \end{bmatrix} \begin{pmatrix} \sigma_{xx} \\ \sigma_{yy} \\ \sigma_{zz} \\ \tau_{yz} \\ \tau_{xz} \\ \tau_{xy} \end{pmatrix}. \tag{1.22}$$

1.2.3 Principal Stresses, Invariants, Mohr's Circles

For each stress state, a special reference system x_h, y_h, z_h, the so-called principal axis system, can be determined. In the sections of this system all stress vectors are parallel to the corresponding unit vectors $\underline{e}_{x,h}, \underline{e}_{y,h}, \underline{e}_{z,h}$ or to the normal vectors of the section surfaces. As a consequence, all shear stresses disappear in this reference frame: $\tau_{yz} = \tau_{xz} = \tau_{xy} = 0$. The normal stresses $\sigma_{xx}, \sigma_{yy}, \sigma_{zz}$ occurring in this specific reference frame are the so-called principal stresses, whereas the designations $\sigma_1, \sigma_2, \sigma_3$ are often used for the principal stresses. In this case, the components of $\underline{t}$ are multiples of the corresponding normal vector multiplied by the corresponding principal stress σ:

$$\underline{t} = \sigma \underline{n}. \tag{1.23}$$

With (1.16) we obtain:

$$\underline{\underline{\sigma}}\,\underline{n} = \sigma \underline{n}, \tag{1.24}$$

which leads to the following eigenvalue problem:

$$\left[\underline{\underline{\sigma}} - \sigma \underline{\underline{I}}\right] \underline{n} = 0. \tag{1.25}$$

Herein, σ are the eigenvalues, and $\underline{n}$ are the associated eigenvectors. The matrix $\underline{\underline{I}}$ is the unit matrix. To avoid the trivial solution it is required that the coefficient determinant in (1.25) becomes zero:

$$\det\left[\underline{\underline{\sigma}} - \sigma \underline{\underline{I}}\right] = 0, \tag{1.26}$$

respectively:

$$\det \begin{bmatrix} \sigma_{xx} - \sigma & \tau_{xy} & \tau_{xz} \\ \tau_{xy} & \sigma_{yy} - \sigma & \tau_{yz} \\ \tau_{xz} & \tau_{yz} & \sigma_{zz} - \sigma \end{bmatrix} = 0. \tag{1.27}$$

This gives the following cubic polynomial whose roots represent the principal stresses $\sigma_1, \sigma_2, \sigma_3$:

$$\sigma^3 - I_1\sigma^2 - I_2\sigma - I_3 = 0, \tag{1.28}$$

wherein:

$$I_1 = \sigma_{xx} + \sigma_{yy} + \sigma_{zz} = \text{spur}\left[\underline{\underline{\sigma}}\right],$$
$$I_2 = \tau_{xy}^2 + \tau_{yz}^2 + \tau_{xz}^2 - \left(\sigma_{xx}\sigma_{yy} + \sigma_{yy}\sigma_{zz} + \sigma_{xx}\sigma_{zz}\right),$$
$$I_3 = \det \underline{\underline{\sigma}}. \tag{1.29}$$

Equation (1.28) always yields real values for the principal stresses $\sigma_1, \sigma_2, \sigma_3$ and is also independent of the reference frame. The quantities I_1, I_2, I_3 are therefore also called the invariants of the stress state.

Equation (1.28) can be discussed in more detail with respect to the principal stresses $\sigma_1, \sigma_2, \sigma_3$. One of the three principal stresses $\sigma_1, \sigma_2, \sigma_3$ is always the maximum normal stress occurring in a body point. It is usually referred to as σ_1. Another principal stress is always the minimum normal stress occurring in the body point, let it be denoted as σ_3. The remaining stress σ_2 takes a value between σ_1 and σ_3. The unit vectors of the principal axis system x_h, y_h, z_h are the eigenvectors of $\underline{\sigma}$ to the respective eigenvalues and always form an orthogonal basis. The determination of the principal stresses and their associated principal axes is called principal axis transformation. The stress tensor in the principal axis system reads:

$$\underline{\underline{\sigma}} = \begin{bmatrix} \sigma_1 & 0 & 0 \\ 0 & \sigma_2 & 0 \\ 0 & 0 & \sigma_3 \end{bmatrix}. \tag{1.30}$$

The invariants I_1, I_2, I_3 read in this case:

$$I_1 = \sigma_1 + \sigma_2 + \sigma_3,$$
$$I_2 = \sigma_1\sigma_2 + \sigma_2\sigma_3 + \sigma_1\sigma_3,$$
$$I_3 = \sigma_1\sigma_2\sigma_3. \tag{1.31}$$

Another axes direction can be identified under which the maximum shear stresses (the so-called principal shear stresses) τ_1, τ_2, τ_3 occur. These arise in sections whose normals are perpendicular to one of the principal axes and form an angle of 45° with the remaining axes. It should be noted that in such sections where the shear stresses become the principal shear stresses, the normal stresses are not necessarily zero. For the principal shear stresses this results in:

$$\tau_1 = \frac{1}{2}\left(\sigma_2 - \sigma_3\right), \quad \tau_2 = \frac{1}{2}\left(\sigma_1 - \sigma_3\right), \quad \tau_3 = \frac{1}{2}\left(\sigma_1 - \sigma_2\right). \tag{1.32}$$

It can be seen that τ_2 is the maximum shear stress.

The spatial stress state can be interpreted very clearly with the help of the so-called Mohr's circles[2] (Fig. 1.6). The circles shown here each denote the stress state in one of the three sections distinguished by the unit vectors $\underline{e}_{x,h}, \underline{e}_{y,h}, \underline{e}_{z,h}$ of the principal

[2] Christian Otto Mohr, 1835–1918, German civil engineer.

Fig. 1.6 Mohr's circles for the spatial stress state with principal normal stresses σ_1, σ_2, σ_3 and principal shear stresses τ_1, τ_2, τ_3

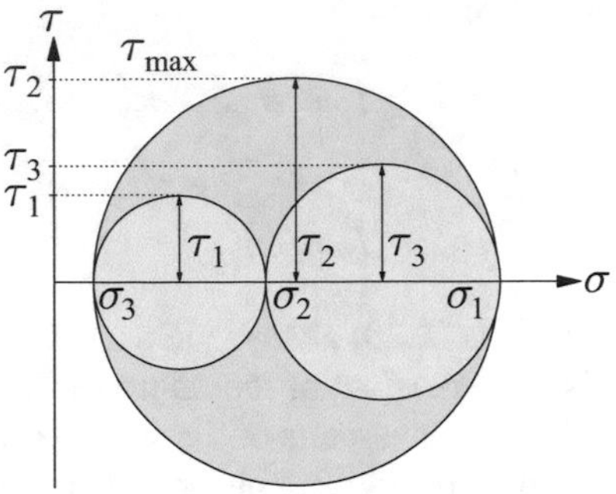

axis system. The three principal stresses are then found at the tangent points of the circles. Any other stress state deviating from the principal axes with basis vectors $\underline{e}_{x,h}$, $\underline{e}_{y,h}$, $\underline{e}_{z,h}$ is found in the area highlighted in dark gray. The circle characterized by σ_1 and σ_3 with the center at the location $\sigma = \frac{1}{2}(\sigma_1 - \sigma_3)$ bounds the stress state on both diagram axes.

1.2.4 Equilibrium Conditions

In this section we want to investigate how the stress states of two infinitesimally adjacent body points are related to each other. This connection is established via the local equilibrium conditions. To derive the equilibrium conditions, we consider an infinitesimal volume element with the edge dimensions $\mathrm{d}x$, $\mathrm{d}y$, $\mathrm{d}z$, which we cut out of the solid body (Fig. 1.7). The volume forces f_x, f_y, f_z are not shown here for reasons of clarity. The stresses at the respective positive and negative cut edges differ by an infinitesimal increment. These increments are denoted as $\mathrm{d}\sigma_{xx}$, $\mathrm{d}\sigma_{yy}$,

Fig. 1.7 Local equilibrium at an infinitesimal volume element

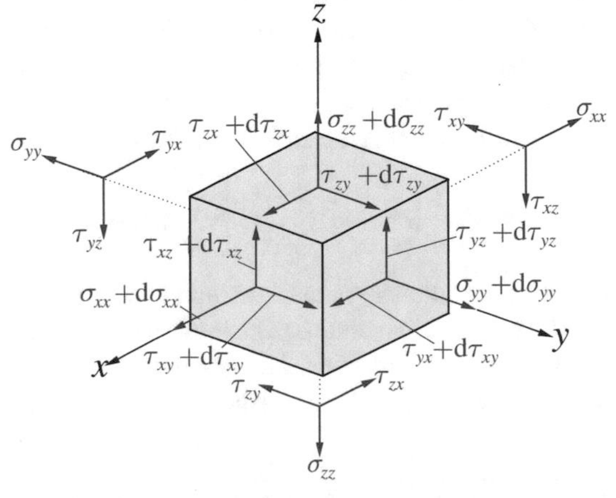

$d\sigma_{zz}$, $d\tau_{xy}$, $d\tau_{xz}$, $d\tau_{yz}$ and can be expanded as Taylor series which are terminated after the first term:

$$
\begin{aligned}
d\sigma_{xx} &= \frac{\partial \sigma_{xx}}{\partial x}dx, & d\sigma_{yy} &= \frac{\partial \sigma_{yy}}{\partial y}dy, & d\sigma_{zz} &= \frac{\partial \sigma_{zz}}{\partial z}dz, \\
d\tau_{xy} &= \frac{\partial \tau_{xy}}{\partial x}dx, & d\tau_{yx} &= \frac{\partial \tau_{yx}}{\partial y}dy, & d\tau_{xz} &= \frac{\partial \tau_{xz}}{\partial x}dx, \\
d\tau_{zx} &= \frac{\partial \tau_{zx}}{\partial z}dz, & d\tau_{yz} &= \frac{\partial \tau_{yz}}{\partial y}dy, & d\tau_{zy} &= \frac{\partial \tau_{zy}}{\partial z}dz.
\end{aligned}
\tag{1.33}
$$

The equilibrium of forces in $x-$ direction is then given as:

$$
\begin{aligned}
&\left(\sigma_{xx} + \frac{\partial \sigma_{xx}}{\partial x}dx\right)dydz - \sigma_{xx}dydz \\
&+ \left(\tau_{yx} + \frac{\partial \tau_{yx}}{\partial y}dy\right)dxdz - \tau_{yx}dxdz \\
&+ \left(\tau_{zx} + \frac{\partial \tau_{zx}}{\partial z}dz\right)dxdy - \tau_{zx}dxdy + f_x dxdydz = 0.
\end{aligned}
\tag{1.34}
$$

Summarizing and observing $\tau_{ij} = \tau_{ji}$ leads to:

$$
\frac{\partial \sigma_{xx}}{\partial x} + \frac{\partial \tau_{xy}}{\partial y} + \frac{\partial \tau_{xz}}{\partial z} + f_x = 0.
\tag{1.35}
$$

In the same way, the equilibrium of forces in $y-$ and in $z-$direction can be formulated, and the local equilibrium conditions are obtained as:

$$
\begin{aligned}
\frac{\partial \sigma_{xx}}{\partial x} + \frac{\partial \tau_{xy}}{\partial y} + \frac{\partial \tau_{xz}}{\partial z} + f_x &= 0, \\
\frac{\partial \tau_{xy}}{\partial x} + \frac{\partial \sigma_{yy}}{\partial y} + \frac{\partial \tau_{yz}}{\partial z} + f_y &= 0, \\
\frac{\partial \tau_{xz}}{\partial x} + \frac{\partial \tau_{yz}}{\partial y} + \frac{\partial \sigma_{zz}}{\partial z} + f_z &= 0.
\end{aligned}
\tag{1.36}
$$

Using the reference system x_1, x_2, x_3 and the notation $\tau_{ij} = \sigma_{ij}$ employing index notation yields:

$$
\frac{\partial \sigma_{ij}}{\partial x_j} + f_i = 0,
\tag{1.37}
$$

with $i, j = 1, 2, 3$. Obviously, the equilibrium conditions are three coupled partial differential equations for six unknown stress components. Thus, any solid body problem is intrinsically statically indeterminate, the existing equilibrium conditions are not sufficient to determine the stress components. Consequently, additional equations must be found to determine the state variables of a solid body.

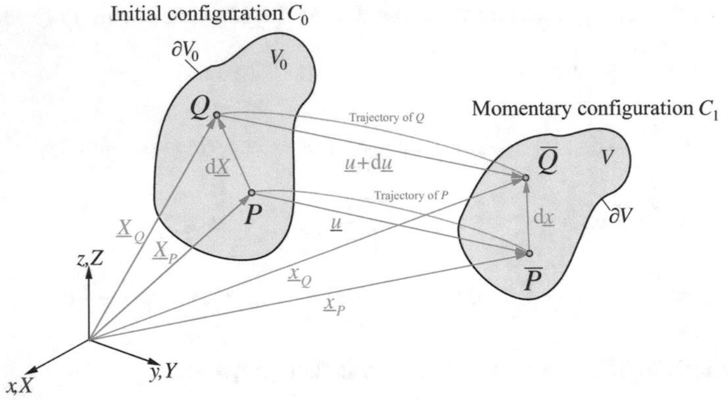

Fig. 1.8 Deformation of a solid

1.3 Deformations and Strains

1.3.1 Introduction

If a solid body is under load, the body points will shift and the displacements u, v, w will occur. This results in the so-called local strains. We summarize both displacements and strains under the term deformation state. For the description of the deformation state we refer to a certain configuration or to a certain deformation state. For this purpose, we distinguish between the undeformed initial state C_0 (the so-called reference configuration) and the current deformed state or momentary configuration C_1 (Fig. 1.8).

We now define a fixed Cartesian coordinate system X, Y, Z in which the position of each body point P in the initial configuration C_0 is given by the vector $\underline{X}$ with the components X, Y, Z (so-called material coordinates or Lagrangian coordinates[3]. The set V_0 of all body points P respectively the set of all location vectors $\underline{X}$ characterizes the undeformed body with the surface ∂V_0.

Now consider the solid body in the deformed state C_1, where each body point P has shifted by $\underline{u}$. Let the position vector in the state C_1 be $\underline{x}$ with the components x, y, z. Thus:

$$\underline{x} = \begin{pmatrix} x \\ y \\ z \end{pmatrix} = \begin{pmatrix} X \\ Y \\ Z \end{pmatrix} + \begin{pmatrix} u \\ v \\ w \end{pmatrix} = \underline{X} + \underline{u}. \tag{1.38}$$

Hence, the deformation of a solid is apparently uniquely characterized by the displacement vector $\underline{u}$:

$$\underline{u} = \underline{x} - \underline{X}. \tag{1.39}$$

[3] Joseph-Louis de Lagrange, 1736–1813, Italian mathematician.

The coordinates x, y, z are also denoted as the Eulerian[4] coordinates. The sum V of all body points $\bar{P}$ in the state C_1 then constitutes the deformed body with the surface ∂V.

For the description of the deformation process in a solid, there are basically two different approaches. On the one hand, there is the so-called Lagrangian approach, in which all state variables are considered as functions of the coordinates X, Y, Z:

$$u = u\,(X, Y, Z)\,, \quad v = v\,(X, Y, Z)\,, \quad w = w\,(X, Y, Z)\,, \tag{1.40}$$

or:

$$\underline{u} = \underline{u}\,(X, Y, Z)\,. \tag{1.41}$$

Thus, it is assumed that the coordinate system X, Y, Z remains at the same position during the deformation process and the deformation event is thus related to the original coordinates.

The other approach is the so-called Eulerian approach, where all state quantities such as displacements u, v, w are described as functions of the spatial coordinates x, y, z

$$u = u\,(x, y, z)\,, \quad v = v\,(x, y, z)\,, \quad w = w\,(x, y, z)\,, \tag{1.42}$$

or:

$$\underline{u} = \underline{u}\,(x, y, z)\,. \tag{1.43}$$

For solid body problems in elasticity theory, the Lagrangian approach is usually employed. We assume that the location $\underline{x}$ of a body point as well as its displacements $\underline{u}$ can be described as a function of the material coordinates X, Y, Z, i.e. $\underline{u} = \underline{u}\,(X, Y, Z)$ and $\underline{x} = \underline{x}\,(X, Y, Z)$.

With the help of Fig. 1.8 we consider the deformation process with respect to the two body points P and Q. Let these two points with location vectors $\underline{X}_P = (X, Y, Z)^T$ and $\underline{X}_Q = (X + \mathrm{d}X, Y + \mathrm{d}Y, Z + \mathrm{d}Z)^T$ be infinitesimally distant from each other in the original state C_0. The difference vector $\mathrm{d}\underline{X}$ is obtained as:

$$\mathrm{d}\underline{X} = \begin{pmatrix} \mathrm{d}X \\ \mathrm{d}Y \\ \mathrm{d}Z \end{pmatrix} = \underline{X}_Q - \underline{X}_P, \tag{1.44}$$

with

$$|\mathrm{d}\underline{X}| = \mathrm{d}S = \sqrt{\mathrm{d}X^2 + \mathrm{d}Y^2 + \mathrm{d}Z^2}. \tag{1.45}$$

In the configuration C_1 the two points P and Q transfer into the points $\bar{P}$ and $\bar{Q}$, with the location vectors $\underline{x}_P = (x, y, z)^T$ and $\underline{x}_Q = (x + \mathrm{d}x, y + \mathrm{d}y, z + \mathrm{d}z)^T$.

[4] Leonhard Euler, 1707–1783, Swiss mathematician.

The difference vector is then in the deformed configuration:

$$\mathrm{d}\underline{x} = \begin{pmatrix} \mathrm{d}x \\ \mathrm{d}y \\ \mathrm{d}z \end{pmatrix} = \underline{x}_Q - \underline{x}_P, \tag{1.46}$$

with:

$$|\mathrm{d}\underline{x}| = \mathrm{d}s = \sqrt{\mathrm{d}x^2 + \mathrm{d}y^2 + \mathrm{d}z^2}. \tag{1.47}$$

From Fig. 1.8 then the following can be deduced:

$$\mathrm{d}\underline{X} + \underline{u} + \mathrm{d}\underline{u} = \underline{u} + \mathrm{d}\underline{x}, \tag{1.48}$$

or:

$$\mathrm{d}\underline{u} = \mathrm{d}\underline{x} - \mathrm{d}\underline{X}. \tag{1.49}$$

In extended form we obtain:

$$\begin{pmatrix} \mathrm{d}u \\ \mathrm{d}v \\ \mathrm{d}w \end{pmatrix} = \begin{pmatrix} \mathrm{d}x \\ \mathrm{d}y \\ \mathrm{d}z \end{pmatrix} - \begin{pmatrix} \mathrm{d}X \\ \mathrm{d}Y \\ \mathrm{d}Z \end{pmatrix}. \tag{1.50}$$

The components $\mathrm{d}u$, $\mathrm{d}v$, $\mathrm{d}w$ of $\mathrm{d}\underline{u}$ can be written as total differentials of the displacement components u, v, w:

$$\begin{aligned} \mathrm{d}u &= \frac{\partial u}{\partial X}\mathrm{d}X + \frac{\partial u}{\partial Y}\mathrm{d}Y + \frac{\partial u}{\partial Z}\mathrm{d}Z, \\ \mathrm{d}v &= \frac{\partial v}{\partial X}\mathrm{d}X + \frac{\partial v}{\partial Y}\mathrm{d}Y + \frac{\partial v}{\partial Z}\mathrm{d}Z, \\ \mathrm{d}w &= \frac{\partial w}{\partial X}\mathrm{d}X + \frac{\partial w}{\partial Y}\mathrm{d}Y + \frac{\partial w}{\partial Z}\mathrm{d}Z. \end{aligned} \tag{1.51}$$

The partial derivatives of the displacements u, v, w with respect to the material coordinates X, Y, Z are summarized in the so-called displacement gradient $\underline{\underline{H}}$, a second-order tensor, as follows:

$$\underline{\underline{H}} = \begin{bmatrix} H_{xx} & H_{xy} & H_{xz} \\ H_{yx} & H_{yy} & H_{yz} \\ H_{zx} & H_{zy} & H_{zz} \end{bmatrix} = \begin{bmatrix} \frac{\partial u}{\partial X} & \frac{\partial u}{\partial Y} & \frac{\partial u}{\partial Z} \\ \frac{\partial v}{\partial X} & \frac{\partial v}{\partial Y} & \frac{\partial v}{\partial Z} \\ \frac{\partial w}{\partial X} & \frac{\partial w}{\partial Y} & \frac{\partial w}{\partial Z} \end{bmatrix}. \tag{1.52}$$

1.3.2 Green-Lagrangian Strain Tensor

The Green[5] -Lagrangian strain tensor is obtained by taking the difference of $\mathrm{d}s^2$ and $\mathrm{d}S^2$:

$$\begin{aligned}
\mathrm{d}s^2 - \mathrm{d}S^2 &= \mathrm{d}x^2 + \mathrm{d}y^2 + \mathrm{d}z^2 - \left(\mathrm{d}X^2 + \mathrm{d}Y^2 + \mathrm{d}Z^2\right) \\
&= (\mathrm{d}u + \mathrm{d}X)^2 + (\mathrm{d}v + \mathrm{d}Y)^2 + (\mathrm{d}w + \mathrm{d}Z)^2 - \left(\mathrm{d}X^2 + \mathrm{d}Y^2 + \mathrm{d}Z^2\right) \\
&= \left(\frac{\partial u}{\partial X}\mathrm{d}X + \frac{\partial u}{\partial Y}\mathrm{d}Y + \frac{\partial u}{\partial Z}\mathrm{d}Z + \mathrm{d}X\right)^2 \\
&\quad + \left(\frac{\partial v}{\partial X}\mathrm{d}X + \frac{\partial v}{\partial Y}\mathrm{d}Y + \frac{\partial v}{\partial Z}\mathrm{d}Z + \mathrm{d}Y\right)^2 \\
&\quad + \left(\frac{\partial w}{\partial X}\mathrm{d}X + \frac{\partial w}{\partial Y}\mathrm{d}Y + \frac{\partial w}{\partial Z}\mathrm{d}Z + \mathrm{d}Z\right)^2 \\
&\quad - \left(\mathrm{d}X^2 + \mathrm{d}Y^2 + \mathrm{d}Z^2\right).
\end{aligned} \tag{1.53}$$

This yields:

$$\begin{aligned}
\mathrm{d}s^2 - \mathrm{d}S^2 = {} & 2E_{xx}\mathrm{d}X^2 + 2E_{yy}\mathrm{d}Y^2 + 2E_{zz}\mathrm{d}Z^2 \\
& + 4E_{xy}\mathrm{d}X\mathrm{d}Y + 4E_{xz}\mathrm{d}X\mathrm{d}Z + 4E_{yz}\mathrm{d}Y\mathrm{d}Z,
\end{aligned} \tag{1.54}$$

or in index notation:

$$\mathrm{d}s^2 - \mathrm{d}S^2 = 2E_{ij}\mathrm{d}X_i\mathrm{d}X_j. \tag{1.55}$$

The quantities E_{xx}, E_{yy}, E_{zz}, E_{xy}, E_{xz}, E_{yz} are the components of the Green-Lagrangian strain tensor $\underline{E}$, defined as:

$$\begin{aligned}
E_{xx} &= \frac{\partial u}{\partial X} + \frac{1}{2}\left(\frac{\partial u}{\partial X}\right)^2 + \frac{1}{2}\left(\frac{\partial v}{\partial X}\right)^2 + \frac{1}{2}\left(\frac{\partial w}{\partial X}\right)^2, \\
E_{yy} &= \frac{\partial v}{\partial Y} + \frac{1}{2}\left(\frac{\partial u}{\partial Y}\right)^2 + \frac{1}{2}\left(\frac{\partial v}{\partial Y}\right)^2 + \frac{1}{2}\left(\frac{\partial w}{\partial Y}\right)^2, \\
E_{zz} &= \frac{\partial w}{\partial Z} + \frac{1}{2}\left(\frac{\partial u}{\partial Z}\right)^2 + \frac{1}{2}\left(\frac{\partial v}{\partial Z}\right)^2 + \frac{1}{2}\left(\frac{\partial w}{\partial Z}\right)^2, \\
E_{xy} &= \frac{1}{2}\left(\frac{\partial u}{\partial Y} + \frac{\partial v}{\partial X} + \frac{\partial u}{\partial X}\frac{\partial u}{\partial Y} + \frac{\partial v}{\partial X}\frac{\partial v}{\partial Y} + \frac{\partial w}{\partial X}\frac{\partial w}{\partial Y}\right), \\
E_{xz} &= \frac{1}{2}\left(\frac{\partial u}{\partial Z} + \frac{\partial w}{\partial X} + \frac{\partial u}{\partial X}\frac{\partial u}{\partial Z} + \frac{\partial v}{\partial X}\frac{\partial v}{\partial Z} + \frac{\partial w}{\partial X}\frac{\partial w}{\partial Z}\right), \\
E_{yz} &= \frac{1}{2}\left(\frac{\partial v}{\partial Z} + \frac{\partial w}{\partial Y} + \frac{\partial u}{\partial Y}\frac{\partial u}{\partial Z} + \frac{\partial v}{\partial Y}\frac{\partial v}{\partial Z} + \frac{\partial w}{\partial Y}\frac{\partial w}{\partial Z}\right).
\end{aligned} \tag{1.56}$$

[5] George Green, 1793–1841, British mathematician and physicist.

Thus, the Green-Lagrangian strain tensor $\underline{\underline{E}}$ describes large (finite) deformations and can be stated as:

$$\underline{\underline{E}} = \begin{bmatrix} E_{xx} & E_{xy} & E_{xz} \\ E_{xy} & E_{yy} & E_{yz} \\ E_{xz} & E_{yz} & E_{zz} \end{bmatrix}. \tag{1.57}$$

In index notation (1.56) yields:

$$E_{ij} = \frac{1}{2}\left(\frac{\partial u_i}{\partial X_j} + \frac{\partial u_j}{\partial X_i} + \frac{\partial u_k}{\partial X_i}\frac{\partial u_k}{\partial X_j}\right). \tag{1.58}$$

The components E_{xx}, E_{yy}, E_{zz} of the Green-Lagrangian strain tensor must not be confused with the elastic moduli of an orthotropic material as will be discussed later in this chapter.

1.3.3 Von-Kármán Strains

In the analysis of plates and shells, the so-called von-Kármán[6] theory is often used. This theory assumes that, although all three displacement components u, v, w can occur in thin-walled plane structures, the plane displacements u and v are are small compared to the deflection w. Consequently, in the Green-Lagrangian strain tensor all terms quadratic in u and v can be omitted and one obtains:

$$\begin{aligned}
E_{xx} &= \frac{\partial u}{\partial x} + \frac{1}{2}\left(\frac{\partial w}{\partial x}\right)^2, \\
E_{yy} &= \frac{\partial v}{\partial y} + \frac{1}{2}\left(\frac{\partial w}{\partial y}\right)^2, \\
E_{zz} &= \frac{\partial w}{\partial z} + \frac{1}{2}\left(\frac{\partial w}{\partial z}\right)^2, \\
E_{xy} &= \frac{1}{2}\left(\frac{\partial u}{\partial y} + \frac{\partial v}{\partial x} + \frac{\partial w}{\partial x}\frac{\partial w}{\partial y}\right), \\
E_{xz} &= \frac{1}{2}\left(\frac{\partial u}{\partial z} + \frac{\partial w}{\partial x} + \frac{\partial w}{\partial x}\frac{\partial w}{\partial z}\right), \\
E_{yz} &= \frac{1}{2}\left(\frac{\partial v}{\partial z} + \frac{\partial w}{\partial y} + \frac{\partial w}{\partial y}\frac{\partial w}{\partial z}\right).
\end{aligned} \tag{1.59}$$

In this case, the distinction between material and Eulerian coordinates is usually no longer made, and the coordinates X, Y, Z are replaced by x, y, z.

[6] Theodore von Kármán, 1881–1963, Austrian-Hungarian engineer and physicist.

1.3.4 Infinitesimal Strain Tensor

In solid body problems where the components of the displacement gradient are very small and $H_{ij} \ll 1$ holds, all product terms and all quadratic terms can be neglected in the Green-Lagrangian strain tensor. The distinction between material coordinates X, Y, Z and spatial coordinates x, y, z is no longer necessary in this case, the Eulerian and Lagrangian approaches coincide, i.e. $x_i \rightarrow X_i$. The displacements u, v, w can then be understood as functions of the spatial coordinates x, y, z: $u = u(x, y, z)$, $v = v(x, y, z)$, $w = w(x, y, z)$. Then, the components of the infinitesimal strain tensor $\underline{\underline{\varepsilon}}$ are obtained as follows:

$$\begin{aligned}
\varepsilon_{xx} &= \frac{\partial u}{\partial x}, \quad \varepsilon_{yy} = \frac{\partial v}{\partial y}, \quad \varepsilon_{zz} = \frac{\partial w}{\partial z}, \\
\gamma_{xy} &= 2\varepsilon_{xy} = \frac{\partial u}{\partial y} + \frac{\partial v}{\partial x}, \\
\gamma_{xz} &= 2\varepsilon_{xz} = \frac{\partial u}{\partial z} + \frac{\partial w}{\partial x}, \\
\gamma_{yz} &= 2\varepsilon_{yz} = \frac{\partial v}{\partial z} + \frac{\partial w}{\partial y}.
\end{aligned} \tag{1.60}$$

Therein, the quantities $\gamma_{xy}, \gamma_{xz}, \gamma_{yz}$ are the so-called technical shear strains. The symmetric infinitesimal strain tensor $\underline{\underline{\varepsilon}}$ is obtained as:

$$\underline{\underline{\varepsilon}} = \begin{bmatrix} \varepsilon_{xx} & \varepsilon_{xy} & \varepsilon_{xz} \\ \varepsilon_{xy} & \varepsilon_{yy} & \varepsilon_{yz} \\ \varepsilon_{xz} & \varepsilon_{yz} & \varepsilon_{zz} \end{bmatrix}. \tag{1.61}$$

The diagonal elements $\varepsilon_{xx}, \varepsilon_{yy}, \varepsilon_{zz}$ are denoted as normal strains. The remaining components $\gamma_{xy} = 2\varepsilon_{xy}, \gamma_{xz} = 2\varepsilon_{xz}, \gamma_{yz} = 2\varepsilon_{yz}$ are the shear strains. Normal strains and shear strains are summarized under the term of strains. In index notation, (1.60) results as:

$$\varepsilon_{ij} = \frac{1}{2}\left(\frac{\partial u_i}{\partial x_j} + \frac{\partial u_j}{\partial x_i}\right). \tag{1.62}$$

The components of the infinitesimal strain tensor are also amenable to a particularly clear geometric interpretation. For this purpose, consider the infinitesimal cube of Fig. 1.9. The normal strain ε_{xx} results as:

$$\varepsilon_{xx} = \frac{\Delta l}{l} = \frac{\left[u(x) + \frac{\partial u}{\partial x}\mathrm{d}x\right] - u(x)}{\mathrm{d}x} = \frac{\partial u}{\partial x}. \tag{1.63}$$

This result corresponds to the expression already derived with (1.60). The same can be done for the two normal strains ε_{yy} and ε_{zz}.

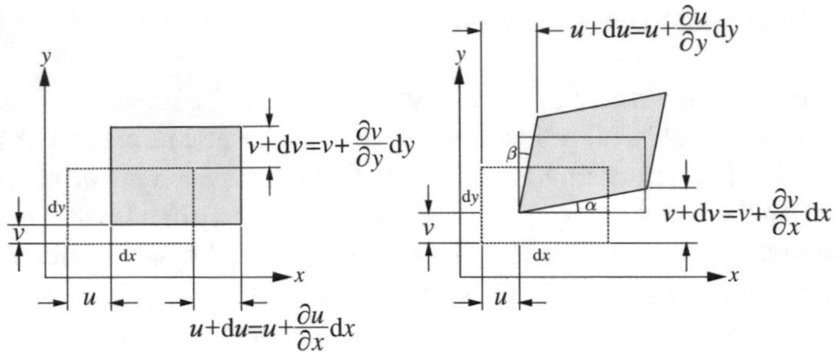

Fig. 1.9 Definition of the infinitesimal normal strains ε_{xx} and ε_{yy} (left) and the infinitesimal shear strain γ_{xy} (right)

The shear strain γ_{xy} in the $xy-$plane can be deduced from Fig. 1.9 as follows:

$$\tan\alpha \simeq \alpha = \frac{\frac{\partial v}{\partial x}\mathrm{d}x}{\mathrm{d}x} = \frac{\partial v}{\partial x}, \quad \tan\beta \simeq \beta = \frac{\frac{\partial u}{\partial y}\mathrm{d}y}{\mathrm{d}y} = \frac{\partial u}{\partial y}. \tag{1.64}$$

The shear strain γ_{xy} thus results as:

$$\gamma_{xy} = \alpha + \beta = \frac{\partial u}{\partial y} + \frac{\partial v}{\partial x}. \tag{1.65}$$

The shear strains γ_{xz} and γ_{yz} can be determined in an analogous way.

The Eq. (1.60) establish relations between displacements u, v, w on the one hand and the strains $\varepsilon_{xx}, \varepsilon_{yy}, \varepsilon_{zz}, \gamma_{xy}, \gamma_{xz}, \gamma_{yz}$ on the other hand. This set of equations is therefore also denoted as the so-called kinematic equations.

As already explained for the stress components, transformation rules can also be derived for the strain components. The corresponding transformation matrix $\underline{\underline{T}}_{\varepsilon}$ that is employed in the transformation rule $\underline{\bar{\varepsilon}} = \underline{\underline{T}}_{\varepsilon}\underline{\varepsilon}$ with $\underline{\varepsilon} = (\varepsilon_{xx}, \varepsilon_{yy}, \varepsilon_{zz}, \gamma_{yz}, \gamma_{xz}, \gamma_{xy})^T$ follows as:

$$\underline{\underline{T}}_{\varepsilon} = \begin{bmatrix} R_{11}^2 & R_{12}^2 & R_{13}^2 & R_{12}R_{13} & R_{11}R_{13} & R_{11}R_{12} \\ R_{21}^2 & R_{22}^2 & R_{23}^2 & R_{22}R_{23} & R_{21}R_{23} & R_{21}R_{22} \\ R_{31}^2 & R_{32}^2 & R_{33}^2 & R_{32}R_{33} & R_{31}R_{33} & R_{31}R_{32} \\ 2R_{21}R_{31} & 2R_{22}R_{32} & 2R_{23}R_{33} & R_{22}R_{33}+R_{23}R_{32} & R_{21}R_{33}+R_{23}R_{31} & R_{21}R_{32}+R_{22}R_{31} \\ 2R_{11}R_{31} & 2R_{12}R_{32} & 2R_{13}R_{33} & R_{12}R_{33}+R_{13}R_{32} & R_{11}R_{33}+R_{13}R_{31} & R_{11}R_{32}+R_{12}R_{31} \\ 2R_{11}R_{21} & 2R_{12}R_{22} & 2R_{13}R_{23} & R_{12}R_{23}+R_{13}R_{22} & R_{11}R_{23}+R_{13}R_{21} & R_{11}R_{22}+R_{12}R_{21} \end{bmatrix}. \tag{1.66}$$

For a pure coordinate rotation around the angle θ with fixed z–axis the transformation rules simplify to:

$$\begin{pmatrix} \bar{\varepsilon}_{xx} \\ \bar{\varepsilon}_{yy} \\ \bar{\varepsilon}_{zz} \\ \bar{\gamma}_{yz} \\ \bar{\gamma}_{xz} \\ \bar{\gamma}_{xy} \end{pmatrix} = \begin{bmatrix} \cos^2\theta & \sin^2\theta & 0 & 0 & 0 & \cos\theta\sin\theta \\ \sin^2\theta & \cos^2\theta & 0 & 0 & 0 & -\cos\theta\sin\theta \\ 0 & 0 & 1 & 0 & 0 & 0 \\ 0 & 0 & 0 & \cos\theta & -\sin\theta & 0 \\ 0 & 0 & 0 & \sin\theta & \cos\theta & 0 \\ -2\cos\theta\sin\theta & 2\cos\theta\sin\theta & 0 & 0 & 0 & \cos^2\theta-\sin^2\theta \end{bmatrix} \begin{pmatrix} \varepsilon_{xx} \\ \varepsilon_{yy} \\ \varepsilon_{zz} \\ \gamma_{yz} \\ \gamma_{xz} \\ \gamma_{xy} \end{pmatrix}. \quad (1.67)$$

The transformation matrices $\underline{\underline{T}}_{\sigma}$ and $\underline{\underline{T}}_{\varepsilon}$ are related as follows:

$$\underline{\underline{T}}_{\sigma}^{-1} = \underline{\underline{T}}_{\varepsilon}^{T}, \quad \underline{\underline{T}}_{\varepsilon}^{-1} = \underline{\underline{T}}_{\sigma}^{T}. \quad (1.68)$$

1.3.5 Compatibility Equations

If the three displacements u, v, w are given, the six components $\varepsilon_{xx}, \varepsilon_{yy}, \varepsilon_{zz}, \gamma_{yz}, \gamma_{xz}, \gamma_{xy}$ of the infinitesimal strain tensor $\underline{\underline{\varepsilon}}$ can be determined by means of the kinematic relations (1.60). Conversely, this also means that if the six strain components are known, there are also a total of six kinematic equations from which the displacements can be determined. The system of equations would be kinematically overdetermined in this case, and as a consequence the components of $\underline{\underline{\varepsilon}}$ cannot be independent of each other. A necessary and sufficient condition for a unique displacement field is that the infinitesimal strains $\varepsilon_{xx}, \varepsilon_{yy}, \varepsilon_{zz}, \gamma_{yz}, \gamma_{xz}, \gamma_{xy}$ satisfy the so-called compatibility conditions. The compatibility conditions are obtained by eliminating the displacements in the kinematic equations:

$$\frac{\partial^2 \varepsilon_{ij}}{\partial x_k \partial x_l} + \frac{\partial^2 \varepsilon_{kl}}{\partial x_i \partial x_j} - \frac{\partial^2 \varepsilon_{ik}}{\partial x_j \partial x_l} - \frac{\partial^2 \varepsilon_{jl}}{\partial x_i \partial x_k} = 0. \quad (1.69)$$

Herein the notation $\varepsilon_{ij} = \frac{1}{2}\gamma_{ij}$ was used for the shear strains, and the reference frame x_1, x_2, x_3 is employed. For the index pairs $(i, j), (k, l) = (1, 1), (2, 2), (3, 3), (1, 2), (1, 3), (2, 3)$ there are six different compatibility conditions, three of which are independent of each other. They result with reference to the reference system x, y, z and by using the technical shear strains as:

$$\frac{\partial^2 \varepsilon_{xx}}{\partial y^2} + \frac{\partial^2 \varepsilon_{yy}}{\partial x^2} - \frac{\partial^2 \gamma_{xy}}{\partial x \partial y} = 0,$$
$$\frac{\partial^2 \varepsilon_{xx}}{\partial z^2} + \frac{\partial^2 \varepsilon_{zz}}{\partial x^2} - \frac{\partial^2 \gamma_{xz}}{\partial x \partial z} = 0,$$
$$\frac{\partial^2 \varepsilon_{yy}}{\partial z^2} + \frac{\partial^2 \varepsilon_{zz}}{\partial y^2} - \frac{\partial^2 \gamma_{yz}}{\partial y \partial z} = 0,$$
$$2\frac{\partial^2 \varepsilon_{xx}}{\partial y \partial z} + \frac{\partial^2 \gamma_{yz}}{\partial x^2} - \frac{\partial^2 \gamma_{xz}}{\partial x \partial y} - \frac{\partial^2 \gamma_{xy}}{\partial x \partial z} = 0,$$

$$2\frac{\partial^2\varepsilon_{yy}}{\partial x\partial z}+\frac{\partial^2\gamma_{xz}}{\partial y^2}-\frac{\partial^2\gamma_{xy}}{\partial y\partial z}-\frac{\partial^2\gamma_{yz}}{\partial x\partial y}=0,$$
$$2\frac{\partial^2\varepsilon_{zz}}{\partial x\partial y}+\frac{\partial^2\gamma_{xy}}{\partial z^2}-\frac{\partial^2\gamma_{yz}}{\partial x\partial z}-\frac{\partial^2\gamma_{xz}}{\partial y\partial z}=0. \tag{1.70}$$

1.4 Constitutive Law

1.4.1 Introduction

The relationships considered so far, i.e. both the equilibrium conditions and the kinematic equations, are independent of the material. Clearly, these relationships are not sufficient to fully describe the mechanical behavior of a three-dimensional solid. Rather, material-specific equations must be found at this point to complete the description of an elastic solid. Such equations, in which the stresses and the strains are related to each other, are called constitutive equations or material equations, and are also referred to as material law or constitutive law.

In all further explanations of this book, we always assume elastic material behavior. Elastic material behavior means that after a complete unloading of the considered solid no permanent deformations remain, the deformation state is completely reversible. Accordingly, the deformation energy or strain energy stored in an elastic solid due to the deformations can be completely recovered and converted into mechanical work. We also assume that the stress state is time-independent and also independent of the loading history and is uniquely determined by the deformation state. We now assume that we are dealing with a geometrically linear problem with small deformations described by the infinitesimal strain tensor $\underline{\underline{\varepsilon}}$. Then we can state a material law or elasticity law in a general form as:

$$\underline{\underline{\sigma}}=\underline{\underline{\sigma}}\left(\underline{\underline{\varepsilon}}\right). \tag{1.71}$$

The actual form of (1.71) depends on the type of material considered and will be concretized below for the case of linear elasticity, i.e. for the case where there is a linear relationship between stresses and strains.

1.4.2 The Generalized Hooke's Law

If there is not only the case of elastic material behavior, but if there is even a linear relationship between stresses and strains, then the material is said to be linearly elastic. In this case, the stresses are proportional to the strains, and the material law is then represented by the generalized Hooke's law[7]. Here we want to distinguish between

[7] Robert Hooke, 1635–1703, English physicist.

isotropic materials (i.e., materials in which the behavior is directionally independent) and anisotropic materials (materials that exhibit directionally dependent behavior). The simplest case of isotropy for the one-dimensional case can be represented as:

$$\sigma = E\varepsilon, \quad \tau = G\gamma, \quad \varepsilon_q = -\nu\varepsilon, \tag{1.72}$$

where σ is the normal stress and τ is the shear stress. The elastic material properties are the modulus of elasticity E, the shear modulus G and the Poisson's[8] ratio ν. The associated strain quantities are the normal strain ε, the transverse strain ε_q, and the shear strain γ. Of course, for the general anisotropic three-dimensional case, Eq. (1.72) are not sufficient. For this purpose, the following general representation in index notation with the reference system x_1, x_2, x_3 as well as the designations σ_{ij} for the shear stresses and $2\varepsilon_{ij} = \gamma_{ij}$ for the shear strains is used, where one speaks of the so-called generalized Hooke's law:

$$\sigma_{ij} = C_{ijkl}\varepsilon_{kl}, \tag{1.73}$$

with $i, j, k, l = 1, 2, 3$. The quantity C_{ijkl} is the elasticity tensor. It is a fourth level tensor, and it has $3^4 = 81$ components by which the material behavior is described.

It is easy to see that the 81 components of the elasticity tensor C_{ijkl} cannot all be independent of each other. Rather, the elasticity tensor exhibits certain symmetry properties. Using $\sigma_{12} = \sigma_{21}$ as an example, we can immediately conclude $C_{12kl} = C_{21kl}$, which must apply analogously to the other stress components. In general it can be concluded:

$$C_{ijkl} = C_{jikl}. \tag{1.74}$$

Moreover, due to $\varepsilon_{ij} = \varepsilon_{ji}$, the infinitesimal strain tensor is also symmetric, i.e.:

$$C_{ijkl} = C_{ijlk}. \tag{1.75}$$

Thus, of the 81 components of the elasticity tensor C_{ijkl}, only 36 are independent of each other. Therefore, the following compact vector-matrix notation is also used in the literature for the representation of the generalized Hooke's law (so-called Voigt notation[9]):

$$\begin{pmatrix} \sigma_{11} \\ \sigma_{22} \\ \sigma_{33} \\ \sigma_{23} \\ \sigma_{13} \\ \sigma_{12} \end{pmatrix} = \begin{bmatrix} C_{11} & C_{12} & C_{13} & C_{14} & C_{15} & C_{16} \\ C_{21} & C_{22} & C_{23} & C_{24} & C_{25} & C_{26} \\ C_{31} & C_{32} & C_{33} & C_{34} & C_{35} & C_{36} \\ C_{41} & C_{42} & C_{43} & C_{44} & C_{45} & C_{46} \\ C_{51} & C_{52} & C_{53} & C_{54} & C_{55} & C_{56} \\ C_{61} & C_{62} & C_{63} & C_{64} & C_{65} & C_{66} \end{bmatrix} \begin{pmatrix} \varepsilon_{11} \\ \varepsilon_{22} \\ \varepsilon_{33} \\ 2\varepsilon_{23} \\ 2\varepsilon_{13} \\ 2\varepsilon_{12} \end{pmatrix}. \tag{1.76}$$

[8] Siméon Denis Poisson, 1781–1840, French physicist and mathematician.

[9] Woldemar Voigt, 1850–1919, German physicist.

The comparison of (1.73) with (1.76) then immediately leads to the relationship between the doubly indexed stiffnesses C_{ij} and the quadruply indexed tensor components C_{ijkl}. For example, obviously $C_{11} = C_{1111}$, $C_{12} = C_{1122}$, $C_{16} = \frac{1}{2}(C_{1112} + C_{1121}) = C_{1112}$ and so forth hold.

In symbolic notation, (1.76) can be given as:

$$\underline{\sigma} = \underline{\underline{C}}\,\underline{\varepsilon}. \tag{1.77}$$

The matrix $\underline{\underline{C}}$ is the so-called stiffness matrix of the material under consideration. Its components C_{ij} $(i, j = 1, 2, 3, 4, 5, 6)$ are the so-called stiffnesses. The elasticity tensor C_{ijkl} and the stiffness matrix $\underline{\underline{C}}$ are both always symmetric as will be proven later on. Hence:

$$C_{ijkl} = C_{klij} \quad \text{and} \quad C_{ij} = C_{ji}. \tag{1.78}$$

Thus, generally anisotropic and linear-elastic material behavior is described by 21 independent material constants.

Completely anisotropic material behavior, i.e. a material with all tensor components C_{ijkl} and stiffnesses C_{ij} not equal to zero, is accompanied by the so-called coupling effects. From (1.76) one can see that any strain component ε_{ij} gives rise to all stress components σ_{ij}. This not only complicates the structural analysis but is also a generally undesired effect of anisotropy.

The generalized Hooke's law can be represented in inverted form as follows:

$$\varepsilon_{ij} = S_{ijkl}\sigma_{kl}, \tag{1.79}$$

where S_{ijkl} is the so-called compliance tensor. It shows the same symmetry properties as the elasticity tensor C_{ijkl} and also has at most 21 independent components:

$$S_{ijkl} = S_{jikl} = S_{ijlk} = S_{klij}. \tag{1.80}$$

In a vector-matrix notation, this representation of Hooke's law reads:

$$\begin{pmatrix} \varepsilon_{11} \\ \varepsilon_{22} \\ \varepsilon_{33} \\ 2\varepsilon_{23} \\ 2\varepsilon_{13} \\ 2\varepsilon_{12} \end{pmatrix} = \begin{bmatrix} S_{11} & S_{12} & S_{13} & S_{14} & S_{15} & S_{16} \\ S_{21} & S_{22} & S_{23} & S_{24} & S_{25} & S_{26} \\ S_{31} & S_{32} & S_{33} & S_{34} & S_{35} & S_{36} \\ S_{41} & S_{42} & S_{43} & S_{44} & S_{45} & S_{46} \\ S_{51} & S_{52} & S_{53} & S_{54} & S_{55} & S_{56} \\ S_{61} & S_{62} & S_{63} & S_{64} & S_{65} & S_{66} \end{bmatrix} \begin{pmatrix} \sigma_{11} \\ \sigma_{22} \\ \sigma_{33} \\ \sigma_{23} \\ \sigma_{13} \\ \sigma_{12} \end{pmatrix}, \tag{1.81}$$

or in symbolic form:

$$\underline{\varepsilon} = \underline{\underline{S}}\,\underline{\sigma}. \tag{1.82}$$

The matrix $\underline{\underline{S}}$ is the so-called compliance matrix, its components are called compliances. It is also symmetric so that $S_{ij} = S_{ji}$ holds. The compliance matrix $\underline{\underline{S}}$ is the inverse of the stiffness matrix $\underline{\underline{C}}$, hence $\underline{\underline{S}} = \underline{\underline{C}}^{-1}$ holds.

1.4.3 Strain Energy

An elastic solid under load develops a state of deformation and strain. The resulting stresses inside the body thus perform work along the displacements. In the case of an elastic body, the work done in this way is stored as energy, which can be fully recovered after the load is removed. This stored energy is the so-called strain energy.

To motivate the notion of strain energy, we consider an infinitesimal volume element with edge lengths dx, dy, dz cut out of an elastic body (Fig. 1.10). We first consider the case that the volume element is under a normal stress σ_{xx} only. The resulting force is $\sigma_{xx}dydz$. An infinitesimal strain $d\varepsilon_{xx}$ now occurs, which leads to the change in length $d\varepsilon_{xx}dx$. The resulting force then performs the work increment $\sigma_{xx}d\varepsilon_{xx}dxdydz$ along this displacement. Relative to the volume $dV = dxdydz$, the following work increment per unit volume is obtained:

$$dU_0 = \sigma_{xx}d\varepsilon_{xx}. \tag{1.83}$$

In the case that all stress components occur, then work is also performed by them along the displacements they cause. The total work increment per unit volume can then be written as:

$$dU_0 = \sigma_{xx}d\varepsilon_{xx} + \sigma_{yy}d\varepsilon_{yy} + \sigma_{zz}d\varepsilon_{zz} + \tau_{yz}d\gamma_{yz} + \tau_{xz}d\gamma_{xz} + \tau_{xy}d\gamma_{xy}. \tag{1.84}$$

In index notation we have:

$$dU_0 = \sigma_{ij}d\varepsilon_{ij}. \tag{1.85}$$

The work done from the strain-free state to the current deformed state with the strains $\varepsilon_{xx}, \varepsilon_{yy}, \varepsilon_{zz}, \gamma_{yz}, \gamma_{xz}, \gamma_{xy}$ is performed by integration over the total work increment dU_0, from the initial strain-free state to the deformed state:

$$U_0 = \int_0^{\varepsilon_{xx}} \sigma_{xx}d\hat{\varepsilon}_{xx} + \int_0^{\varepsilon_{yy}} \sigma_{yy}d\hat{\varepsilon}_{yy} + \int_0^{\varepsilon_{zz}} \sigma_{zz}d\hat{\varepsilon}_{zz} + \int_0^{\gamma_{yz}} \tau_{yz}d\hat{\gamma}_{yz} + \int_0^{\gamma_{xz}} \tau_{xz}d\hat{\gamma}_{xz} + \int_0^{\gamma_{xy}} \tau_{xy}d\hat{\gamma}_{xy}. \tag{1.86}$$

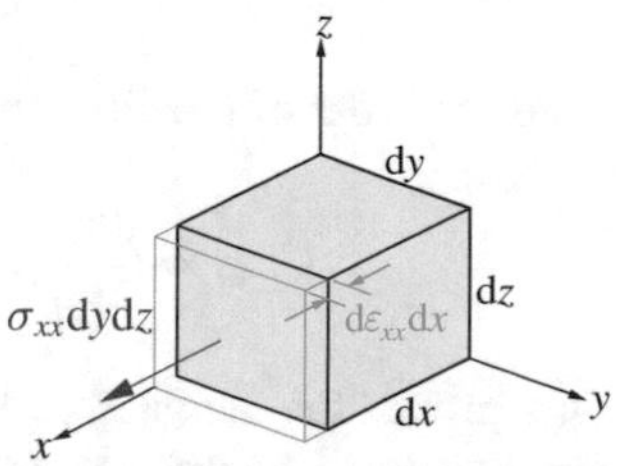

Fig. 1.10 Infinitesimal volume element under normal stress σ_{xx} and the resulting infinitesimal strain $d\varepsilon_{xx}$

Therein, U_0 is the strain energy density. In the case of an elastic material, the strain energy density U_0 does not depend on the deformation history. Rather, it depends exclusively on the current strains $\varepsilon_{xx}, \varepsilon_{yy}, \varepsilon_{zz}, \gamma_{yz}, \gamma_{xz}, \gamma_{xy}$. For this case, the working increment $\mathrm{d}U_0$ is the total differential of the strain energy density U_0:

$$\begin{aligned} &\sigma_{xx}\mathrm{d}\varepsilon_{xx} + \sigma_{yy}\mathrm{d}\varepsilon_{yy} + \sigma_{zz}\mathrm{d}\varepsilon_{zz} + \tau_{yz}\mathrm{d}\gamma_{yz} + \tau_{xz}\mathrm{d}\gamma_{xz} + \tau_{xy}\mathrm{d}\gamma_{xy} \\ &= \frac{\partial U_0}{\partial \varepsilon_{xx}}\mathrm{d}\varepsilon_{xx} + \frac{\partial U_0}{\partial \varepsilon_{yy}}\mathrm{d}\varepsilon_{yy} + \frac{\partial U_0}{\partial \varepsilon_{zz}}\mathrm{d}\varepsilon_{zz} + \frac{\partial U_0}{\partial \gamma_{yz}}\mathrm{d}\gamma_{yz} + \frac{\partial U_0}{\partial \gamma_{xz}}\mathrm{d}\gamma_{xz} + \frac{\partial U_0}{\partial \gamma_{xy}}\mathrm{d}\gamma_{xy}. \end{aligned} \tag{1.87}$$

In index notation we obtain:

$$\sigma_{ij}\mathrm{d}\varepsilon_{ij} = \frac{\partial U_0}{\partial \varepsilon_{ij}}\mathrm{d}\varepsilon_{ij}. \tag{1.88}$$

Thus, the stresses can be determined as the partial derivatives of the potential U_0 with respect to the strains:

$$\begin{aligned} \sigma_{xx} &= \frac{\partial U_0}{\partial \varepsilon_{xx}}, \quad \sigma_{yy} = \frac{\partial U_0}{\partial \varepsilon_{yy}}, \quad \sigma_{zz} = \frac{\partial U_0}{\partial \varepsilon_{zz}}, \\ \tau_{yz} &= \frac{\partial U_0}{\partial \gamma_{yz}}, \quad \tau_{xz} = \frac{\partial U_0}{\partial \gamma_{xz}}, \quad \tau_{xy} = \frac{\partial U_0}{\partial \gamma_{xy}}, \end{aligned} \tag{1.89}$$

or:

$$\sigma_{ij} = \frac{\partial U_0}{\partial \varepsilon_{ij}}. \tag{1.90}$$

In the presence of a linearly elastic material, the stiffnesses C_{ijkl} can be determined as partial derivatives of the stresses σ_{ij} with respect to the strains ε_{kl}:

$$C_{ijkl} = \frac{\partial \sigma_{ij}}{\partial \varepsilon_{kl}}. \tag{1.91}$$

Due to the fact that according to (1.90), the stresses σ_{ij} result from the first partial derivatives of the potential U_0, the stiffnesses C_{ijkl} can be determined as the second partial derivatives of U_0 with respect to the strains, i.e.:

$$C_{ijkl} = \frac{\partial^2 U_0}{\partial \varepsilon_{ij}\partial \varepsilon_{kl}}. \tag{1.92}$$

Here the order of the differentiations is arbitrary, so that:

$$C_{ijkl} = \frac{\partial \sigma_{ij}}{\partial \varepsilon_{kl}} = \frac{\partial^2 U_0}{\partial \varepsilon_{ij}\partial \varepsilon_{kl}} = \frac{\partial^2 U_0}{\partial \varepsilon_{kl}\partial \varepsilon_{ij}} = \frac{\partial \sigma_{kl}}{\partial \varepsilon_{ij}} = C_{klij}. \tag{1.93}$$

This proves the symmetry property $C_{ijkl} = C_{klij}$ of the elasticity tensor and as a consequence also the symmetry of the stiffness matrix $\underline{\underline{C}}$.

In the case of linear elasticity, the strain energy density U_0 is a quadratic form in the strains. In index notation one obtains:

$$U_0 = \int_0^{\varepsilon} \sigma_{ij} \mathrm{d}\hat{\varepsilon}_{ij} = C_{ijkl} \int_0^{\varepsilon} \varepsilon_{kl} \mathrm{d}\hat{\varepsilon}_{ij} = \frac{1}{2} C_{ijkl} \varepsilon_{ij} \varepsilon_{kl} = \frac{1}{2} \sigma_{ij} \varepsilon_{ij}, \tag{1.94}$$

or

$$U_0 = \frac{1}{2} \left(\sigma_{xx} \varepsilon_{xx} + \sigma_{yy} \varepsilon_{yy} + \sigma_{zz} \varepsilon_{zz} + \tau_{yz} \gamma_{yz} + \tau_{xz} \gamma_{xz} + \tau_{xy} \gamma_{xy} \right). \tag{1.95}$$

1.5 Boundary Value Problems

The equations describing a problem of three-dimensional elasticity theory can be summarized for the case of geometric and material linearity as follows. The local equilibrium conditions read:

$$\begin{aligned}
\frac{\partial \sigma_{xx}}{\partial x} + \frac{\partial \tau_{xy}}{\partial y} + \frac{\partial \tau_{xz}}{\partial z} + f_x &= 0, \\
\frac{\partial \tau_{xy}}{\partial x} + \frac{\partial \sigma_{yy}}{\partial y} + \frac{\partial \tau_{yz}}{\partial z} + f_y &= 0, \\
\frac{\partial \tau_{xz}}{\partial x} + \frac{\partial \tau_{yz}}{\partial y} + \frac{\partial \sigma_{zz}}{\partial z} + f_z &= 0,
\end{aligned} \tag{1.96}$$

or in index notation:

$$\frac{\partial \sigma_{ij}}{\partial x_j} + f_i = 0. \tag{1.97}$$

Furthermore, the kinematic equations apply as follows:

$$\begin{aligned}
\varepsilon_{xx} &= \frac{\partial u}{\partial x}, \quad \varepsilon_{yy} = \frac{\partial v}{\partial y}, \quad \varepsilon_{zz} = \frac{\partial w}{\partial z}, \\
\gamma_{xy} &= \frac{\partial u}{\partial y} + \frac{\partial v}{\partial x}, \quad \gamma_{xz} = \frac{\partial u}{\partial z} + \frac{\partial w}{\partial x}, \quad \gamma_{yz} = \frac{\partial v}{\partial z} + \frac{\partial w}{\partial y},
\end{aligned} \tag{1.98}$$

or in index notation:

$$\varepsilon_{ij} = \frac{1}{2} \left(\frac{\partial u_i}{\partial x_j} + \frac{\partial u_j}{\partial x_i} \right). \tag{1.99}$$

Linear elastic material behavior is described by the generalized Hooke's law:

$$\begin{pmatrix} \sigma_{xx} \\ \sigma_{yy} \\ \sigma_{zz} \\ \sigma_{yz} \\ \sigma_{xz} \\ \sigma_{xy} \end{pmatrix} = \begin{bmatrix} C_{11} & C_{12} & C_{13} & C_{14} & C_{15} & C_{16} \\ C_{21} & C_{22} & C_{23} & C_{24} & C_{25} & C_{26} \\ C_{31} & C_{32} & C_{33} & C_{34} & C_{35} & C_{36} \\ C_{41} & C_{42} & C_{43} & C_{44} & C_{45} & C_{46} \\ C_{51} & C_{52} & C_{53} & C_{54} & C_{55} & C_{56} \\ C_{61} & C_{62} & C_{63} & C_{64} & C_{65} & C_{66} \end{bmatrix} \begin{pmatrix} \varepsilon_{xx} \\ \varepsilon_{yy} \\ \varepsilon_{zz} \\ \gamma_{yz} \\ \gamma_{xz} \\ \gamma_{xy} \end{pmatrix}, \tag{1.100}$$

or:

$$\sigma_{ij} = C_{ijkl}\varepsilon_{kl}. \tag{1.101}$$

This set of equations is commonly called field equations, indicating that these equations describe the state and field quantities (displacements u_i, strains ε_{ij} and stresses σ_{ij}) in a solid. There are 15 equations for the 15 unknown field quantities u_i, ε_{ij} and σ_{ij}.

However, the field equations are not sufficient to uniquely describe a given solid problem of three-dimensional elasticity theory. It is necessary to make statements about boundary values for the field quantities and to take them into account in the analysis. Hence, the so-called boundary conditions of a given problem need to be taken into account. If ∂V is the total surface of the solid under consideration, then boundary conditions are usually formulated in such a way that on a part ∂V_t of the body surface ∂V loads are given in the form of the stress vector. Accordingly, $\underline{t} = \underline{t}_0$ is valid on ∂V_t, where $\underline{t}_0$ is the stress vector given on the edge part ∂V_t. On the remaining edge part ∂V_u the displacements are given, i.e. let $u = u_0$, $v = v_0$, $w = w_0$, where u_0, v_0, w_0 are the given displacements. The union of the two partial surfaces ∂V_t and ∂V_u leads to the total body surface ∂V, i.e. $\partial V_t \cup \partial V_u = \partial V$ holds.

A boundary value problem can be stated as follows. On ∂V_t we have:

$$\begin{aligned} \sigma_{xx}n_x + \tau_{xy}n_y + \tau_{xz}n_z &= t_{x0}, \\ \tau_{xy}n_x + \sigma_{yy}n_y + \tau_{yz}n_z &= t_{y0}, \\ \tau_{xz}n_x + \tau_{yz}n_y + \sigma_{zz}n_z &= t_{z0}, \end{aligned} \tag{1.102}$$

or in index notation:

$$\sigma_{ij}n_j = t_{i0}. \tag{1.103}$$

On ∂V_u the following holds:

$$u = u_0, \quad v = v_0, \quad w = w_0. \tag{1.104}$$

Depending on the requirements, the compatibility Eqs. (1.69) and (1.70) are used in addition to the kinematic relations.

If there is a materially and geometrically linear problem, then the superposition principle applies. This means that with two solutions $\sigma_{ij}^{(1)}$, $\varepsilon_{ij}^{(1)}$, $u_i^{(1)}$ and $\sigma_{ij}^{(2)}$, $\varepsilon_{ij}^{(2)}$, $u_i^{(2)}$ of a given boundary value problem also their linear combination $C_1\sigma_{ij}^{(1)} + C_2\sigma_{ij}^{(2)}$, $C_1\varepsilon_{ij}^{(1)} + C_2\varepsilon_{ij}^{(2)}$, $C_1u_i^{(1)} + C_2u_i^{(2)}$ is a solution of the considered problem.

1.6 Material Symmetries

For linear-elastic anisotropic materials with a fully occupied stiffness matrix $\underline{\underline{C}}$ (Eq. (1.76)) pronounced coupling effects occur such that any strain component evokes any stress component. However, many technically relevant materials show a much less complex behavior, and so-called material symmetries are present, meaning that in certain cases certain entries in $\underline{\underline{C}}$ and also in $\underline{\underline{S}}$ can be omitted, even if a direction-dependent behavior remains.

In the following, we consider a material that has three distinct principal directions (the so-called material principal axes), for example a fiber-reinforced plastic as shown in Fig. 1.11.

It is obvious that such a material has pronounced anisotropy properties. Unless otherwise stated, the material principal axes are always denoted by x_1, x_2, x_3. If there is a case with a pronounced principal direction, such as in the case of a unidirectionally fiber-reinforced plastic, then this principal direction is defined as the x_1−axis. Let the displacements belonging to the x_1, x_2, x_3 axis system be u_1, u_2, u_3, and the strain quantities are denoted as $\varepsilon_{11}, \varepsilon_{22}, \varepsilon_{33}$ and $\gamma_{12}, \gamma_{13}, \gamma_{23}$.

1.6.1 Full Anisotropy

In the case where both the stiffness matrix $\underline{\underline{C}}$ and the compliance matrix $\underline{\underline{S}}$ are fully occupied, we speak of full anisotropy. All stiffnesses C_{ij} and compliance S_{ij} are nonzero in this case. The generalized Hooke's law is thus:

$$\begin{pmatrix} \sigma_{11} \\ \sigma_{22} \\ \sigma_{33} \\ \tau_{23} \\ \tau_{13} \\ \tau_{12} \end{pmatrix} = \begin{bmatrix} C_{11} & C_{12} & C_{13} & C_{14} & C_{15} & C_{16} \\ C_{12} & C_{22} & C_{23} & C_{24} & C_{25} & C_{26} \\ C_{13} & C_{23} & C_{33} & C_{34} & C_{35} & C_{36} \\ C_{14} & C_{24} & C_{34} & C_{44} & C_{45} & C_{46} \\ C_{15} & C_{25} & C_{35} & C_{45} & C_{55} & C_{56} \\ C_{16} & C_{26} & C_{36} & C_{46} & C_{56} & C_{66} \end{bmatrix} \begin{pmatrix} \varepsilon_{11} \\ \varepsilon_{22} \\ \varepsilon_{33} \\ \gamma_{23} \\ \gamma_{13} \\ \gamma_{12} \end{pmatrix}, \tag{1.105}$$

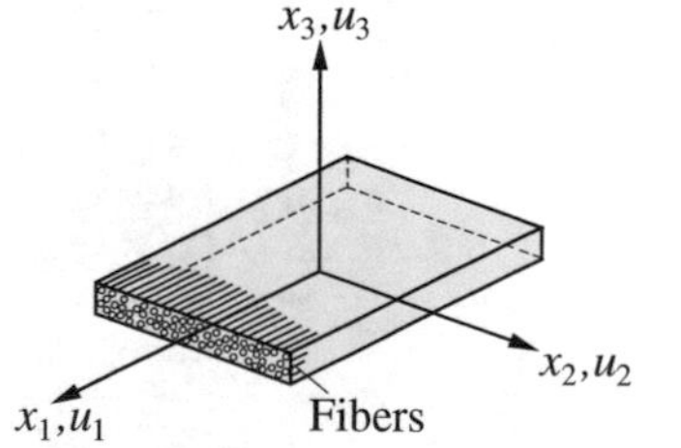

Fig. 1.11 Fiber-reinforced plastic

and

$$\begin{pmatrix} \varepsilon_{11} \\ \varepsilon_{22} \\ \varepsilon_{33} \\ \gamma_{23} \\ \gamma_{13} \\ \gamma_{12} \end{pmatrix} = \begin{bmatrix} S_{11} & S_{12} & S_{13} & S_{14} & S_{15} & S_{16} \\ S_{12} & S_{22} & S_{23} & S_{24} & S_{25} & S_{26} \\ S_{13} & S_{23} & S_{33} & S_{34} & S_{35} & S_{36} \\ S_{14} & S_{24} & S_{34} & S_{44} & S_{45} & S_{46} \\ S_{15} & S_{25} & S_{35} & S_{45} & S_{55} & S_{56} \\ S_{16} & S_{26} & S_{36} & S_{46} & S_{56} & S_{66} \end{bmatrix} \begin{pmatrix} \sigma_{11} \\ \sigma_{22} \\ \sigma_{33} \\ \tau_{23} \\ \tau_{13} \\ \tau_{12} \end{pmatrix}. \tag{1.106}$$

For the shear stresses and shear strains, the designations τ_{ij} and γ_{ij} were used, respectively. Thus, in this case, there are 21 independent material constants. Full anisotropy is thus associated with a very complex relationship between stresses and strains, and pronounced coupling effects are present as shown in Fig. 1.12. The figure shows a cube cut from a fully anisotropic material and stressed by the normal stress σ_{11}. As a consequence of σ_{11} all strain components ε_{11}, ε_{22}, ε_{33}, γ_{23}, γ_{13} and γ_{12} are induced. However, many technically relevant materials exhibit much less complex behavior, which can be attributed to certain symmetry properties as will be discussed in the following. Symmetry properties are associated with rotations and mirrorings of the reference frame, which do not cause any changes in the material behavior, so that the disappearance of certain stiffnesses C_{ij} and compliance S_{ij} can be concluded.

1.6.2 Monotropic Material

Monoclinic or monotropic material has one symmetry plane. For example, let this be the x_1x_2−plane, so that the material properties are mirror symmetric with respect to the plane $x_3 = 0$ (Fig. 1.13). Let the axes x_1, x_2 and x_3 be here not the principal axes of the material, but rather global reference axes which need not coincide with the material principal axes. For example, if the material is a fiber-reinforced plastic

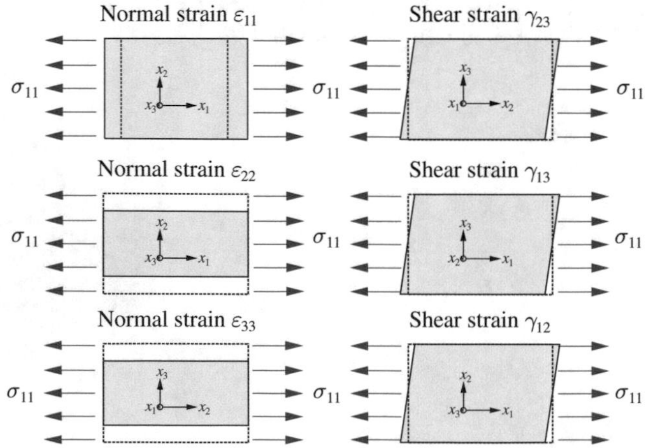

Fig. 1.12 Coupling effects in a fully anisotropic material

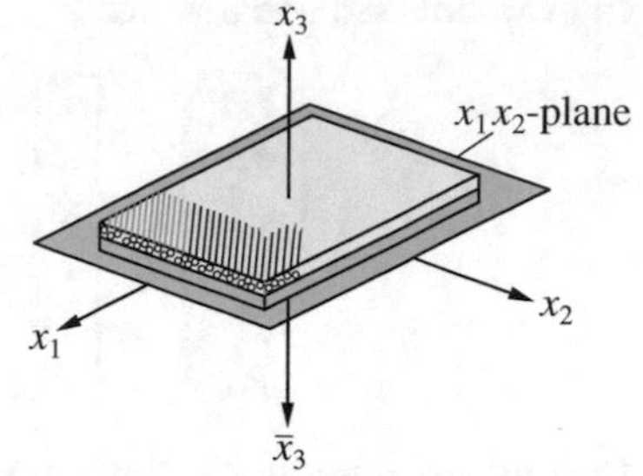

Fig. 1.13 Monotropic material

as shown in Fig. 1.13, then the fibers enclose an arbitrary angle with the x_1-axis at any orientation. The presence of a plane of symmetry means that the material properties must be invariant to a mirroring of the x_3-axis on the x_1x_2-plane, i.e., a mirroring of the form $x_1 = \bar{x}_1, x_2 = \bar{x}_2, x_3 = -\bar{x}_3$. However, this mirroring reverses the signs of the shear strains γ_{23} and γ_{13} and the shear stresses τ_{23} and τ_{13}, thus $\bar{\gamma}_{23} = -\gamma_{23}$, $\bar{\gamma}_{13} = -\gamma_{13}$ and $\bar{\tau}_{23} = -\tau_{23}$, $\bar{\tau}_{13} = -\tau_{13}$. The remaining stress and strain components remain unchanged. One obtains from the fourth and fifth lines of the generalized Hooke's law (1.76):

$$\begin{aligned}\tau_{23} &= C_{14}\varepsilon_{11} + C_{24}\varepsilon_{22} + C_{34}\varepsilon_{33} + C_{44}\gamma_{23} + C_{45}\gamma_{13} + C_{46}\gamma_{12},\\ \tau_{13} &= C_{15}\varepsilon_{11} + C_{25}\varepsilon_{22} + C_{35}\varepsilon_{33} + C_{45}\gamma_{23} + C_{55}\gamma_{13} + C_{56}\gamma_{12}.\end{aligned} \tag{1.107}$$

On the other hand, considering the mirrored reference frame $\bar{x}_1, \bar{x}_2, \bar{x}_3$, we obtain:

$$\begin{aligned}\bar{\tau}_{23} &= -C_{14}\varepsilon_{11} - C_{24}\varepsilon_{22} - C_{34}\varepsilon_{33} + C_{44}\gamma_{23} + C_{45}\gamma_{13} - C_{46}\gamma_{12},\\ \bar{\tau}_{13} &= -C_{15}\varepsilon_{11} - C_{25}\varepsilon_{22} - C_{35}\varepsilon_{33} + C_{45}\gamma_{23} + C_{55}\gamma_{13} - C_{56}\gamma_{12}.\end{aligned} \tag{1.108}$$

However, a constitutive relation must be invariant to a change of the frame of reference. The obvious contradiction between (1.107) and (1.108) can therefore be eliminated only if the following stiffnesses vanish:

$$C_{14} = C_{24} = C_{34} = C_{46} = C_{15} = C_{25} = C_{35} = C_{56} = 0. \tag{1.109}$$

The generalized Hooke's law for a monoclinic material can thus be written as follows:

$$\begin{pmatrix}\sigma_{11}\\ \sigma_{22}\\ \sigma_{33}\\ \tau_{23}\\ \tau_{13}\\ \tau_{12}\end{pmatrix} = \begin{bmatrix} C_{11} & C_{12} & C_{13} & 0 & 0 & C_{16}\\ C_{12} & C_{22} & C_{23} & 0 & 0 & C_{26}\\ C_{13} & C_{23} & C_{33} & 0 & 0 & C_{36}\\ 0 & 0 & 0 & C_{44} & C_{45} & 0\\ 0 & 0 & 0 & C_{45} & C_{55} & 0\\ C_{16} & C_{26} & C_{36} & 0 & 0 & C_{66}\end{bmatrix} \begin{pmatrix}\varepsilon_{11}\\ \varepsilon_{22}\\ \varepsilon_{33}\\ \gamma_{23}\\ \gamma_{13}\\ \gamma_{12}\end{pmatrix}. \tag{1.110}$$

In inverted form one obtains:

$$\begin{pmatrix} \varepsilon_{11} \\ \varepsilon_{22} \\ \varepsilon_{33} \\ \gamma_{23} \\ \gamma_{13} \\ \gamma_{12} \end{pmatrix} = \begin{bmatrix} S_{11} & S_{12} & S_{13} & 0 & 0 & S_{16} \\ S_{12} & S_{22} & S_{23} & 0 & 0 & S_{26} \\ S_{13} & S_{23} & S_{33} & 0 & 0 & S_{36} \\ 0 & 0 & 0 & S_{44} & S_{45} & 0 \\ 0 & 0 & 0 & S_{45} & S_{55} & 0 \\ S_{16} & S_{26} & S_{36} & 0 & 0 & S_{66} \end{bmatrix} \begin{pmatrix} \sigma_{11} \\ \sigma_{22} \\ \sigma_{33} \\ \tau_{23} \\ \tau_{13} \\ \tau_{12} \end{pmatrix}. \tag{1.111}$$

Examples of monotropic material behavior are, as already indicated, fibrous materials like wood or unidirectional fiber-reinforced plastics, if the fiber direction does not coincide with the x_1−axis. Monotropic material is described by 13 material constants. Obviously, certain coupling effects remain in this case. Besides the coupling of the normal stresses σ_{11}, σ_{22} and σ_{33} with the three strains ε_{11}, ε_{22} and ε_{33}, the coupling with the shear strain γ_{12} (so called shear coupling) also remains. Furthermore, there is a coupling between the two shear stresses τ_{23} and τ_{13} with both shear strains γ_{23} and γ_{13}.

1.6.3 Orthogonal Anisotropy/Orthotropy

A technically particularly relevant case of anisotropy is the so-called orthogonal anisotropy (in short: orthotropy), see Fig. 1.14. In the following, let the axes x_1, x_2, x_3 be the material principal axes. An orthotropic material is characterized by mirror symmetry with respect to the planes $x_1 = 0$, $x_2 = 0$, $x_3 = 0$. It can be easily shown that for orthotropic material the following stiffnesses must vanish:

$$\begin{aligned} C_{14} &= C_{24} = C_{34} = C_{46} = C_{15} = C_{25} = C_{35} = C_{56} \\ &= C_{16} = C_{26} = C_{36} = C_{45} = 0. \end{aligned} \tag{1.112}$$

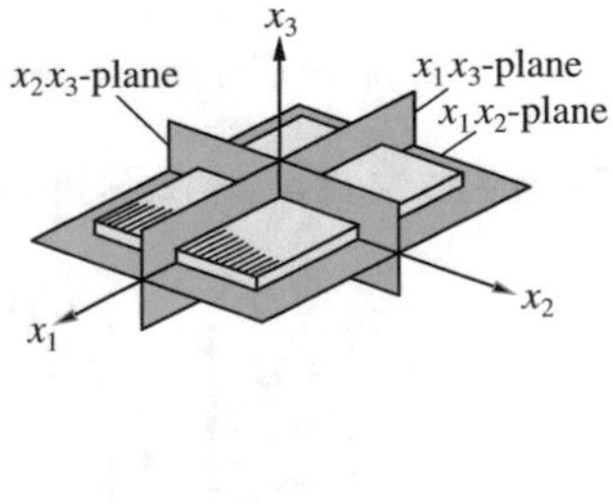

Fig. 1.14 Orthotropic material

The generalized Hooke's law can then be stated as:

$$\begin{pmatrix} \sigma_{11} \\ \sigma_{22} \\ \sigma_{33} \\ \tau_{23} \\ \tau_{13} \\ \tau_{12} \end{pmatrix} = \begin{bmatrix} C_{11} & C_{12} & C_{13} & 0 & 0 & 0 \\ C_{12} & C_{22} & C_{23} & 0 & 0 & 0 \\ C_{13} & C_{23} & C_{33} & 0 & 0 & 0 \\ 0 & 0 & 0 & C_{44} & 0 & 0 \\ 0 & 0 & 0 & 0 & C_{55} & 0 \\ 0 & 0 & 0 & 0 & 0 & C_{66} \end{bmatrix} \begin{pmatrix} \varepsilon_{11} \\ \varepsilon_{22} \\ \varepsilon_{33} \\ \gamma_{23} \\ \gamma_{13} \\ \gamma_{12} \end{pmatrix}. \tag{1.113}$$

In inverted form:

$$\begin{pmatrix} \varepsilon_{11} \\ \varepsilon_{22} \\ \varepsilon_{33} \\ \gamma_{23} \\ \gamma_{13} \\ \gamma_{12} \end{pmatrix} = \begin{bmatrix} S_{11} & S_{12} & S_{13} & 0 & 0 & 0 \\ S_{12} & S_{22} & S_{23} & 0 & 0 & 0 \\ S_{13} & S_{23} & S_{33} & 0 & 0 & 0 \\ 0 & 0 & 0 & S_{44} & 0 & 0 \\ 0 & 0 & 0 & 0 & S_{55} & 0 \\ 0 & 0 & 0 & 0 & 0 & S_{66} \end{bmatrix} \begin{pmatrix} \sigma_{11} \\ \sigma_{22} \\ \sigma_{33} \\ \tau_{23} \\ \tau_{13} \\ \tau_{12} \end{pmatrix}. \tag{1.114}$$

Orthotropic material behavior is thus described by nine independent material parameters. Examples of orthotropic materials are wood or steel-reinforced concrete, but also fiber-reinforced plastics with unidirectional reinforcement, provided that the fiber direction is oriented in the direction of one of the reference axes. Apparently, no coupling effects occur in the case of orthotropy. It should be noted, however, that the remaining nine material constants are different from each other and thus a possibly significant anisotropic behavior is generally present.

In the case of orthotropy, the following relationships exist between the stiffnesses C_{ij} and the compliances S_{ij}:

$$\begin{aligned}
C_{11} &= \frac{S_{23}^2 - S_{22}S_{33}}{S_{11}S_{23}^2 - S_{11}S_{22}S_{33} - 2S_{12}S_{13}S_{23} + S_{22}S_{13}^2 + S_{33}S_{12}^2}, \\
C_{22} &= \frac{S_{13}^2 - S_{11}S_{33}}{S_{11}S_{23}^2 - S_{11}S_{22}S_{33} - 2S_{12}S_{13}S_{23} + S_{22}S_{13}^2 + S_{33}S_{12}^2}, \\
C_{33} &= \frac{S_{12}^2 - S_{11}S_{22}}{S_{11}S_{23}^2 - S_{11}S_{22}S_{33} - 2S_{12}S_{13}S_{23} + S_{22}S_{13}^2 + S_{33}S_{12}^2}, \\
C_{12} &= \frac{S_{12}S_{33} - S_{13}S_{23}}{S_{11}S_{23}^2 - S_{11}S_{22}S_{33} - 2S_{12}S_{13}S_{23} + S_{22}S_{13}^2 + S_{33}S_{12}^2}, \\
C_{13} &= \frac{S_{13}S_{22} - S_{12}S_{23}}{S_{11}S_{23}^2 - S_{11}S_{22}S_{33} - 2S_{12}S_{13}S_{23} + S_{22}S_{13}^2 + S_{33}S_{12}^2}, \\
C_{23} &= \frac{S_{11}S_{23} - S_{12}S_{13}}{S_{11}S_{23}^2 - S_{11}S_{22}S_{33} - 2S_{12}S_{13}S_{23} + S_{22}S_{13}^2 + S_{33}S_{12}^2}, \\
C_{44} &= \frac{1}{S_{44}}, \quad C_{55} = \frac{1}{S_{55}}, \quad C_{66} = \frac{1}{S_{66}}.
\end{aligned} \tag{1.115}$$

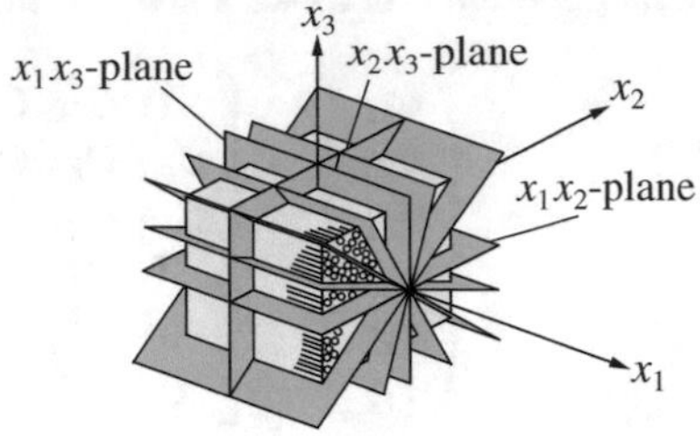

Fig. 1.15 Transversely isotropic material

1.6.4 Transversal Isotropy

Transversal isotropy occurs when a material has a principal direction (here the x_1-axis) and exhibits isotropy with respect to a distinguished plane (here the $x_2 - x_3-$plane). This is illustrated in Fig. 1.15. Thus, for a transversely isotropic material, the material behavior is invariant with respect to a rotation of the reference frame about the principal axis x_1, it behaves identically in any direction perpendicular to the x_1-axis. Accordingly, any plane containing the x_1-axis is a symmetry plane. Transversely isotropic material behavior can be observed, for example, in many unidirectionally reinforced plastics if there is a uniform distribution of fibers. The stiffness matrix $\underline{\underline{C}}$ and the compliance matrix $\underline{\underline{S}}$ are occupied identically to the orthotropic case, but the following identities apply to the stiffnesses C_{ij}:

$$C_{13} = C_{12}, \quad C_{33} = C_{22}, \quad C_{66} = C_{55}, \quad C_{44} = \frac{1}{2}(C_{22} - C_{23}). \tag{1.116}$$

The generalized Hooke's law thus reads:

$$\begin{pmatrix} \sigma_{11} \\ \sigma_{22} \\ \sigma_{33} \\ \tau_{23} \\ \tau_{13} \\ \tau_{12} \end{pmatrix} = \begin{bmatrix} C_{11} & C_{12} & C_{12} & 0 & 0 & 0 \\ C_{12} & C_{22} & C_{23} & 0 & 0 & 0 \\ C_{12} & C_{23} & C_{22} & 0 & 0 & 0 \\ 0 & 0 & 0 & \frac{1}{2}(C_{22} - C_{23}) & 0 & 0 \\ 0 & 0 & 0 & 0 & C_{55} & 0 \\ 0 & 0 & 0 & 0 & 0 & C_{55} \end{bmatrix} \begin{pmatrix} \varepsilon_{11} \\ \varepsilon_{22} \\ \varepsilon_{33} \\ \gamma_{23} \\ \gamma_{13} \\ \gamma_{12} \end{pmatrix}, \tag{1.117}$$

or in inverted form:

$$\begin{pmatrix} \varepsilon_{11} \\ \varepsilon_{22} \\ \varepsilon_{33} \\ \gamma_{23} \\ \gamma_{13} \\ \gamma_{12} \end{pmatrix} = \begin{bmatrix} S_{11} & S_{12} & S_{12} & 0 & 0 & 0 \\ S_{12} & S_{22} & S_{23} & 0 & 0 & 0 \\ S_{12} & S_{23} & S_{22} & 0 & 0 & 0 \\ 0 & 0 & 0 & 2(S_{22} - S_{23}) & 0 & 0 \\ 0 & 0 & 0 & 0 & S_{55} & 0 \\ 0 & 0 & 0 & 0 & 0 & S_{55} \end{bmatrix} \begin{pmatrix} \sigma_{11} \\ \sigma_{22} \\ \sigma_{33} \\ \tau_{23} \\ \tau_{13} \\ \tau_{12} \end{pmatrix}. \tag{1.118}$$

Only five independent material constants remain for transversal isotropy.

1.6.5 Isotropy

The last case of material symmetry is the so-called isotropy. In this case the material behaves identically in each direction, and each axis is a principal axis. Similarly, any plane is a symmetry plane. The generalized Hooke's law is then:

$$\begin{pmatrix}\sigma_{11}\\ \sigma_{22}\\ \sigma_{33}\\ \tau_{23}\\ \tau_{13}\\ \tau_{12}\end{pmatrix}=\begin{bmatrix}C_{11} & C_{12} & C_{12} & 0 & 0 & 0\\ C_{12} & C_{11} & C_{12} & 0 & 0 & 0\\ C_{12} & C_{12} & C_{11} & 0 & 0 & 0\\ 0 & 0 & 0 & \frac{1}{2}(C_{11}-C_{12}) & 0 & 0\\ 0 & 0 & 0 & 0 & \frac{1}{2}(C_{11}-C_{12}) & 0\\ 0 & 0 & 0 & 0 & 0 & \frac{1}{2}(C_{11}-C_{12})\end{bmatrix}\begin{pmatrix}\varepsilon_{11}\\ \varepsilon_{22}\\ \varepsilon_{33}\\ \gamma_{23}\\ \gamma_{13}\\ \gamma_{12}\end{pmatrix}. \tag{1.119}$$

In inverted form:

$$\begin{pmatrix}\varepsilon_{11}\\ \varepsilon_{22}\\ \varepsilon_{33}\\ \gamma_{23}\\ \gamma_{13}\\ \gamma_{12}\end{pmatrix}=\begin{bmatrix}S_{11} & S_{12} & S_{12} & 0 & 0 & 0\\ S_{12} & S_{11} & S_{12} & 0 & 0 & 0\\ S_{12} & S_{12} & S_{11} & 0 & 0 & 0\\ 0 & 0 & 0 & 2(S_{11}-S_{12}) & 0 & 0\\ 0 & 0 & 0 & 0 & 2(S_{11}-S_{12}) & 0\\ 0 & 0 & 0 & 0 & 0 & 2(S_{11}-S_{12})\end{bmatrix}\begin{pmatrix}\sigma_{11}\\ \sigma_{22}\\ \sigma_{33}\\ \tau_{23}\\ \tau_{13}\\ \tau_{12}\end{pmatrix}. \tag{1.120}$$

Thus, only two independent material constants are necessary to describe isotropic material behavior. Many metals and plastics can be assumed to be isotropic.

1.6.6 Representation in Engineering Constants

Besides the description of the linear-elastic material behavior by stiffnesses C_{ij} and compliances S_{ij}, the use of the so-called engineering constants is also very common. For three-dimensional orthotropy there are 12 engineering constants:

- Three generalized elastic moduli E_{11}, E_{22}, E_{33},
- six Poisson's ratios ν_{12}, ν_{13}, ν_{23}, ν_{21}, ν_{31}, ν_{32},
- three generalized shear moduli G_{23}, G_{13}, G_{12}.

The inverted form of the generalized Hooke's law (1.114) can be written as follows when the engineering constants are used:

$$\begin{pmatrix} \varepsilon_{11} \\ \varepsilon_{22} \\ \varepsilon_{33} \\ \gamma_{23} \\ \gamma_{13} \\ \gamma_{12} \end{pmatrix} = \begin{bmatrix} \frac{1}{E_{11}} & -\frac{\nu_{21}}{E_{22}} & -\frac{\nu_{31}}{E_{33}} & 0 & 0 & 0 \\ -\frac{\nu_{12}}{E_{11}} & \frac{1}{E_{22}} & -\frac{\nu_{32}}{E_{33}} & 0 & 0 & 0 \\ -\frac{\nu_{13}}{E_{11}} & -\frac{\nu_{23}}{E_{22}} & \frac{1}{E_{33}} & 0 & 0 & 0 \\ 0 & 0 & 0 & \frac{1}{G_{23}} & 0 & 0 \\ 0 & 0 & 0 & 0 & \frac{1}{G_{13}} & 0 \\ 0 & 0 & 0 & 0 & 0 & \frac{1}{G_{12}} \end{bmatrix} \begin{pmatrix} \sigma_{11} \\ \sigma_{22} \\ \sigma_{33} \\ \tau_{23} \\ \tau_{13} \\ \tau_{12} \end{pmatrix}. \tag{1.121}$$

By comparing (1.114) with (1.121), the following relationships between the engineering constants and compliances S_{ij} are obtained:

$$\begin{aligned} S_{11} &= \frac{1}{E_{11}}, \quad S_{22} = \frac{1}{E_{22}}, \quad S_{33} = \frac{1}{E_{33}}, \\ S_{44} &= \frac{1}{G_{23}}, \quad S_{55} = \frac{1}{G_{13}}, \quad S_{66} = \frac{1}{G_{12}}, \\ S_{12} &= -\frac{\nu_{12}}{E_{11}} = -\frac{\nu_{21}}{E_{22}}, \quad S_{23} = -\frac{\nu_{23}}{E_{22}} = -\frac{\nu_{32}}{E_{33}}, \\ S_{13} &= -\frac{\nu_{31}}{E_{33}} = -\frac{\nu_{13}}{E_{11}}. \end{aligned} \tag{1.122}$$

Due to the symmetry of the compliance matrix $\underline{\underline{S}}$, the following identities can be established:

$$\frac{\nu_{12}}{E_{11}} = \frac{\nu_{21}}{E_{22}}, \quad \frac{\nu_{23}}{E_{22}} = \frac{\nu_{32}}{E_{33}}, \quad \frac{\nu_{31}}{E_{33}} = \frac{\nu_{13}}{E_{11}}. \tag{1.123}$$

Thus, of the twelve engineering constants, only nine are independent of each other.

The stiffnesses C_{ij} can be expressed using the engineering constants as follows:

$$\begin{aligned} C_{11} &= \frac{(1-\nu_{23}\nu_{32})E_{11}}{1-\nu_{12}\nu_{21}-\nu_{23}\nu_{32}-\nu_{31}\nu_{13}-2\nu_{21}\nu_{13}\nu_{32}}, \\ C_{22} &= \frac{(1-\nu_{31}\nu_{13})E_{11}}{1-\nu_{12}\nu_{21}-\nu_{23}\nu_{32}-\nu_{31}\nu_{13}-2\nu_{21}\nu_{13}\nu_{32}}, \\ C_{33} &= \frac{(1-\nu_{21}\nu_{12})E_{33}}{1-\nu_{12}\nu_{21}-\nu_{23}\nu_{32}-\nu_{31}\nu_{13}-2\nu_{21}\nu_{13}\nu_{32}}, \\ C_{44} &= G_{23}, \quad C_{55} = G_{13}, \quad C_{66} = G_{12}, \\ C_{12} &= \frac{(\nu_{12}+\nu_{32}\nu_{13})E_{22}}{1-\nu_{12}\nu_{21}-\nu_{23}\nu_{32}-\nu_{31}\nu_{13}-2\nu_{21}\nu_{13}\nu_{32}} \\ &= \frac{(\nu_{21}+\nu_{31}\nu_{23})E_{11}}{1-\nu_{12}\nu_{21}-\nu_{23}\nu_{32}-\nu_{31}\nu_{13}-2\nu_{21}\nu_{13}\nu_{32}}, \\ C_{13} &= \frac{(\nu_{13}+\nu_{12}\nu_{23})E_{33}}{1-\nu_{12}\nu_{21}-\nu_{23}\nu_{32}-\nu_{31}\nu_{13}-2\nu_{21}\nu_{13}\nu_{32}} \end{aligned}$$

$$
\begin{aligned}
&= \frac{(\nu_{31} + \nu_{21}\nu_{32})E_{11}}{1 - \nu_{12}\nu_{21} - \nu_{23}\nu_{32} - \nu_{31}\nu_{13} - 2\nu_{21}\nu_{13}\nu_{32}}, \\
C_{23} &= \frac{(\nu_{23} + \nu_{21}\nu_{13})E_{33}}{1 - \nu_{12}\nu_{21} - \nu_{23}\nu_{32} - \nu_{31}\nu_{13} - 2\nu_{21}\nu_{13}\nu_{32}} \\
&= \frac{(\nu_{32} + \nu_{12}\nu_{31})E_{11}}{1 - \nu_{12}\nu_{21} - \nu_{23}\nu_{32} - \nu_{31}\nu_{13} - 2\nu_{21}\nu_{13}\nu_{32}},
\end{aligned}
\tag{1.124}
$$

as can be concluded by inverting (1.121).

If the considered material is isotropic, only three engineering constants remain, namely the modulus of elasticity E, the shear modulus G and Poisson's ratio ν. However, of these three engineering constants, only two are independent. The elastic compliances S_{ij} can be given as:

$$
\begin{aligned}
S_{11} = S_{22} = S_{33} &= \frac{1}{E}, \\
S_{12} = S_{13} = S_{23} &= -\frac{\nu}{E}, \\
S_{44} = S_{55} = S_{66} &= \frac{1}{G}.
\end{aligned}
\tag{1.125}
$$

For the elastic stiffnesses C_{ij} one obtains:

$$
\begin{aligned}
C_{11} = C_{22} = C_{33} &= \frac{(1-\nu)\,E}{(1+\nu)\,(1-2\nu)}, \\
C_{12} = C_{13} = C_{23} &= \frac{\nu E}{(1+\nu)\,(1-2\nu)}, \\
C_{44} = C_{55} = C_{66} &= G.
\end{aligned}
\tag{1.126}
$$

For the engineering constants, the following relationship can be deduced:

$$
G = \frac{E}{2(1+\nu)}. \tag{1.127}
$$

Thus, also when using the engineering constants, there are only two independent material constants. The material law (1.121) can then be stated as:

$$
\begin{pmatrix} \varepsilon_{11} \\ \varepsilon_{22} \\ \varepsilon_{33} \\ \gamma_{23} \\ \gamma_{13} \\ \gamma_{12} \end{pmatrix}
=
\begin{bmatrix}
\frac{1}{E} & -\frac{\nu}{E} & -\frac{\nu}{E} & 0 & 0 & 0 \\
-\frac{\nu}{E} & \frac{1}{E} & -\frac{\nu}{E} & 0 & 0 & 0 \\
-\frac{\nu}{E} & -\frac{\nu}{E} & \frac{1}{E} & 0 & 0 & 0 \\
0 & 0 & 0 & \frac{1}{G} & 0 & 0 \\
0 & 0 & 0 & 0 & \frac{1}{G} & 0 \\
0 & 0 & 0 & 0 & 0 & \frac{1}{G}
\end{bmatrix}
\begin{pmatrix} \sigma_{11} \\ \sigma_{22} \\ \sigma_{33} \\ \tau_{23} \\ \tau_{13} \\ \tau_{12} \end{pmatrix}. \tag{1.128}
$$

1.7 Transformation Rules

In this section we want to investigate what changes arise for the material constants S_{ij} and C_{ij} when the considerations are referred to a reference frame other than the material principal axes x_1, x_2, x_3 (the so-called on-axis system). For this purpose the coordinate system x_1, x_2, x_3 is transformed by a pure rotation into an orthogonal coordinate system $\bar{x}_1, \bar{x}_2, \bar{x}_3$ (so called off-axis system), where the origin of both systems is identical. The rotation takes place about the fixed x_3−axis: $x_3 = \bar{x}_3$. Moreover, let us assume that the right angles between the reference axes are preserved also in the transformed state, so there is a pure rotation about the angle θ (see Fig. 1.16). The question thus arises to what extent the elastic properties of an anisotropic material change when such a coordinate transformation is performed, and the compliances and stiffnesses S_{ij} and C_{ij} with respect to the off-axis reference system $\bar{x}_1, \bar{x}_2, \bar{x}_3$ are to be determined. For the transformation of stresses and strains from the reference frame x_1, x_2, x_3 into the coordinate system $\bar{x}_1, \bar{x}_2, \bar{x}_3$ (see Eqs. (1.22) and (1.67)) we obtain:

$$\underline{\bar{\sigma}} = \underline{\underline{T}}_{\sigma}\underline{\sigma}, \quad \underline{\bar{\varepsilon}} = \underline{\underline{T}}_{\varepsilon}\underline{\varepsilon}, \tag{1.129}$$

with

$$\underline{\underline{T}}_{\sigma}^{-1} = \underline{\underline{T}}_{\varepsilon}^{T}, \quad \underline{\underline{T}}_{\varepsilon}^{-1} = \underline{\underline{T}}_{\sigma}^{T}. \tag{1.130}$$

Applying the generalized Hooke's law $\underline{\sigma} = \underline{\underline{C}}\underline{\varepsilon}$ in the first equation in (1.129) gives:

$$\underline{\bar{\sigma}} = \underline{\underline{T}}_{\sigma}\underline{\underline{C}}\underline{\varepsilon}. \tag{1.131}$$

Inserting the relation $\underline{\varepsilon} = \underline{\underline{T}}_{\varepsilon}^{-1}\underline{\bar{\varepsilon}}$, obtained from the inversion of the second equation in (1.129), yields:

$$\underline{\bar{\sigma}} = \underline{\underline{T}}_{\sigma}\underline{\underline{C}}\underline{\underline{T}}_{\varepsilon}^{-1}\underline{\bar{\varepsilon}}, \tag{1.132}$$

or with (1.130):

$$\underline{\bar{\sigma}} = \underline{\underline{T}}_{\sigma}\underline{\underline{C}}\underline{\underline{T}}_{\sigma}^{T}\underline{\bar{\varepsilon}}. \tag{1.133}$$

Thus, with (1.133) there is a relation between the stresses and strains in the off-axis system. In an analogous way one can proceed starting from the second expression

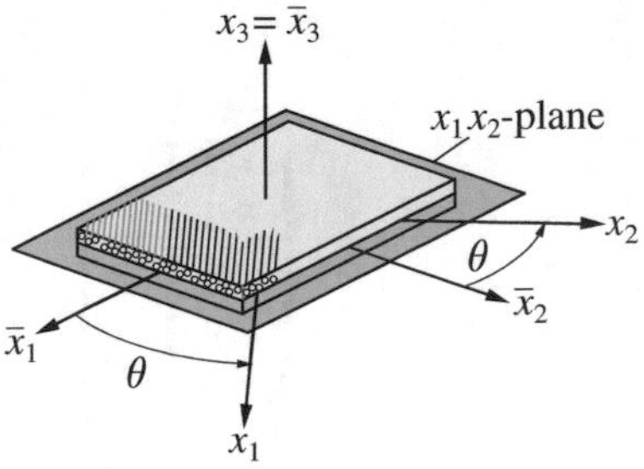

Fig. 1.16 Rotation of the reference system

in (1.129) and then come to the expression inverse to (1.133):

$$\bar{\varepsilon} = \underline{\underline{T}}_{\varepsilon}\underline{\underline{S}}\,\underline{\underline{T}}_{\varepsilon}^{T}\bar{\sigma}. \tag{1.134}$$

The expressions appearing in (1.133) and (1.134) $\underline{\underline{T}}_{\sigma}\underline{\underline{C}}\,\underline{\underline{T}}_{\sigma}^{T}$ and $\underline{\underline{T}}_{\varepsilon}\underline{\underline{S}}\,\underline{\underline{T}}_{\varepsilon}^{T}$ are the transformed stiffness matrix $\underline{\underline{\bar{C}}}$ and the transformed compliance matrix $\underline{\underline{\bar{S}}}$:

$$\begin{aligned}
\underline{\underline{\bar{S}}} &= \underline{\underline{T}}_{\varepsilon}\underline{\underline{S}}\,\underline{\underline{T}}_{\varepsilon}^{T},\\
\underline{\underline{\bar{C}}} &= \underline{\underline{T}}_{\sigma}\underline{\underline{C}}\,\underline{\underline{T}}_{\sigma}^{T}.
\end{aligned} \tag{1.135}$$

With the transformation matrices $\underline{\underline{T}}_{\sigma}$ and $\underline{\underline{T}}_{\varepsilon}$ according to (1.22) and (1.67), the following transformation rules follow for the stiffnesses C_{ij} and the compliances S_{ij}, where orthotropic material is assumed. For the compliances S_{ij} one obtains the following transformation rules:

$$\begin{aligned}
\bar{S}_{11} &= S_{11}\cos^4\theta + S_{22}\sin^4\theta + 2S_{12}\cos^2\theta\sin^2\theta + S_{66}\cos^2\theta\sin^2\theta,\\
\bar{S}_{22} &= S_{11}\sin^4\theta + 2S_{12}\cos^2\theta\sin^2\theta + S_{22}\cos^4\theta + S_{66}\cos^2\theta\sin^2\theta,\\
\bar{S}_{12} &= (S_{11} + S_{22})\cos^2\theta\sin^2\theta + S_{12}\left(\cos^4\theta + \sin^4\theta\right) - S_{66}\cos^2\theta\sin^2\theta,\\
\bar{S}_{66} &= 4\,(S_{11} + S_{22})\cos^2\theta\sin^2\theta - 8S_{12}\cos^2\theta\sin^2\theta + S_{66}\left(\cos^2\theta - \sin^2\theta\right)^2,\\
\bar{S}_{16} &= 2S_{11}\cos^3\theta\sin\theta + 2S_{12}\left(\cos\theta\sin^3\theta - \cos^3\theta\sin\theta\right)\\
&\quad - 2S_{22}\cos\theta\sin^3\theta + S_{66}\left(\cos\theta\sin^3\theta - \cos^3\theta\sin\theta\right),\\
\bar{S}_{26} &= 2S_{11}\cos\theta\sin^3\theta + 2S_{12}\left(\cos^3\theta\sin\theta - \cos\theta\sin^3\theta\right)\\
&\quad - 2S_{22}\cos^3\theta\sin\theta + S_{66}\left(\cos^3\theta\sin\theta - \cos\theta\sin^3\theta\right),\\
\bar{S}_{13} &= S_{13}\cos^2\theta + S_{23}\sin^2\theta,\\
\bar{S}_{23} &= S_{13}\sin^2\theta + S_{23}\cos^2\theta,\\
\bar{S}_{33} &= S_{33},\\
\bar{S}_{36} &= 2S_{13}\cos\theta\sin\theta - 2S_{23}\cos\theta\sin\theta,\\
\bar{S}_{44} &= S_{44}\cos^2\theta + S_{55}\sin^2\theta,\\
\bar{S}_{45} &= S_{55}\cos\theta\sin\theta - S_{44}\cos\theta\sin\theta,\\
\bar{S}_{55} &= S_{44}\sin^2\theta + S_{55}\cos^2\theta.
\end{aligned} \tag{1.136}$$

All other compliances $\bar{S}_{ij}$ not given here vanish in this kind of transformation for orthotropic material. The generalized Hooke's law $\bar{\varepsilon} = \underline{\underline{\bar{S}}}\bar{\sigma}$ in the off-axis system is thus:

$$\begin{pmatrix} \bar{\varepsilon}_{11}\\ \bar{\varepsilon}_{22}\\ \bar{\varepsilon}_{33}\\ \bar{\gamma}_{23}\\ \bar{\gamma}_{13}\\ \bar{\gamma}_{12} \end{pmatrix} = \begin{bmatrix} \bar{S}_{11} & \bar{S}_{12} & \bar{S}_{13} & 0 & 0 & \bar{S}_{16}\\ \bar{S}_{12} & \bar{S}_{22} & \bar{S}_{23} & 0 & 0 & \bar{S}_{26}\\ \bar{S}_{13} & \bar{S}_{23} & \bar{S}_{33} & 0 & 0 & \bar{S}_{36}\\ 0 & 0 & 0 & \bar{S}_{44} & \bar{S}_{45} & 0\\ 0 & 0 & 0 & \bar{S}_{45} & \bar{S}_{55} & 0\\ \bar{S}_{16} & \bar{S}_{26} & \bar{S}_{36} & 0 & 0 & \bar{S}_{66} \end{bmatrix} \begin{pmatrix} \bar{\sigma}_{11}\\ \bar{\sigma}_{22}\\ \bar{\sigma}_{33}\\ \bar{\sigma}_{23}\\ \bar{\sigma}_{13}\\ \bar{\sigma}_{12} \end{pmatrix}. \tag{1.137}$$

Apparently the transformed material law for orthotropic material agrees with that of a monotropic material with a symmetry plane at $x_3 = 0$.

In an analogous way, the transformed stiffnesses $\bar{C}_{ij}$ are obtained:

$$
\begin{aligned}
\bar{C}_{11} &= C_{11}\cos^4\theta + C_{22}\sin^4\theta + 2C_{12}\cos^2\theta\sin^2\theta + 4C_{66}\cos^2\theta\sin^2\theta,\\
\bar{C}_{22} &= C_{11}\sin^4\theta + C_{22}\cos^4\theta + 2C_{12}\cos^2\theta\sin^2\theta + 4C_{66}\cos^2\theta\sin^2\theta,\\
\bar{C}_{12} &= (C_{11}+C_{22})\cos^2\theta\sin^2\theta + C_{12}\left(\cos^4\theta + \sin^4\theta\right) - 4C_{66}\cos^2\theta\sin^2\theta,\\
\bar{C}_{66} &= (C_{11}+C_{22})\cos^2\theta\sin^2\theta - 2C_{12}\cos^2\theta\sin^2\theta + C_{66}\left(\cos^2\theta - \sin^2\theta\right)^2,\\
\bar{C}_{16} &= C_{11}\cos^3\theta\sin\theta + C_{12}\left(\cos\theta\sin^3\theta - \cos^3\theta\sin\theta\right)\\
&\quad - C_{22}\cos\theta\sin^3\theta + 2C_{66}\left(\cos\theta\sin^3\theta - \cos^3\theta\sin\theta\right),\\
\bar{C}_{26} &= C_{11}\cos\theta\sin^3\theta + C_{12}\left(\cos^3\theta\sin\theta - \cos\theta\sin^3\theta\right)\\
&\quad - C_{22}\cos^3\theta\sin\theta + 2C_{66}\left(\cos^3\theta\sin\theta - \cos\theta\sin^3\theta\right),\\
\bar{C}_{13} &= C_{13}\cos^2\theta + C_{23}\sin^2\theta,\\
\bar{C}_{23} &= C_{13}\sin^2\theta + C_{23}\cos^2\theta,\\
\bar{C}_{33} &= C_{33},\\
\bar{C}_{36} &= C_{13}\cos\theta\sin\theta - C_{23}\cos\theta\sin\theta,\\
\bar{C}_{44} &= C_{44}\cos^2\theta + C_{55}\sin^2\theta,\\
\bar{C}_{45} &= C_{55}\cos\theta\sin\theta - C_{44}\cos\theta\sin\theta,\\
\bar{C}_{55} &= C_{44}\sin^2\theta + C_{55}\cos^2\theta.
\end{aligned}
\tag{1.138}
$$

All other transformed stiffnesses $\bar{C}_{ij}$ become zero for orthotropic material for the present type of axis transformation. The transformed material law $\underline{\bar{\sigma}} = \underline{\underline{\bar{C}}}\,\underline{\bar{\varepsilon}}$ in the off-axis system can be stated as:

$$
\begin{pmatrix}\bar{\sigma}_{11}\\ \bar{\sigma}_{22}\\ \bar{\sigma}_{33}\\ \bar{\tau}_{23}\\ \bar{\tau}_{13}\\ \bar{\tau}_{12}\end{pmatrix}
=
\begin{bmatrix}
\bar{C}_{11} & \bar{C}_{12} & \bar{C}_{13} & 0 & 0 & \bar{C}_{16}\\
\bar{C}_{12} & \bar{C}_{22} & \bar{C}_{23} & 0 & 0 & \bar{C}_{26}\\
\bar{C}_{13} & \bar{C}_{23} & \bar{C}_{33} & 0 & 0 & \bar{C}_{36}\\
0 & 0 & 0 & \bar{C}_{44} & \bar{C}_{45} & 0\\
0 & 0 & 0 & \bar{C}_{45} & \bar{C}_{55} & 0\\
\bar{C}_{16} & \bar{C}_{26} & \bar{C}_{36} & 0 & 0 & \bar{C}_{66}
\end{bmatrix}
\begin{pmatrix}\bar{\varepsilon}_{11}\\ \bar{\varepsilon}_{22}\\ \bar{\varepsilon}_{33}\\ \bar{\gamma}_{23}\\ \bar{\gamma}_{13}\\ \bar{\gamma}_{12}\end{pmatrix}.
\tag{1.139}
$$

1.8 Representation of the Basic Equations in Cylindrical Coordinates

In many structural situations, instead of Cartesian coordinates x, y, z, a representation of all relations in cylindrical coordinates r, φ, z is useful (Fig. 1.17, left). To derive the local equilibrium conditions the free body image of Fig. 1.17, right, is considered, in which the stress components $\sigma_{rr}, \sigma_{\varphi\varphi}, \sigma_{zz}, \tau_{r\varphi} = \tau_{\varphi r}, \tau_{z\varphi} = \tau_{\varphi z}, \tau_{rz} = \tau_{zr}$ with

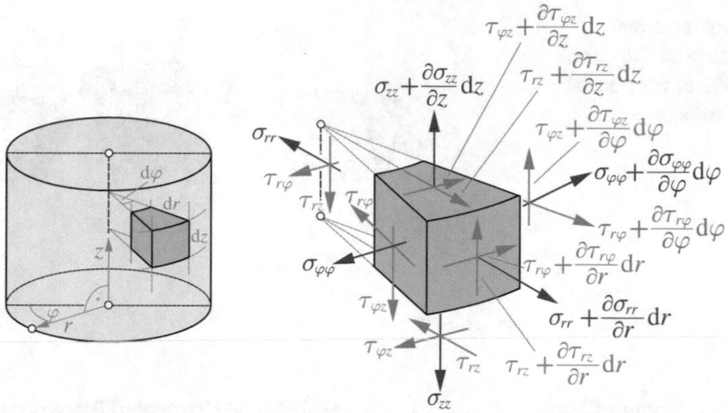

Fig. 1.17 Cylindrical coordinate system (left), free body image of an infinitesimal sectional element (right)

their infinitesimal increments are plotted on the positive cutting edges. For reasons of clarity, the volume forces f_r, f_φ, f_z which may also occur are not shown here. The equilibrium of forces in the radial direction r gives:

$$
\begin{aligned}
&\left(\sigma_{rr} + \frac{\partial \sigma_{rr}}{\partial r}\mathrm{d}r\right)(r + \mathrm{d}r)\,\mathrm{d}\varphi\mathrm{d}z - \sigma_{rr} r\mathrm{d}\varphi\mathrm{d}z \\
&+ \left(\tau_{rz} + \frac{\partial \tau_{rz}}{\partial z}\mathrm{d}z\right) r\mathrm{d}r\mathrm{d}\varphi - \tau_{rz} r\mathrm{d}r\mathrm{d}\varphi \\
&- \left(\sigma_{\varphi\varphi} + \frac{\partial \sigma_{\varphi\varphi}}{\partial \varphi}\mathrm{d}\varphi\right)\mathrm{d}r\mathrm{d}z \sin\left(\frac{\mathrm{d}\varphi}{2}\right) - \sigma_{\varphi\varphi}\mathrm{d}r\mathrm{d}z \sin\left(\frac{\mathrm{d}\varphi}{2}\right) \\
&+ \left(\tau_{r\varphi} + \frac{\partial \tau_{r\varphi}}{\partial \varphi}\mathrm{d}\varphi\right)\mathrm{d}r\mathrm{d}z \cos\left(\frac{\mathrm{d}\varphi}{2}\right) - \tau_{r\varphi}\mathrm{d}r\mathrm{d}z \cos\left(\frac{\mathrm{d}\varphi}{2}\right) + f_r r\mathrm{d}r\mathrm{d}\varphi\mathrm{d}z = 0.
\end{aligned}
\tag{1.140}
$$

If we neglect terms of higher order and consider $\sin\left(\frac{\mathrm{d}\varphi}{2}\right) \simeq \frac{\mathrm{d}\varphi}{2}$ and $\cos\left(\frac{\mathrm{d}\varphi}{2}\right) \simeq 1$, then we obtain:

$$
\frac{\partial \sigma_{rr}}{\partial r} + \frac{1}{r}\frac{\partial \tau_{r\varphi}}{\partial \varphi} + \frac{\partial \tau_{rz}}{\partial z} + \frac{\sigma_{rr} - \sigma_{\varphi\varphi}}{r} + f_r = 0. \tag{1.141}
$$

In the same way, the two remaining equilibrium conditions can be derived:

$$
\begin{aligned}
&\frac{\partial \tau_{r\varphi}}{\partial r} + \frac{1}{r}\frac{\sigma_{\varphi\varphi}}{\partial \varphi} + \frac{\partial \tau_{\varphi z}}{\partial z} + 2\frac{\tau_{r\varphi}}{r} + f_\varphi = 0, \\
&\frac{\partial \tau_{rz}}{\partial r} + \frac{1}{r}\frac{\partial \tau_{\varphi z}}{\partial \varphi} + \frac{\partial \sigma_{zz}}{\partial z} + \frac{\tau_{rz}}{r} + f_z = 0.
\end{aligned}
\tag{1.142}
$$

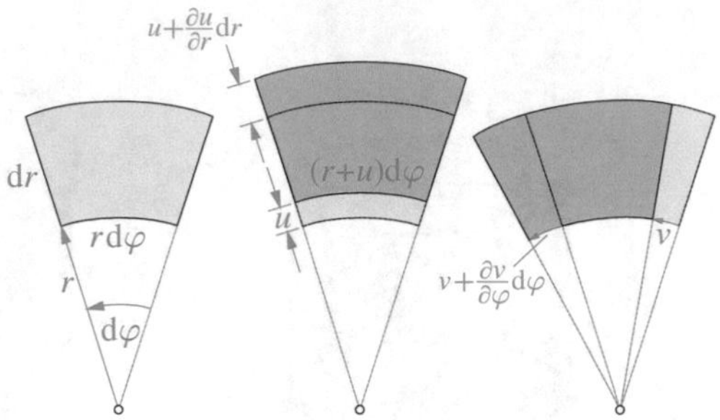

Fig. 1.18 Infinitesimal section element in top view, determination of the normal strains ε_{rr} and $\varepsilon_{\varphi\varphi}$

The strain components $\varepsilon_{rr}, \varepsilon_{\varphi\varphi}, \varepsilon_{zz}, \gamma_{r\varphi}, \gamma_{rz}, \gamma_{\varphi z}$ can be expressed from the kinematic equations by the displacements u, v, w with respect to r, φ, z. We again consider the infinitesimal sectional element in a top view (see Figs. 1.18 and 1.19). The two deformation states shown in Fig. 1.18 lead to the two strains ε_{rr} and $\varepsilon_{\varphi\varphi}$. The radial strain ε_{rr} can be obtained directly from Fig. 1.18, middle, as:

$$\varepsilon_{rr} = \frac{\left(u + \dfrac{\partial u}{\partial r}\mathrm{d}r\right) - u}{\mathrm{d}r} = \frac{\partial u}{\partial r}. \tag{1.143}$$

The strain ε_{rr} is independent of the displacement v.

Analogously, on the basis of Fig. 1.18, middle, for the fraction of tangential strain $\varepsilon_{\varphi\varphi}$ due to the displacement u one obtains:

$$\varepsilon_{\varphi\varphi} = \frac{(r+u)\,\mathrm{d}\varphi - r\mathrm{d}\varphi}{r\mathrm{d}\varphi} = \frac{u}{r}. \tag{1.144}$$

According to Fig. 1.18, right, there is also a contribution due to the tangential displacement v:

$$\varepsilon_{\varphi\varphi} = \frac{\left(v + \dfrac{\partial v}{\partial \varphi}\mathrm{d}\varphi\right) - v}{r\mathrm{d}\varphi} = \frac{1}{r}\frac{\partial v}{\partial \varphi}. \tag{1.145}$$

Thus, for $\varepsilon_{\varphi\varphi}$ we obtain:

$$\varepsilon_{\varphi\varphi} = \frac{1}{r}\frac{\partial v}{\partial \varphi} + \frac{u}{r}. \tag{1.146}$$

The shear strain $\gamma_{r\varphi}$ can be derived from Fig. 1.19. One obtains:

$$\gamma_{r\varphi} = \frac{\partial v}{\partial r} + \frac{1}{r}\frac{\partial u}{\partial \varphi} - \frac{v}{r}. \tag{1.147}$$

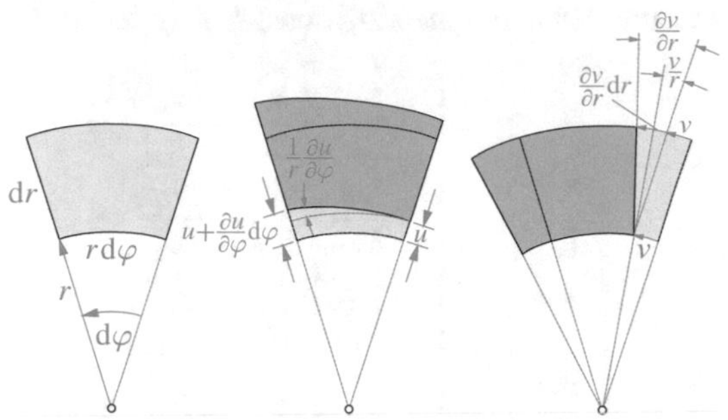

Fig. 1.19 Infinitesimal section element in top view, determination of the shear strain $\gamma_{r\varphi}$

The derivation of the remaining strain components is omitted here. The following kinematic equations in cylindrical coordinates result:

$$\varepsilon_{rr} = \frac{\partial u}{\partial r}, \quad \varepsilon_{\varphi\varphi} = \frac{1}{r}\frac{\partial v}{\partial \varphi} + \frac{u}{r}, \quad \varepsilon_{zz} = \frac{\partial w}{\partial z},$$
$$\gamma_{r\varphi} = \frac{\partial v}{\partial r} + \frac{1}{r}\frac{\partial u}{\partial \varphi} - \frac{v}{r}, \quad \gamma_{rz} = \frac{\partial w}{\partial r} + \frac{\partial u}{\partial z}, \quad \gamma_{\varphi z} = \frac{\partial v}{\partial z} + \frac{1}{r}\frac{\partial w}{\partial \varphi}. \tag{1.148}$$

The generalized Hooke's law can be represented for cylindrical orthotropy (i.e. orthotropy with respect to the three symmetry planes $r\varphi$, rz, φz) in cylindrical coordinates as follows:

$$\begin{pmatrix} \sigma_{rr} \\ \sigma_{\varphi\varphi} \\ \sigma_{zz} \\ \tau_{\varphi z} \\ \tau_{rz} \\ \tau_{r\varphi} \end{pmatrix} = \begin{bmatrix} C_{11} & C_{12} & C_{13} & 0 & 0 & 0 \\ C_{12} & C_{22} & C_{23} & 0 & 0 & 0 \\ C_{13} & C_{23} & C_{33} & 0 & 0 & 0 \\ 0 & 0 & 0 & C_{44} & 0 & 0 \\ 0 & 0 & 0 & 0 & C_{55} & 0 \\ 0 & 0 & 0 & 0 & 0 & C_{66} \end{bmatrix} \begin{pmatrix} \varepsilon_{rr} \\ \varepsilon_{\varphi\varphi} \\ \varepsilon_{zz} \\ \gamma_{\varphi z} \\ \gamma_{rz} \\ \gamma_{r\varphi} \end{pmatrix}. \tag{1.149}$$

The inverted form reads:

$$\begin{pmatrix} \varepsilon_{rr} \\ \varepsilon_{\varphi\varphi} \\ \varepsilon_{zz} \\ \gamma_{\varphi z} \\ \gamma_{rz} \\ \gamma_{r\varphi} \end{pmatrix} = \begin{bmatrix} S_{11} & S_{12} & S_{13} & 0 & 0 & 0 \\ S_{12} & S_{22} & S_{23} & 0 & 0 & 0 \\ S_{13} & S_{23} & S_{33} & 0 & 0 & 0 \\ 0 & 0 & 0 & S_{44} & 0 & 0 \\ 0 & 0 & 0 & 0 & S_{55} & 0 \\ 0 & 0 & 0 & 0 & 0 & S_{66} \end{bmatrix} \begin{pmatrix} \sigma_{rr} \\ \sigma_{\varphi\varphi} \\ \sigma_{zz} \\ \tau_{\varphi z} \\ \tau_{rz} \\ \tau_{r\varphi} \end{pmatrix}. \tag{1.150}$$

Formulated in engineering constants we have:

$$\begin{pmatrix} \varepsilon_{rr} \\ \varepsilon_{\varphi\varphi} \\ \varepsilon_{zz} \\ \gamma_{\varphi z} \\ \gamma_{rz} \\ \gamma_{r\varphi} \end{pmatrix} = \begin{bmatrix} \frac{1}{E_{rr}} & -\frac{\nu_{\varphi r}}{E_{\varphi\varphi}} & -\frac{\nu_{zr}}{E_{zz}} & 0 & 0 & 0 \\ -\frac{\nu_{r\varphi}}{E_{rr}} & \frac{1}{E_{\varphi\varphi}} & -\frac{\nu_{z\varphi}}{E_{zz}} & 0 & 0 & 0 \\ -\frac{\nu_{rz}}{E_{rr}} & -\frac{\nu_{\varphi z}}{E_{\varphi\varphi}} & \frac{1}{E_{zz}} & 0 & 0 & 0 \\ 0 & 0 & 0 & \frac{1}{G_{\varphi z}} & 0 & 0 \\ 0 & 0 & 0 & 0 & \frac{1}{G_{rz}} & 0 \\ 0 & 0 & 0 & 0 & 0 & \frac{1}{G_{r\varphi}} \end{bmatrix} \begin{pmatrix} \sigma_{rr} \\ \sigma_{\varphi\varphi} \\ \sigma_{zz} \\ \tau_{\varphi z} \\ \tau_{rz} \\ \tau_{r\varphi} \end{pmatrix}. \tag{1.151}$$

1.9 Plane Problems

The relations discussed so far have been established for arbitrary three-dimensional problems of elasticity theory. However, many technical applications allow a simplification in such a way that three-dimensional problems can be reduced to two significant dimensions. This is often the case, for example, with the often thin-walled disk, plate and shell structures which are the subject matter of this book. The reduction to two dimensions not only significantly reduces the number of state variables to be determined (displacements, strains, stresses), but also significantly reduces the complexity of the underlying equations and the effort required for analytical or numerical computations.

1.9.1 Plane Strain State

We consider an isotropic and linear-elastic structure where the two displacements u and v occur in $x-$direction and $y-$direction, but the third displacement component w in $z-$direction does not. The two displacements u and v depend only on x and y. Hence:

$$u = u(x, y), \quad v = v(x, y), \quad w = 0. \tag{1.152}$$

In this case, all strain components containing the index z disappear:

$$\gamma_{xz} = \gamma_{yz} = \varepsilon_{zz} = 0. \tag{1.153}$$

Thus, only the plane strain components $\varepsilon_{xx}, \varepsilon_{yy}$ and γ_{xy} remain, which are exclusively functions of x and y. For the described case we thus speak of the so-called plane strain state with respect to the $xy-$plane. The non-vanishing strains are calculated from the kinematic Eq. (1.60) under the condition of geometric linearity as:

$$\varepsilon_{xx}(x,y) = \frac{\partial u}{\partial x}, \quad \varepsilon_{yy}(x,y) = \frac{\partial v}{\partial y}, \quad \gamma_{xy}(x,y) = \frac{\partial u}{\partial y} + \frac{\partial v}{\partial x}. \tag{1.154}$$

Figure 1.20 shows a simple example of a plane strain state at the example of an elastic block under the two normal stresses σ_{xx} and σ_{yy}, which is positioned between two rigid blocks and thus prevented from exhibiting any displacements w.

From Hooke's law (1.119) for isotropic material it can be concluded that in the case of a plane strain state the two shear stresses τ_{yz} and τ_{xz} must become zero. However, the normal stress σ_{zz} in the z−direction generally does not become zero. From the third line in (1.128) we obtain:

$$\varepsilon_{zz} = -\frac{\nu}{E}\sigma_{xx} - \frac{\nu}{E}\sigma_{yy} + \frac{1}{E}\sigma_{zz} = 0. \tag{1.155}$$

From this, the normal stress σ_{zz} can then be determined in the presence of the two normal stresses σ_{xx} and σ_{yy}:

$$\sigma_{zz} = \nu\left(\sigma_{xx} + \sigma_{yy}\right). \tag{1.156}$$

With this expression, σ_{zz} can be eliminated from the constitutive law (1.128), and the following form of the material law for the plane strain state is obtained:

$$\begin{aligned} \varepsilon_{xx} &= \frac{1-\nu^2}{E}\left(\sigma_{xx} - \frac{\nu}{1-\nu}\sigma_{yy}\right), \\ \varepsilon_{yy} &= \frac{1-\nu^2}{E}\left(\sigma_{yy} - \frac{\nu}{1-\nu}\sigma_{xx}\right), \\ \gamma_{xy} &= \frac{2(1+\nu)}{E}\tau_{xy}. \end{aligned} \tag{1.157}$$

It is useful to introduce a substitute elastic modulus $\overline{E}$ and a substitute Poisson's ratio $\overline{\nu}$ as:

$$\overline{E} = \frac{E}{1-\nu^2}, \quad \overline{\nu} = \frac{\nu}{1-\nu}. \tag{1.158}$$

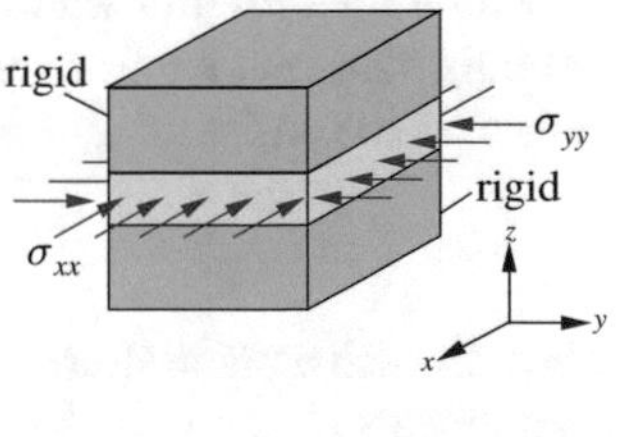

Fig. 1.20 Elastic block between two rigid restraints

Thus one obtains from (1.157):

$$\varepsilon_{xx} = \frac{1}{\overline{E}}\left(\sigma_{xx} - \overline{\nu}\sigma_{yy}\right),$$
$$\varepsilon_{yy} = \frac{1}{\overline{E}}\left(\sigma_{yy} - \overline{\nu}\sigma_{xx}\right),$$
$$\gamma_{xy} = \frac{2\left(1+\overline{\nu}\right)}{\overline{E}}\tau_{xy}. \tag{1.159}$$

Since in the plane state of strain the strains are independent of z, the stresses will also be found to be independent of z. If we also assume that there are no volume forces in the $z-$ direction, then the third equilibrium condition in (1.36) is automatically satisfied. The two remaining conditions in (1.36) are then obtained as:

$$\frac{\partial \sigma_{xx}}{\partial x} + \frac{\partial \tau_{xy}}{\partial y} + f_x = 0,$$
$$\frac{\partial \tau_{xy}}{\partial x} + \frac{\partial \sigma_{yy}}{\partial y} + f_y = 0. \tag{1.160}$$

The following compatibility condition (cf. Eq. (1.70)) remains:

$$\frac{\partial^2 \varepsilon_{xx}}{\partial y^2} + \frac{\partial^2 \varepsilon_{yy}}{\partial x^2} - \frac{\partial^2 \gamma_{xy}}{\partial x \partial y} = 0. \tag{1.161}$$

A problem in the plane strain state is thus fully described by the Eqs. (1.154), (1.157), (1.160) and (1.161) taking into account given boundary conditions. It is worth mentioning here that so far no approximations or simplifications have been introduced that would go beyond the assumption of independence of all state variables from z. Equations (1.154), (1.157), (1.160) and (1.161) thus represent exact governing equations in the framework of the assumed linear elasticity theory for isotropic structures under the assumption of geometric linearity.

1.9.2 Plane Stress State

The assumption of the so-called plane stress state is a simplifying assumption that is very important for thin-walled structures. It assumes that the stresses σ_{zz}, τ_{xz} and τ_{yz} not only vanish at the free surfaces of a given thin-walled structure but also become zero over the entire thickness:

$$\sigma_{zz} = \tau_{xz} = \tau_{yz} = 0. \tag{1.162}$$

Thus, in such a plane stress state with respect to the $xy-$plane, only the plane stress components σ_{xx}, σ_{yy} and τ_{xy} remain, where we want to assume at this point that these remaining stress components are independent of the thickness coordinate z:

$$\sigma_{xx} = \sigma_{xx}(x, y), \quad \sigma_{yy} = \sigma_{yy}(x, y), \quad \tau_{xy} = \tau_{xy}(x, y). \tag{1.163}$$

From the generalized Hooke's law (1.128) one then obtains for the plane stress state:

$$\begin{aligned}
\varepsilon_{xx} &= \frac{1}{E}\left(\sigma_{xx} - \nu\sigma_{yy}\right), \\
\varepsilon_{yy} &= \frac{1}{E}\left(\sigma_{yy} - \nu\sigma_{xx}\right), \\
\varepsilon_{zz} &= -\frac{\nu}{E}\left(\sigma_{xx} + \sigma_{yy}\right), \\
\gamma_{xy} &= \frac{2(1+\nu)}{E}\tau_{xy}, \\
\gamma_{xz} &= \gamma_{yz} = 0.
\end{aligned} \tag{1.164}$$

Comparing these expressions with the constitutive Eq. (1.159) for the plane strain state, it is found that both sets of equations are in perfect agreement except for the substitute quantities $\overline{E}$ and $\overline{\nu}$ in (1.159). Moreover, one notes that in the plane stress state, in contrast to the plane strain state, the normal strain ε_{zz} can also occur in the thickness direction z.

The constitutive law is expressed in terms of the stresses as:

$$\begin{aligned}
\sigma_{xx} &= \frac{E}{1-\nu^2}\left(\varepsilon_{xx} + \nu\varepsilon_{yy}\right), \\
\sigma_{yy} &= \frac{E}{1-\nu^2}\left(\varepsilon_{yy} + \nu\varepsilon_{xx}\right), \\
\tau_{xy} &= G\gamma_{xy}.
\end{aligned} \tag{1.165}$$

The kinematic equations for the plane stress state read:

$$\varepsilon_{xx} = \frac{\partial u}{\partial x}, \quad \varepsilon_{yy} = \frac{\partial v}{\partial y}, \quad \gamma_{xy} = \frac{\partial u}{\partial y} + \frac{\partial v}{\partial x}. \tag{1.166}$$

The equilibrium conditions result as:

$$\begin{aligned}
\frac{\partial \sigma_{xx}}{\partial x} + \frac{\partial \tau_{xy}}{\partial y} + f_x &= 0, \\
\frac{\partial \tau_{xy}}{\partial x} + \frac{\partial \sigma_{yy}}{\partial y} + f_y &= 0.
\end{aligned} \tag{1.167}$$

The remaining compatibility condition can be specified as:

$$\frac{\partial^2 \varepsilon_{xx}}{\partial y^2} + \frac{\partial^2 \varepsilon_{yy}}{\partial x^2} - \frac{\partial^2 \gamma_{xy}}{\partial x \partial y} = 0. \tag{1.168}$$

To describe a plane stress state, given boundary conditions must be taken into account in addition to the basic equations shown above.

The comparison of the equations describing the plane strain state on the one hand and the plane stress state on the other hand shows that all equations with the exception of the elastic constants E and ν or $\overline{E}$ and $\overline{\nu}$ are identical. Thus, for example, once one has determined an solution for a given boundary value problem in the context of the plane strain state, one can use it to determine the corresponding solution for the plane stress state by exchanging the material parameters $\overline{E}$ and $\overline{\nu}$ for E and ν.

1.9.3 Stress Transformation

In this section, we will address the question of how the stress state changes if, instead of considering a section parallel to the global reference axes x, y, z, we examine an arbitrary section through the body point under consideration and thus perform a coordinate transformation as discussed before. Here we use the notations ξ, η for the two planar orthogonal axes rotated by the angle θ about the fixed $z-$axis (Fig. 1.21). This is to express that this transformation can be under any angle and not necessarily a particularly distinguished direction as indicated, for example, in Fig. 1.21. From (1.22) the following relation can be deduced:

$$\begin{pmatrix} \sigma_{\xi\xi} \\ \sigma_{\eta\eta} \\ \tau_{\xi\eta} \end{pmatrix} = \begin{bmatrix} \cos^2\theta & \sin^2\theta & 2\cos\theta\sin\theta \\ \sin^2\theta & \cos^2\theta & -2\cos\theta\sin\theta \\ -\cos\theta\sin\theta & \cos\theta\sin\theta & \cos^2\theta - \sin^2\theta \end{bmatrix} \begin{pmatrix} \sigma_{xx} \\ \sigma_{yy} \\ \tau_{xy} \end{pmatrix}. \tag{1.169}$$

This result is also amenable to a particularly illustrative derivation, as we will briefly explain below. For this purpose, we examine an infinitesimally small sectional element in triangular form, which we have cut out of a plane thin-walled structure (Fig. 1.22) and make the section in such a way that two sectional surfaces parallel to the reference axes x and y and one sectional surface parallel to the rotated axis η result. We denote the sectional surface traversed by the $\xi-$axis as $\mathrm{d}A$, with $\mathrm{d}A = \mathrm{d}\eta h$. The two intersecting surfaces oriented parallel to x and y with lengths $\mathrm{d}x$ and $\mathrm{d}y$ are then obtained as $\mathrm{d}A \sin\theta$ and $\mathrm{d}A \cos\theta$, respectively. The equilibrium of forces in the $\xi-$direction leads to:

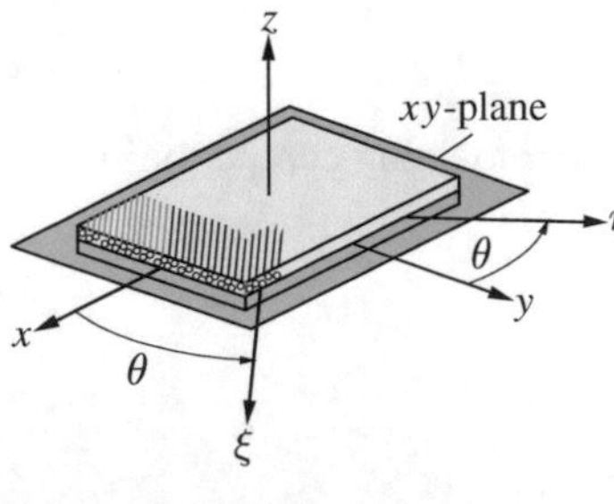

Fig. 1.21 Axis transformation

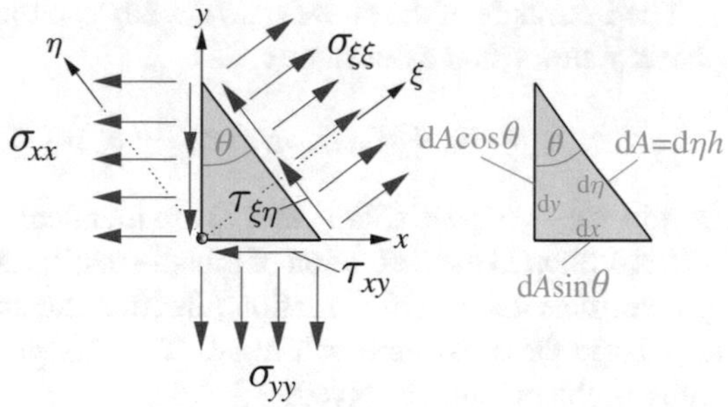

Fig. 1.22 Infinitesimally small sectional element of a thin-walled disk

$$\sigma_{\xi\xi}\mathrm{d}A - \sigma_{xx}\mathrm{d}A\cos\theta\cos\theta - \sigma_{yy}\mathrm{d}A\sin\theta\sin\theta - \tau_{xy}\mathrm{d}A\cos\theta\sin\theta - \tau_{xy}\mathrm{d}A\sin\theta\cos\theta = 0. \tag{1.170}$$

After a short transformation we obtain the normal stress $\sigma_{\xi\xi}$ as:

$$\sigma_{\xi\xi} = \sigma_{xx}\cos^2\theta + \sigma_{yy}\sin^2\theta + 2\tau_{xy}\sin\theta\cos\theta. \tag{1.171}$$

Equilibrium of forces in the $\eta-$ direction yields:

$$\tau_{\xi\eta} = -\sigma_{xx}\sin\theta\cos\theta + \sigma_{yy}\sin\theta\cos\theta + \tau_{xy}\left(\cos^2\theta - \sin^2\theta\right). \tag{1.172}$$

From another sectional view that is not shown here for reasons of brevity and where the $\eta-$axis passes through the sectional surface $\mathrm{d}A$, the normal stress $\sigma_{\eta\eta}$ is obtained as:

$$\sigma_{\eta\eta} = \sigma_{xx}\sin^2\theta + \sigma_{yy}\cos^2\theta - 2\tau_{xy}\sin\theta\cos\theta. \tag{1.173}$$

If one uses

$$\begin{aligned} \cos^2\theta &= \frac{1}{2}\left(1+\cos 2\theta\right), \quad \sin^2\theta = \frac{1}{2}\left(1-\cos 2\theta\right), \\ 2\sin\theta\cos\theta &= \sin 2\theta, \quad \cos^2\theta - \sin^2\theta = \cos 2\theta \end{aligned} \tag{1.174}$$

in (1.171), (1.172) and (1.173) one obtains:

$$\begin{aligned} \sigma_{\xi\xi} &= \frac{1}{2}\left(\sigma_{xx}+\sigma_{yy}\right) + \frac{1}{2}\left(\sigma_{xx}-\sigma_{yy}\right)\cos 2\theta + \tau_{xy}\sin 2\theta, \\ \sigma_{\eta\eta} &= \frac{1}{2}\left(\sigma_{xx}+\sigma_{yy}\right) - \frac{1}{2}\left(\sigma_{xx}-\sigma_{yy}\right)\cos 2\theta - \tau_{xy}\sin 2\theta, \\ \tau_{\xi\eta} &= -\frac{1}{2}\left(\sigma_{xx}-\sigma_{yy}\right)\sin 2\theta + \tau_{xy}\cos 2\theta. \end{aligned} \tag{1.175}$$

These are the transformation equations for the stress components in the case of a plane stress state, from which the stresses $\sigma_{\xi\xi}$, $\sigma_{\eta\eta}$ and $\tau_{\xi\eta}$ in any reference frame ξ, η can be determined from any given stresses σ_{xx}, σ_{yy} and τ_{xy} in the reference frame x, y.

The invariants of the stress state already established with (1.29) are also valid for planar states, whereby currently:

$$I_1 = \sigma_{xx} + \sigma_{yy}, \quad I_2 = \tau_{xy}^2 - \sigma_{xx}\sigma_{yy}, \quad I_3 = \sigma_{xx}\sigma_{yy} - \tau_{xy}^2. \tag{1.176}$$

Herein, the two invariants I_2 and I_3 are identical except for their signs.

In addition to the derivation of transformation equations, it is of particular interest to determine under which section direction the stresses exhibit extreme values and how large these extreme values are. For this purpose, the following two extreme value problems can be solved:

$$\frac{d\sigma_{\xi\xi}}{d\theta} = 0, \quad \frac{d\sigma_{\eta\eta}}{d\theta} = 0. \tag{1.177}$$

It can be seen that both equations lead to the same result for the angle θ_h under which the extremal normal stresses occur:

$$\tan 2\theta_h = \frac{2\tau_{xy}}{\sigma_{xx} - \sigma_{yy}}. \tag{1.178}$$

The direction defined by the angle θ_h is also called the principal direction. The corresponding axes ξ and η are the so-called principal axes. Since the tangent function is a periodic function which is periodic with π, one always obtains from (1.178) two angles θ_h and $\theta_h + \frac{\pi}{2}$ under which the normal stresses become extremal. These two directions are perpendicular to each other and are completely equal. Substituting (1.178) into the transformation Eq. (1.175) yields the following equations for determining the extremal normal stresses, denoted as σ_1 and σ_2:

$$\sigma_{1,2} = \frac{\sigma_{xx} + \sigma_{yy}}{2} \pm \sqrt{\left(\frac{\sigma_{xx} - \sigma_{yy}}{2}\right)^2 + \tau_{xy}^2}. \tag{1.179}$$

The extremal normal stresses σ_1 and σ_2, one of which represents the maximum value and the other the minimum value, are the so-called principal normal stresses.

It is also worth mentioning that if θ_h or $\theta_h + \frac{\pi}{2}$ is substituted into the third transformation equation in (1.175), a vanishing shear stress $\tau_{\xi\eta}$ is obtained. Accordingly, the shear stresses always vanish in sections where extremal normal stresses occur.

In addition to the study of extremal normal stresses, it is of importance to determine the section in which the maximum shear stress occurs and what value it assumes. For this purpose we consider the extremal value problem

$$\frac{d\tau_{\xi\eta}}{d\theta} = 0, \tag{1.180}$$

which leads to the following equation for the corresponding angle θ_h':

$$\tan 2\theta_h' = -\frac{\sigma_{xx} - \sigma_{yy}}{2\tau_{xy}}. \tag{1.181}$$

Here, too, it can be seen that the determination of the principal directions for the extremal shear stresses is based on two equal directions, namely the angle θ_h' on the one hand and the angle $\theta_h' + \frac{\pi}{2}$ on the other hand. The comparison between (1.178) and (1.181) shows furthermore that the relation

$$\tan 2\theta_h' = -\frac{1}{\tan 2\theta_h} \tag{1.182}$$

holds. Consequently, the two directions $2\theta_h'$ and $2\theta_h$ are oriented perpendicular to each other. This means that the direction θ_h' of the principal shear stresses is rotated by 45° with respect to the direction θ_h of the principal normal stresses. The principal shear stress τ_{max} follows by substituting θ_h' according to (1.181) in the transformation Eq. (1.175):

$$\tau_{max} = \pm\sqrt{\left(\frac{\sigma_{xx} - \sigma_{yy}}{2}\right)^2 + \tau_{xy}^2}. \tag{1.183}$$

This can also be expressed using the principal stresses σ_1 and σ_2 as:

$$\tau_{max} = \pm\frac{1}{2}(\sigma_1 - \sigma_2). \tag{1.184}$$

Inserting the angle θ_h' in the transformation Eq. (1.175), we find that both normal stresses at this angle are identical with the value σ_M, which is obtained as follows:

$$\sigma_M = \frac{1}{2}\left(\sigma_{xx} + \sigma_{yy}\right). \tag{1.185}$$

Due to the invariance of the sum of the two normal stresses, this can also be represented as follows:

$$\sigma_M = \frac{1}{2}(\sigma_1 + \sigma_2). \tag{1.186}$$

It follows, therefore, that in those sections where the extremal shear stresses are present, the normal stresses generally do not disappear.

1.9.4 Formulation for Orthotropic Materials

In the following we consider orthotropic material, where the transformed axes ξ and η are the principal axes of the considered orthotropic material (on-axis system), so that here we use the reference axes x_1 and x_2 instead of ξ and η. The axes x, y, z distinguish an arbitrary reference system (off-axis system). The axes x and x_1 as well as y and x_2 are each rotated with respect to each other by the angle θ about the fixed $z-$axis (Fig. 1.23).

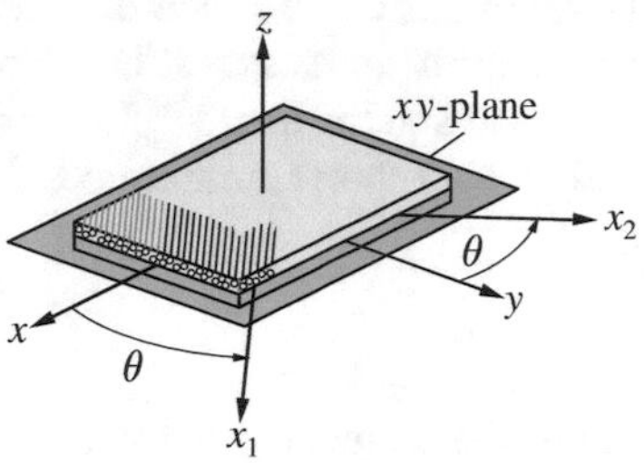

Fig. 1.23 Orthotropic material

The generalized Hooke's law (1.114) can be formulated as follows for an orthotropic thin-walled structure in the principal axis system:

$$\begin{pmatrix} \varepsilon_{11} \\ \varepsilon_{22} \\ \gamma_{12} \end{pmatrix} = \begin{bmatrix} S_{11} & S_{12} & 0 \\ S_{12} & S_{22} & 0 \\ 0 & 0 & S_{66} \end{bmatrix} \begin{pmatrix} \sigma_{11} \\ \sigma_{22} \\ \tau_{12} \end{pmatrix}. \tag{1.187}$$

Formulated in the engineering constants one obtains:

$$\begin{pmatrix} \varepsilon_{11} \\ \varepsilon_{22} \\ \gamma_{12} \end{pmatrix} = \begin{bmatrix} \frac{1}{E_{11}} & -\frac{\nu_{21}}{E_{22}} & 0 \\ -\frac{\nu_{12}}{E_{11}} & \frac{1}{E_{22}} & 0 \\ 0 & 0 & \frac{1}{G_{12}} \end{bmatrix} \begin{pmatrix} \sigma_{11} \\ \sigma_{22} \\ \tau_{12} \end{pmatrix}. \tag{1.188}$$

The normal strain ε_{zz} occurring in the plane stress state can be determined from (1.114) taking into account the requirement $\sigma_{zz} = 0$:

$$\varepsilon_{zz} = S_{13}\sigma_{11} + S_{23}\sigma_{22}. \tag{1.189}$$

The inverse relationship can be stated as follows:

$$\begin{pmatrix} \sigma_{11} \\ \sigma_{22} \\ \tau_{12} \end{pmatrix} = \begin{bmatrix} C_{11} & C_{12} & C_{13} & 0 \\ C_{12} & C_{22} & C_{23} & 0 \\ 0 & 0 & 0 & C_{66} \end{bmatrix} \begin{pmatrix} \varepsilon_{11} \\ \varepsilon_{22} \\ \varepsilon_{zz} \\ \gamma_{12} \end{pmatrix}. \tag{1.190}$$

The normal stress σ_{zz} must vanish in the plane stress state. Thus, we can eliminate the strain ε_{zz} from (1.190). We use the generalized Hooke's law in the form (1.113) for this purpose, and from the third line of (1.113) we obtain with $\sigma_{zz} = 0$:

$$0 = C_{13}\varepsilon_{11} + C_{23}\varepsilon_{22} + C_{33}\varepsilon_{zz}. \tag{1.191}$$

This expression can be solved for the normal strain ε_{zz} as:

$$\varepsilon_{zz} = -\frac{C_{13}}{C_{33}}\varepsilon_{11} - \frac{C_{23}}{C_{33}}\varepsilon_{22}. \tag{1.192}$$

Substituting in the first two equations in (1.190) gives:

$$\sigma_{11} = \left(C_{11} - \frac{C_{13}^2}{C_{33}}\right)\varepsilon_{11} + \left(C_{12} - \frac{C_{13}C_{23}}{C_{33}}\right)\varepsilon_{22},$$
$$\sigma_{22} = \left(C_{12} - \frac{C_{13}C_{23}}{C_{33}}\right)\varepsilon_{11} + \left(C_{22} - \frac{C_{23}^2}{C_{33}}\right)\varepsilon_{22}. \tag{1.193}$$

The third equation in (1.190) remains unchanged.

At this point, the so-called reduced stiffnesses $Q_{11}, Q_{22}, Q_{12}, Q_{66}$ are introduced, so that (1.190) can be represented as follows:

$$\begin{pmatrix} \sigma_{11} \\ \sigma_{22} \\ \tau_{12} \end{pmatrix} = \begin{bmatrix} Q_{11} & Q_{12} & 0 \\ Q_{12} & Q_{22} & 0 \\ 0 & 0 & Q_{66} \end{bmatrix} \begin{pmatrix} \varepsilon_{11} \\ \varepsilon_{22} \\ \gamma_{12} \end{pmatrix}. \tag{1.194}$$

In symbolic form, (1.194) reads:

$$\underline{\sigma} = \underline{\underline{Q}}\,\underline{\varepsilon}. \tag{1.195}$$

Therein, the reduced stiffnesses $Q_{11}, Q_{22}, Q_{12}, Q_{66}$ are defined as:

$$Q_{11} = C_{11} - \frac{C_{13}^2}{C_{33}}, \quad Q_{12} = C_{12} - \frac{C_{13}C_{23}}{C_{33}}, \quad Q_{22} = C_{22} - \frac{C_{23}^2}{C_{33}}, \quad Q_{66} = C_{66}, \tag{1.196}$$

or formulated in the engineering constants:

$$Q_{11} = \frac{E_{11}}{1 - \nu_{12}\nu_{21}}, \quad Q_{22} = \frac{E_{22}}{1 - \nu_{12}\nu_{21}}, \quad Q_{12} = \frac{\nu_{12}E_{22}}{1 - \nu_{12}\nu_{21}}, \quad Q_{66} = G_{12}. \tag{1.197}$$

The engineering constants E_{11}, E_{22}, G_{12}, ν_{12} and ν_{21} are related to the material principal axes.

The relations (1.187), (1.188) and (1.194) are related to the material principal axes of the orthotropic material under consideration. We now consider an orthotropic material oriented under the angle θ in the xy−plane in a plane stress state, and the transformation rules for the stresses and the strains can be taken directly from the previous explanations of this chapter. They read:

$$\begin{pmatrix} \sigma_{xx} \\ \sigma_{yy} \\ \tau_{xy} \end{pmatrix} = \begin{bmatrix} \cos^2\theta & \sin^2\theta & -2\cos\theta\sin\theta \\ \sin^2\theta & \cos^2\theta & 2\cos\theta\sin\theta \\ \cos\theta\sin\theta & -\cos\theta\sin\theta & \cos^2\theta - \sin^2\theta \end{bmatrix} \begin{pmatrix} \sigma_{11} \\ \sigma_{22} \\ \tau_{12} \end{pmatrix}$$

$$= \underline{\underline{T}} \begin{pmatrix} \sigma_{11} \\ \sigma_{22} \\ \tau_{12} \end{pmatrix}, \tag{1.198}$$

and

$$\begin{pmatrix} \varepsilon_{11} \\ \varepsilon_{22} \\ \gamma_{12} \end{pmatrix} = \begin{bmatrix} \cos^2\theta & \sin^2\theta & \cos\theta\sin\theta \\ \sin^2\theta & \cos^2\theta & -\cos\theta\sin\theta \\ -2\cos\theta\sin\theta & 2\cos\theta\sin\theta & \cos^2\theta - \sin^2\theta \end{bmatrix} \begin{pmatrix} \bar{\varepsilon}_{11} \\ \bar{\varepsilon}_{22} \\ \bar{\gamma}_{12} \end{pmatrix}$$

$$= \underline{\underline{T}}^T \begin{pmatrix} \varepsilon_{xx} \\ \varepsilon_{yy} \\ \gamma_{xy} \end{pmatrix}. \tag{1.199}$$

From (1.198) and (1.199), the following relationship between the stresses and the strains with respect to the reference axes x and y can be deduced:

$$\begin{pmatrix} \sigma_{xx} \\ \sigma_{yy} \\ \tau_{xy} \end{pmatrix} = \underline{\underline{T}}\,\underline{\underline{Q}}\,\underline{\underline{T}}^T \begin{pmatrix} \varepsilon_{xx} \\ \varepsilon_{yy} \\ \gamma_{xy} \end{pmatrix} = \underline{\underline{\bar{Q}}} \begin{pmatrix} \varepsilon_{xx} \\ \varepsilon_{yy} \\ \gamma_{xy} \end{pmatrix}. \tag{1.200}$$

Therein, $\underline{\underline{\bar{Q}}} = \underline{\underline{T}}\,\underline{\underline{Q}}\,\underline{\underline{T}}^T$ is the matrix of the so-called transformed reduced stiffnesses. It is generally fully occupied, and it follows:

$$\begin{pmatrix} \sigma_{xx} \\ \sigma_{yy} \\ \tau_{xy} \end{pmatrix} = \begin{bmatrix} \bar{Q}_{11} & \bar{Q}_{12} & \bar{Q}_{16} \\ \bar{Q}_{12} & \bar{Q}_{22} & \bar{Q}_{26} \\ \bar{Q}_{16} & \bar{Q}_{26} & \bar{Q}_{66} \end{bmatrix} \begin{pmatrix} \varepsilon_{xx} \\ \varepsilon_{yy} \\ \gamma_{xy} \end{pmatrix}. \tag{1.201}$$

The quantities $\bar{Q}_{11}, \bar{Q}_{22}, \bar{Q}_{12}, \bar{Q}_{66}, \bar{Q}_{16}, \bar{Q}_{26}$ are the so-called transformed reduced stiffnesses. Evaluating the matrix product $\underline{\underline{T}}\,\underline{\underline{Q}}\,\underline{\underline{T}}^T$ leads to the following transformation rules for the transformed reduced stiffnesses:

$$\begin{aligned}
\bar{Q}_{11} &= Q_{11}\cos^4\theta + 2(Q_{12} + 2Q_{66})\cos^2\theta\sin^2\theta + Q_{22}\sin^4\theta, \\
\bar{Q}_{22} &= Q_{11}\sin^4\theta + 2(Q_{12} + 2Q_{66})\cos^2\theta\sin^2\theta + Q_{22}\cos^4\theta, \\
\bar{Q}_{12} &= (Q_{11} + Q_{22} - 4Q_{66})\cos^2\theta\sin^2\theta + Q_{12}\left(\cos^4\theta + \sin^4\theta\right), \\
\bar{Q}_{66} &= (Q_{11} + Q_{22} - 2Q_{12} - 2Q_{66})\cos^2\theta\sin^2\theta + Q_{66}\left(\cos^4\theta + \sin^4\theta\right), \\
\bar{Q}_{16} &= (Q_{11} - Q_{12} - 2Q_{66})\cos^3\theta\sin\theta + (Q_{12} - Q_{22} + 2Q_{66})\cos\theta\sin^3\theta, \\
\bar{Q}_{26} &= (Q_{11} - Q_{12} - 2Q_{66})\cos\theta\sin^3\theta + (Q_{12} - Q_{22} + 2Q_{66})\cos^3\theta\sin\theta.
\end{aligned} \tag{1.202}$$

In an orthotropic disk in the plane strain state, no displacements w and strains ε_{zz} occur in the thickness direction. The corresponding relations (1.152), (1.153), (1.154), (1.160) and (1.161) are equally valid for an orthotropic disk. It can be easily shown that the generalized Hooke's law in this case can be stated as follows:

$$\begin{pmatrix} \sigma_{11} \\ \sigma_{22} \\ \tau_{12} \end{pmatrix} = \begin{bmatrix} C_{11} & C_{12} & 0 \\ C_{12} & C_{22} & 0 \\ 0 & 0 & C_{66} \end{bmatrix} \begin{pmatrix} \varepsilon_{11} \\ \varepsilon_{22} \\ \gamma_{12} \end{pmatrix}, \tag{1.203}$$

and in inverted form:

$$\begin{pmatrix} \varepsilon_{11} \\ \varepsilon_{22} \\ \gamma_{12} \end{pmatrix} = \begin{bmatrix} R_{11} & R_{12} & 0 \\ R_{12} & R_{22} & 0 \\ 0 & 0 & R_{66} \end{bmatrix} \begin{pmatrix} \sigma_{11} \\ \sigma_{22} \\ \tau_{12} \end{pmatrix}. \tag{1.204}$$

Therein, the quantities R_{ij} $(i, j = 1, 2, 6)$ are the so-called reduced compliances. They can be determined from the compliances S_{ij} as:

$$R_{ij} = S_{ij} - \frac{S_{i3}S_{j3}}{S_{33}}. \tag{1.205}$$

The normal stress σ_{zz} that occurs in a plane strain state can be determined as:

$$\sigma_{zz} = -\frac{1}{S_{33}} \left(S_{13}\sigma_{11} + S_{23}\sigma_{22} \right). \tag{1.206}$$

It is also possible to derive transformation rules for the reduced compliances R_{ij} which, however, is not shown here for reasons of brevity.

1.9.5 Formulation in Polar Coordinates

In many technically relevant cases, it may be advantageous to describe a planar structural mechanics problem by polar coordinates r, φ (Fig. 1.24). The following relationship exists between x, y and r, φ:

$$x = r\cos\varphi, \quad y = r\sin\varphi, \quad r = \sqrt{x^2 + y^2}, \quad \varphi = \arctan\left(\frac{y}{x}\right). \tag{1.207}$$

We first consider a problem in the plane stress state. The equations already derived for the spatial case in cylindrical coordinates can then be directly reused by neglecting all quantities involving the thickness direction z. For the equilibrium conditions (1.141) and (1.142) then remains:

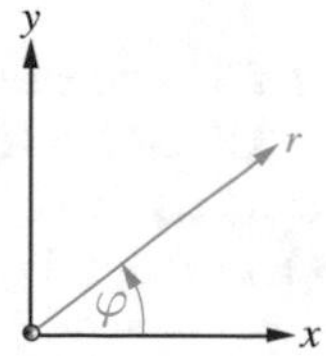

Fig. 1.24 Polar coordinates r, φ

$$\frac{\partial \sigma_{rr}}{\partial r} + \frac{1}{r}\frac{\partial \tau_{r\varphi}}{\partial \varphi} + \frac{\sigma_{rr} - \sigma_{\varphi\varphi}}{r} + f_r = 0,$$
$$\frac{\partial \tau_{r\varphi}}{\partial r} + \frac{1}{r}\frac{\sigma_{\varphi\varphi}}{\partial \varphi} + 2\frac{\tau_{r\varphi}}{r} + f_\varphi = 0. \tag{1.208}$$

The kinematic equations (1.148) then read:

$$\varepsilon_{rr} = \frac{\partial u}{\partial r}, \quad \varepsilon_{\varphi\varphi} = \frac{1}{r}\frac{\partial v}{\partial \varphi} + \frac{u}{r}, \quad \gamma_{r\varphi} = \frac{\partial v}{\partial r} + \frac{1}{r}\frac{\partial u}{\partial \varphi} - \frac{v}{r}, \tag{1.209}$$

and the remaining compatibility condition can be given as:

$$\frac{\partial^2 \varepsilon_{\varphi\varphi}}{\partial r^2} + \frac{1}{r^2}\frac{\partial^2 \varepsilon_{rr}}{\partial \varphi^2} + \frac{2}{r}\frac{\varepsilon_{\varphi\varphi}}{\partial r} - \frac{1}{r}\frac{\partial \varepsilon_{rr}}{\partial r} = \frac{1}{r}\frac{\partial^2 \gamma_{r\varphi}}{\partial r \partial \varphi} + \frac{1}{r^2}\frac{\partial \gamma_{r\varphi}}{\partial \varphi}. \tag{1.210}$$

Hooke's law results in the case of the plane stress state as follows:

$$\begin{pmatrix} \sigma_{rr} \\ \sigma_{\varphi\varphi} \\ \tau_{r\varphi} \end{pmatrix} = \begin{bmatrix} Q_{11} & Q_{12} & 0 \\ Q_{12} & Q_{22} & 0 \\ 0 & 0 & Q_{66} \end{bmatrix} \begin{pmatrix} \varepsilon_{rr} \\ \varepsilon_{\varphi\varphi} \\ \gamma_{r\varphi} \end{pmatrix}, \tag{1.211}$$

or in inverted form:

$$\begin{pmatrix} \varepsilon_{rr} \\ \varepsilon_{\varphi\varphi} \\ \gamma_{r\varphi} \end{pmatrix} = \begin{bmatrix} S_{11} & S_{12} & 0 \\ S_{12} & S_{22} & 0 \\ 0 & 0 & S_{66} \end{bmatrix} \begin{pmatrix} \sigma_{rr} \\ \sigma_{\varphi\varphi} \\ \tau_{r\varphi} \end{pmatrix}. \tag{1.212}$$

For the plane strain state, the reduced compliances R_{ij} $(i, j = 1, 2, 6)$ according to (1.205) have to be used. The normal stress σ_{zz} is obtained analogously to (1.206).

For an isotropic problem, the material law in the plane stress state results as:

$$\begin{pmatrix} \varepsilon_{rr} \\ \varepsilon_{\varphi\varphi} \\ \gamma_{r\varphi} \end{pmatrix} = \begin{bmatrix} \frac{1}{E} & -\frac{\nu}{E} & 0 \\ -\frac{\nu}{E} & \frac{1}{E} & 0 \\ 0 & 0 & \frac{1}{G} \end{bmatrix} \begin{pmatrix} \sigma_{rr} \\ \sigma_{\varphi\varphi} \\ \tau_{r\varphi} \end{pmatrix}. \tag{1.213}$$

In the plane strain state, the equivalent stiffness quantities $\bar{E}$ and $\bar{\nu}$ are to be employed.

Given the stresses σ_{xx}, σ_{yy}, τ_{xy} in Cartesian coordinates, the components σ_{rr}, $\sigma_{\varphi\varphi}$, $\tau_{r\varphi}$ in polar coordinates can be determined from the transformation relations (1.175) as follows:

$$\sigma_{rr} = \frac{1}{2}\left(\sigma_{xx} + \sigma_{yy}\right) + \frac{1}{2}\left(\sigma_{xx} - \sigma_{yy}\right)\cos 2\theta + \tau_{xy}\sin 2\theta,$$
$$\sigma_{\varphi\varphi} = \frac{1}{2}\left(\sigma_{xx} + \sigma_{yy}\right) - \frac{1}{2}\left(\sigma_{xx} - \sigma_{yy}\right)\cos 2\theta - \tau_{xy}\sin 2\theta,$$

$$\tau_{r\varphi} = -\frac{1}{2}\left(\sigma_{xx} - \sigma_{yy}\right)\sin 2\theta + \tau_{xy}\cos 2\theta. \tag{1.214}$$

The sum of the two normal stresses is an invariant, i.e. $\sigma_{xx} + \sigma_{yy} = \sigma_{rr} + \sigma_{\varphi\varphi}$.

References

Altenbach, H., Altenbach, J., Rikards, R.: Einführung in die Mechanik der Laminat- und Sandwichtragwerke. Deutscher Verlag der Grundstoffindustrie Stuttgart, Germany (1996)

Ambartsumyan, S.A.: Theory of Anisotropic Plates. Technomic Publishing Co. Inc., Stamford (1970)

Ashton, J.E., Whitney, J.M.: Theory of Laminated Plates. Technomic Publishing Co. Inc., Stamford (1970)

Becker, W., Gross, D.: Mechanik elastischer Körper und Strukturen. Springer, Berlin (2002)

Eschenauer, H., Schnell, W.: Elastizitätstheorie I, 2nd edn. Bibliographisches Institut, Mannheim (1986)

Frick, A., Klamser, H.: Untersuchung der Verschiebungs- und Spannungsverteilungen an unbelasteten Rändern multidirektionaler Laminate. VDI, Düsseldorf (1990)

Göldner, H., Altenbach, J., Eschke, K., Garz, K.F., Sähn, S.: Lehrbuch Höhere Festigkeitslehre, vol. 1. Physik, Weinheim (1979)

Göldner, H., Altenbach, J., Eschke, K., Garz, K.F., Sähn, S.: Lehrbuch Höhere Festigkeitslehre, vol. 2. Physik, Weinheim (1985)

Hahn, H.G.: Elastizitätstheorie. BG Teubner, Stuttgart (1985)

Jones, R.M.: Mechanics of Composite Materials. Scripta Book Co., Washington, USA (1975)

Lekhnitskii, S.G.: Anisotropic Plates. Gordon and Breach, New York (1968)

Mittelstedt, C., Becker, W.: Strukturmechanik ebener Laminate. Studienbereich Mechanik, Technische Universität, Darmstadt (2016)

Mittelstedt, C.: Rechenmethoden des Leichtbaus. Springer Vieweg, Berlin (2021)

Reddy, J.N.: Mechanics of Laminated Composite Plates and Shells, 2nd edn. CRC Press, Boca Raton (2004)

Tsai, S.W., Hahn, H.T.: Introduction to Composite Materials. Technomic Publishing Co. Inc., Lancaster (1980)

Energy Methods of Elastostatics 2

This chapter is devoted to the presentation of all the necessary basics of energy-based methods which are indispensable for the understanding of the contents of this book. At the beginning, the concepts of work and energy are introduced. This is followed by an introduction to the principle of virtual displacements and its application to rods, beams and continua, before the so-called principle of the stationary value of the total elastic potential is introduced and discussed as a particularly important mechanical principle. The chapter concludes with an introduction to classical approximation methods of elastostatics, namely the Ritz method on the one hand and the Galerkin method on the other hand.

The interested reader can find further information on the energy methods of mechanics in, for example, Becker and Gross (2002), Kossira (1996), Lanczos (1986), Langhaar (2016), Oden and Reddy (1983), Tauchert (1974), or Washizu (1982).

2.1 Work and Energy

In this section, the concepts of mechanical work and energy are introduced and defined for the purposes of this chapter. In the following we always assume so-called conservative forces, i.e. forces for which the work W done by them depends only on the starting and ending point of their motion, but not on the distance covered.

2.1.1 Introduction

To define the concept of mechanical work, we consider the situation of Fig. 2.1. A particle under the force $\underline{F} = (F_x, F_y, F_z)^T$ moving on a spatial trajectory (position vector $\underline{r} = (r_x, r_y, r_z)^T$) from point A to point B is investigated. The position vectors of points A and B are called $\underline{r}_A$ and $\underline{r}_B$, respectively. The force $\underline{F}$ can be

C. Mittelstedt, *Theory of Plates and Shells*,
https://doi.org/10.1007/978-3-662-66805-4_2

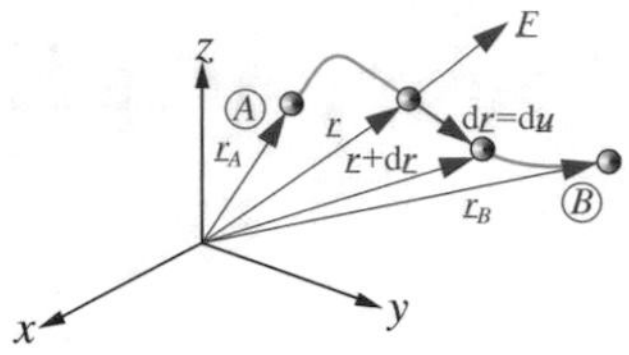

Fig. 2.1 Particle under force $\underline{F}$ on a spatial trajectory between points A and B

location dependent. This means that $\underline{F}$ can have different magnitudes and directions of action at different points of the trajectory. The work increment performed by $\underline{F} = (F_x, F_y, F_z)^T$ along a displacement increment $\mathrm{d}\underline{u} = (\mathrm{d}u, \mathrm{d}v, \mathrm{d}w)^T$ can be given as:

$$\mathrm{d}W = \underline{F}\mathrm{d}\underline{u} = F_x \mathrm{d}u + F_y \mathrm{d}v + F_z \mathrm{d}w. \tag{2.1}$$

The work increment is therefore the scalar product of $\underline{F}$ and $\mathrm{d}\underline{u}$. The total work W done by $\underline{F}$ on the trajectory between point A and point B is the sum of all work increments and can be calculated by integration over the scalar product $\underline{F}\mathrm{d}\underline{u}$:

$$W = \int_A^B \underline{F}\mathrm{d}\underline{u}. \tag{2.2}$$

Work is therefore a scalar quantity. It is given in Newton meters [Nm][1] or in Joules[2] [J], respectively, where 1Nm = 1J. The joule is the SI unit for work and energy:

$$1\mathrm{J} = 1\frac{\mathrm{kg}\cdot\mathrm{m}^2}{\mathrm{s}^2}. \tag{2.3}$$

For the case of a moment $\underline{M} = (M_x, M_y, M_z)^T$ undergoing a rotation $\underline{\varphi} = (\varphi_x, \varphi_y, \varphi_z)^T$, a work increment $\mathrm{d}W$ can be defined as:

$$\mathrm{d}W = \underline{M}\mathrm{d}\underline{\varphi} = M_x \mathrm{d}\varphi_x + M_y \mathrm{d}\varphi_y + M_z \mathrm{d}\varphi_z. \tag{2.4}$$

The total work done between an initial rotation and an end rotation φ_A and φ_B, respectively, is obtained analogously to (2.2) as:

$$W = \int_{\varphi_A}^{\varphi_B} \underline{M}\mathrm{d}\underline{\varphi}. \tag{2.5}$$

As an elementary example, consider a linear-elastic spring with spring stiffness k (Fig. 2.2, left) subjected to a force F and, as a result, to an elongation u. Using the

[1] Isaac Newton, 1642–1726, English scientist.
[2] James Prescott Joule, 1818–1889, English physicist.

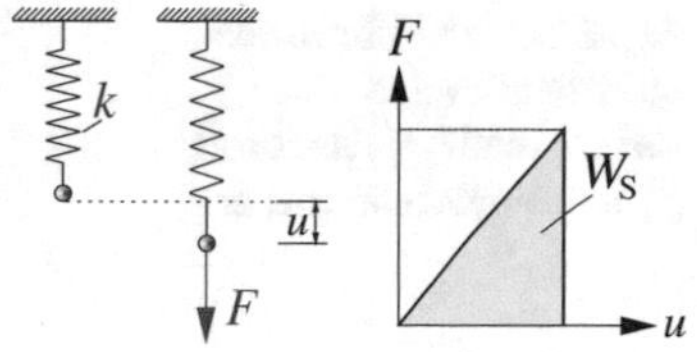

Fig. 2.2 Elongation u of a spring with stiffness k loaded by a force F, corresponding work

linear spring law $F = ku$, the work W_S can be given as:

$$W_S = \int_0^u F\mathrm{d}u = \int_0^u ku\mathrm{d}u = \frac{1}{2}ku^2 = \frac{1}{2}Fu. \tag{2.6}$$

The work performed here thus corresponds to the area below the line in the force-displacement diagram of Fig. 2.2, right.

2.1.2 Inner and Outer Work

In problems of solid mechanics, it is necessary to distinguish fundamentally between different types of work, namely the external work W_a of the applied forces on the one hand and the inner work W_i on the other hand. The external work is the work done by the forces acting on a solid (single forces and moments, line loads, surface loads and volume loads) as a result of the resulting deformations. Taking the linear spring considered above as an example, the work W_S performed as given by (2.6) is an external work W_a.

External forces cause internal forces, i.e. a state of stress. Accordingly, internal work is the work done by the internal forces along the displacements of the individual body points. In the simple example of the linear spring considered above, the external force F produces an internal spring force of the same magnitude. The inner working increment $\mathrm{d}W_i$ is then the product of the inner force F and the displacement increment $\mathrm{d}u$:

$$\mathrm{d}W_i = F\mathrm{d}u. \tag{2.7}$$

The total internal work W_i is then the sum of all work increments $\mathrm{d}W_i$:

$$W_i = \int_0^u F\mathrm{d}u. \tag{2.8}$$

If a linear-elastic spring with the spring law $F = ku$ is given, then the following results:

$$W_i = \int_0^u ku\mathrm{d}u = \frac{1}{2}ku^2 = \frac{1}{2}Fu. \tag{2.9}$$

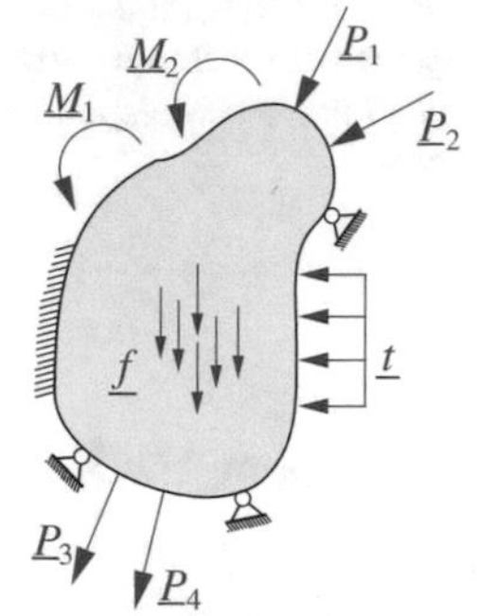

Fig. 2.3 Three-dimensional body under volume load $\underline{f}$, surface load $\underline{t}$, single forces $\underline{P}_i$ and single moments $\underline{M}_j$

Obviously $W_i = W_a$ is valid, i.e. the internal work is identical with the external work. In this context also the term of the so-called strain energy or inner energy is used. If the structure under consideration is elastic or, as a special case, even linear-elastic, as shown in the example of the spring under consideration, then the internal work is stored in the body as internal energy and can be fully recovered when the load is removed from the body.

We now extend the considerations to an arbitrary linear-elastic solid (Fig. 2.3) which is not defined in detail at this point. The internal work W_i will be discussed in detail later. The work W_a results from the given load in the form of concentrated loads, single moments, line loads, surface loads or volume loads due to the deformations resulting from the load.

Let V be the volume of the solid under consideration, and the load be given in terms of m individual forces $\underline{P}_i$ $(i = 1, 2, \ldots, m)$ and n individual moments $\underline{M}_j$ $(j = 1, 2, \ldots., n)$ as well as the volume load $\underline{f}$ and the surface load $\underline{t}$, where $\underline{t}$ acts on the boundary surface ∂V_t. The work W_a can then be given as follows:

$$W_a = \frac{1}{2}\int\limits_V \underline{f}\,\underline{u}\mathrm{d}V + \frac{1}{2}\int\limits_{\partial V_t} \underline{t}\,\underline{u}\mathrm{d}S + \frac{1}{2}\sum_{i=1}^{m} \underline{P}_i \underline{u}_i + \frac{1}{2}\sum_{j=1}^{n} \underline{M}_j \underline{\varphi}_j. \tag{2.10}$$

2.1.3 Principle of Work and Energy and the Law of Conservation of Energy

The principle of work and energy of elastostatics states that the work done by given loads of an elastic solid W_a is completely converted into internal work W_i. It is assumed that the system is conservative. Hence:

$$W_i = W_a. \tag{2.11}$$

This leads to the conclusion that the work done W_a is stored as internal energy in the elastic solid and can be fully recovered when the body is unloaded. The principle of

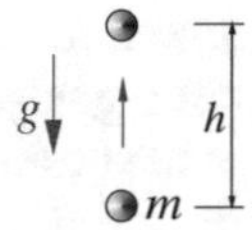

Fig. 2.4 Lifting work using the example of a point mass

work and energy is valid for any elastic closed system. A further conclusion is that in such an elastic system no energy can be lost.

The internal energy or the strain energy, for which there are many symbols in the literature, e.g. Π_i, W_i, U, will be discussed in detail at a later stage.

The concept of energy can be defined quite basically as follows:

Energy is the ability to perform work.

We restrict ourselves in all explanations of this book to static problems. Therefore, the so-called kinetic energy does not play any role in the following. Important at this point, however, is the so-called potential energy. An elementary example is a point mass m (Fig. 2.4), which is lifted against the gravitational field of the earth by the distance h. The work done in this process amounts to $W = mgh$ and at the same time represents the potential energy supplied to the point mass. Obviously, by supplying potential energy, the mass point was put into the position to perform work itself. This also applies analogously to the linear-elastic spring already considered, in which an internal work is also performed by the tensioning or by the work performed, and to which an internal energy was thereby supplied, which in turn can be recovered and converted into work. Potential energy is often simply denoted as potential. The amount of potential must be equal to the amount of work previously performed and required to generate the potential energy.

It is impossible to realize a mechanical system in which the potential energy caused by the performed work is greater than the work itself. Such a so-called perpetuum mobile is a physically impossible structure. However, energy can be transferred from one form to another. For example, the potential energy of a mass m lifted by the height h can be transferred into kinetic energy by releasing it in the gravitational field of the earth. From this, the law of conservation of energy can be motivated:

If we consider a frictionless mechanical and self-contained system without any other influence, its total energy is the same at any time. In this case, energy can neither be lost nor generated, it can only pass from one form to the other.

2.1.4 Strain Energy and Complementary Strain Energy

The Rod

As the first basic structural element, consider the rod of Fig. 2.5 (cross-sectional area A), which is subjected to the normal force N, which causes the uniformly distributed normal stress $\sigma_{xx} = \frac{N}{A}$. Let the rod be elastic, but at this point linear-elastic material behavior is not necessarily assumed. We now examine an infinitesimal element of length $\mathrm{d}x$ and consider its deformation. The displacement u occurs at the negative cutting edge, and the displacement $u + \frac{\mathrm{d}u}{\mathrm{d}x}\mathrm{d}x = u + u'\mathrm{d}x = u + \varepsilon_{xx}\mathrm{d}x$ is present at the positive cutting edge. The normal stress σ_{xx} performs a work increment along the infinitesimal strain $\mathrm{d}\varepsilon_{xx}$, the work increment is $\sigma_{xx}A\mathrm{d}\varepsilon_{xx}\mathrm{d}x$. Considering $\mathrm{d}V = A\mathrm{d}x$, we obtain $\sigma_{xx}\mathrm{d}\varepsilon_{xx}\mathrm{d}V$ for the work increment. Introducing the abbreviation $\mathrm{d}U_0 = \sigma_{xx}\mathrm{d}\varepsilon_{xx}$ then yields the work increment $\mathrm{d}U_0\mathrm{d}V$. The quantity $\mathrm{d}U_0$ is the so-called incremental strain energy density. It is shown in the stress-strain diagram of Fig. 2.6, top left. The strain energy density U_0 is obtained by integrating $\mathrm{d}U_0$ over the entire strain history ε_{xx} up to the desired strain value:

$$U_0 = \int_0^{\varepsilon_{xx}} \sigma_{xx}\mathrm{d}\varepsilon_{xx}. \tag{2.12}$$

The strain energy density is the area below the line of Fig. 2.6, top left.

In the case of linear elasticity, Hooke's law $\sigma_{xx} = E\varepsilon_{xx}$ and the kinematic relation $\varepsilon_{xx} = u'$ give the following expression for the normal stress σ_{xx}:

$$\sigma_{xx} = E\varepsilon_{xx} = Eu'. \tag{2.13}$$

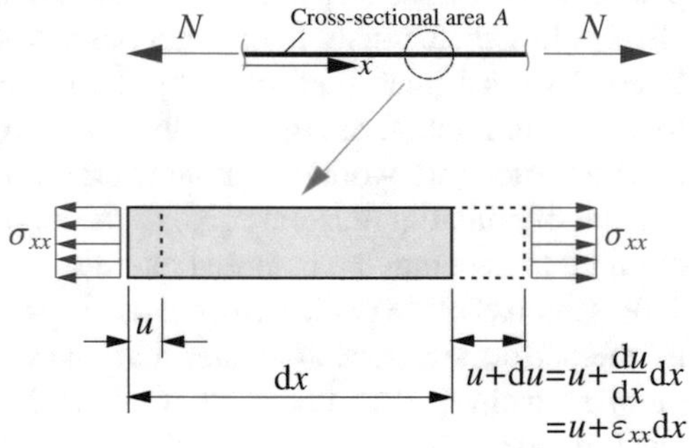

Fig. 2.5 Rod under tension (top), infinitesimal element (bottom) in the undeformed and deformed state

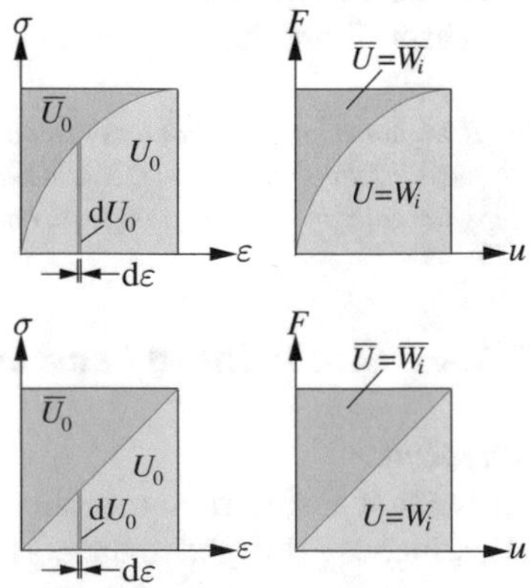

Fig. 2.6 Stress-strain diagram with strain energy density U_0 and complementary strain energy density $\bar{U}_0$ (left), force-displacement diagram with strain energy $U = W_i$ and complementary strain energy $\bar{U} = \bar{W}_i$ (right) at the example of the rod; elastic material behavior (top) and the special case of linear elasticity (bottom)

The normal force N follows from the integration of the normal stress σ_{xx} over the cross section A:

$$N = \int_A \sigma_{xx} \mathrm{d}A = \int_A E u' \mathrm{d}A = E u' \int_A \mathrm{d}A = E A u'. \tag{2.14}$$

Hence:

$$u' = \frac{N}{EA}. \tag{2.15}$$

The normal stress σ_{xx} then follows as:

$$\sigma_{xx} = E u' = \frac{N}{A}. \tag{2.16}$$

The strain energy density U_0 is then obtained as:

$$U_0 = \int_0^{\varepsilon_{xx}} \sigma_{xx} \mathrm{d}\varepsilon_{xx} = E \int_0^{\varepsilon_{xx}} \varepsilon_{xx} \mathrm{d}\varepsilon_{xx} = \frac{1}{2} E \varepsilon_{xx}^2 = \frac{1}{2} \sigma_{xx} \varepsilon_{xx} = \frac{1}{2} E u'^2. \tag{2.17}$$

The work increment $\mathrm{d}W_i$ of the infinitesimal element follows when the strain energy density U_0 is multiplied by the volume $\mathrm{d}V$:

$$\mathrm{d}W_i = \int_0^{\varepsilon_{xx}} \sigma_{xx} \mathrm{d}\varepsilon_{xx} \mathrm{d}V = U_0 \mathrm{d}V. \tag{2.18}$$

The internal energy stored in the rod or the strain energy W_i or U then follows from the integration over the rod volume. With $\mathrm{d}V = A\mathrm{d}x$ it follows:

$$W_i = \int_0^l \int_0^{\varepsilon_{xx}} \sigma_{xx} \mathrm{d}\varepsilon_{xx} A \mathrm{d}x = A \int_0^l U_0 \mathrm{d}x = U. \tag{2.19}$$

The strain energy is shown in the force-displacement diagram of Fig. 2.6, right. It is the surface below the working line.

In the case of linear elasticity, the result for W_i is:

$$W_i = \int_V U_0 \mathrm{d}x = \frac{1}{2} E \int_0^l \int_A u'^2 \mathrm{d}A \mathrm{d}x = \frac{1}{2} E A \int_0^l u'^2 \mathrm{d}x. \tag{2.20}$$

With $u'^2 = \varepsilon_{xx}^2 = \frac{\sigma_{xx}^2}{E^2} = \frac{N^2}{E^2A^2}$ we obtain:

$$W_i = \frac{1}{2}EA\int_0^l \frac{N^2}{E^2A^2}\mathrm{d}x = \frac{1}{2}\int_0^l \frac{N^2}{EA}\mathrm{d}x. \tag{2.21}$$

If the normal force N and the extensional stiffness EA are constant, then it follows:

$$W_i = \frac{N^2 l}{2EA}. \tag{2.22}$$

Quite analogously, the so-called complementary strain energy density can be defined as:

$$\overline{U}_0 = \int_0^{\sigma_{xx}} \varepsilon_{xx}\mathrm{d}\sigma_{xx}. \tag{2.23}$$

From Fig. 2.6, left, it can be concluded that

$$U_0 + \overline{U}_0 = \sigma_{xx}\varepsilon_{xx} \tag{2.24}$$

holds. For the case of linear elasticity one obtains:

$$\overline{U}_0 = \frac{1}{2}\sigma_{xx}\varepsilon_{xx}. \tag{2.25}$$

This result can also be directly deduced from Fig. 2.6, bottom left. For linear elastic material the working line is a straight line. The surface above the working line then represents exactly the value for $\overline{U}_0$, which leads to (2.25).

The complementary strain energy $\overline{W}_i$ or $\overline{U}$ can be determined as follows:

$$\mathrm{d}\overline{W}_i = \int_0^{\sigma_{xx}} \varepsilon_{xx}\mathrm{d}\sigma_{xx}\mathrm{d}V = \overline{U}_0\mathrm{d}V, \tag{2.26}$$

or after integration over the rod volume:

$$\overline{W}_i = \int_0^l \int_0^{\sigma_{xx}} \varepsilon_{xx}\mathrm{d}\sigma_{xx} A\mathrm{d}x = A\int_0^l \overline{U}_0\mathrm{d}x = \overline{U}. \tag{2.27}$$

For the special case of linear elasticity one obtains:

$$\overline{W}_i = \frac{1}{2}\int_0^l \frac{N^2}{EA}\mathrm{d}x. \tag{2.28}$$

In the case of linear elasticity, the expressions for W_i and $\overline{W}_i$ are identical (see also Fig. 2.6, bottom).

The Euler-Bernoulli Beam

In this section, we extend the considerations to the Euler-Bernoulli beam with simultaneous normal force action. We assume the special case of uniaxial bending with the bending moment M and the transverse shear force Q. The prerequisite is that one of the two principal axes of the cross-section under consideration coincides with the $z-$axis. Let the beam be subjected to the arbitrary continuous line load $q(x)$ in the $z-$direction and to the arbitrary continuous axial line load $n(x)$. Let the beam be characterized by the varying but continuous extensional stiffness $EA(x)$ and bending stiffness $EI(x)$. The equilibrium conditions for this beam situation are:

$$N' = -n, \quad Q' = -q, \quad M' = Q. \tag{2.29}$$

The area integrals to be considered here are the moment of inertia I with respect to bending of the beam in $z-$ direction, as well as the static moment $S_y = S$ and the cross-sectional area A:

$$A = \int_A \mathrm{d}A, \quad S = \int_A z\mathrm{d}A, \quad I = \int_A z^2\mathrm{d}A. \tag{2.30}$$

The displacement field of the Euler-Bernoulli beam follows for the case of uniaxial bending under the assumption of the validity of the normal hypothesis and the hypothesis that the cross-sections remain undeformed as:

$$u_P = u - z_P w', \quad w_P = w. \tag{2.31}$$

Here, P is an arbitrary point of the cross section at any location z_P. The displacement quantities u and w are the displacements of the centroid axis in $x-$ and $z-$ direction, respectively.

The only strain component remaining for the Euler-Bernoulli beam is the normal strain ε_{xx}, which can be determined as:

$$\varepsilon_{xx} = \frac{\mathrm{d}u_P}{\mathrm{d}x} = u'_P = u' - z_P w''. \tag{2.32}$$

The normal stress σ_{xx} can then be determined from Hooke's law $\sigma_{xx} = E\varepsilon_{xx}$ as:

$$\sigma_{xx} = E\left(u' - zw''\right), \tag{2.33}$$

where from this point on the indexing concerning the point P is discarded. The normal force N results as the integral of the normal stress σ_{xx} over the cross-sectional area

A:

$$N = \int_A \sigma_{xx} \mathrm{d}A = E \int_A u' \mathrm{d}A - E \int_A z w'' \mathrm{d}A$$
$$= E u' \int_A \mathrm{d}A - E w'' \int_A z \mathrm{d}A. \tag{2.34}$$

The surface integrals occurring here are the cross-sectional area A and the static moment S according to (2.30). The static moment vanishes in the presence of a principal axis system, leaving the following expression for the normal force N:

$$N = EAu'. \tag{2.35}$$

The bending moment M can be determined in the same way:

$$M = \int_A \sigma_{xx} z \mathrm{d}A = E \int_A u' z \mathrm{d}A - E \int_A z^2 w'' \mathrm{d}A$$
$$= E u' \int_A z \mathrm{d}A - E w'' \int_A z^2 \mathrm{d}A. \tag{2.36}$$

With $S = 0$ and the second-order moment of inertia I it follows:

$$M = -EIw''. \tag{2.37}$$

Substituting (2.35) and (2.37) into (2.33) then gives the following expression for the normal stress σ_{xx} for the case of uniaxial bending:

$$\sigma_{xx} = \frac{N}{A} + \frac{M}{I} z. \tag{2.38}$$

The strain energy density U_0 is obtained for the Euler-Bernoulli beam as:

$$U_0 = \int_0^{\varepsilon_{xx}} \sigma_{xx} \mathrm{d}\varepsilon_{xx} = E \int_0^{\varepsilon_{xx}} \varepsilon_{xx} \mathrm{d}\varepsilon_{xx} = \frac{1}{2} E \varepsilon_{xx}^2 = \frac{1}{2} E \left(u' - z w''\right)^2. \tag{2.39}$$

The strain energy W_i can be determined by integration over the beam volume $\mathrm{d}V = \mathrm{d}A\mathrm{d}x$ as:

$$W_i = \int_V U_0 \mathrm{d}V = \frac{1}{2} E \int_0^l \int_A \left(u'^2 - 2zu'w'' + z^2 w''^2\right) \mathrm{d}A\mathrm{d}x. \tag{2.40}$$

Taking into account (2.30) and $S = 0$, we obtain:

$$W_i = \frac{1}{2}EA\int_0^l u'^2 \mathrm{d}x + \frac{1}{2}EI\int_0^l w''^2 \mathrm{d}x. \tag{2.41}$$

The complementary strain energy density follows as:

$$\overline{U}_0 = \int_0^{\sigma_{xx}} \varepsilon_{xx}\mathrm{d}\sigma_{xx} + \int_0^{\tau_{xz}} \gamma_{xz}\mathrm{d}\tau_{xz}. \tag{2.42}$$

It should be noted that the second term appearing here cannot be neglected without further considerations. The shear strain γ_{xz} disappears due to the kinematic assumptions for the Euler-Bernoulli beam, but the shear stress τ_{xz} must occur for equilibrium reasons. From Hooke's law $\sigma_{xx} = E\varepsilon_{xx}$ and $\tau_{xz} = G\gamma_{xz}$ then follows:

$$\overline{U}_0 = \int_0^{\sigma_{xx}} \frac{\sigma_{xx}}{E}\mathrm{d}\sigma_{xx} + \int_0^{\tau_{xz}} \frac{\tau_{xz}}{G}\mathrm{d}\tau_{xz} = \frac{\sigma_{xx}^2}{2E} + \frac{\tau_{xz}^2}{2G}. \tag{2.43}$$

The complementary strain energy $\overline{W}_i$ with $\tau_{xz} = \frac{QS}{Ib}$ can then be determined, e.g., for a rectangular cross section of width b by integration over the volume of the beam:

$$\begin{aligned}
\overline{W}_i &= \int_V \overline{U}_0 \mathrm{d}V = \int_0^l \int_A \left(\frac{\sigma_{xx}^2}{2E} + \frac{\tau_{xz}^2}{2G}\right)\mathrm{d}A\mathrm{d}x \\
&= \int_0^l \int_A \left[\frac{1}{2E}\left(\frac{N}{A} + \frac{M}{I}z\right)^2 + \frac{1}{2G}\left(\frac{QS}{Ib}\right)^2\right]\mathrm{d}A\mathrm{d}x.
\end{aligned} \tag{2.44}$$

After integration it follows:

$$\overline{W}_i = \frac{1}{2}\int_0^l \frac{N^2}{EA}\mathrm{d}x + \frac{1}{2}\int_0^l \frac{M^2}{EI}\mathrm{d}x + \frac{1}{2}\int_0^l \frac{Q^2}{GA_V}\mathrm{d}x. \tag{2.45}$$

In this A_V is defined as:

$$\frac{1}{A_V} = \frac{1}{I^2}\int_A \frac{S^2}{b^2}\mathrm{d}A. \tag{2.46}$$

The quantity A_V is the so-called effective area, which indicates which area of a shear-loaded cross-section is effectively available for load transfer. It can be expressed by means of the so-called shear correction factor K and the actual cross-sectional area

Fig. 2.7 Infinitesimal three-dimensional sectional element

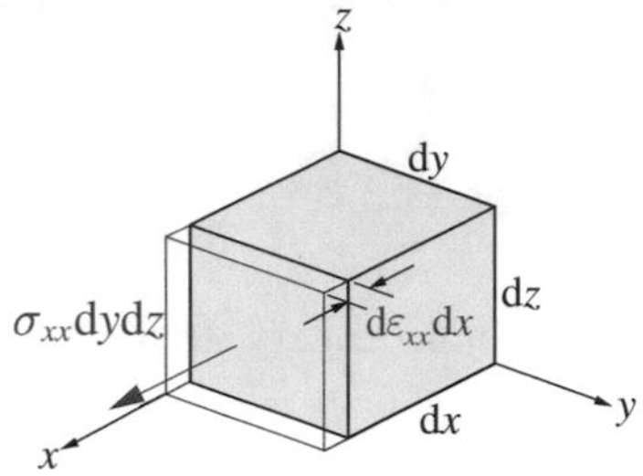

A as:

$$A_V = KA. \tag{2.47}$$

Thus:

$$\overline{W}_i = \frac{1}{2}\int_0^l \frac{N^2}{EA}\mathrm{d}x + \frac{1}{2}\int_0^l \frac{M^2}{EI}\mathrm{d}x + \frac{1}{2}\int_0^l \frac{Q^2}{KGA}\mathrm{d}x. \tag{2.48}$$

Generalization for the Continuum

In this section, we extend our considerations to the three-dimensional continuum by considering the infinitesimal sectional element of a three-dimensional solid shown in Fig. 2.7, on which for the moment only the normal stress σ_{xx} acts. Due to the normal stress σ_{xx}, the working increment $\sigma_{xx}\mathrm{d}y\mathrm{d}z\mathrm{d}\varepsilon_{xx}\mathrm{d}x$ is performed along the infinitesimal strain $\mathrm{d}\varepsilon_{xx}$. With $\mathrm{d}x\mathrm{d}y\mathrm{d}z = \mathrm{d}V$ one obtains for this expression the form $\sigma_{xx}\mathrm{d}\varepsilon_{xx}\mathrm{d}V$ resp. $\mathrm{d}U_0\mathrm{d}V$, so that the working increment $\mathrm{d}W_i$ follows as:

$$\mathrm{d}W_i = \int_0^{\varepsilon_{xx}} \sigma_{xx}\mathrm{d}\varepsilon_{xx}\mathrm{d}V = U_0\mathrm{d}V = \mathrm{d}U. \tag{2.49}$$

Consider now the case that all normal stresses σ_{xx}, σ_{yy}, σ_{zz} and all shear stresses τ_{yz}, τ_{xz}, τ_{xy} occur. Then it follows:

$$\begin{aligned}\mathrm{d}W_i &= \int_0^{\varepsilon_{xx}} \sigma_{xx}\mathrm{d}\varepsilon_{xx}\mathrm{d}V + \int_0^{\varepsilon_{yy}} \sigma_{yy}\mathrm{d}\varepsilon_{yy}\mathrm{d}V + \int_0^{\varepsilon_{zz}} \sigma_{zz}\mathrm{d}\varepsilon_{zz}\mathrm{d}V \\ &\quad + \int_0^{\gamma_{yz}} \tau_{yz}\mathrm{d}\gamma_{yz}\mathrm{d}V + \int_0^{\gamma_{xz}} \tau_{xz}\mathrm{d}\gamma_{xz}\mathrm{d}V + \int_0^{\gamma_{xy}} \tau_{xy}\mathrm{d}\gamma_{xy}\mathrm{d}V \\ &= U_0\mathrm{d}V.\end{aligned} \tag{2.50}$$

In index notation and employing Einstein's summation convention, this expression can be written in the following form:

$$\mathrm{d}W_i = \int_0^{\varepsilon_{ij}} \sigma_{ij} \mathrm{d}\varepsilon_{ij} \mathrm{d}V = U_0 \mathrm{d}V = \mathrm{d}U. \tag{2.51}$$

Analogously, one obtains for $\mathrm{d}\overline{W}_i$:

$$\begin{aligned} \mathrm{d}\overline{W}_i &= \int_0^{\sigma_{xx}} \varepsilon_{xx} \mathrm{d}\sigma_{xx} \mathrm{d}V + \int_0^{\sigma_{yy}} \varepsilon_{yy} \mathrm{d}\sigma_{yy} \mathrm{d}V + \int_0^{\sigma_{zz}} \varepsilon_{zz} \mathrm{d}\sigma_{zz} \mathrm{d}V \\ &\quad + \int_0^{\tau_{yz}} \gamma_{yz} \mathrm{d}\tau_{yz} \mathrm{d}V + \int_0^{\tau_{xz}} \gamma_{xz} \mathrm{d}\tau_{xz} \mathrm{d}V + \int_0^{\tau_{xy}} \gamma_{xy} \mathrm{d}\tau_{xy} \mathrm{d}V \\ &= \overline{U}_0 \mathrm{d}V, \end{aligned} \tag{2.52}$$

or in index notation:

$$\mathrm{d}\overline{W}_i = \int_0^{\sigma_{ij}} \varepsilon_{ij} \mathrm{d}\sigma_{ij} \mathrm{d}V = \overline{U}_0 \mathrm{d}V. \tag{2.53}$$

The total performed work W_i and the complementary work $\overline{W}_i$ is then obtained as:

$$\begin{aligned} W_i &= \int_V U_0 \mathrm{d}V = U, \\ \overline{W}_i &= \int_V \overline{U}_0 \mathrm{d}V = \overline{U}. \end{aligned} \tag{2.54}$$

For the strain energy density U_0 the following holds:

$$\mathrm{d}U_0 = \sigma_{ij} \mathrm{d}\varepsilon_{ij}. \tag{2.55}$$

From this it can be concluded:

$$\sigma_{ij} = \frac{\partial U_0}{\partial \varepsilon_{ij}}. \tag{2.56}$$

Given that the strain energy density is known, a stress component σ_{ij} can be obtained by partial derivation of U_0 according to the associated strain component ε_{ij}. This is quite similar for $\overline{U}_0$ with $\mathrm{d}\overline{U}_0 = \varepsilon_{ij} \mathrm{d}\sigma_{ij}$:

$$\varepsilon_{ij} = \frac{\partial \overline{U}_0}{\partial \sigma_{ij}}. \tag{2.57}$$

2.1.5 General Principle of Work and Energy of Elastostatics

From the equilibrium conditions (1.36) and the kinematic Eq. (1.60), a general principle of work and energy can be derived which establishes a relationship between any equilibrium group of stresses and forces (denoted by the superscript (1)) and any kinematically permissible field of strains and displacements (denoted by the superscript (2)). The equilibrium group and the strain and displacement field do not necessarily have to be a consequence of each other. Now consider an equilibrium group consisting of the three normal stresses $\sigma_{xx}^{(1)}$, $\sigma_{yy}^{(1)}$, $\sigma_{zz}^{(1)}$ and the three shear stresses $\tau_{xy}^{(1)}$, $\tau_{xz}^{(1)}$, $\tau_{yz}^{(1)}$. In addition, the volume forces $f_x^{(1)}$, $f_y^{(1)}$, $f_z^{(1)}$ and the stress vectors $t_x^{(1)}$, $t_y^{(1)}$, $t_z^{(1)}$ are given. The stress field is required to satisfy the equilibrium conditions (1.36) as well as the boundary conditions $t_x = t_{x0}$, $t_y = t_{y0}$, $t_z = t_{z0}$ identically on ∂V_t. The strain and displacement field is composed of the displacements $u^{(2)}$, $v^{(2)}$, $w^{(2)}$ and the strain components $\varepsilon_{xx}^{(2)}$, $\varepsilon_{yy}^{(2)}$, $\varepsilon_{zz}^{(2)}$, $\gamma_{yz}^{(2)}$, $\gamma_{xz}^{(2)}$, $\gamma_{xy}^{(2)}$. Let the displacements and strains be coupled by the kinematic relations (1.60). The displacement field satisfies the given displacement boundary conditions $u = u_0$, $v = v_0$, $w = w_0$ on ∂V_u identically.

We consider the local equilibrium conditions (1.36):

$$\frac{\partial \sigma_{xx}^{(1)}}{\partial x} + \frac{\partial \tau_{xy}^{(1)}}{\partial y} + \frac{\partial \tau_{xz}^{(1)}}{\partial z} + f_x^{(1)} = 0,$$
$$\frac{\partial \tau_{xy}^{(1)}}{\partial x} + \frac{\partial \sigma_{yy}^{(1)}}{\partial y} + \frac{\partial \tau_{yz}^{(1)}}{\partial z} + f_y^{(1)} = 0,$$
$$\frac{\partial \tau_{xz}^{(1)}}{\partial x} + \frac{\partial \tau_{yz}^{(1)}}{\partial y} + \frac{\partial \sigma_{zz}^{(1)}}{\partial z} + f_z^{(1)} = 0. \tag{2.58}$$

In what follows, we make use of the index notation using Einstein's summation convention, and we use the coordinates x_1, x_2, x_3, and the indices i and j (where $i, j = 1, 2, 3$). Thus:

$$\frac{\partial \sigma_{ij}^{(1)}}{\partial x_j} + f_i^{(1)} = 0.$$

These equations are now multiplied by the displacements $u_i^{(2)}$, and an integration over the body volume V is performed:

$$\int_V \frac{\partial \sigma_{ij}^{(1)}}{\partial x_j} u_i^{(2)} \mathrm{d}V + \int_V f_i^{(1)} u_i^{(2)} \mathrm{d}V = 0.$$

We transform the integrand of the first integral using the product rule of differential calculus:

$$\frac{\partial \sigma_{ij}^{(1)}}{\partial x_j} u_i^{(2)} = \frac{\partial}{\partial x_j}\left(\sigma_{ij}^{(1)} u_i^{(2)}\right) - \sigma_{ij}^{(1)} \frac{\partial u_i^{(2)}}{\partial x_j}. \tag{2.59}$$

Thus:

$$\int_V \frac{\partial}{\partial x_j}\left(\sigma_{ij}^{(1)} u_i^{(2)}\right) \mathrm{d}V - \int_V \sigma_{ij}^{(1)} \frac{\partial u_i^{(2)}}{\partial x_j} \mathrm{d}V + \int_V f_i^{(1)} u_i^{(2)} \mathrm{d}V = 0. \tag{2.60}$$

Application of the Gaussian[3] integral theorem to the first integral on the boundary part ∂V yields:

$$\int_V \frac{\partial}{\partial x_j}\left(\sigma_{ij}^{(1)} u_i^{(2)}\right) \mathrm{d}V = \int_{\partial V} \sigma_{ij}^{(1)} u_i^{(2)} n_j \mathrm{d}S. \tag{2.61}$$

Herein, n_j is the component of the normal vector $\underline{n}$ in the direction of the coordinate x_j. It follows:

$$\int_{\partial V} \sigma_{ij}^{(1)} u_i^{(2)} n_j \mathrm{d}S - \int_V \sigma_{ij}^{(1)} \frac{\partial u_i^{(2)}}{\partial x_j} \mathrm{d}V + \int_V f_i^{(1)} u_i^{(2)} \mathrm{d}V = 0. \tag{2.62}$$

Between the internal stresses and the external stress vector the relation $\sigma_{ij}^{(1)} n_j = t_i^{(1)}$ holds. Therein, t_i is the i–th component of the stress vector $\underline{t}$.

With $\sigma_{ij}^{(1)} \frac{\partial u_i^{(2)}}{\partial x_j} = \sigma_{ij}^{(1)} \varepsilon_{ij}^{(2)}$, the general principle of work and energy of elastostatics can be formulated as follows:

$$\int_V \sigma_{ij}^{(1)} \varepsilon_{ij}^{(2)} \mathrm{d}V = \int_V f_i^{(1)} u_i^{(2)} \mathrm{d}V + \int_{\partial V} t_i^{(1)} u_i^{(2)} \mathrm{d}S. \tag{2.63}$$

The surface ∂V of the body can be divided into the parts ∂V_t and ∂V_u. On the part ∂V_t the stresses $\underline{t}_0$ are given, whereas on the part ∂V_u the displacements $\underline{u}_0$ are prescribed. Hence:

$$\int_V \sigma_{ij}^{(1)} \varepsilon_{ij}^{(2)} \mathrm{d}V = \int_V f_i^{(1)} u_i^{(2)} \mathrm{d}V + \int_{\partial V_t} t_{i0}^{(1)} u_i^{(2)} \mathrm{d}S + \int_{\partial V_u} t_i^{(1)} u_{i0}^{(2)} \mathrm{d}S. \tag{2.64}$$

[3] Johann Carl Friedrich Gauß, 1777–1855, German polymath.

Using the reference frame x, y, z, one obtains:

$$\begin{aligned}
&\int\limits_V \left(\sigma_{xx}^{(1)}\varepsilon_{xx}^{(2)} + \sigma_{yy}^{(1)}\varepsilon_{yy}^{(2)} + \sigma_{zz}^{(1)}\varepsilon_{zz}^{(2)} + \tau_{yz}^{(1)}\gamma_{yz}^{(2)} + \tau_{xz}^{(1)}\gamma_{xz}^{(2)} + \tau_{xy}^{(1)}\gamma_{xy}^{(2)}\right) \mathrm{d}V \\
&= \int\limits_V \left(f_x^{(1)}u^{(2)} + f_y^{(1)}v^{(2)} + f_z^{(1)}w^{(2)}\right) \mathrm{d}V \\
&\quad + \int\limits_{\partial V_t} \left(t_{x0}^{(1)}u^{(2)} + t_{y0}^{(1)}v^{(2)} + t_{z0}^{(1)}w^{(2)}\right) \mathrm{d}S \\
&\quad + \int\limits_{\partial V_u} \left(t_x^{(1)}u_0^{(2)} + t_y^{(1)}v_0^{(2)} + t_z^{(1)}w_0^{(2)}\right) \mathrm{d}S. \qquad (2.65)
\end{aligned}$$

The integral of the left side represents the work done by the stresses along the strain components associated with them. It is therefore the internal work. The integral of the right side includes the work performed by the volume forces and by the surface loads. The general principle of work and energy presented here is independent of the type of material behavior and is valid for any material.

2.2 The Principle of Virtual Displacements

The principle of virtual displacements is a fundamental and extremely important principle in the framework of elastostatics, from which a whole series of computational methods in elastostatics can be derived. In this section, we will mainly discuss how the principle of virtual displacements can be used for the derivation of differential equations/equilibrium conditions and boundary conditions for problems in elastostatics.

2.2.1 Virtual Displacements and Virtual Work

To introduce the notion of virtual displacements, let us consider the beam of Fig. 2.8, which is under an arbitrarily distributed line load $q(x)$. The beam has the length l and is simply supported at both ends. Let $w(x)$ be the deflection of the beam in the equilibrium state.

We now introduce an infinitesimal variation $\delta w(x)$ of the deflection from the equilibrium position as indicated in Fig. 2.8. We refer to this variation $\delta w(x)$ as virtual displacement. Virtual displacements have the following properties:

- Virtual displacements are infinitesimally small.
- They are virtual and do not exist in reality.
- Virtual displacements obey the given geometric boundary conditions.

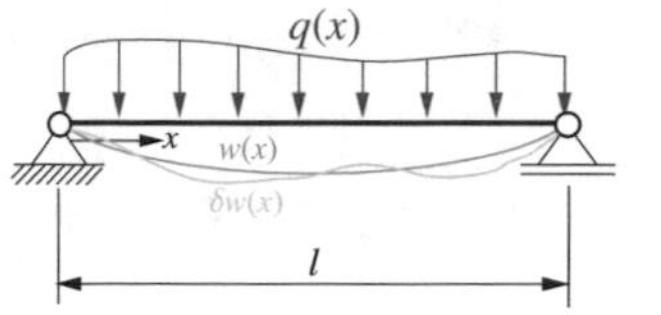

Fig. 2.8 Equilibrium configuration $w(x)$ of a beam and admissible virtual displacements $\delta w(x)$

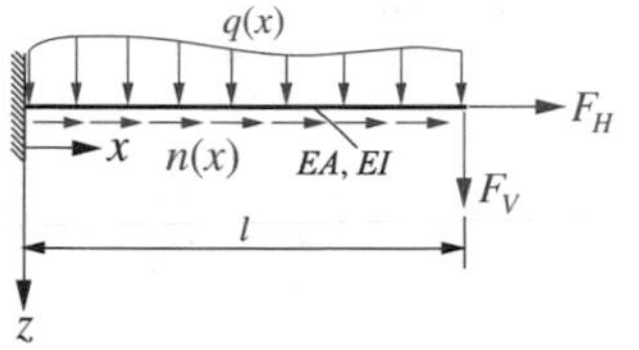

Fig. 2.9 Cantilever beam

From the last requirement it follows, using the example of Fig. 2.8, that at the two support points the virtual displacements $\delta w(x)$ must become zero. Thus $\delta w(x = 0) = 0$ and $\delta w(x = l) = 0$ hold. Any infinitesimal variation $\delta w(x)$ satisfying these conditions is therefore admissible.

In a static system subjected to virtual displacements, virtual work is performed. Since these are virtual displacements from the equilibrium position, the internal and external force and stresses remain unchanged. However, they perform virtual work along the virtual displacements and thus also along the virtual strains. A distinction is made between virtual internal work δW_i and virtual external work δW_a. Using the example of a solid body under a single force F, the virtual work δW_a is the product of the force F and the virtual displacement δu applied at the force application point in the direction of action of the force, i.e. $\delta W_a = F\delta u$, or in vectorial notation $\delta W_a = F\delta \underline{u}$.

As an example, consider the cantilever beam shown in Fig. 2.9, which is loaded by the two line loads $q(x)$ and $n(x)$ and at its free end by the two single forces F_H and F_V. Again, only uniaxial bending is considered. This assumes that y and z are the principal axes of the beam cross-section. The geometrical boundary conditions for the given situation are as follows:

$$w(x = 0) = 0, \quad w'(x = 0) = 0, \quad u(x = 0) = 0. \tag{2.66}$$

The dynamic boundary conditions are:

$$Q(x = l) = F_V, \quad N(x = l) = F_H, \quad M(x = l) = 0. \tag{2.67}$$

We now consider the virtual displacements δu and δw, which are assumed to satisfy the geometric boundary conditions (2.66) identically, i.e. $\delta u(x = 0) = 0$, $\delta w(x = 0) = 0$, $\delta w'(x = 0) = 0$.

Let us first consider the virtual work δW_a. The two single forces F_H and F_V perform virtual work along the virtual displacements $\delta u(x = l)$ and $\delta w(x = l)$. The arbitrary but continuously distributed line load $q(x)$ is multiplied by the virtual displacement $\delta w(x)$ and integrated over the beam length. The same procedure is used for the line load $n(x)$. The virtual work δW_a is then obtained as:

$$\delta W_a = \int_0^l q(x)\delta w \mathrm{d}x + \int_0^l n(x)\delta u \mathrm{d}x + F_V \delta w(x=l) + F_H \delta u(x=l). \quad (2.68)$$

The virtual inner work δW_i results from the virtual strain energy density δU_0:

$$\delta W_i = \int_V \delta U_0 \mathrm{d}V = \int_V \sigma_{ij}\delta\varepsilon_{ij}\mathrm{d}V = \int_0^l \int_A \sigma_{xx}\delta\varepsilon_{xx}\mathrm{d}A\mathrm{d}x. \quad (2.69)$$

With the normal strain ε_{xx} as

$$\varepsilon_{xx} = u' - z w'' \quad (2.70)$$

and the virtual normal strain $\delta\varepsilon_{xx}$

$$\delta\varepsilon_{xx} = \delta u' - z\delta w'' \quad (2.71)$$

the virtual internal work δW_i is given as:

$$\delta W_i = \int_0^l \int_A \sigma_{xx}\left(\delta u' - z\delta w''\right)\mathrm{d}A\mathrm{d}x. \quad (2.72)$$

Integrating this expression over the cross-sectional area A, we obtain with $N = \int_A \sigma_{xx}\mathrm{d}A$ and $M = \int_A \sigma_{xx} z \mathrm{d}A$:

$$\delta W_i = \int_0^l \left(N\delta u' - M\delta w''\right)\mathrm{d}x. \quad (2.73)$$

The relations derived so far are valid independently of the material behavior. In the case of linear elasticity, the constitutive relations $N = EAu'$ and $M = -EIw''$ apply:

$$\delta W_i = \int_0^l EAu'\delta u'\mathrm{d}x + \int_0^l EIw''\delta w''\mathrm{d}x. \quad (2.74)$$

The considerations are now extended to an arbitrary three-dimensional solid with the volume V, which is loaded by the volume forces $\underline{f}$ and m individual forces $\underline{F}_i$. The boundary ∂V is divided into two subdomains ∂V_u and ∂V_t. On the boundary region ∂V_u displacements are given, i.e. the displacement vector $\underline{u}$ is prescribed on ∂V_u with $\underline{u}_0$. On the boundary region ∂V_t let the stress vector $\underline{t}$ be given by $\underline{t}_0$. The body defined in this way is now subjected to virtual displacements $\delta\underline{u}$, which, however, must disappear on ∂V_u. The virtual work δW_a then follows as:

$$\delta W_a = \int_V \underline{f} \delta \underline{u} \mathrm{d}V + \int_{\partial V_t} \underline{t}_0 \delta \underline{u} \mathrm{d}S + \sum_{i=1}^{m} \underline{F}_i \delta \underline{u}_i . \tag{2.75}$$

The virtual strain energy density δU_0 is obtained as:

$$\delta U_0 = \int_0^{\delta \varepsilon_{ij}} \sigma_{ij} \mathrm{d}(\delta \varepsilon_{ij}). \tag{2.76}$$

Since all stress components remain unchanged in the course of virtual displacements, it is possible to write:

$$\delta U_0 = \sigma_{ij} \delta \varepsilon_{ij}. \tag{2.77}$$

The virtual strain energy or the virtual internal work δW_i results from integration of δU_0 over the volume of the solid:

$$\delta W_i = \int_V \delta U_0 \mathrm{d}V = \int_V \sigma_{ij} \delta \varepsilon_{ij} \mathrm{d}V. \tag{2.78}$$

2.2.2 The Principle of Virtual Displacements

The principle of virtual displacements can be verbalized as follows:

> *A body is in equilibrium exactly when, for any permissible virtual displacement from the equilibrium position, the virtual internal work is equal to the virtual external work.*

This means:

$$\delta W_i = \delta W_a. \tag{2.79}$$

The principle of virtual displacements always leads to an equilibrium statement.

2.2.3 Analysis Rules for the Variational Operator δ

The variational operator δ stands for a very small change or variation of a certain quantity, e.g. a displacement u. The varied quantity can depend on one or more variables. One speaks of the so-called first variation δu, if a static system is subjected to a virtual displacement δu. For the variational operator δ certain analysis rules can be established which are very similar to those of differential calculus and which will be applied more frequently in the further course of this book. These rules are briefly discussed below, but detailed mathematical proofs are omitted at this point.

Let the functions f_1, f_2, f_3, ..., f_n be given, which are dependent quantities, e.g. of the displacement u. The following analysis rules apply when dealing with the variational operator δ:

1. The order of variation and differentiation is interchangeable:

$$\delta\,(\nabla u) = \nabla\,(\delta u)\,, \tag{2.80}$$

where ∇ is the Nabla-operator respectively the gradient: $\nabla = \left(\frac{\partial}{\partial x}, \frac{\partial}{\partial y}, \frac{\partial}{\partial z}\right)^T$.

2. The order of integration and variation is interchangeable:

$$\delta\left[\int\limits_V u\mathrm{d}V\right] = \int\limits_V \delta u\mathrm{d}V. \tag{2.81}$$

3. The first variation of the sum of several functions f_1, f_2, f_3, ..., f_n can be formed analogously to the summation rule of differential calculus:

$$\delta\,[f_1 \pm f_2 \pm f_3 \pm \ldots \pm f_n] = \delta f_1 \pm \delta f_2 \pm \delta f_3 \pm \ldots \pm \delta f_n. \tag{2.82}$$

4. The first variation of a product of two functions f_1, f_2 can be formed analogously to the product rule of differential calculus:

$$\delta\,[f_1 f_2] = f_2 \delta f_1 + f_1 \delta f_2. \tag{2.83}$$

5. The first variation of a quotient of two functions f_1, f_2 can be formed analogously to the quotient rule of differential calculus:

$$\delta\left[\frac{f_1}{f_2}\right] = \frac{\delta f_1}{f_2} - \frac{f_1 \delta f_2}{f_2^2}. \tag{2.84}$$

6. The first variation of a function f_1^n can be formed analogously to the chain rule of differential calculus:

$$\delta\left[f_1^n\right] = n f_1^{n-1} \delta f_1. \tag{2.85}$$

7. In the case of a function f that is a function of several dependent variables (e.g., the displacements u, v, w), the total variation δf can be formed from the sum of the partial variations:

$$\delta f\,(u, v, w) = \delta_u f + \delta_v f + \delta_w f. \tag{2.86}$$

The operators δ_u, δ_v, δ_w are the partial variations of u, v, w.

2.2.4 Formulation for the Continuum

In the following, we derive a general formulation of the principle of virtual displacements for the three-dimensional continuum and then discuss and evaluate it for rods and beams. We use the index notation for the moment in conjunction with Einstein's summation convention. The considered solid is in equilibrium and subjected to both surface loads t_i and volume loads f_i. Within the solid, the stress field σ_{ij} arises due to the applied load. Virtual displacements δu_i are imposed, which cause the virtual strains $\delta\varepsilon_{ij}$.

Now the general theorem of work and energy (cf. Sect. 2.1.5) is employed. Herein, the force quantities $\sigma_{ij}^{(1)}, f_i^{(1)}, t_i^{(1)}$ are replaced by the real force quantities σ_{ij}, f_i, t_i. Furthermore, the kinematic quantities $u_i^{(2)}, \varepsilon_{ij}^{(2)}$ are replaced by the virtual kinematic quantities δu_i, $\delta\varepsilon_{ij}$. It should be noted that on those boundary parts ∂V_u where displacements are prescribed, no virtual displacements δu_i can occur. The principle of virtual displacements is then:

$$\int_V \sigma_{ij}\delta\varepsilon_{ij}\mathrm{d}V = \int_V f_i\delta u_i\mathrm{d}V + \int_{\partial V_t} t_{i0}\delta u_i\mathrm{d}S. \tag{2.87}$$

In terms of the reference frame x, y, z it reads:

$$\begin{aligned}&\int_V \left(\sigma_{xx}\delta\varepsilon_{xx} + \sigma_{yy}\delta\varepsilon_{yy} + \sigma_{zz}\delta\varepsilon_{zz} + \tau_{yz}\delta\gamma_{yz} + \tau_{xz}\delta\gamma_{xz} + \tau_{xy}\delta\gamma_{xy}\right)\mathrm{d}V\\ &= \int_V \left(f_x\delta u + f_y\delta v + f_z\delta w\right)\mathrm{d}V + \int_{\partial V_t}\left(t_{x0}\delta u + t_{y0}\delta v + t_{z0}\delta w\right)\mathrm{d}S.\end{aligned} \tag{2.88}$$

Therein, both the inner virtual works δW_i and the external virtual work δW_a occur:

$$\begin{aligned}\delta W_i &= \int_V \sigma_{ij}\delta\varepsilon_{ij}\mathrm{d}V,\\ \delta W_a &= \int_V f_i\delta u_i\mathrm{d}V + \int_{\partial V_t} t_{i0}\delta u_i\mathrm{d}S.\end{aligned} \tag{2.89}$$

The principle of virtual displacements is thus:

$$\delta W_i = \delta W_a. \tag{2.90}$$

The principle of virtual displacements can thus be derived from the general theorem of work and energy of elastostatics. Obviously, the principle of virtual displacements is not only a consequence of the equilibrium conditions, but is even completely equivalent to them. Thus, the principle of virtual displacements is an equilibrium statement formulated over corresponding virtual works. It is valid independently of the constitutive law.

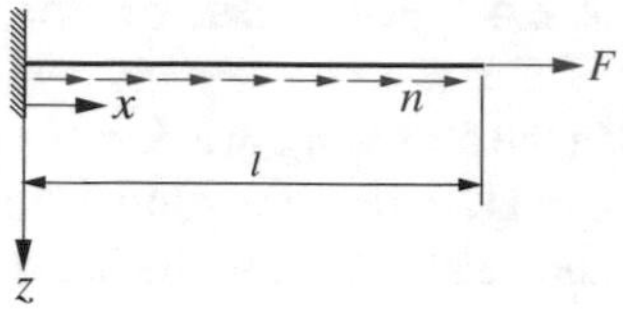

Fig. 2.10 Rod under line load $n(x)$ and single force F

2.2.5 Application to the Rod

The principle of virtual displacements is now used to derive both equilibrium conditions and boundary conditions for a rod. A rod under tension with length l and constant extensional stiffness EA is considered (Fig. 2.10). The bar is subjected to the constant line load $n(x)$ in the axial direction and a single force F at its free end. The equilibrium condition for the rod is:

$$N' = -n. \tag{2.91}$$

For the given rod situation, the boundary conditions can be deduced directly from Fig. 2.10. At the left fixed end $x = 0$ the rod displacement u must become zero. At the free end $x = l$, on the other hand, the normal force N coincides with the applied tensile force F. Thus:

$$u(x = 0) = 0, \quad N(x = l) = F. \tag{2.92}$$

In the following, the relations (2.91) and (2.92) are derived with the help of the principle of virtual displacements. For this purpose, we first consider the virtual internal work δW_i. It is given for the rod as:

$$\delta W_i = \int_V \sigma_{xx} \delta\varepsilon_{xx} \mathrm{d}V. \tag{2.93}$$

The volume integral can be split into a surface integral and an integral with respect to the rod axis x:

$$\delta W_i = \int_0^l \sigma_{xx} A \delta u' \mathrm{d}x = \int_0^l N \delta u' \mathrm{d}x. \tag{2.94}$$

We now perform a partial integration to decrease the degree of the derivative of the virtual displacement δu by one:

$$\delta W_i = N\delta u|_0^l - \int_0^l N' \delta u \mathrm{d}x. \tag{2.95}$$

The virtual external work δW_a consists of two parts, namely a part from the line load $n(x)$ and a part concerning the single force F:

$$\delta W_a = \int_0^l n\delta u \mathrm{d}x + F\delta u(x = l). \tag{2.96}$$

The principle of virtual displacements $\delta W_i = \delta W_a$ then reads for the present case:

$$N\delta u|_0^l - \int_0^l N'\delta u \mathrm{d}x - \int_0^l n\delta u \mathrm{d}x - F\delta u(x = l) = 0. \tag{2.97}$$

With $N\delta u|_0^l = N\delta u(x = l) - N\delta u(x = 0)$ we obtain:

$$-\int_0^l (N' + n)\delta u \mathrm{d}x - N\delta u(x = 0) + (N - F)\,\delta u(x = l) = 0. \tag{2.98}$$

The variations of the displacement u occurring here are not only arbitrary, but also completely independent of each other. Consequently, Eq. (2.98) can only be satisfied if each of the terms occurring here vanishes by itself.

We first consider the integral term in (2.98), which must become zero:

$$\int_0^l \left(N' + n\right)\delta u \mathrm{d}x = 0. \tag{2.99}$$

There are two possible solutions. On the one hand there is the trivial solution $\delta u = 0$. On the other hand, setting the parenthesis term in the integral to zero is possible:

$$N' + n = 0. \tag{2.100}$$

Apparently, this expression corresponds exactly to the equilibrium condition (2.91).

We consider the second term appearing in (2.98). One can interpret this term in such a way that at $x = 0$ either the normal force N becomes zero, or the variation δu vanishes:

$$\text{either} \quad N(x = 0) = 0 \quad \text{or} \quad \delta u(x = 0) = 0. \tag{2.101}$$

The first possibility corresponds to a free unloaded rod end. The second possibility, on the other hand, states that the axial displacement u at the point $x = 0$ must assume a fixed predetermined value $u\,(x = 0) = u_0$ (e.g. the value $u_0 = 0$). For the given example, only the second possibility comes into question, so the boundary condition at $x = 0$ is as follows:

$$u(x = 0) = u_0 = 0. \tag{2.102}$$

Apparently, this expression is identical to the first condition in (2.92).

From the third term appearing in (2.98) it can be concluded that:

$$\text{either} \quad \delta u(x=l) = 0 \quad \text{or} \quad N(x=l) - F = 0. \tag{2.103}$$

Since no displacement u is prescribed for $x = l$, only the second possibility comes into consideration:

$$N(x=l) = F. \tag{2.104}$$

This is the second boundary condition in (2.92).

The principle of virtual displacements not only leads to the equilibrium conditions of a given system, but it also provides in an unambiguous way all potentially possible boundary conditions. The principle of virtual displacements always leads to statements which are formulated in the corresponding force quantities.

2.2.6 Application to the Euler-Bernoulli Beam

As another example of the principle of virtual displacements, consider the cantilever beam of Fig. 2.9. The virtual external work is (see also (2.68)):

$$\delta W_a = \int_0^l q(x)\delta w \mathrm{d}x + \int_0^l n(x)\delta u \mathrm{d}x + F_V \delta w(x=l) + F_H \delta u(x=l). \tag{2.105}$$

The virtual internal work δW_i is given by (2.73):

$$\delta W_i = \int_0^l \left(N\delta u' - M\delta w''\right) \mathrm{d}x. \tag{2.106}$$

The principle of virtual displacements $\delta W_i = \delta W_a$ then results in:

$$\int_0^l \left(N\delta u' - M\delta w''\right) \mathrm{d}x - \int_0^l q(x)\delta w \mathrm{d}x - \int_0^l n(x)\delta u \mathrm{d}x - F_V \delta w(x=l) - F_H \delta u(x=l) = 0. \tag{2.107}$$

The first term of the internal virtual work δW_i is now partially integrated:

$$\int_0^l N\delta u' \mathrm{d}x = N\delta u|_0^l - \int_0^l N'\delta u \mathrm{d}x. \tag{2.108}$$

The second term is partially integrated twice:

$$-\int_0^l M\delta w''\mathrm{d}x = -M\delta w'|_0^l + \int_0^l M'\delta w'\mathrm{d}x = -M\delta w'|_0^l + M'\delta w|_0^l - \int_0^l M''\delta w\mathrm{d}x. \tag{2.109}$$

The principle of virtual displacements is thus:

$$\begin{aligned}
&-\int_0^l \left(N'+n\right)\delta u\mathrm{d}x - \int_0^l \left(M''+q\right)\delta w\mathrm{d}x \\
&- N\delta u(x=0) + (N-F_H)\delta u(x=l) - M'\delta w(x=0) \\
&+ (M'-F_V)\delta w(x=l) + M\delta w'(x=0) - M\delta w'(x=l) = 0.
\end{aligned} \tag{2.110}$$

The two integral terms give the equilibrium conditions of the beam for the special case of uniaxial bending:

$$N' = -n, \quad M'' = -q. \tag{2.111}$$

The boundary terms appearing in (2.110) can be interpreted as follows:

$$\begin{aligned}
&\text{either} \quad N(x=0)=0 \quad \text{or} \quad \delta u(x=0)=0,\\
&\text{either} \quad N(x=l)-F_H=0 \quad \text{or} \quad \delta u(x=l)=0,\\
&\text{either} \quad M'(x=0)=0 \quad \text{or} \quad \delta w(x=0)=0,\\
&\text{either} \quad M'(x=0)-F_V=0 \quad \text{or} \quad \delta w(x=l)-0,\\
&\text{either} \quad M(x=0)=0 \quad \text{or} \quad \delta w'(x=0)=0,\\
&\text{either} \quad M(x=l)=0 \quad \text{or} \quad \delta w'(x=l)=0.
\end{aligned} \tag{2.112}$$

2.3 Principle of the Stationary Value of the Total Elastic Potential

2.3.1 Introduction

The principle of the stationary value of the total elastic potential is the basis of many important approximation methods in elastostatics and can be derived from the principle of virtual displacements. We first consider elastic material behavior, but do not necessarily assume linear elasticity.

For an elastic body, the virtual internal work δW_i corresponds exactly to the virtual change $\delta\Pi_i$ of the total strain energy stored in the body or the internal potential. In index notation we can write:

$$\delta W_i = \int_V \sigma_{ij}\delta\varepsilon_{ij}\mathrm{d}V = \int_V \delta U_0\mathrm{d}V = \delta\Pi_i. \tag{2.113}$$

Consider external forces (here volume forces f_i and surface forces t_{i0}) which have a potential Π_a. Then, it can be assumed that the virtual change $\delta\Pi_a$ of the external potential corresponds to the external virtual work δW_a. However, a negative sign must be used here, because although work has been done, potential energy has been lost:

$$\delta W_a = \int_V f_i \delta u_i \mathrm{d}V + \int_{\partial V_t} t_{i0} \delta u_i \mathrm{d}S = -\delta\Pi_a. \tag{2.114}$$

The total elastic potential Π of the investigated body is composed of the inner potential Π_i and the external potential Π_a of the applied loads:

$$\Pi = \Pi_i + \Pi_a. \tag{2.115}$$

Thus, the principle of virtual displacements is transformed into the principle of virtual total potential:

$$\delta\Pi = \delta\left(\Pi_i + \Pi_a\right) = 0. \tag{2.116}$$

This means that an elastic body is in a state of equilibrium if, for all admissible variations δu_i and $\delta\varepsilon_{ij}$ of the displacements and strains, the change of the total potential vanishes. Accordingly, the first variation of the total potential must vanish for the elastic body under consideration to be in equilibrium. The disappearance of the first variation $\delta\Pi$ of the potential at arbitrary permissible virtual displacements and strains δu_i and $\delta\varepsilon_{ij}$ is equivalent to the elastic total potential Π having an extreme value in the equilibrium state:

$$\Pi = \Pi_i + \Pi_a = \text{Extreme value}. \tag{2.117}$$

This relation is also called the principle of the stationary value of the total elastic potential.

Considering the special case of linear elasticity, the internal potential Π_i is quadratically dependent on the strains. The extreme value required in (2.117) is then a minimum:

$$\Pi = \Pi_i + \Pi_a = \text{Minimum}. \tag{2.118}$$

This relationship is known as the principle of the minimum of the total elastic potential or also as the Green-Dirichlet[4] minimum principle.

2.3.2 Application to the Rod

We consider the rod shown in Fig. 2.11 (length l, constant extensional stiffness EA), which is loaded by the varying but continuous line load $n(x)$. We want to derive the equilibrium conditions and boundary conditions for the given situation using the

[4] Johann Peter Gustav Lejeune Dirichlet, 1805–1859, German mathematician.

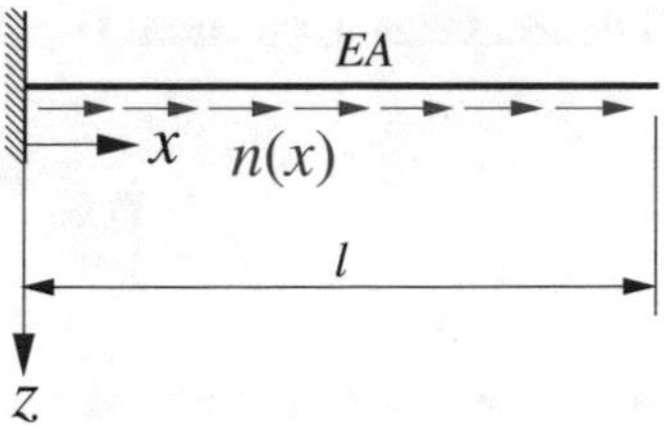

Fig. 2.11 Rod under line load $n(x)$

principle of the minimum of the total potential $\Pi = \Pi_i + \Pi_a =$ minimum. The external potential of the line load $n(x)$ results as:

$$\Pi_a = -\int_0^l nu(x)\mathrm{d}x. \tag{2.119}$$

The inner potential Π_i results here from the acting normal stress σ_{xx} and the normal strain ε_{xx}:

$$\Pi_i = \frac{1}{2}\int_V \sigma_{xx}\varepsilon_{xx}\mathrm{d}V = \frac{1}{2}\int_V \sigma\varepsilon\mathrm{d}V, \tag{2.120}$$

where we can dispense with the indexing of the stress and the strain. We split the volume integral into an integral over the cross-sectional area A and an integral over the rod length l:

$$\Pi_i = \frac{1}{2}\int_A\int_0^l \sigma\varepsilon\mathrm{d}x\mathrm{d}A. \tag{2.121}$$

Integration over the surface A then gives:

$$\Pi_i = \frac{1}{2}\int_0^l \sigma\varepsilon A\mathrm{d}x. \tag{2.122}$$

Using Hooke's law $\sigma = E\varepsilon$ we obtain with $EA =$ const.:

$$\Pi_i = \frac{EA}{2}\int_0^l \varepsilon^2\mathrm{d}x. \tag{2.123}$$

With the kinematic equation $\varepsilon = \frac{\partial u}{\partial x} = u'$ we obtain:

$$\Pi_i = \frac{EA}{2}\int_0^l u'^2\mathrm{d}x. \tag{2.124}$$

The total elastic potential Π can thus be written as:

$$\Pi = \frac{EA}{2} \int_0^l u'^2 \mathrm{d}x - n \int_0^l u(x) \mathrm{d}x, \tag{2.125}$$

where we want to assume from this point on that the line load $n(x)$ is constant along the length of the beam.

We now form the first variation of the total elastic potential:

$$\delta\Pi = \delta \left[\frac{EA}{2} \int_0^l u'^2 \mathrm{d}x - n \int_0^l u(x) \mathrm{d}x \right] = 0. \tag{2.126}$$

The order of variation and integration are interchangeable, so that:

$$\delta\Pi = \frac{EA}{2} \int_0^l \delta\left(u'^2\right) \mathrm{d}x - n \int_0^l \delta u \mathrm{d}x = 0. \tag{2.127}$$

The first integral term is transformed as follows:

$$\delta\left(u'^2\right) = 2u'\delta u'. \tag{2.128}$$

Thus:

$$\delta\Pi = EA \int_0^l u'\delta u' \mathrm{d}x - n \int_0^l \delta u \mathrm{d}x = 0. \tag{2.129}$$

The first term is partially integrated:

$$EA \int_0^l u'\delta u' \mathrm{d}x = EAu'\delta u|_0^l - EA \int_0^l u''\delta u \mathrm{d}x. \tag{2.130}$$

The first variation $\delta\Pi$ of the total elastic potential is thus:

$$\delta\Pi = EAu'\delta u|_0^l - \int_0^l \left(EAu'' + n\right) \delta u \mathrm{d}x = 0. \tag{2.131}$$

For any admissible variation δu this is only feasible if the individual terms in (2.131) vanish for themselves. The integral term in (2.131) delivers:

$$EAu'' = -n. \tag{2.132}$$

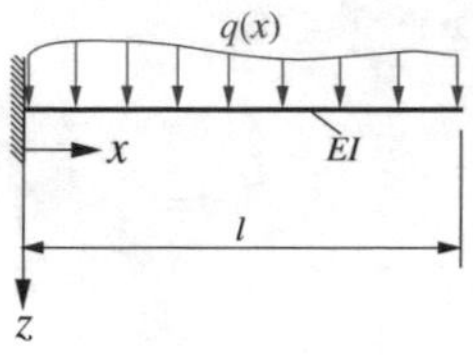

Fig. 2.12 Beam under line load $q(x)$

This represents the differential equation of the rod. The term remaining in (2.131) corresponds to the boundary conditions at $x = 0$ and $x = l$. In contrast to the principle of virtual displacements, here all formulations result in the displacement quantities (in this case the displacement u) and not as in the case of the principle of virtual displacements in the force quantities.

2.3.3 Application to the Euler-Bernoulli Beam

We consider the beam shown in Fig. 2.12, which has the length l and the bending stiffness $EI = EI(x)$ and is loaded by the line load $q(x)$. We assume uniaxial bending in the xz−plane. The beam differential equation in the framework of Euler-Bernoulli theory reads:

$$\left(EIw''\right)'' - q = 0. \tag{2.133}$$

The boundary conditions to be considered here follow from Fig. 2.12 as:

$$w(x=0) = 0, \quad w'(x=0) = 0, \quad M(x=l) = 0, \quad Q(x=l) = 0. \tag{2.134}$$

In the following, we derive the beam differential equation as well as the underlying boundary conditions from the principle of the minimum of the elastic total potential.

The inner potential Π_i reads:

$$\Pi_i = \frac{1}{2}\int_V \sigma\varepsilon \mathrm{d}V. \tag{2.135}$$

Using Hooke's law $\sigma = E\varepsilon$ and splitting the volume integral into an area integral and an integral with respect to the beam longitudinal direction x leads to:

$$\Pi_i = \frac{1}{2}\int_0^l\int_A E\varepsilon^2 \mathrm{d}A\mathrm{d}x. \tag{2.136}$$

With the normal strain ε

$$\varepsilon = u' = -zw'' \tag{2.137}$$

we obtain:

$$\Pi_i = \frac{1}{2}\int_0^l \int_A E\left(-zw''\right)^2 \mathrm{d}A\mathrm{d}x = \frac{1}{2}\int_0^l \int_A Ez^2 w''^2 \mathrm{d}A\mathrm{d}x, \qquad (2.138)$$

or

$$\Pi_i = \frac{1}{2}\int_0^l Ew''^2 \mathrm{d}x \int_A z^2 \mathrm{d}A. \qquad (2.139)$$

With the second-order moment of inertia $\int_A z^2 \mathrm{d}A = I$ we get:

$$\Pi_i = \frac{1}{2}\int_0^l EIw''^2 \mathrm{d}x. \qquad (2.140)$$

The external potential reads for the given situation:

$$\Pi_a = -\int_0^l qw\mathrm{d}x. \qquad (2.141)$$

The total elastic potential $\Pi = \Pi_i + \Pi_a$ can then be given as:

$$\Pi = \frac{1}{2}\int_0^l EIw''^2 \mathrm{d}x - \int_0^l qw\mathrm{d}x. \qquad (2.142)$$

Let it now be required that Π assumes a minimum if $w(x)$ is the exact solution of the problem and all underlying boundary conditions are satisfied. Hence:

$$\delta\Pi = \frac{1}{2}\int_0^l EI\delta(w'')^2 \mathrm{d}x - \int_0^l q\delta w\mathrm{d}x = 0. \qquad (2.143)$$

Applying the chain rule to the first term yields:

$$\int_0^l EIw''\delta w'' \mathrm{d}x - \int_0^l q\delta w\mathrm{d}x = 0. \qquad (2.144)$$

Partial integration of the first term in (2.144) yields:

$$EIw''\delta w'|_0^l - \int_0^l (EIw'')'\delta w'\mathrm{d}x - \int_0^l q\delta w\mathrm{d}x = 0. \qquad (2.145)$$

A further partial integration is performed:

$$\int_0^l \left((EIw'')'' - q\right)\delta w\mathrm{d}x + (EIw'')'\delta w(x=0) - EIw''\delta w'(x=0)$$
$$+EIw''\delta w'(x=l) - (EIw'')'\delta w(x=l) = 0. \qquad (2.146)$$

With the constitutive law of the bar $EIw'' = -M$ or $(EIw'')' = -Q$ one obtains:

$$\int_0^l \left((EIw'')'' - q\right)\delta w\mathrm{d}x - Q\delta w(x=0) + M\delta w'(x=0)$$
$$-M\delta w'(x=l) + Q\delta w(x=l) = 0. \qquad (2.147)$$

Again, for any admissible variations δw and $\delta w'$, this condition can be satisfied only if each term in (2.147) becomes zero. From the integral expression in (2.147) we obtain:

$$(EIw'')'' - q = 0. \qquad (2.148)$$

This is obviously the differential equation of the beam. The remaining terms in (2.147) represent all possible boundary conditions of the given beam problem.

2.4 Approximation Methods of Elastostatics

Among the classical approximation methods of elastostatics are the Ritz method and the Galerkin method, which will be discussed in the following. Both methods have in common that they use suitable approximation approaches for the sought state variables of the system under consideration to solve a given solid problem by using energy formulations. These approximations are often formulated for the displacements of the system under consideration. The resulting strain and stress states can then be determined in a post-processing step. When classifying approximation methods, it is common to distinguish between continuous and discretizing methods. In continuous methods (here the Ritz method and the Galerkin method), the approximation approaches used are formulated on the entire structure under consideration, whereas in discretizing methods (e.g. the finite element method) approximation approaches are used on subdomains. The finite element method (also abbreviated as

FEM) is not part of this book and is only briefly discussed in the context of disk structures.

2.4.1 The Ritz Method

The Ritz[5] method uses the principle of the minimum of the total elastic potential. It assumes that the total potential Π of the structure under consideration can be represented as a function of the displacements u, v, w: $\Pi = \Pi(u, v, w)$. For the displacements u, v, w, approaches of the following form are used:

$$\begin{aligned} u(x, y, z) &\simeq U(x, y, z) = \sum_{i=1}^{i=n_u} A_i U_i(x, y, z), \\ v(x, y, z) &\simeq V(x, y, z) = \sum_{i=1}^{i=n_v} B_i V_i(x, y, z), \\ w(x, y, z) &\simeq W(x, y, z) = \sum_{i=1}^{i=n_w} C_i W_i(x, y, z). \end{aligned} \quad (2.149)$$

The functions $U(x, y, z)$, $V(x, y, z)$, $W(x, y, z)$ consist of the n_u, n_v and n_w functions $U_i(x, y, z)$, $V_i(x, y, z)$, $W_i(x, y, z)$. The quantities A_i, B_i and C_i are constants still to be determined and are the so-called Ritz constants. The approach functions are required to satisfy the geometric boundary conditions of the structure under consideration. Since both the functions $U_i(x, y, z)$, $V_i(x, y, z)$, $W_i(x, y, z)$ and the degrees n_u, n_v, n_w of the series expansions in (2.149) are given by the user, the Ritz constants A_i, B_i, C_i are the actual target of the calculation.

The approach (2.149) is inserted into the total elastic potential Π of the studied structure. The only quantities which are still amenable to variation are the Ritz constants A_i, B_i, C_i. As a consequence, the first variation $\delta\Pi$ of the total elastic potential transforms into the following form:

$$\begin{aligned} \delta\Pi = &\frac{\partial\Pi}{\partial A_1}\delta A_1 + \frac{\partial\Pi}{\partial A_2}\delta A_2 + \frac{\partial\Pi}{\partial A_3}\delta A_3 + \ldots + \frac{\partial\Pi}{\partial A_{n_u}}\delta A_{n_u} \\ &+ \frac{\partial\Pi}{\partial B_1}\delta B_1 + \frac{\partial\Pi}{\partial B_2}\delta B_2 + \frac{\partial\Pi}{\partial B_3}\delta B_3 + \ldots + \frac{\partial\Pi}{\partial B_{n_v}}\delta B_{n_v} \\ &+ \frac{\partial\Pi}{\partial C_1}\delta C_1 + \frac{\partial\Pi}{\partial C_2}\delta C_2 + \frac{\partial\Pi}{\partial C_3}\delta C_3 + \ldots + \frac{\partial\Pi}{\partial C_{n_w}}\delta C_{n_w} = 0. \end{aligned} \quad (2.150)$$

[5] Walter Ritz, 1878–1909, Swiss mathematician and physicist.

The variations of the constants are arbitrary and independent of each other, so that the condition (2.150) can be satisfied only if:

$$\begin{aligned}
&\frac{\partial \Pi}{\partial A_1} = 0, \quad \frac{\partial \Pi}{\partial A_2} = 0, \quad \frac{\partial \Pi}{\partial A_3} = 0, \quad \ldots \quad \frac{\partial \Pi}{\partial A_{n_u}} = 0, \\
&\frac{\partial \Pi}{\partial B_1} = 0, \quad \frac{\partial \Pi}{\partial B_2} = 0, \quad \frac{\partial \Pi}{\partial B_3} = 0, \quad \ldots \quad \frac{\partial \Pi}{\partial B_{n_v}} = 0, \\
&\frac{\partial \Pi}{\partial C_1} = 0, \quad \frac{\partial \Pi}{\partial C_2} = 0, \quad \frac{\partial \Pi}{\partial C_3} = 0, \quad \ldots \quad \frac{\partial \Pi}{\partial C_{n_w}} = 0.
\end{aligned} \qquad (2.151)$$

These are the so-called Ritz equations, from which the constants A_i, B_i, C_i can be determined. If a geometrically linear problem is considered, then this leads to a linear system of equations with $n_u + n_v + n_w$ equations for the $n_u + n_v + n_w$ Ritz constants A_i, B_i, C_i. The approximate displacement field is thus known, and from the kinematic equations, the approximate strain field can be determined. The constitutive law can then be used to calculate the approximate stress field.

The Ritz method is briefly considered below for beam bending problems. For this purpose, consider the beam shown in Fig. 2.13, which is simply supported on both sides and has the length l. The beam has a varying but continuous bending stiffness $EI = EI(x)$ and is loaded by an arbitrary but continuously distributed line load $q = q(x)$. The Ritz method is used to determine an approximate solution for the deflection $w(x)$ of the beam. The total elastic potential of the beam in this case is:

$$\Pi = \Pi_i + \Pi_a = \frac{1}{2} \int_0^l EI(x) w(x)''^2 \mathrm{d}x - \int_0^l q(x) w(x) \mathrm{d}x. \qquad (2.152)$$

For the approximation of the bending line $w(x)$ of the beam an approximation using n approach functions of the following form is used:

$$w(x) \simeq W(x) = \sum_{i=1}^{i=n} C_i W_i(x) = C_1 W_1(x) + C_2 W_2(x) + \ldots + C_n W_n(x). \quad (2.153)$$

Herein it is assumed that the n approach functions $W_i(x)$ satisfy the given geometric boundary conditions $w(x = 0) = w(x = l) = 0$. Substituting the approach (2.153)

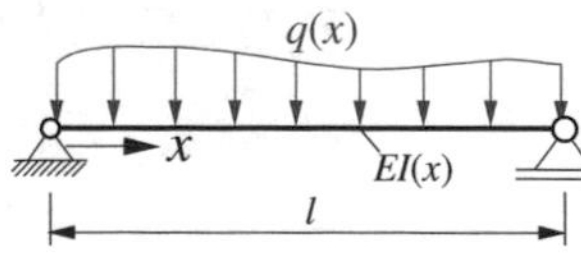

Fig. 2.13 Beam under line load $q(x)$

into the total elastic potential (2.152) yields:

$$\Pi = \frac{1}{2}\int_0^l EI(x)\left(\sum_{i=1}^{i=n} C_i W_i''(x)\right)^2 \mathrm{d}x - \int_0^l q(x)\sum_{i=1}^{i=n} C_i W_i(x)\mathrm{d}x, \tag{2.154}$$

or:

$$\Pi = \frac{1}{2}\sum_{i=1}^{i=n}\sum_{j=1}^{j=n}\int_0^l EI(x)C_iC_jW_i''(x)W_j''(x)\mathrm{d}x - \int_0^l q(x)\sum_{i=1}^{i=n} C_i W_i(x)\mathrm{d}x. \tag{2.155}$$

Written out for n=3, we obtain:

$$\begin{aligned}\Pi &= \frac{1}{2}\int_0^l EI(x)C_1C_1W_1''(x)W_1''(x)\mathrm{d}x + \int_0^l EI(x)C_1C_2W_1''(x)W_2''(x)\mathrm{d}x \\ &+ \int_0^l EI(x)C_1C_3W_1''(x)W_3''(x)\mathrm{d}x + \frac{1}{2}\int_0^l EI(x)C_2C_2W_2''(x)W_2''(x)\mathrm{d}x \\ &+ \int_0^l EI(x)C_2C_3W_2''(x)W_3''(x)\mathrm{d}x + \frac{1}{2}\int_0^l EI(x)C_3C_3W_3''(x)W_3''(x)\mathrm{d}x \\ &- \int_0^l q(x)C_1W_1(x)\mathrm{d}x - \int_0^l q(x)C_2W_2(x)\mathrm{d}x - \int_0^l q(x)C_3W_3(x)\mathrm{d}x. \end{aligned} \tag{2.156}$$

It turns out that systematically recurring integration terms occur, which are provided with abbreviations as follows:

$$K_{ij} = \int_0^l EI(x)W_i''(x)W_j''(x)\mathrm{d}x, \quad F_i = \int_0^l q(x)W_i(x)\mathrm{d}x. \tag{2.157}$$

It then follows from (2.156):

$$\begin{aligned}\Pi &= \frac{1}{2}C_1C_1K_{11} + C_1C_2K_{12} + C_1C_3K_{13} + \frac{1}{2}C_2C_2K_{22} + C_2C_3K_{23} \\ &+ \frac{1}{2}C_3C_3K_{33} - C_1F_1 - C_2F_2 - C_3F_3. \end{aligned} \tag{2.158}$$

The Ritz equations then follow as:

$$\frac{\partial \Pi}{\partial C_1} = C_1K_{11} + C_2K_{12} + C_3K_{13} - F_1 = 0,$$
$$\frac{\partial \Pi}{\partial C_2} = C_1K_{12} + C_2K_{22} + C_3K_{23} - F_2 = 0,$$
$$\frac{\partial \Pi}{\partial C_3} = C_1K_{13} + C_2K_{23} + C_3K_{33} - F_3 = 0. \tag{2.159}$$

In vector-matrix notation, this can be compactly represented as:

$$\begin{bmatrix} K_{11} & K_{12} & K_{13} \\ K_{12} & K_{22} & K_{23} \\ K_{13} & K_{23} & K_{33} \end{bmatrix} \begin{pmatrix} C_1 \\ C_2 \\ C_3 \end{pmatrix} = \begin{pmatrix} F_1 \\ F_2 \\ F_3 \end{pmatrix}. \tag{2.160}$$

For an approach with n functions one obtains:

$$\begin{bmatrix} K_{11} & K_{12} & K_{13} & \dots & K_{1n} \\ K_{12} & K_{22} & K_{23} & \dots & K_{2n} \\ K_{13} & K_{23} & K_{33} & \dots & K_{3n} \\ \vdots & \vdots & \vdots & \ddots & \vdots \\ K_{n1} & K_{n2} & K_{n3} & \dots & K_{nn} \end{bmatrix} \begin{pmatrix} C_1 \\ C_2 \\ C_3 \\ \vdots \\ C_n \end{pmatrix} = \begin{pmatrix} F_1 \\ F_2 \\ F_3 \\ \vdots \\ F_n \end{pmatrix}, \tag{2.161}$$

or in symbolic notation:

$$\underline{\underline{K}}\,\underline{C} = \underline{F}. \tag{2.162}$$

The matrix $\underline{\underline{K}}$ is also called the stiffness matrix. It is always symmetrical and has, when using n approach functions, $n \times n$ rows and columns. The vectors $\underline{C}$ and $\underline{F}$ contain the Ritz constants C_i and the so-called force resultants F_i, respectively. At this point all necessary equations for the determination of the Ritz constants C_i are provided, and the approximated beam deflection $w(x)$ can be completely specified. From this, approximated bending moments, transverse shear forces and stresses can be determined in a post-processing calculation.

2.4.2 The Galerkin Method

The Galerkin method[6] is briefly explained below using the example of a beam of length l with constant bending stiffness EI which is simply supported at both ends. The load is a constant line load q. The Galerkin method starts from the variational statement

[6] Boris Grigorievich Galerkin, 1871–1945, Russian mathematician.

$$\int_0^l (EIw'''' - q)\,\delta w \mathrm{d}x - Q\delta w(x=0) + M\delta w'(x=0)$$
$$-M\delta w'(x=l) + Q\delta w(x=l) = 0, \qquad (2.163)$$

which results, for example, from the principle of virtual displacements. Analogous to the Ritz method, an approximation of the bending line $w(x)$ of the beam of the form

$$w(x) \simeq W(x) = \sum_{i=1}^{i=n} C_i W_i(x) = C_1 W_1(x) + C_2 W_2(x) + \ldots + C_n W_n(x) \quad (2.164)$$

is employed, where now the approach functions are required to satisfy all boundary conditions of the considered problem. For such functions all boundary terms disappear in (2.163) after inserting (2.164), so that the following expression remains:

$$\int_0^l (EIW'''' - q)\,\delta W \mathrm{d}x = 0, \qquad (2.165)$$

or

$$\int_0^l [(EIC_1 W_1'''' + EIC_2 W_2'''' + \ldots + EIC_n W_n'''' - q)$$
$$(\delta C_1 W_1 + \delta C_2 W_2 + \ldots + \delta C_n W_n)]\,\mathrm{d}x = 0. \qquad (2.166)$$

Herein, use has been made of the fact that the only quantities amenable to variation are the constants C_1, C_2,..., C_n. The variations of the constants C_i are independent of each other, so that (2.166) can be satisfied only if:

$$\int_0^l (EIC_1 W_1'''' + EIC_2 W_2'''' + \ldots + EIC_n W_n'''' - q)\, W_1 \mathrm{d}x = 0,$$
$$\int_0^l (EIC_1 W_1'''' + EIC_2 W_2'''' + \ldots + EIC_n W_n'''' - q)\, W_2 \mathrm{d}x = 0,$$
$$\vdots$$
$$\int_0^l (EIC_1 W_1'''' + EIC_2 W_2'''' + \ldots + EIC_n W_n'''' - q)\, W_n \mathrm{d}x = 0. \qquad (2.167)$$

These are the so-called Galerkin equations. After performing the prescribed integrations, they form a linear system of n equations for the n constants $C_1, C_2, \ldots, C_n$.

References

Becker, W., Gross, D.: Mechanik elastischer Körper und Strukturen. Springer, Berlin (2002)
Kossira, H.: Grundlagen des Leichtbaus. Springer, Berlin (1996)
Lanczos, C.: The Variational Principles of Mechanics, 4th edn. Dover, New York (1986)
Langhaar, H.L.: Energy Methods in Applied Mechanics. Dover, New York (2016)
Oden, J.T., Reddy, J.N.: Variational Methods in Theoretical Mechanics, 2nd edn. Springer, Berlin (1983)
Reddy, J.N.: Energy Principles and Variational Methods in Applied Mechanics, 3rd edn. Wiley, New York (1983)
Tauchert, T.R.: Energy Principles in Structural Mechanics. McGraw-Hill, New York (1974)
Washizu, K.: Variational Methods in Elasticity and Plasticity, 3rd edn. Pergamon, New York (1982)

Part II
Disks

Isotropic Disks in Cartesian Coordinates

3

This chapter is devoted to the consideration of isotropic disk structures in Cartesian coordinates. After a short definition of what constitutes a disk, the two basic analytical approaches, namely the displacement method and the force method, are motivated and, for the force method, all basic equations necessary for the description of a disk are compiled. This is followed by an energetic consideration of the disk problem, before the solutions of the disk equation and elementary disk problems are discussed in detail. After that, beam-like disks are considered, i.e. disks which can be regarded as beams in their nature and geometry, but which are no longer amenable to a calculation within the framework of a beam theory. Furthermore, a discussion of St. Venant's principle is given before the isotropic half plane under different boundary loads is investigated. The chapter concludes by introducing the notion of the effective width in beams and load introductions in disk-like structures. Disks in oblique coordinates are treated, for example, in Boresi and Lynn (1974) and Altenbach et al. (2016). Isotropic disks in polar coordinates are treated in Chap. 4. Anisotropic disks are the subject of Chap. 6.

3.1 Introduction

A disk is defined as a plane and thin-walled structure that is loaded exclusively by loads in its plane (Fig. 3.1). The external loads can occur in the form of single forces and line loads, but also as volume loads, where in Fig. 3.1 only single forces are shown. Loads acting across the thickness of the disk must be uniformly distributed along the $z-$direction. In all that follows it is understood that the thickness h of the disk is significantly smaller than the characteristic in-plane dimension l. Let x, y, z be a Cartesian coordinate system. Let the origin of the reference system lie in the middle plane of the disk, which bisects the disk at every point x, y. The middle plane of the disk is spanned by the $x-$ and $y-$axis. Consequently, z is the direction of the

C. Mittelstedt, *Theory of Plates and Shells*,
https://doi.org/10.1007/978-3-662-66805-4_3

Fig. 3.1 Disk

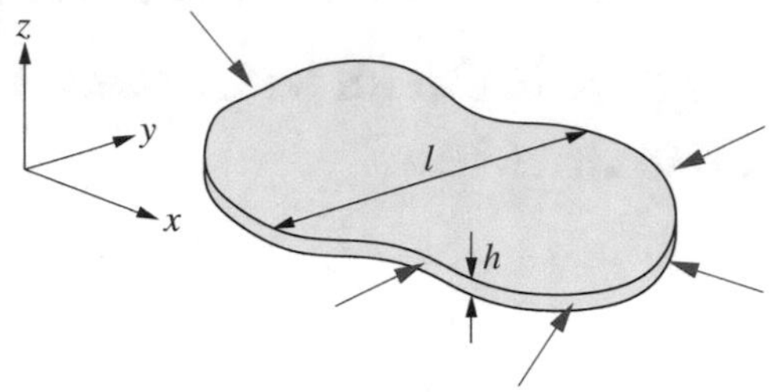

normal of the middle plane of the disk. The disk thickness h can be a function of the two coordinates x and y, i.e. $h = h(x, y)$. It is assumed that the disk is in a plane state of stress with respect to the thickness direction z and that all occurring stresses, but also all loads, are independent of z. Consequently, only the plane stresses σ_{xx}, σ_{yy} and τ_{xy} occur. Then, in the plane stress state, both the transverse shear stresses τ_{xz} and τ_{yz} and the transverse normal stress σ_{zz} vanish. If a plane strain state is considered, then the transverse normal stress σ_{zz} can be determined from the plane stress components in a post-calculation. In any case, the stresses σ_{xx}, σ_{yy} and τ_{xy} will be constant over the thickness h and thus will be independent of z. Therefore:

$$\sigma_{xx} = \sigma_{xx}(x, y), \quad \sigma_{yy} = \sigma_{yy}(x, y), \quad \tau_{xy} = \tau_{xy}(x, y). \tag{3.1}$$

From the plane stress components, resulting force flows can be computed. These can be obtained from the integration of σ_{xx}, σ_{yy} and τ_{xy} over the disk thickness (Fig. 3.2), they act in the middle plane of the disk:

$$\begin{aligned}
N^0_{xx} &= \int_{-\frac{h}{2}}^{+\frac{h}{2}} \sigma_{xx} \mathrm{d}z = \sigma_{xx} h, \\
N^0_{yy} &= \int_{-\frac{h}{2}}^{+\frac{h}{2}} \sigma_{yy} \mathrm{d}z = \sigma_{yy} h, \\
N^0_{xy} &= \int_{-\frac{h}{2}}^{+\frac{h}{2}} \tau_{xy} \mathrm{d}z = \tau_{xy} h.
\end{aligned} \tag{3.2}$$

Thus also the plane force flows N^0_{xx}, N^0_{yy}, N^0_{xy} are functions of x, y only:

$$N^0_{xx} = N^0_{xx}(x, y), \quad N^0_{yy} = N^0_{yy}(x, y), \quad N^0_{xy} = N^0_{xy}(x, y). \tag{3.3}$$

Thus, these are force flows acting in the middle plane of the disk, which also explains the superscript 0, indicating that these are quantities related to the disk's middle plane. The force flows N^0_{xx}, N^0_{yy}, N^0_{xy} thus have the unit of a force per unit length.

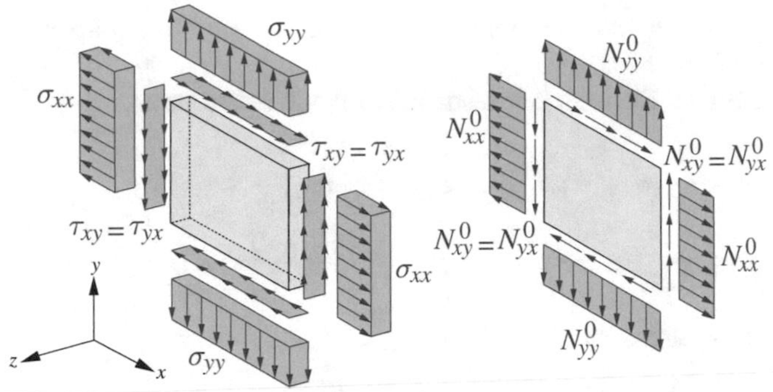

Fig. 3.2 Sectional element of a disk with the corresponding stress components (left), reduction to the middle plane with the corresponding force flows (right)

As further prerequisites we assume that all disk problems treated in this chapter are geometrically and materially linear problems. We thus express the strains by the components of the infinitesimal strain tensor, and Hooke's generalized law is assumed to be valid. Since we consider a disk problem in which the middle plane is exclusively subject to normal and shear strains, but no bending action occurs, the resulting strain quantities ε_{xx}, ε_{yy} and γ_{xy} are also constant over the disk thickness h.

Concerning the way of solving a disk problem, it is in fact irrelevant whether the disk is considered in the plane stress state or in the plane strain state. Once the solution of a given disk problem is available for one of these two states, the solution for the other state can be obtained by applying the corresponding elastic constants. For this, the reader is referred to the explanations in Chap. 1.

The structural analysis of disk structures is discussed in the works by Altenbach et al. (2016), Bauchau and Craig (2009), Becker and Gross (2002), Boresi and Lynn (1974), Chou and Pagano (1967), Eschenauer and Schnell (1993), Eschenauer et al. (1997), Filonenko-Borodich (1967), Girkmann (1974), Göldner et al. (1979), Gross et al. (2014), Hahn (1985), Hake and Meskouris (2007), Leipholz (1974), Massonet (1962), Sadd (2005), Sechler (1952), Timoshenko and Goodier (1951), Wang (1953), and Wiedemann (2006).

3.2 Fundamentals

3.2.1 Basic Equations

We restrict the considerations in this chapter to isotropic disks in the plane stress state, described in Cartesian coordinates. These relations are in the end quite easily transferable to anisotropic and here especially orthotropic disk structures, also in other reference systems, which is the content of Chaps. 4 and 6. The transfer to the plane strain state can then be carried out quite easily by an appropriate use of equivalent stiffnesses.

The generalized Hooke's law reads in the plane stress state:

$$\varepsilon_{xx} = \frac{1}{E}\left(\sigma_{xx} - \nu\sigma_{yy}\right), \quad \varepsilon_{yy} = \frac{1}{E}\left(\sigma_{yy} - \nu\sigma_{xx}\right),$$
$$\varepsilon_{zz} = -\frac{\nu}{E}\left(\sigma_{xx} + \sigma_{yy}\right), \quad \gamma_{xy} = \frac{2\left(1+\nu\right)}{E}\tau_{xy}, \quad \gamma_{xz} = \gamma_{yz} = 0, \tag{3.4}$$

or solved for the stresses:

$$\sigma_{xx} = \frac{E}{1-\nu^2}\left(\varepsilon_{xx} + \nu\varepsilon_{yy}\right), \quad \sigma_{yy} = \frac{E}{1-\nu^2}\left(\varepsilon_{yy} + \nu\varepsilon_{xx}\right), \quad \tau_{xy} = G\gamma_{xy}. \tag{3.5}$$

Due to the assumptions made, the strains of the disk will be constant over z. The set of Eqs. (3.4) and (3.5) can also be formulated in terms of the force flows N^0_{xx}, N^0_{yy}, N^0_{xy} as:

$$\varepsilon_{xx}h = \frac{1}{E}\left(N^0_{xx} - \nu N^0_{yy}\right), \quad \varepsilon_{yy}h = \frac{1}{E}\left(N^0_{yy} - \nu N^0_{xx}\right),$$
$$\varepsilon_{zz}h = -\frac{\nu}{E}\left(N^0_{xx} + N^0_{yy}\right), \quad \gamma_{xy}h = \frac{2\left(1+\nu\right)}{E}N^0_{xy}, \quad \gamma_{xz} = \gamma_{yz} = 0, \tag{3.6}$$

and

$$N^0_{xx} = \frac{Eh}{1-\nu^2}\left(\varepsilon_{xx} + \nu\varepsilon_{yy}\right), \quad N^0_{yy} = \frac{Eh}{1-\nu^2}\left(\varepsilon_{yy} + \nu\varepsilon_{xx}\right), \quad N^0_{xy} = Gh\gamma_{xy}. \tag{3.7}$$

We introduce at this point the so-called disk stiffness A (not to be confused with a cross-sectional area) as follows:

$$A = \frac{Eh}{1-\nu^2}. \tag{3.8}$$

Then (3.7) can be written as:

$$N^0_{xx} = A\left(\varepsilon^0_{xx} + \nu\varepsilon^0_{yy}\right), \quad N^0_{yy} = A\left(\varepsilon^0_{yy} + \nu\varepsilon^0_{xx}\right), \quad N^0_{xy} = A\frac{1-\nu}{2}\gamma^0_{xy}, \tag{3.9}$$

where from here on we also want to assign the strains with a superscript 0 to indicate that these are the strains of the middle plane of the disk.

The kinematic equations are obtained for the disk in the plane state of stress as:

$$\varepsilon^0_{xx} = \frac{\partial u_0}{\partial x}, \quad \varepsilon^0_{yy} = \frac{\partial v_0}{\partial y}, \quad \gamma^0_{xy} = \frac{\partial u_0}{\partial y} + \frac{\partial v_0}{\partial x}. \tag{3.10}$$

Herein, u_0 and v_0 are the displacements of points of the disk middle plane which can be functions of x and y, i.e. $u_0 = u_0(x, y)$ and $v_0 = v_0(x, y)$. The index 0 again indicates that all quantities occurring here are related to the middle plane of the disk.

The equilibrium conditions are in the plane stress state:

$$\frac{\partial \sigma_{xx}}{\partial x} + \frac{\partial \tau_{xy}}{\partial y} + f_x = 0, \quad \frac{\partial \tau_{xy}}{\partial x} + \frac{\partial \sigma_{yy}}{\partial y} + f_y = 0. \tag{3.11}$$

Herein, f_x and f_y are volume loads. We can also write the equilibrium conditions in terms of the force flows N_{xx}^0, N_{yy}^0, N_{xy}^0 as follows:

$$\frac{\partial N_{xx}^0}{\partial x} + \frac{\partial N_{xy}^0}{\partial y} + p_x = 0, \quad \frac{\partial N_{xy}^0}{\partial x} + \frac{\partial N_{yy}^0}{\partial y} + p_y = 0. \tag{3.12}$$

The quantities p_x and p_y are the volume loads multiplied by the disk thickness h, i.e. $p_x = f_x h$ and $p_y = f_y h$.

The following compatibility equation remains in the plane stress state:

$$\frac{\partial^2 \varepsilon_{xx}^0}{\partial y^2} + \frac{\partial^2 \varepsilon_{yy}^0}{\partial x^2} - \frac{\partial^2 \gamma_{xy}^0}{\partial x \partial y} = 0. \tag{3.13}$$

A overview of the basic equations presented here shows that a disk problem in the plane stress state is described by the two displacements u_0, v_0, the three strains ε_{xx}^0, ε_{yy}^0, γ_{xy}^0 and the three stress components σ_{xx}, σ_{yy}, τ_{xy} respectively the three force flows N_{xx}^0, N_{yy}^0, N_{xy}^0. Hooke's law, the equilibrium conditions and the kinematic equations constitute eight equations in total for the eight state variables, so that a disk problem is formulated unambiguously with the specification of boundary conditions. The formulation of boundary conditions will be discussed later.

If a disk problem in a plane strain state is given, then the equations given so far are all valid. Only the elastic constants E and ν have to be replaced by the equivalent stiffnesses $\overline{E}$ and $\overline{\nu}$:

$$\overline{E} = \frac{E}{1-\nu^2}, \quad \overline{\nu} = \frac{\nu}{1-\nu}. \tag{3.14}$$

3.2.2 The Displacement Method

In order to solve a given disk problem analytically in an exact manner (which is unfortunately only possible for rather simple and often strongly idealized problems), there are generally two ways. On the one hand, there is the so-called displacement method, and on the other hand, the so-called force method can be used (Göldner et al. 1979; Altenbach et al. 2016).

In the framework of the displacement method, the set of eight equations is reduced to two differential equations in the displacements u_0 and v_0. Substituting the constitutive relations (3.9) into the equilibrium conditions (3.12), we obtain:

$$\frac{\partial}{\partial x}\left[A\left(\varepsilon_{xx}^{0}+\nu\varepsilon_{yy}^{0}\right)\right]+\frac{\partial}{\partial y}\left[A\frac{1-\nu}{2}\gamma_{xy}^{0}\right]+p_x=0,$$
$$\frac{\partial}{\partial y}\left[A\left(\varepsilon_{yy}^{0}+\nu\varepsilon_{xx}^{0}\right)\right]+\frac{\partial}{\partial x}\left[A\frac{1-\nu}{2}\gamma_{xy}^{0}\right]+p_y=0. \quad (3.15)$$

Using the kinematic Eq. (3.10) then gives the following expressions:

$$\frac{\partial}{\partial x}\left[A\left(\frac{\partial u_0}{\partial x}+\nu\frac{\partial v_0}{\partial y}\right)\right]+\frac{\partial}{\partial y}\left[A\frac{1-\nu}{2}\left(\frac{\partial u_0}{\partial y}+\frac{\partial v_0}{\partial x}\right)\right]+p_x=0,$$
$$\frac{\partial}{\partial y}\left[A\left(\frac{\partial v_0}{\partial y}+\nu\frac{\partial u_0}{\partial x}\right)\right]+\frac{\partial}{\partial x}\left[A\frac{1-\nu}{2}\left(\frac{\partial u_0}{\partial y}+\frac{\partial v_0}{\partial x}\right)\right]+p_y=0. \quad (3.16)$$

These are two coupled second-order inhomogeneous partial differential equations for the two displacements u_0 and v_0 of the disk middle plane. They are valid for an arbitrary disk stiffness $A = A(x, y)$. If, on the other hand, the special case of a constant disk stiffness A is present, then these equations simplify considerably:

$$\frac{\partial^2 u_0}{\partial x^2}+\frac{1-\nu}{2}\frac{\partial^2 u_0}{\partial y^2}+\frac{1+\nu}{2}\frac{\partial^2 v_0}{\partial x\partial y}+\frac{p_x}{A}=0,$$
$$\frac{\partial^2 v_0}{\partial y^2}+\frac{1-\nu}{2}\frac{\partial^2 v_0}{\partial x^2}+\frac{1+\nu}{2}\frac{\partial^2 u_0}{\partial x\partial y}+\frac{p_y}{A}=0. \quad (3.17)$$

In the context of the displacement method, the task is to solve these two differential equations for u_0 and v_0 under consideration of given boundary conditions. Once the displacements are known, the kinematic equations and the constitutive equations provided above can be used to determine all disk strains as well as all disk stresses.

3.2.3 The Force Method

Within the framework of the force method, the solution of a given disk problem is not solved in terms of displacements, but rather a solution for the stress field is sought. Once the stresses are available, the disk strains can be determined, and from this, by integration, the disk displacements can be computed. Starting point are the two equilibrium conditions (3.11) and (3.12), respectively. It is easy to see that these equations are not sufficient to describe the stress state of the disk. Here we make use of the compatibility condition (3.13), in which we replace the strain components by the stresses σ_{xx}, σ_{yy}, τ_{xy} and, assuming constant elastic parameters E and ν, we obtain the following expression:

$$\frac{\partial^2}{\partial x^2}\left(\sigma_{yy}-\nu\sigma_{xx}\right)+\frac{\partial^2}{\partial y^2}\left(\sigma_{xx}-\nu\sigma_{yy}\right)-2\left(1+\nu\right)\frac{\partial^2}{\partial x\partial y}\tau_{xy}=0, \quad (3.18)$$

or formulated in the force flows $N_{xx}^0, N_{yy}^0, N_{xy}^0$:

$$\frac{\partial^2}{\partial x^2}\left(\frac{N_{yy}^0}{h} - \nu\frac{N_{xx}^0}{h}\right) + \frac{\partial^2}{\partial y^2}\left(\frac{N_{xx}^0}{h} - \nu\frac{N_{yy}^0}{h}\right) - 2(1+\nu)\frac{\partial^2}{\partial x \partial y}\left(\frac{N_{xy}^0}{h}\right) = 0. \quad (3.19)$$

If the special case of a constant disk thickness h is given, then this expression simplifies to:

$$\frac{\partial^2}{\partial x^2}\left(N_{yy}^0 - \nu N_{xx}^0\right) + \frac{\partial^2}{\partial y^2}\left(N_{xx}^0 - \nu N_{yy}^0\right) - 2(1+\nu)\frac{\partial^2 N_{xy}^0}{\partial x \partial y} = 0, \quad (3.20)$$

or

$$\frac{\partial^2 N_{xx}^0}{\partial y^2} + \frac{\partial^2 N_{yy}^0}{\partial x^2} - 2\frac{\partial^2 N_{xy}^0}{\partial x \partial y} = \nu\left(\frac{\partial^2 N_{xx}^0}{\partial x^2} + \frac{\partial^2 N_{yy}^0}{\partial y^2} + 2\frac{\partial^2 N_{xy}^0}{\partial x \partial y}\right). \quad (3.21)$$

Forming the partial derivatives of the two equilibrium conditions (3.11) and (3.12) with respect to x and to y, respectively, yields:

$$\frac{\partial^2 \sigma_{xx}}{\partial x^2} + \frac{\partial \tau_{xy}^2}{\partial x \partial y} + \frac{\partial f_x}{\partial x} = 0, \quad \frac{\partial^2 \tau_{xy}}{\partial x \partial y} + \frac{\partial^2 \sigma_{yy}}{\partial y^2} + \frac{\partial f_y}{\partial y} = 0, \quad (3.22)$$

or

$$\frac{\partial^2 N_{xx}^0}{\partial x^2} + \frac{\partial^2 N_{xy}^0}{\partial x \partial y} + \frac{\partial p_x}{\partial x} = 0, \quad \frac{\partial^2 N_{xy}^0}{\partial x \partial y} + \frac{\partial^2 N_{yy}^0}{\partial y^2} + \frac{\partial p_y}{\partial y} = 0. \quad (3.23)$$

We now add the two equilibrium conditions in (3.22) and (3.23), respectively, and obtain:

$$\frac{\partial^2 \sigma_{xx}}{\partial x^2} + \frac{\partial^2 \sigma_{yy}}{\partial y^2} + 2\frac{\partial \tau_{xy}^2}{\partial x \partial y} = -\left(\frac{\partial f_x}{\partial x} + \frac{\partial f_y}{\partial y}\right), \quad (3.24)$$

or

$$\frac{\partial^2 N_{xx}^0}{\partial x^2} + \frac{\partial^2 N_{yy}^0}{\partial y^2} + 2\frac{\partial^2 N_{xy}^0}{\partial x \partial y} = -\left(\frac{\partial p_x}{\partial x} + \frac{\partial p_y}{\partial y}\right). \quad (3.25)$$

Addition of (3.21) and (3.25) yields with the Laplace-Operator $\Delta = \frac{\partial^2}{\partial x^2} + \frac{\partial^2}{\partial y^2}$:

$$\Delta\left(N_{xx}^0 + N_{yy}^0\right) = -(1+\nu)\left(\frac{\partial p_x}{\partial x} + \frac{\partial p_y}{\partial y}\right). \quad (3.26)$$

For the special case of constant quantities p_x and p_y it follows:

$$\Delta\left(N_{xx}^0 + N_{yy}^0\right) = 0, \quad (3.27)$$

or formulated in the stresses:

$$\Delta\left(\sigma_{xx}+\sigma_{yy}\right)=0. \tag{3.28}$$

This provides another equation for determining the forces or stresses of the disk. This equation is also called the disk equation.

In connection with the equilibrium conditions, the disk equation constitutes a differential equation system for the determination of the plane stress components σ_{xx}, σ_{yy}, τ_{xy} or the force flows N^0_{xx}, N^0_{yy}, N^0_{xy}. If volume forces f_x and f_y or their resultants p_x and p_y are present, then this differential equation system is generally inhomogeneous, and the solution for the desired force quantities consists of a homogeneous and a particular solution, where the homogeneous solution corresponds to the case $f_x = f_y = 0$ and $p_x = p_y = 0$, respectively. The particular solution depends mainly on the type of load f_x and f_y or p_x and p_y and shall be assumed to be known in all further details of this chapter. We thus focus our considerations exclusively on the determination of the homogeneous solution. We thus start from the equilibrium conditions in the form

$$\frac{\partial\sigma_{xx}}{\partial x}+\frac{\partial\tau_{xy}}{\partial y}=0,\quad \frac{\partial\tau_{xy}}{\partial x}+\frac{\partial\sigma_{yy}}{\partial y}=0 \tag{3.29}$$

or

$$\frac{\partial N^0_{xx}}{\partial x}+\frac{\partial N^0_{xy}}{\partial y}=0,\quad \frac{\partial N^0_{xy}}{\partial x}+\frac{\partial N^0_{yy}}{\partial y}=0 \tag{3.30}$$

and the disk Eq. (3.27) respectively (3.28).

We obtain a much more convenient representation of the basic equations by introducing the so-called Airy's[1] stress function $F(x, y)$ by which we can express the stresses σ_{xx}, σ_{yy}, τ_{xy} as follows:

$$\sigma_{xx}=\frac{\partial^2 F}{\partial y^2},\quad \sigma_{yy}=\frac{\partial^2 F}{\partial x^2},\quad \tau_{xy}=-\frac{\partial^2 F}{\partial x\partial y}. \tag{3.31}$$

It is easy to see that the above choice of $F(x, y)$ satisfies the equilibrium conditions (3.29) identically. Thus, the task remains to suitably solve the disk Eq. (3.28) in order to solve a given disk problem in the context of the force method. Substituting (3.31) into (3.28) then yields the following expression:

$$\frac{\partial^4 F}{\partial x^4}+2\frac{\partial^4 F}{\partial x^2\partial y^2}+\frac{\partial^4 F}{\partial y^4}=0, \tag{3.32}$$

or

$$\Delta\Delta F=0. \tag{3.33}$$

[1] George Bidell Airy, 1801–1892, English mathematician.

This is a biharmonic differential equation from which the Airy stress function can be determined as a biharmonic function in the case of constant disk thickness h and constant or vanishing volume forces f_x, f_y. Once this solution is found, the stresses can be determined from (3.31). In a post-calculation, strains and displacements can then also be determined.

It should be noted that a representation of the type (3.31) and (3.32) or (3.33) is also possible in the force flows N_{xx}^0, N_{yy}^0, N_{xy}^0. The Airy stress function is then to be introduced as follows[2]:

$$N_{xx}^0 = \frac{\partial^2 F}{\partial y^2}, \quad N_{yy}^0 = \frac{\partial^2 F}{\partial x^2}, \quad N_{xy}^0 = -\frac{\partial^2 F}{\partial x \partial y}. \tag{3.34}$$

It should be noted here that in the presence of e.g. constant volume forces f_x and f_y, Airy's stress function can be applied as follows[3]:

$$\sigma_{xx} = \frac{\partial^2 F}{\partial y^2} - x f_x, \quad \sigma_{yy} = \frac{\partial^2 F}{\partial x^2} - y f_y, \quad \tau_{xy} = -\frac{\partial^2 F}{\partial x \partial y}. \tag{3.35}$$

Here, too, it can be seen that the equilibrium conditions are fulfilled identically in the presence of constant volume forces, so that here, too, the procedure is limited to the solution of the compatibility equation.

To determine the displacements u_0 and v_0 (Altenbach et al. 2016; Girkmann 1974) for given disk stresses σ_{xx}, σ_{yy}, τ_{xy}, we consult the first two constitutive equations in (3.5) and solve them for the displacements:

$$E\frac{\partial u_0}{\partial x} = \sigma_{xx} - \nu\sigma_{yy}, \quad E\frac{\partial v_0}{\partial y} = \sigma_{yy} - \nu\sigma_{xx}. \tag{3.36}$$

Integration with respect to x and y, respectively, then yields the following expressions for the displacements:

$$\begin{aligned} E u_0 &= \int (\sigma_{xx} - \nu\sigma_{yy})\mathrm{d}x + C_1(y), \\ E v_0 &= \int (\sigma_{yy} - \nu\sigma_{xx})\mathrm{d}y + C_2(x). \end{aligned} \tag{3.37}$$

---

[2] Note that F is in the unit of a force in the case of considering the disk stresses, whereas F must be calculated with the unit of a force multiplied by a unit of length when using the internal force flows N_{xx}^0, N_{yy}^0, τ_{xy}^0. We will speak of the Airy stress function F in both cases in the following, and the unit to be used results from the respective context.

[3] Girkmann (1974) adds the volume forces to the definition of the shear stress in the form of $\tau_{xy} = -\frac{\partial^2 F}{\partial x \partial y} - y f_x - x f_y$.

If the stresses are expressed by Airy's stress function, then one obtains:

$$Eu_0 = \int \frac{\partial^2 F}{\partial y^2}\mathrm{d}x - \nu\frac{\partial F}{\partial x} + C_1(y),$$
$$Ev_0 = \int \frac{\partial^2 F}{\partial x^2}\mathrm{d}y - \nu\frac{\partial F}{\partial y} + C_2(x). \tag{3.38}$$

Here $C_1(x)$ and $C_2(y)$ are unknown functions which result from the integrations performed and which are to be determined from given boundary conditions. However, it turns out that these functions are not independent of each other, but are coupled by the kinematic equation $\gamma_{xy} = \frac{\partial u_0}{\partial y} + \frac{\partial v_0}{\partial x} = \frac{\tau_{xy}}{G}$. We obtain:

$$\int \frac{\partial^3 F}{\partial y^3}\mathrm{d}x + \int \frac{\partial^3 F}{\partial x^3}\mathrm{d}y + \frac{\mathrm{d}C_1}{\mathrm{d}y} + \frac{\mathrm{d}C_2}{\mathrm{d}x} = -2\frac{\partial^2 F}{\partial x \partial y}. \tag{3.39}$$

Thus, once the Airy stress function is available, the displacements u_0 and v_0 can be determined from (3.38) taking into account (3.39) and given boundary conditions.

In the case of rectangular disks, stress functions of the form $e^{\pm\alpha y}\cos\alpha x$, $ye^{\pm\alpha y}\cos\alpha x$, $e^{\pm\alpha y}\sin\alpha x$, $ye^{\pm\alpha y}\sin\alpha x$ (where α is an arbitrary constant) occur in many cases. In such cases (3.39) takes on the following form:

$$\frac{\mathrm{d}C_1}{\mathrm{d}y} + \frac{\mathrm{d}C_2}{\mathrm{d}x} = 0. \tag{3.40}$$

This equation, besides $\frac{\mathrm{d}C_1}{\mathrm{d}y} = \frac{\mathrm{d}C_2}{\mathrm{d}x} = 0$, has the following solution:

$$C_2(x) = A_1 - Bx, \quad C_1(y) = A_2 + By, \tag{3.41}$$

where the constant terms describe the rigid body translations $u = \frac{A_2}{E}$, $v = \frac{A_1}{E}$, and the two linear terms describe the rigid body rotation $u = -\frac{By}{E}$, $v = \frac{Bx}{E}$ of the disk.

Besides exact-analytical solutions, which are described by the equations given in this section, all common approximation methods of structural mechanics can be used for disk problems, which will be discussed later.

3.2.4 Boundary Conditions

In addition to the basic equations already discussed, a disk problem is uniquely described by the specification of boundary conditions. Depending on the chosen solution method, different formulations of boundary conditions are necessary.

If the displacement method is used, then displacement boundary conditions are quite easy to realize. On edges with x =const. and y =const. both displacements $u_0 = \hat{u}_0$ and $v_0 = \hat{v}_0$ or stresses $\sigma_{xx} = \hat{\sigma}_{xx}$, $\tau_{xy} = \hat{\tau}_{xy}$ can be given, where the stresses are expressed appropriately via the kinematic and constitutive equations by the displacements u_0 and v_0:

$$\frac{A}{h}\left(\frac{\partial u_0}{\partial x}+\nu\frac{\partial v_0}{\partial y}\right)=\hat{\sigma}_{xx},$$
$$\frac{A}{h}\left(\frac{\partial v_0}{\partial y}+\nu\frac{\partial u_0}{\partial x}\right)=\hat{\sigma}_{yy},$$
$$\frac{A}{h}\frac{1-\nu}{2}\left(\frac{\partial u_0}{\partial y}+\nu\frac{\partial v_0}{\partial x}\right)=\hat{\tau}_{xy}. \tag{3.42}$$

In the context of the force method, stress boundary conditions are particularly easy to specify. If there is an edge with x =const., then the two edge stresses $\hat{\sigma}_{xx}$ and $\hat{\tau}_{xy}$ can be specified. At an edge with y =const., on the other hand, stress boundary conditions of the type $\sigma_{yy}=\hat{\sigma}_{yy}$ and $\tau_{xy}=\hat{\tau}_{xy}$ can be formulated, which can also be expressed by Airy's stress function F. For the case of vanishing volume forces at an edge with x =const. we have:

$$\frac{\partial^2 F}{\partial y^2}=\hat{\sigma}_{xx},\qquad -\frac{\partial^2 F}{\partial x\partial y}=\hat{\tau}_{xy}. \tag{3.43}$$

At an edge with y =const. we get:

$$\frac{\partial^2 F}{\partial x^2}=\hat{\sigma}_{yy},\qquad -\frac{\partial^2 F}{\partial x\partial y}=\hat{\tau}_{xy}. \tag{3.44}$$

Given boundary conditions with respect to displacements u_0 and v_0 can then be satisfied with the displacement solutions (3.38) determined subsequently from the stresses, taking into account (3.39) and (3.41).

The consideration of disks with arbitrary boundary conditions will be discussed later. Further remarks on disk boundary conditions can be found e.g. in Girkmann (1974) or Göldner et al. (1979).

3.3 Energetic Consideration

3.3.1 Strain Energy

For the application of energy-based analysis methods, the strain energy of a disk in the plane state of stress is required. For this purpose, we consider a rectangular disk (length a in $x-$direction, width b in $y-$direction) having constant values for E, ν and h on the whole area under consideration. We then obtain the strain energy as:

$$
\begin{aligned}
\Pi_i &= \frac{1}{2}\int_V \left(\sigma_{xx}\varepsilon_{xx}^0 + \sigma_{yy}\varepsilon_{yy}^0 + \tau_{xy}\gamma_{xy}^0\right) \mathrm{d}V \\
&= \frac{1}{2}\int_{-\frac{h}{2}}^{\frac{h}{2}}\int_0^b\int_0^a \left(\sigma_{xx}\varepsilon_{xx}^0 + \sigma_{yy}\varepsilon_{yy}^0 + \tau_{xy}\gamma_{xy}^0\right) \mathrm{d}x\mathrm{d}y\mathrm{d}z \\
&= \frac{1}{2}\int_0^b\int_0^a \left(N_{xx}^0\varepsilon_{xx}^0 + N_{yy}^0\varepsilon_{yy}^0 + N_{xy}^0\gamma_{xy}^0\right) \mathrm{d}x\mathrm{d}y.
\end{aligned} \tag{3.45}
$$

Depending on the method of analysis to be used, different formulations may be appropriate. It may be useful to express the strain energy Π_i in the force quantities. Thus, in (3.45) we use the constitutive law in the form (3.9) and rearrange this according to the strain quantities and obtain:

$$
\Pi_i = \frac{1}{2Eh}\int_0^b\int_0^a \left[\left(N_{xx}^0\right)^2 + \left(N_{yy}^0\right)^2 - 2\nu N_{xx}^0 N_{yy}^0 + 2(1+\nu)\left(N_{xy}^0\right)^2\right] \mathrm{d}x\mathrm{d}y. \tag{3.46}
$$

This expression can also be formulated in the Airy stress function with (3.34) as:

$$
\Pi_i = \frac{1}{2Eh}\int_0^b\int_0^a \left[\left(\frac{\partial^2 F}{\partial x^2}\right)^2 + \left(\frac{\partial^2 F}{\partial y^2}\right)^2 - 2\nu\frac{\partial^2 F}{\partial x^2}\frac{\partial^2 F}{\partial y^2} + 2(1+\nu)\left(\frac{\partial^2 F}{\partial x \partial y}\right)^2\right] \mathrm{d}x\mathrm{d}y \tag{3.47}
$$

On the other hand, a formulation of the strain energy Π_i in terms of the disk displacements u_0 and v_0 may also be appropriate. Then in (3.45) the force flows N_{xx}^0, N_{yy}^0, N_{xy}^0 and the strains ε_{xx}^0, ε_{yy}^0, γ_{xy}^0 are to be expressed by u_0 and v_0 by means of the kinematic equations and the material law, and we obtain:

$$
\Pi_i = \frac{A}{2}\int_0^b\int_0^a \left[\left(\frac{\partial u_0}{\partial x} + \nu\frac{\partial v_0}{\partial y}\right)\frac{\partial u_0}{\partial x} + \left(\frac{\partial v_0}{\partial y} + \nu\frac{\partial u_0}{\partial x}\right)\frac{\partial v_0}{\partial y} + \frac{1-\nu}{2}\left(\frac{\partial u_0}{\partial y} + \frac{\partial v_0}{\partial x}\right)^2\right] \mathrm{d}x\mathrm{d}y. \tag{3.48}
$$

In the case that both the elastic modulus E and the disk stiffness A are not constant, the corresponding terms in the above equations are to be written inside the integrals.

3.3.2 Energetic Derivation of the Basic Equations

We want to show in this section that with the principle of virtual displacements $\delta W_i = \delta W_a$ we arrive at the already derived set of basic equations and consider again a rectangular disk (length a, width b, constant thickness h), which is now loaded at the edges $x = 0$ and $x = a$ by the boundary loads $\hat{N}^0_{xx}$ and $\hat{N}^0_{xy}$. The boundary loads $\hat{N}^0_{yy}$ and $\hat{N}^0_{xy}$ are present at the two edges $y = 0$ and $y = b$. We can derive the virtual internal work δW_i directly from (3.45) as follows:

$$\delta W_i = \int_0^b \int_0^a \left(N^0_{xx} \delta \varepsilon^0_{xx} + N^0_{yy} \delta \varepsilon^0_{yy} + N^0_{xy} \delta \gamma^0_{xy} \right) \mathrm{d}x \mathrm{d}y$$
$$= \int_0^b \int_0^a \left[N^0_{xx} \frac{\partial \delta u_0}{\partial x} + N^0_{yy} \frac{\partial \delta v_0}{\partial y} + N^0_{xy} \left(\frac{\partial \delta u_0}{\partial y} + \frac{\partial \delta v_0}{\partial x} \right) \right] \mathrm{d}x \mathrm{d}y. \quad (3.49)$$

The virtual external work δW_a presently reads:

$$\delta W_a = \int_0^b \hat{N}^0_{xx} \delta u_0 \Big|_0^a \mathrm{d}y + \int_0^b \hat{N}^0_{xy} \delta v_0 \Big|_0^a \mathrm{d}y + \int_0^a \hat{N}^0_{yy} \delta v_0 \Big|_0^b \mathrm{d}x + \int_0^a \hat{N}^0_{xy} \delta u_0 \Big|_0^b \mathrm{d}x. \quad (3.50)$$

Partial integration of (3.49) gives for the individual terms:

$$\int_0^b \int_0^a N^0_{xx} \frac{\partial \delta u_0}{\partial x} \mathrm{d}x \mathrm{d}y = \int_0^b N^0_{xx} \delta u_0 \Big|_0^a \mathrm{d}y - \int_0^b \int_0^a \frac{\partial N^0_{xx}}{\partial x} \delta u_0 \mathrm{d}x \mathrm{d}y,$$
$$\int_0^b \int_0^a N^0_{yy} \frac{\partial \delta v_0}{\partial y} \mathrm{d}x \mathrm{d}y = \int_0^a N^0_{yy} \delta v_0 \Big|_0^b \mathrm{d}x - \int_0^b \int_0^a \frac{\partial N^0_{yy}}{\partial y} \delta v_0 \mathrm{d}x \mathrm{d}y,$$
$$\int_0^b \int_0^a N^0_{xy} \frac{\partial \delta u_0}{\partial y} \mathrm{d}x \mathrm{d}y = \int_0^a N^0_{xy} \delta u_0 \Big|_0^b \mathrm{d}x - \int_0^b \int_0^a \frac{\partial N^0_{xy}}{\partial y} \delta u_0 \mathrm{d}x \mathrm{d}y,$$
$$\int_0^b \int_0^a N^0_{xy} \frac{\partial \delta v_0}{\partial x} \mathrm{d}x \mathrm{d}y = \int_0^b N^0_{xy} \delta v_0 \Big|_0^a \mathrm{d}y - \int_0^b \int_0^a \frac{\partial N^0_{xy}}{\partial x} \delta v_0 \mathrm{d}x \mathrm{d}y. \quad (3.51)$$

The principle of virtual displacements $\delta W_i = \delta W_a$ then results in:

$$- \int_0^b \int_0^a \left[\left(\frac{\partial N^0_{xx}}{\partial x} + \frac{\partial N^0_{xy}}{\partial y} \right) \delta u_0 + \left(\frac{\partial N^0_{xy}}{\partial x} + \frac{\partial N^0_{yy}}{\partial y} \right) \delta v_0 \right] \mathrm{d}x \mathrm{d}y$$
$$+ \int_0^b \left[\left(N^0_{xx} - \hat{N}^0_{xx} \right) \delta u_0 + \left(N^0_{xy} - \hat{N}^0_{xy} \right) \delta v_0 \right] \Big|_0^a \mathrm{d}y$$

$$+\int_0^a \left[\left(N_{yy}^0-\hat{N}_{yy}^0\right)\delta v_0+\left(N_{xy}^0-\hat{N}_{xy}^0\right)\delta u_0\right]\Bigg|_0^b \mathrm{d}x=0. \tag{3.52}$$

From the first line the equilibrium conditions (3.12) for $p_x=0$ and $p_y=0$ can be deduced. The remaining terms give the associated boundary conditions, according to which either the internal force flows at the relevant edges match the external loads, or the corresponding associated displacements u_0 and v_0 are specified. This equation set can also be written in terms of the stresses σ_{xx}, σ_{yy} and τ_{xy}. However, this remains without representation at this point.

Analogously, the principle of the minimum of the total elastic potential of the disk can be used to derive both the governing differential equations and all potential boundary conditions formulated in terms of the displacement quantities u_0 and v_0. The starting point is the total elastic potential $\Pi=\Pi_i+\Pi_a$, where the internal potential Π_i is already given by (3.48). We assume at this point that in addition to the two volume loads p_x and p_y, the boundary loads $\hat{N}_{xx}^0$, $\hat{N}_{xy}^0$ are also present at the boundaries $x=0, a$, and $\hat{N}_{yy}^0$, $\hat{N}_{xy}^0$ are present at the boundaries $y=0, b$. The external potential is then:

$$\begin{aligned}\Pi_a= & -\int_0^b\int_0^a\left(p_x u_0+p_y v_0\right)\mathrm{d}x\mathrm{d}y \\ & -\int_0^b \hat{N}_{xx}^0 u_0\Big|_0^a \mathrm{d}y-\int_0^b \hat{N}_{xy}^0 v_0\Big|_0^a \mathrm{d}y-\int_0^a \hat{N}_{yy}^0 v_0\Big|_0^b \mathrm{d}x-\int_0^a \hat{N}_{xy}^0 u_0\Big|_0^b \mathrm{d}x.\end{aligned} \tag{3.53}$$

The equations of the given disk problem then follow from the requirement for the vanishing of the first variation $\delta\Pi$ of the total elastic potential $\Pi=\Pi_i+\Pi_a$. For $\delta\Pi_a$ we obtain:

$$\begin{aligned}\delta\Pi_a= & -\int_0^b\int_0^a\left(p_x\delta u_0+p_y\delta v_0\right)\mathrm{d}x\mathrm{d}y \\ & -\int_0^b \hat{N}_{xx}^0\delta u_0\Big|_0^a \mathrm{d}y-\int_0^b \hat{N}_{xy}^0\delta v_0\Big|_0^a \mathrm{d}y-\int_0^a \hat{N}_{yy}^0\delta v_0\Big|_0^b \mathrm{d}x \\ & -\int_0^a \hat{N}_{xy}^0\delta u_0\Big|_0^b \mathrm{d}x.\end{aligned} \tag{3.54}$$

The necessary partial integration of the individual resulting terms in $\delta\Pi_i$ is shown below:

$$A\int_0^b\int_0^a \frac{\partial\delta u_0}{\partial x}\frac{\partial u_0}{\partial x}\mathrm{d}x\mathrm{d}y=A\int_0^b \delta u_0\frac{\partial u_0}{\partial x}\Big|_0^a \mathrm{d}y-A\int_0^b\int_0^a \frac{\partial^2 u_0}{\partial x^2}\delta u_0\mathrm{d}x\mathrm{d}y,$$

$$\nu\frac{A}{2}\int_0^b\int_0^a \frac{\partial\delta u_0}{\partial x}\frac{\partial v_0}{\partial y}\mathrm{d}x\mathrm{d}y=\nu\frac{A}{2}\int_0^b \delta u_0\frac{\partial v_0}{\partial y}\Big|_0^a \mathrm{d}y-\nu\frac{A}{2}\int_0^b\int_0^a \frac{\partial^2 v_0}{\partial x\partial y}\delta u_0\mathrm{d}x\mathrm{d}y,$$

$$\nu\frac{A}{2}\int_0^b\int_0^a \frac{\partial u_0}{\partial x}\frac{\partial \delta v_0}{\partial y}\mathrm{d}x\mathrm{d}y = \nu\frac{A}{2}\int_0^a \delta v_0\frac{\partial u_0}{\partial x}\bigg|_0^b \mathrm{d}x - \nu\frac{A}{2}\int_0^b\int_0^a \frac{\partial^2 u_0}{\partial x\partial y}\delta v_0\mathrm{d}x\mathrm{d}y,$$

$$A\int_0^b\int_0^a \frac{\partial \delta v_0}{\partial y}\frac{\partial v_0}{\partial y}\mathrm{d}x\mathrm{d}y = A\int_0^a \delta v_0\frac{\partial v_0}{\partial y}\bigg|_0^b \mathrm{d}x - A\int_0^b\int_0^a \frac{\partial^2 v_0}{\partial y^2}\delta v_0\mathrm{d}x\mathrm{d}y,$$

$$\frac{A}{2}(1-\nu)\int_0^b\int_0^a \frac{\partial \delta u_0}{\partial y}\frac{\partial u_0}{\partial y}\mathrm{d}x\mathrm{d}y = \frac{A}{2}(1-\nu)\int_0^a \delta u_0\frac{\partial u_0}{\partial y}\bigg|_0^b \mathrm{d}x - \frac{A}{2}(1-\nu)\int_0^b\int_0^a \frac{\partial^2 u_0}{\partial y^2}\delta u_0\mathrm{d}x\mathrm{d}y,$$

$$\frac{A}{2}(1-\nu)\int_0^b\int_0^a \frac{\partial \delta u_0}{\partial y}\frac{\partial v_0}{\partial x}\mathrm{d}x\mathrm{d}y = \frac{A}{2}(1-\nu)\int_0^a \delta u_0\frac{\partial v_0}{\partial x}\bigg|_0^b \mathrm{d}x - \frac{A}{2}(1-\nu)\int_0^b\int_0^a \frac{\partial^2 v_0}{\partial x\partial y}\delta u_0\mathrm{d}x\mathrm{d}y,$$

$$\frac{A}{2}(1-\nu)\int_0^b\int_0^a \frac{\partial u_0}{\partial y}\frac{\partial \delta v_0}{\partial x}\mathrm{d}x\mathrm{d}y = \frac{A}{2}(1-\nu)\int_0^b \delta v_0\frac{\partial u_0}{\partial y}\bigg|_0^a \mathrm{d}y - \frac{A}{2}(1-\nu)\int_0^b\int_0^a \frac{\partial^2 u_0}{\partial x\partial y}\delta v_0\mathrm{d}x\mathrm{d}y,$$

$$\frac{A}{2}(1-\nu)\int_0^b\int_0^a \frac{\partial \delta v_0}{\partial x}\frac{\partial v_0}{\partial x}\mathrm{d}x\mathrm{d}y = \frac{A}{2}(1-\nu)\int_0^b \delta v_0\frac{\partial v_0}{\partial x}\bigg|_0^a \mathrm{d}y - \frac{A}{2}(1-\nu)\int_0^b\int_0^a \frac{\partial^2 v_0}{\partial x^2}\delta v_0\mathrm{d}x\mathrm{d}y. \tag{3.55}$$

From $\delta\Pi = \delta\Pi_i + \delta\Pi_a = 0$ we thus obtain:

$$\begin{aligned}
&-A\int_0^b\int_0^a\left(\frac{\partial^2 u_0}{\partial x^2}+\frac{1-\nu}{2}\frac{\partial^2 u_0}{\partial y^2}+\frac{1+\nu}{2}\frac{\partial^2 v_0}{\partial x\partial y}+\frac{p_x}{A}\right)\delta u_0\mathrm{d}x\mathrm{d}y\\
&-A\int_0^b\int_0^a\left(\frac{\partial^2 v_0}{\partial y^2}+\frac{1-\nu}{2}\frac{\partial^2 v_0}{\partial x^2}+\frac{1+\nu}{2}\frac{\partial^2 u_0}{\partial x\partial y}+\frac{p_y}{A}\right)\delta u_0\mathrm{d}x\mathrm{d}y\\
&+\int_0^b\left[A\left(\frac{\partial u_0}{\partial x}+\nu\frac{\partial v_0}{\partial y}\right)-\hat{N}_{xx}^0\right]\delta u_0\bigg|_0^a\mathrm{d}y\\
&+\int_0^b\left[\frac{A}{2}(1-\nu)\left(\frac{\partial u_0}{\partial y}+\frac{\partial v_0}{\partial x}\right)-\hat{N}_{xy}^0\right]\delta v_0\bigg|_0^a\mathrm{d}y\\
&+\int_0^a\left[A\left(\frac{\partial v_0}{\partial y}+\nu\frac{\partial u_0}{\partial x}\right)-\hat{N}_{yy}^0\right]\delta v_0\bigg|_0^b\mathrm{d}x\\
&+\int_0^a\left[\frac{A}{2}(1-\nu)\left(\frac{\partial u_0}{\partial y}+\frac{\partial v_0}{\partial x}\right)-\hat{N}_{xy}^0\right]\delta u_0\bigg|_0^b\mathrm{d}x = 0.
\end{aligned} \tag{3.56}$$

The differential Eq. (3.17) as well as the associated boundary conditions can be deduced from this expression. The latter have already been provided by (3.52).

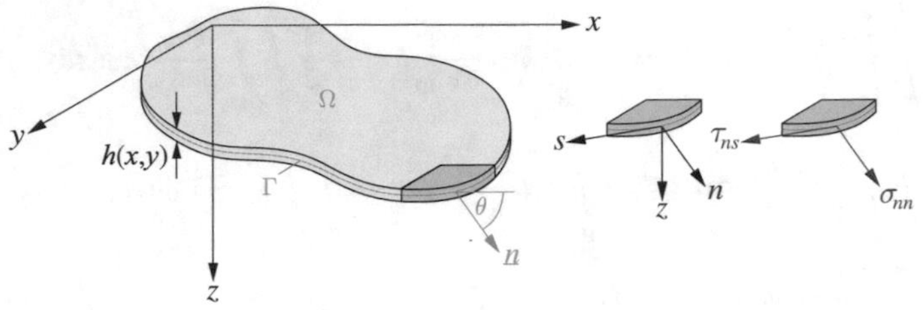

Fig. 3.3 Disk with area Ω and edge Γ of arbitrary contour

3.3.3 Disks with Arbitrary Boundaries

We now consider a disk with the area Ω bounded by the edge Γ of arbitrary contour (Fig. 3.3). Let the boundary be distinguished by the normal direction n and the tangential direction s as well as the normal vector $\underline{n}$, where the normal vector can be given as

$$\underline{n} = n_x \underline{e}_x + n_y \underline{e}_y, \tag{3.57}$$

with $n_x = \cos\theta$ and $n_y = \sin\theta$. Let the boundary stresses with respect to to n and s be the normal stress σ_{nn} and the shear stress τ_{ns}, and let the boundary displacements u_{0n} and u_{0s} be present. Let the boundary be composed of the two parts Γ_u and Γ_σ, where on Γ_σ the stresses σ_{nn} and τ_{ns} are given by the values $\sigma_{nn} = \hat{\sigma}_{nn}$ and $\tau_{ns} = \hat{\tau}_{ns}$. In the following we want to use the principle of virtual displacements $\delta W_i = \delta W_a$ to derive the equilibrium conditions and boundary conditions of an arbitrarily bounded disk.

The virtual internal work δW_i can be written for a disk in a plane stress state as:

$$\delta W_i = \int_\Omega \int_{-\frac{h}{2}}^{+\frac{h}{2}} \left(\sigma_{xx} \delta\varepsilon^0_{xx} + \sigma_{yy} \delta\varepsilon^0_{yy} + \tau_{xy} \delta\gamma^0_{xy} \right) \mathrm{d}z\mathrm{d}\Omega. \tag{3.58}$$

Introduction of the force flows

$$N^0_{xx} = \int_{-\frac{h}{2}}^{+\frac{h}{2}} \sigma_{xx} \mathrm{d}z, \quad N^0_{yy} = \int_{-\frac{h}{2}}^{+\frac{h}{2}} \sigma_{yy} \mathrm{d}z, \quad N^0_{xy} = \int_{-\frac{h}{2}}^{+\frac{h}{2}} \tau_{xy} \mathrm{d}z \tag{3.59}$$

yields

$$\delta W_i = \int_\Omega \left(N^0_{xx} \delta\varepsilon^0_{xx} + N^0_{yy} \delta\varepsilon^0_{yy} + N^0_{xy} \delta\gamma^0_{xy} \right) \mathrm{d}\Omega. \tag{3.60}$$

Let the edge stresses $\hat{\sigma}_{nn}$ and $\hat{\tau}_{ns}$ on Γ_σ be given as the load on the disk, so that the virtual external work results as:

$$\delta W_a = \int_{\Gamma_\sigma} \int_{-\frac{h}{2}}^{+\frac{h}{2}} \left(\hat{\sigma}_{nn} \delta u_{0n} + \hat{\tau}_{ns} \delta u_{0s}\right) \mathrm{d}z\mathrm{d}s. \tag{3.61}$$

With the boundary force flows

$$\hat{N}^0_{nn} = \int_{-\frac{h}{2}}^{+\frac{h}{2}} \hat{\sigma}_{nn} \mathrm{d}z, \quad \hat{N}^0_{ns} = \int_{-\frac{h}{2}}^{+\frac{h}{2}} \hat{\tau}_{ns} \mathrm{d}z \tag{3.62}$$

we obtain for δW_a:

$$\delta W_a = \int_{\Gamma_\sigma} \left(\hat{N}^0_{nn} \delta u_{0n} + \hat{N}^0_{ns} \delta u_{0s}\right) \mathrm{d}s. \tag{3.63}$$

With the virtual strains

$$\delta\varepsilon^0_{xx} = \frac{\partial \delta u_0}{\partial x}, \quad \delta\varepsilon^0_{yy} = \frac{\partial \delta v_0}{\partial y}, \quad \delta\gamma^0_{xy} = \frac{\partial \delta u_0}{\partial y} + \frac{\partial \delta v_0}{\partial x} \tag{3.64}$$

we obtain from the principle of virtual displacements $\delta W_i = \delta W_a$:

$$\int_\Omega \left[N^0_{xx} \delta \frac{\partial \delta u_0}{\partial x} + N^0_{yy} \frac{\partial \delta v_0}{\partial y} + N^0_{xy} \left(\frac{\partial \delta u_0}{\partial y} + \frac{\partial \delta v_0}{\partial x} \right) \right] \mathrm{d}\Omega$$
$$- \int_{\Gamma_\sigma} \left(\hat{N}^0_{nn} \delta u_{0n} + \hat{N}^0_{ns} \delta u_{0s}\right) \mathrm{d}s = 0. \tag{3.65}$$

Partial integration of those terms in (3.65) that involve derivatives of the virtual displacements δu_0 and δv_0 yields:

$$\int_\Omega N^0_{xx} \frac{\partial \delta u_0}{\partial x} \mathrm{d}\Omega = \oint_\Gamma N^0_{xx} n_x \delta u_0 \mathrm{d}s - \int_\Omega \frac{\partial N^0_{xx}}{\partial x} \delta u_0 \mathrm{d}\Omega,$$

$$\int_\Omega N^0_{yy} \frac{\partial \delta v_0}{\partial y} \mathrm{d}\Omega = \oint_\Gamma N^0_{yy} n_y \delta v_0 \mathrm{d}s - \int_\Omega \frac{\partial N^0_{yy}}{\partial y} \delta v_0 \mathrm{d}\Omega,$$

$$\int_\Omega N^0_{xy} \frac{\partial \delta u_0}{\partial y} \mathrm{d}\Omega = \oint_\Gamma N^0_{xy} n_y \delta u_0 \mathrm{d}s - \int_\Omega \frac{\partial N^0_{xy}}{\partial y} \delta u_0 \mathrm{d}\Omega,$$

$$\int_{\Omega} N_{xy}^0 \frac{\partial \delta v_0}{\partial x} \mathrm{d}\Omega = \oint_{\Gamma} N_{xy}^0 n_x \delta v_0 \mathrm{d}s - \int_{\Omega} \frac{\partial N_{xy}^0}{\partial x} \delta v_0 \mathrm{d}\Omega. \tag{3.66}$$

The principle of virtual displacements can then be written as follows noting that on Γ_u the virtual displacements must vanish:

$$\begin{aligned} &\int_{\Omega} \left[-\left(\frac{\partial N_{xx}^0}{\partial x} + \frac{\partial N_{xy}^0}{\partial y} \right) \delta u_0 - \left(\frac{\partial N_{xy}^0}{\partial x} + \frac{\partial N_{yy}^0}{\partial y} \right) \delta v_0 \right] \mathrm{d}\Omega \\ &+ \int_{\Gamma_\sigma} \left[\left(N_{xx}^0 n_x + N_{xy}^0 n_y \right) \delta u_0 + \left(N_{xy}^0 n_x + N_{yy}^0 n_y \right) \delta v_0 \right] \mathrm{d}s \\ &- \int_{\Gamma_\sigma} \left(\hat{N}_{nn}^0 \delta u_{0n} + \hat{N}_{ns}^0 \delta u_{0s} \right) \mathrm{d}s = 0. \end{aligned} \tag{3.67}$$

This equation can be fulfilled only if each of the individual summands becomes zero. The disk equilibrium conditions can thus be read off immediately from the integral expressions with respect to Ω:

$$\frac{\partial N_{xx}^0}{\partial x} + \frac{\partial N_{xy}^0}{\partial y} = 0, \quad \frac{\partial N_{xy}^0}{\partial x} + \frac{\partial N_{yy}^0}{\partial y} = 0. \tag{3.68}$$

To discuss the boundary terms, we establish a relationship between the Cartesian coordinates x, y and the boundary coordinates n, s as:

$$\begin{pmatrix} x \\ y \end{pmatrix} = \begin{bmatrix} \cos\theta & -\sin\theta \\ \sin\theta & \cos\theta \end{bmatrix} \begin{pmatrix} n \\ s \end{pmatrix} = \begin{bmatrix} n_x & -n_y \\ n_y & n_x \end{bmatrix} \begin{pmatrix} n \\ s \end{pmatrix}. \tag{3.69}$$

The transformation of the displacements is performed analogously:

$$\begin{pmatrix} u_0 \\ v_0 \end{pmatrix} = \begin{bmatrix} n_x & -n_y \\ n_y & n_x \end{bmatrix} \begin{pmatrix} u_{0n} \\ u_{0s} \end{pmatrix}. \tag{3.70}$$

The transformation of the stresses σ_{xx}, σ_{yy}, τ_{xy} is performed as:

$$\begin{pmatrix} \sigma_{nn} \\ \tau_{ns} \end{pmatrix} = \begin{bmatrix} n_x^2 & n_y^2 & 2n_x n_y \\ -n_x n_y & n_x n_y & n_x^2 - n_y^2 \end{bmatrix} \begin{pmatrix} \sigma_{xx} \\ \sigma_{yy} \\ \tau_{xy} \end{pmatrix}, \tag{3.71}$$

which is analogously valid for the transformation of the internal force flows:

$$\begin{pmatrix} N_{nn}^0 \\ N_{ns}^0 \end{pmatrix} = \begin{bmatrix} n_x^2 & n_y^2 & 2n_x n_y \\ -n_x n_y & n_x n_y & n_x^2 - n_y^2 \end{bmatrix} \begin{pmatrix} N_{xx}^0 \\ N_{yy}^0 \\ N_{xy}^0 \end{pmatrix}. \tag{3.72}$$

Thus:

$$\left(N_{xx}^{0}n_{x}+N_{xy}^{0}n_{y}\right)\delta u_{0}+\left(N_{xy}^{0}n_{x}+N_{yy}^{0}n_{y}\right)\delta v_{0}=N_{nn}^{0}\delta u_{0n}+N_{ns}^{0}\delta u_{0s}. \quad (3.73)$$

The requirement for the disappearance of the boundary terms in (3.67) can thus be summarized as:

$$\int_{\Gamma_\sigma}\left[\left(N_{nn}^{0}-\hat{N}_{nn}^{0}\right)\delta u_{0n}+\left(N_{ns}^{0}-\hat{N}_{ns}^{0}\right)\delta u_{0s}\right]\mathrm{d}s=0. \quad (3.74)$$

3.4 Elementary Solutions

3.4.1 Solutions of the Disk Equation

In this section we focus on the application of the force method. Starting from the disk equation $\Delta\Delta F = 0$ (3.32) and (3.33), respectively, any number of solutions for the Airy stress function F can be obtained. Therefore, a general and universally applicable solution cannot be given, but rather one will have to choose a suitable solution or combination of solutions from a catalog of solutions which satisfy given boundary conditions. In general, any solution F of the bipotential equation $\Delta\Delta F = 0$ will have an associated stress and hence strain and displacement state. This is then also true for any linear combination of admissible solutions.

Using a polynomial approach of the type (Becker and Gross 2002; Girkmann 1974; Sadd 2005)

$$F(x,y)=\sum_{m=0}^{M}\sum_{n=0}^{N}F_{mn}x^{m}y^{n} \quad (3.75)$$

we can determine the unknown coefficients after inserting them into the disk equation $\Delta\Delta F = 0$ in such a way that the disk equation is exactly fulfilled. We call the resulting polynomials biharmonic polynomials. This is briefly shown below. Inserting the expression (3.75) into the disk Eqs. (3.32) or (3.33) yields:

$$\begin{aligned}&\sum_{m=4}^{M}\sum_{n=0}^{N}m(m-1)(m-2)(m-3)F_{mn}x^{m-4}y^{n}\\&+2\sum_{m=2}^{M}\sum_{n=2}^{N}m(m-1)n(n-1)F_{mn}x^{m-2}y^{n-2}\\&+\sum_{m=0}^{M}\sum_{n=4}^{N}n(n-1)(n-2)(n-3)F_{mn}x^{m}y^{n-4}=0.\end{aligned} \quad (3.76)$$

It should be noted that those terms in (3.75) for which $m + n \leq 1$ holds have no influence on the stress field due to the definition (3.31) and thus can be neglected. The expression (3.76) can be put into the following form:

$$\sum_{m=2}^{M}\sum_{n=2}^{N}\big[(m+2)(m+1)m(m-1)F_{m+2,n-2} + 2m(m-1)n(n-1)F_{mn} + (n+2)(n+1)n(n-1)F_{m-2,n+2}\big]x^{m-2}y^{n-2} = 0. \tag{3.77}$$

This expression must become zero independently of x and y, so that:

$$(m+2)(m+1)m(m-1)F_{m+2,n-2} + 2m(m-1)n(n-1)F_{mn} + (n+2)(n+1)n(n-1)F_{m-2,n+2} = 0. \tag{3.78}$$

This is a recursive formula for the determination of the series coefficients F_{mn}. Possible terms of biharmonic polynomials resulting from this are (Becker and Gross 2002; Gross et al. 2014)

$$\begin{aligned} F(x,y) = C,\quad & x,\quad x^2,\quad x^3,\quad xy,\quad x^2y,\quad x^3y,\quad x^4-3x^2y^2,\\ & x^4y - x^2y^3,\quad x^5 - 5x^3y^2,\quad x^5y - \frac{5}{3}x^3y^3,\\ & x^6 - 10x^4y^2 + 5x^2y^4, \ldots, \end{aligned} \tag{3.79}$$

where x and y are interchangeable.

Further solutions for the disk equation $\Delta\Delta F = 0$ can be obtained (Altenbach et al. 2016) by setting $\Delta F = \bar{F}$ and requiring that $\Delta\bar{F} = 0$ holds. Thus, any solution of the potential equation $\Delta\bar{F} = 0$ is at the same time a solution of the bipotential equation $\Delta\bar{F} = 0$. Now a separation approach of the following kind is used (Altenbach et al. 2016; see also Sadd 2005):

$$\bar{F} = F_1(x)F_2(y). \tag{3.80}$$

From the potential equation $\Delta\bar{F} = 0$ then follows:

$$\Delta\bar{F} = \frac{\mathrm{d}^2F_1}{\mathrm{d}x^2}F_2 + F_1\frac{\mathrm{d}^2F_2}{\mathrm{d}y^2} = 0, \tag{3.81}$$

which can be transformed to:

$$\frac{1}{F_1}\frac{\mathrm{d}^2F_1}{\mathrm{d}x^2} = -\frac{1}{F_2}\frac{\mathrm{d}^2F_2}{\mathrm{d}y^2}. \tag{3.82}$$

It becomes clear that the left side of this equation depends exclusively on x, whereas on the right side only the variable y appears. Obviously, (3.82) can be satisfied only if both expressions of this equation are constant for themselves and, for example, the left side corresponds to the value $-\lambda^2$ and the right side to the value λ^2, i.e.:

$$\frac{1}{F_1}\frac{\mathrm{d}^2 F_1}{\mathrm{d}x^2} = -\frac{1}{F_2}\frac{\mathrm{d}^2 F_2}{\mathrm{d}y^2} = -\lambda^2. \tag{3.83}$$

These are two similar differential equations of the form

$$\frac{\mathrm{d}^2 F_1}{\mathrm{d}x^2} + F_1\lambda^2 = 0, \quad \frac{\mathrm{d}^2 F_2}{\mathrm{d}y^2} - F_2\lambda^2 = 0, \tag{3.84}$$

where λ^2 can take arbitrary positive or negative values or even the value zero. Thus, the types of solutions in (3.84) are explicitly controlled by the value λ^2. For $F_1(x)$, we obtain the following solution for $\lambda^2 > 0$:

$$F_1(x) = C_1 \cos\lambda x + C_2 \sin\lambda x. \tag{3.85}$$

If λ^2 takes the value zero, then:

$$F_1(x) = C_1 x + C_2. \tag{3.86}$$

In the case of $\lambda^2 < 0$ it follows:

$$F_1(x) = C_1 \cosh\bar{\lambda} x + C_2 \sinh\bar{\lambda} x, \tag{3.87}$$

where $\lambda = i\bar{\lambda}$. This can also be written as follows when using exponential functions:

$$F_1(x) = C_1 e^{\bar{\lambda} x} + C_2 e^{\;\bar{\lambda} x}. \tag{3.88}$$

Analogous solutions can then be given for $F_2(y)$, noting the opposite sign of λ^2.

These functions F_1 and F_2 then form according to (3.80) the potential function $\bar{F}$ as follows. If the case $\lambda^2 > 0$ is present, then:

$$\bar{F} = (C_1 \cos\lambda x + C_2 \sin\lambda x)\left(D_1 e^{\lambda y} + D_2 e^{-\lambda y}\right), \tag{3.89}$$

where we have chosen the formulation with exponential functions for $F_2(y)$.

For $\lambda^2 = 0$ we have:

$$\bar{F} = (C_1 x + C_2)(D_1 y + D_2). \tag{3.90}$$

In the case of $\lambda^2 < 0$ we obtain:

$$\bar{F} = \left(C_1 e^{\bar{\lambda} x} + C_2 e^{-\bar{\lambda} x}\right)\left(D_1 \cos\bar{\lambda} y + D_2 \sin\bar{\lambda} y\right), \tag{3.91}$$

where again the notation using exponential functions was used.

These solutions are valid for arbitrary values of λ and can be combined linearly. Finally, a complete set of solutions is obtained from the requirement $\Delta F = \bar{F}$, which is an inhomogeneous partial differential equation for the Airy stress function.

Gross et al. (2014) also give combinations of exponential functions and trigonometric functions:

$$F(x, y) = e^{\pm\lambda y} \cos \lambda x, \quad x e^{\pm\lambda y} \cos \lambda x, \tag{3.92}$$

where x and y are also interchangeable and the cosine function can be replaced by a sine function (see also Göldner et al. 1979). Functions of the form

$$F(x, y) = \sinh \alpha y \sin \alpha x, \quad y \sinh \alpha y \sin \alpha x, \quad x \sinh \alpha y \sin \alpha x \tag{3.93}$$

are equivalent (Hake und Meskouris 2007, a formal derivation of these terms can be found in Sadd 2005), where the sin and sinh functions can be replaced by corresponding cosine terms, and y and x are interchangeable. Girkmann (1974) and Hake and Meskouris (2007) also provide logarithmic functions which satisfy the disk equation:

$$\begin{aligned} F(x, y) = \ln(x^2 + y^2), \quad (x^2 + y^2) \ln(x^2 + y^2), \quad (ax + by) \ln(x^2 + y^2), \\ \ln[(x + c)^2 + y^2], \quad (x + c) \ln[(x + c)^2 + y^2], \ldots, \end{aligned} \tag{3.94}$$

wherein c is a constant.

3.4.2 Elementary Cases

In this section we demonstrate the application of the force method to some elementary examples of isotropic disks in Cartesian coordinates. We begin the elaborations by considering a rectangular isotropic disk (length a, width b, constant thickness h) and expose this disk to some elementary load cases (Fig. 3.4). For the case of uniaxial tensile load $\sigma_{xx} = \sigma_0$ (Fig. 3.4, top right), we first consider the stress boundary conditions, which for the given situation are as follows. At the disk edges $x = 0$ and $x = a$ we have:

$$\sigma_{xx} = \sigma_0, \quad \tau_{xy} = 0. \tag{3.95}$$

At the two edges $y = 0$ and $y = b$ on the other hand the following holds:

$$\sigma_{yy} = 0, \quad \tau_{xy} = 0. \tag{3.96}$$

Since the boundary stresses are constant, the Airy stress function must be chosen in the form of a quadratic polynomial:

$$F = C_1 x^2 + C_2 y^2 + C_3 xy + C_4 x + C_5 y + C_6. \tag{3.97}$$

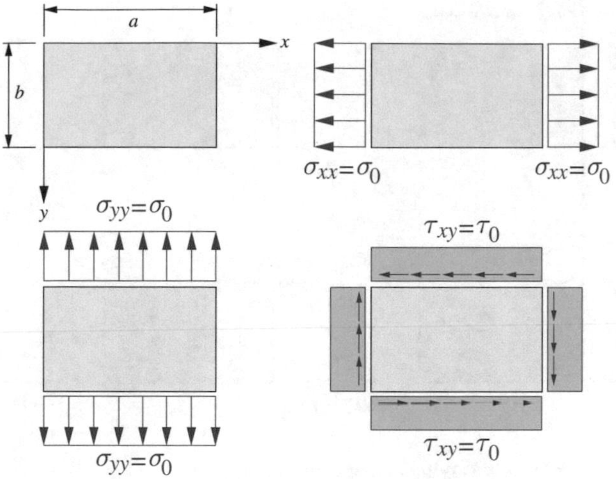

Fig. 3.4 Disk under elementary load cases

Therein, the quantities $C_1, \ldots, C_6$ are constants that are fitted to the given boundary conditions. The stress components are obtained from this as:

$$\sigma_{xx} = \frac{\partial^2 F}{\partial y^2} = 2C_2, \quad \sigma_{yy} = \frac{\partial^2 F}{\partial x^2} = 2C_1, \quad \tau_{xy} = -\frac{\partial^2 F}{\partial x \partial y} = -C_3. \quad (3.98)$$

Obviously, the constants C_4, C_5 and C_6 drop out of the stress calculation, they have no influence on the stresses.

With the given stress state, the remaining constants C_1, C_2 and C_3 are obtained as:

$$C_1 = 0, \quad C_2 = \frac{\sigma_0}{2}, \quad C_3 = 0. \quad (3.99)$$

Thus, a formulation for the stress function F of the type $F = \frac{1}{2}\sigma_0 y^2$ is suitable to describe the given problem. The stress state in the disk is then given by (3.31):

$$\sigma_{xx} = \frac{\partial^2 F}{\partial y^2} = \sigma_0, \quad \sigma_{yy} = \frac{\partial^2 F}{\partial x^2} = 0, \quad \tau_{xy} = -\frac{\partial^2 F}{\partial x \partial y} = 0. \quad (3.100)$$

If, on the other hand, the case of a uniaxial tensile stress $\sigma_{yy} = \sigma_0$ is present, then the stress function $F = \frac{1}{2}\sigma_0 x^2$ leads quite analogously to the following stress field in the disk:

$$\sigma_{xx} = \frac{\partial^2 F}{\partial y^2} = 0, \quad \sigma_{yy} = \frac{\partial^2 F}{\partial x^2} = \sigma_0, \quad \tau_{xy} = -\frac{\partial^2 F}{\partial x \partial y} = 0. \quad (3.101)$$

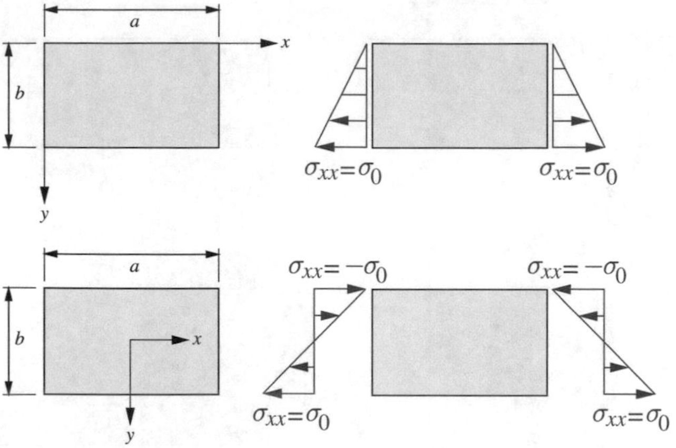

Fig. 3.5 Disks under linearly varying boundary stress σ_{xx}

In the case of a uniform shear load $\tau_{xy} = \tau_0$, a formulation for the stress function F is suitable with $F = -\tau_0 xy$, so that:

$$\sigma_{xx} = \frac{\partial^2 F}{\partial y^2} = 0, \quad \sigma_{yy} = \frac{\partial^2 F}{\partial x^2} = 0, \quad \tau_{xy} = -\frac{\partial^2 F}{\partial x \partial y} = \tau_0. \tag{3.102}$$

Analogously, further elementary stress states can be constructed by a suitable choice of the Airy stress function F. It is also shown that by a convenient choice of the reference frame with the same formulation for F different stress states can be represented, which we will show by the example of Fig. 3.5. Figure 3.5, top, shows a disk under the linearly varying boundary stress σ_{xx}, which takes the value $\sigma_{xx} = 0$ at $y = 0$ and $\sigma_{xx} = \sigma_0$ at $y = b$. In the given coordinate system with origin in the upper left corner of the disk, a formulation für F of the type $F = \frac{1}{6b}\sigma_0 y^3$ is suitable to represent this stress state:

$$\sigma_{xx} = \frac{\partial^2 F}{\partial y^2} = \sigma_0 \frac{y}{b}, \quad \sigma_{yy} = \frac{\partial^2 F}{\partial x^2} = 0, \quad \tau_{xy} = -\frac{\partial^2 F}{\partial x \partial y} = 0. \tag{3.103}$$

The case of a linearly varying boundary load with boundary values $\sigma_{xx} = \pm\sigma_0$ for $y = \pm\frac{b}{2}$ (Fig. 3.5, bottom), on the other hand, can be represented by the stress function $F = \frac{1}{3b}\sigma_0 y^3$. The following stress field is then obtained:

$$\sigma_{xx} = \frac{\partial^2 F}{\partial y^2} = 2\sigma_0 \frac{y}{b}, \quad \sigma_{yy} = \frac{\partial^2 F}{\partial x^2} = 0, \quad \tau_{xy} = -\frac{\partial^2 F}{\partial x \partial y} = 0. \tag{3.104}$$

Apparently, identical formulations for the Airy stress function were used in both cases considered here, differing only by the constant factors $\frac{\sigma_0}{6b}$ and $\frac{\sigma_0}{3b}$.

3.5 Beam-type Disks

The computational rules presented for the analysis of disks are also particularly useful for deriving improved computational models for beam structures, especially in cases where both the Euler-Bernoulli beam theory and higher beam theories give inconclusive results. For illustration, consider a cantilever beam as shown in Fig. 3.6, which has length a, height $2b$, and thickness h (Göldner et al. 1979; Eschenauer and Schnell 1993; Becker and Gross 2002). Let the cantilever be isotropic and homogeneous and also loaded by the surface load p_0 at $y = b$. The left end of the cantilever at $x = 0$ is free, while the right end at $x = a$ is clamped.

The following approach for the Airy stress function F in terms of a biharmonic polynomial is appropriate in this case:

$$F = C_1 x^2 + C_2 x^2 y + C_3 y^3 + C_4 \left(y^5 - 5x^2 y^3 \right). \tag{3.105}$$

From this, the stress components follow as:

$$\begin{aligned} \sigma_{xx} &= \frac{\partial^2 F}{\partial y^2} = 6C_3 y + C_4 \left(20y^3 - 30x^2 y \right), \\ \sigma_{yy} &= \frac{\partial^2 F}{\partial x^2} = 2C_1 + 2C_2 y - 10C_4 y^3, \\ \tau_{xy} &= -\frac{\partial^2 F}{\partial x \partial y} = -2C_2 x + 30C_4 x y^2. \end{aligned} \tag{3.106}$$

The constants C_1, C_2, C_3, C_4 are now adjusted to the given boundary conditions, which can be given in terms of the stresses as follows:

$$\begin{aligned} &\sigma_{yy} (y = b) = -p_0, \quad \tau_{xy} (y = b) = 0, \\ &\sigma_{yy} (y = -b) = 0, \quad \tau_{xy} (y = -b) = 0, \\ &\sigma_{xx} (x = 0) = 0, \quad \tau_{xy} (x = 0) = 0. \end{aligned} \tag{3.107}$$

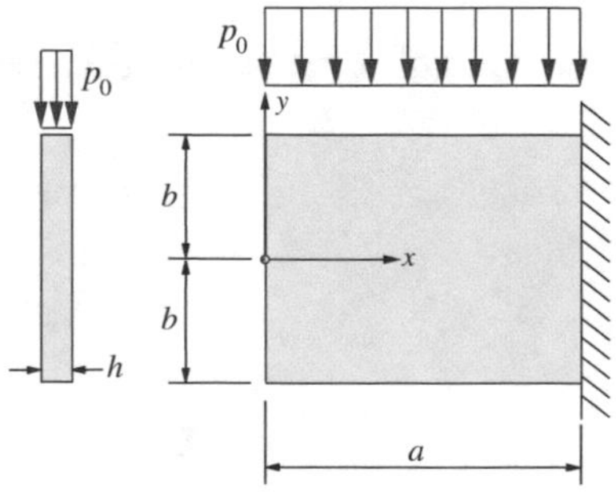

Fig. 3.6 Cantilever disk

We first address the boundary conditions at the top edge at $y = b$. The first two boundary conditions in (3.107) result in the following expressions using (3.106):

$$\begin{aligned} 2C_1 + 2C_2 b - 10C_4 b^3 &= -p_0, \\ -2C_2 x + 30C_4 x b^2 &= 0. \end{aligned} \tag{3.108}$$

With respect to the boundary conditions at the bottom of the disk at $y = -b$, it can be seen that the chosen approach (3.105) yields a shear stress distribution τ_{xy}, which identically satisfies the condition $\tau_{xy}(y = -b) = 0$. Evaluating the boundary condition $\sigma_{yy}(y = -b) = 0$, on the other hand, yields:

$$2C_1 - 2C_2 b + 10C_4 b^3 = 0. \tag{3.109}$$

In addition, the conditions at the free end of the cantilever at $x = 0$, i.e. the requirements $\sigma_{xx} = 0$ and $\tau_{xy} = 0$, have to be fulfilled. However, it turns out that these conditions cannot be exactly satisfied point by point, so here we resort to a formulation that satisfies these conditions in an integral sense and requires that the stress resultants resulting from σ_{xx} and τ_{xy} become zero:

$$N = h\int_{-b}^{b} \sigma_{xx}\mathrm{d}y = 0, \quad M = h\int_{-b}^{b} \sigma_{xx} y\mathrm{d}y = 0, \quad Q = h\int_{-b}^{b} \tau_{xy}\mathrm{d}y = 0. \tag{3.110}$$

This reveals that the resultants N and Q with the chosen approach (3.105) automatically become zero at the edge $x = 0$, whereas the requirement $M = 0$ yields the following:

$$4C_3 b^3 + 8C_4 b^5 = 0. \tag{3.111}$$

Thus, there are four equations from which the four constants C_1, C_2, C_3, C_4 can be determined. They follow as:

$$C_1 = -\frac{p_0}{4}, \quad C_2 = -\frac{3p_0}{8b}, \quad C_3 = \frac{p_0}{20b}, \quad C_4 = -\frac{p_0}{40b^3}. \tag{3.112}$$

The stress distributions in the disk are thus given as:

$$\begin{aligned} \sigma_{xx} &= p_0 \frac{y}{b}\left(\frac{3}{10} + \frac{3x^2}{4b^2} - \frac{y^2}{2b^2}\right), \\ \sigma_{yy} &= p_0\left(-\frac{1}{2} - \frac{3y}{4b} + \frac{y^3}{4b^3}\right), \\ \tau_{xy} &= p_0 \frac{x}{b}\left(\frac{3}{4} - \frac{3y^2}{4b^2}\right). \end{aligned} \tag{3.113}$$

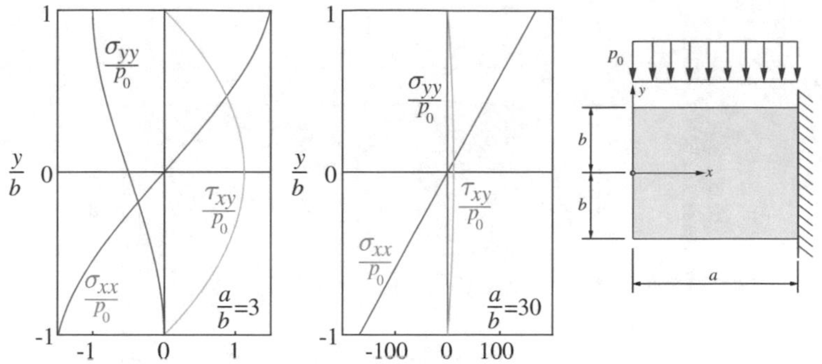

Fig. 3.7 Stress distributions in the cantilever disk for $\frac{a}{b} = 3$ (left) and $\frac{a}{b} = 30$ (right), evaluated at $x = \frac{a}{2}$

This provides a closed-form solution for the stress components in the disk. The resulting strain and displacement components can be easily determined from this, but this remains without representation here. A discussion (see Fig. 3.7) shows that in the case of slender beams with $a >> 2b$ at sufficient distance from the free end (i.e., for $x >> b$), the disk stress σ_{xx} essentially corresponds to the bending stress according to Euler-Bernoulli theory. The normal stress σ_{yy} remains small compared to σ_{xx} despite the transverse load p_0 in the $y-$direction, provided it is a slender disk. The distribution of the shear stress τ_{xy} agrees with the formulation according to the Euler-Bernoulli beam theory. On the other hand, a fundamentally different picture emerges for rather compact disks. In particular, the pronounced nonlinearity of σ_{xx} and the occurrence of significant values for the transverse normal stress σ_{yy} should be emphasized.

Another example of a beam-like disk is the simply supported beam of Fig. 3.8 (length $2a$, height $2b$, thickness h, surface load p_0 at the top at $y = b$, coordinate origin as shown, see Gross et al. 2014; also solved by Massonet 1962; Eschenauer et al. 1997; Sadd 2005; Göldner et al. 1979). For this disk situation, due to the symmetry properties of the given problem, an approach for the Airy stress function F is suitable which is symmetric with respect to x and non-symmetric with respect to y and also satisfies the disk equation:

$$F = C_1 x^2 + C_2 x^2 y + C_3 y^3 + C_4 \left(x^4 y - x^2 y^3\right) + C_5 \left(y^5 - 5y^3 x^2\right). \tag{3.114}$$

The stress components of the disk then follow as:

$$\sigma_{xx} = \frac{\partial^2 F}{\partial y^2} = 6C_3 y - 6C_4 x^2 y + C_5 \left(20y^3 - 30x^2 y\right),$$

$$\sigma_{yy} = \frac{\partial^2 F}{\partial x^2} = 2C_1 + 2C_2 y + C_4 \left(12x^2 y - 2y^3\right) - 10C_5 y^3,$$

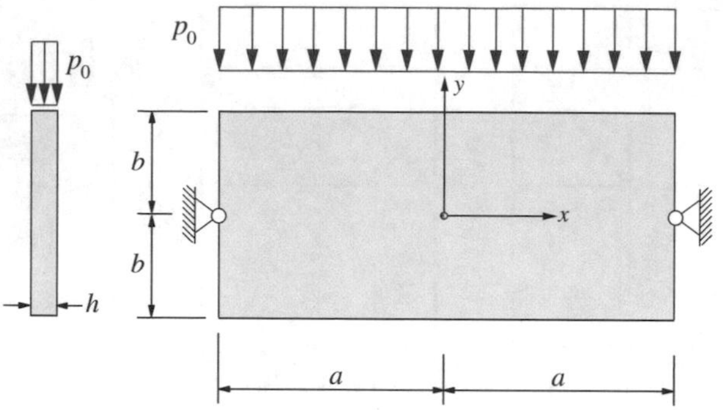

Fig. 3.8 Simply supported disk

$$\tau_{xy} = -\frac{\partial^2 F}{\partial x \partial y} = -2C_2 x - C_4\left(4x^3 - 6xy^2\right) + 30C_5 xy^2. \tag{3.115}$$

The constants $C_1, \ldots, C_5$ are again determined from the stress boundary conditions. Again, we want to require that they are exactly satisfied at the two edges at $y = \pm b$, i.e.:

$$\sigma_{yy}(y = b) = -p_0, \quad \tau_{xy}(y = b) = 0, \quad \sigma_{yy}(y = -b) = 0, \quad \tau_{xy}(y = -b) = 0. \tag{3.116}$$

This gives $C_4 = 0$ and two equations for the constants $C_1, \ldots, C_5$:

$$2C_1 - 2C_2 b + 10C_5 b^3 = 0, \quad -2C_2 + 30C_5 b^2 = 0. \tag{3.117}$$

At the two disk edges $x = \pm a$, as in the previous example, we want to make demands on the stress resultants as follows:

$$N(x = \pm a) = h \int_{-b}^{b} \sigma_{xx} \mathrm{d}y = 0,$$

$$Q(x = \pm a) = h \int_{-b}^{b} \tau_{xy} \mathrm{d}y = \pm p_0 a h,$$

$$M(x = \pm a) = h \int_{-b}^{b} \sigma_{xx} y \mathrm{d}y = 0. \tag{3.118}$$

It can be seen that the requirement for the disappearance of the normal force N is fulfilled identically. From the remaining requirements two further equations for the constants $C_1, \ldots, C_5$ result as follows:

$$-4C_2 b + 20C_5 b^3 = p_0, \quad 2C_3 b^3 + C_5\left(4b^5 - 10b^3 a^2\right) = 0. \tag{3.119}$$

This provides four equations for determining C_1, C_2, C_3, and C_5 (C_4 has already been determined as $C_4 = 0$), leading to:

$$C_1 = -\frac{p_0}{4}, \quad C_2 = -\frac{3p_0}{8b}, \quad C_3 = -\frac{p_0}{8b^3}\left(a^2 - \frac{2b^2}{5}\right), \quad C_5 = -\frac{p_0}{40b^3}. \tag{3.120}$$

The disk stresses can then be expressed as:

$$\begin{aligned} \sigma_{xx} &= -\frac{3p_0a^2y}{4b^3}\left(1 - \frac{x^2}{a^2} + \frac{2y^2}{3a^2} - \frac{2b^2}{5a^2}\right), \\ \sigma_{yy} &= -p_0\left(\frac{1}{2} + \frac{3y}{4b} - \frac{y^3}{4b^3}\right), \\ \tau_{xy} &= \frac{3p_0x}{4b}\left(1 - \frac{y^2}{b^2}\right). \end{aligned} \tag{3.121}$$

An illustration of the distribution of the stress components is given in Fig. 3.9. It can be seen that the distribution of the normal stress σ_{xx} deviates from the results according to elementary beam theory. The disk solution predicts a nonlinear distribution over y, whereas the beam theory predicts a linear distribution. The distribution of τ_{xy} corresponds again to what would result according to the Euler-Bernoulli beam theory. The transverse normal stress σ_{yy} remains usually quite small compared to σ_{xx} and can be neglected for very slender beam-like disks with $a >> b$. When interpreting the results, it should be noted that they are accurate if the support situation is designed so that the support reactions are introduced by distributed boundary loads in such a way that they correspond exactly to the boundary stresses. If the support situation differs from this, then edge stresses result at the ends of the disk due to the introduction of forces, which usually decay rapidly over x and are relieved after about one disk height. This must be taken into account in the design of such structures, and the subject of force introduction analysis will be addressed later.

An iterative solution for beam-like disks under lateral load is described in Altenbach et al. (2016). Girkmann (1974) and Wang (1953) consider beam-like disks

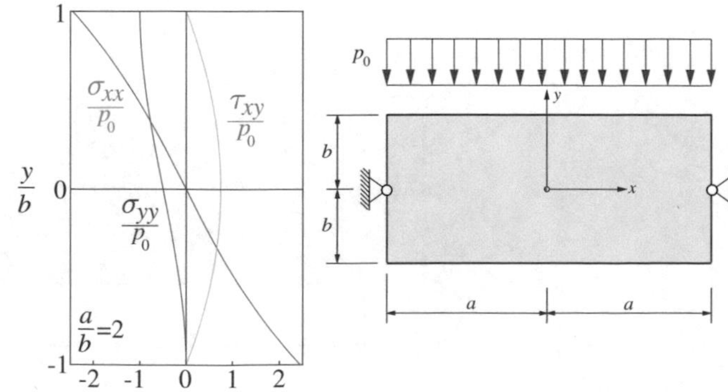

Fig. 3.9 Stress distributions for the simply supported disk evaluated at $x = \frac{a}{2}$

under concentrated loads and sinusoidally distributed surface loads. Eschenauer and Schnell (1993) and Hake and Meskouris (2007) address a beam-like disk loaded by a sinusoidal boundary load (Fig. 3.10). This analysis procedure is briefly outlined below. The disk under consideration is again supported in such a way that the external load $p_x(x)$ is introduced by shear into the two vertical edges. The disk has the dimensions a and b, the coordinate system is introduced as indicated.

We assume that the given load $p_x(x)$ can be represented in terms of a Fourier series as follows:

$$p(x) = \sum_{n=1}^{\infty} p_n \sin\left(\frac{n\pi x}{a}\right). \tag{3.122}$$

We perform the calculation below in general form for the series term n, i.e. for

$$p(x) = p_n \sin \alpha_n x, \quad \alpha_n = \frac{n\pi}{a}. \tag{3.123}$$

We use a product approach for the Airy stress function as follows:

$$F(x, y) = f(x)g(y). \tag{3.124}$$

From this we obtain the stress components $\sigma_{xx}, \sigma_{yy}, \tau_{xy}$ as:

$$\sigma_{xx} = \frac{\partial^2 F}{\partial y^2} = f\frac{\mathrm{d}^2 g}{\mathrm{d}y^2}, \quad \sigma_{yy} = \frac{\partial^2 F}{\partial x^2} = \frac{\mathrm{d}^2 f}{\mathrm{d}x^2} g, \quad \tau_{xy} = -\frac{\partial^2 F}{\partial x \partial y} = -\frac{\mathrm{d}f}{\mathrm{d}x}\frac{\mathrm{d}g}{\mathrm{d}y}. \tag{3.125}$$

We first consider the boundary condition that at the loaded disk edge $y = b$ the normal stress σ_{yy} must be in equilibrium with the boundary stress, i.e.:

$$\sigma_{yy}(y = b) = \frac{\mathrm{d}^2 f}{\mathrm{d}x^2} g = -\frac{p(x)}{h} = -\frac{1}{h} p_n \sin \alpha_n x. \tag{3.126}$$

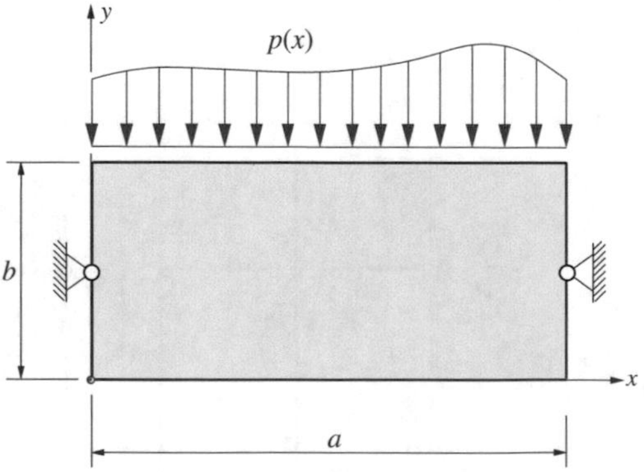

Fig. 3.10 Beam-type disk under edge load

It follows that the expression $\frac{d^2 f}{dx^2}$ is of the form

$$\frac{d^2 f}{dx^2} = C \sin \alpha_n x. \tag{3.127}$$

Through integration we obtain:

$$f = -\frac{C}{\alpha_n} \sin \alpha_n x. \tag{3.128}$$

For the Airy stress function, an expression of the form

$$F = \frac{1}{\alpha_n^2} [A_n \cosh \alpha_n y + \alpha_n y B_n \sinh \alpha_n y + C_n \sinh \alpha_n y + \alpha_n y D_n \cosh \alpha_n y] \sin \alpha_n x \tag{3.129}$$

is employed. Thus, the stress components can be expressed as:

$$\begin{aligned}
\sigma_{xx} &= [(A_n + 2B_n) \cosh \alpha_n y + B_n \alpha_n y \sinh \alpha_n y \\
&\quad + (C_n + 2D_n) \sinh \alpha_n y + D_n \alpha_n y \cosh \alpha_n y] \sin \alpha_n x, \\
\sigma_{yy} &= -[A_n \cosh \alpha_n y + B_n \alpha_n y \sinh \alpha_n y \\
&\quad + C_n \sinh \alpha_n y + D_n \alpha_n y \cosh \alpha_n y] \sin \alpha_n x, \\
\tau_{xy} &= -[(A_n + B_n) \sinh \alpha_n y + B_n \alpha_n y \cosh \alpha_n y \\
&\quad + (C_n + D_n) \cosh \alpha_n y + D_n \alpha_n y \sinh \alpha_n y] \cos \alpha_n x.
\end{aligned} \tag{3.130}$$

The constants A_n, B_n, C_n, D_n appearing here are adapted to the given boundary conditions. Here it turns out that the conditions $\sigma_{xx}(x = 0)$ and $\sigma_{xx}(x = a)$ are satisfied for each y. From the conditions

$$\sigma_{yy}(y = 0) = 0, \quad \sigma_{yy}(y = b) = -\frac{p(x)}{h}, \quad \tau_{xy}(y = 0), \quad \tau_{xy}(y = b) = 0 \tag{3.131}$$

the following constants A_n, B_n, C_n, D_n result after inserting (3.130):

$$\begin{aligned}
A_n &= 0, \quad B_n = \frac{p_n}{h} \frac{\alpha_n b \sinh \alpha_n b}{\sinh^2 \alpha_n b - \alpha_n^2 b^2}, \\
C_n &= -D_n = \frac{p_n}{h} \frac{\sinh \alpha_n b + \alpha_n b \cosh \alpha_n b}{\sinh^2 \alpha_n b - \alpha_n^2 b^2}.
\end{aligned} \tag{3.132}$$

Thus, the total stress field of the disk can be fully specified at this point, which can be simplified as follows due to $A_n = 0$ and $C_n = -D_n$:

$$\begin{aligned}
\sigma_{xx} &= [B_n (2 \cosh \alpha_n y + \alpha_n y \sinh \alpha_n y) - C_n (\sinh \alpha_n y + \alpha_n y \cosh \alpha_n y)] \sin \alpha_n x, \\
\sigma_{yy} &= -[B_n \alpha_n y \sinh \alpha_n y + C_n (\sinh \alpha_n y - \alpha_n y \cosh \alpha_n y)] \sin \alpha_n x,
\end{aligned}$$

$$\tau_{xy} = -\left[B_n \left(\sinh \alpha_n y + \alpha_n y \cosh \alpha_n y\right) + C_n \alpha_n y \sinh \alpha_n y\right] \cos \alpha_n x. \tag{3.133}$$

In Timoshenko and Goodier (1951) an analogous formulation is found for the beam-like disk supported on both sides and loaded on two edges by a periodic load.

Massonet (1962), Girkmann (1974), and Eschenauer and Schnell (1993) consider the disk strip (height b) infinitely extended in the $x-$direction under the periodic boundary loads p and $\bar{p}$ (Fig. 3.11) (period l), which form an equilibrium group. If the boundary loads are in the form of even functions of x, then the boundary load is developed in terms of the following series:

$$p(x) = a_0 + \sum_{n=1}^{\infty} a_n \cos \alpha_n x, \quad \bar{p}(x) = a_0 + \sum_{n=1}^{\infty} \bar{a}_n \cos \alpha_n x, \tag{3.134}$$

with $\alpha_n = \frac{2n\pi}{l}$. Since the two boundary loads form an equilibrium group, the term a_0 must be identical in both series expansions. The boundary conditions given at the edge $y = +\frac{b}{2}$ are:

$$\sigma_{yy} = \frac{p}{h}, \quad \tau_{xy} = 0. \tag{3.135}$$

On the other hand, at the edge $y = -\frac{b}{2}$ holds:

$$\sigma_{yy} = \frac{\bar{p}}{h}, \quad \tau_{xy} = 0. \tag{3.136}$$

The Airy stress function is also developed in the form of a Fourier series as follows:

$$F(x, y) = \sum_{n=1}^{\infty} F_n(y) \cos \alpha_n x. \tag{3.137}$$

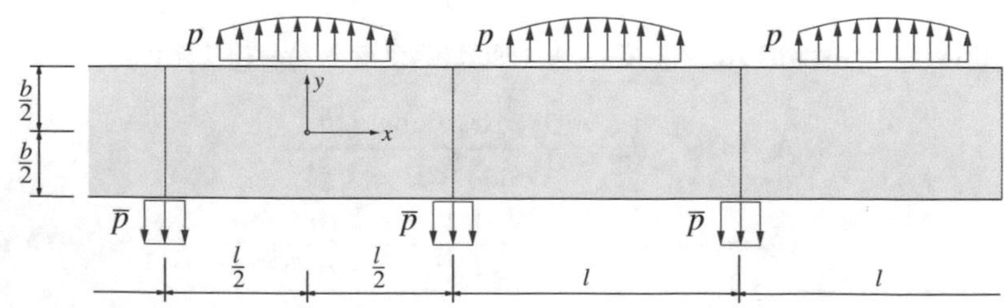

Fig. 3.11 Disk strip under periodic load

Substitution into the disk equation $\Delta\Delta F = 0$ then leads to the following differential equation:

$$\frac{\mathrm{d}^4 F_n}{\mathrm{d}y^4} - 2\alpha_n^2 \frac{\mathrm{d}^2 F_n}{\mathrm{d}y^2} + \alpha_n^4 F_n = 0. \tag{3.138}$$

For $n \geq 1$ the general solution is as follows:

$$F_n = \frac{1}{\alpha_n^2}\left(A_n \cosh\alpha_n y + B_n \alpha_n y \cosh\alpha_n y + C_n \sinh\alpha_n y + D_n \alpha_n y \sinh\alpha_n y\right), \tag{3.139}$$

whereas for $n = 0$ $F_0 = A_0 x^2$ holds. The stress components $\sigma_{xx}, \sigma_{yy}, \tau_{xy}$ are then:

$$\begin{aligned}
\sigma_{xx} &= \sum_{n=1}^{\infty} [(A_n + 2D_n)\cosh\alpha_n y + B_n \alpha_n y \cosh\alpha_n y \\
&\quad + (2B_n + C_n)\sinh\alpha_n y + D_n \alpha_n y \sinh\alpha_n y]\cos\alpha_n x, \\
\sigma_{yy} &= 2A_0 - \sum_{n=1}^{\infty} (A_n \cosh\alpha_n y + B_n \alpha_n y \cosh\alpha_n y \\
&\quad + C_n \sinh\alpha_n y + D_n \alpha_n y \sinh\alpha_n y)\cos\alpha_n x, \\
\tau_{xy} &= \sum_{n=1}^{\infty} [(A_n + D_n)\sinh\alpha_n y + B_n \alpha_n y \sinh\alpha_n y \\
&\quad + (B_n + C_n)\cosh\alpha_n y + D_n \alpha_n y \cosh\alpha_n y]\sin\alpha_n x.
\end{aligned} \tag{3.140}$$

From the boundary conditions (3.135) and (3.136), the constants A_n, B_n, C_n, D_n are obtained as:

$$\begin{aligned}
A_n &= -\frac{a_n + \bar{a}_n}{h} \frac{\sinh\left(\frac{\alpha_n b}{2}\right) + \frac{\alpha_n b}{2}\cosh\left(\frac{\alpha_n b}{2}\right)}{\sinh\alpha_n b + \alpha_n b}, \\
B_n &= \frac{a_n - \bar{a}_n}{h} \frac{\cosh\left(\frac{\alpha_n b}{2}\right)}{\sinh\alpha_n b + \alpha_n b}, \\
C_n &= -\frac{a_n - \bar{a}_n}{h} \frac{\cosh\left(\frac{\alpha_n b}{2}\right) + \frac{\alpha_n b}{2}\sinh\left(\frac{\alpha_n b}{2}\right)}{\sinh\alpha_n b + \alpha_n b}, \\
D_n &= \frac{a_n + \bar{a}_n}{h} \frac{\sinh\left(\frac{\alpha_n b}{2}\right)}{\sinh\alpha_n b + \alpha_n b}.
\end{aligned} \tag{3.141}$$

In the case of an antimetric load, the series development of the loads is performed as follows:

$$p(x) = \sum_{n=1}^{\infty} a_n \sin\alpha_n x, \quad \bar{p}(x) = \sum_{n=1}^{\infty} \bar{a}_n \sin\alpha_n x. \tag{3.142}$$

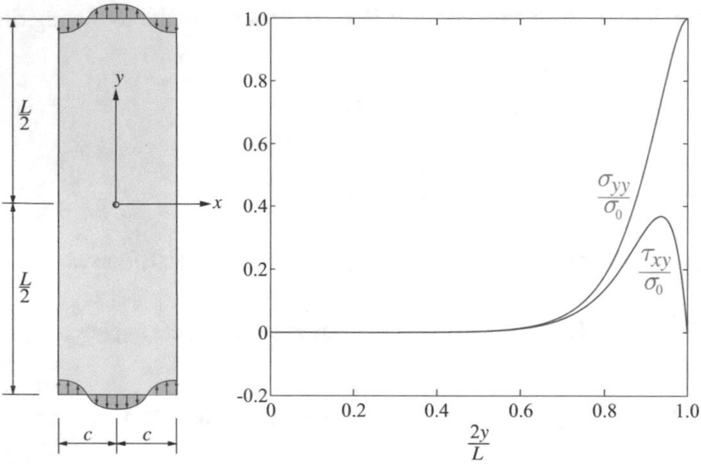

Fig. 3.12 Bar-type disk under cosine boundary load (left), distributions of stresses σ_{yy} and τ_{xy} (right) over disk height for $\frac{L}{c} = 10$

The Airy stress function in this case is:

$$F(x, y) = \sum_{n=1}^{\infty} F_n(y) \sin \alpha_n x, \tag{3.143}$$

with F_n according to (3.139).

3.6 St. Venant's Principle

An important principle used in the analysis of disks, but also in other types of thin-walled structures, is the so-called principle of St. Venant[4] (see, for example, Timoshenko and Goodier 1951; Filonenko-Borodich 1967; Sechler 1952). For illustration, consider the disk situation of Fig. 3.12, where a bar-type disk (height L, width $2c$, thickness h) is subjected to a cosine boundary load

$$\sigma_{yy} = \sigma_0 \cos \alpha x \tag{3.144}$$

($\alpha = \frac{\pi}{c}$). St. Venant's principle states that in the case of a load in equilibrium with itself, only a very local perturbation of the stress state due to the load introduction occurs and the remaining disk area remains unperturbed. It is irrelevant what form the

[4] Adhémar Jean Claude Barré de Saint-Venant, 1797–1886, French mathematician and engineer.

load introduction actually takes, and one load group can be replaced by another load group in this local boundary region as long as the load types are statically equivalent to each other. The stress state of the disk is then always concentrated in a very small area in the immediate vicinity of the load introduction, the propagation of which depends on the dimensions of the area on which the load is applied.

St. Venant's principle has no formal proof, but can be illustrated quite clearly by simple examples. Sechler (1952) uses the bar-shaped disk of Fig. 3.12 under the given cosine-shaped load on the two edges $y = \pm\frac{L}{2}$. This boundary load has no resultant force in the $y-$direction, and the associated moment at the ends of the disk also vanishes. From the illustration, it can be expected that a fairly localized stress state will develop at the horizontal edges of this disk, but it will decay very rapidly until all stress components vanish at some edge distance. For the calculation, a separation approach of the form

$$F = \bar{F}(y)\cos\alpha x \tag{3.145}$$

is employed. Insertion into the disk equation $\Delta\Delta F = 0$, after a short transformation, yields the following fourth order linear homogeneous differential equation for the still unknown function $\bar{F}(y)$:

$$\frac{\mathrm{d}^4\bar{F}}{\mathrm{d}y^4} - 2\alpha^2\frac{\mathrm{d}^2\bar{F}}{\mathrm{d}y^2} + \alpha^4\bar{F} = 0. \tag{3.146}$$

Its general solution is:

$$\bar{F} = C_1\cosh\alpha y + C_2\sinh\alpha y + C_3 y\cosh\alpha y + C_4 y\sinh\alpha y. \tag{3.147}$$

Thus, with (3.144), the stress components can be specified as:

$$\begin{aligned}
\sigma_{xx} &= \cos\alpha x\,[\,C_1\alpha^2\cosh\alpha y + C_2\alpha^2\sinh\alpha y \\
&\quad + C_3\left(2\alpha\sinh\alpha y + \alpha^2 y\cosh\alpha y\right) + C_4\left(2\alpha\cosh\alpha y + \alpha^2 y\sinh\alpha y\right)\big],\\
\sigma_{yy} &= -\alpha^2\cos\alpha x\left(C_1\cosh\alpha y + C_2\sinh\alpha y + C_3 y\cosh\alpha y + C_4 y\sinh\alpha y\right),\\
\tau_{xy} &= \alpha\sin\alpha x\,[C_1\alpha\sinh\alpha y + C_2\alpha\cosh\alpha y \\
&\quad + C_3\left(\cosh\alpha y + \alpha y\sinh\alpha y\right) + C_4\left(\sinh\alpha y + \alpha y\cosh\alpha y\right)].
\end{aligned} \tag{3.148}$$

From the requirement that at the loaded ends $y = \frac{L}{2}$ the normal stress σ_{yy} must correspond to the applied boundary load, i.e. $\sigma_{yy} = \sigma_0\cos\alpha x$, the constants C_1, C_2, C_3, C_4 can be determined as follows:

$$C_1 = -\frac{2\sigma_0}{\alpha^2}\frac{\sinh\beta + \beta\cosh\beta}{2\beta + \sinh 2\beta},\quad C_2 = C_3 = 0,\quad C_4 = \frac{2\sigma_0}{\alpha}\frac{\sinh\beta}{2\beta + \sinh 2\beta}. \tag{3.149}$$

Thus, the stresses can be specified as:

$$\sigma_{xx} = 2\sigma_0\frac{(\sinh\beta - \beta\cosh\beta)\cosh\alpha y + \alpha y\sinh\beta\sinh\alpha y}{2\beta + \sinh 2\beta}\cos\alpha x,$$

$$\sigma_{yy} = 2\sigma_0 \frac{(\sinh\beta + \beta\cosh\beta)\cosh\alpha y - \alpha y \sinh\beta \sinh\alpha y}{2\beta + \sinh 2\beta} \cos\alpha x,$$
$$\tau_{xy} = 2\sigma_0 \frac{\beta\cosh\beta \sinh\alpha y - \alpha y \sinh\beta \cosh\alpha y}{2\beta + \sinh 2\beta} \sin\alpha x, \tag{3.150}$$

with $\beta = \frac{\pi L}{2c}$. If L is significantly larger than c, it may be assumed that $\sinh\beta = \cosh\beta$. The stresses can then be written as:

$$\sigma_{xx} = \sigma_0 \cos\alpha x \frac{(1-\beta)\cosh\alpha y + \alpha y \sinh\alpha y}{\beta + \sinh^2\beta} \sinh\beta,$$
$$\sigma_{yy} = \sigma_0 \cos\alpha x \frac{(1+\beta)\cosh\alpha y - \alpha y \sinh\alpha y}{\beta + \sinh^2\beta} \sinh\beta,$$
$$\tau_{xy} = \sigma_0 \sin\alpha x \frac{\beta\sinh\alpha y - \alpha y \cosh\alpha y}{\beta + \sinh^2\beta} \sinh\beta. \tag{3.151}$$

Since for sufficiently large values for β the relation $\beta << \sinh\beta$ holds, (3.151) can also be written as follows:

$$\sigma_{xx} = \sigma_0 \cos\alpha x \frac{(1-\beta)\cosh\alpha y + \alpha y \sinh\alpha y}{\sinh\beta},$$
$$\sigma_{yy} = \sigma_0 \cos\alpha x \frac{(1+\beta)\cosh\alpha y - \alpha y \sinh\alpha y}{\sinh\beta},$$
$$\tau_{xy} = \sigma_0 \sin\alpha x \frac{\beta\sinh\alpha y - \alpha y \cosh\alpha y}{\sinh\beta}. \tag{3.152}$$

A plot of the distribution (3.151) of the normal stress σ_{yy} for $x = 0$ and the shear stress für τ_{xy} at $x = \frac{c}{2}$ is given in Fig. 3.12, right. It is clear that these stress components decay very rapidly as the edge distance increases, and are concentrated in a region that is approximately of the extent of the loaded edge.

It should be noted here that the stress solution shown exactly satisfies the boundary conditions at the loaded edges, but no statements have yet been made about boundary conditions at the unloaded edges at $x = \pm c$. For this example, further details can be found, e.g., in Sechler (1952).

3.7 The Isotropic Half-Plane

3.7.1 Decay Behaviour of Boundary Perturbations

In this section we want to take a closer look at the problem of boundary perturbations at disk structures and assume a theoretically infinitely wide disk (the so-called half-plane with infinite length with respect to the positive $x-$ direction), which is subjected to a harmonic boundary stress at the position $x = 0$ (see also Wiedemann (2006) and Girkmann (1974); Fig. 3.13). The disk also has an infinite width with respect to the $y-$direction. It is of particular interest here not only to present a closed-form analytical solution of the disk equation, but also to use this solution to study the fundamental decay behavior of such edge perturbations. The knowledge about this has a high practical relevance, since edge stresses always have to be accommodated by appropriate strengthening measures for the usually thin-walled disk structures, e.g. in applications of lightweight construction.

We consider a harmonically distributed boundary stress, which can be boundary stresses in the form of normal or shear stresses that are self-equilibrating. Regardless of the concrete case, an approach for the Airy stress function can be sought in the following form:

$$F = \bar{F}_n \sin\left(\frac{n\pi y}{b}\right). \tag{3.153}$$

Therein, n is the half-wave number of the used sinusoidal approach and $\bar{F}_n$ is an unknown function of x. Substituting (3.153) into the disk Eqs. (3.32) and (3.33), respectively, yields the following ordinary fourth-order linear homogeneous differential equation for $\bar{F}_n$:

$$\frac{\mathrm{d}^4 \bar{F}_n}{\mathrm{d}x^4} - 2\left(\frac{n\pi}{b}\right)^2 \frac{\mathrm{d}^2 \bar{F}_n}{\mathrm{d}x^2} + \left(\frac{n\pi}{b}\right)^4 \bar{F}_n = 0. \tag{3.154}$$

The general solution is given as:

$$\bar{F}_n = \left(\frac{b}{n\pi}\right)^2 \left[\left(C_{1n} + C_{2n}\frac{n\pi x}{b}\right) e^{\frac{n\pi x}{b}} + \left(C_{3n} + C_{4n}\frac{n\pi x}{b}\right) e^{\frac{n\pi x}{b}}\right]. \tag{3.155}$$

Since we may assume that edge perturbations due to the load introduction completely decay with increasing distance to the disk edge in an infinitely extended disk, we set the constants C_{3n} and C_{4n} to zero. This leaves:

$$\bar{F}_n = \left(\frac{b}{n\pi}\right)^2 \left(C_{1n} + C_{2n}\frac{n\pi x}{b}\right) e^{-\frac{n\pi x}{b}}. \tag{3.156}$$

The corresponding stresses of the disk are then:

$$\begin{aligned}
\sigma_{xx,n} &= -\left[C_{1n} + C_{2n}\frac{n\pi x}{b}\right] e^{-\frac{n\pi x}{b}} \sin\left(\frac{n\pi y}{b}\right), \\
\sigma_{yy,n} &= \left[C_{1n} + C_{2n}\left(\frac{n\pi x}{b} - 2\right)\right] e^{-\frac{n\pi x}{b}} \sin\left(\frac{n\pi y}{b}\right), \\
\tau_{xy,n} &= \left[C_{1n} + C_{2n}\left(\frac{n\pi x}{b} - 1\right)\right] e^{-\frac{n\pi x}{b}} \cos\left(\frac{n\pi y}{b}\right).
\end{aligned} \tag{3.157}$$

The constants $C_{1,n}$ and $C_{2,n}$ can be determined from the given boundary conditions.

The advantage of this representation of a boundary stress is that any boundary stress can be approximated in terms of Fourier series whose individual members each satisfy the disk equation. The solutions shown below can then be superposed arbitrarily for different values n to represent a given boundary load.

We now consider the case of an infinite disk, which is under the edge load $\sigma_{xx,n} = \sigma_{0,n}\sin\left(\frac{n\pi y}{b}\right)$ (amplitude $\sigma_{0,n}$) at the edge $x = 0$. Let the edge of the half-plane also be free of any shear stress τ_{xy}. The constants $C_{1,n}$ and $C_{2,n}$ can thus be determined as:

$$C_{1n} = C_{2n} = -\sigma_{0,n}. \tag{3.158}$$

Thus:

$$\begin{aligned}
\sigma_{xx,n} &= \sigma_{0,n}\left[1 + \frac{n\pi x}{b}\right]e^{-\frac{n\pi x}{b}}\sin\left(\frac{n\pi y}{b}\right),\\
\sigma_{yy,n} &= -\sigma_{0,n}\left[\frac{n\pi x}{b} - 1\right]e^{-\frac{n\pi x}{b}}\sin\left(\frac{n\pi y}{b}\right),\\
\tau_{xy,n} &= -\sigma_{0,n}\frac{n\pi x}{b}e^{-\frac{n\pi x}{b}}\cos\left(\frac{n\pi y}{b}\right).
\end{aligned} \tag{3.159}$$

A discussion of the result shows (Fig. 3.13) that the region in which these edge perturbations arise and show significant values is narrowed down to a range of about the order of the half-wavelength $b_n = \frac{b}{n}$ with respect to the $x-$direction.

It can be shown, as discussed in Wiedemann (2006), that the decay behavior is also significantly controlled by the transverse strain behavior of the disk. In particular, the decay of the edge strain is accelerated if the transverse strain ε_{yy}^0 is hindered. Currently, we want to assume that ε_{yy}^0 is zero at the point $x = 0$. This can be realized by an edge rib with sufficiently high extensional stiffness. Hence, in this case $\varepsilon_{yy}^0(x = 0) = 0$, and the transverse stress $\sigma_{yy,n}$ is obtained as $\sigma_{yy,n} = \nu\sigma_{xx,n}$, therefore with $\nu_1 = \frac{1+\nu}{2}$:

$$C_{1n} = -\sigma_{0,n}, \quad C_{2n} = -\nu_1\sigma_{0,n}. \tag{3.160}$$

Thus:

$$\begin{aligned}
\sigma_{xx,n} &= \sigma_{0,n}\left[1 + \nu_1\frac{n\pi x}{b}\right]e^{-\frac{n\pi x}{b}}\sin\left(\frac{n\pi y}{b}\right),\\
\sigma_{yy,n} &= -\sigma_{0,n}\left[1 + \nu_1\left(\frac{n\pi x}{b} - 2\right)\right]e^{-\frac{n\pi x}{b}}\sin\left(\frac{n\pi y}{b}\right),\\
\tau_{xy,n} &= -\sigma_{0,n}\left[1 + \nu_1\left(\frac{n\pi x}{b} - 1\right)\right]e^{-\frac{n\pi x}{b}}\cos\left(\frac{n\pi y}{b}\right).
\end{aligned} \tag{3.161}$$

A plot of the determined stress distributions with respect to the $x-$direction in the disk loaded by a harmonic boundary stress with $n = 1$ and $\sigma_0 = 1$ with $\nu = 0.3$ is given in Fig. 3.13. Here σ_{xx} and σ_{yy} were evaluated at the position $y = \frac{b}{2}$, whereas the evaluation for τ_{xy} takes place at $y = 0$. The decay behavior of the individual stress components with increasing edge distance can be observed particularly well here.

If it is possible to restrain the transverse strain ε_{yy}^0 across the entire width of the disk, for example, by continuously arranging stiffeners in the $y-$direction, which can be thought of as computationally smeared with the disk, then for our computational purposes we can assume that ε_{yy}^0 becomes zero for all x. The compatibility Eq. (3.13)

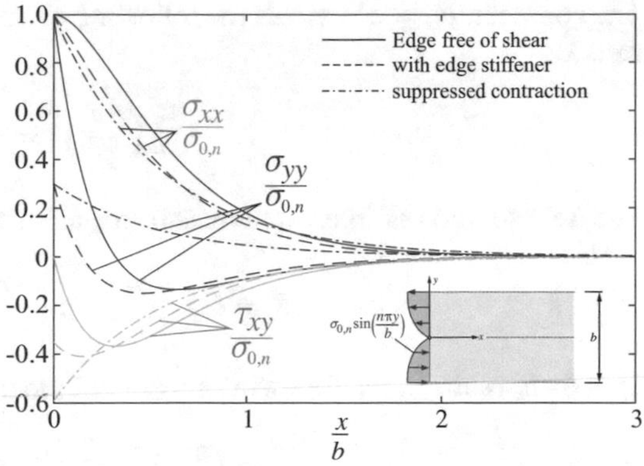

Fig. 3.13 Stress distributions in the half-plane under harmonic boundary stress $\sigma_{xx,n} = \sigma_{0,n} \sin\left(\frac{n\pi y}{b}\right)$ for $n = 1$ and $\nu = 0.3$

is then:

$$\frac{\partial^2 \varepsilon_{xx}^0}{\partial y^2} - \frac{\partial^2 \gamma_{xy}^0}{\partial x \partial y} = 0. \tag{3.162}$$

Moreover, from the material law (3.5) it follows with $\varepsilon_{yy}^0 = 0$:

$$\sigma_{yy} = \nu \sigma_{xx}. \tag{3.163}$$

For the strain ε_{xx}^0 we thus obtain:

$$\varepsilon_{xx}^0 = \frac{1}{E} \sigma_{xx} (1 - \nu^2). \tag{3.164}$$

The compatibility condition is then with $\nu_2^2 = \frac{1-\nu}{2}$ and $\varepsilon_{yy}^0 = 0$:

$$\nu_2^2 \frac{\partial^4 F}{\partial y^4} + \frac{\partial^4 F}{\partial x^2 \partial y^2} = 0. \tag{3.165}$$

For the Airy stress function, an approach of the form (3.153) is again used:

$$F = \bar{F}_n \sin\left(\frac{n\pi y}{b}\right), \tag{3.166}$$

which, after substitution into the compatibility condition (3.165), leads to the following ordinary differential equation:

$$\frac{\mathrm{d}^2 \bar{F}_n}{\mathrm{d}x^2} - \nu_2^2 \left(\frac{n\pi}{b}\right)^2 \bar{F}_n = 0. \tag{3.167}$$

The approach $\bar{F}_n = e^{\lambda x}$ yields the following characteristic polynomial for the exponent λ:

$$\lambda^2 - \nu_2^2 \left(\frac{n\pi}{b}\right)^2 = 0. \tag{3.168}$$

The only solution of interest here is the negative exponent $\lambda = -\nu_2 \frac{n\pi}{b}$, so that:

$$F = Ce^{-\nu_2 \frac{n\pi x}{b}} \sin\left(\frac{n\pi y}{b}\right). \tag{3.169}$$

This then results in the following stress components:

$$\begin{aligned}
\sigma_{xx} &= -C\left(\frac{n\pi}{b}\right)^2 e^{-\nu_2 \frac{n\pi x}{b}} \sin\left(\frac{n\pi y}{b}\right),\\
\sigma_{yy} &= -C\nu\left(\frac{n\pi}{b}\right)^2 e^{-\nu_2 \frac{n\pi x}{b}} \sin\left(\frac{n\pi y}{b}\right),\\
\tau_{xy} &= \nu_2 C\left(\frac{n\pi}{b}\right)^2 e^{-\nu_2 \frac{n\pi x}{b}} \cos\left(\frac{n\pi y}{b}\right).
\end{aligned} \tag{3.170}$$

The constant C is determined from the boundary condition that the normal stress σ_{xx} at the disk edge $x = 0$ is equal to the applied boundary stress, which leads to $C = -\left(\frac{b}{n\pi}\right)^2 \sigma_{0,n}$. Then:

$$\begin{aligned}
\sigma_{xx} &= \sigma_{0,n} e^{-\nu_2 \frac{n\pi x}{b}} \sin\left(\frac{n\pi y}{b}\right),\\
\sigma_{yy} &= \sigma_{0,n} \nu e^{-\nu_2 \frac{n\pi x}{b}} \sin\left(\frac{n\pi y}{b}\right),\\
\tau_{xy} &= -\nu_2 \sigma_{0,n} e^{-\nu_2 \frac{n\pi x}{b}} \cos\left(\frac{n\pi y}{b}\right).
\end{aligned} \tag{3.171}$$

A discussion of this solution (Fig. 3.13) indicates that the full transverse strain restraint accelerates the decay of the edge stresses, but weakens compared to the previous solution approaches for increasing values of x.

As a further technically significant case we want to consider the cosine edge stress in the form of an edge shear stress $\tau_{xy} = \tau_{0,n} \cos\left(\frac{n\pi y}{b}\right)$. A normal stress σ_{xx} is not present in this case. The constants $C_{1,n}$ and $C_{2,n}$ then follow as:

$$C_{1n} = 0, \quad C_{2n} = -\tau_{0,n}. \tag{3.172}$$

Thus we obtain the stress field of the infinitely extended disk as:

$$\begin{aligned}
\sigma_{xx,n} &= \tau_{0,n} \frac{n\pi x}{b} e^{-\frac{n\pi x}{b}} \sin\left(\frac{n\pi y}{b}\right),\\
\sigma_{yy,n} &= -\tau_{0,n} \left(\frac{n\pi x}{b} - 2\right) e^{-\frac{n\pi x}{b}} \sin\left(\frac{n\pi y}{b}\right),\\
\tau_{xy,n} &= -\tau_{0,n} \left(\frac{n\pi x}{b} - 1\right) e^{-\frac{n\pi x}{b}} \cos\left(\frac{n\pi y}{b}\right).
\end{aligned} \tag{3.173}$$

For the case of boundary stress in the form of a harmonically distributed boundary shear stress, a rapid decay over x is again observed. A plot of the distribution of the stress components with respect to the $x-$direction is shown in Fig. 3.14.

3.7.2 The Half-Plane Under Periodic Boundary Load

The isotropic half-plane under boundary loads is of special technical importance, e.g., for lightweight construction, since this situation can be regarded as representative for load introductions into thin-walled lightweight structures. We consider the half-plane infinitely extended with respect to x, which is bounded by the straight edge $y = 0$ (shown in Fig. 3.15 as loaded by an edge load p). We discuss in this section some load situations of the isotropic half-plane, where we will assume periodic loads.

We start with an example which can also be found in this form in Girkmann (1974) and Altenbach et al. (2016). A half plane (Fig. 3.16) is considered, which is supported at regular intervals $2a$ on the partial length $2c$ and loaded by the surface load p. The surface loads p_1 are therefore to be understood as support reactions. The period of

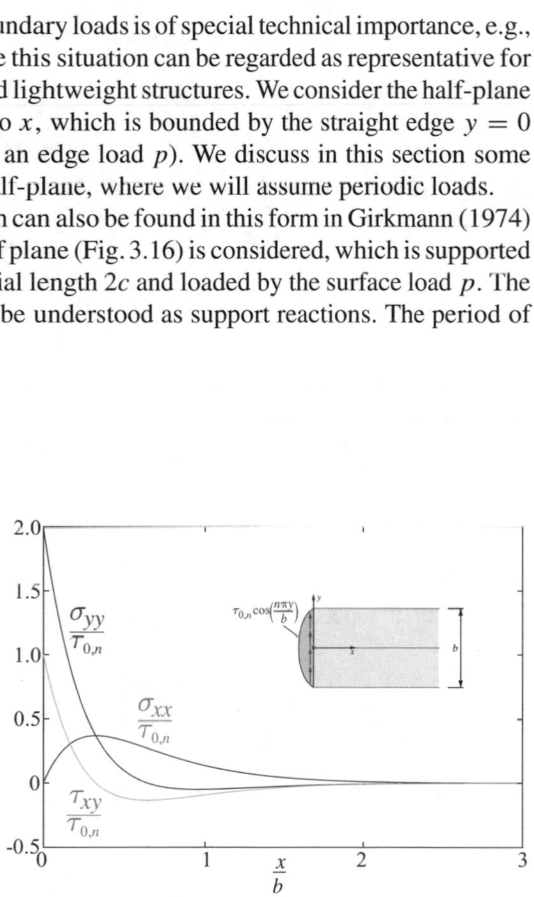

Fig. 3.14 Stress distributions in the half-plane under harmonic boundary stress $\tau_{xx,n} = \tau_{0,n} \cos\left(\frac{n\pi y}{b}\right)$ for $n = 1$

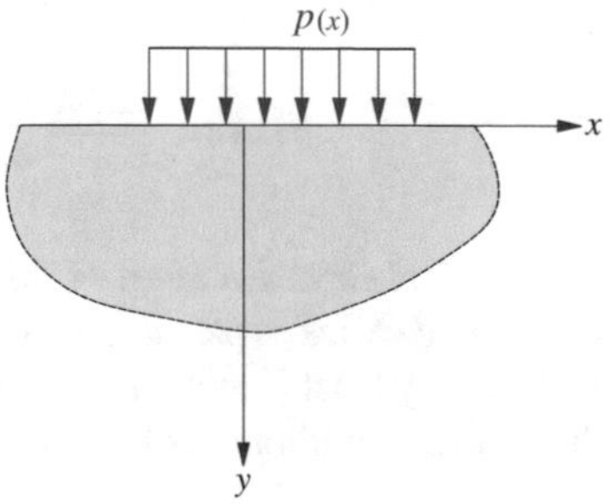

Fig. 3.15 Half-plane under boundary load

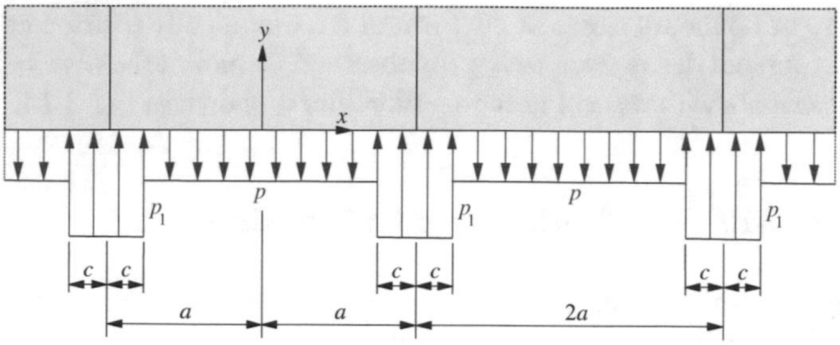

Fig. 3.16 Half-plane loaded by periodic edge load

this load situation is obviously $L = 2a$. For reasons of equilibrium, the following must hold:

$$p_1 = p\frac{a-c}{c}. \tag{3.174}$$

In the following we assume a surface $y = 0$ that is free of shear stresses. At this boundary $\sigma_{yy} = \frac{p(x)}{h}$ is valid.

The given load situation is represented by a Fourier series:

$$p(x) = \sum_{n=1}^{\infty} a_n \cos\left(\frac{n\pi x}{a}\right). \tag{3.175}$$

For the coefficients a_n the following results:

$$a_n = \frac{2}{a}\left[p\int_0^{a-c} \cos\left(\frac{n\pi x}{a}\right) \mathrm{d}x - p_1 \int_{a-c}^{a} \cos\left(\frac{n\pi x}{a}\right) \mathrm{d}x\right], \tag{3.176}$$

which leads to the following result with (3.174):

$$a_n = -\frac{2pa}{\pi c}\frac{1}{n} \sin\left(\frac{n\pi c}{a}\right) \cos(n\pi). \tag{3.177}$$

With $\cos(n\pi) = (-1)^n$ the series expansion follows as:

$$p(x) = -\frac{2pa}{\pi c}\sum_{n=1}^{\infty} \frac{(-1)^n}{n} \sin\left(\frac{n\pi c}{a}\right) \cos\left(\frac{n\pi x}{a}\right). \tag{3.178}$$

A graphical representation of this series expansion for different length ratios $\frac{c}{a}$ and degrees of series expansion N is shown in Fig. 3.17. In Girkmann (1974) it is shown how this type of loading can also be developed as an odd function in x. However, this remains without further consideration at this point.

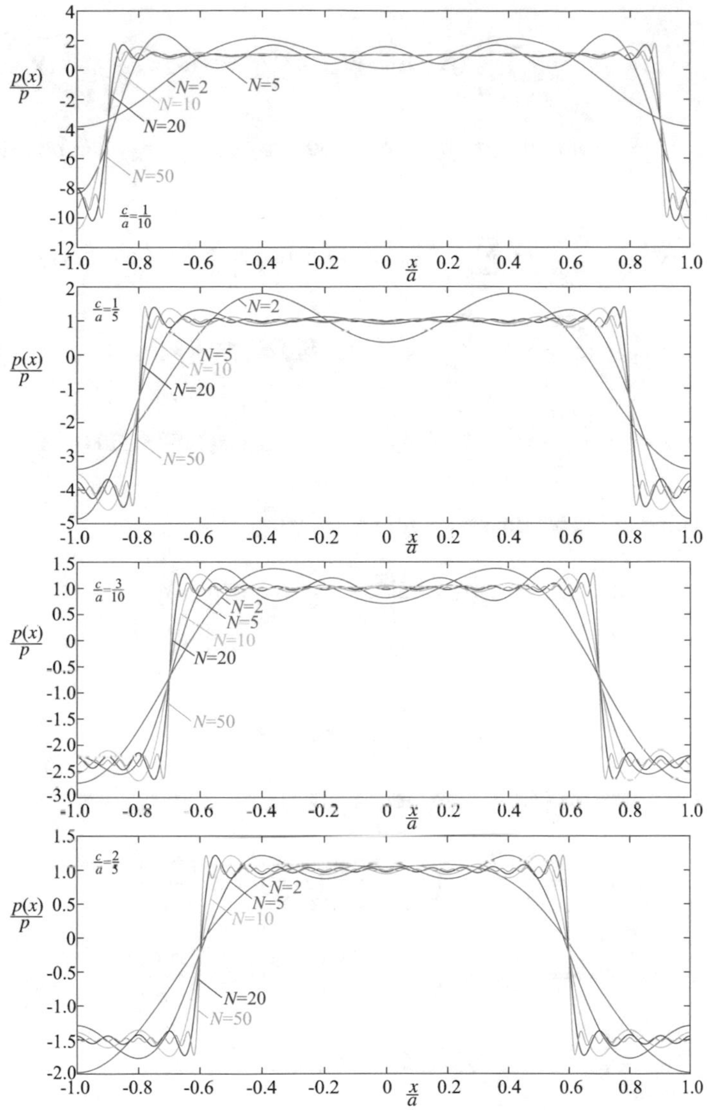

Fig. 3.17 Approximation of the boundary load as a periodically recurring function over the period length $L = 2a$ for different length ratios $\frac{c}{a}$ and degrees of series expansion N

We assume that the Airy stress function can also be represented in the form of a series expansion, whereby we choose only those from the selection of solution functions which show a decaying behavior from the edge. The following approach is chosen for this purpose:

$$F = \sum_{n=1}^{\infty} \frac{1}{\alpha_n^2} (A_n + \alpha_n y B_n)\, e^{-\alpha_n y} \cos(\alpha_n x), \tag{3.179}$$

where $\alpha_n = \frac{n\pi}{a}$. From this, the stress components $\sigma_{xx}, \sigma_{yy}, \tau_{xy}$ can be determined as:

$$\begin{aligned}
\sigma_{xx} &= \frac{\partial^2 F}{\partial y^2} = \sum_{n=1}^{\infty} [(A_n - 2B_n) + \alpha_n y B_n]\, e^{-\alpha_n y} \cos(\alpha_n x), \\
\sigma_{yy} &= \frac{\partial^2 F}{\partial x^2} = -\sum_{n=1}^{\infty} (A_n + \alpha_n y B_n)\, e^{-\alpha_n y} \cos(\alpha_n x), \\
\tau_{xy} &= -\frac{\partial^2 F}{\partial x \partial y} = -\sum_{n=1}^{\infty} [(A_n - B_n) + \alpha_n y B_n]\, e^{-\alpha_n y} \sin(\alpha_n x).
\end{aligned} \tag{3.180}$$

The evaluation of the boundary conditions

$$\sigma_{yy}(y=0) = \frac{p(x)}{h}, \quad \tau_{xy}(y=0) = 0 \tag{3.181}$$

yields:

$$A_n = -\frac{1}{h} a_n, \quad B_n = A_n. \tag{3.182}$$

The stress components (3.180) can then be given as:

$$\begin{aligned}
\sigma_{xx} &= \frac{1}{h} \sum_{n=1}^{\infty} a_n \left(1 - \frac{n\pi y}{a}\right) e^{-\frac{n\pi y}{a}} \cos\left(\frac{n\pi x}{a}\right), \\
\sigma_{yy} &= \frac{1}{h} \sum_{n=1}^{\infty} a_n \left(1 + \frac{n\pi y}{a}\right) e^{-\frac{n\pi y}{a}} \cos\left(\frac{n\pi x}{a}\right), \\
\tau_{xy} &= \frac{1}{h} \sum_{n=1}^{\infty} a_n \frac{n\pi y}{a} e^{-\frac{n\pi y}{a}} \sin\left(\frac{n\pi x}{a}\right).
\end{aligned} \tag{3.183}$$

A graphical representation of the distribution of stress components $\sigma_{xx}, \sigma_{yy}, \tau_{xy}$ for $-a \le x \le +a$, $0 \le y \le 2a$ for the disk with unit thickness $h = 1$ and aspect ratio $\frac{c}{a} = \frac{2}{5}$ for $N = 50$ is shown in Figs. 3.18, 3.19 and 3.20. The pronounced decay behavior of the stresses in such an edge-loaded disk is particularly evident here. The situation of the isotropic half plane under boundary normal load can also be summarized in general terms as follows (see e.g. Altenbach et al. 2016). Let a periodic boundary load be given, which can be expressed as a Fourier series as

$$p(x) = \frac{1}{2} a_0 + \sum_{n=1}^{\infty} a_n \cos\left(\frac{2n\pi x}{L}\right) + \sum_{n=1}^{\infty} b_n \sin\left(\frac{2n\pi x}{L}\right). \tag{3.184}$$

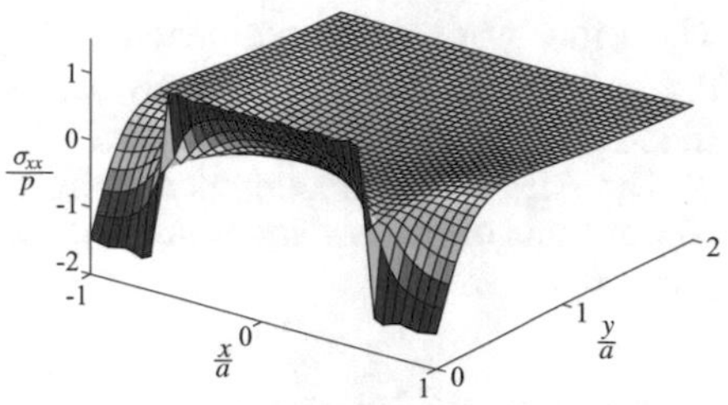

Fig. 3.18 Normal stress σ_{xx} for the disk with periodic edge load with unit thickness $h = 1$ and aspect ratio $\frac{c}{a} = \frac{2}{5}$ for $N = 50$

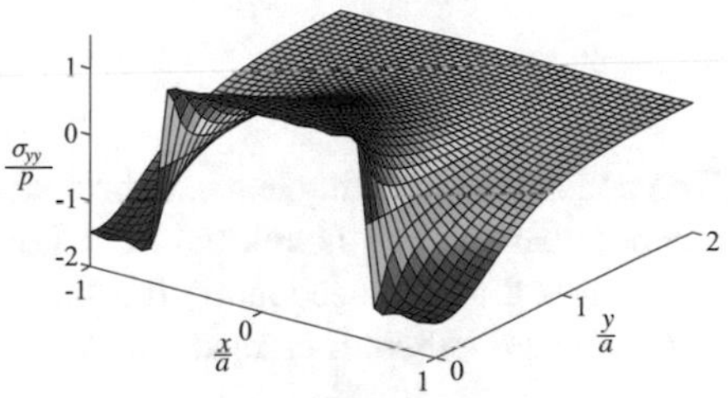

Fig. 3.19 Normal stress σ_{yy} for the disk with periodic edge load with unit thickness $h = 1$ and aspect ratio $\frac{c}{a} = \frac{2}{5}$ for $N = 50$

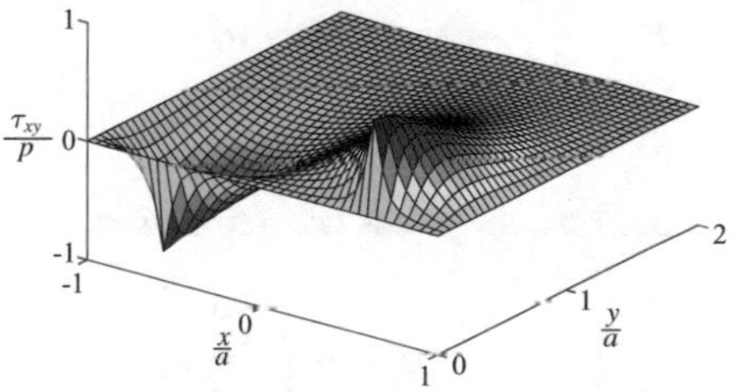

Fig. 3.20 Shear stress τ_{xy} for the disk with periodic edge load with unit thickness $h = 1$ and aspect ratio $\frac{c}{a} = \frac{2}{5}$ for $N = 50$

where this type of representation is periodic with period L. The coefficients a_n and b_n follow as:

$$a_n = \frac{2}{L} \int_{-\frac{L}{2}}^{+\frac{L}{2}} p \cos\left(\frac{2n\pi x}{L}\right) \mathrm{d}x = \frac{2}{L} \int_{0}^{L} p \cos\left(\frac{2n\pi x}{L}\right) \mathrm{d}x,$$

$$b_n = \frac{2}{L} \int_{-\frac{L}{2}}^{+\frac{L}{2}} p \sin\left(\frac{2n\pi x}{L}\right) \mathrm{d}x = \frac{2}{L} \int_{0}^{L} p \sin\left(\frac{2n\pi x}{L}\right) \mathrm{d}x. \tag{3.185}$$

Also:

$$a_0 = \frac{2}{L} \int_{-\frac{L}{2}}^{+\frac{L}{2}} p \mathrm{d}x = \frac{2}{L} \int_{0}^{L} p \mathrm{d}x. \tag{3.186}$$

The value $\frac{1}{2}a_0$ can be interpreted as the average value of the function $f_x(x)$ over the period length L. Accordingly, a_0 vanishes whenever the average value of the function $f(x)$ over L becomes zero.

The Airy stress function is now also developed in the form of a Fourier series, where again only those terms are considered which show a decaying behavior. With $\alpha_n = \frac{2n\pi}{L}$:

$$F = A_0 x^2 + \sum_{n=1}^{\infty} (A_n + B_n \alpha_n y) e^{-\alpha_n y} \cos(\alpha_n x) + \sum_{n=1}^{\infty} (C_n + D_n \alpha_n y) e^{-\alpha_n y} \sin(\alpha_n x) . \tag{3.187}$$

The constants occurring here can be concluded from the boundary conditions that at the boundary $y = 0$ the normal force flow must correspond to the applied boundary normal load and consequently the shear force flow at this boundary must become zero. From coefficient comparison we obtain:

$$F = \frac{a_0 x^2}{4h} - \sum_{n=1}^{\infty} \frac{a_n}{\alpha_n^2 h} (1 + \alpha_n y) e^{-\alpha_n y} \cos(\alpha_n x) - \sum_{n=1}^{\infty} \frac{b_n}{\alpha_n^2 h} (1 + \alpha_n y) e^{-\alpha_n y} \sin(\alpha_n x) . \tag{3.188}$$

From this, the stress components σ_{xx}, σ_{yy}, τ_{xy} result as follows:

$$\begin{aligned}
\sigma_{xx} &= \frac{1}{h} \sum_{n=1}^{\infty} a_n (1 - \alpha_n y) e^{-\alpha_n y} \cos(\alpha_n x) \\
&\quad + \frac{1}{h} \sum_{n=1}^{\infty} b_n (1 - \alpha_n y) e^{-\alpha_n y} \sin(\alpha_n x) , \\
\sigma_{yy} &= \frac{a_0}{2h} + \frac{1}{h} \sum_{n=1}^{\infty} a_n (1 + \alpha_n y) e^{-\alpha_n y} \cos(\alpha_n x) \\
&\quad + \frac{1}{h} \sum_{n=1}^{\infty} b_n (1 + \alpha_n y) e^{-\alpha_n y} \sin(\alpha_n x) , \\
\tau_{xy} &= \frac{1}{h} \sum_{n=1}^{\infty} a_n \alpha_n y e^{-\alpha_n y} \sin(\alpha_n x) \\
&\quad - \frac{1}{h} \sum_{n=1}^{\infty} b_n \alpha_n y e^{-\alpha_n y} \cos(\alpha_n x) .
\end{aligned} \tag{3.189}$$

If the case of a half-plane under a boundary shear load is considered, the procedure is identical to that described so far. Further explanations can thus be omitted at this point.

3.7.3 The Half-Plane Under Non-periodic Load

A very useful way to represent boundary loads in the context of disk theory is to formulate them as Fourier integrals. For this purpose, we consider a load given on an edge, where both the load itself and its first derivative are continuous. Starting point of the considerations is the Eqs. (3.184) with (3.185) and (3.186), which we present here once more for the sake of better readability and here first assume a very general denotation of a function $f(x)$ which approximates a given function $f_x(x)$:

$$f(x) = \frac{1}{2}a_0 + \sum_{n=1}^{\infty} a_n \cos\left(\frac{2n\pi x}{L}\right) + \sum_{n=1}^{N} b_n \sin\left(\frac{2n\pi x}{L}\right),$$

$$a_0 = \frac{2}{L}\int_{-\frac{L}{2}}^{+\frac{L}{2}} f_x \mathrm{d}x, \quad a_n = \frac{2}{L}\int_{-\frac{L}{2}}^{+\frac{L}{2}} f_x \cos\left(\frac{2n\pi x}{L}\right)\mathrm{d}x, \quad b_n = \frac{2}{L}\int_{-\frac{L}{2}}^{+\frac{L}{2}} f_x \sin\left(\frac{2n\pi x}{L}\right)\mathrm{d}x. \tag{3.190}$$

We now insert the coefficients a_n and b_n into the formulation for $f(x)$ and assume the period length $L = 2l$. Let $x = \lambda$:

$$f(x) = \frac{1}{2l}\int_{-l}^{+l} f_\lambda \mathrm{d}\lambda + \frac{1}{l}\sum_{n=1}^{\infty}\int_{-l}^{+l} f_\lambda \cos\left(\frac{n\pi\lambda}{l}\right)\mathrm{d}\lambda \cos\left(\frac{n\pi x}{l}\right)$$
$$+ \frac{1}{l}\sum_{n=1}^{\infty}\int_{-l}^{+l} f_\lambda \sin\left(\frac{n\pi\lambda}{l}\right)\mathrm{d}\lambda \sin\left(\frac{n\pi x}{l}\right). \tag{3.191}$$

At this point we introduce the auxiliary quantity $\alpha_n = \frac{n\pi}{l}$ and the difference $\delta\alpha = \frac{\pi}{l}$. We expand the second and third terms in (3.191) with $\frac{\pi}{l}$, truncate l and complete the boundary transition $l \to \infty$. Hereby, the quantity $\delta\alpha$ transitions to the differential quantity $\mathrm{d}\alpha$ and assumes all values between zero and infinity. Moreover, the sums in (3.191) transcend into integrals extending from $\alpha = 0$ to $\alpha = \infty$. If we further assume that the first integral expression in (3.191) (i.e., the expression $\int_{-\infty}^{+\infty} f_x \mathrm{d}x$) results in a finite value, then the first term in (3.191) vanishes. The result of the limit $l \to \infty$ is thus:

$$f(x) = \frac{1}{\pi}\int_{0}^{\infty} \cos(\alpha x)\,\mathrm{d}\alpha \int_{-\infty}^{+\infty} f_\lambda \cos(\alpha\lambda)\,\mathrm{d}\lambda$$
$$+ \frac{1}{\pi}\int_{0}^{\infty} \sin(\alpha x)\,\mathrm{d}\alpha \int_{-\infty}^{+\infty} f_\lambda \sin(\alpha\lambda)\,\mathrm{d}\lambda. \tag{3.192}$$

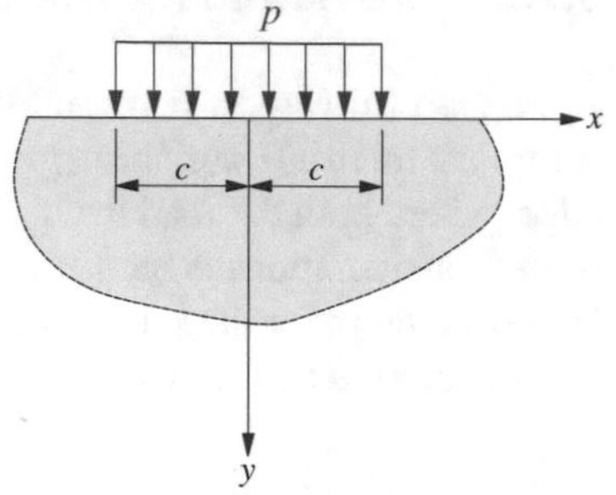

Fig. 3.21 Half-plane under edge load p

If f_x is an even function in x, then:

$$f(x) = \frac{2}{\pi} \int_0^\infty \cos(\alpha x)\, \mathrm{d}\alpha \int_0^\infty f_\lambda \cos(\alpha\lambda)\, \mathrm{d}\lambda. \tag{3.193}$$

If, on the other hand, f_x is an odd function in x, then we obtain:

$$f(x) = \frac{2}{\pi} \int_0^\infty \sin(\alpha x)\, \mathrm{d}\alpha \int_0^\infty f_\lambda \sin(\alpha\lambda)\, \mathrm{d}\lambda. \tag{3.194}$$

In the context of the solution of disk problems, recurrent integral formulas occur frequently, which in some cases allow an elementary closed-analytical solution.

As an introduction we consider the example of Fig. 3.21. Let there be a boundary load p acting on a segment of the edge with length $2c$. Let the half-plane be infinitely extended in the $x-$ direction and have a rectilinear boundary.

We first represent the boundary load p as a Fourier integral. Since the load is symmetric in y, we use the following representation:

$$p(x) = \frac{2}{\pi} \int_0^\infty \cos\alpha x \mathrm{d}\alpha \int_0^c p_x \cos\alpha\lambda \mathrm{d}\lambda = \frac{2p_x}{\pi} \int_0^\infty \frac{\sin\alpha c}{\alpha} \cos\alpha x \mathrm{d}\alpha. \tag{3.195}$$

For the Airy stress function F the following approach is used:

$$F = \int_0^\infty \frac{1}{\alpha^2} (A + \alpha y B)\, e^{-\alpha y} \cos\alpha x \mathrm{d}\alpha, \tag{3.196}$$

wherein A and B are functions of α. From this, the following stress components can then be determined:

$$\sigma_{xx} = \frac{\partial^2 F}{\partial y^2} = \int_0^\infty [(A - 2B) + \alpha y B]\, e^{-\alpha y} \cos\alpha x \mathrm{d}\alpha,$$

$$\sigma_{yy} = \frac{\partial^2 F}{\partial x^2} = -\int_0^\infty (A + \alpha y B)\, e^{-\alpha y} \cos \alpha x \mathrm{d}\alpha,$$

$$\sigma_{yy} = -\frac{\partial^2 F}{\partial x \partial} = -\int_0^\infty [(A - B) + \alpha y B]\, e^{-\alpha y} \sin \alpha x \mathrm{d}\alpha. \tag{3.197}$$

The boundary conditions to be fulfilled here require that the normal stress σ_{yy} at the boundary $y = 0$ must correspond to the negative boundary load $p(x)$ for each x. Moreover, at this boundary the shear stress τ_{xy} vanishes. From this follows:

$$A = B = \frac{2p}{\pi h} \frac{\sin \alpha c}{\alpha}. \tag{3.198}$$

Thus, the stress components follow as:

$$\sigma_{xx} = -\frac{2p}{\pi h} \int_0^\infty \frac{\sin \alpha c}{\alpha} (1 - \alpha y)\, e^{-\alpha y} \cos \alpha x \mathrm{d}\alpha,$$

$$\sigma_{yy} = -\frac{2p}{\pi h} \int_0^\infty \frac{\sin \alpha c}{\alpha} (1 + \alpha y)\, e^{-\alpha y} \cos \alpha x \mathrm{d}\alpha,$$

$$\sigma_{yy} = -\frac{2p}{\pi h} \int_0^\infty \frac{\sin \alpha c}{\alpha} \alpha y e^{-\alpha y} \sin \alpha x \mathrm{d}\alpha. \tag{3.199}$$

The integrals appearing here are easily solved. Using

$$\sin \alpha c \cos \alpha x = \frac{1}{2} [\sin \alpha (x + c) - \sin \alpha (x - c)] \tag{3.200}$$

and the integrals

$$\int_0^\infty \frac{\sin \alpha (x \pm c)}{\alpha} e^{-\alpha y} \mathrm{d}\alpha = \arctan\left(\frac{x \pm c}{y}\right),$$

$$\int_0^\infty \sin \alpha (x \pm c) e^{-\alpha y} \mathrm{d}\alpha = \frac{x \pm c}{(x \pm c)^2 + y^2} \tag{3.201}$$

the stress components can be expressed as:

$$\sigma_{xx} = -\frac{p}{\pi h}\left[\arctan\left(\frac{x+c}{y}\right) - \arctan\left(\frac{x-c}{y}\right) - y\left(\frac{x+c}{(x+c)^2 + y^2} - \frac{x-c}{(x-c)^2 + y^2}\right)\right],$$

$$\sigma_{yy} = -\frac{p}{\pi h}\left[\arctan\left(\frac{x+c}{y}\right) - \arctan\left(\frac{x-c}{y}\right) + y\left(\frac{x+c}{(x+c)^2 + y^2} - \frac{x-c}{(x-c)^2 + y^2}\right)\right],$$

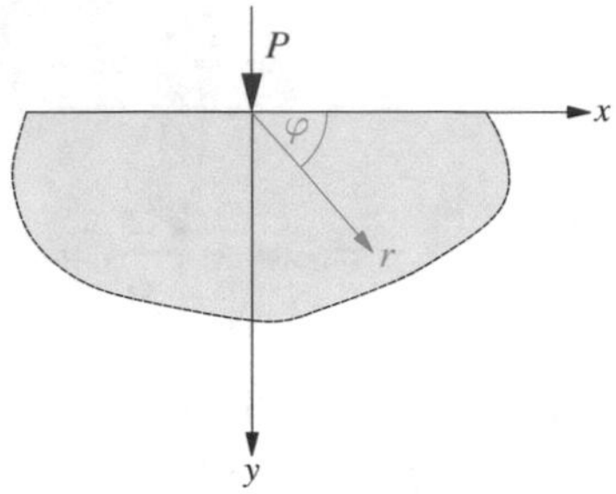

Fig. 3.22 Half-plane under edge normal force P

$$\tau_{xy} = -\frac{py}{\pi h}\left[\frac{y}{(x-c)^2+y^2} - \frac{y}{(x+c)^2+y^2}\right]. \tag{3.202}$$

We consider as another example the half-plane under the boundary normal force P as shown in Fig. 3.22. This case can be derived from the previous situation by considering the boundary transition $c \to 0$ and by setting $P = 2pc$. It follows for the stress components:

$$\sigma_{xx} = -\frac{P}{\pi h}\int_0^\infty (1-\alpha y)\, e^{-\alpha y} \cos\alpha x \,\mathrm{d}\alpha,$$

$$\sigma_{yy} = -\frac{P}{\pi h}\int_0^\infty (1+\alpha y)\, e^{-\alpha y} \cos\alpha x \,\mathrm{d}\alpha,$$

$$\sigma_{yy} = -\frac{P}{\pi h}\int_0^\infty \alpha y e^{-\alpha y} \sin\alpha x \,\mathrm{d}\alpha. \tag{3.203}$$

The integrals appearing here can be solved elementarily, and it follows:

$$\frac{\sigma_{xx} bh}{P} = -\frac{2}{\pi}\frac{\left(\frac{x}{b}\right)^2\left(\frac{y}{b}\right)}{\left[\left(\frac{x}{b}\right)^2+\left(\frac{y}{b}\right)^2\right]^2},$$

$$\frac{\sigma_{yy} bh}{P} = -\frac{2}{\pi}\frac{\left(\frac{y}{b}\right)^3}{\left[\left(\frac{x}{b}\right)^2+\left(\frac{y}{b}\right)^2\right]^2},$$

$$\frac{\tau_{xy} bh}{P} = -\frac{2}{\pi}\frac{\left(\frac{x}{b}\right)\left(\frac{y}{b}\right)^2}{\left[\left(\frac{x}{b}\right)^2+\left(\frac{y}{b}\right)^2\right]^2}. \tag{3.204}$$

The two stress components σ_{yy} and τ_{xy} are shown in Fig. 3.23. The strongly local effect of the edge normal force P on the stress state of the half-plane can be clearly

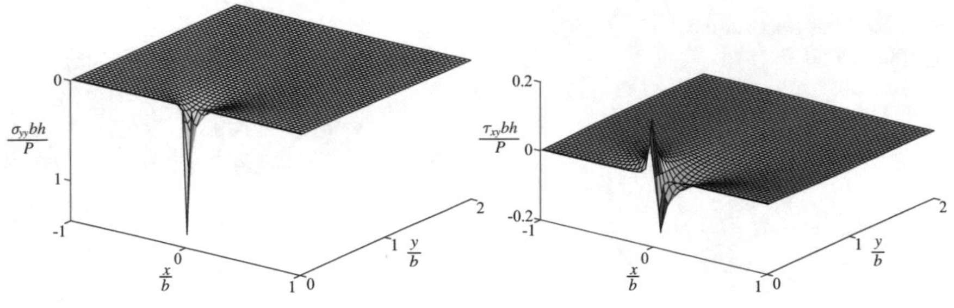

Fig. 3.23 Stresses σ_{yy} and τ_{xy} in a half-plane under edge normal load P

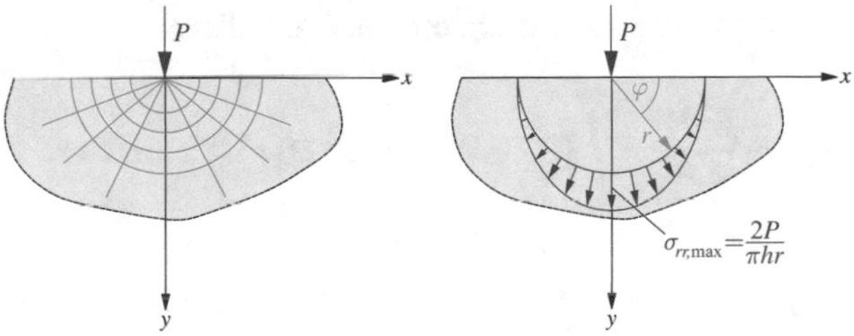

Fig. 3.24 Principal stress lines (left) and stress distribution at r = const. (right) of the half-plane loaded by an edge normal force

seen here; the edge stresses decay very quickly. At the point $x = 0$, $y = 0$ a stress singularity occurs.

The transition to polar coordinates r, φ results from (3.204):

$$\frac{\sigma_{rr}Rh}{P} = -\frac{2}{\pi}\frac{\sin\varphi}{\frac{r}{R}}, \quad \frac{\sigma_{\varphi\varphi}Rh}{P} = 0, \quad \frac{\tau_{xy}Rh}{P} = 0. \tag{3.205}$$

Accordingly, in this reference system only the radial stress σ_{rr} is present, and the main principal stress lines that arise are formed by rays emerging from the load application point and concentric circles arranged therein. Along a concentric circle $r = const.$ the distribution of σ_{rr} is sinusoidal with the maximum value $\sigma_{rr} = -\frac{2P}{\pi hr}$ (Fig. 3.24).

Finally, we consider the case of the half-plane loaded by an edge shear load (Fig. 3.25). Since the resulting stress state will be antimetric with respect to x, the following approach is used for the Airy stress function:

$$F = \int_0^\infty \frac{1}{\alpha^2}(C + \alpha y D)e^{-\alpha y} \sin\alpha x \mathrm{d}\alpha. \tag{3.206}$$

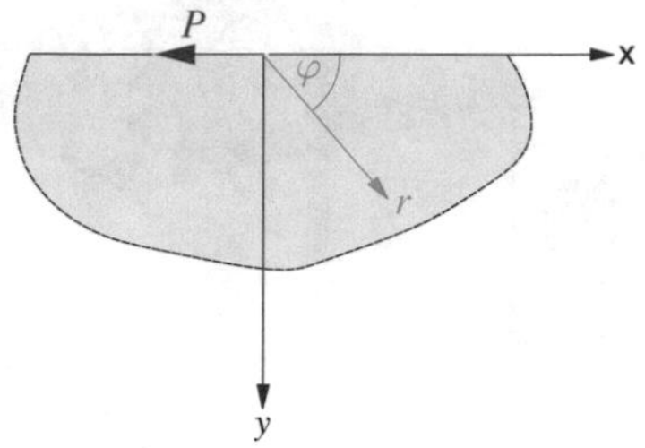

Fig. 3.25 Half-plane under edge shear load P

Herein C and D are functions of α determined from the given boundary conditions. With (3.206), the stress components are obtained as follows:

$$\sigma_{xx} = \frac{\partial^2 F}{\partial y^2} = \int_0^\infty [(C - 2D) + \alpha y D]\, e^{-\alpha y} \sin \alpha x \mathrm{d}\alpha,$$

$$\sigma_{yy} = \frac{\partial^2 F}{\partial x^2} = -\int_0^\infty (C + \alpha y D)\, e^{-\alpha y} \sin \alpha x \mathrm{d}\alpha,$$

$$\tau_{xy} = -\frac{\partial^2 F}{\partial x \partial y} = \int_0^\infty [(C - D) + \alpha y D]\, e^{-\alpha y} \cos \alpha x \mathrm{d}\alpha. \tag{3.207}$$

The fulfillment of the boundary conditions $\sigma_{yy}(y = 0) = 0$ and $\tau_{xy}(y = 0) = \frac{p(x)}{h}$ results in:

$$C = 0, \quad D = -\frac{P}{\pi h}. \tag{3.208}$$

Hence:

$$\sigma_{xx} = \frac{P}{\pi h} \int_0^\infty (2 - \alpha y)\, e^{-\alpha y} \sin \alpha x \mathrm{d}\alpha,$$

$$\sigma_{yy} = \frac{P}{\pi h} \int_0^\infty \alpha y e^{-\alpha y} \sin \alpha x \mathrm{d}\alpha,$$

$$\tau_{xy} = \frac{P}{\pi h} \int_0^\infty (1 - \alpha y) e^{-\alpha y} \cos \alpha x \mathrm{d}\alpha. \tag{3.209}$$

The integrals occurring here can be solved in an exact closed-form analytical manner, and it follows:

$$\sigma_{xx} = \frac{2P}{\pi h} \frac{x^3}{\left(x^2 + y^2\right)^2}, \quad \sigma_{yy} = \frac{2P}{\pi h} \frac{x y^2}{\left(x^2 + y^2\right)^2}, \quad \tau_{xy} = \frac{2P}{\pi h} \frac{x^2 y}{\left(x^2 + y^2\right)^2}. \tag{3.210}$$

The transition to polar coordinates leads to:

$$\sigma_{rr} = \frac{2P}{\pi h}\frac{\cos\varphi}{r}, \quad \sigma_{\varphi\varphi} = 0, \quad \tau_{r\varphi} = 0. \tag{3.211}$$

In this case, the stress state again exhibits a singularity at $r = 0$. At the locations $\varphi = 0$ and $\varphi = \pi$ with fixed r, the maximum and minimum stress values are obtained with $\sigma_{rr} = \pm\frac{2P}{\pi hr}$, respectively.

3.8 The Effective Width

3.8.1 Effective Width of Flanges of Beams Under Bending

For applications e.g. in civil engineering or lightweight construction, the estimation of the so-called effective width b_m of beams is technically particularly relevant. Here we want to consider the case of a beam under a moment load M_y (see Fig. 3.26, top) and treat the double-symmetric I-beam (height h, flange width b, wall thickness t as indicated) and also assume, for simplicity, that the bending moment is carried exclusively by the flanges of the beam (see Fig. Fig. 3.26, bottom left). Consider now a beam which can no longer be suitably treated by means of a beam theory and, in particular, can no longer be classified as sufficiently slender. In such a case, a stress state will develop which is in contrast to the predictions of, for example, the Euler-Bernoulli beam theory, but also the Timoshenko beam theory. In particular, it will be seen that the flanges of such a beam will no longer carry across their full width, but the bending stresses will be unevenly distributed across the flange width, as shown in Fig. 3.26, bottom left. This effect will be more pronounced the wider the flange of the beam becomes. The concept of effective width b_m is helpful at this point: Here it is assumed that the bending stress in the flange caused by the bending moment M_y is distributed uniformly over a reduced width, namely the effective width b_m of the flange, and thus only this width b_m can be used for stress transfer (Fig. 3.26, bottom right). The value of the bending stress σ_f applied over b_m then corresponds to the value of the stress at the intersection point between flange and web with the amount $\sigma_f = \sigma_{\max}$. It should be noted that b_m depends not only on the type of loading of the bending beam under consideration, but also on its cross-sectional shape and the underlying boundary conditions.

In order to determine the effective width b_m by means of a disk analysis, Wiedemann (2006) considers the flanges as disks and first considers a double-symmetric I-beam of length l, which is simply supported at both ends and subjected to the constant line load q_z (Fig. 3.27, left). Furthermore, it is assumed that the transverse shear stress only occurs in the web and may be assumed to be constant (Fig. 3.27, bottom right). The shear stress $\tau_{xz,s}$ in the web, which according to an exact calculation is parabolic in the web and linear in the flanges (Fig. 3.27, top right), can then be determined approximately from the transverse shear force $Q(x)$ divided by the height

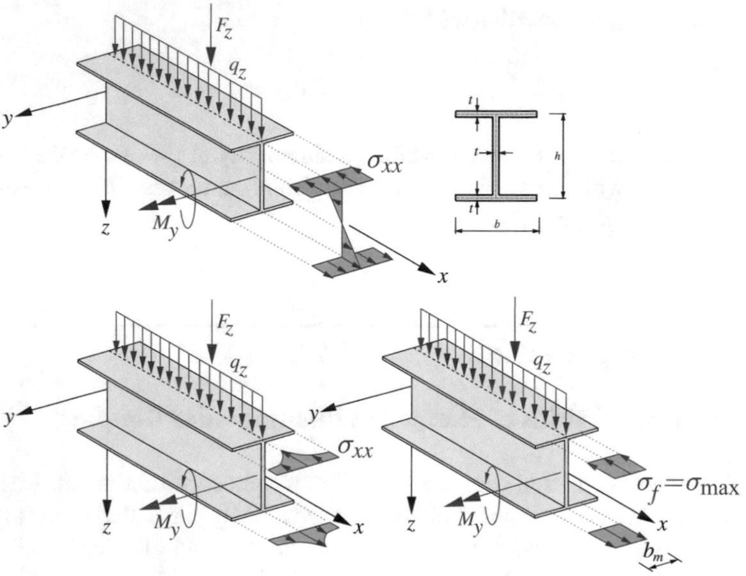

Fig. 3.26 Beam under bending (top), simplifying assumption of load transfer exclusively through the flanges and nonlinear stress distribution of σ_{xx} across the flange width (bottom left), introduction of the effective width b_m (bottom right)

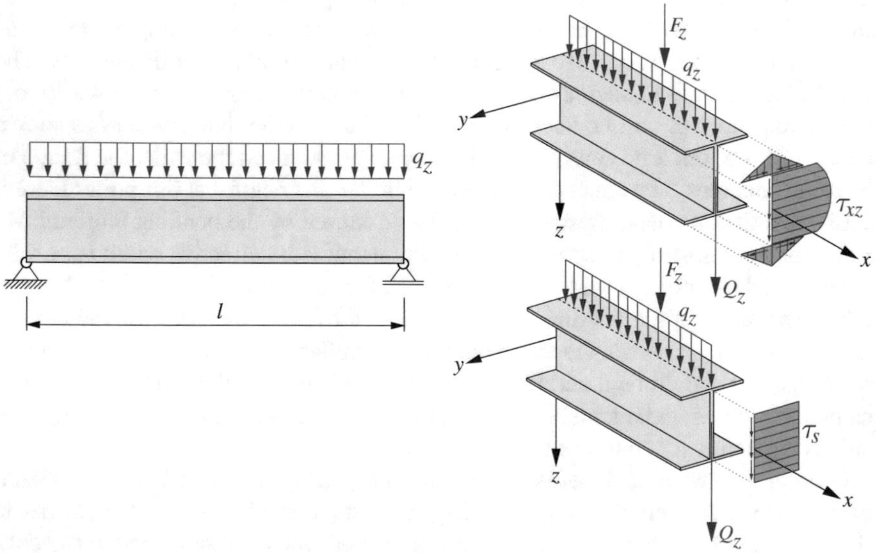

Fig. 3.27 Simply supported I-beam under constant line load q_z (left), actual shear stress distribution (top right), simplified assumption of shear being taken up solely by the web (bottom right) and assumption of constant shear stress τ_s

$h(x)$ and the thickness $t(x)$ of the web. By developing the flange shear stresses as a Fourier series, assuming the flanges to be infinitely extended in the y–direction, Wiedemann arrives at the following expression for the effective width b_m with the requirement for the flange and web shear stresses to coincide at the point $y = 0$ and for the intersection edge between web and flange disks to remain straight:

$$\frac{b_m}{l} = \frac{\mu\pi^2\left[\left(\frac{x}{l}\right)^2 - \frac{x}{l}\right]}{4\sum_{m=1}^{M}\frac{1}{m^2}\sin\left(\frac{m\pi x}{l}\right)}, \tag{3.212}$$

wherein $\mu = \sqrt{\frac{1-\nu}{2}}$. For a finite width flange plate it is assumed that the free flange edge is free of shear stresses. For this case, Wiedemann gives the following expression for the effective width b_m with a reference frame shifted by $\frac{b}{2}$:

$$\frac{b_m}{l} = \frac{\mu\pi^2\left[\left(\frac{x}{l}\right)^2 - \frac{x}{l}\right]}{4\sum_{m=1}^{M}\frac{1}{m^2}\frac{\sin\left(\frac{m\pi x}{l}\right)}{\tanh\left(\frac{m\pi b}{2\mu l}\right)}}. \tag{3.213}$$

A graphical representation for the effective width is shown in Fig. 3.28 for a simply supported beam under a constant line load (Fig. 3.28, left) and a concentrated load at midspan (Fig. 3.28, right). A discussion of the results of Fig. 3.28, left, shows that the flange width is available for load transfer of the bending load only up to a certain extent. In the case of an infinitely extended flange disk, the effective width b_m results in about 40 percent of the beam length. If, on the other hand, there is a beam profile with a narrow flange, then this can be assumed to be fully load-bearing: $b_m = b$. Note, however, that the effective width b_m is a function of the beam axis x. Thus, it will take different values at the support points than at midspan. The effective width b_m turns out to be particularly small at the support points and increases significantly in the direction of the center of the beam at $x = \frac{l}{2}$. The reason is to be found in the force transmission of the support forces via the web plate, and the load transmission between web and flanges can thus only occur at a certain distance from the support points.

A similar computational approach is employed by Girkmann (1974) (also found in modified form in Hake and Meskouris 2007), who considers two flanges of a double-symmetric I-profile hinged to a web (Fig. 3.29), where the assumption is made that the flanges, which are assumed to extend infinitely in the y–direction, carry exclusively as disks and thus their thickness h is significantly smaller than the width. Let the considered beam be simply supported at the locations $x = 0$ and $x = l$, so that at these locations the bending moment M_y vanishes. The underlying load case is not yet specified at this point, and the bending moment curve over the beam length l is represented as a Fourier series in general as follows:

$$M_y(x) = \sum_{n=1}^{\infty} m_n \sin\alpha_n x, \tag{3.214}$$

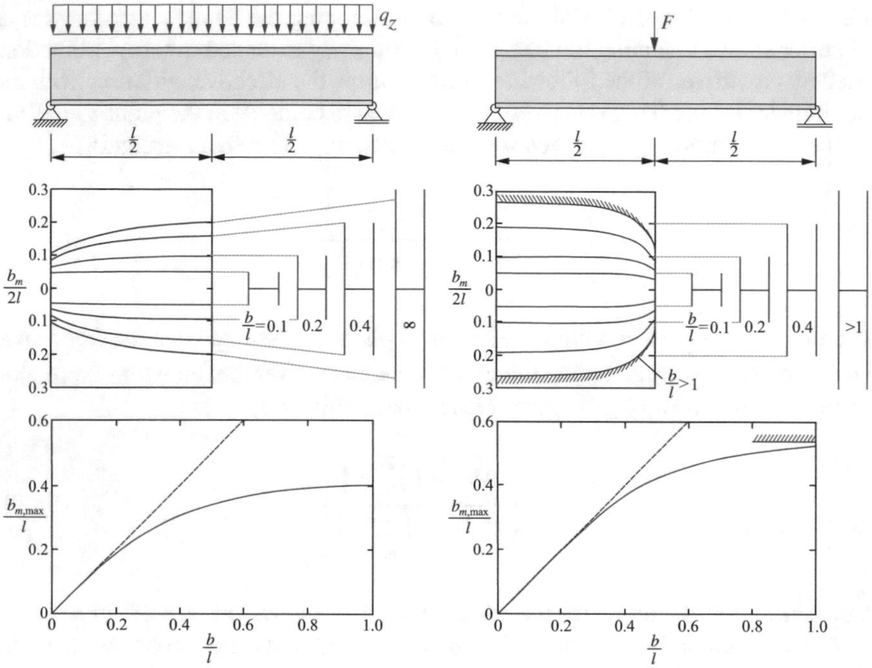

Fig. 3.28 Effective width for a simply supported I-beam; uniform line load (left), single force at midspan (right), after Wiedemann (2006)

with $\alpha_n = \frac{n\pi}{l}$.

To represent the stress components in the flange of the beam under consideration, an approach of the following form is used for the Airy stress function:

$$F = \sum_{n=1}^{\infty} \frac{1}{\alpha_n^2} \left(A_n + \alpha_n y B_n\right) e^{-\alpha_n y} \sin \alpha_n x. \tag{3.215}$$

This allows the stress components in the flanges to be represented as:

$$\sigma_{xx} = \sum_{n=1}^{\infty} \left[\left(A_n - 2B_n\right) + \alpha_n y B_n\right] e^{-\alpha_n y} \sin \alpha_n x,$$

$$\sigma_{yy} = -\sum_{n=1}^{\infty} \left(A_n + \alpha_n y B_n\right) e^{-\alpha_n y} \sin \alpha_n x,$$

$$\tau_{xy} = \sum_{n=1}^{\infty} \left[\left(A_n - B_n\right) + \alpha_n y B_n\right] e^{-\alpha_n y} \cos \alpha_n x. \tag{3.216}$$

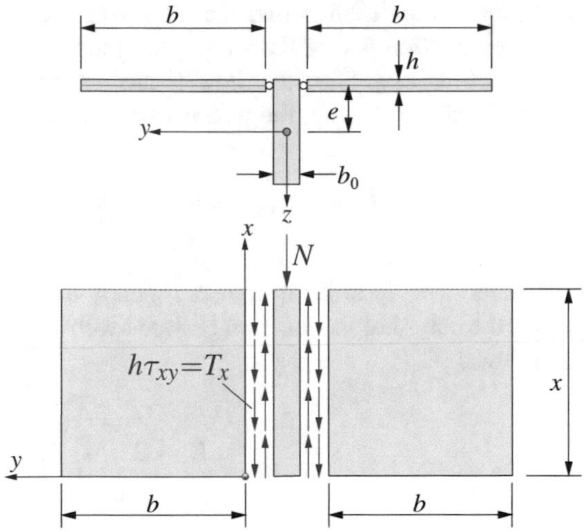

Fig. 3.29 Mechanical model for determining the effective flange width for the simply supported I-beam, according to Girkmann (1974)

For the solution of the given problem, the longitudinal strain ε_{xx}^0 of the flange disks and the transverse displacement v are also required. From

$$\varepsilon_{xx}^0 = \frac{1}{E}\left(\sigma_{xx} - \nu\sigma_{yy}\right) \tag{3.217}$$

with (3.216) one obtains:

$$E\varepsilon_{xx}^0 = \sum_{n=1}^{\infty}\{(1+\nu)A_n - [2-(1+\nu)\alpha_n y]\,B_n\}\,e^{-\alpha_n y}\sin\alpha_n x. \tag{3.218}$$

For the transverse displacement v we obtain:

$$Ev_0 = \int\frac{\partial^2 F}{\partial x^2}\mathrm{d}y - \nu\frac{\partial F}{\partial y} + C_2(x), \tag{3.219}$$

where $C_2(x) = D_1 - D_2 x$. With (3.215) and σ_{yy} (3.216):

$$\begin{aligned} Ev_0 = {} & (1+\nu)\sum_{i=1}^{\infty}\frac{1}{\alpha_n}\left[\left(A_n + \frac{1-\nu}{1+\nu}B_n\right) + \alpha_n y B_n\right]e^{-\alpha_n y}\sin\alpha_n x \\ & + D_1 - D_2 x. \end{aligned} \tag{3.220}$$

The first boundary condition to be formulated is that at the point $y = 0$ the displacement v_0 of the flange disk becomes zero: $v_0(y = 0) = 0$. This is satisfied for any x if $D_1 = D_2 = 0$ and

$$A_n = -\frac{1-\nu}{1+\nu}B_n. \tag{3.221}$$

The second condition is that the elongation ε_{xx}^0 of the flange disk must be equal to the elongation $\bar{\varepsilon}_{xx}$ of the web at the junction between the web and the flanges: $\varepsilon_{xx}^0(y=0) = \bar{\varepsilon}_{xx}$. The shear flow T acts between the web and the flanges and must match the shear stress of the web multiplied by h:

$$T_x = h\tau_{xy}(y=0) = h\sum_{n=1}^{\infty}(A_n - B_n)\cos\alpha_n x. \tag{3.222}$$

The shear flow T_x is in equilibrium with a normal force N in the web, which also acts at the height of the junction between web and flanges and can thus be calculated as follows:

$$N = 2\int_0^x T_x \mathrm{d}x. \tag{3.223}$$

With positive shear flow T_x, N is a compressive force (Fig. 3.29). Thus:

$$N = 2h\sum_{n=1}^{\infty}\left[(A_n - B_n)\int_0^x \cos\alpha_n x \mathrm{d}x\right] = 2h\sum_{n=1}^{\infty}(A_n - B_n)\frac{\sin\alpha_n x}{\alpha_n}. \tag{3.224}$$

Due to the excentricity of the normal force N the following normal stress $\bar{\sigma}_{xx}$ arises in the web, acting at the junction of web and flanges:

$$\bar{\sigma}_{xx} = -\frac{N}{A} - \frac{M_y + Ne}{I_{yy}}e, \tag{3.225}$$

where I_{yy} and A are the corresponding second-order moment of inertia and the cross-sectional area of the web, respectively. Introducing the radius of gyration $i^2 = \frac{I}{A}$ gives:

$$\bar{\sigma}_{xx} = -\frac{M_y e + N\left(e^2 + i^2\right)}{I_{yy}}. \tag{3.226}$$

Thus, the strain $\bar{\varepsilon}_{xx}$ of the beam can be represented as:

$$\bar{\varepsilon}_{xx} = \frac{\bar{\sigma}_{xx}}{E} = -\frac{M_y e + N\left(e^2 + i^2\right)}{EI_{yy}}. \tag{3.227}$$

Substituting the terms (3.224) and (3.214) for normal force N and bending moment M_y yields:

$$E\bar{\varepsilon}_{xx} = -\frac{1}{I_{yy}}\sum_{n=1}^{\infty}\left[em_n + \frac{2h}{\alpha_n}\left(e^2 + i^2\right)(A_n - B_n)\right]\sin\alpha_n x. \tag{3.228}$$

Thus, the second necessary condition, namely the matching of the longitudinal strain $\varepsilon^0_{xx}(y = 0)$ in the flange disk at the junction to the web with the longitudinal strain $\bar{\varepsilon}_{xx}$ of the web at $z = -e$ for all x, can be formulated. The following equation is obtained:

$$A_n\left[(1+\nu)+\frac{2h}{\alpha_n I_{yy}}\left(e^2+i^2\right)\right]-B_n\left[2+\frac{2h}{\alpha_n I_{yy}}\left(e^2+i^2\right)\right]=-\frac{em_n}{I_{yy}}. \quad (3.229)$$

Solving with (3.221) gives the following constants A_n and B_n:

$$A_n = -\frac{(1-\nu)em_n}{(3-\nu)(1+\nu)I_{yy}+4\frac{h}{\alpha_n}\left(e^2+i^2\right)},$$

$$B_n = \frac{(1+\nu)em_n}{(3-\nu)(1+\nu)I_{yy}+4\frac{h}{\alpha_n}\left(e^2+i^2\right)}. \quad (3.230)$$

Thus, all stress components in the flange disks can be determined with (3.216). The effective width is determined according to this calculation model by requiring that the beam suffers the same curvature at any point x according to the present model as the same beam under elementary bending theory. Further elaboration is omitted at this point; Fig. 3.30 contains selected results for the effective width b_m (after Girkmann 1974, for $b \to \infty$ and $A = 0.1hl$). The similarity of the results with the effective width given in Fig. 3.28 for the beam under constant line load and single force is apparent. Furthermore, it is shown that in the presence of a sinusoidal load of the type $p(x) = p_0 \sin\left(\frac{n\pi x}{l}\right)$, a constant value $b_m = 0.1815\frac{l}{n}$ results.

3.8.2 Effective Width for Load Introductions

A further technically relevant load introduction situation, discussed in detail in Wiedemann (2006), consists in the concentrated introduction of normal stresses σ_0 occurring only in certain areas into a disk / a half-plane, as shown in Fig. 3.31. Here, the boundary stress $\sigma_{xx} = \sigma_0$ is introduced on the width b_{m0} in an infinitely extended disk, this force introduction being made at regular intervals b. This situation can also be characterized again by an effective width b_m. In case of such a load introduction, the distribution of the normal stress σ_{xx} in the half-plane / the disk will change significantly over x. At the location $x = 0$, i.e. at the edge of the disk, the stress σ_{xx} will exactly correspond to the value σ_0 and, moreover, will only form on the width b_{m0}. However, moving away from the edge of the disk, the stress σ_{xx} will become more and more uniform until, at a certain distance from the edge, a uniform stress distribution with the value $\sigma_{xx} = \sigma^\infty_{xx}$ is obtained. This value is calculated as:

$$\sigma^\infty_{xx} = \sigma_0\frac{b_{m0}}{b}. \quad (3.231)$$

Fig. 3.30 Effective flange width for the I-beam simply supported at both ends under different load cases, after Girkmann (1974)

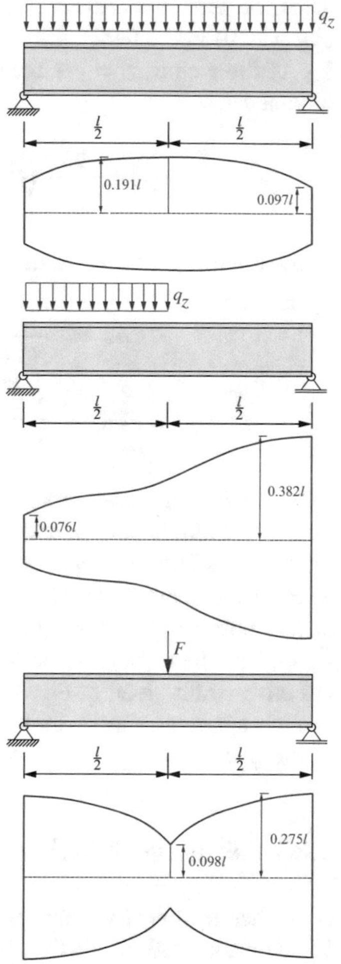

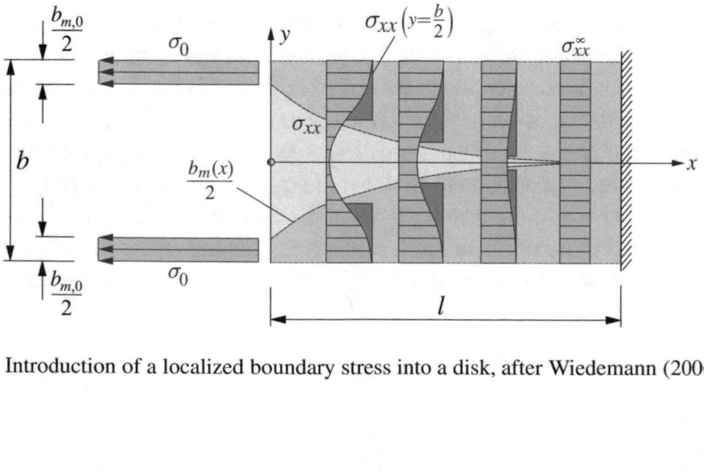

Fig. 3.31 Introduction of a localized boundary stress into a disk, after Wiedemann (2006)

Obviously, the effective width b_m takes exactly the value $b_m = b_{m0}$ at the point $x = 0$ and increases further with increasing values for x, until finally the value $b_m = b$ is reached in the presence of the uniform stress distribution $\sigma_{xx} = \sigma_{xx}^{\infty}$ and here the disk carries the load along over its entire width. Indicated here again is the procedure of replacing the nonlinear stress distribution of σ_{xx} with respect to y by constant stresss distributed over the effective width b_m and in which one assumes a uniform stress distribution with the value of the maximum boundary value $\sigma_{xx}\left(y = \frac{b}{2}\right)$.

With respect to the y−direction, the stress σ_{xx} can now be split into the constant value σ_{xx}^{∞} and a variable part (Wiedemann 2006), where a Fourier series is applied for the variable part, so that ($n = 1, 3, 5, \ldots, N$):

$$\frac{\sigma_{xx}(y)}{\sigma_0} = \frac{b_{m0}}{b} - \frac{2}{\pi}\sum_{n=1}^{N}\frac{1}{n}\sin\left(\frac{n\pi b_{m0}}{b}\right)\cos\left(2\frac{n\pi y}{b}\right). \tag{3.232}$$

This harmonically distributed boundary perturbation and the associated stresses inside the disk can be approached computationally by the means of the previous sections. We assume here the case of a completely vanishing strain ε_{yy}^{0} (which can be ensured, e.g., by the arrangement of stiffening elements) and also assume that at a certain edge distance at the point $x = l$ there is a rigid restraint of the disk, requiring that the shear stress τ_{xy} vanishes there. The solution is composed of a part which does not change over x and a part which decays in accordance with (3.171). Wiedemann gives the following expression for the effective width b_m:

$$\frac{b_m}{b} = \frac{\sigma_{xx}^{\infty}}{\sigma_{\max}} = \frac{1}{1 - \frac{2}{\pi}\frac{b}{b_{m0}}\sum_{n=1}^{N}\frac{(-1)^n}{n}\sin\left(\frac{n\pi b_{m0}}{b}\right)\frac{\cosh\left(\frac{2\mu n\pi\bar{x}}{b}\right)}{\cosh\left(\frac{2\mu n\pi l}{b}\right)}}, \tag{3.233}$$

with $\mu^2 = \frac{1-\nu}{2}$. Here $\sigma_{\max}$ is the boundary value of the normal stress σ_{xx} at the location $y = \frac{b}{2}$ and $\bar{x} = l - x$. A qualitative representation of the distribution of the stress σ_{xx} over x as well as the associated variation of the effective width b_m is shown in Fig. 3.31.

If, in contrast to the edge stress σ_0 distributed over the width b_{m0}, concentrated individual forces F are to be introduced into the disk via stiffeners at $y = \pm\frac{b}{2}$ by shear (Fig. 3.32), then the effective width b_m according to Wiedemann (2006) can be given as:

$$\frac{b_m}{b} = 1 - \frac{8}{\pi^2}\sum_{n=1}^{N}\frac{1}{n^2}\frac{\cosh\left(\frac{\mu n\pi\bar{x}}{b}\right)}{\sinh\left(\frac{\mu n\pi l}{b}\right)}, \tag{3.234}$$

which can be simplified for large lengths $l > 2b$ to ($n = 1, 3, 5, \ldots, N$):

$$\frac{b_m}{b} = 1 - \frac{8}{\pi^2}\sum_{n=1}^{N}\frac{1}{n^2}e^{-\frac{\mu n\pi x}{b}}. \tag{3.235}$$

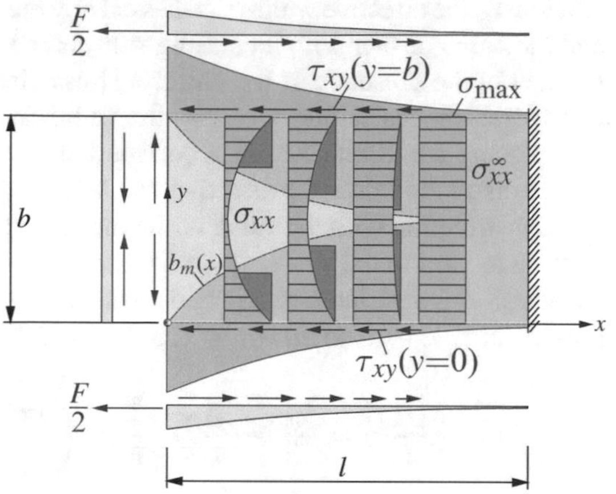

Fig. 3.32 Introduction of an edge load F into a disk using edge stiffeners, after Wiedemann (2006)

Simplified, this formula is:

$$\frac{b_m}{b} = 1 - e^{-\frac{\mu\pi x}{b}}. \tag{3.236}$$

A prerequisite for the validity of these results is that the load is applied by means of stiffeners which are rigidly connected to the disk and whose cross sections vary over x in such a way that their elongation corresponds to the elongation of the disk at any point x. In addition, a stiffener is required at the transverse edge $x = 0$ of the disk. A qualitative illustration is also given in Fig. 3.32.

The qualitative results of Fig. 3.32 show that the variation of the effective width b_m has quite similar characteristics as in the case of the edge load σ_0 distributed over a certain edge width b_{m0}. The single forces F introduced into the edge stiffeners are transferred by shear into the disk. The shear flow between the stiffener and the disk decreases continuously, so that the width b_m increases continuously as well. At a certain edge distance, a uniform distribution of the normal stress σ_{xx}^{∞} is achieved. It is remarkable that the maximum value σ_{max} is constant, independent of the position on the longitudinal axis x.

A final elementary case of load introduction, which can be handled by the means of disk analysis discussed so far, is the introduction of a bending moment represented by the pair of forces F introduced into the two edge stiffeners shown in Fig. 3.33. For this problem, the definition of an effective width is not meaningful, but the problem is closely related to what has already been discussed in this section. The load-bearing effect between stiffener and disk is quite analogous to what has already been discussed for the normal force introduction, so that more detailed elaborations can be omitted at this point. Further details can be found in Wiedemann (2006). In the present case, at a sufficient edge distance, a linear stress distribution σ_{xx}^{∞} can be expected as follows:

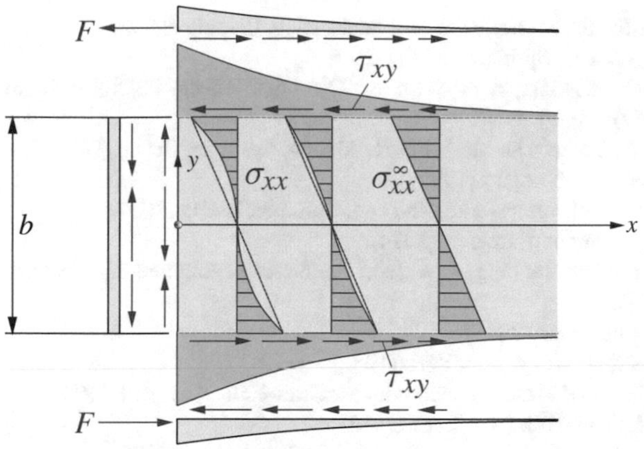

Fig. 3.33 Introduction of a pair of forces into a disk by means of edge stiffeners, after Wiedemann (2006)

$$\sigma_{xx}^{\infty} = \frac{2\sigma_{\max} y}{b}, \tag{3.237}$$

where $\sigma_{\max}$ is the maximum value of the normal stress σ_{xx} at the position $y = \frac{b}{2}$. Since the pair of forces is again introduced into the disk by shear, a form of σ_{xx} that deviates significantly from (3.237) will occur near the edge, which according to Wiedemann can be described as follows:

$$\frac{\sigma_{xx}}{\sigma_{\max}} = \frac{2y}{b} + \frac{2}{\pi} \sum_{n=1}^{N} \frac{(-1)^n}{n} e^{-\frac{2\mu n \pi x}{b}} \sin\left(\frac{2n\pi y}{b}\right). \tag{3.238}$$

A qualitative evaluation of this result is shown in Fig. 3.33.

References

Altenbach, H., Altenbach, J., Naumenko, K.: Ebene Flächentragwerke, 2nd edn. Springer, Berlin (2016)

Bauchau, O.A., Craig, J.I.: Structural Analysis. Springer, Dordrecht (2009)

Becker, W., Gross, D.: Mechanik elastischer Körper und Strukturen. Springer, Berlin (2002)

Boresi, A.P., Lynn, P.P.: Elasticity in Engineering Mechanics. Prentice-Hall, Eaglewood Cliffs (1974)

Chou, P.C., Pagano, N.J.: Elasticity. Van Nostrand, Princeton (1967)

Eschenauer, H., Schnell, W.: Elastizitätstheorie, 3rd edn. BI Wissenschaftsverlag, Mannheim (1993)

Eschenauer, H., Olhoff, N., Schnell, W.: Applied Structural Mechanics. Springer, Berlin (1997)

Filonenko-Boroditsch, M.M.: Elastizitätstheorie. VEB Fachbuchverlag, Leipzig (1967)
Girkmann, K.: Flächentragwerke, 6th edn. Springer, Wien (1974)
Göldner, H., Altenbach, J., Eschke, K., Garz, K.F., Sähn, S.: Lehrbuch Höhere Festigkeitslehre Band 1. Physik, Weinheim (1979)
Gross, D., Hauger, W., Wriggers, P.: Technische Mechanik 4, 9th edn. Springer, Berlin (2014)
Hahn, H.G.: Elastizitätstheorie. Teubner, Stuttgart (1985)
Hake, E., Meskouris, K.: Statik der Flächentragwerke, 2nd edn. Springer, Berlin (2007)
Leipholz, H.: Theory of Elasticity. Noordhoff, Leiden (1974)
Massonet, C.: Two-dimensional problems. In: Flügge, W. (ed.) Handbook of Engineering Mechanics. McGraw-Hill, New York (1962)
Sadd, M.H.: Elasticity. Elsevier, Amsterdam (2005)
Sechler, E.E.: Elasticity in Engineering. Wiley, New York (1952)
Timoshenko, S., Goodier, J.N.: Theory of Elasticity, 2nd edn. McGraw-Hill, New York (1951)
Wang, C.T.: Applied Elasticity. McGraw-Hill, New York (1953)
Wiedemann, J.: Leichtbau, Elemente und Konstruktion, 3rd edn. Springer, Berlin (2006)

Isotropic Disks in Polar Coordinates

4

This chapter deals with isotropic disks in polar coordinates. After a short presentation of all necessary basic equations, as already in the case of Cartesian coordinates, the force method is discussed in detail, where besides the derivation of the disk equation and solution approaches, also an energetic consideration is presented. In addition to elementary basic cases, selected technically relevant cases are discussed, namely rotationally symmetric disks as well as non-rotationally symmetric circular arc disks and wedge-shaped disks, before this chapter concludes with the consideration of disks containing circular holes.

4.1 Fundamentals

4.1.1 Basic Equations

In this chapter, we will extend our considerations to isotropic disk problems that can be advantageously described in polar coordinates r, φ. Isotropic disk problems in polar coordinates are discussed in Altenbach et al. (2016), Bauchau and Craig (2009), Becker and Gross (2002), Boresi and Lynn (1974), Chou and Pagano (1967), Eschenauer and Schnell (1993), Eschenauer et al. (1997), Filonenko-Borodich (1967), Girkmann (1974), Göldner et al. (1979), Gross et al. (2014), Hahn (1985), Hake and Meskouris (2007), Leipholz (1974), Massonet (1962), Sadd (2005), Sechler (1952), Timoshenko and Goodier (1951), Wang (1953), among others.

The following relationships exist between the Cartesian coordinates x and y and the polar coordinates r and φ (Fig. 4.1):

$$x = r\cos\varphi, \quad y = r\sin\varphi, \quad r = \sqrt{x^2 + y^2}, \quad \varphi = \arctan\left(\frac{y}{x}\right). \tag{4.1}$$

C. Mittelstedt, *Theory of Plates and Shells*,
https://doi.org/10.1007/978-3-662-66805-4_4

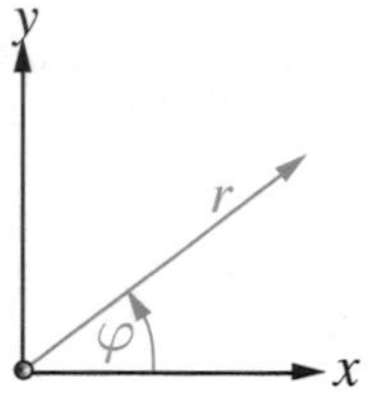

Fig. 4.1 Polar coordinates r, φ

Let the plane stress components σ_{rr}, $\sigma_{\varphi\varphi}$ and $\tau_{r\varphi}$ as well as the plane strain components ε^0_{rr}, $\varepsilon^0_{\varphi\varphi}$ and $\gamma^0_{r\varphi}$ be given. Whether a normal stress σ_{zz} or a normal strain ε_{zz} occurs is determined by the respective conditions. All of the mentioned quantities are constant over the thickness h of the disk and depend only on r and φ.

The equilibrium conditions are:

$$\begin{aligned}\frac{\partial\sigma_{rr}}{\partial r}+\frac{1}{r}\frac{\partial\tau_{r\varphi}}{\partial\varphi}+\frac{\sigma_{rr}-\sigma_{\varphi\varphi}}{r}+f_r &= 0,\\ \frac{\partial\tau_{r\varphi}}{\partial r}+\frac{1}{r}\frac{\sigma_{\varphi\varphi}}{\partial\varphi}+2\frac{\tau_{r\varphi}}{r}+f_\varphi &= 0.\end{aligned} \tag{4.2}$$

Again, we want to use corresponding force flows instead of stresses, which we define as follows:

$$N^0_{rr}=\int_{-\frac{h}{2}}^{\frac{h}{2}}\sigma_{rr}\mathrm{d}z=\sigma_{rr}h,\ N^0_{\varphi\varphi}=\int_{-\frac{h}{2}}^{\frac{h}{2}}\sigma_{\varphi\varphi}\mathrm{d}z=\sigma_{\varphi\varphi}h,\ N^0_{r\varphi}=\int_{-\frac{h}{2}}^{\frac{h}{2}}\tau_{r\varphi}\mathrm{d}z=\tau_{r\varphi}h. \tag{4.3}$$

Thus, we can also write the equilibrium conditions as:

$$\begin{aligned}\frac{\partial N^0_{rr}}{\partial r}+\frac{1}{r}\frac{\partial N^0_{r\varphi}}{\partial\varphi}+\frac{N^0_{rr}-N^0_{\varphi\varphi}}{r}+p_r &= 0,\\ \frac{\partial N^0_{r\varphi}}{\partial r}+\frac{1}{r}\frac{N^0_{\varphi\varphi}}{\partial\varphi}+2\frac{N^0_{r\varphi}}{r}+p_\varphi &= 0,\end{aligned} \tag{4.4}$$

with $p_r = f_r h$ and $p_\varphi = f_\varphi h$.

The kinematic equations result in:

$$\varepsilon^0_{rr}=\frac{\partial u_0}{\partial r},\quad \varepsilon^0_{\varphi\varphi}=\frac{1}{r}\frac{\partial v_0}{\partial\varphi}+\frac{u_0}{r},\quad \gamma^0_{r\varphi}=\frac{\partial v_0}{\partial r}+\frac{1}{r}\frac{\partial u_0}{\partial\varphi}-\frac{v_0}{r}, \tag{4.5}$$

where u_0 is the radial displacement and v_0 is the tangential displacement. The compatibility condition is given here as follows:

$$\frac{\partial^2 \varepsilon_{\varphi\varphi}^0}{\partial r^2} + \frac{1}{r^2}\frac{\partial^2 \varepsilon_{rr}^0}{\partial \varphi^2} + \frac{2}{r}\frac{\partial \varepsilon_{\varphi\varphi}^0}{\partial r} - \frac{1}{r}\frac{\partial \varepsilon_{rr}^0}{\partial r} = \frac{1}{r}\frac{\partial^2 \gamma_{r\varphi}^0}{\partial r \partial \varphi} + \frac{1}{r^2}\frac{\partial \gamma_{r\varphi}^0}{\partial \varphi}. \tag{4.6}$$

Hooke's law is given in the case of plane state of stress as:

$$\begin{pmatrix} \varepsilon_{rr}^0 \\ \varepsilon_{\varphi\varphi}^0 \\ \gamma_{r\varphi}^0 \end{pmatrix} = \begin{bmatrix} \frac{1}{E} & -\frac{\nu}{E} & 0 \\ -\frac{\nu}{E} & \frac{1}{E} & 0 \\ 0 & 0 & \frac{1}{G} \end{bmatrix} \begin{pmatrix} \sigma_{rr} \\ \sigma_{\varphi\varphi} \\ \tau_{r\varphi} \end{pmatrix}. \tag{4.7}$$

If a plane strain state is present, then the equivalent stiffness quantities $\bar{E}$ and $\bar{\nu}$ are to be used instead of E and ν. Formulated in terms of the force flows N_{rr}^0, $N_{\varphi\varphi}^0$, $N_{r\varphi}^0$, Eq. (4.7) reads:

$$\begin{pmatrix} \varepsilon_{rr}^0 \\ \varepsilon_{\varphi\varphi}^0 \\ \gamma_{r\varphi}^0 \end{pmatrix} = \begin{bmatrix} \frac{1}{Eh} & -\frac{\nu}{Eh} & 0 \\ -\frac{\nu}{Eh} & \frac{1}{Eh} & 0 \\ 0 & 0 & \frac{1}{Gh} \end{bmatrix} \begin{pmatrix} N_{rr}^0 \\ N_{\varphi\varphi}^0 \\ N_{r\varphi}^0 \end{pmatrix}, \tag{4.8}$$

or in its inverted form:

$$\begin{pmatrix} N_{rr}^0 \\ N_{\varphi\varphi}^0 \\ N_{r\varphi}^0 \end{pmatrix} = \begin{bmatrix} A & A\nu & 0 \\ A\nu & A & 0 \\ 0 & 0 & A\frac{1-\nu}{2} \end{bmatrix} \begin{pmatrix} \varepsilon_{rr}^0 \\ \varepsilon_{\varphi\varphi}^0 \\ \gamma_{r\varphi}^0 \end{pmatrix}, \tag{4.9}$$

with the disk stiffness $A = \frac{Eh}{1-\nu^2}$.

In the case of a rotationally symmetric disk situation, where all state variables are a function of the radial coordinate r, the given basic equations are significantly reduced. All partial derivatives with respect to φ vanish, and for the equilibrium conditions we obtain $\tau_{r\varphi} = 0$:

$$\frac{\partial \sigma_{rr}}{\partial r} + \frac{\sigma_{rr} - \sigma_{\varphi\varphi}}{r} + f_r = 0, \tag{4.10}$$

or

$$\frac{\partial N_{rr}^0}{\partial r} + \frac{N_{rr}^0 - N_{\varphi\varphi}^0}{r} + p_r = 0. \tag{4.11}$$

The kinematic equations reduce to:

$$\varepsilon_{rr}^0 = \frac{\partial u}{\partial r}, \quad \varepsilon_{\varphi\varphi}^0 = \frac{u}{r}, \quad \gamma_{r\varphi}^0 = 0. \tag{4.12}$$

The following compatibility condition remains:

$$\frac{\partial^2 \varepsilon_{\varphi\varphi}^0}{\partial r^2} + \frac{2}{r} \frac{\varepsilon_{\varphi\varphi}^0}{\partial r} - \frac{1}{r} \frac{\partial \varepsilon_{rr}^0}{\partial r} = 0. \tag{4.13}$$

Hooke's law then reads:

$$\begin{pmatrix} \varepsilon_{rr}^0 \\ \varepsilon_{\varphi\varphi}^0 \end{pmatrix} = \begin{bmatrix} \frac{1}{E} & -\frac{\nu}{E} \\ -\frac{\nu}{E} & \frac{1}{E} \end{bmatrix} \begin{pmatrix} \sigma_{rr} \\ \sigma_{\varphi\varphi} \end{pmatrix}, \tag{4.14}$$

or

$$\begin{pmatrix} \varepsilon_{rr}^0 \\ \varepsilon_{\varphi\varphi}^0 \end{pmatrix} = \begin{bmatrix} \frac{1}{Eh} & -\frac{\nu}{Eh} \\ -\frac{\nu}{Eh} & \frac{1}{Eh} \end{bmatrix} \begin{pmatrix} N_{rr}^0 \\ N_{\varphi\varphi}^0 \end{pmatrix}, \tag{4.15}$$

or in its inverted form:

$$\begin{pmatrix} N_{rr}^0 \\ N_{\varphi\varphi}^0 \end{pmatrix} = \begin{bmatrix} A & A\nu \\ A\nu & A \end{bmatrix} \begin{pmatrix} \varepsilon_{rr}^0 \\ \varepsilon_{\varphi\varphi}^0 \end{pmatrix}. \tag{4.16}$$

4.1.2 The Displacement Method

As already shown for isotropic disks in Cartesian coordinates x, y, also when using polar coordinates r, φ basically two methods of solution can be used, namely on the one hand the displacement method and on the other hand the force method. The equations of the displacement method are obtained, if in the constitutive relations the strains ε_{rr}^0, $\varepsilon_{\varphi\varphi}^0$, $\gamma_{r\varphi}^0$ are expressed by the displacements u_0, v_0, and the resulting expressions for the stresses σ_{rr}, $\sigma_{\varphi\varphi}$, $\tau_{r\varphi}$, respectively the force flows N_{rr}^0, $N_{\varphi\varphi}^0$, $N_{r\varphi}^0$ into the equilibrium conditions. One obtains the following two differential equations in the displacements u_0 and v_0 (see, e.g., Göldner et al. 1979):

$$\begin{aligned} &\frac{\partial^2 u_0}{\partial r^2} + \frac{1}{r}\frac{\partial u_0}{\partial r} - \frac{u_0}{r^2} + \frac{1-\nu}{2r^2}\frac{\partial^2 u_0}{\partial \varphi^2} \\ &- \frac{3-\nu}{2r^2}\frac{\partial v_0}{\partial \varphi} + \frac{1+\nu}{2r}\frac{\partial^2 v_0}{\partial r \partial \varphi} + \frac{1-\nu^2}{E} f_r = 0, \\ &\frac{\partial^2 v_0}{\partial r^2} + \frac{1}{r}\frac{\partial v_0}{\partial r} - \frac{v_0}{r^2} + \frac{2}{1-\nu}\frac{1}{r^2}\frac{\partial^2 v_0}{\partial \varphi^2} \\ &+ \frac{1+\nu}{1-\nu}\frac{1}{r}\frac{\partial^2 u_0}{\partial r \partial \varphi} + \frac{3-\nu}{1-\nu}\frac{1}{r^2}\frac{\partial u_0}{\partial \varphi} + \frac{2(1+\nu)}{E} f_\varphi = 0. \end{aligned} \tag{4.17}$$

4.1.3 The Force Method

We exploit the fact that the bipotential equation $\Delta\Delta F = 0$ is valid independently of the coordinate system under consideration. When moving from Cartesian coordinates x, y to polar coordinates r, φ one obtains the following relations for partial derivatives:

$$\frac{\partial}{\partial x} = \frac{\partial}{\partial r}\frac{\partial r}{\partial x} + \frac{\partial}{\partial \varphi}\frac{\partial \varphi}{\partial x}, \quad \frac{\partial}{\partial y} = \frac{\partial}{\partial r}\frac{\partial r}{\partial y} + \frac{\partial}{\partial \varphi}\frac{\partial \varphi}{\partial y}. \tag{4.18}$$

Then with (4.1):

$$\begin{aligned}
\frac{\partial x}{\partial r} &= \cos\varphi, \quad \frac{\partial x}{\partial \varphi} = -r\sin\varphi, \quad \frac{\partial y}{\partial r} = \sin\varphi, \quad \frac{\partial y}{\partial \varphi} = r\cos\varphi, \\
\frac{\partial r}{\partial x} &= \cos\varphi, \quad \frac{\partial r}{\partial y} = \sin\varphi, \quad \frac{\partial \varphi}{\partial x} = -\frac{1}{r}\sin\varphi, \quad \frac{\partial \varphi}{\partial y} = \frac{1}{r}\cos\varphi.
\end{aligned} \tag{4.19}$$

Thus it follows from (4.18):

$$\begin{aligned}
\frac{\partial}{\partial x} &= \cos\varphi\frac{\partial}{\partial r} - \frac{1}{r}\sin\varphi\frac{\partial}{\partial \varphi}, \\
\frac{\partial}{\partial y} &= \sin\varphi\frac{\partial}{\partial r} + \frac{1}{r}\cos\varphi\frac{\partial}{\partial \varphi},
\end{aligned} \tag{4.20}$$

and the partial derivatives of the Airy stress function F take the following form:

$$\begin{aligned}
\frac{\partial F}{\partial x} &= \cos\varphi\frac{\partial F}{\partial r} - \frac{1}{r}\sin\varphi\frac{\partial F}{\partial \varphi}, \\
\frac{\partial F}{\partial y} &= \sin\varphi\frac{\partial F}{\partial r} + \frac{1}{r}\cos\varphi\frac{\partial F}{\partial \varphi}.
\end{aligned} \tag{4.21}$$

For the second derivatives one obtains:

$$\begin{aligned}
\frac{\partial^2 F}{\partial x^2} &= \frac{\partial^2 F}{\partial r^2}\cos^2\varphi - 2\left(\frac{1}{r}\frac{\partial^2 F}{\partial r\partial\varphi} - \frac{1}{r^2}\frac{\partial F}{\partial \varphi}\right)\sin\varphi\cos\varphi \\
&\quad + \left(\frac{1}{r^2}\frac{\partial^2 F}{\partial \varphi^2} + \frac{1}{r}\frac{\partial F}{\partial r}\right)\sin^2\varphi, \\
\frac{\partial^2 F}{\partial y^2} &= \frac{\partial^2 F}{\partial r^2}\sin^2\varphi + 2\left(\frac{1}{r}\frac{\partial^2 F}{\partial r\partial\varphi} - \frac{1}{r^2}\frac{\partial F}{\partial \varphi}\right)\sin\varphi\cos\varphi \\
&\quad + \left(\frac{1}{r^2}\frac{\partial^2 F}{\partial \varphi^2} + \frac{1}{r}\frac{\partial F}{\partial r}\right)\cos^2\varphi, \\
\frac{\partial^2 F}{\partial x\partial y} &= \left(\frac{\partial^2 F}{\partial r^2} - \frac{1}{r}\frac{\partial F}{\partial r} - \frac{1}{r^2}\frac{\partial^2 F}{\partial \varphi^2}\right)\sin\varphi\cos\varphi \\
&\quad + \left(\frac{1}{r^2}\frac{\partial F}{\partial \varphi} - \frac{1}{r}\frac{\partial^2 F}{\partial r\partial\varphi}\right)\left(\sin^2\varphi - \cos^2\varphi\right).
\end{aligned} \tag{4.22}$$

In polar coordinates the Laplace operator Δ is thus:

$$\Delta = \frac{\partial^2}{\partial r^2} + \frac{1}{r^2}\frac{\partial^2}{\partial \varphi^2} + \frac{1}{r}\frac{\partial}{\partial r}, \tag{4.23}$$

so that the disk equation assumes the following form for a constant disk thickness and constant volume forces (Girkmann 1974):

$$\Delta\Delta F = \left(\frac{\partial^2}{\partial r^2} + \frac{1}{r^2}\frac{\partial^2}{\partial \varphi^2} + \frac{1}{r}\frac{\partial}{\partial r}\right)\left(\frac{\partial^2 F}{\partial r^2} + \frac{1}{r^2}\frac{\partial^2 F}{\partial \varphi^2} + \frac{1}{r}\frac{\partial F}{\partial r}\right) = 0. \tag{4.24}$$

The stresses $\sigma_{rr}, \sigma_{\varphi\varphi}, \tau_{r\varphi}$ are to be derived from the Airy stress function as follows:

$$\sigma_{rr} = \frac{1}{r^2}\frac{\partial^2 F}{\partial \varphi^2} + \frac{1}{r}\frac{\partial F}{\partial r}, \quad \sigma_{\varphi\varphi} = \frac{\partial^2 F}{\partial r^2}, \quad \tau_{r\varphi} = -\frac{\partial}{\partial r}\left(\frac{1}{r}\frac{\partial F}{\partial \varphi}\right). \tag{4.25}$$

In the case of a rotationally symmetric disk situation, the disk equation reduces significantly to:

$$\frac{\mathrm{d}^4 F}{\mathrm{d}r^4} + \frac{2}{r}\frac{\mathrm{d}^3 F}{\mathrm{d}r^3} - \frac{1}{r^2}\frac{\mathrm{d}^2 F}{\mathrm{d}r^2} + \frac{1}{r^3}\frac{\mathrm{d}F}{\mathrm{d}r} = 0. \tag{4.26}$$

This is an Eulerian differential equation, the solution of which will be discussed at a later point.

Solutions of the disk equation (4.24) can be obtained quite analogously to disk problems in Cartesian coordinates. We first consider again the potential equation $\Delta\bar{F} = 0$ such that $\Delta F = \bar{F}$ and use the approach $\bar{F} = F_1(r)F_2(\varphi)$ (Altenbach et al. 2016). Then from

$$\Delta\bar{F} = \frac{\partial^2 \bar{F}}{\partial r^2} + \frac{1}{r^2}\frac{\partial^2 \bar{F}}{\partial \varphi^2} + \frac{1}{r}\frac{\partial \bar{F}}{\partial r} = 0 \tag{4.27}$$

after multiplication by $\frac{r^2}{F_1 F_2}$ the following representation results:

$$r^2\frac{1}{F_1}\frac{\mathrm{d}^2 F_1}{\mathrm{d}r^2} + r\frac{1}{F_1}\frac{\mathrm{d}F_1}{\mathrm{d}r} + \frac{1}{F_2}\frac{\mathrm{d}^2 F_2}{\mathrm{d}\varphi^2} = 0, \tag{4.28}$$

which we can also write as follows:

$$r^2\frac{1}{F_1}\frac{\mathrm{d}^2 F_1}{\mathrm{d}r^2} + r\frac{1}{F_1}\frac{\mathrm{d}F_1}{\mathrm{d}r} = -\frac{1}{F_2}\frac{\mathrm{d}^2 F_2}{\mathrm{d}\varphi^2} = -\lambda^2. \tag{4.29}$$

This describes two differential equations for the two unknown functions $F_1(r)$ and $F_2(\varphi)$. The differential equation concerning F_1 can be expressed as:

$$r^2 \frac{\mathrm{d}^2 F_1}{\mathrm{d}r^2} + r \frac{\mathrm{d}F_1}{\mathrm{d}r} + \lambda^2 F_1 = 0. \tag{4.30}$$

This is an Eulerian differential equation whose general solution for $\lambda^2 > 0$ is as follows:

$$F_1(r) = C_1 r^{\lambda} + C_2 r^{-\lambda}. \tag{4.31}$$

For $\lambda^2 = 0$, on the other hand, we obtain:

$$F_1(r) = C_1 + C_2 \ln r. \tag{4.32}$$

The linear differential equation appearing in (4.29) concerning $F_2(\varphi)$ is:

$$\frac{\mathrm{d}^2 F_2}{\mathrm{d}\varphi^2} - \lambda^2 F_2 = 0. \tag{4.33}$$

Their general solution is in the case $\lambda^2 > 0$:

$$F_2(\varphi) = D_1 \cos \lambda\varphi + D_2 \sin \lambda\varphi. \tag{4.34}$$

If $\lambda^2 = 0$, then:

$$F_2(\varphi) = D_1 + D_2 \varphi. \tag{4.35}$$

Thus the function $\bar{F}$ can be given for different values of λ^2. These solutions can be linearly combined with each other arbitrarily. A complete set of solutions is then obtained from the requirement $\Delta F = \bar{F}$, which leads to an inhomogeneous partial differential equation for the Airy stress function.

In Girkmann (1974) the following functions are given for which the disk equation is satisfied:

$$r^2, \quad \ln r, \quad r^2 \ln r, \quad \varphi, \quad \varphi^2, \quad \varphi^3, \quad r^2\varphi, \quad \varphi \ln r, \quad r^2 \varphi \ln r, \quad \sin 2\varphi, \ \cos 2\varphi,$$
$$r\varphi \sin \varphi, \quad r\varphi \cos \varphi, \quad r \ln r \sin \varphi, \quad r \ln r \cos \varphi, \quad \cos(n \ln r) \cosh n\varphi,$$
$$r^2 \cos(n \ln r) \cosh n\varphi, \ldots \tag{4.36}$$

Girkmann also suggests the following way to form solutions of the disk equations, using the fact that any biharmonic function F can be represented as:

$$F = \bar{F}_1 + r^2 \bar{F}_2, \tag{4.37}$$

where $\bar{F}_1$ and $\bar{F}_2$ are harmonic functions, i.e. they satisfy the condition $\Delta \bar{F} = 0$. In polar coordinates the following representation is valid for harmonic functions:

$$\bar{F} = r^n \left(C_1 \cos n\varphi + C_2 \sin n\varphi \right). \tag{4.38}$$

Thus, the following solutions of the disk equation can be formed:

$$\begin{aligned} F &= \sum_n \left(A_n r^n + B_n r^{-n} + C_n r^{n+2} + D_n r^{-n+2} \right) \cos n\varphi \\ &\quad + \sum_n \left(\bar{A}_n r^n + \bar{B}_n r^{-n} + \bar{C}_n r^{n+2} + \bar{D}_n r^{-n+2} \right) \sin n\varphi, \end{aligned} \tag{4.39}$$

where $n = 2, 3, 4, \ldots$. For $n = 1$ the terms multiplied by A_1 and D_1 or $\bar{A}_1$ and $\bar{D}_1$ match, so the following formulation is used:

$$\begin{aligned} F_1 &= \left(A_1 r + B_1 r^{-1} + C_1 r^3 + D_1 r \ln r \right) \cos n\varphi \\ &\quad + \left(\bar{A}_1 r + \bar{B}_1 r^{-1} + \bar{C}_1 r^3 + \bar{D}_1 r \ln r \right) \sin n\varphi. \end{aligned} \tag{4.40}$$

Further solutions of the disk equation can be constructed by also considering the case where all state quantities depend exclusively on φ. The disk equation can then be stated as:

$$\frac{\mathrm{d}^4 F}{\mathrm{d}\varphi^4} + 4 \frac{\mathrm{d}^2 F}{\mathrm{d}\varphi^2} = 0. \tag{4.41}$$

The solution of this ordinary fourth-order homogeneous linear differential equation can be given as:

$$F(\varphi) = D_1 \varphi + D_2 \varphi^2 + D_3 \cos 2\varphi + D_4 \sin 2\varphi. \tag{4.42}$$

Göldner et al. (1979) give other classes of solutions for disk problems in polar coordinates. In the case of a rotationally symmetric disk situation where all state variables depend exclusively on r, the general solution of the reduced form of the disk equation (4.26) can be given as:

$$F(r) = C_1 + C_2 \ln r + C_3 r^2 + C_4 r^2 \ln r. \tag{4.43}$$

4.2 Energetic Consideration

4.2.1 Strain Energy

For the application of energy-based computational methods the strain energy of the isotropic disk must be provided also in polar coordinates. We consider a disk of thickness h bounded by the radial coordinates $r = R_i$ and $r = R_a$ and the two

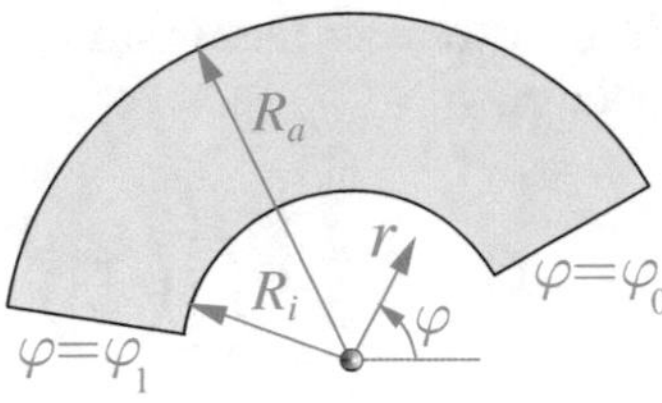

Fig. 4.2 Circular disk

circumferential coordinates $\varphi = \varphi_0$ and $\varphi = \varphi_1$ (see Fig. 4.2). The strain energy Π_i can be given in general terms for this case as:

$$\Pi_i = \frac{1}{2}\int_V \left(\sigma_{rr}\varepsilon_{rr}^0 + \sigma_{\varphi\varphi}\varepsilon_{\varphi\varphi}^0 + \tau_{r\varphi}\gamma_{r\varphi}^0\right) \mathrm{d}V. \tag{4.44}$$

With $\mathrm{d}V = r\mathrm{d}\varphi\mathrm{d}r\mathrm{d}z$ we obtain:

$$\Pi_i = \frac{1}{2}\int_{-\frac{h}{2}}^{+\frac{h}{2}}\int_{R_i}^{R_a}\int_{\varphi_0}^{\varphi_1} \left(\sigma_{rr}\varepsilon_{rr}^0 + \sigma_{\varphi\varphi}\varepsilon_{\varphi\varphi}^0 + \tau_{r\varphi}\gamma_{r\varphi}^0\right) r\mathrm{d}\varphi\mathrm{d}r\mathrm{d}z. \tag{4.45}$$

The integration with respect to the z-direction results in a representation in the force flows N_{rr}^0, $N_{\varphi\varphi}^0$, $N_{r\varphi}^0$:

$$\Pi_i = \frac{1}{2}\int_{R_i}^{R_a}\int_{\varphi_0}^{\varphi_1} \left(N_{rr}^0\varepsilon_{rr}^0 + N_{\varphi\varphi}^0\varepsilon_{\varphi\varphi}^0 + N_{r\varphi}^0\gamma_{r\varphi}^0\right) r\mathrm{d}\varphi\mathrm{d}r. \tag{4.46}$$

Replacing the strain quantities with the force flows N_{rr}^0, $N_{\varphi\varphi}^0$, $N_{r\varphi}^0$, we obtain the following representation:

$$\begin{aligned}\Pi_i &= \frac{1}{2Eh}\int_{R_i}^{R_a}\int_{\varphi_0}^{\varphi_1} \Big[\left(N_{rr}^0\right)^2 + \left(N_{\varphi\varphi}^0\right)^2 - 2\nu N_{rr}^0 N_{\varphi\varphi}^0 \\ &\quad + 2\left(1+\nu\right)\left(N_{r\varphi}^0\right)^2\Big] r\mathrm{d}\varphi\mathrm{d}r.\end{aligned} \tag{4.47}$$

Similarly, in (4.46) the force flows N_{rr}^0, $N_{\varphi\varphi}^0$, $N_{r\varphi}^0$ can be expressed by the strains ε_{rr}^0, $\varepsilon_{\varphi\varphi}^0$, $\gamma_{r\varphi}^0$:

$$\Pi_i = \frac{A}{2}\int_{R_i}^{R_a}\int_{\varphi_0}^{\varphi_1} \left[\left(\varepsilon_{rr}^0\right)^2 + \left(\varepsilon_{\varphi\varphi}^0\right)^2 + 2\nu\varepsilon_{rr}^0\varepsilon_{\varphi\varphi}^0 + \frac{1-\nu}{2}\left(\gamma_{r\varphi}^0\right)^2\right] r\mathrm{d}\varphi\mathrm{d}r. \tag{4.48}$$

If one expresses the strains ε_{rr}^0, $\varepsilon_{\varphi\varphi}^0$, $\gamma_{r\varphi}^0$ by the displacements u_0 and v_0, then one obtains:

$$\begin{aligned}\Pi_i &= \frac{A}{2}\int_{R_i}^{R_a}\int_{\varphi_0}^{\varphi_1}\Bigg[\left(\frac{\partial u_0}{\partial r}\right)^2+\frac{1}{r^2}\left(\frac{\partial v_0}{\partial \varphi}\right)^2+\frac{2}{r^2}u_0\frac{\partial v_0}{\partial \varphi}+\left(\frac{u_0}{r}\right)^2+\frac{2\nu}{r}\frac{\partial u_0}{\partial r}\frac{\partial v_0}{\partial \varphi}\\ &+2\nu\frac{u_0}{r}\frac{\partial u_0}{\partial r}+\frac{1-\nu}{2}\left(\frac{\partial v_0}{\partial r}\right)^2+\frac{1-\nu}{2}\frac{1}{r^2}\left(\frac{\partial u_0}{\partial \varphi}\right)^2+\frac{1-\nu}{2}\left(\frac{v_0}{r}\right)^2\\ &-(1-\nu)\frac{1}{r}\frac{\partial v_0}{\partial r}\frac{\partial u_0}{\partial \varphi}-(1-\nu)\frac{v_0}{r}\frac{\partial v_0}{\partial r}-(1-\nu)\frac{v_0}{r^2}\frac{\partial u_0}{\partial \varphi}\Bigg]r\mathrm{d}\varphi\mathrm{d}r. \qquad (4.49)\end{aligned}$$

4.2.2 Energetic Derivation of the Basic Equations

We use the principle of virtual displacements $\delta W_i = \delta W_a$ and first consider the virtual internal work δW_i, which can be given as follows:

$$\delta W_i = \int_{R_i}^{R_a}\int_{\varphi_0}^{\varphi_1}\left(N_{rr}^0\delta\varepsilon_{rr}^0+N_{\varphi\varphi}^0\delta\varepsilon_{\varphi\varphi}^0+N_{r\varphi}^0\delta\gamma_{r\varphi}^0\right)r\mathrm{d}\varphi\mathrm{d}r. \qquad (4.50)$$

The virtual strain quantities $\delta\varepsilon_{rr}^0$, $\delta\varepsilon_{\varphi\varphi}^0$, $\delta\gamma_{r\varphi}^0$ can be specified as:

$$\delta\varepsilon_{rr}^0=\frac{\partial\delta u_0}{\partial r},\quad \delta\varepsilon_{\varphi\varphi}^0=\frac{1}{r}\frac{\partial\delta v_0}{\partial\varphi}+\frac{\delta u_0}{r},\quad \delta\gamma_{r\varphi}^0=\frac{\partial\delta v_0}{\partial r}+\frac{1}{r}\frac{\partial\delta u_0}{\partial\varphi}-\frac{\delta v_0}{r}. \qquad (4.51)$$

Partial integration of the individual terms in (4.50) with (4.51) yields:

$$\begin{aligned}\int_{R_i}^{R_a}\int_{\varphi_0}^{\varphi_1} N_{rr}^0\frac{\partial\delta u_0}{\partial r}r\mathrm{d}\varphi\mathrm{d}r &= \int_{\varphi_0}^{\varphi_1} N_{rr}^0 r\delta u_0\Big|_{R_i}^{R_a}\mathrm{d}\varphi-\int_{R_i}^{R_a}\int_{\varphi_0}^{\varphi_1}\frac{\partial}{\partial r}\left(N_{rr}^0 r\right)\delta u_0\mathrm{d}\varphi\mathrm{d}r,\\ \int_{R_i}^{R_a}\int_{\varphi_0}^{\varphi_1} N_{\varphi\varphi}^0\frac{\partial\delta v_0}{\partial \varphi}\mathrm{d}\varphi\mathrm{d}r &= \int_{R_i}^{R_a} N_{\varphi\varphi}^0\delta v_0\Big|_{\varphi_0}^{\varphi_1}\mathrm{d}r-\int_{R_i}^{R_a}\int_{\varphi_0}^{\varphi_1}\frac{\partial N_{\varphi\varphi}^0}{\partial \varphi}\delta v_0\mathrm{d}\varphi\mathrm{d}r,\\ \int_{R_i}^{R_a}\int_{\varphi_0}^{\varphi_1} \frac{\partial\delta v_0}{\partial r}rN_{r\varphi}^0\mathrm{d}\varphi\mathrm{d}r &= \int_{\varphi_0}^{\varphi_1} rN_{r\varphi}^0\delta v_0\Big|_{R_i}^{R_a}\mathrm{d}\varphi-\int_{R_i}^{R_a}\int_{\varphi_0}^{\varphi_1}\frac{\partial}{\partial r}\left(rN_{r\varphi}^0\right)\delta v_0\mathrm{d}\varphi\mathrm{d}r,\\ \int_{R_i}^{R_a}\int_{\varphi_0}^{\varphi_1} N_{r\varphi}^0\frac{\partial\delta u_0}{\partial \varphi}\mathrm{d}\varphi\mathrm{d}r &= \int_{R_i}^{R_a} N_{r\varphi}^0\delta u_0\Big|_{\varphi_0}^{\varphi_1}\mathrm{d}r-\int_{R_i}^{R_a}\int_{\varphi_0}^{\varphi_1}\frac{\partial N_{r\varphi}^0}{\partial \varphi}\delta u_0\mathrm{d}\varphi\mathrm{d}r. \qquad (4.52)\end{aligned}$$

Now the boundary force flows $\hat{N}_{rr}^0$ and $\hat{N}_{\varphi\varphi}^0$ are prescribed on the edges $r = R_i$, $r = R_a$ and $\varphi = \varphi_0$, $\varphi = \varphi_1$, respectively. In addition, the boundary shear force flows $\hat{N}_{r\varphi}^0$ are prescribed on these edges. Likewise, the volume forces p_r and $p_{r\varphi}$ are present. Then for the virtual work δW_a we obtain:

$$
\begin{aligned}
\delta W_a &= \int_{R_i}^{R_a} \hat{N}^0_{\varphi\varphi}\delta v_0\Big|_{\varphi_0}^{\varphi_1}\mathrm{d}r + \int_{R_i}^{R_a} \hat{N}^0_{r\varphi}\delta u_0\Big|_{\varphi_0}^{\varphi_1}\mathrm{d}r + \int_{\varphi_0}^{\varphi_1} r\hat{N}^0_{rr}\delta u_0\Big|_{R_i}^{R_a}\mathrm{d}\varphi \\
&+ \int_{\varphi_0}^{\varphi_1} r\hat{N}^0_{r\varphi}\delta v_0\Big|_{R_i}^{R_a}\mathrm{d}\varphi \\
&+ \int_{R_i}^{R_a}\int_{\varphi_0}^{\varphi_1} p_r\delta u_0 r\mathrm{d}\varphi\mathrm{d}r + \int_{R_i}^{R_a}\int_{\varphi_0}^{\varphi_1} p_\varphi\delta v_0 r\mathrm{d}\varphi\mathrm{d}r.
\end{aligned}
\tag{4.53}
$$

Sorting the respective terms by the variations δu_0 and δv_0 yields:

$$
\begin{aligned}
&-\int_{R_i}^{R_a}\int_{\varphi_0}^{\varphi_1}\left(\frac{\partial N^0_{rr}}{\partial r} + \frac{1}{r}\frac{\partial N^0_{r\varphi}}{\partial\varphi} + \frac{N^0_{rr}-N^0_{\varphi\varphi}}{r} + p_r\right)\delta u_0 r\mathrm{d}\varphi\mathrm{d}r \\
&-\int_{R_i}^{R_a}\int_{\varphi_0}^{\varphi_1}\left(\frac{\partial N^0_{r\varphi}}{\partial r} + \frac{1}{r}\frac{\partial N^0_{\varphi\varphi}}{\partial\varphi} + \frac{2}{r}N^0_{r\varphi} + p_\varphi\right)\delta v_0 r\mathrm{d}\varphi\mathrm{d}r \\
&+\int_{R_i}^{R_a}\left(N^0_{\varphi\varphi}-\hat{N}^0_{\varphi\varphi}\right)\delta v_0\Big|_{\varphi_0}^{\varphi_1}\mathrm{d}r + \int_{R_i}^{R_a}\left(N^0_{r\varphi}-\hat{N}^0_{r\varphi}\right)\delta u_0\Big|_{\varphi_0}^{\varphi_1}\mathrm{d}r \\
&+\int_{\varphi_0}^{\varphi_1} r\left(N^0_{rr}-\hat{N}^0_{rr}\right)\delta u_0\Big|_{R_i}^{R_a}\mathrm{d}\varphi + \int_{\varphi_0}^{\varphi_1} r\left(N^0_{r\varphi}-\hat{N}^0_{r\varphi}\right)\delta v_0\Big|_{R_i}^{R_a}\mathrm{d}\varphi = 0.
\end{aligned}
\tag{4.54}
$$

From the double integral terms in (4.54), the equilibrium conditions (4.4) can be confirmed. The boundary terms, on the other hand, express that either the boundary loads $\hat{N}^0_{rr}$, $\hat{N}^0_{\varphi\varphi}$, $\hat{N}^0_{r\varphi}$ on the boundaries in question coincide with the internal force flows N^0_{rr}, $N^0_{\varphi\varphi}$, $N^0_{r\varphi}$, or that the virtual displacements become zero.

The equations derived with the principle of virtual displacements are formulated in the force quantities N^0_{rr}, $N^0_{\varphi\varphi}$, $N^0_{r\varphi}$ and the corresponding loads. A formulation in the displacements u_0 and v_0 is obtained by applying the principle of the minimum of the total elastic potential. The inner potential Π_i has already been provided by (4.49). If now again the boundary loads $\hat{N}^0_{rr}$, $\hat{N}^0_{\varphi\varphi}$, $\hat{N}^0_{r\varphi}$ as well as the volume loads p_r and p_φ are present, then the external potential Π_a can be represented as:

$$
\begin{aligned}
\Pi_a &= -\int_{R_i}^{R_a}\int_{\varphi_0}^{\varphi_1} p_r u_0 r\mathrm{d}\varphi\mathrm{d}r - \int_{R_i}^{R_a}\int_{\varphi_0}^{\varphi_1} p_\varphi v_0 r\mathrm{d}\varphi\mathrm{d}r \\
&- \int_{R_i}^{R_a} \hat{N}^0_{\varphi\varphi} v_0\Big|_{\varphi_0}^{\varphi_1}\mathrm{d}r - \int_{R_i}^{R_a} \hat{N}^0_{r\varphi} u_0\Big|_{\varphi_0}^{\varphi_1}\mathrm{d}r - \int_{\varphi_0}^{\varphi_1} r\hat{N}^0_{rr} u_0\Big|_{R_i}^{R_a}\mathrm{d}\varphi
\end{aligned}
$$

$$-\int_{\varphi_0}^{\varphi_1} r\hat{N}^0_{r\varphi} v_0 \Big|_{R_i}^{R_a} \mathrm{d}\varphi. \tag{4.55}$$

The principle of the minimum of the total elastic potential $\delta\Pi = \delta\Pi_i + \delta\Pi_a = 0$ requires forming the first variation $\delta\Pi$ of the total elastic potential. For $\delta\Pi_a$ we obtain from (4.55):

$$\begin{aligned}
\delta\Pi_a = & -\int_{R_i}^{R_a}\int_{\varphi_0}^{\varphi_1} p_r \delta u_0 r \mathrm{d}\varphi \mathrm{d}r - \int_{R_i}^{R_a}\int_{\varphi_0}^{\varphi_1} p_\varphi \delta v_0 r \mathrm{d}\varphi \mathrm{d}r \\
& -\int_{R_i}^{R_a} \hat{N}^0_{\varphi\varphi} \delta v_0 \Big|_{\varphi_0}^{\varphi_1} \mathrm{d}r - \int_{R_i}^{R_a} \hat{N}^0_{r\varphi} \delta u_0 \Big|_{\varphi_0}^{\varphi_1} \mathrm{d}r - \int_{\varphi_0}^{\varphi_1} r\hat{N}^0_{rr} \delta u_0 \Big|_{R_i}^{R_a} \mathrm{d}\varphi \\
& -\int_{\varphi_0}^{\varphi_1} r\hat{N}^0_{r\varphi} \delta v_0 \Big|_{R_i}^{R_a} \mathrm{d}\varphi.
\end{aligned} \tag{4.56}$$

For forming $\delta\Pi_i$ we perform term for term variation as well as partial integration if necessary to reduce the degree of the derivative for δu_0 and δv_0. This results in:

$$\begin{aligned}
\frac{A}{2}\delta \int_{R_i}^{R_a}\int_{\varphi_0}^{\varphi_1} \left(\frac{\partial u_0}{\partial r}\right)^2 r \mathrm{d}\varphi \mathrm{d}r &= \frac{A}{2}\int_{R_i}^{R_a}\int_{\varphi_0}^{\varphi_1} 2\frac{\partial u_0}{\partial r}\frac{\partial \delta u_0}{\partial r} r \mathrm{d}\varphi \mathrm{d}r \\
&= \frac{A}{2}\int_{\varphi_0}^{\varphi_1} 2r\frac{\partial u_0}{\partial r}\delta u_0 \Big|_{R_i}^{R_a} \mathrm{d}\varphi \\
&\quad -\frac{A}{2}\int_{R_i}^{R_a}\int_{\varphi_0}^{\varphi_1} 2\frac{\partial}{\partial r}\left(r\frac{\partial u_0}{\partial r}\right)\delta u_0 \mathrm{d}\varphi \mathrm{d}r, \\
\frac{A}{2}\delta \int_{R_i}^{R_a}\int_{\varphi_0}^{\varphi_1} \frac{1}{r^2}\left(\frac{\partial v_0}{\partial \varphi}\right)^2 r \mathrm{d}\varphi \mathrm{d}r &= \frac{A}{2}\int_{R_i}^{R_a}\int_{\varphi_0}^{\varphi_1} \frac{2}{r}\frac{\partial v_0}{\partial \varphi}\frac{\partial \delta v_0}{\partial \varphi} \mathrm{d}\varphi \mathrm{d}r \\
&= \frac{A}{2}\int_{R_i}^{R_a} \frac{2}{r}\frac{\partial v_0}{\partial \varphi}\delta v_0 \Big|_{\varphi_0}^{\varphi_1} \mathrm{d}r \\
&\quad -\frac{A}{2}\int_{R_i}^{R_a}\int_{\varphi_0}^{\varphi_1} \frac{2}{r}\frac{\partial^2 v_0}{\partial \varphi^2}\delta v_0 \mathrm{d}\varphi \mathrm{d}r,
\end{aligned}$$

$$
\begin{aligned}
\frac{A}{2}\delta\int_{R_i}^{R_a}\int_{\varphi_0}^{\varphi_1}\frac{2}{r}u_0\frac{\partial v_0}{\partial\varphi}\mathrm{d}\varphi\mathrm{d}r &= \frac{A}{2}\int_{R_i}^{R_a}\int_{\varphi_0}^{\varphi_1}\frac{2}{r}\delta u_0\frac{\partial v_0}{\partial\varphi}\mathrm{d}\varphi\mathrm{d}r \\
&\quad+\frac{A}{2}\int_{R_i}^{R_a}\int_{\varphi_0}^{\varphi_1}\frac{2}{r}u_0\frac{\partial\delta v_0}{\partial\varphi}\mathrm{d}\varphi\mathrm{d}r \\
&= \frac{A}{2}\int_{R_i}^{R_a}\int_{\varphi_0}^{\varphi_1}\frac{2}{r}\frac{\partial v_0}{\partial\varphi}\delta u_0\mathrm{d}\varphi\mathrm{d}r \\
&\quad+\frac{A}{2}\int_{R_i}^{R_a}\frac{2}{r}u_0\delta v_0\Big|_{\varphi_0}^{\varphi_1}\mathrm{d}\varphi\mathrm{d}r \\
&\quad-\frac{A}{2}\int_{R_i}^{R_a}\int_{\varphi_0}^{\varphi_1}\frac{2}{r}\frac{\partial u_0}{\partial\varphi}\delta v_0\mathrm{d}\varphi\mathrm{d}r,
\end{aligned}
$$

$$
\frac{A}{2}\delta\int_{R_i}^{R_a}\int_{\varphi_0}^{\varphi_1}\left(\frac{u_0}{r}\right)^2 r\mathrm{d}\varphi\mathrm{d}r = \frac{A}{2}\int_{R_i}^{R_a}\int_{\varphi_0}^{\varphi_1}\frac{2u_0}{r}\delta u_0\mathrm{d}\varphi\mathrm{d}r,
$$

$$
\begin{aligned}
\frac{A}{2}\delta\int_{R_i}^{R_a}\int_{\varphi_0}^{\varphi_1}\frac{2\nu}{r}\frac{\partial u_0}{\partial r}\frac{\partial v_0}{\partial\varphi}r\mathrm{d}\varphi\mathrm{d}r &= \frac{A}{2}\int_{R_i}^{R_a}\int_{\varphi_0}^{\varphi_1}2\nu\frac{\partial\delta u_0}{\partial r}\frac{\partial v_0}{\partial\varphi}\mathrm{d}\varphi\mathrm{d}r \\
&\quad+\frac{A}{2}\int_{R_i}^{R_a}\int_{\varphi_0}^{\varphi_1}2\nu\frac{\partial u_0}{\partial r}\frac{\partial\delta v_0}{\partial\varphi}\mathrm{d}\varphi\mathrm{d}r \\
&= \frac{A}{2}\int_{\varphi_0}^{\varphi_1}2\nu\frac{\partial v_0}{\partial\varphi}\delta u_0\Big|_{R_i}^{R_a}\mathrm{d}\varphi \\
&\quad-\frac{A}{2}\int_{R_i}^{R_a}\int_{\varphi_0}^{\varphi_1}2\nu\frac{\partial^2 v_0}{\partial r\partial\varphi}\delta u_0\mathrm{d}\varphi\mathrm{d}r \\
&\quad+\frac{A}{2}\int_{R_i}^{R_a}2\nu\frac{\partial u_0}{\partial r}\delta v_0\Big|_{\varphi_0}^{\varphi_1}\mathrm{d}r \\
&\quad-\frac{A}{2}\int_{R_i}^{R_a}\int_{\varphi_0}^{\varphi_1}2\nu\frac{\partial^2 u_0}{\partial r\partial\varphi}\delta v_0\mathrm{d}\varphi\mathrm{d}r,
\end{aligned}
$$

$$
\begin{aligned}
\frac{A}{2}\delta\int_{R_i}^{R_a}\int_{\varphi_0}^{\varphi_1} 2\nu\frac{u_0}{r}\frac{\partial u_0}{\partial r} r\mathrm{d}\varphi\mathrm{d}r &= \frac{A}{2}\int_{R_i}^{R_a}\int_{\varphi_0}^{\varphi_1} 2\nu\frac{\partial u_0}{\partial r}\delta u_0\mathrm{d}\varphi\mathrm{d}r \\
&+\frac{A}{2}\int_{R_i}^{R_a}\int_{\varphi_0}^{\varphi_1} 2\nu u_0\frac{\partial \delta u_0}{\partial r}\mathrm{d}\varphi\mathrm{d}r \\
&= \frac{A}{2}\int_{\varphi_0}^{\varphi_1} 2\nu u_0\delta u_0|_{R_i}^{R_a}\,\mathrm{d}\varphi,
\end{aligned}
$$

$$
\begin{aligned}
\frac{A}{2}\delta\int_{R_i}^{R_a}\int_{\varphi_0}^{\varphi_1} \frac{1-\nu}{2}\left(\frac{\partial v_0}{\partial r}\right)^2 r\mathrm{d}\varphi\mathrm{d}r &= \frac{A}{2}\int_{R_i}^{R_a}\int_{\varphi_0}^{\varphi_1} (1-\nu)\,\frac{\partial v_0}{\partial r}\frac{\partial \delta v_0}{\partial r} r\mathrm{d}\varphi\mathrm{d}r \\
&= \frac{A}{2}\int_{\varphi_0}^{\varphi_1} (1-\nu)\,r\frac{\partial v_0}{\partial r}\delta v_0\bigg|_{R_i}^{R_a}\,\mathrm{d}\varphi \\
&-\frac{A}{2}\int_{R_i}^{R_a}\int_{\varphi_0}^{\varphi_1} (1-\nu)\,\frac{\partial}{\partial r}\left(r\frac{\partial v_0}{\partial r}\right)\delta v_0\mathrm{d}\varphi\mathrm{d}r,
\end{aligned}
$$

$$
\begin{aligned}
\frac{A}{2}\delta\int_{R_i}^{R_a}\int_{\varphi_0}^{\varphi_1} \frac{1-\nu}{2}\frac{1}{r^2}\left(\frac{\partial u_0}{\partial \varphi}\right)^2 r\mathrm{d}\varphi\mathrm{d}r &= \frac{A}{2}\int_{R_i}^{R_a}\int_{\varphi_0}^{\varphi_1} (1-\nu)\,\frac{1}{r}\frac{\partial u_0}{\partial \varphi}\frac{\partial \delta u_0}{\partial \varphi}\mathrm{d}\varphi\mathrm{d}r \\
&= \frac{A}{2}\int_{R_i}^{R_a} (1-\nu)\,\frac{1}{r}\frac{\partial u_0}{\partial \varphi}\delta u_0\bigg|_{\varphi_0}^{\varphi_1}\,\mathrm{d}r \\
&-\frac{A}{2}\int_{R_i}^{R_a}\int_{\varphi_0}^{\varphi_1} (1-\nu)\,\frac{1}{r}\frac{\partial^2 u_0}{\partial \varphi^2}\delta u_0\mathrm{d}\varphi\mathrm{d}r,
\end{aligned}
$$

$$
\frac{A}{2}\delta\int_{R_i}^{R_a}\int_{\varphi_0}^{\varphi_1} \frac{1-\nu}{2}\left(\frac{v_0}{r}\right)^2 r\mathrm{d}\varphi\mathrm{d}r = \frac{A}{2}\int_{R_i}^{R_a}\int_{\varphi_0}^{\varphi_1} (1-\nu)\,\frac{v_0}{r}\delta v_0\mathrm{d}\varphi\mathrm{d}r,
$$

$$
\begin{aligned}
\frac{A}{2}\delta\int_{R_i}^{R_a}\int_{\varphi_0}^{\varphi_1} (1-\nu)\,\frac{1}{r}\frac{\partial v_0}{\partial r}\frac{\partial u_0}{\partial \varphi} r\mathrm{d}\varphi\mathrm{d}r &= \frac{A}{2}\int_{R_i}^{R_a}\int_{\varphi_0}^{\varphi_1} (1-\nu)\,\frac{\partial \delta v_0}{\partial r}\frac{\partial u_0}{\partial \varphi}\mathrm{d}\varphi\mathrm{d}r \\
+\frac{A}{2}\int_{R_i}^{R_a}\int_{\varphi_0}^{\varphi_1} (1-\nu)\,\frac{\partial v_0}{\partial r}\frac{\partial \delta u_0}{\partial \varphi}\mathrm{d}\varphi\mathrm{d}r &= \frac{A}{2}\int_{\varphi_0}^{\varphi_1} (1-\nu)\,\frac{\partial u_0}{\partial \varphi}\delta v_0\bigg|_{R_i}^{R_a}\,\mathrm{d}\varphi
\end{aligned}
$$

$$
-\frac{A}{2}\int_{R_i}^{R_a}\int_{\varphi_0}^{\varphi_1}(1-\nu)\frac{\partial^2 u_0}{\partial r\partial\varphi}\delta v_0 \mathrm{d}\varphi\mathrm{d}r \qquad +\frac{A}{2}\int_{R_i}^{R_a}(1-\nu)\frac{\partial v_0}{\partial r}\delta u_0\Big|_{\varphi_0}^{\varphi_1}\mathrm{d}\varphi\mathrm{d}r
$$

$$
-\frac{A}{2}\int_{R_i}^{R_a}\int_{\varphi_0}^{\varphi_1}(1-\nu)\frac{\partial^2 v_0}{\partial r\partial\varphi}\delta u_0 \mathrm{d}\varphi\mathrm{d}r,
$$

$$
-\frac{A}{2}\delta\int_{R_i}^{R_a}\int_{\varphi_0}^{\varphi_1}(1-\nu)\frac{v_0}{r}\frac{\partial v_0}{\partial r}r\mathrm{d}\varphi\mathrm{d}r = -\frac{A}{2}\int_{R_i}^{R_a}\int_{\varphi_0}^{\varphi_1}(1-\nu)\frac{\partial v_0}{\partial r}\delta v_0 \mathrm{d}\varphi\mathrm{d}r
$$

$$
-\frac{A}{2}\int_{R_i}^{R_a}\int_{\varphi_0}^{\varphi_1}(1-\nu)\,v_0\frac{\partial\delta v_0}{\partial r}\mathrm{d}\varphi\mathrm{d}r = -\frac{A}{2}\int_{\varphi_0}^{\varphi_1}(1-\nu)\,v_0\delta v_0|_{R_i}^{R_a}\,\mathrm{d}\varphi,
$$

$$
-\frac{A}{2}\delta\int_{R_i}^{R_a}\int_{\varphi_0}^{\varphi_1}(1-\nu)\frac{v_0}{r}\frac{\partial u_0}{\partial\varphi}\mathrm{d}\varphi\mathrm{d}r = -\frac{A}{2}\int_{R_i}^{R_a}\int_{\varphi_0}^{\varphi_1}(1-\nu)\frac{\delta v_0}{r}\frac{\partial u_0}{\partial\varphi}\mathrm{d}\varphi\mathrm{d}r
$$

$$
-\frac{A}{2}\int_{R_i}^{R_a}\int_{\varphi_0}^{\varphi_1}(1-\nu)\frac{v_0}{r}\frac{\partial\delta u_0}{\partial\varphi}\mathrm{d}\varphi\mathrm{d}r = -\frac{A}{2}\int_{R_i}^{R_a}\int_{\varphi_0}^{\varphi_1}(1-\nu)\frac{\delta v_0}{r}\frac{\partial u_0}{\partial\varphi}\mathrm{d}\varphi\mathrm{d}r
$$

$$
\frac{A}{2}\int_{R_i}^{R_a}(1-\nu)\frac{v_0}{r}\delta u_0\Big|_{\varphi_0}^{\varphi_1}\mathrm{d}r \qquad +\frac{A}{2}\int_{R_i}^{R_a}\int_{\varphi_0}^{\varphi_1}(1-\nu)\frac{1}{r}\frac{\partial v_0}{\partial\varphi}\delta u_0\mathrm{d}\varphi\mathrm{d}r. \quad (4.57)
$$

In summary, $\delta\Pi = \delta\Pi_i + \delta\Pi_a = 0$ gives the following expression:

$$
-A\int_{R_i}^{R_a}\int_{\varphi_0}^{\varphi_1}\left(\frac{\partial^2 u_0}{\partial r^2}+\frac{1}{r}\frac{\partial u_0}{\partial r}-\frac{u_0}{r^2}+\frac{1-\nu}{2r^2}\frac{\partial^2 u_0}{\partial\varphi^2}-\frac{3-\nu}{2r^2}\frac{\partial v_0}{\partial\varphi}\right.
$$

$$
\left.+\frac{1+\nu}{2r}\frac{\partial^2 v_0}{\partial r\partial\varphi}+\frac{1-\nu^2}{E}f_r\right)\delta u_0 r\mathrm{d}\varphi\mathrm{d}r
$$

$$
-\frac{A}{2}\int_{R_i}^{R_a}\int_{\varphi_0}^{\varphi_1}(1-\nu)\left[\frac{\partial^2 v_0}{\partial r^2}+\frac{1}{r}\frac{\partial v_0}{\partial r}-\frac{v_0}{r^2}+\frac{2}{1-\nu}\frac{1}{r^2}\frac{\partial^2 v_0}{\partial^2\varphi}+\frac{1+\nu}{1-\nu}\frac{1}{r}\frac{\partial^2 u_0}{\partial r\partial\varphi}\right.
$$

$$
\left.+\frac{3-\nu}{1-\nu}\frac{1}{r^2}\frac{\partial u_0}{\partial\varphi}+\frac{2(1+\nu)}{E}f_\varphi\right]\delta v_0 r\mathrm{d}\varphi\mathrm{d}r
$$

$$
+\int_{\varphi_0}^{\varphi_1}\left\{A\left[\frac{\partial u_0}{\partial r}+\nu\left(\frac{1}{r}\frac{\partial v_0}{\partial\varphi}+\frac{u_0}{r}\right)\right]-\hat{N}_{rr}^0\right\}r\delta u_0\Big|_{R_i}^{R_a}\mathrm{d}\varphi
$$

$$+\int_{\varphi_0}^{\varphi_1}\left[\frac{A}{2}(1-\nu)\left(\frac{\partial v_0}{\partial r}+\frac{1}{r}\frac{\partial u_0}{\partial \varphi}-\frac{v_0}{r}\right)-\hat{N}^0_{r\varphi}\right]r\delta v_0\Bigg|_{R_i}^{R_a}\mathrm{d}\varphi$$

$$+\int_{R_i}^{R_a}\left[A\left(\frac{1}{r}\frac{\partial v_0}{\partial r}+\frac{u_0}{r}+\nu\frac{\partial u_0}{\partial r}\right)-\hat{N}^0_{\varphi\varphi}\right]\delta v_0\Bigg|_{\varphi_0}^{\varphi_1}\mathrm{d}r$$

$$+\int_{R_i}^{R_a}\left[\frac{A}{2}(1-\nu)\left(\frac{\partial v_0}{\partial r}+\frac{1}{r}\frac{\partial u_0}{\partial \varphi}-\frac{v_0}{r}\right)-\hat{N}^0_{r\varphi}\right]\delta u_0\Bigg|_{\varphi_0}^{\varphi_1}\mathrm{d}r=0. \tag{4.58}$$

The differential equations (4.17) of the displacement method can be read from the double integral terms. The remaining terms again describe the boundary conditions of the current disk situation.

4.3 Elementary Cases

The first simple case is the circular disk (radius R) under constant tensile edge stress σ_0 (Fig. 4.3). Using the approach $F = \frac{1}{2}\sigma_0 r^2$, (4.25) gives the following stress components of the circular disk loaded in this way:

$$\sigma_{rr}=\sigma_0,\quad \sigma_{\varphi\varphi}=\sigma_0,\quad \tau_{r\varphi}=0. \tag{4.59}$$

Obviously, in this situation there is a uniform normal stress state which is free from any shear stress.

Another quite elementary case is given with the situation of Fig. 4.4. Consider an infinitely extended disk with a circular hole (radius R), where a constant internal pressure σ_0 is present in the circular opening. Obviously, this is a rotationally symmetric situation, so that all state quantities will turn out to be independent of φ. An

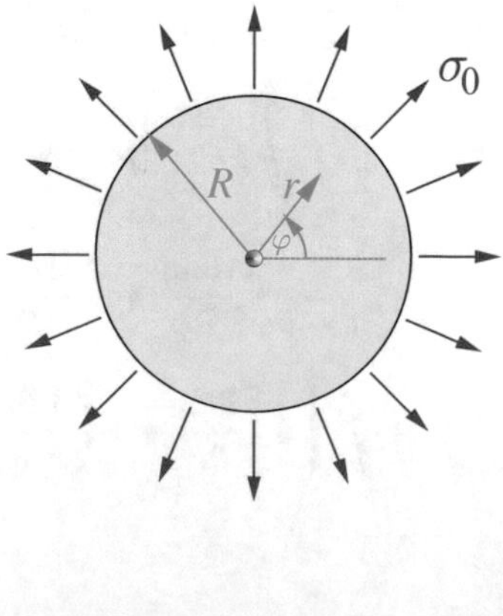

Fig. 4.3 Circular disk under constant tensile edge stress σ_0

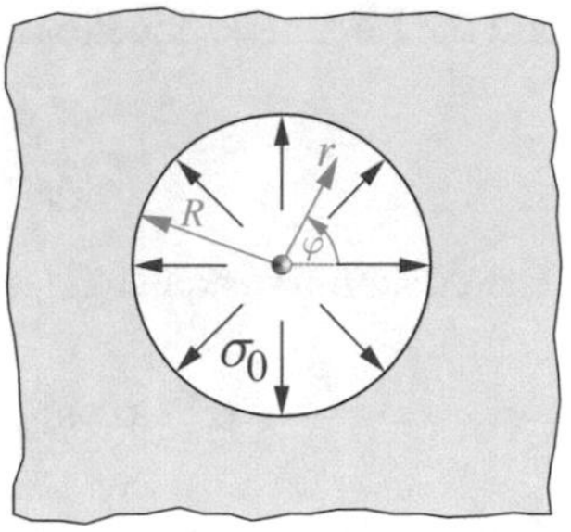

Fig. 4.4 Infinitely extended disk with circular hole under constant internal pressure σ_0

approach of the form

$$F(r) = -\sigma_0 R^2 \ln\left(\frac{r}{R}\right) \tag{4.60}$$

for the Airy stress function proves expedient here and yields the following stress components with (4.25):

$$\sigma_{rr} = -\sigma_0 \frac{R^2}{r^2}, \quad \sigma_{\varphi\varphi} = \sigma_0 \frac{R^2}{r^2}, \quad \tau_{r\varphi} = 0. \tag{4.61}$$

Thus, there is a compressive stress σ_{rr} in the radial direction, whereas the tangential stress $\sigma_{\varphi\varphi}$ turns out to be a tensile stress. Due to the rotationally symmetric situation, the shear stress $\tau_{r\varphi}$ vanishes at every point of the disk. The resulting expressions for σ_{rr} and $\sigma_{\varphi\varphi}$ show that due to their proportionality to $\frac{1}{r^2}$ they decay rapidly with increasing distance from the circular hole edge and tend to zero at a sufficient distance.

4.4 Rotationally Symmetric Disks

We now want to focus the considerations on the investigation of such disk situations with constant disk thickness h and vanishing volume forces, where the geometry as well as load and boundary conditions are rotationally symmetric and therefore all state variables are decoupled from the tangential coordinate φ. For this purpose, one can derive the displacement differential equations in a rather simple way, which we will discuss in the following (see e.g. Altenbach et al. 2016).

The equilibrium conditions are:

$$\frac{\mathrm{d}\sigma_{rr}}{\mathrm{d}r} + \frac{\sigma_{rr} - \sigma_{\varphi\varphi}}{r} = 0, \quad \frac{\mathrm{d}\tau_{r\varphi}}{\mathrm{d}r} + \frac{2}{r}\tau_{r\varphi} = 0, \tag{4.62}$$

and the kinematic equations to be applied here result in:

$$\varepsilon_{rr}^0 = \frac{\mathrm{d}u_0}{\mathrm{d}r}, \quad \varepsilon_{\varphi\varphi}^0 = \frac{u_0}{r}, \quad \gamma_{r\varphi}^0 = \frac{\mathrm{d}v_0}{\mathrm{d}r} - \frac{v_0}{r}. \tag{4.63}$$

The constitutive equations remain unchanged and read as follows:

$$\sigma_{rr} = \frac{E}{1-\nu^2}\left(\varepsilon_{rr}^0 + \nu\varepsilon_{\varphi\varphi}^0\right), \quad \sigma_{\varphi\varphi} = \frac{E}{1-\nu^2}\left(\varepsilon_{\varphi\varphi}^0 + \nu\varepsilon_{rr}^0\right), \quad \tau_{r\varphi} = G\gamma_{r\varphi}^0. \tag{4.64}$$

The governing equations show that this disk problem decomposes into two subproblems, namely, first, a problem involving the quantities u_0, ε_{rr}^0, $\varepsilon_{\varphi\varphi}^0$, σ_{rr}, $\sigma_{\varphi\varphi}$, and a subproblem involving the state variables v_0, $\gamma_{r\varphi}^0$, $\tau_{r\varphi}$.

The first subproblem is described by the following displacement differential equation, which can be derived quite easily from the preceding equations:

$$\frac{\mathrm{d}^2 u_0}{\mathrm{d}r^2} + \frac{1}{r}\frac{\mathrm{d}u_0}{\mathrm{d}r} - \frac{u_0}{r^2} = 0. \tag{4.65}$$

This equation also follows from (4.17), if there the volume force f_r is set to zero and all derivatives concerning the circumferential coordinate are omitted. Its general solution is:

$$u_0 = C_1 r + \frac{C_2}{r}. \tag{4.66}$$

Herein, C_1 and C_2 are constants which are adjusted to given boundary conditions. Using the displacement solution (4.66), the stress components $\sigma_{rr}, \sigma_{\varphi\varphi}$ can be written as:

$$\begin{aligned} \sigma_{rr} &= \frac{E}{1-\nu^2}\left[(1+\nu)\,C_1 - (1-\nu)\,\frac{C_2}{r^2}\right], \\ \sigma_{\varphi\varphi} &= \frac{E}{1-\nu^2}\left[(1+\nu)\,C_1 + (1-\nu)\,\frac{C_2}{r^2}\right]. \end{aligned} \tag{4.67}$$

The second subproblem is described by the following differential equation, which can also be derived from (4.17):

$$\frac{\mathrm{d}^2 v_0}{\mathrm{d}r^2} + \frac{1}{r}\frac{\mathrm{d}v_0}{\mathrm{d}r} - \frac{v_0}{r^2} = 0, \tag{4.68}$$

with the general solution

$$v_0 = C_3 r + \frac{C_4}{r}. \tag{4.69}$$

Thus, the shear stress $\tau_{r\varphi}$ is obtained as:

$$\tau_{r\varphi} = -G\frac{2C_4}{r^2}. \tag{4.70}$$

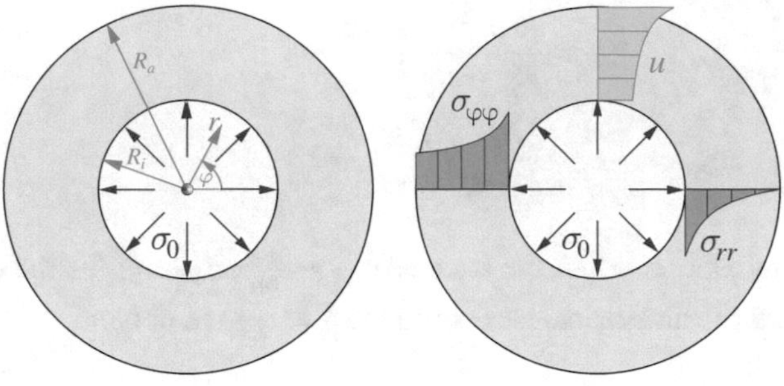

Fig. 4.5 Circular ring disk under internal pressure σ_0 (left), qualitative sketch of the state variables (right)

To illustrate the first subproblem, let us consider again the situation of Fig. 4.3. Due to the stress singularity at $r = 0$, the constant C_2 is set to zero, so that:

$$u_0 = C_1 r, \quad \sigma_{rr} = \sigma_{\varphi\varphi} = \frac{E}{1-\nu} C_1. \tag{4.71}$$

From the requirement that at the boundary $r = R$ the normal stress σ_{rr} must agree with the applied stress σ_0 it follows:

$$C_1 = \frac{\sigma_0}{E}(1-\nu). \tag{4.72}$$

From this then follow the stresses σ_{rr}, $\sigma_{\varphi\varphi}$ and the displacement u_0 as:

$$u_0 = \frac{\sigma_0 r}{E}(1-\nu), \quad \sigma_{rr} = \sigma_{\varphi\varphi} = \sigma_0. \tag{4.73}$$

Also consider the disk of Fig. 4.5, left. Given here is a circular ring disk with the inner radius R_i and the outer radius R_a, where at the point $r = R_i$ a constant pressure σ_0 is applied. The boundary conditions to be considered here are:

$$\sigma_{rr}(r = R_i) = -\sigma_0, \quad \sigma_{rr}(r = R_a) = 0. \tag{4.74}$$

The constants C_1 and C_2 result from this as:

$$C_1 = \frac{\sigma_0}{E}(1-\nu)\frac{R_i^2}{R_a^2 - R_i^2}, \quad C_2 = \frac{\sigma_0}{E}(1+\nu)\frac{R_i^2 R_a^2}{R_a^2 - R_i^2}. \tag{4.75}$$

Thus, the displacement u_0 and the two stress components σ_{rr}, $\sigma_{\varphi\varphi}$ can be written as:

$$u_0 = \frac{\sigma_0 r}{E}\frac{R_i^2}{R_a^2 - R_i^2}\left[(1-\nu) + \frac{R_a^2}{r^2}(1+\nu)\right],$$

$$\sigma_{rr} = -\sigma_0 \frac{R_i^2}{R_a^2 - R_i^2}\left(\frac{R_a^2}{r^2} - 1\right),$$
$$\sigma_{\varphi\varphi} = \sigma_0 \frac{R_i^2}{R_a^2 - R_i^2}\left(\frac{R_a^2}{r^2} + 1\right). \quad (4.76)$$

With the radius ratio $\rho = \frac{R_i}{R_a}$, the stresses $\bar{\sigma}_{rr} = \frac{\sigma_{rr}}{\sigma_0}$, $\bar{\sigma}_{\varphi\varphi} = \frac{\sigma_{\varphi\varphi}}{\sigma_0}$, the displacement $\bar{u}_0 = \frac{u_0}{R_a}\frac{E}{\sigma_0}$ and the dimensionless coordinate $\xi = \frac{r}{R_a}$ we obtain:

$$\bar{u}_0 = \frac{\rho^2 \xi}{1-\rho^2}\left[(1-\nu) + \frac{1}{\xi^2}(1+\nu)\right],$$
$$\bar{\sigma}_{rr} = -\frac{\rho^2}{1-\rho^2}\left(\frac{1}{\xi^2} - 1\right), \quad \bar{\sigma}_{\varphi\varphi} = \frac{\rho^2}{1-\rho^2}\left(\frac{1}{\xi^2} + 1\right). \quad (4.77)$$

A graphical representation of the two dimensionless stresses $\bar{\sigma}_{rr}$ and $\bar{\sigma}_{\varphi\varphi}$ for different values ρ is given in Fig. 4.6.

It is also possible to derive a differential equation for the stress component σ_{rr} for the first subproblem by eliminating the displacement u_0 and its first derivative $\frac{\mathrm{d}u_0}{\mathrm{d}r}$ in the constitutive relations. This gives the following equation:

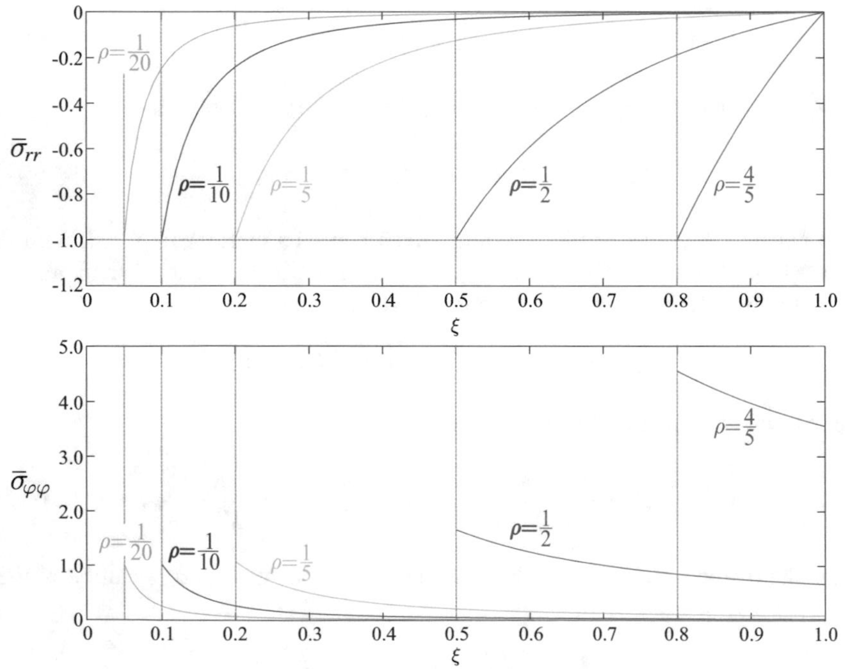

Fig. 4.6 Dimensionless stresses $\bar{\sigma}_{rr}$ and $\bar{\sigma}_{\varphi\varphi}$ of the circular ring disk under internal pressure σ_0

$$(1+\nu)\left(\sigma_{rr}-\sigma_{\varphi\varphi}\right)-r\left(\frac{\mathrm{d}\sigma_{\varphi\varphi}}{\mathrm{d}r}-\nu\frac{\sigma_{rr}}{\mathrm{d}r}\right)=0. \tag{4.78}$$

If, by means of the equilibrium condition (4.62), one also eliminates the circumferential stress $\sigma_{\varphi\varphi}$ by means of

$$\sigma_{\varphi\varphi}=r\frac{\mathrm{d}\sigma_{rr}}{\mathrm{d}r}+\sigma_{rr},\qquad \frac{\mathrm{d}\sigma_{\varphi\varphi}}{\mathrm{d}r}=2\frac{\mathrm{d}\sigma_{rr}}{\mathrm{d}r}+r\frac{\mathrm{d}^2\sigma_{rr}}{\mathrm{d}r^2}, \tag{4.79}$$

then finally one obtains the following equation:

$$\frac{\mathrm{d}^2\sigma_{rr}}{\mathrm{d}r^2}+\frac{3}{r}\frac{\mathrm{d}\sigma_{rr}}{\mathrm{d}r}=0. \tag{4.80}$$

This is a homogeneous Euler differential equation for the normal stress σ_{rr}, its general solution is:

$$\sigma_{rr}=D_1+\frac{1}{r^2}D_2. \tag{4.81}$$

Once σ_{rr} is present, then from the first equation in (4.79) the normal stress $\sigma_{\varphi\varphi}$ can also be determined as:

$$\sigma_{\varphi\varphi}=r\frac{\mathrm{d}\sigma_{rr}}{\mathrm{d}r}+\sigma_{rr}=D_1-\frac{1}{r^2}D_2. \tag{4.82}$$

If there is also interest in the displacement u_0 (the tangential displacement v_0 vanishes due to the assumed rotational symmetry of the given disk situation), then it can be easily determined from the kinematic equations as:

$$u_0(r)=\frac{r}{E}\left[(1-\nu)\,D_1-\frac{1}{r^2}\,(1+\nu)\,D_2\right]. \tag{4.83}$$

As an example, let us consider again the situation of Fig. 4.3. Since this is a solid disk, which with the solution (4.81) would have a stress singularity for $r\rightarrow 0$ due to the term $\frac{1}{r^2}$, the corresponding term is neglected and it remains:

$$\sigma_{rr}=D_1. \tag{4.84}$$

This reduces the formulation (4.82) for the circumferential stress $\sigma_{\varphi\varphi}$ to:

$$\sigma_{\varphi\varphi}=D_1. \tag{4.85}$$

Obviously, both stresses σ_{rr} and $\sigma_{\varphi\varphi}$ not only assume constant values, but also turn out to be identical. We determine the constant D_1 from the condition that the normal stress σ_{rr} coincides with the applied load σ_0, which leads to $D_1=\sigma_0$. Thus:

$$\sigma_{rr}=\sigma_{\varphi\varphi}=\sigma_0. \tag{4.86}$$

The displacement u_0 can then be determined as:

$$u_0 = \frac{r}{E}(1-\nu)\,\sigma_0. \tag{4.87}$$

It thus exhibits a linear distribution over r. The results obtained in this way agree with the findings already provided.

For further illustration, let us consider the situation of Fig. 4.5, left, again. Evaluating the boundary conditions leads to the two constants D_1 and D_2:

$$D_1 = \sigma_0 \frac{R_i^2}{R_a^2 - R_i^2}, \quad D_2 = \sigma_0 \frac{R_i^2 R_a^2}{R_i^2 - R_a^2}, \tag{4.88}$$

so that the stress components can be written as:

$$\sigma_{rr} = -\sigma_0 \frac{R_i^2}{R_a^2 - R_i^2}\left(\frac{R_a^2}{r^2} - 1\right), \quad \sigma_{\varphi\varphi} = \sigma_0 \frac{R_i^2}{R_a^2 - R_i^2}\left(\frac{R_a^2}{r^2} + 1\right). \tag{4.89}$$

The displacement u_0 is then given as:

$$u_0 = \frac{r\sigma_0}{E}\frac{R_i^2}{R_a^2 - R_i^2}\left[(1-\nu) + (1+\nu)\frac{R_a^2}{r^2}\right]. \tag{4.90}$$

These results agree with the already determined expressions (4.76) for the stresses and displacement.

Another example is the circular ring segment shown in Fig. 4.7, left (inner radius R_i, outer radius R_a, arbitrary opening angle), under the moment M_0, where M_0 is in the unit of a bending moment. This is a task solved by various authors, see e.g. Girkmann (1974), Altenbach et al. (2016) and Gross et al. (2009). Regardless of the concrete opening angle of the disk, this is in any case a rotationally symmetric situation concerning the stress components (the moment distribution is independent of φ, so this necessarily follows for the stresses σ_{rr} and $\sigma_{\varphi\varphi}$ as well), so we can use

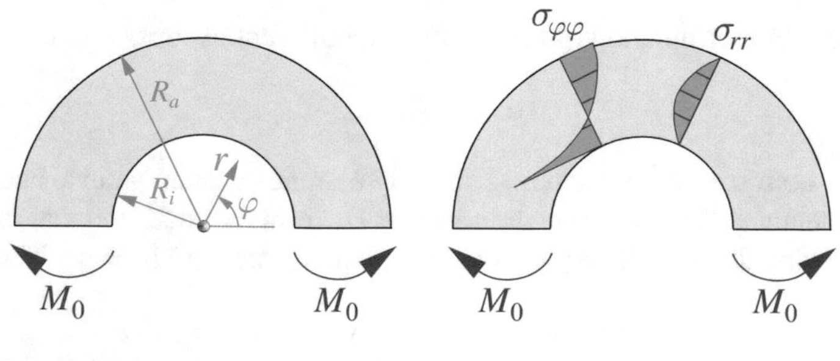

Fig. 4.7 Circular disk under moment load M_0

the solution (4.43) for the disk equation and the resulting stress components. For the determination of the constants C_2, C_3, C_4 the conditions are provided that the radial stress σ_{rr} must vanish for $r = R_i$ as well as for $r = R_a$. Moreover, it is required that the resultant of the circumferential stress $\sigma_{\varphi\varphi}$ coincides with the applied moment M_0. This can be formulated as follows:

$$h \int_{R_i}^{R_a} \sigma_{\varphi\varphi} r \, \mathrm{d}r = -M_0. \tag{4.91}$$

This provides three equations for determining the constants C_2, C_3, C_4, and it follows:

$$\begin{aligned}
C_2 &= -\frac{4M_0}{\kappa} R_i^2 R_a^2 \ln\left(\frac{R_a}{R_i}\right), \\
C_3 &= \frac{M_0}{\kappa}\left[R_a^2 - R_i^2 + 2\left(R_a^2 \ln R_a - R_i^2 \ln R_i\right)\right], \\
C_4 &= -\frac{2M_0}{\kappa}\left(R_a^2 - R_i^2\right),
\end{aligned} \tag{4.92}$$

wherein $\kappa = h\left[\left(R_a^2 - R_i^2\right)^2 - 4R_i^2 R_a^2 \ln^2\left(\frac{R_a}{R_i}\right)\right]$. For the stress components we then obtain:

$$\begin{aligned}
\sigma_{rr} &= -\frac{4M_0}{\kappa}\left[-R_i^2 \ln\left(\frac{r}{R_i}\right) + R_a^2 \ln\left(\frac{r}{R_a}\right) + \frac{R_i^2 R_a^2}{r^2} \ln\left(\frac{R_a}{R_i}\right)\right], \\
\sigma_{\varphi\varphi} &= -\frac{4M_0}{\kappa}\left[\left(R_a^2 - R_i^2\right) - R_i^2 \ln\left(\frac{r}{R_i}\right) + R_a^2 \ln\left(\frac{r}{R_a}\right) - \frac{R_i^2 R_a^2}{r^2} \ln\left(\frac{R_a}{R_i}\right)\right], \\
\tau_{r\varphi} &= 0.
\end{aligned} \tag{4.93}$$

A qualitative representation of the stress distributions is given in Fig. 4.7, right. A graphical representation of the normal stresses σ_{rr}, $\sigma_{\varphi\varphi}$ for different radius ratios $\rho = \frac{R_i}{R_a}$ is shown in Fig. 4.8. The nonlinear character of the circumferential stress $\sigma_{\varphi\varphi}$ is clearly evident here, which can be interpreted in this example similar to a bending stress of a curved beam. Its maximum value occurs at the point $r = R_i$, and as a result of the stress profile, the neutral fiber is displaced from the center towards the inner edge. It should also be noted here that although a rotationally symmetric stress state is established in this concrete disk situation, the displacement state will not be rotationally symmetric. We do not specify the displacement components at this point, they can be determined from the kinematic equations if the stresses are given.

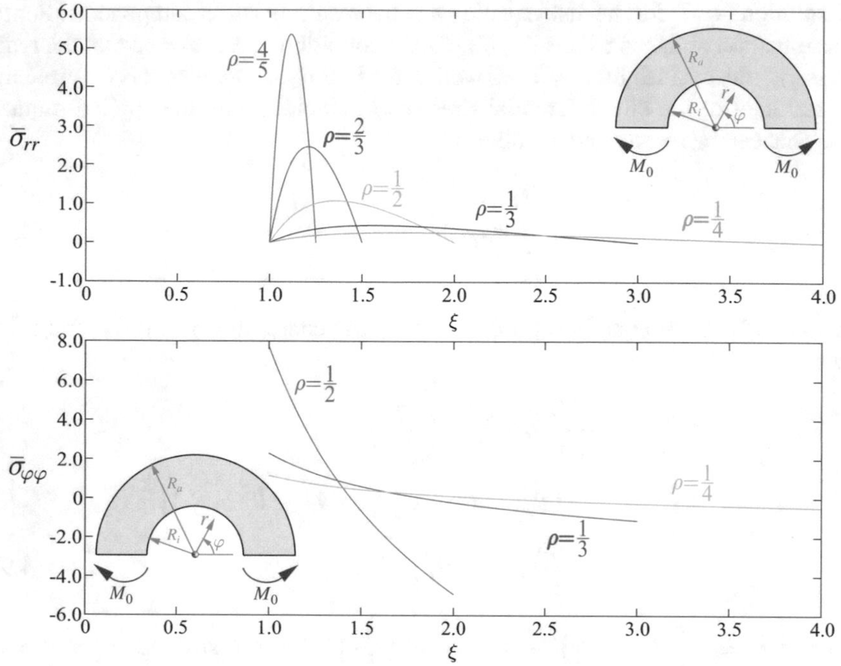

Fig. 4.8 Normal stresses $\bar{\sigma}_{rr} = \frac{\sigma_{rr} h R_i^2}{M_0}$, $\bar{\sigma}_{\varphi\varphi} = \frac{\sigma_{\varphi\varphi} h R_i^2}{M_0}$ of the circular disk under moment load M_0 for different radius ratios $\rho = \frac{R_i}{R_a}$ for $M_0 = 1\,\text{Nmm}$, $h = 1\,\text{mm}$, $R_i = 1\,\text{mm}$, plotted over $\xi = \frac{r}{R_i}$

4.5 Non-rotationally Symmetric Circular Disks

A quite easily accessible example for the treatment of non-rotationally symmetric disk problems (considered by several authors, see e.g. Göldner et al. 1979) is the clamped circular disk with the two radii R_i and R_a, which is under the two force flows V_0 and H_0 and the moment flow M_0 at the free end (Fig. 4.9). To determine the stresses in this disk, the following approach for the Airy stress function is used,

Fig. 4.9 Circular disk clamped on one side under boundary loads

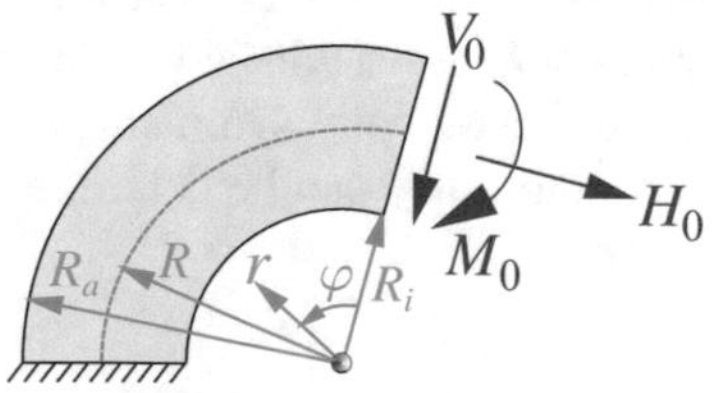

which consists of one part independent of φ and two parts dependent on φ:

$$\begin{aligned} F &= C_{01} + C_{02} \ln\left(\frac{r}{R_i}\right) + C_{03} r^2 + C_{04} r^2 \ln\left(\frac{r}{R_i}\right) \\ &+ \left[C_{1c} r + \frac{1}{r} C_{2c} + C_{3c} r^3 + C_{4c} r \ln\left(\frac{r}{R_i}\right)\right] \cos\varphi \\ &+ \left[C_{1s} r + \frac{1}{r} C_{2s} + C_{3s} r^3 + C_{4s} r \ln\left(\frac{r}{R_i}\right)\right] \sin\varphi. \end{aligned} \tag{4.94}$$

However, the constants C_{01}, C_{1c}, C_{1s} do not contribute to the stress state, so we can set them to zero without loss of generality. Thus:

$$\begin{aligned} F &= C_{02} \ln\left(\frac{r}{R_i}\right) + C_{03} r^2 + C_{04} r^2 \ln\left(\frac{r}{R_i}\right) \\ &+ \left[\frac{1}{r} C_{2c} + C_{3c} r^3 + C_{4c} r \ln\left(\frac{r}{R_i}\right)\right] \cos\varphi \\ &+ \left[\frac{1}{r} C_{2s} + C_{3s} r^3 + C_{4s} r \ln\left(\frac{r}{R_i}\right)\right] \sin\varphi. \end{aligned} \tag{4.95}$$

Using (4.25) we obtain the stress components as:

$$\begin{aligned} \sigma_{rr} &= \frac{1}{r^2} C_{02} + 2C_{03} + C_{04}\left[1 + 2\ln\left(\frac{r}{R_i}\right)\right] \\ &+ \left(-\frac{2}{r^3} C_{2c} + 2C_{3c} r + \frac{1}{r} C_{4c}\right) \cos\varphi + \left(-\frac{2}{r^3} C_{2s} + 2C_{3s} r + \frac{1}{r} C_{4s}\right) \sin\varphi, \\ \sigma_{\varphi\varphi} &= -\frac{1}{r^2} C_{02} + 2C_{03} + C_{04}\left[2\ln\left(\frac{r}{R_i}\right) + 3\right] \\ &+ \left(\frac{2}{r^3} C_{2c} + 6C_{3c} r + \frac{1}{r} C_{4c}\right) \cos\varphi + \left(\frac{2}{r^3} C_{2s} + 6C_{3s} r + \frac{1}{r} C_{4s}\right) \sin\varphi, \\ \tau_{r\varphi} &= \left(-\frac{2}{r^3} C_{2c} + 2C_{3c} r + \frac{1}{r} C_{4c}\right) \sin\varphi + \left(\frac{2}{r^3} C_{2s} - 2C_{3s} r - \frac{1}{r} C_{4s}\right) \cos\varphi. \end{aligned} \tag{4.96}$$

For the determination of the free constants, the following boundary conditions are available at the unloaded edges at $r = R_i$ and $r = R_a$:

$$\sigma_{rr}(r = R_i) = 0, \ \sigma_{rr}(r = R_a) = 0, \ \tau_{r\varphi}(r = R_i) = 0, \ \tau_{r\varphi}(r = R_a) = 0. \tag{4.97}$$

This leads to the following expressions concerning the requirements for σ_{rr}:

$$\begin{aligned} &\frac{1}{R_i^2} C_{02} + 2C_{03} + C_{04} \\ &+ \left(-\frac{2}{R_i^3} C_{2c} + 2C_{3c} R_i + \frac{1}{R_i} C_{4c}\right) \cos\varphi \end{aligned}$$

$$
\begin{aligned}
&+\left(-\frac{2}{R_i^3}C_{2s}+2C_{3s}R_i+\frac{1}{R_i}C_{4s}\right)\sin\varphi=0,\\
&\frac{1}{R_a^2}C_{02}+2C_{03}+C_{04}\left[1+2\ln\left(\frac{R_a}{R_i}\right)\right]\\
&+\left(-\frac{2}{R_a^3}C_{2c}+2C_{3c}R_a+\frac{1}{R_a}C_{4c}\right)\cos\varphi\\
&+\left(-\frac{2}{R_a^3}C_{2s}+2C_{3s}R_a+\frac{1}{R_a}C_{4s}\right)\sin\varphi=0.
\end{aligned}
\tag{4.98}
$$

For any angle φ this can only be fulfilled if the individual terms vanish for themselves. Therefore it holds:

$$
\begin{aligned}
\frac{1}{R_i^2}C_{02}+2C_{03}+C_{04} &= 0,\\
-\frac{2}{R_i^3}C_{2c}+2C_{3c}R_i+\frac{1}{R_i}C_{4c} &= 0,\\
-\frac{2}{R_i^3}C_{2s}+2C_{3s}R_i+\frac{1}{R_i}C_{4s} &= 0,\\
\frac{1}{R_a^2}C_{02}+2C_{03}+C_{04}\left[1+2\ln\left(\frac{R_a}{R_i}\right)\right] &= 0,\\
-\frac{2}{R_a^3}C_{2c}+2C_{3c}R_a+\frac{1}{R_a}C_{4c} &= 0,\\
-\frac{2}{R_a^3}C_{2s}+2C_{3s}R_a+\frac{1}{R_a}C_{4s} &= 0.
\end{aligned}
\tag{4.99}
$$

The boundary conditions concerning the shear stress $\tau_{r\varphi}$ result as:

$$
\begin{aligned}
&\left(-\frac{2}{R_i^3}C_{2c}+2C_{3c}R_i+\frac{1}{R_i}C_{4c}\right)\sin\varphi\\
&+\left(\frac{2}{R_i^3}C_{2s}-2C_{3s}R_i-\frac{1}{R_i}C_{4s}\right)\cos\varphi=0,\\
&\left(-\frac{2}{R_a^3}C_{2c}+2C_{3c}R_a+\frac{1}{R_a}C_{4c}\right)\sin\varphi\\
&+\left(\frac{2}{R_a^3}C_{2s}-2C_{3s}R_a-\frac{1}{R_a}C_{4s}\right)\cos\varphi=0.
\end{aligned}
\tag{4.100}
$$

Again, the satisfaction of the conditions for arbitrary angles φ can be ensured only if the individual bracket terms vanish. Obviously, the resulting terms do not provide any new information compared to (4.99), so that with (4.99) a total of six conditions are available to determine the free constants.

At the loaded edge of the disk, it is required that the stress components $\sigma_{\varphi\varphi}$ and $\tau_{r\varphi}$ are in equilibrium with the respective external loads. Since this cannot be fulfilled in an exact manner, resultants of the stresses are employed. At point $\varphi = 0$ we have:

$$\int_{R_i}^{R_a} \sigma_{\varphi\varphi} \mathrm{d}r = H_0, \quad \int_{R_i}^{R_a} \tau_{r\varphi} \mathrm{d}r = V_0, \quad \int_{R_i}^{R_a} \sigma_{\varphi\varphi} r \mathrm{d}r = M_0 + H_0 R. \tag{4.101}$$

This results in the following conditions:

$$\begin{aligned}
\frac{R_i - R_a}{R_i R_a} C_{02} + 2C_{03}\left(R_a - R_i\right) + C_{04}\left[R_a - R_i + 2R_a \ln\left(\frac{R_a}{R_i}\right)\right] & \\
+ \frac{R_a^2 - R_i^2}{R_i^2 R_a^2} C_{2c} + 3C_{3c}\left(R_a^2 - R_i^2\right) + C_{4c} \ln\left(\frac{R_a}{R_i}\right) &= H_0, \\
\frac{R_i^2 - R_a^2}{R_i^2 R_a^2} C_{2s} + C_{3s}\left(R_a^2 - R_i^2\right) + C_{4s} \ln\left(\frac{R_a}{R_i}\right) &= V_0, \\
2\frac{R_a - R_i}{R_i R_a} C_{2c} + 2C_{3c}\left(R_a^3 - R_i^3\right) + C_{4c}\left(R_a - R_i\right) - C_{02} \ln\left(\frac{R_a}{R_i}\right) & \\
+ C_{03}\left(R_a^2 - R_i^2\right) + C_{04}\left[R_a^2 \ln\left(\frac{R_a}{R_i}\right) + R_a^2 - R_i^2\right] &= M_0 + H_0 R.
\end{aligned} \tag{4.102}$$

Thus, a total of nine equations are available for the unknown constants. This eventually results in the following:

$$\begin{aligned}
C_{02} &= \frac{4R_i^2 R_a^2 \ln\left(\frac{R_a}{R_i}\right)\left(M_0 + H_0 R\right)}{\left(R_a^2 - R_i^2\right)^2 - 4R_i^2 R_a^2 \ln^2\left(\frac{R_a}{R_i}\right)}, \\
C_{03} &= \frac{\left[R_i^2 - R_a^2 - 2R_a^2 \ln\left(\frac{R_a}{R_i}\right)\right]\left(M_0 + H_0 R\right)}{\left(R_a^2 - R_i^2\right)^2 - 4R_i^2 R_a^2 \ln^2\left(\frac{R_a}{R_i}\right)}, \\
C_{04} &= \frac{2\left(R_a^2 - R_i^2\right)\left(M_0 + H_0 R\right)}{\left(R_a^2 - R_i^2\right)^2 - 4R_i^2 R_a^2 \ln^2\left(\frac{R_a}{R_i}\right)}, \\
C_{2c} &= \frac{R_i^2 R_a^2 H_0}{2\left[R_i^2 - R_a^2 + \left(R_i^2 + R_a^2\right) \ln\left(\frac{R_a}{R_i}\right)\right]}, \\
C_{3c} &= \frac{H_0}{2\left[R_a^2 - R_i^2 + \left(R_i^2 + R_a^2\right) \ln\left(\frac{R_a}{R_i}\right)\right]}, \\
C_{4c} &= \frac{\left(R_a^2 + R_i^2\right) H_0}{R_i^2 - R_a^2 + \left(R_i^2 + R_a^2\right) \ln\left(\frac{R_a}{R_i}\right)}, \\
C_{2s} &= -\frac{R_i^2 R_a^2 V_0}{2\left[\left(R_i^2 + R_a^2\right) \ln\left(\frac{R_a}{R_i}\right) + R_i^2 - R_a^2\right]},
\end{aligned}$$

$$
\begin{aligned}
C_{3s} &= \frac{V_0}{2\left[\left(R_i^2+R_a^2\right)\ln\left(\frac{R_a}{R_i}\right)+R_i^2-R_a^2\right]}, \\
C_{4s} &= -\frac{\left(R_i^2+R_a^2\right)V_0}{\left(R_i^2+R_a^2\right)\ln\left(\frac{R_a}{R_i}\right)+R_i^2-R_a^2}.
\end{aligned} \tag{4.103}
$$

Thus, the stress state in the circular disk under consideration can be determined at any point r, φ. The displacements can then be obtained by integration from the constitutive equations taking into account the kinematics.

If the special case of a disk with $M_0 = 0$ and $H_0 = 0$ is present, the constants C_{2s}, C_{3s}, C_{4s} remaining here result as:

$$
C_{2s} = -\frac{V_0 R_i^2 R_a^2}{2\kappa}, \quad C_{3s} = \frac{V_0}{2\kappa}, \quad C_{4s} = -\frac{V_0}{\kappa}\left(R_i^2+R_a^2\right), \tag{4.104}
$$

wherein:

$$
\kappa = R_i^2 - R_a^2 + \left(R_i^2+R_a^2\right)\ln\left(\frac{R_a}{R_i}\right). \tag{4.105}
$$

The resulting stress field then reads with (4.96):

$$
\begin{aligned}
\sigma_{rr} &= \frac{V_0}{\kappa}\left(\frac{R_i^2 R_a^2}{r^3}+r-\frac{R_i^2+R_a^2}{r}\right)\sin\varphi, \\
\sigma_{\varphi\varphi} &= \frac{V_0}{\kappa}\left(-\frac{R_i^2 R_a^2}{r^3}+3r-\frac{R_i^2+R_a^2}{r}\right)\sin\varphi, \\
\tau_{r\varphi} &= \frac{V_0}{\kappa}\left(-\frac{R_i^2 R_a^2}{r^3}-r+\frac{R_i^2+R_a^2}{r}\right)\cos\varphi.
\end{aligned} \tag{4.106}
$$

Analogous expressions can be derived for the case $M_0 = 0$, $V_0 = 0$, which is not elaborated on further at this point.

4.6 Wedge-shaped Disks

We consider the situation of Fig. 4.10, left. Let a wedge-shaped disk (thickness h) with the two straight edges $\varphi = \pm\alpha$ be given, which is loaded at its tip by the two force flows H_0 and V_0 and by the moment flow M_0 (cf. see also Göldner et al. 1979; Altenbach et al. 2016; Becker and Gross 2002; Eschenauer and Schnell 1993, and numerous other authors). Here V_0 and M_0 will cause antimetric components in the stress components with respect to the axis $\varphi = 0$, whereas H_0 will cause symmetric components. A corresponding approach for the Airy stress function is:

$$
F = C_1 r\varphi\sin\varphi + C_2 r\varphi\cos\varphi + C_3\varphi + C_4\sin 2\varphi, \tag{4.107}
$$

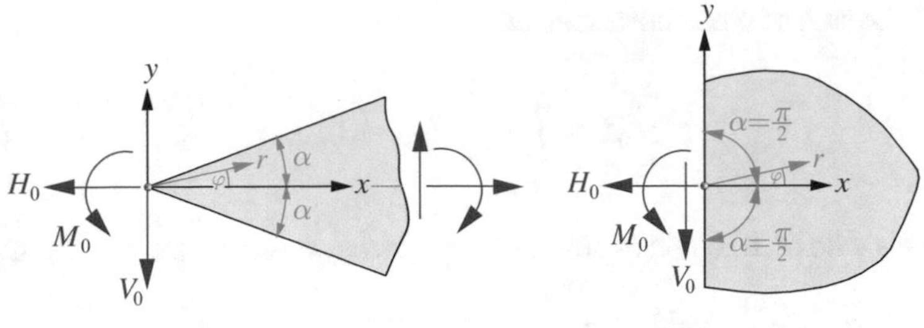

Fig. 4.10 Wedge-shaped disk (left), special case of a half-plane (right)

with constants C_1, C_2, C_3, C_4 to be determined from suitable conditions. The stress components in the wedge-shaped disk are thus:

$$\sigma_{rr} = \frac{2C_1}{r}\cos\varphi - \frac{2C_2}{r}\sin\varphi - \frac{4C_4}{r^2}\sin 2\varphi, \quad \sigma_{\varphi\varphi} = 0, \quad \tau_{r\varphi} = \frac{C_3}{r^2} + \frac{2C_4}{r^2}\cos 2\varphi. \tag{4.108}$$

Obviously, this approach leads to a vanishing circumferential stress $\sigma_{\varphi\varphi}$.

The boundary conditions to be fulfilled here are as follows. First, the circumferential stress $\sigma_{\varphi\varphi}$ at the two straight edges $\varphi = \pm\alpha$ must vanish for each r. This condition is already automatically fulfilled with the chosen approach. Similarly, the shear stress $\tau_{r\varphi}$ must vanish at these edges, i.e. $\tau_{r\varphi}(\varphi = \pm\alpha) = 0$ for each r. This gives the following condition with (4.108):

$$C_3 = -2C_4\cos 2\alpha. \tag{4.109}$$

We obtain further conditions for the determination of the constants C_1, C_2, C_3, C_4 by making a cut through the disk at an arbitrary point r and observing the equilibrium at the resulting free body image with the released stresses σ_{rr} and $\tau_{r\varphi}$. The sum of forces in x-direction gives:

$$H_0 h + \int_{-\alpha}^{+\alpha} \left(\tau_{r\varphi}\sin\varphi - \sigma_{rr}\cos\varphi\right) hr\mathrm{d}\varphi = 0. \tag{4.110}$$

The vertical sum of forces follows as:

$$V_0 h - \int_{-\alpha}^{+\alpha} \left(\tau_{r\varphi}\cos\varphi + \sigma_{rr}\sin\varphi\right) hr\mathrm{d}\varphi = 0. \tag{4.111}$$

Finally, the sum of moments results in:

$$M_0 h + \int_{-\alpha}^{+\alpha} \tau_{r\varphi} h r^2 \mathrm{d}\varphi = 0. \tag{4.112}$$

This leads with (4.108) and (4.109) to the following constants C_1, C_2, C_3, C_4:

$$\begin{aligned} C_1 &= \frac{H_0}{2\alpha + \sin 2\alpha}, \quad C_2 = \frac{V_0}{\sin 2\alpha - 2\alpha}, \\ C_3 &= -\frac{M_0 \cos 2\alpha}{2\alpha \cos 2\alpha - \sin 2\alpha}, \quad C_4 = \frac{M_0}{4\alpha \cos 2\alpha - 2 \sin 2\alpha}. \end{aligned} \tag{4.113}$$

An important special case arises for $\alpha = \frac{\pi}{2}$ (Fig. 4.10, right). In this case, the wedge-shaped disk transitions into a half-plane loaded by the force flows H_0 and V_0 and the moment flow M_0. The constants are then given as:

$$C_1 = \frac{H_0}{\pi}, \quad C_2 = -\frac{V_0}{\pi}, \quad C_3 = -\frac{M_0}{\pi}, \quad C_4 = -\frac{M_0}{2\pi}. \tag{4.114}$$

4.7 Disks with Circular Holes

If a disk is weakened by a hole, this leads to stress concentrations in the vicinity of the hole, which must be taken into account in the analysis and design accordingly. We first consider the situation of Fig. 4.11 (a special situation considered by a good number of authors, cf. e.g., Wang 1953; Göldner et al. 1979; Hahn 1985; Timoshenko and Goodier 1951; Ugural and Fenster 2003; Bauchau and Craig 2009). Consider an infinitely extended disk of constant thickness h, which is weakened by a circular hole of radius R. At a sufficiently large distance from the circular hole the constant normal stress $\sigma_{xx}^{\infty} = \sigma_0$ is applied, where the infinity symbol indicates that this load application occurs at a theoretically infinite distance and therefore there is no interaction of the hole with the loaded boundary. We can thus establish the following

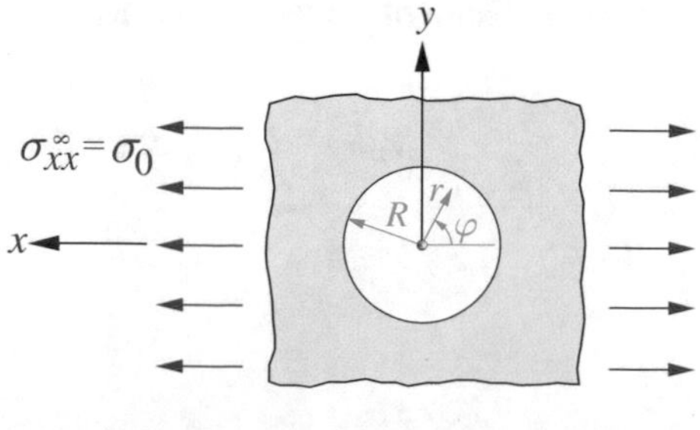

Fig. 4.11 Disk with a circular hole under uniaxial tension

boundary conditions. At the unloaded hole edge, both the radial normal stress σ_{rr} as well as the shear stress $\tau_{r\varphi}$ must vanish:

$$\sigma_{rr}(r = R) = 0, \quad \tau_{r\varphi}(r = R) = 0. \tag{4.115}$$

At a sufficient distance from the edge of the hole (i.e. for theoretically infinite values of r), the stress state of the unweakened disk has to be restored so that:

$$\sigma_{xx}^{\infty} = \sigma_0, \quad \sigma_{yy}^{\infty} = 0, \quad \tau_{xy}^{\infty} = 0. \tag{4.116}$$

This elementary stress state can be described by the following Airy stress function:

$$F^{\infty} = \frac{1}{2}\sigma_0 y^2. \tag{4.117}$$

The description of the stress state near the hole, on the other hand, can be advantageously carried out in polar coordinates, so that we also want to describe the stress state at infinity in r and φ. This results in:

$$\sigma_{rr}^{\infty} = \frac{1}{2}\sigma_0\left(1 + \cos 2\varphi\right), \sigma_{\varphi\varphi}^{\infty} = \frac{1}{2}\sigma_0\left(1 - \cos 2\varphi\right), \tau_{r\varphi} = -\frac{1}{2}\sigma_0 \sin 2\varphi. \tag{4.118}$$

Therefore, to describe the stress state in the vicinity of the hole, we use a form for the Airy stress function composed of two parts, where the first part must depend exclusively on r and where the second part is a function of both r and φ:

$$F = f_1(r) + f_2(r)\cos 2\varphi. \tag{4.119}$$

Substituting this expression into the disk equation and taking into account that the resulting expression must be satisfied for any φ, this allows to write two equations of determination for the two functions f_1 and f_2:

$$\begin{aligned} \left(\frac{\mathrm{d}^2}{\mathrm{d}r^2} + \frac{1}{r}\frac{\mathrm{d}}{\mathrm{d}r}\right)\left(\frac{\mathrm{d}^2 f_1}{\mathrm{d}r^2} + \frac{1}{r}\frac{\mathrm{d}f_1}{\mathrm{d}r}\right) &= 0, \\ \left(\frac{\mathrm{d}^2}{\mathrm{d}r^2} + \frac{1}{r}\frac{\mathrm{d}}{\mathrm{d}r} - \frac{4}{r^2}\right)\left(\frac{\mathrm{d}^2 f_2}{\mathrm{d}r^2} + \frac{1}{r}\frac{\mathrm{d}f_2}{\mathrm{d}r} - \frac{4f_2}{r^2}\right) &= 0. \end{aligned} \tag{4.120}$$

For these two differential equations the following general solutions can be given:

$$\begin{aligned} f_1 &= C_1 r^2 \ln r + C_2 r^2 + C_3 \ln r + C_4, \\ f_2 &= D_1 r^2 + D_2 r^4 + \frac{1}{r^2} D_3 + D_4. \end{aligned} \tag{4.121}$$

This allows the stresses in the disk to be written as:

$$\begin{aligned}\sigma_{rr} &= C_1(1+2\ln r)+2C_2+\frac{1}{r^2}C_3-\left(2D_1+\frac{6}{r^4}D_3+\frac{4}{r^2}D_4\right)\cos 2\varphi,\\ \sigma_{\varphi\varphi} &= C_1(3+2\ln r)+2C_2-\frac{1}{r^2}C_3+\left(2D_1+12D_2r^2+\frac{6}{r^4}D_3+\frac{4}{r^2}D_4\right)\cos 2\varphi,\\ \tau_{r\varphi} &= \left(2D_1+6D_2r^2-\frac{6}{r^4}D_3-\frac{2}{r^2}D_4\right)\sin 2\varphi.\end{aligned} \tag{4.122}$$

The constant C_4 does not appear in these equations and therefore obviously has no influence on the stress state of the perforated disk. Furthermore, we may already set the two constants C_1 and D_2 to zero at this point - the resulting stress state at a sufficient distance from the hole edge must assume finite values for each stress component.

We first evaluate the boundary conditions at the hole edge by using (4.115). It follows by observing that the equations have to hold again for arbitrary φ:

$$\begin{aligned}2C_2+\frac{1}{R^2}C_3 &= 0,\ 2D_1+\frac{6}{R^4}D_3+\frac{4}{R^2}D_4=0,\\ 2D_1-\frac{6}{R^4}D_3-\frac{2}{R^2}D_4 &= 0.\end{aligned} \tag{4.123}$$

Moreover, from the requirement that at infinity the stress state of the unweakened disk is restored:

$$C_2=\frac{1}{4}\sigma_0,\quad D_1=-\frac{1}{4}\sigma_0. \tag{4.124}$$

Solving the Eqs. (4.123) with (4.124) finally yields:

$$C_3=-\frac{R^2}{2}\sigma_0,\quad D_3=-\frac{R^4}{4}\sigma_0,\quad D_4=\frac{R^2}{2}\sigma_0. \tag{4.125}$$

Thus, all stress components can be specified as:

$$\begin{aligned}\sigma_{rr} &= \frac{\sigma_0}{2}\left[\left(1-\frac{R^2}{r^2}\right)+\left(1+\frac{3R^4}{r^4}-\frac{4R^2}{r^2}\right)\cos 2\varphi\right],\\ \sigma_{\varphi\varphi} &= \frac{\sigma_0}{2}\left[\left(1+\frac{R^2}{r^2}\right)-\left(1+\frac{3R^4}{r^4}\right)\cos 2\varphi\right],\\ \tau_{r\varphi} &= -\frac{\sigma_0}{2}\left(1-\frac{3R^4}{r^4}+\frac{2R^2}{r^2}\right)\sin 2\varphi.\end{aligned} \tag{4.126}$$

With the dimensionless coordinate $\xi=\frac{r}{R}$ and the relative stress $\bar{\sigma}_{\varphi\varphi}=\frac{\sigma_{\varphi\varphi}}{\sigma_0}$, the circumferential stress can be given as:

$$\bar{\sigma}_{\varphi\varphi}=\frac{1}{2}\left[1+\frac{1}{\xi^2}-\left(1+\frac{3}{\xi^4}\right)\cos 2\varphi\right]. \tag{4.127}$$

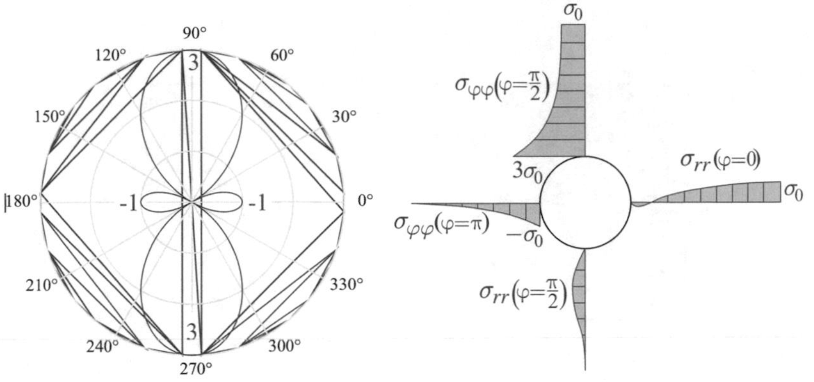

Fig. 4.12 Stresses σ_{rr} and $\sigma_{\varphi\varphi}$ near the hole: Polar diagram for $\bar{\sigma}_{\varphi\varphi}$ at the hole edge $\xi = 1$ (left), qualitative distributions of σ_{rr} and $\sigma_{\varphi\varphi}$ over r at selected locations φ (right)

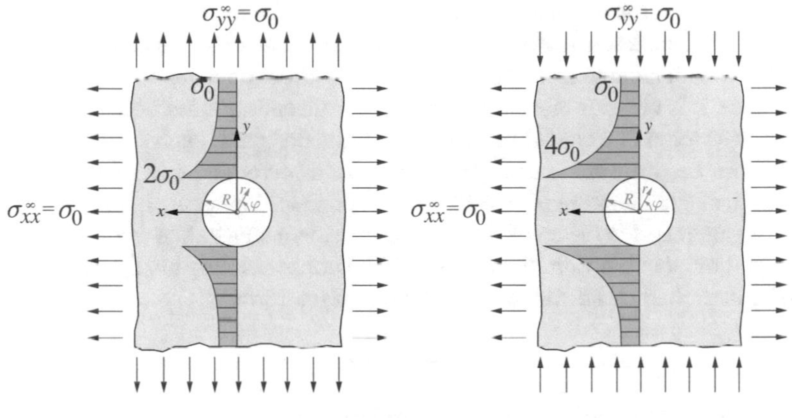

Fig. 4.13 Perforated disk under biaxial load

At the hole edge $\xi = 1$ then results:

$$\bar{\sigma}_{\varphi\varphi} = 1 - 2\cos 2\varphi. \tag{4.128}$$

A graphical representation of the stress components σ_{rr} and $\sigma_{\varphi\varphi}$ is shown in Fig. 4.12. From the polar diagram in Fig. 4.12, left, it can be seen that the stress component $\sigma_{\varphi\varphi}$ at the point $\varphi = \frac{\pi}{2}$ assumes its maximum value with the magnitude $3\sigma_0$. This ratio $\frac{\sigma_{max}}{\sigma_0}$ of the maximum stress to the stress in the undisturbed region is also addressed as the so-called stress concentration factor. On the other hand, $\sigma_{\varphi\varphi}$ at locations $\varphi = 0$ and $\varphi = \pi$ is present as a compressive stress with magnitude σ_0. The diagrams of Fig. 4.12, right, show the decay behavior of the hole-edge stresses over r at selected locations φ.

By simple superposition, the case of biaxial loading can also be treated, as shown in Fig. 4.13 for two selected cases. We will skip the detailed way of calculation at

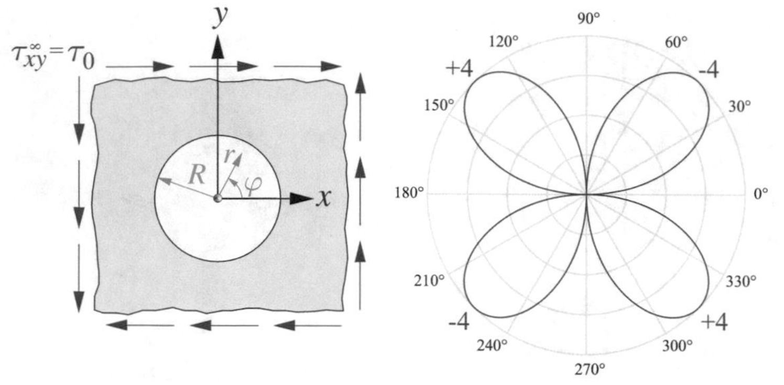

Fig. 4.14 Perforated disk under shear (left), polar diagram for $\bar{\sigma}_{\varphi\varphi}$ at the hole edge $\xi = 1$ (right)

this point, but it turns out that in case of a biaxial tensile load with identical amounts the stress concentration factor is reduced to the value 2, but in case of tension in x-direction with simultaneous compression in y-direction a factor $\frac{\sigma_{max}}{\sigma_0} = 4$ results. In any case, it is obvious that such local stress concentrations at hole edges must be taken into account and cannot be neglected in the design of such structures.

Furthermore, we want to consider the case of a perforated disk loaded at infinity by a constant shear load τ_0 (Fig. 4.14, left, cf. e.g. Göldner et al. 1979). In this case, we will again resort to the consideration that the elementary state of the unperturbed disk is perturbed in the vicinity of the hole. The corresponding unperturbed state can again be described by an Airy stress function of the form

$$F^{\infty} = -\tau_0 xy \tag{4.129}$$

which can be represented in polar coordinates as

$$F^{\infty} = -\frac{1}{2}\tau_0 r^2 \sin 2\varphi. \tag{4.130}$$

An approach to the Airy stress function can be constructed analogous to the previous example, and to describe the stress state in the perforated disk, the Airy stress function is chosen as follows:

$$F = \tau_0 \left(R^2 - \frac{1}{2}r^2 - \frac{R^4}{2r^2} \right) \sin 2\varphi. \tag{4.131}$$

The resulting stress components are then:

$$\sigma_{rr} = \tau_0 \left[1 + 3\left(\frac{R}{r}\right)^4 - 4\left(\frac{R}{r}\right)^2 \right] \sin 2\varphi,$$

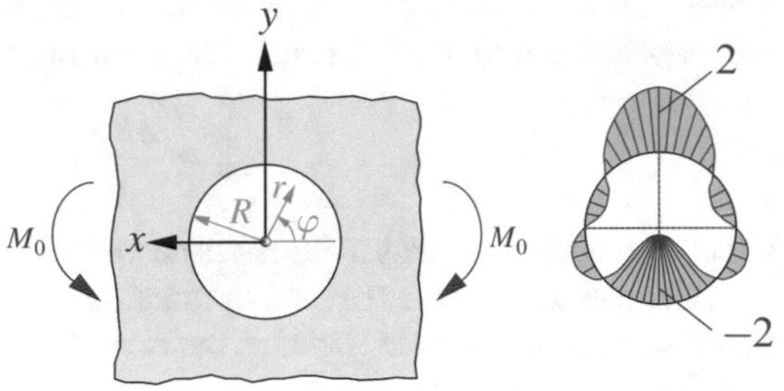

Fig. 4.15 Perforated disk under bending (left), polar diagram for $\sigma_{\varphi\varphi}$ at the hole edge $r = R$ (right), after Göldner et al. (1979)

$$\sigma_{\varphi\varphi} = \tau_0 \left[-1 - 3\left(\frac{R}{r}\right)^4 \right] \sin 2\varphi,$$

$$\tau_{r\varphi} = \tau_0 \left[1 - 3\left(\frac{R}{r}\right)^4 + 2\left(\frac{R}{r}\right)^2 \right] \cos 2\varphi. \tag{4.132}$$

The circumferential stress $\sigma_{\varphi\varphi}$ can be given in dimensionless form with $\bar{\sigma}_{\varphi\varphi} = \frac{\sigma_{\varphi\varphi}}{\tau_0}$ and $\xi = \frac{r}{R}$ as:

$$\bar{\sigma}_{\varphi\varphi} = -\left(1 + \frac{3}{\xi^4}\right) \sin 2\varphi, \tag{4.133}$$

and at the hole edge $\xi = 1$:

$$\bar{\sigma}_{\varphi\varphi} = -4 \sin 2\varphi. \tag{4.134}$$

A polar diagram for $\bar{\sigma}_{\varphi\varphi}$ along the hole edge at location $r = R$ or $\xi = 1$ is given in Fig. 4.14, right. The stress concentration factor here is apparently $\left|\frac{\tau_{max}}{\tau_0}\right| = 4$, occurring at four different locations along the hole edge. It should be noted here that the same result is obtained if the situation of Fig. 4.13, right, transformed by the angle 45°, is considered with $\sigma_0 = \tau_0$.

A final example for the stress state in a perforated plate is shown in Fig. 4.15, left. Let a disk be given here which contains a circular hole (radius R) and which is under bending M_0. Let it be assumed that the dimension of the disk in x-direction is sufficiently large so that no interaction of the boundary stress state due to the load introduction with the state in the hole region occurs. Furthermore, it is assumed that the disk is also sufficiently extended in the y-direction (i.e., this dimension is significantly larger than R), so that no interaction of the free disk edge with the hole stress state occurs. The undisturbed stress state that will occur in such a disk due to bending can be described by the Airy stress function as follows:

$$F^{\infty} = \frac{M_0 y^3}{6I}, \tag{4.135}$$

where I is the second-order moment of inertia with respect to bending about the z-axis pointing into the disk plane. The resulting normal stress is then

$$\sigma_{xx}^{\infty} = \frac{M_0 y}{I}, \tag{4.136}$$

which corresponds exactly to the result according to the Euler-Bernoulli beam theory.

An expression for the Airy stress function that suitably describes the stress state in the vicinity of the hole, satisfies the boundary conditions at the hole edges, and transitions to the solution of the undisturbed state at a sufficient distance from the hole edge can be given as follows:

$$F = \frac{M_0}{I}\left[\left(\frac{r^3}{8} - \frac{R^2 r}{4} + \frac{R^4}{8r}\right)\sin\varphi + \left(-\frac{r^3}{24} + \frac{R^4}{8r} - \frac{R^6}{12r^3}\right)\sin 3\varphi\right]. \tag{4.137}$$

From this, the stress components σ_{rr}, $\sigma_{\varphi\varphi}$, and $\tau_{r\varphi}$ can be determined as:

$$\begin{aligned}
\sigma_{rr} &= \frac{M_0}{I}\left[\left(\frac{r}{4} - \frac{R^4}{4r^3}\right)\sin\varphi + \left(\frac{r}{4} - \frac{5R^4}{4r^3} + \frac{R^6}{r^5}\right)\sin 3\varphi\right], \\
\sigma_{\varphi\varphi} &= \frac{M_0}{I}\left[\left(\frac{3r}{4} + \frac{R^4}{4r^3}\right)\sin\varphi + \left(-\frac{r}{4} + \frac{R^4}{4r^3} - \frac{R^6}{r^5}\right)\sin 3\varphi\right], \\
\tau_{r\varphi} &= \frac{M_0}{I}\left[\left(-\frac{r}{4} + \frac{R^4}{4r^3}\right)\cos\varphi + \left(\frac{r}{4} + \frac{3R^4}{4r^3} - \frac{R^6}{r^5}\right)\cos 3\varphi\right].
\end{aligned} \tag{4.138}$$

The resulting stress state is shown in Fig. 4.15, right, in the form of a polar diagram (see Göldner et al. 1979) for the circumferential stress $\sigma_{\varphi\varphi}$ at the hole edge $r = R$, where here a normalization of the results to the maximum stress $\sigma_{xx}^{\max} = \frac{M_0 R}{I}$ of the undisturbed state was performed. The stress concentration amounts to the value 2.

References

Altenbach, H., Altenbach, J., Naumenko, K.: Ebene Flächentragwerke, 2nd edn. Springer, Berlin (2016)

Bauchau, O.A., Craig, J.I.: Structural Analysis. Springer, Dordrecht (2009)

Becker, W., Gross, D.: Mechanik elastischer Körper und Strukturen. Springer, Berlin (2002)

Boresi, A.P., Lynn, P.P.: Elasticity in Engineering Mechanics. Prentice-Hall, Eaglewood Cliffs (1974)

Chou, P.C., Pagano, N.J.: Elasticity. Van Nostrand, Princeton (1967)

Eschenauer, H., Schnell, W.: Elastizitätstheorie, 3rd edn. BI Wissenschaftsverlag, Mannheim (1993)

Eschenauer, H., Olhoff, N., Schnell, W.: Applied Structural Mechanics. Springer, Berlin (1997)

Filonenko-Boroditsch, M.M.: Elastizitätstheorie. VEB Fachbuchverlag, Leipzig (1967)

Girkmann, K.: Flächentragwerke, 6th edn. Springer, Wien (1974)

Göldner, H., Altenbach, J., Eschke, K., Garz, K.F., Sähn, S.: Lehrbuch Höhere Festigkeitslehre Band 1. Physik, Weinheim (1979)

Gross, D., Hauger, W., Wriggers, P.: Technische Mechanik 4, 9th edn. Springer, Berlin (2014)
Hahn, H.G.: Elastizitätstheorie. Teubner, Stuttgart (1985)
Hake, E., Meskouris, K.: Statik der Flächentragwerke, 2nd edn. Springer, Berlin (2007)
Leipholz, H.: Theory of Elasticity. Noordhoff, Leyden (1974)
Massonet, C.: Two-dimensional problems. In: Flügge, W. (ed.) Handbook of Engineering Mechanics. McGraw-Hill, New York (1962)
Sadd, M.H.: Elasticity. Elsevier, Amsterdam (2005)
Sechler, E.E.: Elasticity in Engineering. Wiley, New York (1952)
Timoshenko, S., Goodier, J.N.: Theory of Elasticity, 2nd edn. McGraw-Hill, New York (1951)
Ugural, A.C., Fenster, S.K.: Advanced Strength and Applied Elasticity. Prentice Hall, Hoboken (2003)
Wang, C.T.: Applied Elasticity. McGraw-Hill, New York (1953)

Approximation Methods for Isotropic Disks

5

In this chapter we will discuss approximation methods of structural mechanics for application to isotropic disk structures. First of all, we will discuss the classical Ritz method, where we will consider two versions, namely a displacement-based formulation on the one hand, and a force-based formulation on the other hand. Furthermore, we will discuss the finite element method (FEM). Approximate analysis methods, as they are relevant in the context of this chapter, are described e.g. in Becker and Gross (2002), Göldner et al. (1979, 1985), Kossira (1996), Lanczos (1986), Langhaar (2016), Mittelstedt (2021), Reddy (2017), Tauchert (1974) and Washizu (1982). A wealth of further information on the finite element method can be found in the works of Altenbach and Sakharov (1982), Bathe (1996), Betten (2003, 2004), Crisfield (1991, 1997), Oden and Reddy (2011), Reddy (2005), Zienkiewicz et al. (2013) and Zienkiewicz and Taylor (2013). For a discussion of the application of the finite difference method to disk structures see Girkmann (1974). Altenbach et al. (2016) discuss other classical approximation methods for disk structures.

5.1 The Displacement-Based Ritz Method

Starting point is the principle of the minimum of the total elastic potential $\Pi = \Pi_i + \Pi_a$, where Π_i is the inner potential/the strain energy of the structure under consideration, and Π_a is the potential of the applied loads. Here we will consider the Ritz method for disks in Cartesian coordinates in the plane stress state, the transition to the plane strain state does not present any difficulties. The inner potential Π of a disk has already been given in Chap. 3 with Eqs. (3.46), (3.47) and (3.48). Here we will turn to the Ritz method in the form of approaches for the displacements u_0 and v_0 of the disk. For better readability, the strain energy for this case for a rectangular disk (length a, width b) is given again here:

C. Mittelstedt, *Theory of Plates and Shells*,
https://doi.org/10.1007/978-3-662-66805-4_5

$$\Pi_i = \frac{A}{2}\int_0^b\int_0^a\left[\left(\frac{\partial u_0}{\partial x}+\nu\frac{\partial v_0}{\partial y}\right)\frac{\partial u_0}{\partial x}+\left(\frac{\partial v_0}{\partial y}+\nu\frac{\partial u_0}{\partial x}\right)\frac{\partial v_0}{\partial y}\right.$$
$$\left.+\frac{1-\nu}{2}\left(\frac{\partial u_0}{\partial y}+\frac{\partial v_0}{\partial x}\right)^2\right]\mathrm{d}x\mathrm{d}y. \tag{5.1}$$

The quantity A is the stiffness of the disk (see Eq. (3.8)), assumed as constant, and must not be confused with a surface or an area. We now introduce approximations in the form of separation approaches for the displacements u_0 and v_0 as follows:

$$u_0 = \sum_{m=1}^{M}\sum_{n=1}^{N} U_{mn}u_{1,m}(x)u_{2,n}(y),$$
$$v_0 = \sum_{m=1}^{K}\sum_{n=1}^{L} V_{mn}v_{1,m}(x)v_{2,n}(y). \tag{5.2}$$

The quantities U_{mn} and V_{mn} are unknown coefficients and are the actual target of the calculation. The functions $u_{1,m}(x)$, $u_{2,n}(y)$, $v_{1,m}(x)$, $v_{2,n}(y)$ are one-dimensional approach functions, which are used to approximate the displacement field of the disk. They are required to satisfy the underlying geometric boundary conditions of the considered disk situation and to be continuously differentiable at least once. The degrees of approach, i.e. the numbers of approach functions, are described by the integers K, L, M and N. Substituting the approach (5.2) into the inner potential Π_i (5.1), we get:

$$\Pi_i = \frac{A}{2}\int_0^b\int_0^a\left[\sum_{m=1}^{M}\sum_{n=1}^{N}\sum_{p=1}^{M}\sum_{q=1}^{N} U_{mn}U_{pq}\frac{\mathrm{d}u_{1,m}}{\mathrm{d}x}\frac{\mathrm{d}u_{1,p}}{\mathrm{d}x}u_{2,n}u_{2,q}\right.$$
$$+2\nu\sum_{m=1}^{M}\sum_{n=1}^{N}\sum_{p=1}^{K}\sum_{q=1}^{L} U_{mn}V_{pq}\frac{\mathrm{d}u_{1,m}}{\mathrm{d}x}v_{1,p}u_{2,n}\frac{\mathrm{d}v_{2,q}}{\mathrm{d}y}$$
$$+\sum_{m=1}^{K}\sum_{n=1}^{L}\sum_{p=1}^{K}\sum_{q=1}^{L} V_{mn}V_{pq}v_{1,m}v_{1,p}\frac{\mathrm{d}v_{2,n}}{\mathrm{d}y}\frac{\mathrm{d}v_{2,q}}{\mathrm{d}y}$$
$$+\frac{1-\nu}{2}\sum_{m=1}^{M}\sum_{n=1}^{N}\sum_{p=1}^{M}\sum_{q=1}^{N} U_{mn}U_{pq}u_{1,m}u_{1,p}\frac{\mathrm{d}u_{2,n}}{\mathrm{d}y}\frac{\mathrm{d}u_{2,q}}{\mathrm{d}y}$$
$$+(1-\nu)\sum_{m=1}^{M}\sum_{n=1}^{N}\sum_{p=1}^{K}\sum_{q=1}^{L} U_{mn}V_{pq}u_{1,m}\frac{\mathrm{d}v_{1,p}}{\mathrm{d}x}\frac{\mathrm{d}u_{2,n}}{\mathrm{d}y}v_{2,q}$$
$$\left.+\frac{1-\nu}{2}\sum_{m=1}^{K}\sum_{n=1}^{L}\sum_{p=1}^{K}\sum_{q=1}^{L} V_{mn}V_{pq}\frac{\mathrm{d}v_{1,m}}{\mathrm{d}x}\frac{\mathrm{d}v_{1,p}}{\mathrm{d}x}v_{2,n}v_{2,q}\right]\mathrm{d}x\mathrm{d}y. \tag{5.3}$$

We introduce the following abbreviations at this point:

$$\begin{aligned}
\Omega_{ij}^{mp,1} &= \int_0^a \frac{\mathrm{d}^i u_{1,m}}{\mathrm{d}x^i}\frac{\mathrm{d}^j u_{1,p}}{\mathrm{d}x^j}\mathrm{d}x, \quad \Omega_{ij}^{mp,2} = \int_0^a \frac{\mathrm{d}^i v_{1,m}}{\mathrm{d}x^i}\frac{\mathrm{d}^j v_{1,p}}{\mathrm{d}x^j}\mathrm{d}x,\\
\Omega_{ij}^{mp,3} &= \int_0^a \frac{\mathrm{d}^i u_{1,m}}{\mathrm{d}x^i}\frac{\mathrm{d}^j v_{1,p}}{\mathrm{d}x^j}\mathrm{d}x, \quad \Delta_{ij}^{nq,1} = \int_0^b \frac{\mathrm{d}^i u_{2,n}}{\mathrm{d}y^i}\frac{\mathrm{d}^j u_{2,q}}{\mathrm{d}y^j}\mathrm{d}y,\\
\Delta_{ij}^{nq,2} &= \int_0^b \frac{\mathrm{d}^i v_{2,n}}{\mathrm{d}y^i}\frac{\mathrm{d}^j v_{2,q}}{\mathrm{d}y^j}\mathrm{d}y, \quad \Delta_{ij}^{nq,3} = \int_0^b \frac{\mathrm{d}^i u_{2,n}}{\mathrm{d}y^i}\frac{\mathrm{d}^j v_{2,q}}{\mathrm{d}y^j}\mathrm{d}y. \qquad (5.4)
\end{aligned}$$

Thus (5.3) can also be written as:

$$\begin{aligned}
\Pi_i &= \frac{A}{2}\left[\sum_{m=1}^{M}\sum_{n=1}^{N}\sum_{p=1}^{M}\sum_{q=1}^{N} U_{mn}U_{pq}\left(\Omega_{11}^{mp,1}\Delta_{00}^{nq,1} + \frac{1-\nu}{2}\Omega_{00}^{mp,1}\Delta_{11}^{nq,1}\right)\right.\\
&+ \sum_{m=1}^{M}\sum_{n=1}^{N}\sum_{p=1}^{K}\sum_{q=1}^{L} U_{mn}V_{pq}\left(2\nu\Omega_{10}^{mp,3}\Delta_{01}^{nq,3} + (1-\nu)\,\Omega_{01}^{mp,3}\Delta_{10}^{nq,3}\right)\\
&\left.+ \sum_{m=1}^{K}\sum_{n=1}^{L}\sum_{p=1}^{K}\sum_{q=1}^{L} V_{mn}V_{pq}\left(\Omega_{00}^{mp,2}\Delta_{11}^{nq,2} + \frac{1-\nu}{2}\Omega_{11}^{mp,2}\Delta_{00}^{nq,2}\right)\right] \qquad (5.5)
\end{aligned}$$

With the further abbreviations

$$\begin{aligned}
\Gamma_1^{mnpq} &= \Omega_{11}^{mp,1}\Delta_{00}^{nq,1} + \frac{1-\nu}{2}\Omega_{00}^{mp,1}\Delta_{11}^{nq,1},\\
\Gamma_2^{mnpq} &= 2\nu\Omega_{10}^{mp,3}\Delta_{01}^{nq,3} + (1-\nu)\,\Omega_{01}^{mp,3}\Delta_{10}^{nq,3},\\
\Gamma_3^{mnpq} &= \Omega_{00}^{mp,2}\Delta_{11}^{nq,2} + \frac{1-\nu}{2}\Omega_{11}^{mp,2}\Delta_{00}^{nq,2} \qquad (5.6)
\end{aligned}$$

we obtain (5.5) in the following form:

$$\begin{aligned}
\Pi_i &= \frac{A}{2}\left[\sum_{m=1}^{M}\sum_{n=1}^{N}\sum_{p=1}^{M}\sum_{q=1}^{N} U_{mn}U_{pq}\Gamma_1^{mnpq}\right.\\
&+ \sum_{m=1}^{M}\sum_{n=1}^{N}\sum_{p=1}^{K}\sum_{q=1}^{L} U_{mn}V_{pq}\Gamma_2^{mnpq}\\
&\left.+ \sum_{m=1}^{K}\sum_{n=1}^{L}\sum_{p=1}^{K}\sum_{q=1}^{L} V_{mn}V_{pq}\Gamma_3^{mnpq}\right]. \qquad (5.7)
\end{aligned}$$

Furthermore, the external potential Π_a is needed, where for the sake of clarity we will consider here the disk situation of Fig. 5.1. Let a rectangular disk (length a, width b) be given, which is fixed at the edge $x = 0$ and free of any support at all other edges. At the boundary $x = a$ it is loaded by an arbitrary but continuously

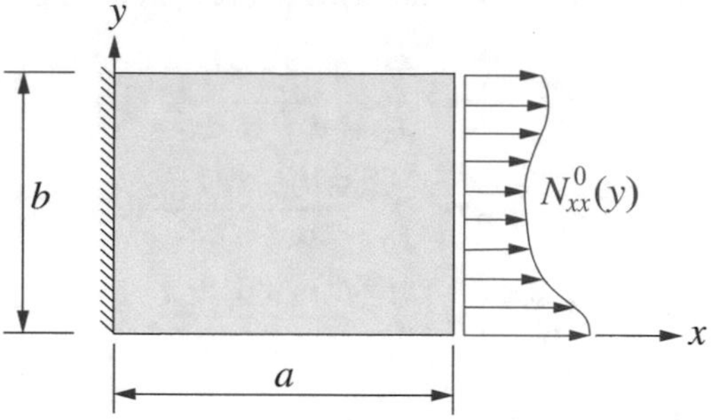

Fig. 5.1 Clamped rectangular disk under arbitrarily varying boundary load $N_{xx}^0(y)$

changing boundary force flow $N_{xx}^0(y)$. The external potential Π_a is then:

$$\Pi_a = -\int_0^b N_{xx}^0(y) \sum_{m=1}^{M} \sum_{n=1}^{N} U_{mn} u_{1,m}(x=a) u_{2,n}(y) \mathrm{d}y. \tag{5.8}$$

The still unknown coefficients U_{mn} and V_{mn} can be determined from the Ritz equations if the total elastic potential $\Pi = \Pi_i + \Pi_a$ is available:

$$\frac{\partial \Pi}{\partial U_{mn}} = 0, \quad \frac{\partial \Pi}{\partial V_{mn}} = 0. \tag{5.9}$$

This leads to a linear system of equations with $M \cdot N + K \cdot L$ equations for the $M \cdot N + K \cdot L$ coefficients U_{mn} and V_{mn}, which can be solved with any algorithm for solving such systems of equations. The displacement approach is then complete, and approximate strains and stresses can be determined using the kinematic equations of the disk and the constitutive equations.

Let us illustrate the Ritz method using the series expansion (5.2) with $M = N = K = L = 2$:

$$\begin{aligned} u_0 &= U_{11} u_{1,1}(x) u_{2,1}(y) + U_{12} u_{1,1}(x) u_{2,2}(y) + U_{21} u_{1,2}(x) u_{2,1}(y) \\ &\quad + U_{22} u_{1,2}(x) u_{2,2}(y), \\ v_0 &= V_{11} v_{1,1}(x) v_{2,1}(y) + V_{12} v_{1,1}(x) v_{2,2}(y) + V_{21} v_{1,2}(x) v_{2,1}(y) \\ &\quad + V_{22} v_{1,2}(x) v_{2,2}(y). \end{aligned} \tag{5.10}$$

The internal potential (5.7) can be written as:

$$\begin{aligned} \Pi_i &= \frac{A}{2} \Big[U_{11}^2 \Gamma_1^{1111} + U_{11} U_{12} \left(\Gamma_1^{1112} + \Gamma_1^{1211} \right) + U_{11} U_{21} \left(\Gamma_1^{1121} + \Gamma_1^{2111} \right) \\ &\quad + U_{11} U_{22} \left(\Gamma_1^{1122} + \Gamma_1^{2211} \right) + U_{12}^2 \Gamma_1^{1212} + U_{12} U_{21} \left(\Gamma_1^{1221} + \Gamma_1^{2112} \right) \\ &\quad + U_{12} U_{22} \left(\Gamma_1^{1222} + \Gamma_1^{2212} \right) + U_{21}^2 \Gamma_1^{2121} + U_{21} U_{22} \left(\Gamma_1^{2122} + \Gamma_1^{2221} \right) \\ &\quad + U_{22}^2 \Gamma_1^{2222} \end{aligned}$$

$$
\begin{aligned}
&+U_{11}V_{11}\Gamma_2^{1111}+U_{11}V_{12}\Gamma_2^{1112}+U_{11}V_{21}\Gamma_2^{1121}+U_{11}V_{22}\Gamma_2^{1122}\\
&+U_{12}V_{11}\Gamma_2^{1211}+U_{12}V_{12}\Gamma_2^{1212}+U_{12}V_{21}\Gamma_2^{1221}+U_{12}V_{22}\Gamma_2^{1222}\\
&+U_{21}V_{11}\Gamma_2^{2111}+U_{21}V_{12}\Gamma_2^{2112}+U_{21}V_{21}\Gamma_2^{2121}+U_{21}V_{22}\Gamma_2^{2122}\\
&+U_{22}V_{11}\Gamma_2^{2211}+U_{22}V_{12}\Gamma_2^{2212}+U_{22}V_{21}\Gamma_2^{2221}+U_{22}V_{22}\Gamma_2^{2222}\\
&+V_{11}^2\Gamma_3^{1111}+V_{11}V_{12}\left(\Gamma_3^{1112}+\Gamma_3^{1211}\right)+V_{11}V_{21}\left(\Gamma_3^{1121}+\Gamma_3^{2111}\right)\\
&+V_{11}V_{22}\left(\Gamma_3^{1122}+\Gamma_3^{2211}\right)+V_{12}^2\Gamma_3^{1212}+V_{12}V_{21}\left(\Gamma_3^{1221}+\Gamma_3^{2112}\right)\\
&+V_{12}V_{22}\left(\Gamma_3^{1222}+\Gamma_3^{2212}\right)+V_{21}^2\Gamma_3^{2121}+V_{21}V_{22}\left(\Gamma_3^{2122}+\Gamma_3^{2221}\right)\\
&+V_{22}^2\Gamma_3^{2222}\Big].
\end{aligned}
\tag{5.11}
$$

The Ritz equations (5.9) are then:

$$
\begin{aligned}
\frac{\partial\Pi}{\partial U_{11}}&=0, \quad \frac{\partial\Pi}{\partial U_{12}}=0, \quad \frac{\partial\Pi}{\partial U_{21}}=0, \quad \frac{\partial\Pi}{\partial U_{22}}=0,\\
\frac{\partial\Pi}{\partial V_{11}}&=0, \quad \frac{\partial\Pi}{\partial V_{12}}=0, \quad \frac{\partial\Pi}{\partial V_{21}}=0, \quad \frac{\partial\Pi}{\partial V_{22}}=0.
\end{aligned}
\tag{5.12}
$$

Performing the necessary differentiations finally yields a linear system of equations of order $M \times N + K \times L$ for the coefficients $U_{11}, U_{12}, U_{21}, U_{22}, V_{11}, V_{12}, V_{21}, V_{22}$, which is specifically given as follows:

$$
\frac{A}{2}\begin{bmatrix} \underline{\underline{K}}_1 & \underline{\underline{K}}_2 \\ \underline{\underline{K}}_2^T & \underline{\underline{K}}_3 \end{bmatrix}\begin{pmatrix} \underline{U} \\ \underline{V} \end{pmatrix} = \underline{F}, \tag{5.13}
$$

where $\underline{U} = (U_{11}, U_{12}, U_{21}, U_{22})^T$ and $\underline{V} = (V_{11}, V_{12}, V_{21}, V_{22})^T$. The symmetric submatrices $\underline{\underline{K}}_1$, $\underline{\underline{K}}_2$ and $\underline{\underline{K}}_3$, which together with the factor $\frac{A}{2}$ form the stiffness matrix of the given disk problem, can be given as:

$$
\begin{aligned}
\underline{\underline{K}}_1 &= \begin{bmatrix}
2\Gamma_1^{1111} & \Gamma_1^{1112}+\Gamma_1^{1211} & \Gamma_1^{1121}+\Gamma_1^{2111} & \Gamma_1^{1122}+\Gamma_1^{2211}\\
\Gamma_1^{1112}+\Gamma_1^{1211} & 2\Gamma_1^{1212} & \Gamma_1^{1221}+\Gamma_1^{2112} & \Gamma_1^{1222}+\Gamma_1^{2212}\\
\Gamma_1^{1121}+\Gamma_1^{2111} & \Gamma_1^{1221}+\Gamma_1^{2112} & 2\Gamma_1^{2121} & \Gamma_1^{2122}+\Gamma_1^{2221}\\
\Gamma_1^{1122}+\Gamma_1^{2211} & \Gamma_1^{1222}+\Gamma_1^{2212} & \Gamma_1^{2122}+\Gamma_1^{2221} & 2\Gamma_1^{2222}
\end{bmatrix},\\
\underline{\underline{K}}_2 &= \begin{bmatrix}
\Gamma_2^{1111} & \Gamma_2^{1112} & \Gamma_2^{1121} & \Gamma_2^{1122}\\
\Gamma_2^{1211} & \Gamma_2^{1212} & \Gamma_2^{1221} & \Gamma_2^{1222}\\
\Gamma_2^{2111} & \Gamma_2^{2112} & \Gamma_2^{2121} & \Gamma_2^{2122}\\
\Gamma_2^{2211} & \Gamma_2^{2212} & \Gamma_2^{2221} & \Gamma_2^{2222}
\end{bmatrix},\\
\underline{\underline{K}}_3 &= \begin{bmatrix}
2\Gamma_3^{1111} & \Gamma_3^{1112}+\Gamma_3^{1211} & \Gamma_3^{1121}+\Gamma_3^{2111} & \Gamma_3^{1122}+\Gamma_3^{2211}\\
\Gamma_3^{1112}+\Gamma_3^{1211} & 2\Gamma_3^{1212} & \Gamma_3^{1221}+\Gamma_3^{2112} & \Gamma_3^{1222}+\Gamma_3^{2212}\\
\Gamma_3^{1121}+\Gamma_3^{2111} & \Gamma_3^{1221}+\Gamma_3^{2112} & 2\Gamma_3^{2121} & \Gamma_3^{2122}+\Gamma_3^{2221}\\
\Gamma_3^{1122}+\Gamma_3^{2211} & \Gamma_3^{1222}+\Gamma_3^{2212} & \Gamma_3^{2122}+\Gamma_3^{2221} & 2\Gamma_3^{2222}
\end{bmatrix}.
\end{aligned}
\tag{5.14}
$$

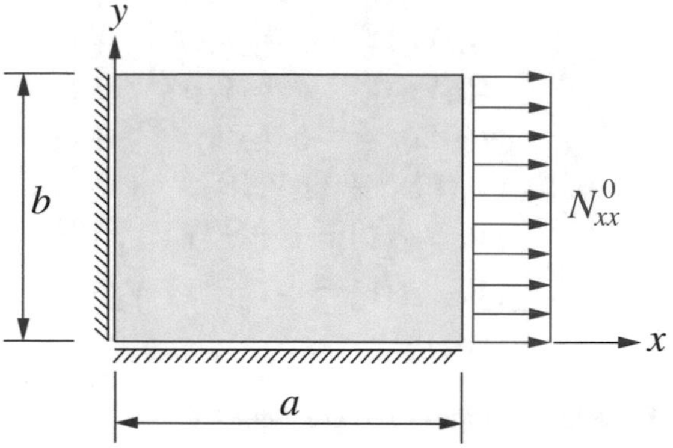

Fig. 5.2 Rectangular disk under constant boundary load N_{xx}^0

Accordingly, these matrices can also be written for higher degrees of series expansion M, N, K, L. The force vector $\underline{F}$ is not specified at this point since it depends on the concrete load case.

In order to illustrate the Ritz method, we consider the elementary simple disk situation of Fig. 5.2. Given is a rectangular disk, which is supported at the two edges $x = 0$ and $y = 0$ with respect to the displacements u_0 and v_0. All other displacements are unobstructed at these edges, the remaining two edges are free of any bearing. The disk is loaded at its edge at $x = a$ by the constant edge force flow N_{xx}^0. We choose for this very simple situation an approach of the form (5.2), where we fix the degrees of series expansion all with the value 1, i.e.:

$$u_0 = U u_1(x) u_2(y), \quad v_0 = V v_1(x) v_2(y). \tag{5.15}$$

It is clear that the displacement field of the considered disk will be such that the longitudinal displacement is a linear function of x and independent of y. Analogously, this is also true for the transverse displacement v_0, so that the following approaches are used:

$$u_0 = Ux, \quad v_0 = Vy. \tag{5.16}$$

Substituting (5.16) into the strain energy Π_i (5.1) then gives:

$$\Pi_i = \frac{A}{2} ab \left(U^2 + V^2 + 2\nu UV \right). \tag{5.17}$$

The external potential (5.8) then takes on the following form:

$$\Pi_a = -N_{xx}^0 Uab. \tag{5.18}$$

The two Ritz equations to be evaluated here are then:

$$\begin{aligned}\frac{\partial \Pi}{\partial U} = Aab\,(U + \nu V) - N^0_{xx}ab &= 0,\\ \frac{\partial \Pi}{\partial V} = Aab\,(V + \nu U) &= 0.\end{aligned} \tag{5.19}$$

Solving gives with $A = \frac{Eh}{1-\nu^2}$:

$$U = \frac{N^0_{xx}}{Eh}, \quad V = -\nu\frac{N^0_{xx}}{Eh}, \tag{5.20}$$

and thus the approximate disk displacements u_0 and v_0 can be formulated as:

$$u_0 = \frac{N^0_{xx}x}{Eh}, \quad v_0 = -\nu\frac{N^0_{xx}y}{Eh}. \tag{5.21}$$

For comparison, we consider the exact disk solution that can be derived using an approach to Airy's stress function of the form $F = \frac{1}{2}\sigma_0 y^2$ and the associated constant normal stress $\sigma_{xx} = \sigma_0 = \frac{N^0_{xx}}{h}$. The strains ε_{xx} and ε_{yy} are obtained as:

$$\begin{aligned}\varepsilon_{xx} &= \frac{1}{E}\left(\sigma_{xx} - \nu\sigma_{yy}\right) = \frac{N^0_{xx}}{Eh} = \frac{\partial u_0}{\partial x},\\ \varepsilon_{yy} &= \frac{1}{E}\left(\sigma_{yy} - \nu\sigma_{xx}\right) = -\frac{\nu N^0_{xx}}{Eh} = \frac{\partial v_0}{\partial y}.\end{aligned} \tag{5.22}$$

The displacements of the disk can then be determined by integrating the strains:

$$\begin{aligned}u_0 &= \int \varepsilon_{xx}\mathrm{d}x = \frac{N^0_{xx}x}{Eh},\\ v_0 &= \int \varepsilon_{yy}\mathrm{d}y = -\nu\frac{N^0_{xx}y}{Eh}.\end{aligned} \tag{5.23}$$

Obviously, for this simple example, the Ritz solution (5.21) and the exact solution (5.23) match each other. The reason is that for the Ritz method the exact displacement functions were used as approximation approaches.

We also consider the disk situation as given in Fig. 5.3. This is the same situation as discussed in the previous example, but now the edge at $x = 0$ is additionally restrained concerning the displacement v_0. For the approximate analysis in the framework of the Ritz method, we will assume that the displacements u_0 evolve linearly over x and do not change at all over y. For the transverse displacement v_0, on the

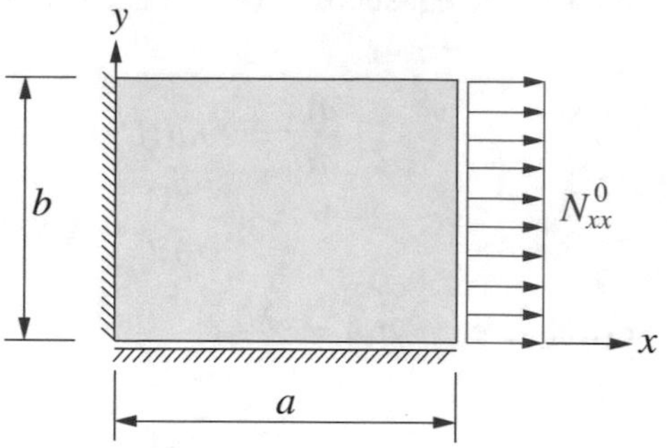

Fig. 5.3 Rectangular disk under constant boundary load N^0_{xx}

other hand, we assume that it evolves parabolically over x and linearly over y. We thus employ the following functions:

$$u_0 = Ux, \quad v_0 = V_1 xy + V_2 x^2 y. \tag{5.24}$$

Substituting in the strain energy (5.1) yields after performing the necessary differentiations:

$$\begin{aligned}\Pi_i &= \frac{A}{2}\int_0^b\int_0^a \Big\{\left[U + \nu\left(V_1 x + V_2 x^2\right)\right]U + \left(V_1 x + V_2 x^2 + \nu U\right)\left(V_1 x + V_2 x^2\right) \\ &\quad + \frac{1-\nu}{2}\left(V_1 y + 2V_2 xy\right)^2\Big\}\,\mathrm{d}x\mathrm{d}y.\end{aligned} \tag{5.25}$$

Performing the prescribed integration results in:

$$\begin{aligned}\Pi_i &= \frac{A}{2}\Big[U^2 ab + \nu U V_1 a^2 b + \frac{2}{3}\nu U V_2 a^3 b + \frac{1}{3}V_1^2 a^3 b + \frac{1}{2}V_1 V_2 a^4 b + \frac{1}{5}V_2^2 a^5 b \\ &\quad + \frac{1-\nu}{6}V_1^2 ab^3 + \frac{1-\nu}{3}V_1 V_2 a^2 b^3 + \frac{2}{9}(1-\nu)\,V_2^2 a^3 b^3\Big].\end{aligned} \tag{5.26}$$

Since for this example the choice of the function u_0 is identical to the previously discussed disk situation, the external potential Π_a remains unchanged with respect to (5.18), i.e. $\Pi_a = -N^0_{xx} Uab$. Thus, we can evaluate the Ritz equations, which are currently as follows:

$$\frac{\partial \Pi}{\partial U} = 0, \quad \frac{\partial \Pi}{\partial V_1} = 0, \quad \frac{\partial \Pi}{\partial V_2} = 0. \tag{5.27}$$

After some elementary arithmetic operations, the following linear equation system for the still unknown coefficients U, V_1, V_2 is obtained:

$$\begin{aligned} U + \frac{\nu}{2} V_1 a + \frac{\nu}{3} V_2 a^2 &= \frac{N_{xx}^0}{A}, \\ \nu U a + \frac{1}{3}\left[2a^2 + (1-\nu)\, b^2\right] V_1 + \left(\frac{a^2}{2} + \frac{1-\nu}{3} b^2\right) a V_2 &= 0, \\ \frac{2}{3}\nu U a + \left(\frac{a^2}{2} + \frac{1-\nu}{3} b^2\right) V_1 + 2a \left[\frac{a^2}{5} + \frac{2}{9}(1-\nu)\, b^2\right] V_2 &= 0. \end{aligned} \tag{5.28}$$

From this, the constants U, V_1, V_2 can be determined. The result, however, is not represented here. If the approach coefficients are available, then the completely determined approximate displacements can also be used to determine approximate strains and thus also approximate stresses. However, this remains here without further presentation.

We now want to illustrate this example a little more concretely by not limiting the degrees M, N, K, L and using polynomials as approach functions. The series expansion (5.2) would thus be:

$$\begin{aligned} u_0 &= \sum_{m=1}^{M} \sum_{n=1}^{N} U_{mn} x^m y^n, \\ v_0 &= \sum_{m=1}^{K} \sum_{n=1}^{L} V_{mn} x^m y^n, \end{aligned} \tag{5.29}$$

where the geometric boundary conditions $u_0(x = 0, y) = v_0(x = 0, y) = 0$ and $v_0(x, y = 0)$ are satisfied a priori. The abbreviations (5.4) are then with (5.29):

$$\begin{aligned} \Omega_{00}^{mp,1} &= \int_0^a x^m x^p \mathrm{d}x = \frac{a^{m+p+1}}{m+p+1}, \\ \Omega_{01}^{mp,1} &= p \int_0^a x^m x^{p-1} \mathrm{d}x = \frac{p a^{m+p}}{m+p}, \\ \Omega_{10}^{mp,1} &= m \int_0^a x^{m-1} x^p \mathrm{d}x = \frac{m a^{m+p}}{m+p}, \\ \Omega_{11}^{mp,1} &= mp \int_0^a x^{m-1} x^{p-1} \mathrm{d}x = \frac{mp a^{m+p-1}}{m+p-1}. \end{aligned} \tag{5.30}$$

Since in the specific polynomial approach used here (5.29) the employed functions are identical, it can be concluded that

$$\Omega_{ij}^{mp,1} = \Omega_{ij}^{mp,2} = \Omega_{ij}^{mp,3} \tag{5.31}$$

holds. The auxiliary quantities $\Delta_{ij}^{nq,1} = \Delta_{ij}^{nq,2} = \Delta_{ij}^{nq,3}$ can be obtained from (5.31) by exchanging the indices m and p for n and q. The disk length a is then also to be exchanged for the width b. The inner potential Π_i can then be written as:

$$\begin{aligned}
\Pi_i = \frac{A}{2}\Bigg\{ & \sum_{m=2}^{M}\sum_{n=1}^{N}\sum_{p=2}^{M}\sum_{q=1}^{N} U_{mn}U_{pq}a^{m+p-1}b^{n+q-1} \\
& \times\left[\frac{mpb^2}{(m+p-1)(n+q+1)} + \frac{1-\nu}{2}\frac{nqa^2}{(m+p+1)(n+q-1)}\right] \\
& + \sum_{m=2}^{M}\sum_{n=1}^{N}\sum_{p=2}^{K}\sum_{q=2}^{L} U_{mn}V_{pq}\frac{a^{m+p}b^{n+q}}{(m+p)(n+q)}(2\nu mq + (1-\nu)np) \\
& + \sum_{m=2}^{K}\sum_{n=2}^{L}\sum_{p=2}^{K}\sum_{q=2}^{L} V_{mn}V_{pq}a^{m+p-1}b^{n+q-1} \\
& \times\left[\frac{nqa^2}{(m+p+1)(n+q-1)}\right. \\
& \left.+\frac{1-\nu}{2}\frac{mpb^2}{(m+p-1)(n+q+1)}\right]\Bigg\}.
\end{aligned} \tag{5.32}$$

The external potential Π_a is given for the present case after performing the necessary integrations:

$$\Pi_a = -N_{xx}^0 \sum_{m=1}^{M}\sum_{n=1}^{N} U_{mn}\frac{a^m b^{n+1}}{n+1}. \tag{5.33}$$

It should be noted that the consideration of additional displacement terms of the type

$$u_0 = \sum_{m=1}^{M} U_{m0}x^m, \tag{5.34}$$

which are constant over y and variable in x and which also fulfill the given boundary conditions has been neglected. These can, if necessary, be incorporated into the above formulation, but this is not shown here for reasons of brevity.

5.2 The Force-Based Ritz Method

In addition to a displacement-based formulation of the Ritz method for disk structures, a formulation with approximation approaches for the Airy stress function F can be given, where the basis of the considerations is the strain energy in the formulation (3.47), which we give here again for better readability:

$$\begin{aligned} \Pi_i &= \frac{1}{2Eh}\int_0^b \int_0^a \left[\left(\frac{\partial^2 F}{\partial x^2}\right)^2 + \left(\frac{\partial^2 F}{\partial y^2}\right)^2 - 2\nu \frac{\partial^2 F}{\partial x^2}\frac{\partial^2 F}{\partial y^2} \right. \\ &\quad \left. +2(1+\nu)\left(\frac{\partial^2 F}{\partial x \partial y}\right)^2 \right] \mathrm{d}x\mathrm{d}y. \end{aligned} \tag{5.35}$$

Consider a rectangular disk of dimensions a and b, which is currently under an arbitrary load, i.e. the arbitrary but continuously distributed force flows $\hat{N}_{xx}^0$, $\hat{N}_{xy}^0$ and $\hat{N}_{yy}^0$, $\hat{N}_{xy}^0$ are given on the edges $x = 0$, $x = a$ and $y = 0$, $y = b$, respectively. We formulate an approach for the Airy stress function F as follows:

$$F(x, y) = F_0(x, y) + \sum_{m=1}^{M}\sum_{n=1}^{N} C_{mn} F_{mx}(x) F_{ny}(y). \tag{5.36}$$

The approach for the Airy stress function is required to be twofold continuously differentiable and to satisfy the corresponding boundary conditions. The element F_0 is chosen in such a way that all underlying boundary conditions are fulfilled. Then the functions $F_{mx}(x)$ and $F_{ny}(y)$ can be chosen in such a way that they satisfy homogeneous boundary conditions. The approximate force flows N_{xx}^0, N_{yy}^0, N_{xy}^0 of the disk are then obtained as:

$$\begin{aligned} N_{xx}^0 &= \frac{\partial^2 F_0}{\partial y^2} + \sum_{m=1}^{M}\sum_{n=1}^{N} C_{mn} F_{mx} \frac{\mathrm{d}^2 F_{ny}}{\mathrm{d}y^2}, \\ N_{yy}^0 &= \frac{\partial^2 F_0}{\partial x^2} + \sum_{m=1}^{M}\sum_{n=1}^{N} C_{mn} \frac{\mathrm{d}^2 F_{mx}}{\mathrm{d}x^2} F_{ny}, \\ N_{xy}^0 &= -\frac{\partial^2 F_0}{\partial x \partial y} - \sum_{m=1}^{M}\sum_{n=1}^{N} C_{mn} \frac{\mathrm{d}F_{mx}}{\mathrm{d}x}\frac{\mathrm{d}F_{ny}}{\mathrm{d}y}. \end{aligned} \tag{5.37}$$

From this, approximate strains and displacements of the considered disk can be calculated. Substituting the approach (5.36) into (5.35) yields:

$$
\begin{aligned}
\Pi_i \;=\; & \frac{1}{2Eh}\int_0^b\int_0^a\Bigg[\left(\frac{\partial^2 F_0}{\partial x^2}\right)^2+\left(\frac{\partial^2 F_0}{\partial y^2}\right)^2+2\frac{\partial^2 F_0}{\partial x^2}\sum_{m=1}^{M}\sum_{n=1}^{N}C_{mn}\frac{\mathrm{d}^2 F_{mx}}{\mathrm{d}x^2}F_{ny}\\
&+2\frac{\partial^2 F_0}{\partial y^2}\sum_{m=1}^{M}\sum_{n=1}^{N}C_{mn}F_{mx}\frac{\mathrm{d}^2 F_{ny}}{\mathrm{d}y^2}\\
&+\sum_{m=1}^{M}\sum_{n=1}^{N}\sum_{p=1}^{M}\sum_{q=1}^{N}C_{mn}C_{pq}\frac{\mathrm{d}^2 F_{mx}}{\mathrm{d}x^2}\frac{\mathrm{d}^2 F_{px}}{\mathrm{d}x^2}F_{ny}F_{qy}\\
&+\sum_{m=1}^{M}\sum_{n=1}^{N}\sum_{p=1}^{M}\sum_{q=1}^{N}C_{mn}C_{pq}F_{mx}F_{px}\frac{\mathrm{d}^2 F_{ny}}{\mathrm{d}y^2}\frac{\mathrm{d}^2 F_{qy}}{\mathrm{d}y^2}\Bigg]\mathrm{d}x\mathrm{d}y\\
&-\frac{\nu}{Eh}\int_0^b\int_0^a\Bigg[\frac{\partial^2 F_0}{\partial x^2}\frac{\partial^2 F_0}{\partial y^2}+\frac{\partial^2 F_0}{\partial x^2}\sum_{m=1}^{M}\sum_{n=1}^{N}C_{mn}F_{mx}\frac{\mathrm{d}^2 F_{ny}}{\mathrm{d}y^2}\\
&+\frac{\partial^2 F_0}{\partial y^2}\sum_{m=1}^{M}\sum_{n=1}^{N}C_{mn}\frac{\mathrm{d}^2 F_{mx}}{\mathrm{d}x^2}F_{ny}\\
&+\sum_{m=1}^{M}\sum_{n=1}^{N}\sum_{p=1}^{M}\sum_{q=1}^{N}C_{mn}C_{pq}\frac{\mathrm{d}^2 F_{mx}}{\mathrm{d}x^2}F_{px}F_{ny}\frac{\mathrm{d}^2 F_{qy}}{\mathrm{d}y^2}\Bigg]\mathrm{d}x\mathrm{d}y\\
&+\frac{1+\nu}{Eh}\int_0^b\int_0^a\Bigg[\left(\frac{\partial^2 F_0}{\partial x\partial y}\right)^2+2\frac{\partial^2 F_0}{\partial x\partial y}\sum_{m=1}^{M}\sum_{n=1}^{N}C_{mn}\frac{\mathrm{d}F_{mx}}{\mathrm{d}x}\frac{\mathrm{d}F_{ny}}{\mathrm{d}y}\\
&+\sum_{m=1}^{M}\sum_{n=1}^{N}\sum_{p=1}^{M}\sum_{q=1}^{N}C_{mn}C_{pq}\frac{\mathrm{d}F_{mx}}{\mathrm{d}x}\frac{\mathrm{d}F_{px}}{\mathrm{d}x}\frac{\mathrm{d}F_{nq}}{\mathrm{d}y}\frac{\mathrm{d}F_{qy}}{\mathrm{d}y}\Bigg]\mathrm{d}x\mathrm{d}y. \qquad (5.38)
\end{aligned}
$$

At this point, the following abbreviations prove to be useful:

$$
\Phi_1=\int_0^b\int_0^a\left(\frac{\partial^2 F_0}{\partial x^2}\right)^2\mathrm{d}x\mathrm{d}y,\quad \Phi_2=\int_0^b\int_0^a\left(\frac{\partial^2 F_0}{\partial y^2}\right)^2\mathrm{d}x\mathrm{d}y,
$$

$$
\Phi_3=\int_0^b\int_0^a\frac{\partial^2 F_0}{\partial x^2}\frac{\partial^2 F_0}{\partial y^2}\mathrm{d}x\mathrm{d}y,\quad \Phi_4=\int_0^b\int_0^a\left(\frac{\partial^2 F_0}{\partial x\partial y}\right)^2\mathrm{d}x\mathrm{d}y, \qquad (5.39)
$$

and

$$\Lambda_1^{mn} = 2\int_0^b\int_0^a \frac{\partial^2 F_0}{\partial x^2}\frac{\mathrm{d}^2 F_{mx}}{\mathrm{d}x^2}F_{ny}\mathrm{d}x\mathrm{d}y, \quad \Lambda_2^{mn} = 2\int_0^b\int_0^a \frac{\partial^2 F_0}{\partial y^2}F_{mx}\frac{\mathrm{d}^2 F_{ny}}{\mathrm{d}y^2}\mathrm{d}x\mathrm{d}y,$$

$$\Lambda_3^{mn} = \int_0^b\int_0^a \frac{\partial^2 F_0}{\partial x^2}F_{mx}\frac{\mathrm{d}^2 F_{ny}}{\mathrm{d}y^2}\mathrm{d}x\mathrm{d}y, \quad \Lambda_4^{mn} = \int_0^b\int_0^a \frac{\partial^2 F_0}{\partial y^2}\frac{\mathrm{d}^2 F_{mx}}{\mathrm{d}x^2}F_{ny}\mathrm{d}x\mathrm{d}y,$$

$$\Lambda_5^{mn} = 2\int_0^b\int_0^a \frac{\partial^2 F_0}{\partial x\partial y}\frac{\mathrm{d}F_{mx}}{\mathrm{d}x}\frac{\mathrm{d}F_{ny}}{\mathrm{d}y}\mathrm{d}x\mathrm{d}y \tag{5.40}$$

and as well

$$\Omega_{ij}^{mp} = \int_0^a \frac{\mathrm{d}^i F_{mx}}{\mathrm{d}x^i}\frac{\mathrm{d}^j F_{px}}{\mathrm{d}x^j}\mathrm{d}x, \quad \Delta_{ij}^{nq} = \int_0^b \frac{\mathrm{d}^i F_{ny}}{\mathrm{d}y^i}\frac{\mathrm{d}^j F_{qy}}{\mathrm{d}y^j}\mathrm{d}y. \tag{5.41}$$

Thus (5.38) can be represented as:

$$\begin{aligned}
\Pi_i &= \frac{1}{2Eh}\{\Phi_1 + \Phi_2 - 2\nu\Phi_3 + 2(1+\nu)\Phi_4 \\
&\quad + \sum_{m=1}^{M}\sum_{n=1}^{N} C_{mn}\left[\Lambda_1^{mn} + \Lambda_2^{mn} - 2\nu\left(\Lambda_3^{mn} + \Lambda_4^{mn}\right) + 2(1+\nu)\Lambda_5^{mn}\right] \\
&\quad + \sum_{m=1}^{M}\sum_{n=1}^{N}\sum_{p=1}^{M}\sum_{q=1}^{N} C_{mn}C_{pq}\left[\Omega_{22}^{mp}\Delta_{00}^{nq} + \Omega_{00}^{mp}\Delta_{22}^{nq}\right. \\
&\quad \left. - 2\nu\Omega_{20}^{mp}\Delta_{02}^{nq} + 2(1+\nu)\Omega_{11}^{mp}\Delta_{11}^{nq}\right]\}.
\end{aligned} \tag{5.42}$$

With the abbreviations

$$\begin{aligned}
\Gamma_0 &= \frac{1}{2Eh}\left[\Phi_1 + \Phi_2 - 2\nu\Phi_3 + 2(1+\nu)\Phi_4\right], \\
\Gamma_1^{mn} &= \frac{1}{2Eh}\left[\Lambda_1^{mn} + \Lambda_2^{mn} - 2\nu\left(\Lambda_3^{mn} + \Lambda_4^{mn}\right) + 2(1+\nu)\Lambda_5^{mn}\right], \\
\Gamma_2^{mnpq} &= \frac{1}{2Eh}\left[\Omega_{22}^{mp}\Delta_{00}^{nq} + \Omega_{00}^{mp}\Delta_{22}^{nq}\right. \\
&\quad \left. -2\nu\Omega_{20}^{mp}\Delta_{02}^{nq} + 2(1+\nu)\Omega_{11}^{mp}\Delta_{11}^{nq}\right]
\end{aligned} \tag{5.43}$$

this leads to:

$$\Pi_i = \Gamma_0 + \sum_{m=1}^{M}\sum_{n=1}^{N} C_{mn}\Gamma_1^{mn} + \sum_{m=1}^{M}\sum_{n=1}^{N}\sum_{p=1}^{M}\sum_{q=1}^{N} C_{mn}C_{pq}\Gamma_2^{mnpq}. \tag{5.44}$$

The Ritz equations

$$\frac{\partial \Pi_i}{\partial C_{mn}} = 0 \tag{5.45}$$

then lead to the equations necessary for the determination of the constants C_{mn}. Further details on the force-based Ritz method can be found in Altenbach et al. (2016) for the case where the resulting stress state is found to be independent of the Poisson's ratio ν.

5.3 Finite Elements for Disks

In addition to the analytical solutions already discussed and the approximation solutions using the Ritz method, purely numerical methods such as the finite element method are used when the aforementioned methods reach their limits. This will always be the case when either the disk geometry and/or the underlying load and boundary conditions are too complex for an analytical treatment. We will therefore discuss in this section the finite element analysis of plane isotropic disks in the plane stress state.

When applying the finite element method, a vector-matrix notation is very convenient. We first consider the kinematic relations:

$$\varepsilon_{xx}^0 = \frac{\partial u_0}{\partial x}, \quad \varepsilon_{yy}^0 = \frac{\partial v_0}{\partial y}, \quad \gamma_{xy}^0 = \frac{\partial u_0}{\partial y} + \frac{\partial v_0}{\partial x}. \tag{5.46}$$

In vector-matrix notation, this reads:

$$\begin{pmatrix} \varepsilon_{xx}^0 \\ \varepsilon_{yy}^0 \\ \gamma_{xy}^0 \end{pmatrix} = \begin{bmatrix} \frac{\partial}{\partial x} & 0 \\ 0 & \frac{\partial}{\partial y} \\ \frac{\partial}{\partial y} & \frac{\partial}{\partial x} \end{bmatrix} \begin{pmatrix} u_0 \\ v_0 \end{pmatrix}, \tag{5.47}$$

or in symbolic form:

$$\underline{\varepsilon}^0 = \underline{\underline{B}}\,\underline{u}. \tag{5.48}$$

Here $\underline{\varepsilon}^0$ is the strain vector, $\underline{\underline{B}}$ the so-called operator matrix and $\underline{u}$ the displacement vector. Hooke's law for the plane stress state of an isotropic disk can be formulated quite similarly, which we also state again here:

$$\sigma_{xx} = \frac{E}{1-\nu^2}\left(\varepsilon_{xx}^0 + \nu\varepsilon_{yy}^0\right), \sigma_{yy} = \frac{E}{1-\nu^2}\left(\varepsilon_{yy}^0 + \nu\varepsilon_{xx}^0\right), \tau_{xy} = G\gamma_{xy}^0. \quad (5.49)$$

With $G = \frac{E}{2(1+\nu)}$ this can be expressed in vector-matrix form as:

$$\begin{pmatrix} \sigma_{xx} \\ \sigma_{yy} \\ \tau_{xy} \end{pmatrix} = \frac{E}{1-\nu^2}\begin{bmatrix} 1 & \nu & 0 \\ \nu & 1 & 0 \\ 0 & 0 & \frac{1-\nu}{2} \end{bmatrix}\begin{pmatrix} \varepsilon_{xx}^0 \\ \varepsilon_{yy}^0 \\ \gamma_{xy}^0 \end{pmatrix}, \quad (5.50)$$

or in symbolic form:

$$\underline{\sigma} = \underline{\underline{Q}}\varepsilon^0. \quad (5.51)$$

In this, $\underline{\sigma}$ is the stress vector and $\underline{\underline{Q}}$ is the stiffness matrix. It contains the reduced stiffnesses of the plane stress state.

Now consider a disk structure discretized by disk elements, for example with four nodes, where the nodes are arranged in the element corners (Fig. 5.4). It should be noted that, depending on the requirements, one can also use elements with other numbers of nodes. The volume V of the disk is thus meshed by n elements, where the volume of the element e has the value V_e. Thus, the disk is geometrically approximated by the union of all elements, i.e.:

$$V \simeq \bigcup_{e-1}^{e=n} V_e. \quad (5.52)$$

Each node i of the element e has the degree of freedom $u_{i,e}$, i.e. the displacement of the node i of the element e in x-direction, and the degree of freedom $v_{i,e}$ (displacement of the node i of the element e in y-direction). The nodal degrees of freedom are summarized in the vector $\underline{v}_{i,e} = \begin{pmatrix} u_{i,e} & v_{i,e} \end{pmatrix}^T$. From the nodal displacements, which are still unknown at this point, an approximate displacement field $\underline{\tilde{u}}_e(x, y)$ in element e can be obtained using the shape functions $N_{i,e}(x, y)$:

$$\underline{\tilde{u}}_e(x, y) = \sum_{i=1}^{i=n_e} N_{i,e}(x, y)\underline{v}_{i,e}, \quad (5.53)$$

where n_e is the number of nodes of the considered disk element. Using the example of a four-node disk element, we get:

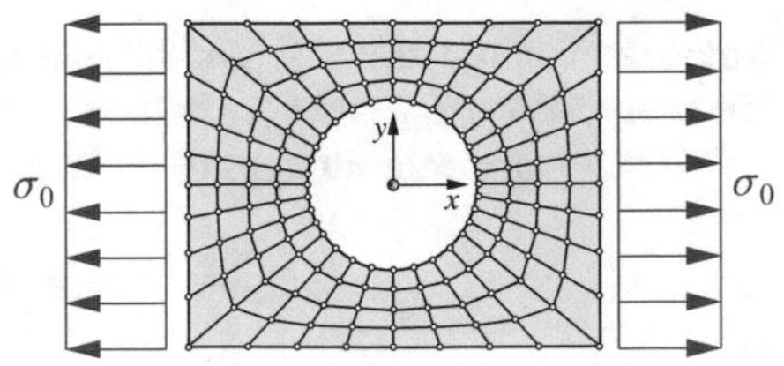

Fig. 5.4 Perforated disk, discretized with disk elements with four nodes

$$\underline{\tilde{u}}_e = \begin{pmatrix} \tilde{u}_e \\ \tilde{v}_e \end{pmatrix} = N_{1,e} \begin{pmatrix} u_{1,e} \\ v_{1,e} \end{pmatrix} + N_{2,e} \begin{pmatrix} u_{2,e} \\ v_{2,e} \end{pmatrix} + N_{3,e} \begin{pmatrix} u_{3,e} \\ v_{3,e} \end{pmatrix} + N_{4,e} \begin{pmatrix} u_{4,e} \\ v_{4,e} \end{pmatrix}. \tag{5.54}$$

The shape functions $N_{i,e}$ are formulated in such a way that they assume the value 1 in node i. In all further nodes they become zero. Typical for the formulation of the shape functions is the use of polynomials.

From the kinematic relations, the approximate strain field in element e can then be determined using the chosen displacement approach:

$$\underline{\tilde{\varepsilon}}_e = \begin{pmatrix} \tilde{\varepsilon}_{xx,e} \\ \tilde{\varepsilon}_{yy,e} \\ \tilde{\gamma}_{xy,e} \end{pmatrix} = \sum_{i=1}^{i=n_e} \begin{bmatrix} \frac{\partial N_{i,e}}{\partial x} & 0 \\ 0 & \frac{\partial N_{i,e}}{\partial y} \\ \frac{\partial N_{i,e}}{\partial y} & \frac{\partial N_{i,e}}{\partial x} \end{bmatrix} \begin{pmatrix} u_{i,e} \\ v_{i,e} \end{pmatrix} = \sum_{i=1}^{i=n_e} \underline{\underline{B}}_{i,e} \underline{v}_{i,e}. \tag{5.55}$$

The introduction of the following matrices turns out to be convenient:

$$\underline{\underline{N}}_{i,e} = \begin{bmatrix} N_{i,e} & 0 \\ 0 & N_{i,e} \end{bmatrix}, \quad \underline{\underline{N}}_e = \begin{bmatrix} \underline{\underline{N}}_{1,e} & \underline{\underline{N}}_{2,e} & \underline{\underline{N}}_{3,e} & \underline{\underline{N}}_{4,e} \end{bmatrix},$$

$$\underline{\underline{B}}_e = \begin{bmatrix} \underline{\underline{B}}_{1,e} & \underline{\underline{B}}_{2,e} & \underline{\underline{B}}_{3,e} & \underline{\underline{B}}_{4,e} \end{bmatrix}. \tag{5.56}$$

We also summarize the nodal displacements $u_{1,e}$, $v_{1,e}$,...,$u_{4,e}$, $v_{4,e}$ in the vector $\underline{v}_e$:

$$\underline{v}_e = \begin{pmatrix} u_{1,e} & v_{1,e} & u_{2,e} & v_{2,e} & u_{3,e} & v_{3,e} & u_{4,e} & v_{4,e} \end{pmatrix}^T. \tag{5.57}$$

We can then formulate the approximated displacements and strains of the element as:

$$\underline{\tilde{u}}_e = \underline{\underline{N}}_e \underline{v}_e, \quad \underline{\tilde{\varepsilon}}_e = \underline{\underline{B}}_e \underline{v}_e. \tag{5.58}$$

For the derivation of the element equations, the principle of the minimum of the total elastic potential is used. Substitution of the approximate displacements and strains of the element then gives:

$$\delta\Pi = \int_{V_e} \underline{\sigma}_e \delta\underline{\tilde{\varepsilon}}_e \mathrm{d}V_e - \int_{V_e} \underline{f}_e \delta\underline{\tilde{u}}_e \mathrm{d}V_e - \int_{S_{t,e}} \underline{t}_{0,e} \delta\underline{\tilde{u}}_e \mathrm{d}S_e = 0, \tag{5.59}$$

where the volume forces $f_{x,e}$ and $f_{y,e}$ were combined in the vector $\underline{f}_e$ and the given boundary loads $t_{x0,e}$ and $t_{y0,e}$ in the vector $\underline{t}_{0,e}$.

We summarize the stresses of the element e in the vector

$$\underline{\sigma}_e = \begin{pmatrix} \sigma_{xx,e} & \sigma_{yy,e} & \tau_{xy,e} \end{pmatrix}^T. \tag{5.60}$$

They can be determined from the strains $\underline{\tilde{\varepsilon}}_e$ using Hooke's law as:

$$\underline{\sigma}_e = \begin{pmatrix} \sigma_{xx,e} \\ \sigma_{yy,e} \\ \tau_{xy,e} \end{pmatrix} = \begin{bmatrix} \frac{E_e}{1-\nu_e^2} & \frac{\nu_e E_e}{1-\nu_e^2} & 0 \\ \frac{\nu_e E_e}{1-\nu_e^2} & \frac{E_e}{1-\nu_e^2} & 0 \\ 0 & 0 & G_e \end{bmatrix} \begin{pmatrix} \tilde{\varepsilon}_{xx,e} \\ \tilde{\varepsilon}_{yy,e} \\ \tilde{\gamma}_{xy,e} \end{pmatrix} = \underline{\underline{Q}}_e \, \underline{\tilde{\varepsilon}}_e. \tag{5.61}$$

Finally, the first variation of the total elastic potential of the disk discretized by n elements reads:

$$\sum_{e=1}^{e=n} \left\{ \int_{V_e} \left[\underline{\underline{Q}}_e \underline{\underline{B}}_e \underline{v}_e \right]^T \underline{\underline{B}}_e \mathrm{d}V_e - \int_{V_e} \underline{\underline{N}}_e^T \underline{f}_e \mathrm{d}V_e - \int_{S_{t,e}} \underline{\underline{N}}_e^T \underline{t}_{0,e} \mathrm{d}S_e \right\} \delta \underline{v}_e = 0. \tag{5.62}$$

The variation $\delta \underline{v}_e$ of the element node displacements is arbitrary within the given geometric boundary conditions, so that only the vanishing of the bracket term can be considered as a solution for (5.62):

$$\int_{V_e} \left[\underline{\underline{Q}}_e \underline{\underline{B}}_e \underline{v}_e \right]^T \underline{\underline{B}}_e \mathrm{d}V_e - \int_{V_e} \underline{\underline{N}}_e^T \underline{f}_e \mathrm{d}V_e - \int_{S_{t,e}} \underline{\underline{N}}_e^T \underline{t}_{0,e} \mathrm{d}S_e = 0. \tag{5.63}$$

At this point we introduce the following quantities:

$$\underline{\underline{K}}_e = \int_{V_e} \underline{\underline{B}}_e^T \underline{\underline{Q}}_e \underline{\underline{B}}_e \mathrm{d}V_e, \quad \underline{F}_e = \int_{V_e} \underline{\underline{N}}_e^T \underline{f}_e \mathrm{d}V_e + \int_{S_{t,e}} \underline{\underline{N}}_e^T \underline{t}_{0,e} \mathrm{d}S_e. \tag{5.64}$$

Thus (5.63) passes into the following compact form:

$$\underline{\underline{K}}_e \underline{v}_e = \underline{F}_e. \tag{5.65}$$

Here $\underline{\underline{K}}_e$ is the element stiffness matrix. It has as many rows and columns as there are nodal degrees of freedom in element e. The vector $\underline{F}_e$ is the element load vector. This expression is a linear system of equations for the unknown nodal displacements $\underline{v}_e$. The integrations prescribed in (5.64) cannot be performed exactly-analytically in many cases, so that numerical integration methods such as Gauß integration are often used.

Assembling the contributions of all elements into a single system of equations, we obtain a system of equations of the type

$$\underline{\underline{K}} \underline{v} = \underline{F}. \tag{5.66}$$

The matrix $\underline{\underline{K}}$ is thus the stiffness matrix of the entire system discretized with disk elements. Its size corresponds to the number of all nodal degrees of freedom of the discretized disk structure. This is analogously true for the force vector $\underline{F}$. Formally, this procedure corresponds to the summation of all element-wise contributions to the potential of the whole structure.

References

Altenbach, J., Sacharov, A.S.: Die Methode der Finiten Elemente in der Festkörpermechanik. Hanser, München (1982)

Altenbach, H., Altenbach, J., Naumenko, K.: Ebene Flächentragwerke, 2nd edn. Springer, Berlin (2016)

Bathe, K.J.: Finite Element Procedures. Prentice Hall, New Jersey (1996)

Becker, W., Gross, D.: Mechanik elastischer Körper und Strukturen. Springer, Berlin (2002)

Betten, J.: Finite Elemente für Ingenieure Teil 1: Grundlagen, Matrixmethoden, Elastisches Kontinuum, 2nd edn. Springer, Berlin (2003)

Betten, J.: Finite Elemente für Ingenieure Teil 2: Variationsrechnung, Energiemethoden, Näherungsverfahren, Nichtlinearitäten, numerische Integrationen, 2nd edn. Springer, Berlin (2004)

Crisfield, M.A.: Non-linear Finite Element Analysis of Solids and Structures, Volume 1: Essentials. Wiley, Chichester (1991)

Crisfield, M.A.: Non-linear Finite Element Analysis of Solids and Structures, Volume 2: Advanced Topics. Wiley, Chichester (1997)

Girkmann, K.: Flächentragwerke, 6th edn. Springer, Wien (1974)

Göldner, H., Altenbach, J., Eschke, K., Garz, K.F., Sähn, S.: Lehrbuch Höhere Festigkeitslehre, vol. 1. Physik, Weinheim (1979)

Göldner, H., Altenbach, J., Eschke, K., Garz, K.F., Sähn, S.: Lehrbuch Höhere Festigkeitslehre, vol. 2, 3rd edn. Fachbuchverlag Leipzig, Germany (1985)

Kossira, H.: Grundlagen des Leichtbaus. Springer, Berlin (1996)

Lanczos, C.: The Variational Principles of Mechanics, 4th edn. Dover, New York (1986)

Langhaar, H.L.: Energy Methods in Applied Mechanics. Dover, New York (2016)

Mittelstedt, C.: Rechenmethoden des Leichtbaus. Springer, Heidelberg (2021)

Oden, J.T., Reddy, J.N.: An Introduction to the Mathematical Theory of Finite Elements. Dover, New York (2011)

Reddy, J.N.: An Introduction to the Finite Element Method, 3rd edn. Mcgraw Hill, New York (2005)

Reddy, J.N.: Energy Principles and Variational Methods in Applied Mechanics, 3rd edn. Wiley, New York (2017)

Tauchert, T.R.: Energy Principles in Structural Mechanics. McGraw-Hill, New York (1974)

Washizu, K.: Variational Methods in Elasticity and Plasticity, 3rd edn. Pergamon, New York (1982)

Zienkiewicz, O.C., Taylor, R.L.: The Finite Element Method for Solid and Structural Mechanics, 7th edn. Butterworth Heinemann, Oxford (2013)

Zienkiewicz, O.C., Taylor, R.L., Zhu, J.Z.: The Finite Element Method: Its Basis and Fundamentals, 7th edn. Butterworth Heinemann, Oxford (2013)

Anisotropic Disks

6

In this chapter we want to extend the considerations made so far with respect to isotropic disks to anisotropic disk structures, whereby we will mainly restrict ourselves to orthotropic disks due to their high practical relevance. All considerations will focus on the plane stress state, the transfer of the results to the plane strain state is then very easy to accomplish by replacing the elastic constants (see also Chap. 1). Let the assumed anisotropic materials be such that they have at least one symmetry plane coincident with the disk middle plane. Otherwise, all the same conditions apply as have already been discussed for isotropic disks. We discuss at the beginning all equations necessary for the treatment of anisotropic disks, both in Cartesian and in polar coordinates. This is followed by a consideration of elementary fundamental equations and a discussion of beam-like disks, before treating the decay behavior of boundary perturbations in infinitely extended orthotropic disks. The chapter concludes with a discussion of both homogeneous and layered circular ring and arc disks.

6.1 Basic Equations

6.1.1 Cartesian Coordinates

We start with anisotropic disks which can be advantageously described in Cartesian coordinates x, y, z, where z is the thickness coordinate and the disk middle plane is spanned by the x- and y-axes. Let the disk thickness be h. The basic equations describing the disk behavior are given here again for clarity.

C. Mittelstedt, *Theory of Plates and Shells*,
https://doi.org/10.1007/978-3-662-66805-4_6

In addition to the equilibrium conditions in the absence of volume forces

$$\frac{\partial \sigma_{xx}}{\partial x} + \frac{\partial \tau_{xy}}{\partial y} = 0, \quad \frac{\partial \tau_{xy}}{\partial x} + \frac{\partial \sigma_{yy}}{\partial y} = 0 \tag{6.1}$$

and the kinematic relations

$$\varepsilon^0_{xx} = \frac{\partial u_0}{\partial x}, \quad \varepsilon^0_{yy} = \frac{\partial v_0}{\partial y}, \quad \gamma^0_{xy} = \frac{\partial u_0}{\partial y} + \frac{\partial v_0}{\partial x} \tag{6.2}$$

a constitutive relation is also required, which we state here in the form of Hooke's law for a monoclinic disk in the plane stress state:

$$\begin{pmatrix} \varepsilon^0_{xx} \\ \varepsilon^0_{yy} \\ \gamma^0_{xy} \end{pmatrix} = \begin{bmatrix} S_{11} & S_{12} & S_{16} \\ S_{12} & S_{22} & S_{26} \\ S_{16} & S_{26} & S_{66} \end{bmatrix} \begin{pmatrix} \sigma_{xx} \\ \sigma_{yy} \\ \tau_{xy} \end{pmatrix}. \tag{6.3}$$

In the special case of orthotropy, the shear coupling terms S_{16} and S_{26} vanish, and we obtain:

$$\begin{pmatrix} \varepsilon^0_{xx} \\ \varepsilon^0_{yy} \\ \gamma^0_{xy} \end{pmatrix} = \begin{bmatrix} S_{11} & S_{12} & 0 \\ S_{12} & S_{22} & 0 \\ 0 & 0 & S_{66} \end{bmatrix} \begin{pmatrix} \sigma_{xx} \\ \sigma_{yy} \\ \tau_{xy} \end{pmatrix}, \tag{6.4}$$

or formulated in engineering constants:

$$\begin{pmatrix} \varepsilon^0_{xx} \\ \varepsilon^0_{yy} \\ \gamma^0_{xy} \end{pmatrix} = \begin{bmatrix} \frac{1}{E_{xx}} & -\frac{\nu_{yx}}{E_{yy}} & 0 \\ -\frac{\nu_{xy}}{E_{xx}} & \frac{1}{E_{yy}} & 0 \\ 0 & 0 & \frac{1}{G_{xy}} \end{bmatrix} \begin{pmatrix} \sigma_{xx} \\ \sigma_{yy} \\ \tau_{xy} \end{pmatrix}. \tag{6.5}$$

In addition, the compatibility condition is required. It currently reads:

$$\frac{\partial^2 \varepsilon^0_{xx}}{\partial y^2} + \frac{\partial^2 \varepsilon^0_{yy}}{\partial x^2} - \frac{\partial^2 \gamma^0_{xy}}{\partial x \partial y} = 0. \tag{6.6}$$

It should be noted that in the plane stress state a strain ε_{zz} in the thickness direction also occurs, which can be calculated as in a post-calculation as follows:

$$\varepsilon_{zz} = S_{13}\sigma_{xx} + S_{23}\sigma_{yy}. \tag{6.7}$$

The equation describing the behavior of a disk in the plane state of stress is also a disk equation in the case of the anisotropic disk, which we obtain after defining the Airy stress function F for the case of vanishing volume forces as

$$\sigma_{xx} = \frac{\partial^2 F}{\partial y^2}, \quad \sigma_{yy} = \frac{\partial^2 F}{\partial x^2}, \quad \tau_{xy} = -\frac{\partial^2 F}{\partial x \partial y} \tag{6.8}$$

(see e.g. Lekhnitskii 1968; Rand and Rovenski 2005; Karttunen and von Hertzen 2016):

$$S_{22}\frac{\partial^4 F}{\partial x^4} + (2S_{12} + S_{66})\frac{\partial^4 F}{\partial x^2 \partial y^2} + S_{11}\frac{\partial^4 F}{\partial y^4}$$
$$-2S_{16}\frac{\partial^4 F}{\partial x \partial y^3} - 2S_{26}\frac{\partial^4 F}{\partial x^3 \partial y} = 0. \tag{6.9}$$

For the case of orthotropy, this equation reduces to (Hashin 1967; Blumer 1979; von Roth and Epple 1981):

$$S_{22}\frac{\partial^4 F}{\partial x^4} + (2S_{12} + S_{66})\frac{\partial^4 F}{\partial x^2 \partial y^2} + S_{11}\frac{\partial^4 F}{\partial y^4} = 0, \tag{6.10}$$

or formulated in the corresponding engineering constants:

$$\frac{1}{E_{yy}}\frac{\partial^4 F}{\partial x^4} + \left(\frac{1}{G_{xy}} - \frac{2\nu_{xy}}{E_{xx}}\right)\frac{\partial^4 F}{\partial x^2 \partial y^2} + \frac{1}{E_{xx}}\frac{\partial^4 F}{\partial y^4} = 0. \tag{6.11}$$

This form of the disk equation reduces to the special case of isotropy for the case $E_{xx} = E_{yy} = E$, $G_{xy} = G = \frac{E}{2(1+\nu)}$ and $\nu_{12} = \nu$.

An alternative form of the disk equation is given by Blumer (1979). With the two orthotropy parameters

$$s^2 = \frac{E_{xx}}{E_{yy}}, \quad k^2 = \frac{E_{xx}}{G_{xy}} \tag{6.12}$$

it follows from (6.11):

$$s^2\frac{\partial^4 F}{\partial x^4} + \left(k^2 - 2\nu_{xy}\right)\frac{\partial^4 F}{\partial x^2 \partial y^2} + \frac{\partial^4 F}{\partial y^4} = 0. \tag{6.13}$$

For the special case of isotropy $s^2 = 1$ and $k^2 = 2(1 + \nu)$ holds.

Solutions of the disk equation (6.9) can again be found in the form of polynomial functions as in the case of the isotropic disk (cf. Hashin 1967; Jayne and Tang 1970). Starting from the disk equation (6.9), an approach of the form

$$F = \sum_{m=0}^{M}\sum_{n=0}^{N} F_{mn} x^m y^n \tag{6.14}$$

after substitution into (6.9) leads to the following expression:

$$\sum_{m=2}^{M}\sum_{n=2}^{N}\left[S_{22}(m+2)(m+1)m(m-1)F_{m+2,n-2}\right.$$
$$-2S_{26}(m+1)m(m-1)(n-1)F_{m+1,n-1}$$

$$
\begin{aligned}
&+(2S_{12}+S_{66})m(m-1)n(n-1)F_{mn}\\
&-2S_{16}(m-1)(n+1)n(n-1)F_{m-1,n+1}\\
&+S_{11}(n+2)(n+1)n(n-1)]\,x^{m-2}y^{n-2}=0.
\end{aligned}
\tag{6.15}
$$

For any x and y, this can only be satisfied if the bracket expression becomes zero, i.e.:

$$
\begin{aligned}
&S_{22}(m+2)(m+1)m(m-1)F_{m+2,n-2}-2S_{26}(m+1)m(m-1)(n-1)F_{m+1,n-1}\\
&+(2S_{12}+S_{66})m(m-1)n(n-1)F_{mn}-2S_{16}(m-1)(n+1)n(n-1)F_{m-1,n+1}\\
&+S_{11}(n+2)(n+1)n(n-1)=0.
\end{aligned}
\tag{6.16}
$$

The special case of orthotropy follows with $S_{16}=S_{26}=0$ as:

$$
\begin{aligned}
&S_{22}(m+2)(m+1)m(m-1)F_{m+2,n-2}+(2S_{12}+S_{66})m(m-1)n(n-1)F_{mn}\\
&+S_{11}(n+2)(n+1)n(n-1)=0.
\end{aligned}
\tag{6.17}
$$

As in the case of the isotropic disk, combinations of exponential and trigonometric functions can be used for the orthotropic disk. First of all, an approach of the form

$$
F=C_m e^{\lambda_m\omega_m y}\cos\omega_m x
\tag{6.18}
$$

(see also Blumer 1979 or von Roth and Epple 1981), with $\omega_m=\frac{m\pi}{l}$ is employed. This, after insertion into the disk equation (6.13), leads to the following fourth-order characteristic polynomial for the determination of λ_m:

$$
\lambda_m^4-\left(k^2-2\nu_{xy}\right)\lambda_m^2+s^2=0.
\tag{6.19}
$$

The four roots are:

$$
\lambda_{m,1,2,3,4}=\pm\sqrt{\frac{k^2-2\nu_{xy}}{2}\pm\sqrt{\frac{\left(k^2-2\nu_{xy}\right)^2}{4}-s^2}},
\tag{6.20}
$$

which leads to the following expression for F:

$$
\begin{aligned}
F\;=\;&\sum_{m=1}^{M}\left(C_{m1}e^{\lambda_{m1}\omega_m y}+C_{m2}e^{-\lambda_{m1}\omega_m y}+C_{m3}e^{\lambda_{m2}\omega_m y}\right.\\
&\left.+C_{m4}e^{-\lambda_{m2}\omega_m y}\right)\cos\omega_m x.
\end{aligned}
\tag{6.21}
$$

In Blumer (1979) an expression equivalent to this can be found as:

$$
\begin{aligned}
F\;=\;&\sum_{m=1}^{M}(C_{m1}\cosh\lambda_{m1}\omega_m y+C_{m2}\sinh\lambda_{m1}\omega_m y\\
&+C_{m3}\cosh\lambda_{m2}\omega_m y+C_{m4}\sinh\lambda_{m2}\omega_m y)\cos\omega_m x.
\end{aligned}
\tag{6.22}
$$

In the approach (6.18), the cosine function may be replaced by a sine term, resulting in the following solution:

$$F = \sum_{m=1}^{M} \left(C_{m1} e^{\lambda_{m1}\omega_m y} + C_{m2} e^{-\lambda_{m1}\omega_m y} + C_{m3} e^{\lambda_{m2}\omega_m y} + C_{m4} e^{-\lambda_{m2}\omega_m y} \right) \sin \omega_m x. \tag{6.23}$$

The solutions for λ_m are always real-valued if both $s \geq 0$ and $s \leq \nu_{xy} - \frac{k^2}{2}$ are satisfied.

6.1.2 Polar Coordinates

The formulation of anisotropic disk problems in polar coordinates in the absence of volume loads, as already shown for Cartesian coordinates, is also possible by using the basic equations already derived in Chap. 1 in connection with the corresponding material law, as discussed in Chap. 1. With the equilibrium conditions

$$\begin{aligned} \frac{\partial \sigma_{rr}}{\partial r} + \frac{1}{r}\frac{\partial \tau_{r\varphi}}{\partial \varphi} + \frac{\sigma_{rr} - \sigma_{\varphi\varphi}}{r} &= 0, \\ \frac{\partial \tau_{r\varphi}}{\partial r} + \frac{1}{r}\frac{\sigma_{\varphi\varphi}}{\partial \varphi} + 2\frac{\tau_{r\varphi}}{r} &= 0, \end{aligned} \tag{6.24}$$

the kinematic conditions

$$\varepsilon^0_{rr} = \frac{\partial u_0}{\partial r}, \quad \varepsilon^0_{\varphi\varphi} = \frac{1}{r}\frac{\partial v_0}{\partial \varphi} + \frac{u_0}{r}, \quad \gamma^0_{r\varphi} = \frac{\partial v_0}{\partial r} + \frac{1}{r}\frac{\partial u_0}{\partial \varphi} - \frac{v_0}{r} \tag{6.25}$$

and the compatibility condition

$$\frac{\partial^2 \varepsilon^0_{\varphi\varphi}}{\partial r^2} + \frac{1}{r^2}\frac{\partial^2 \varepsilon^0_{rr}}{\partial \varphi^2} + \frac{2}{r}\frac{\varepsilon^0_{\varphi\varphi}}{\partial r} - \frac{1}{r}\frac{\partial \varepsilon^0_{rr}}{\partial r} = \frac{1}{r}\frac{\partial^2 \gamma^0_{r\varphi}}{\partial r \partial \varphi} + \frac{1}{r^2}\frac{\partial \gamma^0_{r\varphi}}{\partial \varphi} \tag{6.26}$$

follows the disk equation with the anisotropic material law

$$\begin{pmatrix} \varepsilon^0_{rr} \\ \varepsilon^0_{\varphi\varphi} \\ \gamma^0_{r\varphi} \end{pmatrix} = \begin{bmatrix} S_{11} & S_{12} & S_{16} \\ S_{12} & S_{22} & S_{26} \\ S_{16} & S_{26} & S_{66} \end{bmatrix} \begin{pmatrix} \sigma_{rr} \\ \sigma_{\varphi\varphi} \\ \tau_{r\varphi} \end{pmatrix} \tag{6.27}$$

and the definition of the Airy stress function F

$$\sigma_{rr} = \frac{1}{r^2}\frac{\partial^2 F}{\partial \varphi^2} + \frac{1}{r}\frac{\partial F}{\partial r}, \quad \sigma_{\varphi\varphi} = \frac{\partial^2 F}{\partial r^2}, \quad \tau_{r\varphi} = -\frac{\partial}{\partial r}\left(\frac{1}{r}\frac{\partial F}{\partial \varphi}\right) \tag{6.28}$$

as shown below for the case of vanishing volume forces (Lekhnitskii (1968)):

$$\begin{aligned} & S_{22}\left(\frac{\partial^4 F}{\partial r^4}+\frac{2}{r}\frac{\partial^3 F}{\partial r^3}\right) \\ + & S_{11}\left(\frac{1}{r^4}\frac{\partial^4 F}{\partial \varphi^4}-\frac{1}{r^2}\frac{\partial^2 F}{\partial r^2}+\frac{2}{r^4}\frac{\partial^2 F}{\partial \varphi^2}+\frac{1}{r^3}\frac{\partial F}{\partial r}\right) \\ + & (2S_{12}+S_{66})\left(\frac{1}{r^2}\frac{\partial^4 F}{\partial r^2\partial \varphi^2}-\frac{1}{r^3}\frac{\partial^3 F}{\partial r\partial \varphi^2}+\frac{1}{r^4}\frac{\partial^2 F}{\partial \varphi^2}\right) \\ + & 2S_{16}\left(-\frac{1}{r^3}\frac{\partial^4 F}{\partial r\partial \varphi^3}+\frac{1}{r^4}\frac{\partial^3 F}{\partial \varphi^3}-\frac{1}{r^3}\frac{\partial^2 F}{\partial r\partial \varphi}+\frac{1}{r^4}\frac{\partial F}{\partial \varphi}\right) \\ + & 2S_{26}\left(-\frac{1}{r}\frac{\partial^4 F}{\partial r^3\partial \varphi}-\frac{1}{r^3}\frac{\partial^2 F}{\partial r\partial \varphi}+\frac{1}{r^4}\frac{\partial F}{\partial \varphi}\right)=0. \end{aligned} \tag{6.29}$$

For the special case of cylindrical orthotropy with $S_{16} = S_{26} = 0$ this equation reduces to:

$$\begin{aligned} & S_{22}\left(\frac{\partial^4 F}{\partial r^4}+\frac{2}{r}\frac{\partial^3 F}{\partial r^3}\right) \\ + & S_{11}\left(\frac{1}{r^4}\frac{\partial^4 F}{\partial \varphi^4}-\frac{1}{r^2}\frac{\partial^2 F}{\partial r^2}+\frac{2}{r^4}\frac{\partial^2 F}{\partial \varphi^2}+\frac{1}{r^3}\frac{\partial F}{\partial r}\right) \\ + & (2S_{12}+S_{66})\left(\frac{1}{r^2}\frac{\partial^4 F}{\partial r^2\partial \varphi^2}-\frac{1}{r^3}\frac{\partial^3 F}{\partial r\partial \varphi^2}+\frac{1}{r^4}\frac{\partial^2 F}{\partial \varphi^2}\right)=0. \end{aligned} \tag{6.30}$$

Formulated in terms of the engineering constants E_{rr}, $E_{\varphi\varphi}$, $G_{r\varphi}$, and $\nu_{r\varphi}$, the disk equation (6.30) is:

$$\begin{aligned} & \frac{1}{E_{\varphi\varphi}}\frac{\partial^4 F}{\partial r^4}+\left(\frac{1}{G_{r\varphi}}-\frac{2\nu_{\varphi r}}{E_{\varphi\varphi}}\right)\frac{1}{r^2}\frac{\partial^4 F}{\partial r^2\partial \varphi^2}+\frac{1}{E_{rr}}\frac{1}{r^4}\frac{\partial^4 F}{\partial \varphi^4} \\ + & \frac{2}{E_{\varphi\varphi}}\frac{1}{r}\frac{\partial^3 F}{\partial r^3}-\left(\frac{1}{G_{r\varphi}}-\frac{2\nu_{\varphi r}}{E_{\varphi\varphi}}\right)\frac{1}{r^3}\frac{\partial^3 F}{\partial r\partial \varphi^2}-\frac{1}{E_{rr}}\frac{1}{r^2}\frac{\partial^2 F}{\partial r^2} \\ + & \left(2\frac{1-\nu_{r\varphi}}{E_{rr}}+\frac{1}{G_{r\varphi}}\right)\frac{1}{r^4}\frac{\partial^2 F}{\partial \varphi^2}+\frac{1}{E_{rr}}\frac{1}{r^3}\frac{\partial F}{\partial r}=0. \end{aligned} \tag{6.31}$$

It should be noted that in the presence of a plane stress state, additionally the strain ε_{zz} occurs in the thickness direction of the disk, which can be determined in a post-calculation from the plane stress components. A formulation of the above equations for the plane strain state is easily achieved by applying the corresponding elastic constants (see Chap. 1).

Blumer (1979) introduces at this point the two orthotropy parameters

$$s^2=\frac{E_{\varphi\varphi}}{E_{rr}},\quad k^2=\frac{E_{\varphi\varphi}}{G_{r\varphi}}, \tag{6.32}$$

with which the following representation of the disk equation (6.31) results:

$$\frac{\partial^4 F}{\partial r^4}+\frac{2}{r}\frac{\partial^3 F}{\partial r^3}-\frac{s^2}{r^2}\frac{\partial^2 F}{\partial r^2}+\frac{s^2}{r^3}\frac{\partial F}{\partial r}+\frac{s^2}{r^4}\frac{\partial^4 F}{\partial \varphi^4}+\frac{k^2-2\nu_{\varphi r}}{r^2}\frac{\partial^4 F}{\partial r^2 \partial \varphi^2}$$
$$-\frac{k^2-2\nu_{\varphi r}}{r^3}\frac{\partial^3 F}{\partial r \partial \varphi^2}+\frac{k^2+2s^2-2\nu_{\varphi r}}{r^4}\frac{\partial^2 F}{\partial \varphi^2}=0. \tag{6.33}$$

The special case of isotropy follows with $s^2 = 1$ and $k^2 = 2(1 + \nu)$.

A solution approach for the Airy stress function can be found in the following form (Blumer 1979; von Roth and Epple 1981; Noack and von Roth 1972):

$$F = C r^{\lambda_m} \cos m\varphi. \tag{6.34}$$

Substituting into the disk equation (6.33) yields the following polynomial for determining the exponent λ_m (Blumer 1979):

$$\lambda_m^4 - 4\lambda_m^3 + \left(5 - s^2 - m^2k^2 + 2m^2\nu_{\varphi r}\right)\lambda_m^2$$
$$+\left(2s^2 + 2m^2k^2 - 4m^2\nu_{\varphi r} - 2\right)\lambda_m$$
$$+\left(m^4s^2 - 2s^2m^2 - m^2k^2 + 2m^2\nu_{\varphi r}\right) = 0. \tag{6.35}$$

With the substitution $\lambda_m = x_m + 1$ this gives the following expression:

$$x_m^4 - 2q_m^2x_m^2 + s^2\left(m^2-1\right)^2, \tag{6.36}$$

with

$$q_m^2 = \frac{1}{2}\left[1 + s^2 + m^2\left(k^2 - 2\nu_{\varphi r}\right)\right]. \tag{6.37}$$

The roots of (6.36) are:

$$x_{m,1,2,3,4} = \pm\sqrt{q_m^2 \pm \sqrt{q_m^4 - s^2\left(m^2-1\right)^2}} = \lambda_{m,1,2,3,4} - 1. \tag{6.38}$$

Thus, the following disk solution can be given:

$$F = \sum_{m=2}^{M}\left(C_{m1}r^{\lambda_{m1}} + C_{m2}r^{\lambda_{m2}} + C_{m3}r^{\lambda_{m3}} + C_{m4}r^{\lambda_{m4}}\right)\cos m\varphi. \tag{6.39}$$

Therein, the cosine function may be replaced by a sine function, so that the following expression is also a solution of the disk equation (6.33):

$$F = \sum_{m=2}^{M}\left(C_{m1}r^{\lambda_{m1}} + C_{m2}r^{\lambda_{m2}} + C_{m3}r^{\lambda_{m3}} + C_{m4}r^{\lambda_{m4}}\right)\sin m\varphi. \tag{6.40}$$

The special case $m = 1$ leads with $q_m^2 = \frac{1}{2}\left[1 + s^2 + k^2 - 2\nu_{\varphi r}\right]$ to the following roots:

$$\lambda_{11} = \lambda_{12} = 1, \quad \lambda_{13,4} = 1 \pm \sqrt{1 + s^2 + k^2 - 2\nu_{\varphi r}} = 1 \pm n. \tag{6.41}$$

Thus, the solution F is given as:

$$F = \left(C_{11}r + C_{12}r \ln r + C_{13}r^{1+n} + C_{14}r^{1-n}\right) \cos \varphi. \tag{6.42}$$

The expression

$$F = \left(C_{11}r + C_{12}r \ln r + C_{13}r^{1+n} + C_{14}r^{1-n}\right) \sin \varphi \tag{6.43}$$

is also a solution of the disk equation.

A special case is given when the Airy stress function F turns out to be independent of the angle φ. Then the disk equation reduces to:

$$\frac{\mathrm{d}^4 F}{\mathrm{d}r^4} + \frac{2}{r}\frac{\mathrm{d}^3 F}{\mathrm{d}r^3} + \frac{S_{11}}{S_{22}}\left(-\frac{1}{r^2}\frac{\mathrm{d}^2 F}{\mathrm{d}r^2} + \frac{1}{r^3}\frac{\mathrm{d}F}{\mathrm{d}r}\right) = 0. \tag{6.44}$$

With $S_{11} = E_{rr}^{-1}$ and $S_{22} = E_{\varphi\varphi}^{-1}$ and also with $s^2 = \frac{E_{\varphi\varphi}}{E_{rr}}$ it follows:

$$\frac{\mathrm{d}^4 F}{\mathrm{d}r^4} + \frac{2}{r}\frac{\mathrm{d}^3 F}{\mathrm{d}r^3} - \frac{s^2}{r^2}\frac{\mathrm{d}^2 F}{\mathrm{d}r^2} + \frac{s^2}{r^3}\frac{\mathrm{d}F}{\mathrm{d}r} = 0. \tag{6.45}$$

For a solution we can go back directly to the roots (6.38) with $m = 0$, and it follows:

$$\lambda_{01} = 0, \quad \lambda_{02} = 2, \quad \lambda_{03} = 1 + s, \quad \lambda_{04} = 1 - s. \tag{6.46}$$

The stress function in this case is:

$$F = C_{01} + C_{02}r^2 + C_{03}r^{1+s} + C_{04}r^{1-s}. \tag{6.47}$$

6.2 Elementary Cases

As already shown for isotropic disks, also in the case of anisotropic disks a number of exact-analytical solutions for some elementary simple disk situations can be postulated. We will discuss a selection of such solutions in more detail in this section. First, we consider the disk situation as already discussed in Chap. 3 using the example of Fig. 3.4. Let a rectangular disk of length a and width b be given here, which is loaded by the uniaxial stress σ_0 in the x-direction. The disk obeys the material law (6.3), i.e. it behaves monoclinically. With an approach of the kind

$$F = \frac{1}{2}\sigma_0 y^2 \tag{6.48}$$

for the Airy stress function, the stress components in the disk are given by (6.8) as $\sigma_{xx} = \sigma_0$, $\sigma_{yy} = 0$, $\tau_{xy} = 0$. Obviously, also the monoclinic case present here does not change anything in the stress balance of the disk in comparison with the isotropic disk (see Chap. 3, Eq. (3.100)). On the other hand, it can be seen that the deformation behavior is much more complex in the case of the monoclinic disk. If in the case of the isotropic disk there is a strain state in which only the two normal strains ε_{xx}^0 and ε_{yy}^0 occur, the monoclinic case considered here additionally evokes the shear strain γ_{xy}^0, as can be easily understood from Eq. (6.3). We have:

$$\begin{pmatrix} \varepsilon_{xx}^0 \\ \varepsilon_{yy}^0 \\ \gamma_{xy}^0 \end{pmatrix} = \begin{pmatrix} S_{11}\sigma_0 \\ S_{12}\sigma_0 \\ S_{16}\sigma_0 \end{pmatrix}. \tag{6.49}$$

Obviously, the term S_{16} causes a shear strain γ_{xy}^0 of the disk even under pure tensile loading. This phenomenon is called shear coupling (see also Chap. 1).

Using the approach $F = \frac{1}{2}\sigma_0 x^2$, the analogous case of a uniaxial constant tensile load σ_0 in the y-direction can be represented (cf. Eq. (3.101)):

$$\sigma_{xx} = \frac{\partial^2 F}{\partial y^2} = 0, \quad \sigma_{yy} = \frac{\partial^2 F}{\partial x^2} = \sigma_0, \quad \tau_{xy} = -\frac{\partial^2 F}{\partial x \partial y} = 0. \tag{6.50}$$

The strain field is then obtained as:

$$\begin{pmatrix} \varepsilon_{xx}^0 \\ \varepsilon_{yy}^0 \\ \gamma_{xy}^0 \end{pmatrix} = \begin{pmatrix} S_{12}\sigma_0 \\ S_{22}\sigma_0 \\ S_{26}\sigma_0 \end{pmatrix}. \tag{6.51}$$

Here, the term S_{26} invokes the shear coupling.

A pure shear load τ_0 can be represented via the approach $F = -\tau_0 xy$, which leads to the following stress field of the disk (see also Eq. (3.102)):

$$\sigma_{xx} = \frac{\partial^2 F}{\partial y^2} = 0, \quad \sigma_{yy} = \frac{\partial^2 F}{\partial x^2} = 0, \quad \tau_{xy} = -\frac{\partial^2 F}{\partial x \partial y} = \tau_0. \tag{6.52}$$

The strain field is then obtained as:

$$\begin{pmatrix} \varepsilon_{xx}^0 \\ \varepsilon_{yy}^0 \\ \gamma_{xy}^0 \end{pmatrix} = \begin{pmatrix} S_{16}\tau_0 \\ S_{26}\tau_0 \\ S_{66}\tau_0 \end{pmatrix}. \tag{6.53}$$

It is shown that a constant shear loading of a monoclinic disk gives rise not only to the (intuitively expected) shear strain γ_{xy}^0, but also to the two normal strains ε_{xx}^0 and ε_{yy}^0 due to the two shear coupling terms S_{16} and S_{26}.

If, on the other hand, a purely orthotropic disk with $S_{16} = S_{26} = 0$ is present, which is subjected to the elementary load cases discussed here, then the strain field simplifies considerably and then corresponds to the state that would be intuitively expected: In the case of tensile loading σ_0, only the two normal strains ε^0_{xx} and ε^0_{yy} occur. Under the shear load τ_0, on the other hand, only the shear strain γ^0_{xy} is induced.

It turns out that even in the case of monoclinic or orthotropic disks, in the presence of such elementary stress cases, the approach to finding the Airy stress function is largely analogous to that for isotropic disks (Chap. 3).

6.3 Beam-type Disks

As already shown in Chap. 3 for monoclinic disks, solutions for beam-type structures will be investigated. We first consider again the rectangular monoclinic disk (length $2a$, width $2b$), which is supported at the points $x = \pm a$ in such a way that the applied constant surface load p_0 is transferred by shear at the ends $x = \pm a$ (cf. also Fig. 3.8). This example was also treated by Lekhnitskii (1968) and Hashin (1967). The objective here is a formulation for Airy's stress function in terms of a polynomial of the fifth degree as follows:

$$F = \sum_{m=0}^{2}\sum_{n=0}^{5} F_{mn} x^m y^n, \tag{6.54}$$

with $m + n \leq 5$. The resulting stress field after satisfying the underlying boundary conditions is finally given as:

$$\begin{aligned}
\sigma_{xx} &= \frac{3p_0}{4b^3}\left(x^2 - a^2\right) y + \frac{3p_0\eta}{b}\left(\frac{y^2}{b^2} - \frac{1}{3}\right) x + \frac{p_0\beta}{2b}\left(\frac{y^2}{b^2} - \frac{3}{5}\right) y, \\
\sigma_{yy} &= \frac{p_0}{4}\left(\frac{y^3}{b^3} - \frac{3y}{b} - 2\right), \\
\tau_{xy} &= \frac{p_0}{4b}\left(4\eta y + 3x\right)\left(1 - \frac{y^2}{b^2}\right),
\end{aligned} \tag{6.55}$$

with the abbreviations

$$\eta = \frac{S_{16}}{2S_{11}}, \quad \beta = 2\left(\frac{S_{16}}{S_{11}}\right)^2 - \frac{2S_{12} + S_{66}}{2S_{11}}. \tag{6.56}$$

Obviously, it results in the case of the monoclinic beam-type disk that the stress field depends on the elastic properties of the disk in contrast to the isotropic disk. An exception is the normal stress σ_{yy}, which also proves to be material-independent in the present case.

If the special case of orthotropy is present, where the shear coupling term S_{16} vanishes, then the solution (6.55) simplifies as follows:

$$
\begin{aligned}
\sigma_{xx} &= \frac{3p_0}{4b^3}\left(x^2 - a^2\right) y + \frac{p_0\beta}{2b}\left(\frac{y^2}{b^2} - \frac{3}{5}\right) y, \\
\sigma_{yy} &= \frac{p_0}{4}\left(\frac{y^3}{b^3} - \frac{3y}{b} - 2\right), \\
\tau_{xy} &= \frac{3p_0 x}{4b}\left(1 - \frac{y^2}{b^2}\right).
\end{aligned}
\tag{6.57}
$$

It can be seen that now also the shear stress distribution is independent of the degree of orthotropy. It can be easily shown that the coefficient β takes the value $\beta = -1$ in the case of isotropy. The stress distribution in the disk then corresponds exactly to the results (3.121), Chap. 3.

The benefit of the solution shown here is that it allows the influence of anisotropy on the stress balance of the monoclinic disc to be evaluated against an orthotropic or isotropic disc. This is an aspect of the design and construction of such disk-type structures that is usually not negligible. A study for the special case of orthotropy ($\eta = 0$) for the normal stress σ_{xx} is shown in Fig. 6.1. Here, a dimensionless representation in the form of the relative normal stress $\bar{\sigma}_{xx}$ was assumed as

$$
\bar{\sigma}_{xx} = \frac{\sigma_{xx}}{p_0} = \frac{3}{4}\left(\xi 1^2 - \rho^2\right)\xi_2 + \frac{\beta}{2}\left(\xi_2^2 - \frac{3}{5}\right)\xi_2,
\tag{6.58}
$$

where the quantities $\xi_1 = \frac{x}{b}$, $\xi_2 = \frac{y}{b}$, $\rho = \frac{a}{b}$ were used. The parameter β was varied here between $\beta = -1$ (isotropy) and $\beta = -30$ (typical CFRP single layer). Obviously, a strongly nonlinear distribution of the normal stress over the cross-

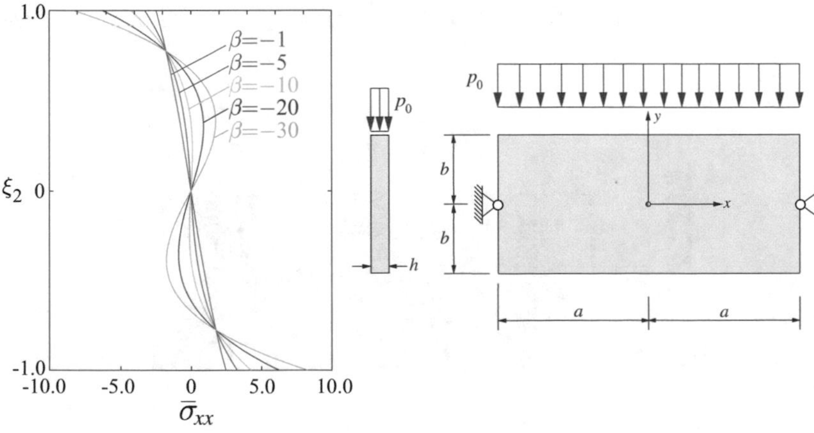

Fig. 6.1 Relative normal stress $\bar{\sigma}_{xx}$ of the rectangular orthotropic disk under surface load p_0 for different values β, evaluated for $\rho = 2$ at $\xi_1 = \frac{\rho}{2}$

section thickness results, whereby this effect becomes more pronounced the more the parameter β deviates from the isotropy value $\beta = -1$. For particularly high degrees of orthotropy, there are even several zero crossings of the normal stress $\bar{\sigma}_{xx}$, which is an effect not known from isotropic structures and which becomes more pronounced the smaller the aspect ratio ρ is.

In the case that the considered rectangular disk is under a more complex load (Fig. 6.2, see Lekhnitskii 1968), then it is convenient to represent the load as a Fourier series:

$$p(x) = p_0 + \sum_{m=1}^{\infty} p_m \cos\left(\frac{m\pi x}{a}\right), \tag{6.59}$$

wherein

$$p_0 = \frac{1}{a}\int_0^a p\mathrm{d}x, \quad p_m = \frac{2}{a}\int_0^a p\cos\left(\frac{m\pi x}{a}\right)\mathrm{d}x. \tag{6.60}$$

The boundary conditions at the loaded edge $y = b$ are:

$$\sigma_{yy} = -p, \quad \tau_{xy} = 0. \tag{6.61}$$

On the other hand, at the unloaded longitudinal edge $y = -b$ we have:

$$\sigma_{yy} = 0, \quad \tau_{xy} = 0. \tag{6.62}$$

The stress distribution due to the constant load term p_0 is known from the previous example. For the variable part, the following approach for the Airy stress function is made for the part affine to $p_m \cos\left(\frac{m\pi x}{a}\right)$:

$$F = F_m(y)\cos\left(\frac{m\pi x}{a}\right). \tag{6.63}$$

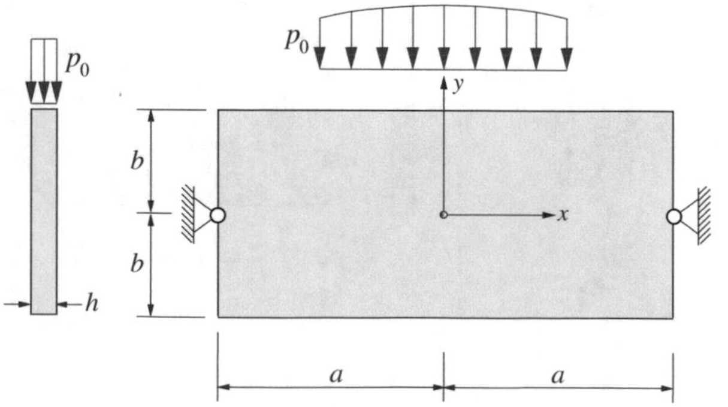

Fig. 6.2 Beam-type disk under arbitrary surface load $p(x)$

Substituting in the disk equation (6.11) leads to the following differential equation to determine $F_m(y)$:

$$\frac{1}{E_{xx}}\frac{\mathrm{d}^4 F_m}{\mathrm{d}y^4} - \left(\frac{1}{G_{xy}} - \frac{2\nu_{xy}}{E_{xx}}\right)\left(\frac{m\pi}{a}\right)^2 \frac{\mathrm{d}^2 F_m}{\mathrm{d}y^2} + \frac{1}{E_{yy}}\left(\frac{m\pi}{a}\right)^4 F_m = 0. \quad (6.64)$$

The approach

$$F_m = Ce^{\lambda \frac{m\pi y}{a}} \quad (6.65)$$

leads after substitution into (6.64) to the following characteristic polynomial to determine the exponent λ:

$$\lambda^4 - \left(k^2 - 2\nu_{xy}\right)\lambda^2 + s^2 = 0, \quad (6.66)$$

with $s^2 = \frac{E_{xx}}{E_{yy}}$ and $k^2 = \frac{E_{xx}}{G_{xy}}$.

There are three possible cases for the solution of (6.66). The first case is with unequal real-valued solutions $\pm\lambda_1, \pm\lambda_2$ ($\lambda_1 > 0, \lambda_2 > 0$), so that the expression for the function F_m is as follows:

$$\begin{aligned} F_m &= \left[C_{m1}\cosh\left(\frac{\lambda_1 m\pi y}{a}\right) + C_{m2}\sinh\left(\frac{\lambda_1 m\pi y}{a}\right) + C_{m3}\cosh\left(\frac{\lambda_2 m\pi y}{a}\right)\right. \\ &\quad \left. + C_{m4}\sinh\left(\frac{\lambda_2 m\pi y}{a}\right)\right]\cos\left(\frac{m\pi x}{a}\right). \end{aligned} \quad (6.67)$$

The second case is when there are identical real-valued zeros $\pm\lambda$ ($\lambda > 0$). Then it follows:

$$\begin{aligned} F_m &= \left[(C_{m1} + C_{m2}y)\cosh\left(\frac{\lambda m\pi y}{a}\right) + (C_{m3} + C_{m4}y)\sinh\left(\frac{\lambda m\pi y}{a}\right)\right] \\ &\quad \times \cos\left(\frac{m\pi x}{a}\right). \end{aligned} \quad (6.68)$$

The third case leads to complex zeros $\lambda_{1,2} = s \pm ti$, $\lambda_{3,4} = -s \pm ti$ ($s > 0, t > 0$) such that:

$$\begin{aligned} F_m &= \left\{\left[C_{m1}\cosh\left(\frac{sm\pi y}{a}\right) + C_{m2}\sinh\left(\frac{sm\pi y}{a}\right)\right]\cos\left(\frac{tm\pi y}{a}\right)\right. \\ &\quad \left. + \left[C_{m3}\cosh\left(\frac{sm\pi y}{a}\right) + C_{m4}\sinh\left(\frac{sm\pi y}{a}\right)\right]\sin\left(\frac{tm\pi y}{a}\right)\right\} \\ &\quad \times \cos\left(\frac{m\pi x}{a}\right). \end{aligned} \quad (6.69)$$

The constants appearing in the solutions (6.67)–(6.69) are adapted for each m to the underlying boundary conditions at the two longitudinal edges $y = \pm b$, and the stress components in the disk can then be determined by summation. A detailed description is not given at this point.

Further examples of rectangular anisotropic disks under different boundary conditions and loads can be found, for example, in Lekhnitskii (1968), Jayne and Tang (1970), Karttunen and von Hertzen (2016) and Ding et al. (2006).

6.4 Decay Behaviour of Edge Perturbations

We consider the situation of the orthotropic half-plane (Wiedemann 2006), which is infinitely extended in the x-direction and bounded by the y-axis. The half-plane is loaded at $x = 0$ by a harmonic boundary load of the form

$$\sigma_{xx,n} = \sigma_{0,n} \sin\left(\frac{n\pi y}{b}\right), \tag{6.70}$$

where $\sigma_{0,n}$ is the amplitude of the edge load with n waves on the edge segment with length b.

The following approach for the Airy stress function F is used:

$$F = F_n(x) \sin\left(\frac{n\pi y}{b}\right). \tag{6.71}$$

Insertion into the disk equation

$$\frac{1}{E_{yy}}\frac{\partial^4 F}{\partial x^4} + \left(\frac{1}{G_{xy}} - \frac{2\nu_{xy}}{E_{xx}}\right)\frac{\partial^4 F}{\partial x^2 \partial y^2} + \frac{1}{E_{xx}}\frac{\partial^4 F}{\partial y^4} = 0 \tag{6.72}$$

leads to the following ordinary differential equation:

$$\frac{1}{E_{yy}}\frac{\mathrm{d}^4 F_n}{\partial x^4} - \left(\frac{1}{G_{xy}} - \frac{2\nu_{xy}}{E_{xx}}\right)\left(\frac{n\pi}{b}\right)^2 \frac{\mathrm{d}^2 F_n}{\mathrm{d}x^2} + \frac{1}{E_{xx}}\left(\frac{n\pi}{b}\right)^4 F_n = 0. \tag{6.73}$$

Introduction of a scaled coordinate

$$\xi = x\sqrt[4]{\frac{S_{11}}{S_{22}}} = x\sqrt[4]{\frac{E_{yy}}{E_{xx}}} \tag{6.74}$$

with

$$\frac{\mathrm{d}^2}{\mathrm{d}x^2} = \sqrt{\frac{E_{yy}}{E_{xx}}}\frac{\mathrm{d}^2}{\mathrm{d}\xi^2}, \quad \frac{\mathrm{d}^4}{\mathrm{d}x^4} = \frac{E_{yy}}{E_{xx}}\frac{\mathrm{d}^4}{\mathrm{d}\xi^4} \tag{6.75}$$

leads to the following representation of (6.73):

$$\frac{\mathrm{d}^4 F_n}{\partial \xi^4} - \frac{k^2 - 2\nu_{xy}}{s}\left(\frac{n\pi}{b}\right)^2 \frac{\mathrm{d}^2 F_n}{\mathrm{d}\xi^2} + \left(\frac{n\pi}{b}\right)^4 F_n = 0. \tag{6.76}$$

The following approach for F_n is used:

$$F_n = -C_n \left(\frac{b}{n\pi} \right)^2 e^{\frac{\lambda n \pi \xi}{b}}. \tag{6.77}$$

Therein λ is an exponent still to be determined. Substituting (6.77) into (6.76) then yields after some transformations the following characteristic polynomial for the determination of λ:

$$\lambda^4 - 2\zeta\lambda^2 + 1 = 0. \tag{6.78}$$

Therein ζ is the so-called shear number of the considered orthotropic half-plane:

$$\zeta = \frac{S_{12} + 2S_{66}}{2\sqrt{S_{11}S_{22}}} = \frac{k^2 - 2\nu_{xy}}{2s}. \tag{6.79}$$

For the solution of (6.78) a case distinction has to be made. If $\zeta > 1$, then (6.78) gives four real and distinct solutions for the exponent λ:

$$\lambda_{1,2,3,4} = \pm\sqrt{\zeta \pm \sqrt{\zeta^2 - 1}}. \tag{6.80}$$

If, on the other hand, the case $\zeta < 1$ is present, then complex roots result as follows:

$$\lambda_{1,2,3,4} = \pm \left(\xi_1 \pm i\xi_2 \right), \tag{6.81}$$

with

$$\xi_{1,2} = \sqrt{\frac{1 \pm \zeta}{2}}. \tag{6.82}$$

In the following we assume that $\zeta > 1$ holds. The case $\zeta < 1$ can be treated analogously, but remains here without representation. With (6.80) the solution for the function $\bar{F}_n$ is given:

$$\begin{aligned} F_n = & -C_{n1} \left(\frac{b}{n\pi} \right)^2 e^{\frac{\lambda_1 n\pi\xi}{b}} - C_{n2} \left(\frac{b}{n\pi} \right)^2 e^{\frac{\lambda_2 n\pi\xi}{b}} - C_{n3} \left(\frac{b}{n\pi} \right)^2 e^{\frac{\lambda_3 n\pi\xi}{b}} \\ & -C_{n4} \left(\frac{b}{n\pi} \right)^2 e^{\frac{\lambda_4 n\pi\xi}{b}}. \end{aligned} \tag{6.83}$$

The normal stress distribution $\sigma_{xx,n}$ occurring due to the harmonic boundary stress is given as:

$$\sigma_{xx,n} = \left(C_{n1} e^{\frac{\lambda_1 n\pi\xi}{b}} + C_{n2} e^{\frac{\lambda_2 n\pi\xi}{b}} + C_{n3} e^{\frac{\lambda_3 n\pi\xi}{b}} + C_{n4} e^{\frac{\lambda_4 n\pi\xi}{b}} \right) \sin\left(\frac{n\pi y}{b} \right). \tag{6.84}$$

If we assume that we are dealing with a disk infinitely extended in the x-direction, then the two exponential functions increasing with x can be neglected. Let these be the two functions with exponents λ_1 and λ_3. Thus:

$$\sigma_{xx,n} = \left(C_{n2}e^{\frac{\lambda_2 n\pi\xi}{b}} + C_{n4}e^{\frac{\lambda_4 n\pi\xi}{b}}\right)\sin\left(\frac{n\pi y}{b}\right). \tag{6.85}$$

The two remaining constants C_{n2} and C_{n4} are determined from the two boundary conditions that the loaded edge is free of any shear stress on the one hand, and that the normal stress $\sigma_{xx,n}$ at the loaded edge matches the applied harmonic edge stress. Thus:

$$\tau_{xy,n}(\xi = 0) = 0, \quad \sigma_{xx,n}(\xi = 0) = \sigma_{0,n}\sin\left(\frac{n\pi y}{b}\right). \tag{6.86}$$

This yields:

$$C_{n2} = \frac{\sigma_{0,n}\lambda_4}{\lambda_4 - \lambda_2}, \quad C_{n4} = -\frac{\sigma_{0,n}\lambda_2}{\lambda_4 - \lambda_2}. \tag{6.87}$$

This gives the distribution of the normal stress σ_{xx} for the boundary stress problem at hand. A graphical representation of the relative normal stress $\bar{\sigma}_{xx,n} = \frac{\sigma_{xx,n}}{\sigma_{0,n}}$ is shown in Fig. 6.3 for different values of the shear number ζ. For the case $\zeta = 1$ the isotropic half plane (see Fig. 3.13) is included as a special case. Basically, the decay becomes slower with increasing shear number ζ. Moreover, for $\zeta < 1$ zero crossings occur. The case $\zeta = -1$ is a borderline case in the sense that no decay can be observed here, but rather the stress curve fluctuates harmonically over ξ. All in all, the degree of orthotropy has a significant influence on the decay behavior of an edge stress in an orthotropic half-plane.

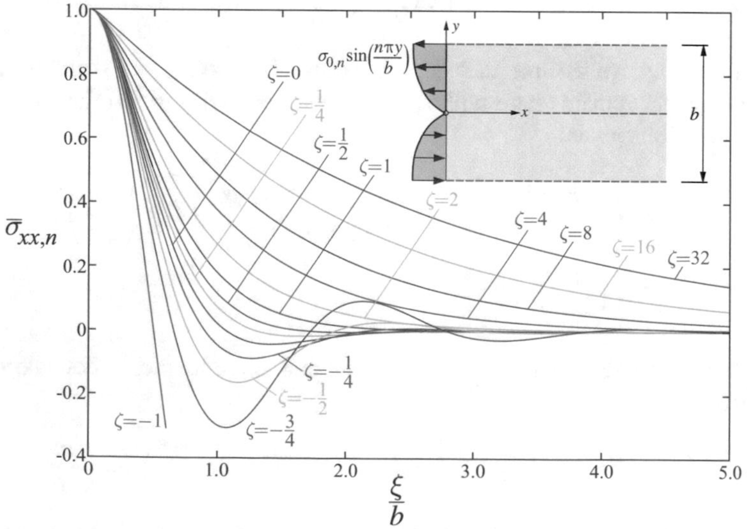

Fig. 6.3 Decay behavior of a harmonic boundary load

6.5 Orthotropic Circular Ring Disks

In this section, we consider the circular ring disk of Fig. 6.4 under constant internal pressure σ_0. Polar orthotropy is assumed here, with the pole of orthotropy found at the coordinate origin. The circular ring disk is bounded by the two edges at the positions $r = R_i$ and $r = R_a$.

Due to the rotational symmetry of the given situation, the approach (6.47) is used, i.e.:

$$F = C_{01} + C_{02}r^2 + C_{03}r^{1+s} + C_{04}r^{1-s}. \tag{6.88}$$

After fulfilling the boundary conditions

$$\sigma_{rr}(r = R_i) = -\sigma_0, \quad \sigma_{rr}(r = R_a) = 0 \tag{6.89}$$

the stresses can be determined as (Lekhnitskii 1968):

$$\begin{aligned} \sigma_{rr} &= \frac{\sigma_0 \rho^{s+1}}{1-\rho^{2s}} \left(\frac{r}{R_a}\right)^{s-1} - \frac{\sigma_0}{1-\rho^{2s}} \rho^{s+1} \left(\frac{R_a}{r}\right)^{s+1}, \\ \sigma_{\varphi\varphi} &= \frac{\sigma_0 \rho^{s+1}}{1-\rho^{2s}} s \left(\frac{r}{R_a}\right)^{s-1} + \frac{\sigma_0}{1-\rho^{2s}} s \rho^{s+1} \left(\frac{R_a}{r}\right)^{s+1}, \\ \tau_{r\varphi} &= 0. \end{aligned} \tag{6.90}$$

where $\rho = \frac{R_i}{R_a}$. For the special case of isotropy with $s = 1$ these expressions take the form (4.76).

A graphical representation of the stress distribution of $\bar{\sigma}_{rr} = \frac{\sigma_{rr}}{\sigma_0}$ and $\bar{\sigma}_{\varphi\varphi} = \frac{\sigma_{\varphi\varphi}}{\sigma_0}$ over the dimensionless radial coordinate $\xi = \frac{r}{R_a}$ for different values of s is shown in Fig. 6.5. The influence of orthotropy on the stress distributions is clearly evident: the radial stress σ_{rr} decays faster with increasing s, whereas the circumferential stress $\sigma_{\varphi\varphi}$ shows an increasing concentration at the inner edge $r = R_i$.

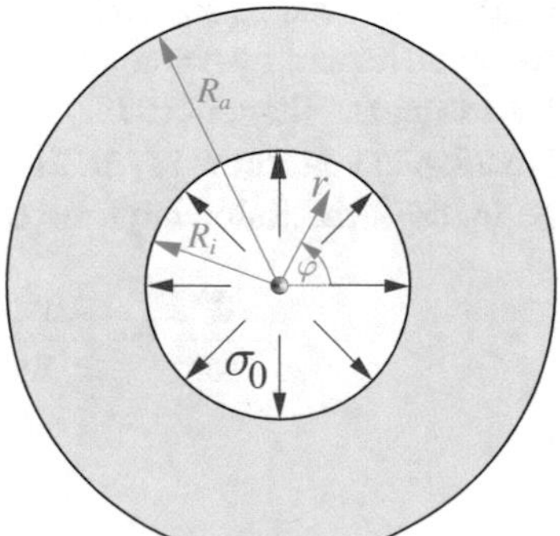

Fig. 6.4 Circular ring disk under internal pressure σ_0

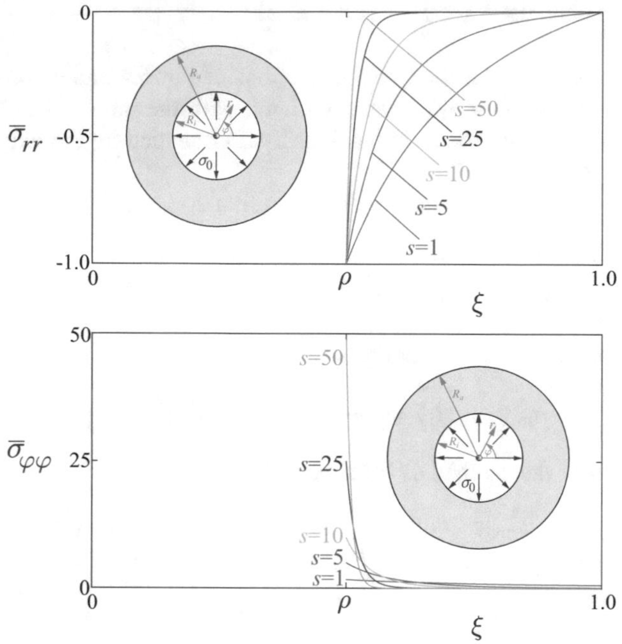

Fig. 6.5 Stress distributions $\bar{\sigma}_{rr} = \frac{\sigma_{rr}}{\sigma_0}$ and $\bar{\sigma}_{\varphi\varphi} = \frac{\sigma_{\varphi\varphi}}{\sigma_0}$ of the circular ring disk under internal pressure σ_0 for different values of s

6.6 Orthotropic Circular Arc Disks

A technically significant disk situation is the orthotropic circular arc disk under boundary loads. We first consider the situation of Fig. 6.6. Consider a polar orthotropic circular arc disk of arbitrary opening angle which is loaded by the boundary moments M_0. Let the boundary of the disk be distinguished by the inner radius R_i and the outer radius R_a, and let the ratio of R_i and R_a be called $\rho = \frac{R_i}{R_a}$. This problem has been treated by a number of authors, and reference is made here to the works of, for example, Blumer (1979), Buchmann (1969), Ko (1988), Lekhnitskii (1968) and Noack and von Roth (1972). Since this is a rotationally symmetric situation, use can be made of the following form of the disk equation:

$$\frac{\mathrm{d}^4 F}{\mathrm{d}r^4} + \frac{2}{r}\frac{\mathrm{d}^3 F}{\mathrm{d}r^3} - \frac{s^2}{r^2}\frac{\mathrm{d}^2 F}{\mathrm{d}r^2} + \frac{s^2}{r^3}\frac{\mathrm{d}F}{\mathrm{d}r} = 0, \tag{6.91}$$

with

$$s = \sqrt{\frac{E_{\varphi\varphi}}{E_{rr}}}. \tag{6.92}$$

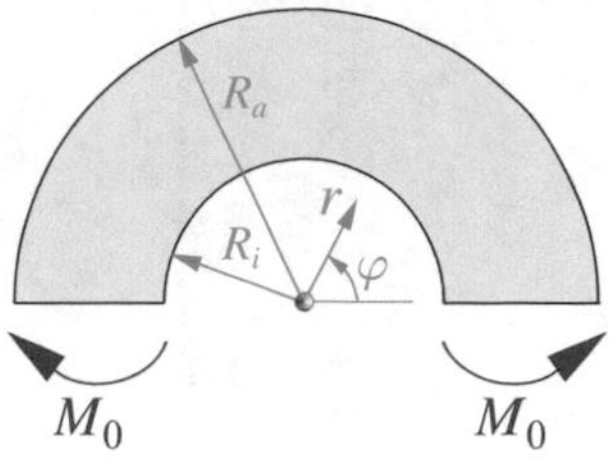

Fig. 6.6 Orthotropic circular arc disk under boundary moments

The solution of the disk equation (6.91) is:

$$F = C_0 + C_1 r^2 + C_2 r^{1+s} + C_3 r^{1-s}. \tag{6.93}$$

From this, the stresses in the disk can be calculated as follows:

$$\begin{aligned}\sigma_{rr} &= 2C_1 + (1+s)C_2 r^{s-1} + (1-s)C_3 r^{-s-1},\\ \sigma_{\varphi\varphi} &= 2C_1 + (1+s)sC_2 r^{s-1} - (1-s)C_3 r^{-s-1},\\ \tau_{r\varphi} &= 0.\end{aligned} \tag{6.94}$$

The constants C_1, C_2, C_3 can be obtained from the following boundary conditions:

$$\begin{aligned}\sigma_{rr}(r = R_i) &= 0,\\ \sigma_{rr}(r = R_a) &= 0,\\ h\int_{R_i}^{R_a} \sigma_{\varphi\varphi} r\,\mathrm{d}r &= -M_0.\end{aligned} \tag{6.95}$$

This ultimately results in the following stress components:

$$\begin{aligned}\sigma_{rr} &= -\frac{M_0}{R_a^2 h\kappa}\left[1 - \frac{1-\rho^{s+1}}{1-\rho^{2s}}\left(\frac{r}{R_a}\right)^{s-1} - \frac{1-\rho^{s-1}}{1-\rho^{2s}}\rho^{s+1}\left(\frac{R_a}{r}\right)^{s+1}\right],\\ \sigma_{\varphi\varphi} &= -\frac{M_0}{R_a^2 h\kappa}\left[1 - \frac{1-\rho^{s+1}}{1-\rho^{2s}}s\left(\frac{r}{R_a}\right)^{s-1} + \frac{1-\rho^{s-1}}{1-\rho^{2s}}s\rho^{s+1}\left(\frac{R_a}{r}\right)^{s+1}\right],\\ \tau_{r\varphi} &= 0,\end{aligned} \tag{6.96}$$

wherein $\rho = \frac{R_i}{R_a}$, and

$$\kappa = \frac{1-\rho^2}{2} - \frac{k}{k+1}\frac{\left(1-\rho^{s+1}\right)^2}{1-\rho^{2s}} + \frac{s\rho^2}{s-1}\frac{\left(1-\rho^{s-1}\right)^2}{1-\rho^{2s}}. \tag{6.97}$$

A graphical representation of the stress components $\bar{\sigma}_{rr}$ and $\bar{\sigma}_{\varphi\varphi}$ for different parameters s with $\rho = \frac{1}{2}$ over the disk height over $\xi = \frac{r}{R_a}$ is shown in Fig. 6.7. The influence of the degree of orthotropy s becomes evident from the diagrams shown

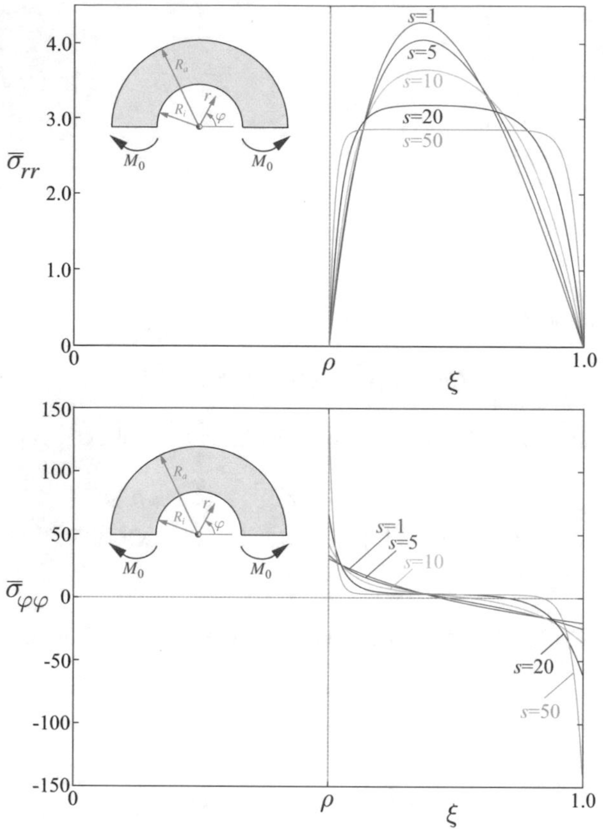

Fig. 6.7 Stress distributions $\bar{\sigma}_{rr}$ and $\bar{\sigma}_{\varphi\varphi}$ for the orthotropic circular arc disk under boundary moments M_0 for different parameters s with $\rho = \frac{1}{2}$

here. In the case of the relative radial stress $\bar{\sigma}_{rr}$ it can be seen that with increasing s the stress distribution becomes more even with a simultaneous decrease of the maximum value. For the relative tangential stress $\bar{\sigma}_{\varphi\varphi}$, on the other hand, there is a pronounced increase in the boundary values at $\xi = \rho$ and $\xi = 1$. All in all, these results show that the effects of orthotropy can be quite serious and therefore have to be taken into account in the design and analysis of such structures.

The more general case of the circular arc disk under combined loading by boundary forces N_0, Q_0 and moments M_0 (Fig. 6.8) is discussed by Blumer (1979), Lekhnitskii (1968), Noack and von Roth (1972), among others. This solution is briefly outlined here. For this disk situation, the following approach for the Airy stress function is chosen:

$$F = F_0 + F_1 \sin\varphi + F_2 \cos\varphi, \tag{6.98}$$

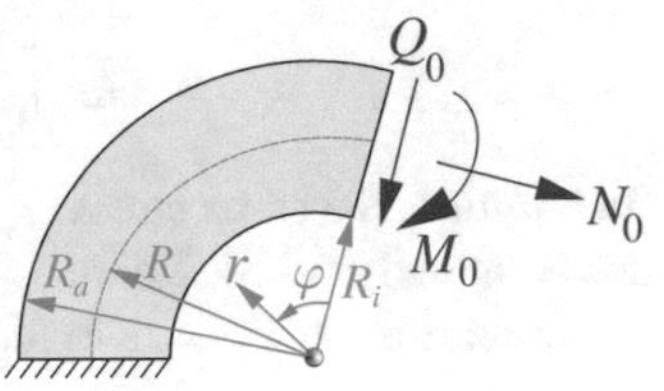

Fig. 6.8 Circular arc disk under boundary moments and forces

wherein the functions F_0, F_1 and F_2 are defined as follows:

$$\begin{aligned} F_0 &= C_{01} + C_{02}r^2 + C_{03}r^{1-s} + C_{04}r^{1+s}, \\ F_1 &= C_{11}r^{1-n} + C_{12}r^{1+n} + C_{13}r + C_{14}r\ln r, \\ F_2 &= D_{11}r^{1-n} + D_{12}r^{1+n} + D_{13}r + D_{14}r\ln r. \end{aligned} \tag{6.99}$$

The stresses in the disk can thus be determined as:

$$\begin{aligned} \sigma_{rr} &= \frac{1}{r}\frac{\mathrm{d}F_0}{\mathrm{d}r} + \left(\frac{1}{r}\frac{\mathrm{d}F_1}{\mathrm{d}r} - \frac{1}{r^2}F_1\right)\sin\varphi + \left(\frac{1}{r}\frac{\mathrm{d}F_2}{\mathrm{d}r} - \frac{1}{r^2}F_2\right)\cos\varphi, \\ \sigma_{\varphi\varphi} &= \frac{\mathrm{d}^2F_0}{\mathrm{d}r^2} + \frac{\mathrm{d}^2F_1}{\mathrm{d}r^2}\sin\varphi + \frac{\mathrm{d}^2F_2}{\mathrm{d}r^2}\cos\varphi, \\ \tau_{r\varphi} &= \left(\frac{1}{r}\frac{\mathrm{d}F_2}{\mathrm{d}r} - \frac{1}{r^2}F_2\right)\sin\varphi - \left(\frac{1}{r}\frac{\mathrm{d}F_1}{\mathrm{d}r} - \frac{1}{r^2}F_1\right)\cos\varphi. \end{aligned} \tag{6.100}$$

The constants appearing in (6.99) are determined from the following boundary conditions, whereby it is to be noted here that the constants C_{01}, C_{13} and D_{13} drop out of the stress expressions and thus are disregarded in all further considerations. Thus, nine boundary conditions for nine constants have to be established.

It is required that the radial stress σ_{rr} at the inner edge $r = R_i$ becomes zero: $\sigma_{rr}(r = R_i) = 0$. From this follows:

$$\begin{aligned} &\frac{1}{R_i}\left.\frac{\mathrm{d}F_0}{\mathrm{d}r}\right|_{r=R_i} + \left(\frac{1}{R_i}\left.\frac{\mathrm{d}F_1}{\mathrm{d}r}\right|_{r=R_i} - \frac{1}{R_i^2}F_1(r=R_i)\right)\sin\varphi \\ &+ \left(\frac{1}{R_i}\left.\frac{\mathrm{d}F_2}{\mathrm{d}r}\right|_{r=R_i} - \frac{1}{R_i^2}F_2(r=R_i)\right)\cos\varphi = 0. \end{aligned} \tag{6.101}$$

This can be satisfied for any angle φ only if each of the three terms appearing here vanishes by itself:

$$\begin{aligned} \left.\frac{\mathrm{d}F_0}{\mathrm{d}r}\right|_{r=R_i} &= 0, \\ \left.\frac{\mathrm{d}F_1}{\mathrm{d}r}\right|_{r=R_i} - \frac{1}{R_i}F_1(r=R_i) &= 0, \end{aligned}$$

$$\left.\frac{\mathrm{d}F_2}{\mathrm{d}r}\right|_{r=R_i} - \frac{1}{R_i}F_2(r=R_i) = 0. \tag{6.102}$$

In addition, the shear stress $\tau_{r\varphi}$ must disappear at the edge $r = R_i$. This condition is automatically fulfilled here.

Conditions analogous to (6.102) must hold at the outer edge $r = R_a$:

$$\begin{aligned} \left.\frac{\mathrm{d}F_0}{\mathrm{d}r}\right|_{r=R_a} &= 0, \\ \left.\frac{\mathrm{d}F_1}{\mathrm{d}r}\right|_{r=R_a} - \frac{1}{R_a}F_1(r=R_a) &= 0, \\ \left.\frac{\mathrm{d}F_2}{\mathrm{d}r}\right|_{r=R_a} - \frac{1}{R_a}F_2(r=R_a) &= 0. \end{aligned} \tag{6.103}$$

At the boundary $\varphi = 0$, the forces and moments N_0, Q_0, M_0 must be in equilibrium with the respective stress components. The following relation between N_0 and the tangential stress $\sigma_{\varphi\varphi}$ holds:

$$\int_{R_i}^{R_a} \sigma_{\varphi\varphi}\mathrm{d}r = \frac{N_0}{h}, \tag{6.104}$$

or

$$\int_{R_i}^{R_a} \left(\frac{\mathrm{d}^2F_0}{\mathrm{d}r^2} + \frac{\mathrm{d}^2F_2}{\mathrm{d}r^2}\right) = \frac{N_0}{h}. \tag{6.105}$$

With $\left.\frac{\mathrm{d}F_0}{\mathrm{d}r}\right|_{r=R_i} = 0$ and $\left.\frac{\mathrm{d}F_0}{\mathrm{d}r}\right|_{r=R_a} = 0$ it follows:

$$\left.\frac{\mathrm{d}F_2}{\mathrm{d}r}\right|_{r=R_a} - \left.\frac{\mathrm{d}F_2}{\mathrm{d}r}\right|_{r=R_i} = \frac{N_0}{h}. \tag{6.106}$$

The moment M_0 must also be in equilibrium with the resulting moment of the tangential stress $\sigma_{\varphi\varphi}$, i.e.:

$$\int_{R_i}^{R_a} \sigma_{\varphi\varphi} r\mathrm{d}r = \frac{M_0 + N_0 R}{h}, \tag{6.107}$$

or

$$\int_{R_i}^{R_a} \left(r\frac{\mathrm{d}^2F_0}{\mathrm{d}r^2} + r\frac{\mathrm{d}^2F_2}{\mathrm{d}r^2}\right)\mathrm{d}r = \frac{M_0 + N_0 R}{h}, \tag{6.108}$$

which after partial integration leads to the following expression:

$$F_0(r = R_a) - F_0(r = R_i) = -\frac{M_0 + N_0 R}{h}. \tag{6.109}$$

Finally, the shear force Q_0 must be in equilibrium with the shear stress $\tau_{r\varphi}$:

$$\int_{R_i}^{R_a} \tau_{r\varphi} \mathrm{d}r = \frac{Q_0}{h}, \tag{6.110}$$

or

$$\int_{R_i}^{R_a} \left(\frac{1}{r}\frac{\mathrm{d}F_1}{\mathrm{d}r} - \frac{1}{r^2} F_1 \right) \mathrm{d}r = \frac{Q_0}{h}. \tag{6.111}$$

After partial integration this expression changes into:

$$\frac{1}{R_a} F_1(r = R_a) - \frac{1}{R_i} F_1(r = R_i) = -\frac{Q_0}{h}. \tag{6.112}$$

Thus, with (6.102), (6.103), (6.106), (6.109) and (6.112) there are nine conditions for determining the nine unknown constants in (6.99) and (6.100), respectively. Evaluating these conditions leads to the following linear system of equations for their determination:

$$\begin{aligned}
2R_i C_{02} + (1-s) C_{03} R_i^{-s} + (1+s) C_{04} R_i^{s} &= 0, \\
-n R_i^{-n} C_{11} + n R_i^{n} C_{12} + C_{14} &= 0, \\
-n R_i^{-n} D_{11} + n R_i^{n} D_{12} + D_{14} &= 0, \\
2R_a C_{02} + (1-s) C_{03} R_a^{-s} + (1+s) C_{04} R_a^{s} &= 0, \\
-n R_a^{-n} C_{11} + n R_a^{n} C_{12} + C_{14} &= 0, \\
-n R_a^{-n} D_{11} + n R_a^{n} D_{12} + D_{14} &= 0, \\
(1-n)\left(R_a^{-n} - R_i^{-n}\right) D_{11} + (1+n)\left(R_a^{n} - R_i^{n}\right) D_{12} + \ln\left(\frac{R_a}{R_i}\right) D_{14} &= \frac{N_0}{h}, \\
\left(R_a^2 - R_i^2\right) C_{02} + \left(R_a^{1-s} - R_i^{1-s}\right) C_{03} + \left(R_a^{1+s} - R_i^{1+s}\right) C_{04} &= -\frac{M_0 + N_0 R}{h}, \\
(1-n)\left(R_a^{-n} - R_i^{-n}\right) C_{11} + (1+n)\left(R_a^{n} - R_i^{n}\right) C_{12} + \ln\left(\frac{R_a}{R_i}\right) C_{14} &= -\frac{Q_0}{h}.
\end{aligned} \tag{6.113}$$

With the abbreviations

$$\begin{aligned}
\kappa_0 &= \frac{1}{\left(R_a^2 - R_i^2\right) - \frac{2}{1+s}\frac{\left(R_a^{s+1} - R_i^{s+1}\right)^2}{R_a^{2s} - R_i^{2s}} + \frac{2}{1-s}\frac{\left(R_a^{s-1} - R_i^{s-1}\right)^2}{R_a^{2s} - R_i^{2s}} R_i^2 R_a^2}, \\
\kappa_1 &= \frac{n\left(R_a^n + R_i^n\right)}{\ln\left(\frac{R_a}{R_i}\right) n \left(R_a^n + R_i^n\right) - 2R_a\left(R_a^n - R_i^n\right)}
\end{aligned} \tag{6.114}$$

the constants result as:

$$
\begin{aligned}
C_{02} &= -\kappa_0 \frac{M_0 + N_0 R}{h}, \\
C_{03} &= \frac{2}{1-s} \frac{R_a^{s-1} - R_i^{s-1}}{R_a^{2s} - R_i^{2s}} R_i^{s+1} R_a^{1+s} \kappa_0 \frac{M_0 + N_0 R}{h}, \\
C_{04} &= \frac{2}{1+s} \frac{R_a^{s+1} - R_i^{s+1}}{R_a^{2s} - R_i^{2s}} \kappa_0 \frac{M_0 + N_0 R}{h}, \\
C_{11} &= -\frac{R_i^n R_a^n}{n\left(R_a^n + R_i^n\right)} \kappa_1 \frac{Q_0}{h}, \\
C_{12} &= \frac{1}{n\left(R_a^n + R_i^n\right)} \kappa_1 \frac{Q_0}{h}, \\
C_{14} &= -\kappa_1 \frac{Q_0}{h}, \\
D_{11} &= \frac{R_i^n R_a^n}{n\left(R_a^n + R_i^n\right)} \kappa_1 \frac{N_0}{h}, \\
D_{12} &= -\frac{1}{n\left(R_a^n + R_i^n\right)} \kappa_1 \frac{N_0}{h}, \\
D_{14} &= \kappa_1 \frac{N_0}{h}.
\end{aligned} \tag{6.115}
$$

From this, using (6.99) and (6.100), the stresses in the disk can then be determined as follows:

$$
\begin{aligned}
\sigma_{rr} &= \left[-2 + 2r^{s-1} \frac{R_a^{s+1} - R_i^{s+1}}{R_a^{2s} - R_i^{2s}} + \frac{2}{r^{s+1}} \frac{R_a^{s-1} - R_i^{s-1}}{R_a^{2s} - R_i^{2s}} R_i^{s+1} R_a^{s+1}\right] \kappa_0 \frac{M_0 + N_0 R}{h} \\
&+ \left[\frac{1}{r} - \frac{r^{n-1}}{R_a^n + R_i^n} - \frac{R_i^n R_a^n}{r^{n+1}\left(R_a^n + R_i^n\right)}\right] \kappa_1 \cos\varphi \frac{N_0}{h} \\
&+ \left[-\frac{1}{r} + \frac{r^{n-1}}{R_a^n + R_i^n} + \frac{R_i^n R_a^n}{r^{n+1}\left(R_a^n + R_i^n\right)}\right] \kappa_1 \sin\varphi \frac{Q_0}{h}, \\
\sigma_{\varphi\varphi} &= \left[-2 + 2s r^{s-1} \frac{R_a^{s+1} - R_i^{s+1}}{R_a^{2s} - R_i^{2s}} - \frac{2s}{r^{s+1}} \frac{R_a^{s-1} - R_i^{s-1}}{R_a^{2s} - R_i^{2s}} R_i^{s+1} R_a^{s+1}\right] \kappa_0 \frac{M_0 + N_0 R}{h} \\
&+ \left[\frac{1}{r} - (1+n) \frac{r^{n-1}}{R_a^n + R_i^n} - (1-n) \frac{R_i^n R_a^n}{r^{n+1}\left(R_a^n + R_i^n\right)}\right] \kappa_1 \cos\varphi \frac{N_0}{h} \\
&+ \left[-\frac{1}{r} + (1+n) \frac{r^{n-1}}{R_a^n + R_i^n} + (1-n) \frac{R_i^n R_a^n}{r^{n+1}\left(R_a^n + R_i^n\right)}\right] \kappa_1 \sin\varphi \frac{Q_0}{h}, \\
\tau_{r\varphi} &= \left[\frac{1}{r} - \frac{r^{n-1}}{R_a^n + R_i^n} - \frac{R_i^n R_a^n}{r^{n+1}\left(R_a^n + R_i^n\right)}\right] \kappa_1 \sin\varphi \frac{N_0}{h} \\
&+ \left[\frac{1}{r} - \frac{r^{n-1}}{R_a^n + R_i^n} - \frac{R_i^n R_a^n}{r^{n+1}\left(R_a^n + R_i^n\right)}\right] \kappa_1 \cos\varphi \frac{Q_0}{h}.
\end{aligned} \tag{6.116}
$$

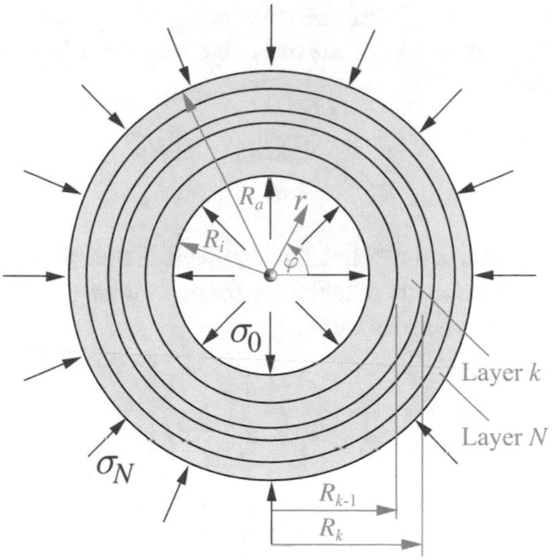

Fig. 6.9 Layered circular ring disk

6.7 Layered Circular Ring Disks

Layered disks are technically significant structures. We begin the elaborations by considering the situation of Fig. 6.9. Let a circular ring disk be given, which consists of N polar orthotropic layers (Lekhnitskii 1968). The respective layer thicknesses are arbitrary. The disk is loaded by the internal pressure σ_0 and the external pressure σ_N. We assume perfect bonding of the layers. Let the layer k be bounded by the two radial coordinates $r = R_{k-1}$ and $r = R_k$, where $R_0 = R_i$ and $R_N = R_a$. The problem is described by polar coordinates r, φ. Let $\sigma_{rr,k}$ and $\sigma_{\varphi\varphi,k}$ be the two normal stresses in the layer k. Due to the rotational symmetry of the given situation, the shear stress $\tau_{r\varphi,k}$ is absent in each layer. The same is true for the circumferential displacement v.

We introduce the two layerwise quantities:

$$\rho_k = \frac{R_{k-1}}{R_k}, \quad s_k = \sqrt{\frac{E_{\varphi\varphi,k}}{E_{rr,k}}}. \tag{6.117}$$

The boundary conditions to be fulfilled here are at the disk surfaces:

$$\sigma_{rr,1}(r = R_i) = -\sigma_0, \quad \sigma_{rr,N}(r = R_a) = -\sigma_0. \tag{6.118}$$

In addition, there are continuity conditions in the boundary surfaces between the individual layers. At the point $r = R_{k-1}$ between the two layers $k - 1$ and k must hold:

$$\begin{aligned} \sigma_{rr,k-1}(r = R_{k-1}) &= \sigma_{rr,k}(r = R_{k-1}), \\ u_{k-1}(r = R_{k-1}) &= u_k(r = R_{k-1}), \end{aligned} \tag{6.119}$$

where u is the radial displacement. Extending those expressions already found for the homogeneous polar orthotropic annular disk, the layer-wise stresses and displacements are as follows:

$$\begin{aligned} \sigma_{rr,k} &= \frac{\sigma_{k-1}\rho_k^{s_k+1}}{1-\rho_k^{2s_k}}\left[\left(\frac{r}{R_k}\right)^{s_k-1} - \left(\frac{R_k}{r}\right)^{s_k+1}\right] \\ &\quad + \frac{\sigma_k}{1-\rho_k^{2s_k}}\left[-\left(\frac{r}{R_k}\right)^{s_k-1} + \rho_k^{2s_k}\left(\frac{R_k}{r}\right)^{s_k+1}\right], \\ \sigma_{\varphi\varphi,k} &= \frac{\sigma_{k-1}\rho_k^{s_k+1}s_k}{1-\rho_k^{2s_k}}\left[\left(\frac{r}{R_k}\right)^{s_k-1} + \left(\frac{R_k}{r}\right)^{s_k+1}\right] \\ &\quad - \frac{\sigma_k s_k}{1-\rho_k^{2s_k}}\left[\left(\frac{r}{R_k}\right)^{s_k-1} + \rho_k^{2s_k}\left(\frac{R_k}{r}\right)^{s_k+1}\right], \\ u_k &= \frac{\sigma_{k-1}R_k\rho_k^{s_k+1}}{E_{\varphi\varphi,k}\left(1-\rho_k^{2s_k}\right)}\left[\left(s_k - \nu_{\varphi r,k}\right)\left(\frac{r}{R_k}\right)^{s_k} + \left(s_k - \nu_{\varphi r,k}\right)\left(\frac{R_k}{r}\right)^{s_k}\right] \\ &\quad - \frac{\sigma_k R_k}{E_{\varphi\varphi,k}\left(1-\rho_k^{2s_k}\right)}\left[\left(s_k - \nu_{\varphi r,k}\right)\left(\frac{r}{R_k}\right)^{s_k}\right. \\ &\quad \left. + \left(s_k + \nu_{\varphi r,k}\right)\rho_k^{2s_k}\left(\frac{R_k}{r}\right)^{s_k}\right]. \end{aligned} \tag{6.120}$$

Herein, σ_{k-1} and σ_k are the radial stresses acting in the boundary surface of layer k at $r = R_{k-1}$ and $r = R_k$, u_k is the radial displacement of layer k.

The fulfillment of the boundary and continuity conditions (6.118) and (6.119) leads to the following equation, which is to be evaluated for $k = 1, 2, ..., N - 1$, taking into account that the stresses σ_0 and σ_N are given:

$$\sigma_{k+1}R_{k+1}\alpha_{k+1} + \sigma_k R_k \beta_k + \sigma_{k-1}R_{k-1}\alpha_{k-1} = 0, \tag{6.121}$$

with the abbreviations:

$$\begin{aligned} \alpha_k &= \frac{2s_k}{E_{\varphi\varphi,k}}\frac{\rho_k^{s_k}}{1-\rho_k^{2s_k}}, \\ \beta_k &= \frac{1}{E_{\varphi\varphi,k}}\left(\nu_{\varphi r,k} - s_k\frac{1+\rho_k^{2s_k}}{1-\rho_k^{2s_k}}\right) \end{aligned}$$

$$-\frac{1}{E_{\varphi\varphi,k+1}}\left(\nu_{\varphi r,k+1}+s_{k+1}\frac{1+\rho_{k+1}^{2s_{k+1}}}{1-\rho_{k+1}^{2s_{k+1}}}\right). \tag{6.122}$$

With this, the radial stresses σ_k in the boundary surfaces of the individual layers can be calculated, with which the layer-by-layer stresses can then be determind by means of (6.120).

6.8 Layered Circular Arc Disks

In this section, we consider the situation of a layered circular arc disk under moment load M_0 (cf. Ko and Jackson 1989; Schnabel et al. 2017), as shown in Fig. 6.10. To determine the stress field in this layered disk, the Airy stress function in each layer is formulated as follows ($k = 1, 2, ..., N$):

$$F_k = C_{01,k} + C_{02,k}r^2 + C_{03,k}r^{1+s_k} + C_{04,k}r^{1-s_k}, \tag{6.123}$$

with

$$s_k = \sqrt{\frac{E_{\varphi\varphi,k}}{E_{rr,k}}}. \tag{6.124}$$

The layerwise constants $C_{02,k}$, $C_{03,k}$, $C_{04,k}$ are determined from boundary and continuity conditions for stresses and displacements. The constant $C_{01,k}$ plays no role in the determination of the state quantities of the disk and can be neglected. The stress field is then:

$$\begin{aligned}
\sigma_{rr,k} &= 2C_{02,k} + C_{03,k}(s_k+1)\,r^{s_k-1} + C_{04,k}(1-s_k)\,r^{-s_k-1},\\
\sigma_{\varphi\varphi,k} &= 2C_{02,k} + C_{03,k}s_k(s_k+1)\,r^{s_k-1} - C_{04,k}s_k(1-s_k)\,r^{-s_k-1},\\
\tau_{r\varphi} &= 0.
\end{aligned} \tag{6.125}$$

Once the stresses have been determined, the strains $\varepsilon_{rr,k}$, $\varepsilon_{\varphi\varphi,k}$, $\gamma_{r\varphi,k}$ (with $i = 1, ..., N$) in each layer can be calculated. The displacements in each layer then follow

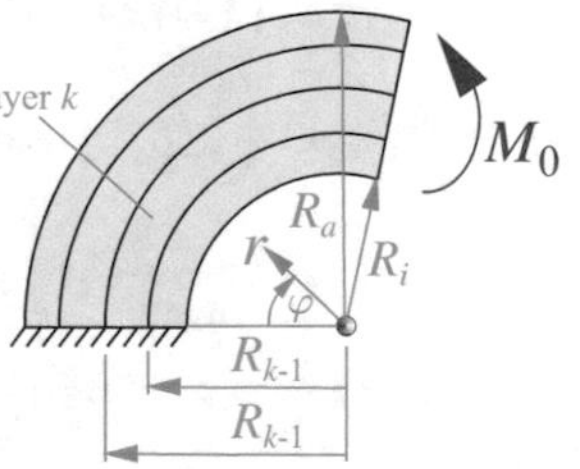

Fig. 6.10 Layered circular arc disk under boundary moment load

from the kinematic equations (4.5). For the displacements we have:

$$
\begin{aligned}
u_k &= 2C_{02,k}\left(S_{11,k}+S_{12,k}\right)r + C_{03,k}\left(\frac{1}{s_k}S_{11,k}+S_{12,k}\right)(s_k+1)\,r^{s_k} \\
&\quad -C_{04,k}\left(\frac{1}{s_k}S_{11,k}-S_{12,k}\right)(1-s_k)\,r^{-s_k}, \\
v_k &= 2C_{02,k}\left(S_{22,k}-S_{11,k}\right)r\varphi.
\end{aligned}
\tag{6.126}
$$

The following boundary conditions can be postulated. First, the radial normal stress σ_{rr} at the inner and outer edge of the disk must vanish for each φ:

$$
\sigma_{rr,1}(r=R_i)=0, \quad \sigma_{rr,N}(r=R_a)=0. \tag{6.127}
$$

This leads to the following two conditions:

$$
\begin{aligned}
2C_{02,1}+C_{03,1}\,(s_1+1)\,R_i^{s_1-1}+C_{04,1}\,(1-s_1)\,R_i^{-s_1-1} &= 0, \\
2C_{02,N}+C_{03,N}\,(s_N+1)\,R_a^{s_N-1}+C_{04,N}\,(1-s_N)\,R_a^{-s_N-1} &= 0.
\end{aligned}
\tag{6.128}
$$

In the interfaces between the individual layers, the continuity of the displacements u and v as well as the radial stress σ_{rr} must be ensured. It holds:

$$
\begin{aligned}
\sigma_{rr,k}\,(r=R_k) &= \sigma_{rr,k+1}\,(r=R_k), \\
u_k\,(r=R_k) &= u_{k+1}\,(r=R_k), \\
v_k\,(r=R_k) &= v_{k+1}\,(r=R_k),
\end{aligned}
\tag{6.129}
$$

with $i = 1, 2, ..., N-1$. This leads to the following expressions:

$$
\begin{aligned}
&2C_{02,k}+C_{03,k}\,(s_k+1)\,R_k^{s_k-1}+C_{04,k}\,(1-s_k)\,R_k^{-s_k-1} \\
-\;&2C_{03,k+1}-C_{03,k+1}\,(s_{k+1}+1)\,R_k^{s_{k+1}-1}-C_{04,k+1}\,(1-s_{k+1})\,R_k^{-s_{k+1}-1}=0, \\
&2C_{02,k}\left(S_{11,k}+S_{12,k}\right)R_k+C_{03,k}\left(\frac{1}{s_k}S_{11,k}+S_{12,k}\right)(s_k+1)\,R_k^{s_k} \\
-\;&C_{04,k}\left(\frac{1}{s_k}S_{11,k}-S_{12,k}\right)(1-s_k)\,R_k^{-s_k} \\
-\;&2C_{02,k+1}\left(S_{11,k+1}+S_{12,k+1}\right)R_k \\
-\;&C_{03,k+1}\left(\frac{1}{s_{k+1}}S_{11,k+1}+S_{12,k+1}\right)(s_{k+1}+1)\,R_k^{s_{k+1}} \\
+\;&C_{04,k+1}\left(\frac{1}{s_{k+1}}S_{11,k+1}-S_{12,k+1}\right)(1-s_{k+1})\,R_k^{-s_{k+1}}=0, \\
&C_{02,k}\left(S_{22,k}-S_{11,k}\right)-C_{02,k+1}\left(S_{22,k+1}-S_{11,k+1}\right)=0.
\end{aligned}
\tag{6.130}
$$

At the loaded edge of the disk, the resultant of the tangential normal stress $\sigma_{\varphi\varphi}$ multiplied by the respective lever arm must correspond to the applied moment M_0.

The contributions of the individual layers must be summed up:

$$\sum_{k=1}^{N} \int_{R_{k-1}}^{R_k} \sigma_{\varphi\varphi,k} r \mathrm{d}r = -M_0. \tag{6.131}$$

This results in the following condition:

$$\sum_{k=1}^{N} \Big[C_{02,k} \left(R_k^2 - R_{k-1}^2 \right) + C_{03,k} s_k \left(R_k^{s_k+1} - R_{k-1}^{s_k+1} \right) - C_{04,k} s_k \left(R_k^{-s_k+1} - R_{k-1}^{-s_k+1} \right) \Big] = -M_0. \tag{6.132}$$

This provides $3N$ conditions for the $3N$ constants $C_{02,k}$, $C_{03,k}$, $C_{04,k}$.

The stresses that can be determined with the described solution method are evaluated below for a typical composite material (see Table 6.1). A comparison of the results for an asymmetric cross-ply laminate [0°/90°] and a symmetric cross-ply laminate $[0°/90°]_S$ with finite element calculations (see Schnabel et al. 2017) is given in Figs. 6.11 and 6.12. The coordinate ξ is measured starting from the inner

Table 6.1 Material properties of a typical composite material (layer thickness $t = 0{,}125$ mm)

0°-layer			90°-layer			
$E_{\varphi\varphi}$	132.000	[MPa]	$E_{\varphi\varphi}$	10.800	[MPa]	Tangential modulus of elasticity
E_{rr}	10.800	[MPa]	E_{rr}	10.800	[MPa]	Radial modulus of elasticity
E_{zz}	10.800	[MPa]	E_{zz}	132.000	[MPa]	Transversal modulus of elasticity
$\nu_{r\varphi}$	0,0195	[–]	$\nu_{r\varphi}$	0,238	[–]	Poisson's ratio
ν_{rz}	0,238	[–]	ν_{rz}	0,0195	[–]	Poisson's ratio
$\nu_{\varphi z}$	0,238	[–]	$\nu_{\varphi z}$	0,0195	[–]	Poisson's ratio
$G_{r\varphi}$	5650	[MPa]	$G_{r\varphi}$	3360	[MPa]	Shear modulus
G_{rz}	3360	[MPa]	G_{rz}	5650	[MPa]	Shear modulus
$G_{\varphi z}$	5650	[MPa]	$G_{\varphi z}$	5650	[MPa]	Shear modulus

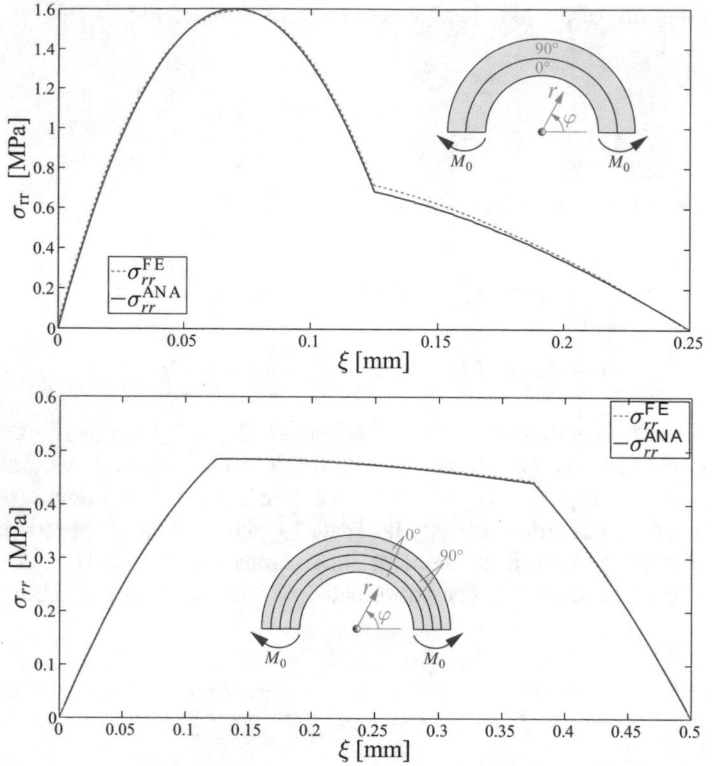

Fig. 6.11 Radial stress σ_{rr} in an asymmetric cross-ply laminate $[0°/90°]$ (top) and in a symmetric cross-ply laminate $[0°/90°]_S$ (bottom)

edge in the direction of the outer edge, i.e. $\xi = r - R_i$. The layerwise character of the stress field can be clearly observed. Due to the different properties of the individual layers of the disk, in the case of the tangential stress $\sigma_{\varphi\varphi}$ there are discontinuities in the stress distribution at the interfaces between the 0°- and the 90°-layers, whereas in the case of the radial stress σ_{rr} there are discontinuities in the slope over ξ. The comparison with the accompanying FEM simulations is very satisfactory, and the analytical disk solution presented here gives almost identical results, but with a fraction of the computational effort required for the numerical calculation.

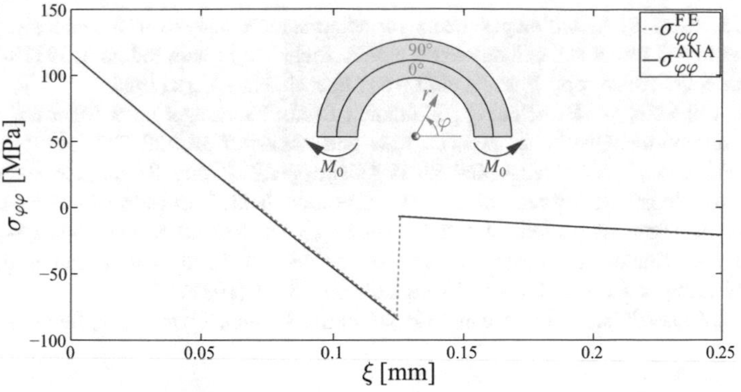

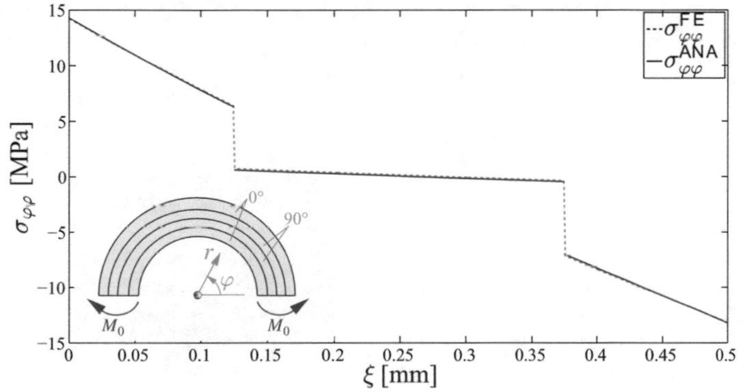

Fig. 6.12 Tangential stress $\sigma_{\varphi\varphi}$ in an asymmetric cross-ply laminate [0°/90°] (top) and in a symmetric cross-ply laminate $[0°/90°]_S$ (bottom)

References

Blumer, H.: Spannungsberechnungen an anisotropen Kreisbogenscheiben und Satteltraegern konstanter Dicke. Research report, Universität (TH) Karlsruhe, Germany (1979)

Buchmann, W.: Berechnung polarorthotroper Kreisbogenscheiben konstanter Dicke unter reiner Biegebeanspruchung. Die Bautechnik **1**, 27–32 (1969)

Ding, H.J., Huang, D.J., Wang, H.M.: Analytical calculation for fixed-fixed anisotropic beam subjected to uniform load. Appl. Math. Mech. **27**, 1305–1310 (2006)

Hashin, Z.: Plane anisotropic beams. J. Appl. Mech. **34**, 257–262 (1967)

Jayne, B.A., Tang, R.C.: Power series stress function for anisotropic and orthotropic beams. Wood Fiber Sci. **2** (1970)

Karttunen, A.T., von Hertzen, R.: On the foundations of anisotropic interior beam theories. Compos. Part B Eng. **87**, 299–310 (2016)

Ko, W.L.: Delamination stresses in semicircular laminated composite bars. NASA Technical Memorandum 4026 (1988)

Ko, W.L., Jackson, R.H.: Multilayer theory for delamination analysis of a composite curved bar subjected to end forces and end moments. NASA Technical Memorandum 4139 (1989)
Lekhnitskii, S.G.: Anisotropic Plates. Gordon and Breach, New York (1968)
Noack, D., von Roth, W.: Berechnung gekrümmter Brettschichtträger unter Belastung durch Momente, Normal- und Querkräfte. Holz als Roh- und Werkstoff **30**, 220–233 (1972)
Rand, O., Rovenski, V.: Analytical Methods in Anisotropic Elasticity. Birkhäuser, Boston (2005)
Schnabel, J.E., Yousfi, M., Mittelstedt, C.: Free-edge stress fields in cylindrically curved symmetric and unsymmetric cross-ply laminates under bending load. Compos. Struct. **180**, 862–875 (2017)
von Roth, W., Epple, A.: Vergleichende isotrope und orthotrope Berechnung gekrümmter Brettschichtträger. Holz als Roh- und Werkstoff **39**, 25–31 (1981)
Wiedemann, J.: Leichtbau, Elemente und Konstruktion, 3rd edn. Springer, Berlin (2006)

Part III
Plates

Kirchhoff Plate Theory in Cartesian Coordinates

7

This chapter is dedicated to the so-called Kirchhoff plate theory in Cartesian coordinates. Starting from the assumptions of this elementary important plate theory, both the displacement field and the strain field of the Kirchhoff plate are derived and discussed in detail, from which finally the stress field of the plate can be determined. The internal forces and moments of the Kirchhoff plate then follow directly from the stresses. The effective properties of plates for various special cases are then discussed before the basic equations of plate bending are derived. An important part is the discussion of the so-called equivalent shear forces and the plate boundary conditions. After elementary solutions of the plate equation, boundary value problems in the form of the bending of plate strips as well as Navier-type solutions and Lévy-type solutions are treated. This is followed by the energetic treatment of plate problems, and both the principle of virtual displacements and the principle of the stationary value of the total elastic potential are used to derive the plate equation and boundary conditions. Furthermore, plates with arbitrary boundaries are considered as well. The discussion of two interesting special cases, namely the plate on an elastic foundation and the membrane, completes the present chapter.

7.1 Introduction

A plate (see Fig. 7.1) is a thin planar surface structure which, in contrast to the disk, is not loaded in its plane but exclusively perpendicular to it, for example by surface or line loads or single forces, but also by moments or other loads which lead to a curvature and/or a twist of the plate. The thickness h of the plate is significantly smaller than the characteristic dimensions of the plate in its plane and may generally be a continuous function of the two coordinates x and y. Let $h << l$ hold if l is a characteristic dimension in the plane of the plate. We want to start the considerations using a Cartesian coordinate system x, y, z, and the plate is bisected at each point

C. Mittelstedt, *Theory of Plates and Shells*,
https://doi.org/10.1007/978-3-662-66805-4_7

x, y by its so-called plate middle plane, which is spanned by the coordinate axes x and y. Thus the origin of the coordinates is in the xy-plane, and $-\frac{h}{2} \leq z \leq \frac{h}{2}$. Hence, in fact, disks and plates are identical with respect to their geometrical properties, they differ only in terms of their loading.

The mechanics of isotropic and anisotropic plate structures are discussed, for example, in the works of Altenbach et al. (2016), Ambartsumyan (1970), Becker and Gross (2002), Czerwenka and Schnell (1970), Eschenauer and Schnell (1993), Eschenauer et al. (1997), Girkmann (1974), Göldner et al. (1979), Gross et al. (2014), Hake and Meskouris (2007), Lekhnitskii (1968), Mansfield (1989), Marguerre and Woernle (1975), Mittelstedt and Becker (2016), Reddy (2006), Timoshenko and Woinowsky-Krieger (1964), Turner (1965), Ugural (1981), Vinson (1974), Wang (1953), and Wiedemann (2006), among others.

The disk and plate behavior can be considered separately in the case of small deformations, even if loads are present simultaneously in the plane of the plate structure and perpendicular to it. In this case, the stress fields can be superposed.

Typical and technically significant plate shapes are plates with rectangular or circular planform. The present chapter is devoted to the treatment of plate problems in Cartesian coordinates, which is particularly useful in the treatment of rectangular plates. Altenbach et al. (2016) also provide details on the analytical treatment of oblique plates.

7.2 The Kirchhoff Plate Theory

7.2.1 Assumptions, Kinematics and Displacement Field

Consider a plate of thickness h as shown in Fig. 7.1. The Kirchhoff plate theory[1] is a very widely used theory for the analysis of thin-walled plate structures. It is an important prerequisite that the plate structure under consideration is thin, although a more precise definition of this term is not easy to establish. At this point, we will assume that the plate thickness h is significantly smaller than the characteristic dimensions of the plate under consideration in the xy-plane.

Kirchhoff's plate theory is based on the following assumptions:

- We assume linear-elastic material behavior and assume that Hooke's law is valid. In all further explanations we will assume orthotropic plates, where the principal material directions coincide with the reference axes x, y and z. Let the material under consideration be homogeneous.
- We want to assume that the cross-section of the plate remains plane even in the deformed state and thus no cross-sectional warpings occur. This corresponds to

[1] Gustav Robert Kirchhoff, 1824–1887, German physicist.

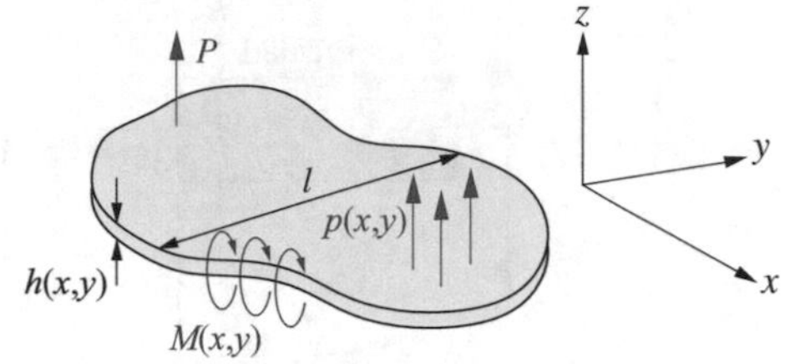

Fig. 7.1 Plate

the hypothesis of the cross-sections remaining plane, as it is also used for the Euler-Bernoulli beam.

- We further assume that the cross section is perpendicular to the middle plane of the plate even in the deformed state. This is called the normal hypothesis.
- We assume geometric linearity and therefore assume that the deformations are small compared to the dimensions of the plate.
- The normal stress σ_{zz} in the thickness direction of the plate is assumed to be negligibly small. The two transverse shear stresses τ_{xz} and τ_{yz} cannot be determined from constitutive relations, but must be determined in a post-calculation from equilibrium considerations.
- The thickness h of the plate does not change in the deformed state.

With these assumptions it is clear that the Kirchhoff plate theory is an extension of the Euler-Bernoulli beam theory for plate structures. Therefore, we also speak of the so-called shear-rigid plate.

We now consider the sectional element of a Kirchhoff plate shown in Fig. 7.2, both in the undeformed and in the deformed state. We will first examine a point which is located on the middle plane of the plate at $z = 0$ and exhibits the displacement or deflection w_0 in z-direction. The subscript 0 indicates that this is a quantity that refers to the middle plane of the plate and thus is only a function of the two plane coordinate directions x and y: $w_0 = w_0(x, y)$. Also associated with the deflection w_0 of the plate midplane is a slope $\frac{\partial w_0}{\partial x}$ as indicated. The slope $\frac{\partial w_0}{\partial y}$ occurs as well, but is not representable here in the given perspective. The displacements u_0 and v_0 in the plate middle plane do not occur in pure bending.

In addition to the deflection w_0 and the two inclinations $\frac{\partial w_0}{\partial x}$ and $\frac{\partial w_0}{\partial y}$, the two bending angles ψ_x and ψ_y also occur, which describe the rotation of the cross-section about the y-axis and the x-axis, respectively. The bending angle ψ_x is shown in Fig. 7.2 and is negative in the given situation. Since, according to the assumptions made at the beginning of this section, we assume that the cross section remains plane in the deformed state, the two bending angles are exclusively functions of the two plane coordinates x and y, i.e. $\psi_x = \psi_x(x, y)$ and $\psi_y = \psi_y(x, y)$ since the cross-section of the plate remains plane.

We now consider the displacement of the point P as shown in Fig. 7.2. This point has the distance z_P from the plate middle plane. It will undergo the deflection w and

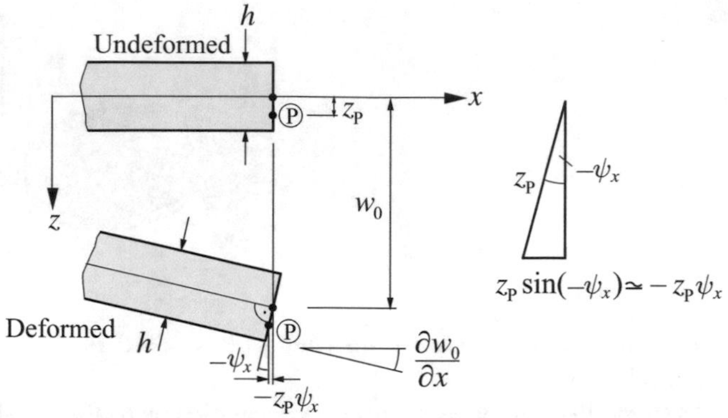

Fig. 7.2 Kinematics of a plate element in the context of Kirchhoff's plate theory

the horizontal displacement u_P during deformation. We can deduce from Fig. 7.2:

$$u_P = -z_P \sin(-\psi_x) = z_P \sin \psi_x. \tag{7.1}$$

We assume small deformations according to above assumptions, which is also true for the bending angle ψ_x, and $\sin(\psi_x) \simeq \psi_x$ is valid:

$$u_P = z_P \psi_x. \tag{7.2}$$

This relation holds for any point outside the plate midplane, so we want to drop the index P from here on:

$$u = z\psi_x. \tag{7.3}$$

The displacement v of any point can be formulated in the same way:

$$v = z\psi_y. \tag{7.4}$$

Since we have assumed that the thickness h of the plate does not change due to the deformation, we can also equate the deflection w of any point with the deflection w_0 of the plate center plane:

$$w = w_0. \tag{7.5}$$

We also bring into play the normal hypothesis, according to which cross-sections of the plate are perpendicular to the plate midplane even in the deformed state, i.e. the right angle marked in Fig. 7.2 is preserved. Thus, a relation between the two bending angles ψ_x and ψ_y and the deflection $w = w_0$ of the plate can be established immediately as follows:

$$-\psi_x = \frac{\partial w_0}{\partial x}, \quad -\psi_y = \frac{\partial w_0}{\partial y}. \tag{7.6}$$

With the now established relations we can write down the displacement field according to Kirchhoff's plate theory as:

$$u(x, y, z) = -z\frac{\partial w_0(x, y)}{\partial x}, \quad v(x, y, z) = -z\frac{\partial w_0(x, y)}{\partial y}, \quad w(x, y) = w_0(x, y). \tag{7.7}$$

7.2.2 Strain and Stress Field

Using the displacement field derived with (7.7), the strain field of the Kirchhoff plate can be derived using the kinematic relations (1.60). It follows:

$$\begin{aligned}
\varepsilon_{xx} &= \frac{\partial u}{\partial x} = -z\frac{\partial^2 w_0}{\partial x^2}, \\
\varepsilon_{yy} &= \frac{\partial v}{\partial y} = -z\frac{\partial^2 w_0}{\partial y^2}, \\
\varepsilon_{zz} &= \frac{\partial w}{\partial z} = 0, \\
\gamma_{xy} &= \frac{\partial u}{\partial y} + \frac{\partial v}{\partial x} = -2z\frac{\partial^2 w_0}{\partial x \partial y}, \\
\gamma_{xz} &= \frac{\partial u}{\partial z} + \frac{\partial w}{\partial x} = 0, \\
\gamma_{yz} &= \frac{\partial v}{\partial z} + \frac{\partial w}{\partial y} = 0.
\end{aligned} \tag{7.8}$$

Consistent with the assumptions made, both the strain ε_{zz} as a consequence of the assumption of the invariability of the plate thickness h and the two shear strains γ_{xz} and γ_{yz} as a consequence of the normal hypothesis and the hypothesis of the flatness of the cross sections disappear. Therefore, only the two strains ε_{xx} and ε_{yy} and the shear strain γ_{xy} remain, which are shown to vary linearly over z.

We introduce at this point the so-called curvatures κ_{xx}^0 and κ_{yy}^0 as well as the so-called twist κ_{xy}^0 of the midplane of the plate as follows:

$$\kappa_{xx}^0 = -\frac{\partial^2 w_0}{\partial x^2}, \quad \kappa_{yy}^0 = -\frac{\partial^2 w_0}{\partial y^2}, \quad \kappa_{xy}^0 = -2\frac{\partial^2 w_0}{\partial x \partial y}. \tag{7.9}$$

They are shown schematically in Fig. 7.3 and are summarized in the vector $\underline{\kappa}^0$ as follows:

$$\underline{\kappa}^0 = \begin{pmatrix} \kappa_{xx}^0 \\ \kappa_{yy}^0 \\ \kappa_{xy}^0 \end{pmatrix} = -\begin{pmatrix} \frac{\partial^2 w_0}{\partial x^2} \\ \frac{\partial^2 w_0}{\partial y^2} \\ 2\frac{\partial^2 w_0}{\partial x \partial y} \end{pmatrix}. \tag{7.10}$$

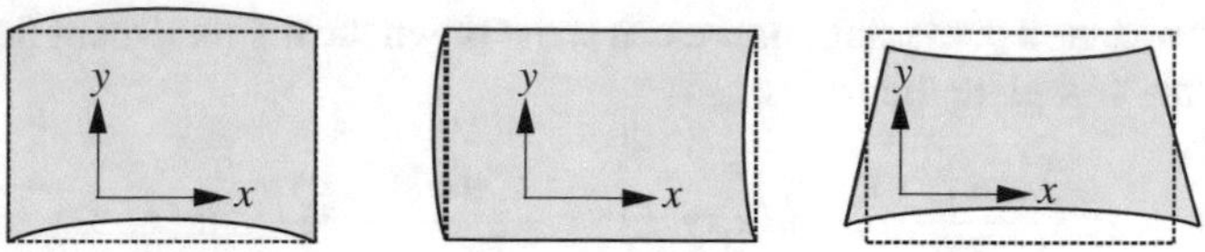

Fig. 7.3 Curvatures and twist of the plate middle plane

The strain field of the Kirchhoff plate theory, consisting of the two plane strains ε_{xx} and ε_{yy} and the plane shear strain γ_{xy}, can then be given in a vector-matrix notation as:

$$\begin{pmatrix} \varepsilon_{xx} \\ \varepsilon_{yy} \\ \gamma_{xy} \end{pmatrix} = z\underline{\kappa}^0 = z \begin{pmatrix} \kappa_{xx}^0 \\ \kappa_{yy}^0 \\ \kappa_{xy}^0 \end{pmatrix} = -z \begin{pmatrix} \frac{\partial^2 w_0}{\partial x^2} \\ \frac{\partial^2 w_0}{\partial y^2} \\ 2\frac{\partial^2 w_0}{\partial x \partial y} \end{pmatrix}. \tag{7.11}$$

Once the strain field, here expressed by the curvatures κ_{xx}^0 and κ_{yy}^0 and the twist κ_{xy}^0, is given, then the stress field of the plate can be determined by (1.194) as follows:

$$\begin{aligned} \begin{pmatrix} \sigma_{xx} \\ \sigma_{yy} \\ \tau_{xy} \end{pmatrix} &= \begin{bmatrix} Q_{11} & Q_{12} & 0 \\ Q_{12} & Q_{22} & 0 \\ 0 & 0 & Q_{66} \end{bmatrix} \begin{pmatrix} \varepsilon_{xx} \\ \varepsilon_{yy} \\ \gamma_{xy} \end{pmatrix} \\ &= -\begin{bmatrix} Q_{11} & Q_{12} & 0 \\ Q_{12} & Q_{22} & 0 \\ 0 & 0 & Q_{66} \end{bmatrix} z \begin{pmatrix} \frac{\partial^2 w_0}{\partial x^2} \\ \frac{\partial^2 w_0}{\partial y^2} \\ 2\frac{\partial^2 w_0}{\partial x \partial y} \end{pmatrix}. \end{aligned} \tag{7.12}$$

In expanded form we have:

$$\begin{aligned} \sigma_{xx} &= -Q_{11} z \frac{\partial^2 w_0}{\partial x^2} - Q_{12} z \frac{\partial^2 w_0}{\partial y^2}, \\ \sigma_{yy} &= -Q_{12} z \frac{\partial^2 w_0}{\partial x^2} - Q_{22} z \frac{\partial^2 w_0}{\partial y^2}, \\ \tau_{xy} &= -2Q_{66} z \frac{\partial^2 w_0}{\partial x \partial y}. \end{aligned} \tag{7.13}$$

Here, instead of the reduced stiffnesses Q_{11}, Q_{22}, Q_{12} and Q_{66}, we use the engineering constants E_{xx}, E_{yy}, ν_{xy}, ν_{yx} and G_{xy} of the orthotropic plate material, and we obtain:

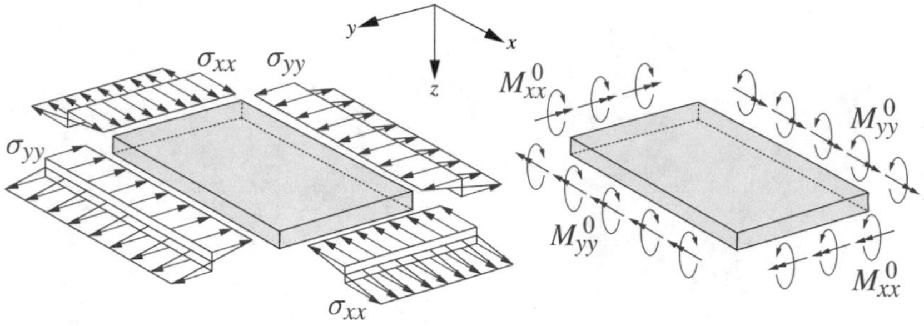

Fig. 7.4 Normal stresses (left) and resultant bending moment flows (right) of the plate

$$
\begin{aligned}
\sigma_{xx} &= -\frac{z}{1-\nu_{xy}\nu_{yx}}\left(E_{xx}\frac{\partial^2 w_0}{\partial x^2}+\nu_{xy}E_{yy}\frac{\partial^2 w_0}{\partial y^2}\right),\\
\sigma_{yy} &= -\frac{z}{1-\nu_{xy}\nu_{yx}}\left(\nu_{xy}E_{yy}\frac{\partial^2 w_0}{\partial x^2}+E_{yy}\frac{\partial^2 w_0}{\partial y^2}\right),\\
\tau_{xy} &= -2G_{xy}z\frac{\partial^2 w_0}{\partial x\partial y}.
\end{aligned}
\tag{7.14}
$$

If, on the other hand, an isotropic plate is present, then we obtain:

$$
\begin{aligned}
\sigma_{xx} &= -\frac{Ez}{1-\nu^2}\left(\frac{\partial^2 w_0}{\partial x^2}+\nu\frac{\partial^2 w_0}{\partial y^2}\right),\\
\sigma_{yy} &= -\frac{Ez}{1-\nu^2}\left(\nu\frac{\partial^2 w_0}{\partial x^2}+\frac{\partial^2 w_0}{\partial y^2}\right),\\
\tau_{xy} &= -2Gz\frac{\partial^2 w_0}{\partial x\partial y}=-\frac{Ez}{1+\nu}\frac{\partial^2 w_0}{\partial x\partial y}.
\end{aligned}
\tag{7.15}
$$

Obviously, the distribution of the stresses σ_{xx}, σ_{yy}, τ_{xy} varies linearly over z, analogous to the Euler-Bernoulli beam. The individual stress components with respect to the z-direction are shown in Figs. 7.4, 7.5 and 7.6 together with the associated force and moment flows.

At this point some contradictions of Kirchhoff's plate theory should be pointed out. On the one hand, we assume that the thickness h of the plate does not change during the deformation process, which leads to $\varepsilon_{zz} = 0$ (see Eq. (7.8)). This corresponds to a plane strain condition with respect to the z-direction. On the other hand, we require the presence of a plane stress state, which leads to the vanishing of the normal stress σ_{zz}. However, it should be noted that these two requirements are mutually exclusive. Another contradiction is that we use the hypothesis of plane cross sections and also the normal hypothesis. Both hypotheses enforce a disappearance of the two shear strains γ_{xz} and γ_{yz}, so we have no possibility to calculate the shear stresses τ_{xz} and τ_{yz} from a constitutive relation. From these two stresses, we can calculate the shear

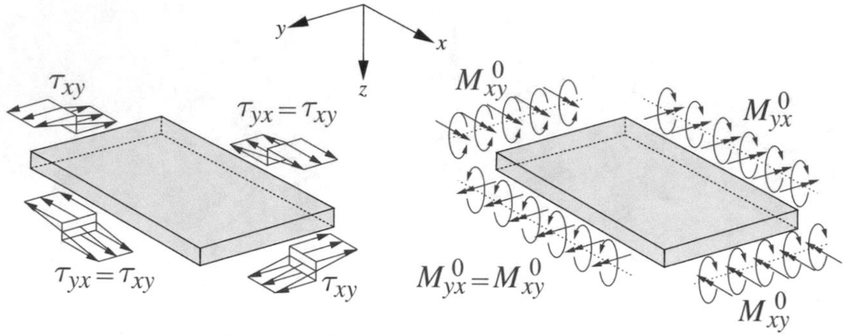

Fig. 7.5 Shear stresses (left) and resultant twisting moment flows (right) of the plate

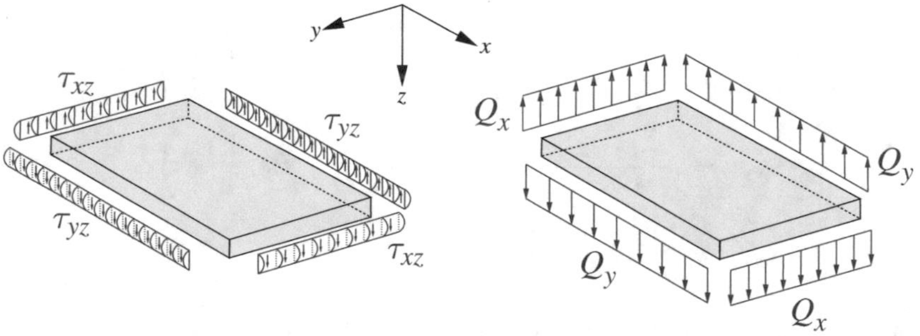

Fig. 7.6 Transverse shear stresses (left) and resultant transverse shear force flows (right) of the plate

force flows Q_x and Q_y (Fig. 7.6) of the plate, which are essential for equilibrium demands and therefore must also occur in a plate. We will discuss this in more detail later. All in all, it remains to be said that Kirchhoff's plate theory contains contradictions which have to be taken into account. Practical experience shows, however, that these contradictions and the resulting errors usually remain negligibly small when sufficiently thin plate structures are considered.

7.2.3 Force and Moment Flows, Constitutive Law

In a Kirchhoff plate, the two bending moment flows M_{xx}^0, M_{yy}^0 (see Fig. 7.4, right) and the twisting moment flow M_{xy}^0 (Fig. 7.5, right) occur. They can be interpreted as the resultants of the plane stress components σ_{xx}, σ_{yy}, and τ_{xy}, and thus have the unit of a moment per unit length, in SI units [Nm/m] = [N]. They can be determined from the integration of the stress components σ_{xx}, σ_{yy} and τ_{xy}, multiplied by the lever arm z, over the plate thickness h as:

$$\begin{pmatrix} M^0_{xx} \\ M^0_{yy} \\ M^0_{xy} \end{pmatrix} = \int_{-\frac{h}{2}}^{\frac{h}{2}} \begin{pmatrix} \sigma_{xx} \\ \sigma_{yy} \\ \tau_{xy} \end{pmatrix} z\mathrm{d}z. \tag{7.16}$$

We can replace the stresses by (7.12) and obtain:

$$\begin{pmatrix} M^0_{xx} \\ M^0_{yy} \\ M^0_{xy} \end{pmatrix} = -\int_{-\frac{h}{2}}^{\frac{h}{2}} \begin{bmatrix} Q_{11} & Q_{12} & 0 \\ Q_{12} & Q_{22} & 0 \\ 0 & 0 & Q_{66} \end{bmatrix} \begin{pmatrix} \frac{\partial^2 w_0}{\partial x^2} \\ \frac{\partial^2 w_0}{\partial y^2} \\ 2\frac{\partial^2 w_0}{\partial x \partial y} \end{pmatrix} z^2\mathrm{d}z. \tag{7.17}$$

Performing the prescribed integration results in:

$$\begin{pmatrix} M^0_{xx} \\ M^0_{yy} \\ M^0_{xy} \end{pmatrix} = -\begin{bmatrix} D_{11} & D_{12} & 0 \\ D_{12} & D_{22} & 0 \\ 0 & 0 & D_{66} \end{bmatrix} \begin{pmatrix} \frac{\partial^2 w_0}{\partial x^2} \\ \frac{\partial^2 w_0}{\partial y^2} \\ 2\frac{\partial^2 w_0}{\partial x \partial y} \end{pmatrix}. \tag{7.18}$$

For the sake of convenience, we have introduced the abbreviation

$$D_{ij} = \int_{\frac{h}{2}}^{\frac{h}{2}} Q_{ij} z^2 \mathrm{d}z \tag{7.19}$$

at this point $(i, j = 1, 2, 6)$. The quantities D_{ij} are the so-called plate stiffnesses. They establish a relation between the moment flows on the one hand and the curvatures and the twist on the other hand. They can therefore be interpreted in a similar way as the bending stiffness EI of the Euler-Bernoulli beam. Thus, Eq. (7.18) is the constitutive law of the Kirchhoff plate. The matrix of plate stiffnesses

$$\underline{\underline{D}} = \begin{bmatrix} D_{11} & D_{12} & 0 \\ D_{12} & D_{22} & 0 \\ 0 & 0 & D_{66} \end{bmatrix} \tag{7.20}$$

is denoted as the stiffness matrix of the plate. Thus, we can also write the constitutive law symbolically as:

$$\underline{M}^0 = \underline{\underline{D}}\underline{\kappa}^0, \tag{7.21}$$

where the vector $\underline{M}^0 = (M^0_{xx}, M^0_{yy}, M^0_{xy})^T$ contains the moment flows of the plate.

If the elastic properties of the plate do not depend on the z-coordinate, then the plate stiffnesses D_{ij} are:

$$D_{ij} = \frac{Q_{ij} h^3}{12}. \tag{7.22}$$

Equation (7.18) yields:

$$\begin{aligned}
M_{xx}^0 &= -D_{11}\frac{\partial^2 w_0}{\partial x^2} - D_{12}\frac{\partial^2 w_0}{\partial y^2},\\
M_{yy}^0 &= -D_{12}\frac{\partial^2 w_0}{\partial x^2} - D_{22}\frac{\partial^2 w_0}{\partial y^2},\\
M_{xy}^0 &= -2D_{66}\frac{\partial^2 w_0}{\partial x \partial y},
\end{aligned} \tag{7.23}$$

respectively expressed in the reduced stiffnesses Q_{ij} $(i, j = 1, 2, 6)$ of the plate:

$$\begin{aligned}
M_{xx}^0 &= -\frac{Q_{11}h^3}{12}\frac{\partial^2 w_0}{\partial x^2} - \frac{Q_{12}h^3}{12}\frac{\partial^2 w_0}{\partial y^2},\\
M_{yy}^0 &= -\frac{Q_{12}h^3}{12}\frac{\partial^2 w_0}{\partial x^2} - \frac{Q_{22}h^3}{12}\frac{\partial^2 w_0}{\partial y^2},\\
M_{xy}^0 &= -\frac{Q_{66}h^3}{6}\frac{\partial^2 w_0}{\partial x \partial y}.
\end{aligned} \tag{7.24}$$

If we use the engineering constants for orthotropic material, we obtain:

$$\begin{aligned}
M_{xx}^0 &= -\frac{E_{xx}h^3}{12(1-\nu_{xy}\nu_{yx})}\left(\frac{\partial^2 w_0}{\partial x^2} + \nu_{yx}\frac{\partial^2 w_0}{\partial y^2}\right),\\
M_{yy}^0 &= -\frac{E_{yy}h^3}{12(1-\nu_{xy}\nu_{yx})}\left(\nu_{xy}\frac{\partial^2 w_0}{\partial x^2} + \frac{\partial^2 w_0}{\partial y^2}\right),\\
M_{xy}^0 &= -\frac{G_{xy}h^3}{6}\frac{\partial^2 w_0}{\partial x \partial y}.
\end{aligned} \tag{7.25}$$

In the case of an isotropic plate, the result is:

$$\begin{aligned}
M_{xx}^0 &= -\frac{Eh^3}{12(1-\nu^2)}\left(\frac{\partial^2 w_0}{\partial x^2} + \nu\frac{\partial^2 w_0}{\partial y^2}\right),\\
M_{yy}^0 &= -\frac{Eh^3}{12(1-\nu^2)}\left(\nu\frac{\partial^2 w_0}{\partial x^2} + \frac{\partial^2 w_0}{\partial y^2}\right),\\
M_{xy}^0 &= -\frac{Eh^3}{12(1+\nu)}\frac{\partial^2 w_0}{\partial x \partial y}.
\end{aligned} \tag{7.26}$$

Using the isotropic plate stiffness

$$D = \frac{Eh^3}{12(1-\nu^2)} \tag{7.27}$$

then yields:

$$
\begin{aligned}
M_{xx}^{0} &= -D\left(\frac{\partial^2 w_0}{\partial x^2} + \nu \frac{\partial^2 w_0}{\partial y^2}\right), \\
M_{yy}^{0} &= -D\left(\nu \frac{\partial^2 w_0}{\partial x^2} + \frac{\partial^2 w_0}{\partial y^2}\right), \\
M_{xy}^{0} &= -D(1-\nu)\frac{\partial^2 w_0}{\partial x \partial y}.
\end{aligned} \tag{7.28}
$$

It can be seen that the integrations performed with respect to the plate thickness have reduced the actual three-dimensional plate structure to its middle plane. Thus, Kirchhoff's plate theory is a consideration of a two-dimensional structure with effective properties represented by the plate stiffnesses. Such a way of looking at a plate is also called single-layer theory.

It is important to emphasize that the indexing of moment flows in the case of a plate is performed differently than it is used to treat beams. In the case of the plate, the moment flows are given the indices of the stress that causes them. For example, the moment flow M_{xx}^{0} is a consequence of the normal stress σ_{xx}, so it has a direction of rotation around the y-axis. On the other hand, when calculating beam structures, the indexing is usually done in such a way that the index of the bending moment indicates the axis about which the moment rotates the beam. As an example, consider the bending moment M_y of a beam, which is caused by the normal stress σ_{xx}, but rotates the cross-section about the y-axis.

The transverse shear force flows Q_x and Q_y, which are necessary for equilibrium reasons, are given in the unit of a force per unit length, e.g. [N/m], and can be obtained from the transverse shear stresses τ_{xz} and τ_{yz} by integration over the plate thickness:

$$
\begin{pmatrix} Q_x \\ Q_y \end{pmatrix} = \int_{-\frac{h}{2}}^{\frac{h}{2}} \begin{pmatrix} \tau_{xz} \\ \tau_{yz} \end{pmatrix} \mathrm{d}z. \tag{7.29}
$$

However, within the framework of Kirchhoff's plate theory it is not possible to determine the transverse shear forces by a constitutive relation, since the corresponding shear strains γ_{xz} and γ_{yz} disappear by definition due to the hypothesis of the cross sections remaining plane and the normal hypothesis. Therefore, the calculation of the transverse forces from the transverse shear stresses does not lead to any result. However, the transverse shear stresses can be obtained, similar to the Euler-Bernoulli beam, from a post-calculation of the plane stress components, which will be discussed later.

7.2.4 Transformation Rules

For the moment flows M^0_{xx}, M^0_{yy}, M^0_{xy} transformation equations can be given similar to the plane stress components (see Chap. 1). It turns out that the moment flows can be transformed identically to the stress components, i.e. for the moment flows $M^0_{\xi\xi}$, $M^0_{\eta\eta}$, $M^0_{\xi\eta}$ in the coordinate system rotated by the angle θ (Fig. 7.7) we obtain:

$$\begin{aligned} M^0_{\xi\xi} &= M^0_{xx}\cos^2\theta + M^0_{yy}\sin^2\theta + 2M^0_{xy}\cos\theta\sin\theta, \\ M^0_{\eta\eta} &= M^0_{xx}\sin^2\theta + M^0_{yy}\cos^2\theta - 2M^0_{xy}\cos\theta\sin\theta, \\ M^0_{\xi\eta} &= \left(M^0_{yy} - M^0_{xx}\right)\sin\theta\cos\theta + M^0_{xy}\left(\cos^2\theta - \sin^2\theta\right). \end{aligned} \tag{7.30}$$

Consequently, the same invariants occur here as for the stress components (Sect. 1.2.3), i.e. it holds:

$$M^0_{xx} + M^0_{yy} = M^0_{\xi\xi} + M^0_{\eta\eta},\ M^0_{xx}M^0_{yy} - \left(M^0_{xy}\right)^2 = M^0_{\xi\xi}M^0_{\eta\eta} - \left(M^0_{\xi\eta}\right)^2. \tag{7.31}$$

Analogous to the plane stress state, a coordinate system can also be determined for the moment flows of the Kirchhoff plate, in which the bending moment flows become extremal and the twisting moment flow disappears. Let θ_h be the corresponding principal axis angle, which can be determined as follows:

$$\tan 2\theta_h = \frac{2M^0_{xy}}{M^0_{yy} - M^0_{xx}}, \tag{7.32}$$

and the corresponding extremal principal moment flows M_1, M_2 are given as:

$$M_{1,2} = \frac{1}{2}\left(M^0_{xx} + M^0_{yy}\right) \pm \sqrt{\left(\frac{M^0_{xx} - M^0_{yy}}{2}\right)^2 + \left(M^0_{xy}\right)^2}. \tag{7.33}$$

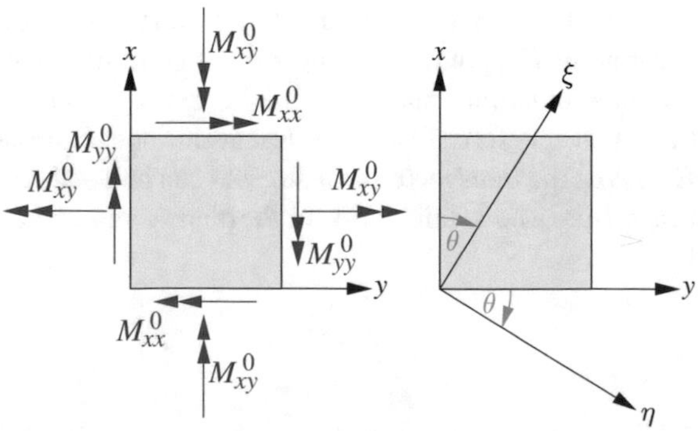

Fig. 7.7 Transformation of the moment flows

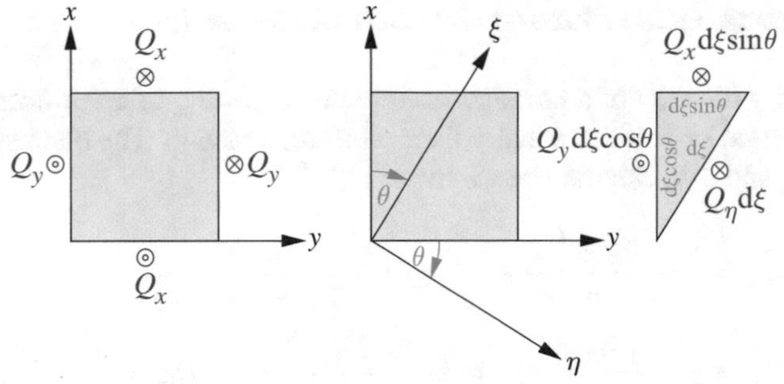

Fig. 7.8 Transformation of the transverse shear force flows

Similarly, a distinguished coordinate direction $\theta_h' = \theta_h + \frac{\pi}{4}$ can be given under which the extremal twisting moment flow $M^0_{\xi\eta,\max}$ is obtained as:

$$M^0_{\xi\eta,\max} = \sqrt{\left(\frac{M^0_{xx} - M^0_{yy}}{2}\right)^2 + \left(M^0_{xy}\right)^2}. \tag{7.34}$$

The transformation of the transverse shear force flows Q_x, Q_y is shown in Fig. 7.8. From the free body image of Fig. 7.8, right, the shear force sum with the shear force flow Q_η acting on the intersection edge of the lenth $\mathrm{d}\xi$ is given as:

$$Q_\eta \mathrm{d}\xi + Q_x \mathrm{d}\xi \sin\theta - Q_y \mathrm{d}\xi \cos\theta = 0, \tag{7.35}$$

which can be transformed to:

$$Q_\eta = -Q_x \sin\theta + Q_y \cos\theta. \tag{7.36}$$

Analogously, the following expression for the transverse shear force flow Q_ξ can be derived:

$$Q_\xi = Q_x \cos\theta + Q_y \sin\theta. \tag{7.37}$$

7.3 Effective Stiffnesses for Selected Plate Structures

In this section we discuss the calculation of the plate stiffnesses D_{11}, D_{22}, D_{12}, and D_{66} for selected plate structures as relevant in engineering practice.

7.3.1 Homogeneous Plate of Orthotropic Material

We first consider the case of a homogeneous plate consisting of an orthotropic material such as wood or unidirectionally fiber reinforced plastic. The plate stiffnesses D_{11}, D_{22}, D_{12} and D_{66} can then be given as:

$$D_{11} = \frac{E_{xx}h^3}{12(1-\nu_{xy}\nu_{yx})}, \quad D_{22} = \frac{E_{yy}h^3}{12(1-\nu_{xy}\nu_{yx})},$$
$$D_{12} = \frac{\nu_{yx}E_{xx}h^3}{12(1-\nu_{xy}\nu_{yx})} = \frac{\nu_{xy}E_{yy}h^3}{12(1-\nu_{xy}\nu_{yx})}, \quad D_{66} = \frac{G_{xy}h^3}{6}. \tag{7.38}$$

7.3.2 Homogeneous Plate of Isotropic Material

For a homogeneous isotropic plate we obtain:

$$D_{11} = D_{22} = \frac{Eh^3}{12(1-\nu^2)} = D, \quad D_{12} = \frac{\nu Eh^3}{12(1-\nu^2)}, \quad D_{66} = \frac{Eh^3}{12(1+\nu)}. \tag{7.39}$$

7.3.3 Reinforced Concrete Plate

We consider a plate made of concrete reinforced by steel reinforcements in x- and y-directions. A schematic representation is given in Fig. 7.9. The plate stiffnesses can then be determined as follows:

$$D_{11} = \frac{E_B}{1-\nu_B^2}\left[I_{Bx} + \left(\frac{E_S}{E_B} - 1\right) I_{Sx}\right],$$
$$D_{22} = \frac{E_B}{1-\nu_B^2}\left[I_{By} + \left(\frac{E_S}{E_B} - 1\right) I_{Sy}\right],$$
$$D_{12} = \nu_B\sqrt{D_{11}D_{22}}, \quad D_{66} = \frac{1-\nu_B}{2}\sqrt{D_{11}D_{22}}. \tag{7.40}$$

Therein we have:

- ν_B=Poisson's ratio of the unreinforced concrete,
- E_B=Modulus of elasticity of the unreinforced concrete,
- E_S=Modulus of elasticity of the reinforcing steel,
- I_{Bx}, I_{By}=Moments of inertia of the unreinforced concrete with respect to the midplane of the plate,
- I_{Sx}, I_{Sy}=Moments of inertia of the steel cross-section with respect to the midplane of the plate.

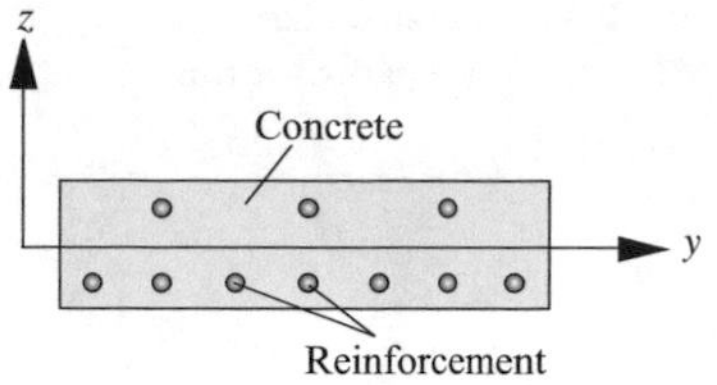

Fig. 7.9 Reinforced concrete plate

7.3.4 Isotropic Plate Reinforced by Equidistant Stiffeners

An isotropic plate with the elastic properties E and ν is considered, which is stiffened by equidistant stiffeners with distance s in the direction of the x-axis (Fig. 7.10). The stiffeners are described by their effective properties E_S (modulus of elasticity) and I_S (moment of inertia of the stiffener with respect to the plate midplane). The plate stiffnesses can then be given as follows:

$$D_{11} = \frac{Eh^3}{12(1-\nu^2)} + \frac{E_S I_S}{s}, \quad D_{22} = \frac{Eh^3}{12(1-\nu^2)}, \quad D_{12} = \nu D_{11}. \tag{7.41}$$

The plate stiffness D_{66} can be calculated as in the case of a homogeneous isotropic plate.

7.3.5 Isotropic Plate Reinforced by Equidistant Ribs

We further consider an isotropic plate stiffened by equidistant ribs with spacing s (Fig. 7.11). Let the width of the stiffener cross-section be t, and the height of the stiffeners be given by the value H. Let I_S be the moment of inertia and GI_T the torsional stiffness of the highlighted section. Let the size GI_{TP} be the torsional stiffness of the plate. Then the plate stiffnesses are as follows:

$$D_{11} = \frac{Eh^3 s}{12\left[s - t + t\left(\frac{h}{H}\right)^3\right]}, \quad D_{22} = \frac{EI}{s}, \quad 2D_{66} = 2GI_{TP} + \frac{GI_T}{s}. \tag{7.42}$$

The plate stiffness D_{12} may be taken as $D_{12} = 0$ for the case $\nu = 0$.

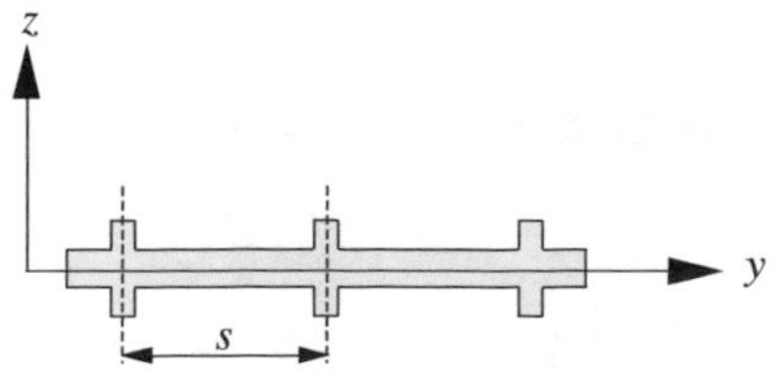

Fig. 7.10 Isotropic plate reinforced by equidistant stiffeners

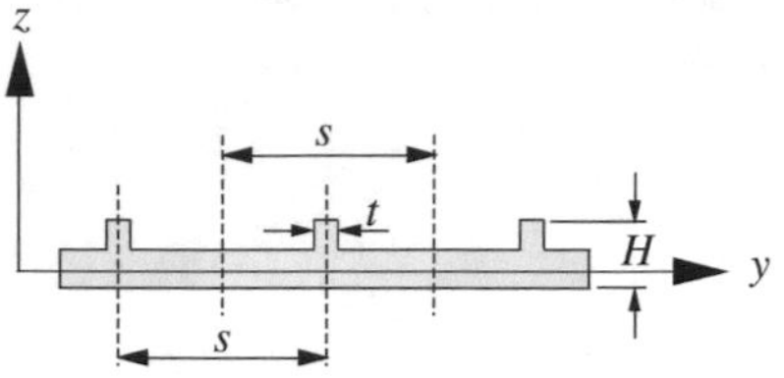

Fig. 7.11 Isotropic plate reinforced by equidistant ribs

7.3.6 Corrugated Metal Sheet

A corrugated sheet of thickness h is iven, the shape of which can be described by a sinusoidal function with half-wavelength s and amplitude H (Fig. 7.12). The plate stiffnesses can be given for this case as:

$$D_{11} = \frac{s}{\mu} \frac{Eh^3}{12(1-\nu^2)}, \quad D_{22} = EI, \quad 2D_{66} = \frac{s}{\mu} \frac{Eh^3}{12(1+\nu)}, \quad D_{12} = 0. \tag{7.43}$$

Herein, the following auxiliary quantities were used:

$$\mu = s\left(1 + \frac{\pi^2 H^2}{4s^2}\right), \quad I = \frac{H^2 h}{2}\left[1 - \frac{0{,}81}{1 + 2{,}5\left(\frac{H}{2s}\right)^2}\right]. \tag{7.44}$$

7.3.7 Symmetrical Cross-Ply Composite Laminate

Laminates are layered structures (see Fig. 7.13) used in a variety of engineering applications. Fiber-reinforced plastics, but also wood-based materials such as plywood boards and the like are worth mentioning here. The calculation of laminate structures will be discussed in detail at a later point in this book, but it is useful to point out the close relationship between the calculation of such layered structures and the analysis of plates. A laminate is composed of an arbitrary number of individual layers, which we will assume here to be orthotropic. If, for example, the laminate is a fiber-reinforced plastic in which unidirectional fibers are inserted into a plastic, then it is basically possible to arrange the fiber directions of the individual layers arbitrarily in such a way that the given requirements for the component under consideration are met to the best possible extent. At this point, we will restrict our considerations to

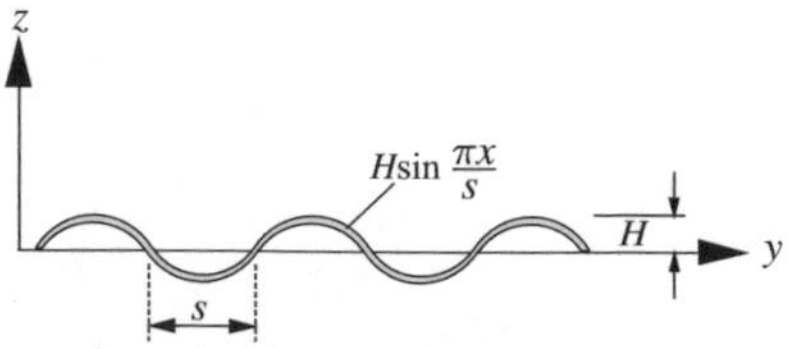

Fig. 7.12 Corrugated metal sheet

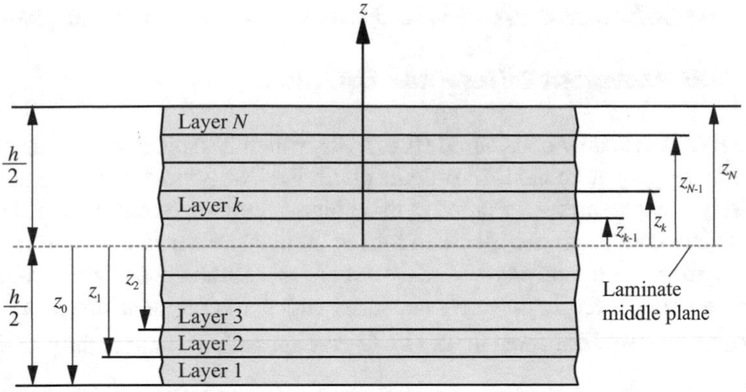

Fig. 7.13 Section of a laminate

the case where the fiber directions of the individual layers are either parallel to the x-axis or parallel to the y-axis. Such laminates are called cross-ply laminates. We refer to a layer with the fiber orientation in the x-direction as a 0°-layer. If the fiber orientation runs in the direction of the y-axis, then we speak of a so-called 90°-layer, since the fiber orientation of this layer is rotated by 90° with respect to the x-axis. Accordingly, the orientations of the fibers of the individual layers are crosswise and perpendicular to each other, which explains the notion of a cross-ply laminate. Furthermore, we want to assume here that the cross-ply laminate under consideration is symmetrical with respect to its center plane, i.e. has identical layer sequences below and above the laminate middle plane.

Let M be the number of 0°-layers and N be the number of 90°-layers. Let the layer k be bounded by the coordinates z_{k-1} and z_k. The plate stiffnesses of the laminate can then be calculated with the layerwise constant orthotropic properties $E_{xx,k}$, $E_{yy,k}$, $\nu_{xy,k}$, $\nu_{yx,k}$ and $G_{xy,k}$ as:

$$
\begin{aligned}
D_{11} &= \frac{1}{3}\sum_{k=1}^{M}\frac{E_{xx,k}}{1-\nu_{xy,k}\nu_{yx,k}}\left(z_k^3-z_{k-1}^3\right)+\frac{1}{3}\sum_{k=1}^{N}\frac{E_{yy,k}}{1-\nu_{xy,k}\nu_{yx,k}}\left(z_k^3-z_{k-1}^3\right),\\
D_{22} &= \frac{1}{3}\sum_{k=1}^{M}\frac{E_{yy,k}}{1-\nu_{xy,k}\nu_{yx,k}}\left(z_k^3-z_{k-1}^3\right)+\frac{1}{3}\sum_{k=1}^{N}\frac{E_{xx,k}}{1-\nu_{xy,k}\nu_{yx,k}}\left(z_k^3-z_{k-1}^3\right),\\
D_{12} &= \frac{1}{3}\sum_{k=1}^{M}\frac{\nu_{xy,k}E_{yy,k}}{1-\nu_{xy,k}\nu_{yx,k}}\left(z_k^3-z_{k-1}^3\right)+\frac{1}{3}\sum_{k=1}^{N}\frac{\nu_{xy,k}E_{yy,k}}{1-\nu_{xy,k}\nu_{yx,k}}\left(z_k^3-z_{k-1}^3\right),\\
D_{66} &= \frac{1}{3}\sum_{k=1}^{M}G_{xy,k}\left(z_k^3-z_{k-1}^3\right)+\frac{1}{3}\sum_{k=1}^{N}G_{xy,k}\left(z_k^3-z_{k-1}^3\right).
\end{aligned}
\tag{7.45}
$$

7.4 Basic Equations of Plate Bending in Cartesian Coordinates

7.4.1 Displacement Differential Equation

A plate structure is now considered (Fig. 7.14), which is subjected to a surface load $p(x, y)$ acting perpendicular to the middle plane. Basically, the distribution of $p(x, y)$ is arbitrary, but not further specified at this point. Likewise, no concrete dimensions and boundary conditions are specified for the moment. We now cut an infinitesimal section element of dimensions $\mathrm{d}x$ and $\mathrm{d}y$ out of the plate and consider the local equilibrium conditions. At the negative sectional edges, the moment flows M_{xx}^0, M_{yy}^0, M_{xy}^0 and transverse shear force flows Q_x, Q_y are applied, whereas at the positive sections the quantities with the respective infinitesimal increments, i.e. $M_{xx}^0 + \frac{\partial M_{xx}^0}{\partial x}\mathrm{d}x$, $M_{yy}^0 + \frac{\partial M_{yy}^0}{\partial y}\mathrm{d}y$, $M_{xy}^0 + \frac{\partial M_{xy}^0}{\partial x}\mathrm{d}x$, $M_{xy}^0 + \frac{\partial M_{xy}^0}{\partial y}\mathrm{d}y$, $Q_x + \frac{\partial Q_x}{\partial x}\mathrm{d}x$ and $Q_y + \frac{\partial Q_y}{\partial y}\mathrm{d}y$, are applied. From the sum of transverse forces we get:

$$\left(Q_x + \frac{\partial Q_x}{\partial x}\mathrm{d}x\right)\mathrm{d}y + \left(Q_y + \frac{\partial Q_y}{\partial y}\mathrm{d}y\right)\mathrm{d}x - Q_x\mathrm{d}y - Q_y\mathrm{d}x + p\mathrm{d}x\mathrm{d}y = 0, \quad (7.46)$$

or after division by $\mathrm{d}A = \mathrm{d}x\mathrm{d}y$:

$$\frac{\partial Q_x}{\partial x} + \frac{\partial Q_y}{\partial y} + p = 0. \quad (7.47)$$

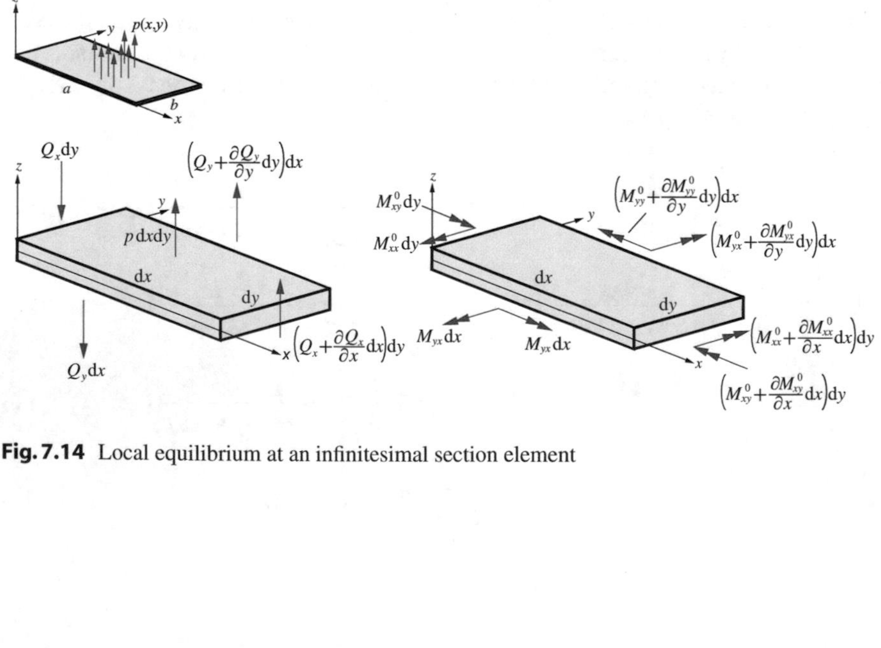

Fig. 7.14 Local equilibrium at an infinitesimal section element

The moment equilibrium about the y-axis with reference point in the center of gravity of the infinitesimal section element results in:

$$\left(M_{xx}^0 + \frac{\partial M_{xx}^0}{\partial x}\mathrm{d}x\right)\mathrm{d}y + \left(M_{xy}^0 + \frac{\partial M_{yx}^0}{\partial y}\mathrm{d}y\right)\mathrm{d}x - M_{xx}^0\mathrm{d}y - M_{yx}^0\mathrm{d}x$$
$$-\left(Q_x + \frac{\partial Q_x}{\partial x}\mathrm{d}x\right)\mathrm{d}y\frac{\mathrm{d}x}{2} - Q_x\mathrm{d}y\frac{\mathrm{d}x}{2} = 0, \tag{7.48}$$

or:

$$\frac{\partial M_{xx}^0}{\partial x}\mathrm{d}x\mathrm{d}y + \frac{\partial M_{xy}^0}{\partial y}\mathrm{d}x\mathrm{d}y - Q_x\mathrm{d}x\mathrm{d}y - \frac{\partial Q_x}{\partial x}\frac{\mathrm{d}x^2\mathrm{d}y}{2} = 0. \tag{7.49}$$

The last term may be assumed to be small of higher order when compared to the other terms, leaving the following expression:

$$\frac{\partial M_{xx}^0}{\partial x} + \frac{\partial M_{xy}^0}{\partial y} = Q_x. \tag{7.50}$$

The moment equilibrium around the x-axis yields analogously:

$$\frac{\partial M_{xy}^0}{\partial x} + \frac{\partial M_{yy}^0}{\partial y} = Q_y. \tag{7.51}$$

The three equilibrium conditions used here are called plate equilibrium. From the two moment equilibrium relations (7.50) and (7.51), the transverse shear force flows Q_x and Q_y of the plate can then also be calculated if the moment flows are known.

The three equilibrium conditions (7.47), (7.50) and (7.51) can be put into a more compact form. If the first moment equilibrium (7.50) is partially differentiated with respect to x as well as the second moment equilibrium (7.51) is partially differentiated with respect to y and this is substituted into the sum of transverse forces (7.47), then the so-called condensed plate equilibrium is obtained as follows:

$$\frac{\partial^2 M_{xx}^0}{\partial x^2} + 2\frac{\partial^2 M_{xy}^0}{\partial x \partial y} + \frac{\partial^2 M_{yy}^0}{\partial y^2} + p = 0. \tag{7.52}$$

Once the moment flows have been determined, the transverse shear forces from the two equations (7.50) and (7.51) can be calculated quite easily in a post-calculation.

The relationships presented so far are independent of the concrete material behavior. If we assume linear-elastic and orthotropic material and use the constitutive relations (7.23) for the moment flows M_{xx}^0, M_{yy}^0 and M_{xy}^0, then we obtain the following equation after insertion into (7.52):

$$D_{11}\frac{\partial^4 w_0}{\partial x^4} + 2\left(D_{12} + 2D_{66}\right)\frac{\partial^4 w_0}{\partial x^2 \partial y^2} + D_{22}\frac{\partial^4 w_0}{\partial y^4} = p. \tag{7.53}$$

This is a fourth-order linear partial imhomogeneous differential equation with constant coefficients for the deflection $w_0(x, y)$ of the plate midplane. It is also called plate equation. If a solution of this differential equation can be found which satisfies the given boundary conditions, then the moment and shear force flows M_{xx}^0, M_{yy}^0, M_{xy}^0, Q_x, Q_y as well as the stresses σ_{xx}, σ_{yy}, τ_{xy} can be determined. We will discuss various exact solutions in detail later in this chapter.

In the case of an isotropic plate, (7.53) takes the following form:

$$\frac{\partial^4 w_0}{\partial x^4} + 2\frac{\partial^4 w_0}{\partial x^2 \partial y^2} + \frac{\partial^4 w_0}{\partial y^4} = \frac{p}{D}. \tag{7.54}$$

This equation can be advantageously represented using the Laplace operator $\Delta = \frac{\partial^2}{\partial x^2} + \frac{\partial^2}{\partial y^2}$ as:

$$\Delta\Delta w_0 = \frac{p}{D}. \tag{7.55}$$

7.4.2 Equivalent Transverse Shear Forces

A boundary value problem of the Kirchhoff plate theory is, besides the provided basic equations, furthermore uniquely described by the respective boundary conditions. We distinguish between the so-called geometric/kinematic boundary conditions and the so-called dynamic/physical boundary conditions. The former make demands on kinematic boundary conditions (i.e. generally displacements and rotations), whereas the latter require the fulfillment of boundary conditions for force quantities (i.e. transverse shear forces and moments). The plate equation (7.53) or (7.54) is a fourth-order differential equation, therefore two boundary conditions must be satisfied at each boundary.

We consider a plate boundary at the location $x = const.$ Here, with respect to the kinematic boundary conditions, both a given boundary displacement $\widehat{w}_0$ and a given boundary slope $\widehat{\frac{\partial w_0}{\partial x}}$ can be considered, where the symbolism $\widehat{...}$ denotes a given quantity. In the same way, at an edge with $y = const.$ a plate deflection $\widehat{w}_0$ or an edge rotation $\widehat{\frac{\partial w_0}{\partial y}}$ can be given.

With respect to the dynamic boundary conditions, at a boundary $x = const.$ requirements can be given concerning the boundary force quantities Q_x, M_{xx}^0, M_{xy}^0. At an edge $y = const.$, analogously, statements concerning the quantities Q_y, M_{yy}^0, M_{xy}^0 can be given. However, in the framework of Kirchhoff plate theory, only two boundary force quantities can be given at a time. Therefore, the procedure of combining the transverse shear force flows Q_x, Q_y and the twisting moment flows M_{xy}^0 to the so-called equivalent transverse shear forces has been established. The procedure is shown in Fig. 7.15 using the example of the twisting moment M_{xy}^0 for an edge at the position $x = const.$

We now consider the twisting moment M_{xy}^0 at the location $y = y_1$. Furthermore, we consider the twisting moment at a point $y = y_2$ infinitesimally adjacent on the y-axis as indicated. Here the twisting moment will occur in addition to an infinitesimal

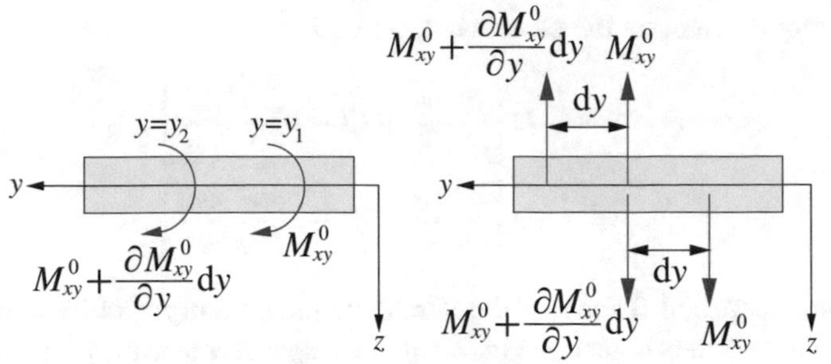

Fig. 7.15 Equivalent transverse shear forces at a boundary with $x = \text{const.}$

increment, i.e. $M^0_{xy} + \frac{\partial M^0_{xy}}{\partial y}\mathrm{d}y$. The twisting moment can now be understood as a pair of forces with the lever arm $\mathrm{d}y$. Accordingly, a couple M^0_{xy} and a couple $M^0_{xy} + \frac{\partial M^0_{xy}}{\partial y}\mathrm{d}y$ occur at the position $y = y_1$. Obviously, these forces partially cancel each other out, and the only remaining force component in the z-direction is the infinitesimal increment $\frac{\partial M^0_{xy}}{\partial y}\mathrm{d}y$. If we relate this force component to the length $\mathrm{d}y$ and add this to the transverse shear force Q_x, then we obtain the so-called Kirchhoff equivalent transverse shear force $\bar{Q}_x$ as follows:

$$\bar{Q}_x = Q_x + \frac{\partial M^0_{xy}}{\partial y}. \tag{7.56}$$

Analogously, we obtain for the second equivalent transverse shear force:

$$\bar{Q}_y = Q_y + \frac{\partial M^0_{xy}}{\partial x}. \tag{7.57}$$

From the equilibrium conditions (7.50) and (7.51) we obtain:

$$\begin{aligned}
\bar{Q}_x &= \frac{\partial M^0_{xx}}{\partial x} + 2\frac{\partial M^0_{xy}}{\partial y}, \\
\bar{Q}_y &= 2\frac{\partial M^0_{xy}}{\partial x} + \frac{\partial M^0_{yy}}{\partial y}.
\end{aligned} \tag{7.58}$$

Using the constitutive law (7.18), we get:

$$\begin{aligned}
\bar{Q}_x &= -D_{11}\frac{\partial^3 w_0}{\partial x^3} - (D_{12} + 4D_{66})\frac{\partial^3 w_0}{\partial x \partial y^2}, \\
\bar{Q}_y &= -(D_{12} + 4D_{66})\frac{\partial^3 w_0}{\partial x^2 \partial y} - D_{22}\frac{\partial^3 w_0}{\partial y^3}.
\end{aligned} \tag{7.59}$$

For the isotropic plate, on the other hand, we obtain:

$$\bar{Q}_x = -D\left[\frac{\partial^3 w_0}{\partial x^3} + (2-\nu)\frac{\partial^3 w_0}{\partial x \partial y^2}\right],$$
$$\bar{Q}_y = -D\left[\frac{\partial^3 w_0}{\partial y^3} + (2-\nu)\frac{\partial^3 w_0}{\partial x^2 \partial y}\right]. \quad (7.60)$$

At this point, a special feature of the Kirchhoff plate theory should be pointed out which results at corners of plates (Fig. 7.16). If we again replace the twisting moments by equivalent transverse shear forces, it can be deduced that at plate corners single transverse forces in z-direction result, which, however, do not cancel each other in their effect. The resulting force is also called corner force and can be calculated in the case of a right-angled corner of an orthotropic plate as:

$$F = -2M_{xy}^0 = -4D_{66}\frac{\partial^2 w_0}{\partial x \partial y}. \quad (7.61)$$

For the isotropic plate we obtain:

$$F = 2D(1-\nu)\frac{\partial^2 w_0}{\partial x \partial y}. \quad (7.62)$$

Resulting single forces also arise at locations that have discontinuities in the twisting moment flows, as can be seen from Fig. 7.17.

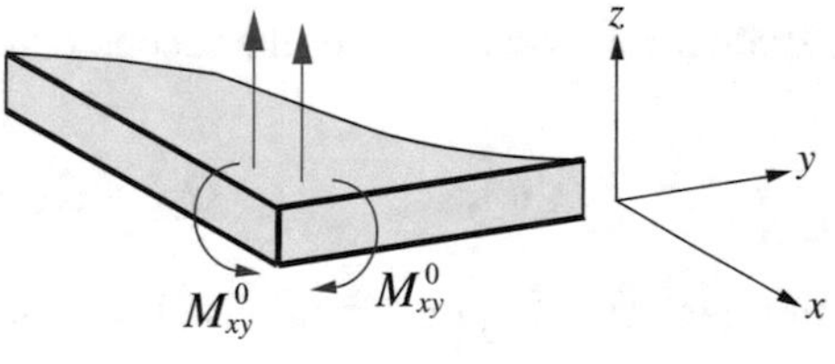

Fig. 7.16 Corner forces

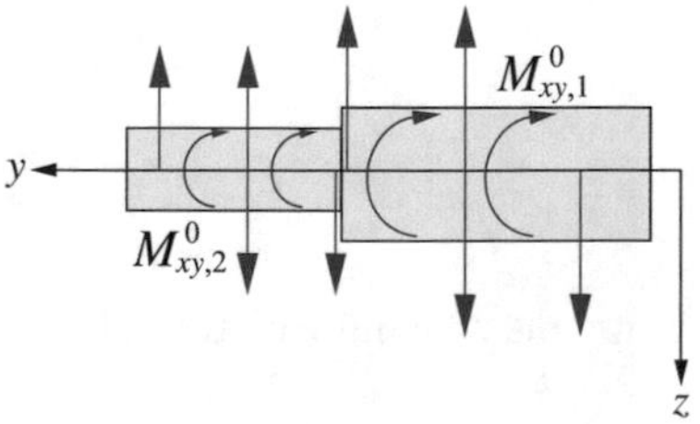

Fig. 7.17 Resulting transverse shear forces at discontinuity points of the twisting moment flow

7.4.3 Boundary Conditions

The unambiguous solution of a static plate bending problem requires the clear specification of boundary conditions. Typical boundary conditions in the case of the Kirchhoff plate will be discussed below.

First, let an edge be given at the point $x = const.$ which is free of any support. In this case, both the Kirchhoff equivalent transverse shear force flow $\bar{Q}_x$ and the bending moment flow M^0_{xx} must vanish at this boundary:

$$\bar{Q}_x = 0, \quad M^0_{xx} = 0. \tag{7.63}$$

Using the constitutive law (7.18) results in:

$$\begin{aligned} D_{11}\frac{\partial^2 w_0}{\partial x^2} + D_{12}\frac{\partial^2 w_0}{\partial y^2} &= 0, \\ D_{11}\frac{\partial^3 w_0}{\partial x^3} + (D_{12} + 4D_{66})\frac{\partial^2 w_0}{\partial x \partial y^2} &= 0. \end{aligned} \tag{7.64}$$

In the case of an isotropic plate we obtain:

$$\frac{\partial^3 w_0}{\partial x^3} + (2-\nu)\frac{\partial^3 w_0}{\partial x \partial y^2} = 0, \quad \frac{\partial^2 w_0}{\partial x^2} + \nu\frac{\partial^2 w_0}{\partial y^2} = 0. \tag{7.65}$$

If there is a plate edge at the position $x = const.$, which is freely rotatable, then both the deflection w_0 and the bending moment flow M^0_{xx} must become zero:

$$w_0 = 0, \quad M^0_{xx} = 0. \tag{7.66}$$

We can also write the second condition in (7.66) using (7.18):

$$D_{11}\frac{\partial^2 w_0}{\partial x^2} + D_{12}\frac{\partial^2 w_0}{\partial y^2} = 0. \tag{7.67}$$

However, at a freely rotating edge at the position $x = const.$ no twist and no curvatures about the x-axis can take place, so we can omit the second term in (7.67):

$$\frac{\partial^2 w_0}{\partial x^2} = 0. \tag{7.68}$$

In the case of an isotropic plate we get

$$\frac{\partial^2 w_0}{\partial x^2} + \nu\frac{\partial^2 w_0}{\partial y^2} = 0, \tag{7.69}$$

where of course also here the second term disappears by definition, so that also for the isotropic plate the expression (7.68) remains.

If the boundary at the point $x = const.$ is fully clamped, then both the deflection w_0 and the rotation $\frac{\partial w_0}{\partial x}$ must become zero:

$$w_0 = 0, \quad \frac{\partial w_0}{\partial x} = 0. \tag{7.70}$$

In addition, no twisting moments M_{xy}^0 occur at clamped plate edges, since the derivatives $\frac{\partial^2 w_0}{\partial y^2}$ and $\frac{\partial^2 w_0}{\partial x \partial y}$ must also become zero.

Attention must also be paid to the corners of plates. First, we consider the case where two rigidly clamped edges meet in a right-angled corner of a plate. There the following partial derivative of the deflection w_0 becomes zero:

$$\frac{\partial^2 w_0}{\partial x \partial y} = 0. \tag{7.71}$$

As a consequence, at such a corner both the twisting moment flow M_{xy}^0 and the corner force F disappear.

If there is a right-angled plate corner where two simply supported plate edges meet, then both the deflection w_0 and the bending moment flows M_{xx}^0 and M_{yy}^0 must vanish there:

$$\begin{aligned} w_0 &= 0, \\ D_{11}\frac{\partial^2 w_0}{\partial x^2} + D_{12}\frac{\partial^2 w_0}{\partial y^2} &= 0, \\ D_{12}\frac{\partial^2 w_0}{\partial x^2} + D_{22}\frac{\partial^2 w_0}{\partial y^2} &= 0, \end{aligned} \tag{7.72}$$

or in the case of isotropic plates:

$$\begin{aligned} w_0 &= 0, \\ \frac{\partial^2 w_0}{\partial x^2} - \nu\frac{\partial^2 w_0}{\partial y^2} &= 0, \\ \nu\frac{\partial^2 w_0}{\partial x^2} - \frac{\partial^2 w_0}{\partial y^2} &= 0. \end{aligned} \tag{7.73}$$

As a further case, consider a free right-angled corner where two free plate edges meet. Here, the corner force F must vanish, so that it must hold:

$$\frac{\partial^2 w_0}{\partial x \partial y} = 0. \tag{7.74}$$

Of particular practical relevance are elastically supported plate edges (see Fig. 7.18), which are either supported by spring elements or stiffened by edge stiffeners. Considering first the case of elastically supported edges (stiffness $k = k(y)$, Fig. 7.18,

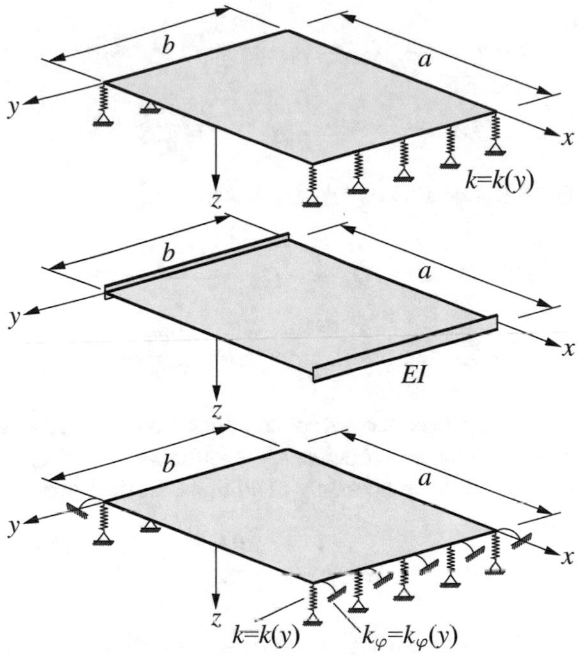

Fig. 7.18 Elastically supported plate edges

top) at $x = 0$ and $x = a$, then the boundary conditions for $x = 0$, $0 \leq y \leq b$ are obtained as:

$$
\begin{aligned}
kw_0 &= \bar{Q}_x = Q_x + \frac{\partial M_{xy}^0}{\partial y} = -D_{11}\frac{\partial^3 w_0}{\partial x^3} - (D_{12} + 4D_{66})\frac{\partial^3 w_0}{\partial x \partial y^2}, \\
M_{xx}^0 &= -D_{11}\frac{\partial^2 w_o}{\partial x^2} - D_{12}\frac{\partial w_0}{\partial y^2} = 0,
\end{aligned} \tag{7.75}
$$

whereas at the edge $x = a$, $0 \leq y \leq b$ we have:

$$
\begin{aligned}
kw_0 &= -\bar{Q}_x = -Q_x - \frac{\partial M_{xy}^0}{\partial y} = D_{11}\frac{\partial^3 w_0}{\partial x^3} + (D_{12} + 4D_{66})\frac{\partial^3 w_0}{\partial x \partial y^2}, \\
M_{xx}^0 &= -D_{11}\frac{\partial^2 w_o}{\partial x^2} - D_{12}\frac{\partial w_0}{\partial y^2} = 0.
\end{aligned} \tag{7.76}
$$

If, on the other hand, edge stiffeners with negligible torsional stiffness $GI_T \simeq 0$ and with the bending stiffness EI are present at the plate edges $x = 0$, $x = a$ (Fig. 7.18, middle), then for $x = 0$, $0 \leq y \leq b$:

$$
\begin{aligned}
EIw'''' &= \bar{Q}_x = Q_x + \frac{\partial M_{xy}^0}{\partial y} = -D_{11}\frac{\partial^3 w_0}{\partial x^3} - (D_{12} + 4D_{66})\frac{\partial^3 w_0}{\partial x \partial y^2}, \\
M_{xx}^0 &= -D_{11}\frac{\partial^2 w_o}{\partial x^2} - D_{12}\frac{\partial w_0}{\partial y^2} = 0.
\end{aligned}
\tag{7.77}
$$

For the edge $x = a, 0 \leq y \leq b$ we have:

$$
\begin{aligned}
EIw'''' &= -\bar{Q}_x = -Q_x - \frac{\partial M_{xy}^0}{\partial y} = D_{11}\frac{\partial^3 w_0}{\partial x^3} + (D_{12} + 4D_{66})\frac{\partial^3 w_0}{\partial x \partial y^2}, \\
M_{xx}^0 &= -D_{11}\frac{\partial^2 w_o}{\partial x^2} - D_{12}\frac{\partial w_0}{\partial y^2} = 0.
\end{aligned}
\tag{7.78}
$$

The case where the two plate edges are elastically supported with respect to both deflection w_0 and rotation $\frac{\partial w_0}{\partial x}$ (stiffnesses k and k_φ, Fig. 7.18, bottom) is described by the following boundary conditions. At the boundary $x = 0, 0 \leq y \leq b$ we have:

$$
\begin{aligned}
kw_0 &= \bar{Q}_x = Q_x + \frac{\partial M_{xy}^0}{\partial y} = -D_{11}\frac{\partial^3 w_0}{\partial x^3} - (D_{12} + 4D_{66})\frac{\partial^3 w_0}{\partial x \partial y^2}, \\
k_\varphi \frac{\partial w_0}{\partial x} &= M_{xx}^0 = -D_{11}\frac{\partial^2 w_o}{\partial x^2} - D_{12}\frac{\partial w_0}{\partial y^2} = 0.
\end{aligned}
\tag{7.79}
$$

For the edge $x = a, 0 \leq y \leq b$ the following conditions hold:

$$
\begin{aligned}
kw_0 &= -\bar{Q}_x = -Q_x - \frac{\partial M_{xy}^0}{\partial y} = D_{11}\frac{\partial^3 w_0}{\partial x^3} + (D_{12} + 4D_{66})\frac{\partial^3 w_0}{\partial x \partial y^2}, \\
k_\varphi \frac{\partial w_0}{\partial x} &= -M_{xx}^0 = D_{11}\frac{\partial^2 w_o}{\partial x^2} + -D_{12}\frac{\partial w_0}{\partial y^2} = 0.
\end{aligned}
\tag{7.80}
$$

The boundary conditions for plates with arbitrary boundary will be discussed in detail later.

7.5 Elementary Solutions of the Plate Equation

The plate equation (7.53) or (7.55) is an inhomogeneous partial differential equation. Accordingly, the solution for the deflection w_0 is composed of a homogeneous solution and a particular solution, i.e.:

$$
w_0 = w_{0,h} + w_{0,p}. \tag{7.81}
$$

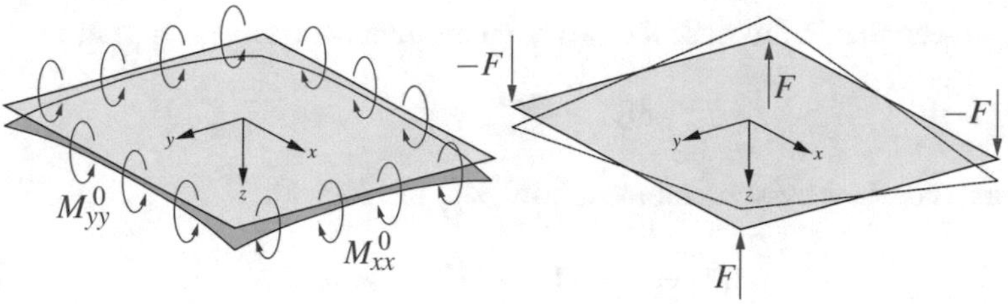

Fig. 7.19 Plate under bending moment flow (left), plate under corner forces (right)

Let all further explanations of this section be related to the Eq. (7.55) of the isotropic plate, and we study the homogeneous plate equation

$$D\Delta\Delta w_0 = 0. \tag{7.82}$$

Due to the structure of this equation in comparison with the disk equation $\Delta\Delta F = 0$ (Eq. (3.33)), it is obvious that we can quite fundamentally fall back on the explanations for the isotropic disk (Becker and Gross 2002; Altenbach et al. 2016). There is no general solution to this problem, but a catalog of solutions can be used. For example, the expressions

$$\begin{aligned} w_0 \;=\; & C,\;\; x^2,\;\; y^2,\;\; xy,\;\; x^2+y^2, \\ & \ln\left(x^2+y^2\right),\;\; x\ln\left(x^2+y^2\right),\;\; e^{\lambda x}\sin\lambda y,\;\; \ldots \end{aligned} \tag{7.83}$$

are solutions of the plate equation (7.82). From these elementary solutions of the plate equation, the force and moment quantities of the plate can be determined. This is shown here for two cases (Fig. 7.19). The first case (Fig. 7.19, left) consists of a parabolic deformation of the plate, in the following form:

$$w_0 = C\left(x^2+y^2\right). \tag{7.84}$$

From this, the bending moment flows M^0_{xx}, M^0_{yy}, M^0_{xy} can be determined as:

$$M^0_{xx} = M^0_{yy} = -2D(1+\nu)C = \text{const.}, \quad M^0_{xy} = 0. \tag{7.85}$$

This case of plate deformation thus corresponds to a constant bending moment flow on all sides, as shown in Fig. 7.19, left. The shear force flows Q_x, Q_y as well as the equivalent transverse shear forces $\bar{Q}_x$, $\bar{Q}_y$ result to zero in this case:

$$Q_x = Q_y = 0, \quad \bar{Q}_x = \bar{Q}_y = 0. \tag{7.86}$$

Let the second case (Fig. 7.19, right) be a plate deformation of the type

$$w_0 = Cxy. \tag{7.87}$$

It can be seen that in this case the two bending moment flows disappear, i.e.

$$M_{xx}^0 = M_{yy}^0 = 0, \tag{7.88}$$

whereas a constant twisting moment flow M_{xy}^0 results with

$$M_{xy}^0 = -D(1-\nu)\,C = \text{const.} \tag{7.89}$$

This results in corner forces $F = \pm 2M_{xy}^0 = \pm 2D(1-\nu)\,C$ at the plate corners. The transverse shear force flows Q_x, Q_y and the equivalent transverse shear force flows $\bar{Q}_x$, $\bar{Q}_y$ also follow to zero in this case:

$$Q_x = Q_y = 0, \quad \bar{Q}_x = \bar{Q}_y = 0. \tag{7.90}$$

In the same way, further elementary deformation states can be constructed. We do not give a description here, but refer to Altenbach et al. (2016) for further details.

7.6 Bending of Plate Strips

Some plate problems can be traced back to the analysis of so-called plate stripes and thus to a one-dimensional problem. On the one hand, this is the case when dealing with a beam-like structure, i.e. a very long plate is assumed, where the length a is significantly larger than the plate width b (Fig. 7.20). On the other hand, a uniaxial problem can be assumed when the plate under consideration is subjected to cylindrical bending. This is the case when the width b of the plate is significantly larger than the length a (Fig. 7.21). In both cases, it is assumed that the load p is exclusively a function of the longitudinal coordinate x, i.e. $p = p(x)$. Both situations described also have in common that the deflection w_0 of the plate is exclusively a function of x. Furthermore, all derivatives with respect to y become zero. It is easy to understand that the presence of a beam-like structure is a problem of the plane stress state, whereas the consideration of an infinitely wide plate leads to the consideration of a plate strip in the plane strain state.

If such a plate situation is given, then the following expression remains for the plate equation (7.53) due to the disappearance of all derivatives of w_0 with respect

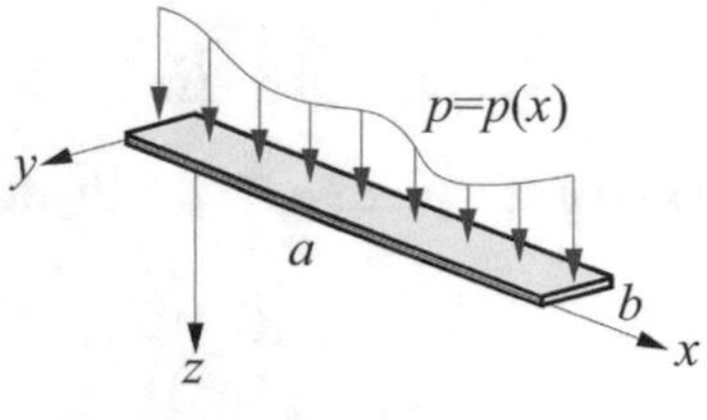

Fig. 7.20 Long plate, idealization as a plate strip

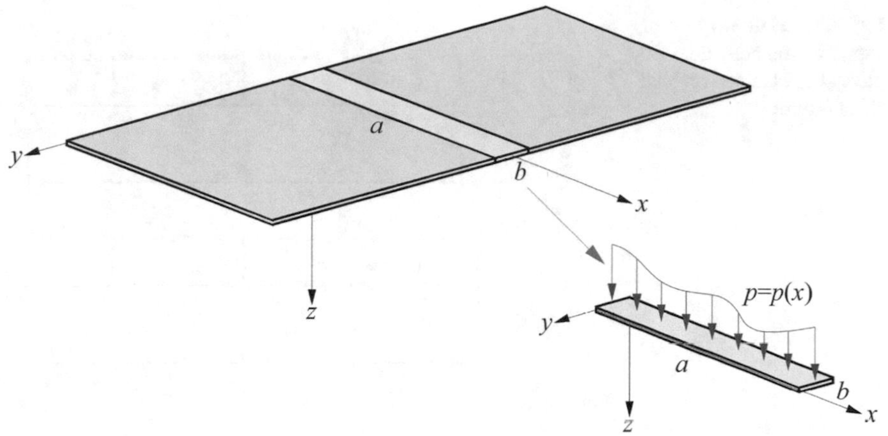

Fig. 7.21 Wide plate, idealization as a plate strip

to y:

$$D_{11}\frac{\mathrm{d}^4 w_0}{\mathrm{d}x^4} = p. \tag{7.91}$$

Both the deflection w_0 and the load p are functions of the coordinate x. This equation and all following relations are also valid in the case of isotropic plates, if the orthotropic plate stiffness D_{11} is replaced by the isotropic plate stiffness D according to (7.27).

The expression (7.91) corresponds to the differential equation of uniaxial bending of the Euler-Bernoulli beam, where here the bending stiffness EI is to be exchanged for the plate stiffness D_{11}.

From Eq. (7.91), the deflection $w_0(x)$ can be obtained by quadruple integration:

$$\begin{aligned}
D_{11}\frac{\mathrm{d}^4 w_0}{\mathrm{d}x^4} &= p,\\
D_{11}\frac{\mathrm{d}^3 w_0}{\mathrm{d}x^3} &= \int p\mathrm{d}x + C_1,\\
D_{11}\frac{\mathrm{d}^2 w_0}{\mathrm{d}x^2} &= \int\int p\mathrm{d}x\mathrm{d}x + C_1 x + C_2,\\
D_{11}\frac{\mathrm{d}w_0}{\mathrm{d}x} &= \int\int\int p\mathrm{d}x\mathrm{d}x\mathrm{d}x + \frac{1}{2}C_1 x^2 + C_2 x + C_3,\\
D_{11} w_0 &= \int\int\int\int p\mathrm{d}x\mathrm{d}x\mathrm{d}x\mathrm{d}x + \frac{1}{6}C_1 x^3 + \frac{1}{2}C_2 x^2 + C_3 x + C_4.
\end{aligned} \tag{7.92}$$

The integration constants C_1, C_2, C_3 and C_4 are to be determined from the given boundary conditions.

We consider for illustration the two plate strips of Fig. 7.22. First, we consider a plate strip (Fig. 7.22, top) of length a which is simply supported on both sides and

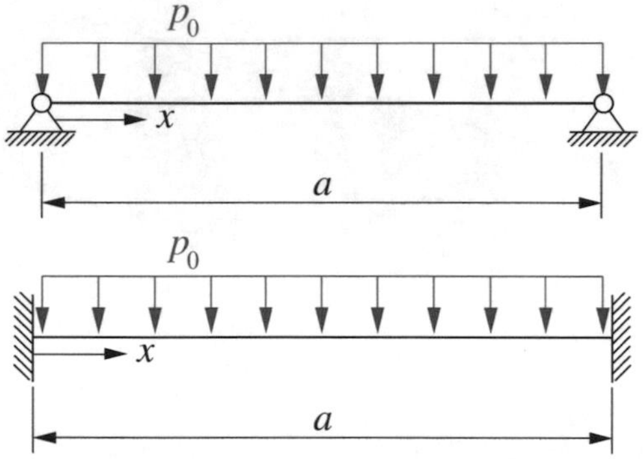

Fig. 7.22 Plate strip simply supported on both sides (top), plate strip clamped on both sides (bottom)

subjected to a constant surface load $p = p_0$. The relations (7.92) take the following form:

$$\begin{aligned} D_{11}\frac{\mathrm{d}^4 w_0}{\mathrm{d}x^4} &= p_0, \\ D_{11}\frac{\mathrm{d}^3 w_0}{\mathrm{d}x^3} &= p_0 x + C_1, \\ D_{11}\frac{\mathrm{d}^2 w_0}{\mathrm{d}x^2} &= \frac{1}{2}p_0 x^2 + C_1 x + C_2, \\ D_{11}\frac{\mathrm{d}w_0}{\mathrm{d}x} &= \frac{1}{6}p_0 x^3 + \frac{1}{2}C_1 x^2 + C_2 x + C_3, \\ D_{11} w_0 &= \frac{1}{24}p_0 x^4 + \frac{1}{6}C_1 x^3 + \frac{1}{2}C_2 x^2 + C_3 x + C_4. \end{aligned} \tag{7.93}$$

The boundary conditions to be applied here to determine the constants C_1, C_2, C_3 and C_4 are:

$$w_0(x=0) = 0, \quad \left.\frac{\mathrm{d}^2 w_0}{\mathrm{d}x^2}\right|_{x=0} = 0, \quad w_0(x=a) = 0, \quad \left.\frac{\mathrm{d}^2 w_0}{\mathrm{d}x^2}\right|_{x=a} = 0. \tag{7.94}$$

Evaluation yields with (7.93):

$$C_1 = -\frac{1}{2}p_0 a, \quad C_2 = 0, \quad C_3 = \frac{1}{24}p_0 a^3, \quad C_4 = 0. \tag{7.95}$$

Thus, the deflection $w_0(x)$ as well as the moment $M_{xx}^0(x)$ can be given as:

$$\begin{aligned} w_0(x) &= \frac{p_0 a^4}{24 D_{11}}\left[\left(\frac{x}{a}\right)^4 - 2\left(\frac{x}{a}\right)^3 + \left(\frac{x}{a}\right)\right], \\ M_{xx}^0(x) &= -\frac{p_0 a^2}{2}\left[\left(\frac{x}{a}\right)^2 - \left(\frac{x}{a}\right)\right]. \end{aligned} \tag{7.96}$$

As a further example, consider the plate strip clamped on both sides (Fig. 7.22, bottom). The boundary conditions to be applied here are:

$$w_0(x=0)=0,\quad \left.\frac{\mathrm{d}w_0}{\mathrm{d}x}\right|_{x=0}=0,\quad w_0(x=a)=0,\quad \left.\frac{\mathrm{d}w_0}{\mathrm{d}x}\right|_{x=a}=0. \tag{7.97}$$

From this, the constants C_1, C_2, C_3 and C_4 can be determined as:

$$C_1=-\frac{1}{2}p_0a,\quad C_2=\frac{p_0a^2}{12},\quad C_3=0,\quad C_4=0. \tag{7.98}$$

The deflection $w_0(x)$ and the moment $M_{xx}^0(x)$ can then be given as follows:

$$\begin{aligned} w_0(x) &= \frac{p_0a^4}{24D_{11}}\left[\left(\frac{x}{a}\right)^4-2\left(\frac{x}{a}\right)^3+\left(\frac{x}{a}\right)^2\right],\\ M_{xx}^0(x) &= -\frac{p_0a^2}{12}\left[6\left(\frac{x}{a}\right)^2-6\left(\frac{x}{a}\right)+1\right]. \end{aligned} \tag{7.99}$$

7.7 Navier Solution for Static Plate Bending Problems

7.7.1 Determination of the Plate Deflection

Another class of problems for which exact closed-form analytical solutions for static plate bending problems can be derived are the so-called Navier plate solutions[2]. A Navier plate is a rectanular plate whose edges are all simply supported. We consider orthotropic plates with plate stiffnesses D_{11}, D_{22}, D_{12} and D_{66} with length a and width b, and the loading is in the form of an arbitrarily distributed surface load $p=p(x,y)$. The considered situation is shown in Fig. 7.23. The surface load $p(x,y)$ is represented in the framework of Navier's solution in the form of a Fourier series:

$$p(x,y)=\sum_{m=1}^{m=\infty}\sum_{n=1}^{n=\infty}P_{mn}\sin\left(\frac{m\pi x}{a}\right)\sin\left(\frac{n\pi y}{b}\right). \tag{7.100}$$

The corresponding coefficients P_{mn} can be determined as follows:

$$P_{mn}=\frac{4}{ab}\int_0^b\int_0^a p(x,y)\sin\left(\frac{m\pi x}{a}\right)\sin\left(\frac{n\pi y}{b}\right)\mathrm{d}x\mathrm{d}y. \tag{7.101}$$

We will discuss the determination of the coefficients P_{mn} in detail later in this section.

[2] Claude Louis Marie Henri Navier, 1785–1836, French mathematician.

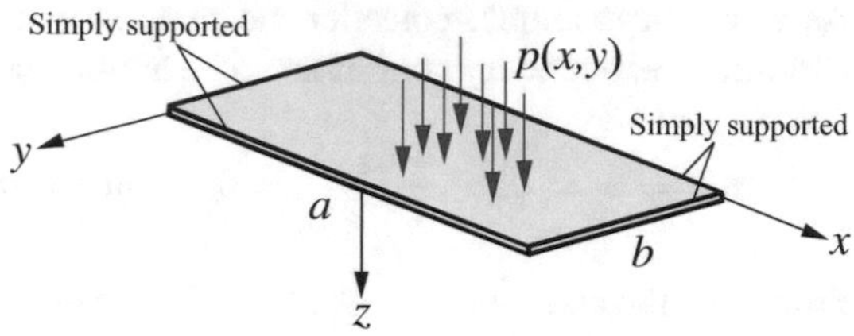

Fig. 7.23 Navier plate

To solve the given plate problem we use the plate equation (7.53), which we state again here for better clarity:

$$D_{11}\frac{\partial^4 w_0}{\partial x^4}+2\left(D_{12}+2D_{66}\right)\frac{\partial^4 w_0}{\partial x^2 \partial y^2}+D_{22}\frac{\partial^4 w_0}{\partial y^4}-p=0. \tag{7.102}$$

The boundary conditions are as follows for a rectangular plate simply supported at all edges. At the edges at $x = 0$ and $x = a$ for $0 \leq y \leq b$ both the deflection w_0 of the plate and the bending moment flow M^0_{xx} must vanish:

$$w_0 = 0, \quad M^0_{xx} = -D_{11}\frac{\partial^2 w_0}{\partial x^2} - D_{12}\frac{\partial^2 w_0}{\partial y^2} = 0. \tag{7.103}$$

Similarly, at the edges $y = 0$ and $y = b$ für $0 \leq x \leq a$ both the deflection w_0 and the bending moment flow M^0_{yy} vanish:

$$w_0 = 0, \quad M^0_{yy} = -D_{12}\frac{\partial^2 w_0}{\partial x^2} - D_{22}\frac{\partial^2 w_0}{\partial y^2} = 0. \tag{7.104}$$

We use a displacement approach of the form

$$w_0 = \sum_{m=1}^{m=\infty}\sum_{n=1}^{n=\infty} W_{mn}\sin\left(\frac{m\pi x}{a}\right)\sin\left(\frac{n\pi y}{b}\right). \tag{7.105}$$

It can be easily shown that this formulation satisfies all the underlying boundary conditions (7.103) and (7.104) identically. We substitute this approach into the plate equation (7.102) and obtain:

$$\sum_{m=1}^{m=\infty}\sum_{n=1}^{n=\infty}\left\{-W_{mn}\left[D_{11}\left(\frac{m\pi}{a}\right)^4+2\left(D_{12}+2D_{66}\right)\left(\frac{m\pi}{a}\right)^2\left(\frac{n\pi}{b}\right)^2\right.\right.$$
$$\left.\left.+D_{22}\left(\frac{n\pi}{b}\right)^4\right]+P_{mn}\right\}\sin\left(\frac{m\pi x}{a}\right)\sin\left(\frac{n\pi y}{b}\right)=0. \tag{7.106}$$

An non-trivial solution is obtained by setting the expression in brackets to zero. One can solve this directly for the coefficient W_{mn} as follows:

$$W_{mn} = \frac{P_{mn}}{\pi^4 \left[D_{11} \frac{m^4}{a^4} + 2\left(D_{12} + 2D_{66}\right) \frac{m^2 n^2}{a^2 b^2} + D_{22} \frac{n^4}{b^4} \right]}. \tag{7.107}$$

The series expansion (7.105) can thus be represented as:

$$w_0 = \sum_{m=1}^{m=\infty} \sum_{n=1}^{n=\infty} \frac{P_{mn} \sin\left(\frac{m\pi x}{a}\right) \sin\left(\frac{n\pi y}{b}\right)}{\pi^4 \left[D_{11} \frac{m^4}{a^4} + 2\left(D_{12} + 2D_{66}\right) \frac{m^2 n^2}{a^2 b^2} + D_{22} \frac{n^4}{b^4} \right]}. \tag{7.108}$$

It is important to note that the form of Eq. (7.108) suggests that it is an approximation of the plate deflection due to its double-sum representation. However, Eq. (7.108) is a consequence of the evaluation of the plate equation (7.102), and all underlying boundary conditions are identically satisfied by the chosen approach (7.105). Thus, the solution (7.108) is exact in the sense of the underlying Kirchhoff plate theory. However, care must be taken that the series expansion in (7.108) is carried out with a sufficiently large number of series elements, since the quality of the solution always increases with increasing degree of expansion. Usually the first few elements of the series are sufficient to obtain sufficiently accurate results, because the coefficients W_{mn} will quickly become small with increasing degrees of approximation due to the terms m^4 and n^4 appearing in the denominator. This must be ensured in any case by convergence studies. However, some caution is required concerning the convergence of the force and moment flows of the plate. Moment and shear force flows are, after all, quantities arising from second and third order partial derivatives of the plate deflection w_0, respectively. Therefore, the convergence behavior of moment and transverse force flows is characterized by $\frac{1}{m^2}, \frac{1}{n^2}$ and $\frac{1}{m}, \frac{1}{m}$, respectively.

In the case of an isotropic plate, the coefficients W_{mn} are obtained as:

$$W_{mn} = \frac{P_{mn}}{D} \frac{a^4 b^4}{\pi^4 \left(m^2 b^2 + n^2 a^2\right)^2}. \tag{7.109}$$

The plate deflection $w_0(x, y)$ is then:

$$w_0 = \sum_{m=1}^{m=\infty} \sum_{n=1}^{n=\infty} \frac{P_{mn}}{D} \frac{a^4 b^4}{\pi^4 \left(m^2 b^2 + n^2 a^2\right)^2} \sin\left(\frac{m\pi x}{a}\right) \sin\left(\frac{n\pi y}{b}\right). \tag{7.110}$$

7.7.2 Moments, Forces and Stresses of the Plate

With the available solution for the plate deflection w_0, the moment flows M^0_{xx}, M^0_{yy} and M^0_{xy} can be determined:

$$
\begin{aligned}
M^0_{xx} &= -D_{11}\frac{\partial^2 w_0}{\partial x^2} - D_{12}\frac{\partial^2 w_0}{\partial y^2} \\
&= \sum_{m=1}^{m=\infty}\sum_{n=1}^{n=\infty}\left(D_{11}\frac{m^2\pi^2}{a^2} + D_{12}\frac{n^2\pi^2}{b^2}\right) W_{mn}\sin\left(\frac{m\pi x}{a}\right)\sin\left(\frac{n\pi y}{b}\right), \\
M^0_{yy} &= -D_{12}\frac{\partial^2 w_0}{\partial x^2} - D_{22}\frac{\partial^2 w_0}{\partial y^2} \\
&= \sum_{m=1}^{m=\infty}\sum_{n=1}^{n=\infty}\left(D_{12}\frac{m^2\pi^2}{a^2} + D_{22}\frac{n^2\pi^2}{b^2}\right) W_{mn}\sin\left(\frac{m\pi x}{a}\right)\sin\left(\frac{n\pi y}{b}\right), \\
M^0_{xy} &= -2D_{66}\frac{\partial^2 w_0}{\partial x\partial y} \\
&= -2\sum_{m=1}^{m=\infty}\sum_{n=1}^{n=\infty}\frac{m\pi}{a}\frac{n\pi}{b}D_{66}W_{mn}\cos\left(\frac{m\pi x}{a}\right)\cos\left(\frac{n\pi y}{b}\right).
\end{aligned}
\tag{7.111}
$$

By means of the moment equilibria (7.50) and (7.51) the shear force flows Q_x and Q_y can be determined, but this is not shown here.

The stresses σ_{xx}, σ_{yy} and τ_{xy} can be calculated from Hooke's law (7.12) as:

$$
\begin{aligned}
\begin{pmatrix}\sigma_{xx}\\ \sigma_{yy}\\ \tau_{xy}\end{pmatrix} &= -z\begin{bmatrix}Q_{11} & Q_{12} & 0\\ Q_{12} & Q_{22} & 0\\ 0 & 0 & Q_{66}\end{bmatrix}\begin{pmatrix}\frac{\partial^2 w_0}{\partial x^2}\\ \frac{\partial^2 w_0}{\partial y^2}\\ 2\frac{\partial^2 w_0}{\partial x\partial y}\end{pmatrix} \\
&= z\sum_{m=1}^{m=\infty}\sum_{n=1}^{n=\infty}W_{mn}\begin{bmatrix}Q_{11} & Q_{12} & 0\\ Q_{12} & Q_{22} & 0\\ 0 & 0 & Q_{66}\end{bmatrix} \\
&\quad\times\begin{pmatrix}\frac{m^2\pi^2}{a^2}\sin\left(\frac{m\pi x}{a}\right)\sin\left(\frac{n\pi y}{b}\right)\\ \frac{n^2\pi^2}{b^2}\sin\left(\frac{m\pi x}{a}\right)\sin\left(\frac{n\pi y}{b}\right)\\ -2\frac{m\pi}{a}\frac{n\pi}{b}\cos\left(\frac{m\pi x}{a}\right)\cos\left(\frac{n\pi y}{b}\right)\end{pmatrix} \\
&= z\sum_{m=1}^{m=\infty}\sum_{n=1}^{n=\infty}W_{mn}\begin{pmatrix}\left(Q_{11}\frac{m^2\pi^2}{a^2} + Q_{12}\frac{n^2\pi^2}{b^2}\right)\sin\left(\frac{m\pi x}{a}\right)\sin\left(\frac{n\pi y}{b}\right)\\ \left(Q_{12}\frac{m^2\pi^2}{a^2} + Q_{22}\frac{n^2\pi^2}{b^2}\right)\sin\left(\frac{m\pi x}{a}\right)\sin\left(\frac{n\pi y}{b}\right)\\ -2Q_{66}\frac{m\pi}{a}\frac{n\pi}{b}\cos\left(\frac{m\pi x}{a}\right)\cos\left(\frac{n\pi y}{b}\right)\end{pmatrix}.
\end{aligned}
\tag{7.112}
$$

As expected, all stress components are always linear across the plate thickness, regardless of the underlying load case. For certain selected load cases, some general statements can also be made. In the case of a uniform surface load $p(x, y) = p_0$, the maximum deflection of the plate will always occur in the middle of the plate. In addition, the maximum bending stresses will always occur at this point, namely on both surfaces of the plate at $z = \pm\frac{h}{2}$. The maximum shear stresses, on the other hand, will always be found in the corners of the plate.

For the transverse shear stresses τ_{xz}, τ_{yz} no statement has been made so far. They cannot be calculated from constitutive relations within the framework of the Kirchhoff plate theory. This is analogous to the Euler-Bernoulli beam theory, where the shear stresses must be determined from the normal stresses in a post-calculation. The normal stress σ_{zz} cannot be calculated either, but will not play a role in structural behavior for sufficiently thin plates, unless highly concentrated loads perpendicular to the plate midplane are to be introduced.

The distributions of the transverse shear stresses can be obtained in a post-calculation by applying the three-dimensional local equilibrium conditions (see Eq. (1.36)).

$$
\begin{aligned}
\frac{\partial \sigma_{xx}}{\partial x} + \frac{\partial \tau_{xy}}{\partial y} + \frac{\partial \tau_{xz}}{\partial z} &= 0, \\
\frac{\partial \tau_{xy}}{\partial x} + \frac{\partial \sigma_{yy}}{\partial y} + \frac{\partial \tau_{yz}}{\partial z} &= 0.
\end{aligned}
\tag{7.113}
$$

Integration over the thickness assuming that there are no volume forces yields:

$$
\begin{aligned}
\tau_{xz} &= -\int_{-\frac{h}{2}}^{z} \left(\frac{\partial \sigma_{xx}}{\partial x} + \frac{\partial \tau_{xy}}{\partial y} \right) \mathrm{d}z + F_1(x, y), \\
\tau_{yz} &= -\int_{-\frac{h}{2}}^{z} \left(\frac{\partial \tau_{xy}}{\partial x} + \frac{\partial \sigma_{yy}}{\partial y} \right) \mathrm{d}z + F_2(x, y).
\end{aligned}
\tag{7.114}
$$

Here $F_1(x, y)$ and $F_2(x, y)$ are the initial values of the stresses σ_{xx}, σ_{yy} and τ_{xy} at the locations $z = -\frac{h}{2}$. Evaluating finally results in:

$$
\begin{aligned}
\tau_{xz} &= \frac{h^2}{8}\left[1 - \left(\frac{2z}{h}\right)^2\right] \sum_{m=1}^{m=\infty} \sum_{n=1}^{n=\infty} \\
&\quad \times \left[\left(\frac{m\pi}{a}\right)^3 Q_{11} + \left(\frac{m\pi}{a}\right)\left(\frac{n\pi}{b}\right)^2 (Q_{12} + 2Q_{66}) \right] \\
&\quad \times W_{mn} \cos\left(\frac{m\pi x}{a}\right) \sin\left(\frac{n\pi y}{b}\right), \\
\tau_{yz} &= \frac{h^2}{8}\left[1 - \left(\frac{2z}{h}\right)^2\right] \sum_{m=1}^{m=\infty} \sum_{n=1}^{n=\infty} \\
&\quad \times \left[\left(\frac{n\pi}{b}\right)^3 Q_{22} + \left(\frac{m\pi}{a}\right)^2\left(\frac{n\pi}{b}\right) (Q_{12} + 2Q_{66}) \right] \\
&\quad \times W_{mn} \sin\left(\frac{m\pi x}{a}\right) \cos\left(\frac{n\pi y}{b}\right).
\end{aligned}
\tag{7.115}
$$

As expected, the transverse shear stresses are parabolic over z and vanish at the plate surfaces at $z = \pm\frac{h}{2}$.

7.7.3 Special Load Cases

We want to shed some more light on the determination of the load coefficients P_{mn} (7.101) in (7.100) and consider some selected cases of loads $p(x, y)$ (see Reddy 2006).

If the load $p(x, y)$ is in the form of a uniform surface load $p(x, y) = p_0$ (Fig. 7.24), then the Fourier coefficients P_{mn} are obtained as:

$$P_{mn} = \frac{16 p_0}{mn\pi^2} \quad (m, n = 1, 3, 5, ...). \tag{7.116}$$

The deflection w_0 of the plate can then be written as:

$$w_0 = \frac{16 p_0}{\pi^6} \sum_{m=1,3,5,...}^{m=\infty} \sum_{n=1,3,5,...}^{n=\infty} \frac{\frac{1}{mn} \sin\left(\frac{m\pi x}{a}\right) \sin\left(\frac{n\pi y}{b}\right)}{D_{11}\frac{m^4}{a^4} + 2\left(D_{12} + 2D_{66}\right)\frac{m^2 n^2}{a^2 b^2} + D_{22}\frac{n^4}{b^4}}. \tag{7.117}$$

In the case of a constant line load $p(x, y) = p_0$ acting at the point x_0 and running over the entire width b of the plate (Fig. 7.25), we obtain:

$$P_{mn} = \frac{8 p_0}{an\pi} \sin\left(\frac{m\pi x_0}{a}\right) \quad (m = 1, 3, 5, ...; n = 1, 2, 3, ...). \tag{7.118}$$

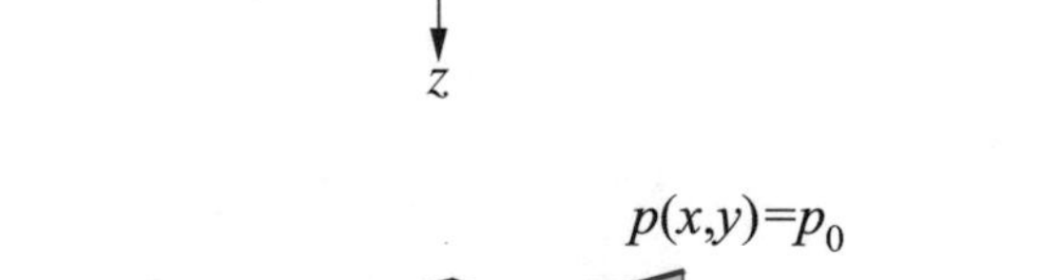

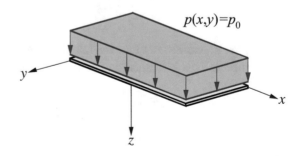

Fig. 7.24 Plate under uniform surface load $p(x, y) = p_0$

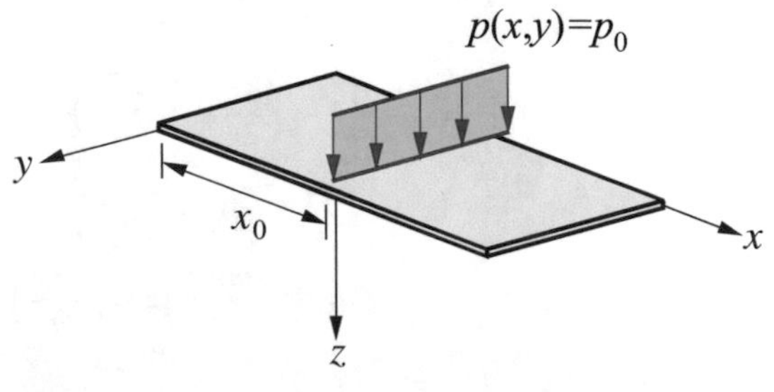

Fig. 7.25 Plate under uniform line load $p(x, y) = p_0$

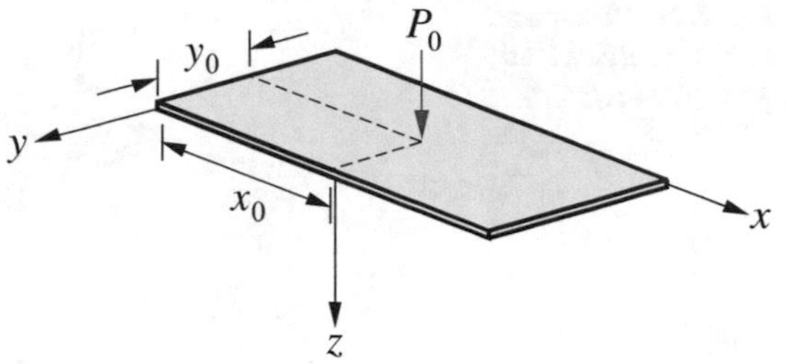

Fig. 7.26 Plate under point load P_0

On the other hand, if a single force P is present at the location $x = x_0$ and $y = y_0$ (Fig. 7.26), then the load coefficients P_{mn} are obtained as:

$$P_{mn} = \frac{4P_0}{ab} \sin\left(\frac{m\pi x_0}{a}\right) \sin\left(\frac{n\pi y_0}{b}\right) \quad (m, n = 1, 2, 3, ...). \tag{7.119}$$

An interesting special case is when the load $p(x, y)$ is in the form of a sinusoidal distribution of the type

$$p(x, y) = p_0 \sin\left(\frac{\pi x}{a}\right) \sin\left(\frac{\pi y}{b}\right). \tag{7.120}$$

It is then shown that a series expansion for the plate deflection w_0 is not necessary, but a fully closed-form analytical solution is obtained for $m = 1$ and $n = 1$. All other load coefficients P_{mn} are zero, and the only non-zero coefficient is $P_{11} = p_0$. Then the only remaining coefficient W_{11} of deflection is:

$$W_{11} = \frac{p_0}{\pi^4 \left[\frac{D_{11}}{a^4} + \frac{2}{a^2b^2}(D_{12} + 2D_{66}) + \frac{D_{22}}{b^4}\right]}. \tag{7.121}$$

The exact solution for the deflection w_0 of the plate can then be written as:

$$w_0 = \frac{p_0 \sin\left(\frac{\pi x}{a}\right) \sin\left(\frac{\pi y}{b}\right)}{\pi^4 \left[\frac{D_{11}}{a^4} + \frac{2}{a^2b^2}(D_{12} + 2D_{66}) + \frac{D_{22}}{b^4}\right]}. \tag{7.122}$$

Furthermore, let a plate be given which is under a constant surface load $p(x, y) = p_0$ acting on a subregion of the plate with length $2c$ and width $2d$. The center of gravity of this subregion is located at $x = x_0$ and $y = y_0$ (Fig. 7.27). Equation (7.101) then leads to the following expression for the load coefficients P_{mn}:

$$P_{mn} = \frac{4p_0}{ab} \int_{y_0-d}^{y_0+d} \int_{x_0-c}^{x_0+c} \sin\left(\frac{m\pi x}{a}\right) \sin\left(\frac{n\pi y}{b}\right) \mathrm{d}x\mathrm{d}y. \tag{7.123}$$

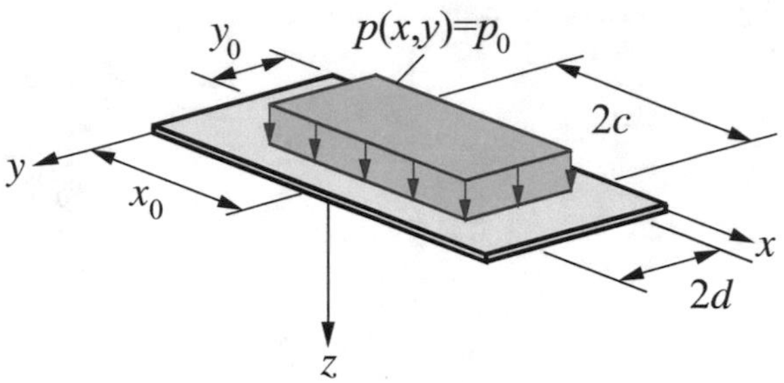

Fig. 7.27 Plate under uniform surface load $p(x, y) = p_0$

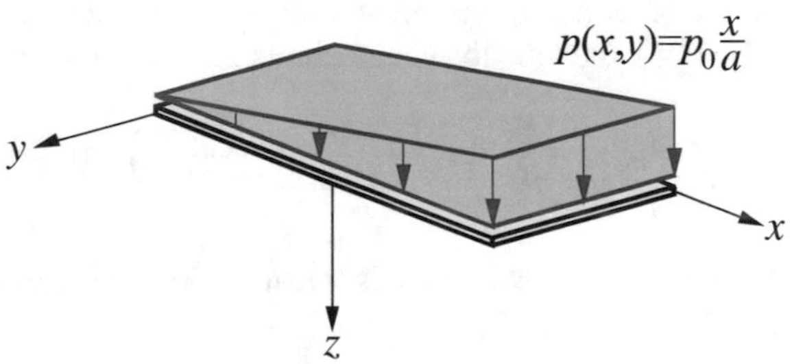

Fig. 7.28 Plate under linearly varying surface load $p(x, y) = p_0 \frac{x}{a}$

The integration prescribed here is easily carried out, and it results:

$$P_{mn} = \frac{16 p_0}{\pi^2 mn} \sin\left(\frac{m\pi x_0}{a}\right) \sin\left(\frac{m\pi c}{a}\right) \sin\left(\frac{n\pi y_0}{b}\right) \sin\left(\frac{n\pi d}{b}\right). \tag{7.124}$$

The plate deflection then results as:

$$w_0 = \frac{16 p_0}{\pi^6} \sum_{m=1}^{m=\infty} \sum_{n=1}^{n=\infty} \frac{\sin\left(\frac{m\pi x_0}{a}\right) \sin\left(\frac{m\pi c}{a}\right) \sin\left(\frac{n\pi y_0}{b}\right) \sin\left(\frac{n\pi d}{b}\right) \sin\left(\frac{m\pi x}{a}\right) \sin\left(\frac{n\pi y}{b}\right)}{D_{11}\frac{m^4}{a^4} + 2\left(D_{12} + 2D_{66}\right)\frac{m^2 n^2}{a^2 b^2} + D_{22}\frac{n^4}{b^4}}. \tag{7.125}$$

Finally, we want to investigate the case where the surface load $p(x, y)$ is in a linear form (Fig. 7.28). We then obtain the Fourier coefficients as:

$$P_{mn} = \frac{4 p_0}{a^2 b} \int_0^b \int_0^a x \sin\left(\frac{m\pi x}{a}\right) \sin\left(\frac{n\pi y}{b}\right) \mathrm{d}x\mathrm{d}y. \tag{7.126}$$

This leads to:

$$P_{mn} = \frac{4 p_0}{mn\pi^2} \left[(-1)^n - 1\right] (-1)^m. \tag{7.127}$$

For even values for n this expression becomes zero, and for odd values remains:

$$P_{mn} = \frac{8 p_0}{mn\pi^2} (-1)^{m+1}. \tag{7.128}$$

7.8 Lévy-type solutions for static plate bending problems

7.8.1 Introduction

A further important class of exact closed-analytic solutions are the so-called Lévy-type[3] solutions. Let again an orthotropic plate with dimensions a and b be given under an arbitrary static surface load $p(x, y)$, where we require that the plate be simply supported at two opposing edges. These are the two edges at $x = 0$ and $x = a$. The two plate edges at $y = 0$ and $y = b$ can be subject to any boundary conditions, e.g. free, simply supported, clamped, elastically restrained, reinforced by stiffeners, etc. Let the plate stiffnesses be given by D_{11}, D_{22}, D_{12} and D_{66}. Figure 7.29 shows the considered situation. We will show that by a suitable solution approach in the x-direction we can reduce the plate differential equation to an ordinary differential equation, which can then be solved in an exact closed-form analytical manner for elementary cases.

7.8.2 Orthotropic Plates

The starting point of the considerations is again the plate equation (7.53), which is presented again here for better readability:

$$D_{11}\frac{\partial^4 w_0}{\partial x^4}+D_{22}\frac{\partial^4 w_0}{\partial y^4}+2\left(D_{12}+2D_{66}\right)\frac{\partial^4 w_0}{\partial x^2\partial y^2}-p=0. \tag{7.129}$$

For the deflection w_0 of the plate we use an approach of the form

$$w_0=\sum_{m=1}^{m=\infty}\sin\left(\frac{m\pi x}{a}\right)W_m\left(y\right). \tag{7.130}$$

It is shown that the approach (7.130) is identical with the displacement approach (7.105) in the framework of Navier's solution with respect to the x-direction due to the conditions of simply supported edges at $x = 0$ and $x = a$. The plate deflection is described with respect to the y-direction by the functions $W_m\left(y\right)$, which are to be determined.

As in the framework of Navier's solution, we develop the external load in the form of a series expansion as follows:

$$p=\sum_{m=1}^{m=\infty}\sin\left(\frac{m\pi x}{a}\right)P_m\left(y\right), \tag{7.131}$$

[3] Maurice Lévy, 1838–1910, French engineer.

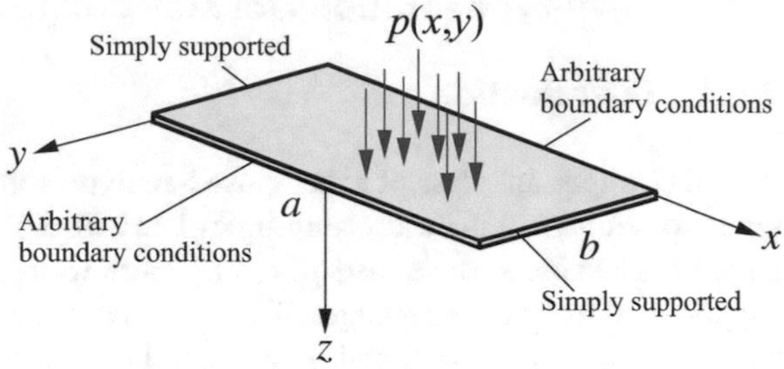

Fig. 7.29 Plate in the framework of a Lévy-type solution

with:

$$P_m(y) = \frac{2}{a} \int_0^a p(x, y) \sin\left(\frac{m\pi x}{a}\right) \mathrm{d}x. \tag{7.132}$$

Substituting (7.130) and (7.131) into the plate equation (7.129) then gives:

$$\sum_{m=1}^{m=\infty} \left[D_{22} \frac{\partial^4 W_m}{\partial y^4} - 2(D_{12} + 2D_{66}) \left(\frac{m\pi}{a}\right)^2 \frac{\partial^2 W_m}{\partial y^2} + D_{11} \left(\frac{m\pi}{a}\right)^4 W_m - P_m \right] \sin\left(\frac{m\pi x}{a}\right) = 0. \tag{7.133}$$

This condition must be satisfied regardless of the value of x, so we require that the term in the square brackets disappears:

$$D_{22} \frac{\mathrm{d}^4 W_m}{\mathrm{d}y^4} - 2(D_{12} + 2D_{66}) \left(\frac{m\pi}{a}\right)^2 \frac{\mathrm{d}^2 W_m}{\mathrm{d}y^2} + D_{11} \left(\frac{m\pi}{a}\right)^4 W_m = P_m. \tag{7.134}$$

This is an ordinary fourth order linear inhomogeneous differential equation with constant coefficients from which the displacement function W_m can be determined. Using the approaches (7.130) and (7.131) thus reduces the plate equation (7.129) to an ordinary differential equation which is straightforward in its solution. Since this is an inhomogeneous differential equation, the solution W_m will be composed of a homogeneous solution $W_{m,h}$ and a particular solution $W_{m,p}$.

Let us first turn to the determination of the homogeneous solution $W_{m,h}$. An approach of the form

$$W_{m,h} = C_m e^{\lambda_m y} \tag{7.135}$$

leads after substitution into (7.134) to the following fourth-order polynomial for the exponent λ:

$$D_{22} \lambda_m^4 - 2(D_{12} + 2D_{66}) \left(\frac{m\pi}{a}\right)^2 \lambda_m^2 + D_{11} \left(\frac{m\pi}{a}\right)^4 = 0. \tag{7.136}$$

This equation gives four solutions for λ as follows:

$$\lambda_{m,1,2,3,4} = \pm \frac{m\pi}{a\sqrt{D_{22}}} \sqrt{D_{12} + 2D_{66} \pm \sqrt{(D_{12} + 2D_{66})^2 - D_{11}D_{22}}}. \quad (7.137)$$

To uniquely specify the solution for the plate deflection w_0, a case distinction is performed, where the term under the second root in (7.137) is decisive. The following three cases can be distinguished.

If $(D_{12} + 2D_{66})^2 > D_{11}D_{22}$ holds, then all solutions for λ_m are real-valued and different from each other. Also, $\lambda_{m,1} = -\lambda_{m,2}$ and $\lambda_{m,3} = -\lambda_{m,4}$ hold. The displacement function $W_{m,h}$ is then:

$$W_{m,h} = C_{m,1}e^{\lambda_{m,1}y} + C_{m,2}e^{\lambda_{m,2}y} + C_{m,3}e^{\lambda_{m,3}y} + C_{m,4}e^{\lambda_{m,4}y}. \quad (7.138)$$

An alternative representation is

$$\begin{aligned} W_{m,h} &= C_{m,1}\cosh\left(\lambda_{m,1}y\right) + C_{m,2}\sinh\left(\lambda_{m,1}y\right) \\ &+ C_{m,3}\cosh\left(\lambda_{m,3}y\right) + C_{m,4}\sinh\left(\lambda_{m,3}y\right), \end{aligned} \quad (7.139)$$

it is equally common in the literature.

If $(D_{12} + 2D_{66})^2 = D_{11}D_{22}$ holds, then the exponents λ_m are real-valued and result as

$$\lambda_m = \pm \frac{m\pi}{a\sqrt{D_{22}}} \sqrt{D_{12} + 2D_{66}}, \quad (7.140)$$

they are pairwise identical, i.e. $\lambda_{m,1} = \lambda_{m,2} = -\lambda_{m,3} = -\lambda_{m,4} = \lambda_m$. The displacement function $W_{m,h}$ is then:

$$W_{m,h} = \left(C_{m,1} + C_{m,2}y\right)\cosh\left(\lambda y_m\right) + \left(C_{m,3} + C_{m,4}y\right)\sinh\left(\lambda y_m\right). \quad (7.141)$$

If, on the other hand, the case $(D_{12} + 2D_{66})^2 < D_{11}D_{22}$ is present, then the exponents λ_m are complex-valued and exist as complex conjugate pairs, i.e. $\pm\lambda_{m,2} \pm i\lambda_{m,2}$. The displacement function $W_{m,h}$ can then be written as:

$$\begin{aligned} W_{m,h} &= \left[C_{m,1}\cos\left(\lambda_{m,2}y\right) + C_{m,2}\sin\left(\lambda_{m,2}y\right)\right]\cosh\left(\lambda_{m,1}y\right) \\ &+ \left[C_{m,3}\cos\left(\lambda_{m,2}y\right) + C_{m,4}\sin\left(\lambda_{m,2}y\right)\right]\sinh\left(\lambda_{m,1}y\right). \end{aligned} \quad (7.142)$$

In all three of the above cases, the constants C_m are determined from the boundary conditions at the edges at $y = 0$ and $y = b$.

The discussed homogeneous solutions are always valid independently of the underlying load case. On the other hand, the particular solution $W_{m,p}$ cannot be given in general form, but must be determined depending on the concrete mathematical form of the term P_m. Usually, the particular solution $W_{m,p}$ can be determined quite easily with right-hand side type approaches and coefficient comparison. This is briefly shown in general for the case of a constant term P_m. The approach for $W_{m,p}$ in this case is chosen conveniently in the form of a constant k, which after insertion into (7.134) and coefficient comparison yields the following particular solution:

$$W_{m,p} = \frac{P_m}{D_{11}} \left(\frac{a}{m\pi} \right)^4. \tag{7.143}$$

We will illustrate the procedure using the Lévy-type solution approach by considering the deflection of a rectangular orthotropic plate under a constant surface load $p(x, y) = p_0$. We assume that $(D_{12} + 2D_{66})^2 > D_{11} D_{22}$ holds. The total solution w_0 for the plate deflection is then given as:

$$\begin{aligned} w_0 &= \sum_{m=1}^{m=\infty} \left\{ \sin\left(\frac{m\pi x}{a} \right) \left[C_{m,1} e^{\lambda_{m,1} y} + C_{m,2} e^{\lambda_{m,2} y} + C_{m,3} e^{\lambda_{m,3} y} \right.\right. \\ &\quad \left.\left. + C_{m,4} e^{\lambda_{m,4} y} + \frac{P_m}{D_{11}} \left(\frac{a}{m\pi} \right)^4 \right] \right\}. \end{aligned} \tag{7.144}$$

Let us now consider the case where the edges at $x = 0$ and $x = a$ are simply supported, and where the two edges at $y = 0$ and $y = b$ are clamped. The boundary conditions at the edges $y = 0$ and $y = b$ are then:

$$w_0 (y = 0) = 0, \quad w_0 (y = b) = 0, \quad \left. \frac{\partial w_0}{\partial y} \right|_{y=0} = 0, \quad \left. \frac{\partial w_0}{\partial y} \right|_{y=b} = 0. \tag{7.145}$$

Substituting the solution (7.144) into the given boundary conditions then leads to the following system of four linear equations for the constants C_m:

$$\begin{aligned} & C_{m,1} + C_{m,2} + C_{m,3} + C_{m,4} = -\frac{P_m}{D_{11}} \left(\frac{a}{m\pi} \right)^4, \\ & C_{m,1} e^{\lambda_{m,1} b} + C_{m,2} e^{\lambda_{m,2} b} + C_{m,3} e^{\lambda_{m,3} b} + C_{m,4} e^{\lambda_{m,4} b} = -\frac{P_m}{D_{11}} \left(\frac{a}{m\pi} \right)^4, \\ & \lambda_{m,1} C_{m,1} + \lambda_{m,2} C_{m,2} + \lambda_{m,3} C_{m,3} + \lambda_{m,4} C_{m,4} = 0, \\ & \lambda_{m,1} C_{m,1} e^{\lambda_{m,1} b} + \lambda_{m,2} C_{m,2} e^{\lambda_{m,2} b} + \lambda_{m,3} C_{m,3} e^{\lambda_{m,3} b} + \lambda_{m,4} C_{m,4} e^{\lambda_{m,4} b} \\ & \quad = 0. \end{aligned} \tag{7.146}$$

This system of equations can then be quite easily solved for the constants C_m. The solution for the plate deflection w_0 is then obtained by summing up all individual solutions for $m = 1$ up to a certain degree $m = M$, which has to be determined by a convergence analysis.

We obtain a more compact representation by using the form (7.139) instead of the solution representation (7.144), thus:

$$\begin{aligned} w_0 &= \sum_{m=1}^{m=\infty} \left\{ \sin\left(\frac{m\pi x}{a} \right) \left[C_{m,1} \cosh\left(\lambda_{m,1} y \right) + C_{m,2} \sinh\left(\lambda_{m,1} y \right) \right.\right. \\ &\quad + C_{m,3} \cosh\left(\lambda_{m,3} y \right) + C_{m,4} \sinh\left(\lambda_{m,3} y \right) \\ &\quad \left.\left. + \frac{P_m}{D_{11}} \left(\frac{a}{m\pi} \right)^4 \right] \right\}. \end{aligned} \tag{7.147}$$

Evaluating the given boundary conditions then yields the following system of equations:

$$
\begin{aligned}
C_{m,1} + C_{m,3} &= -\frac{P_m}{D_{11}}\left(\frac{a}{m\pi}\right)^4, \\
C_{m,1}\cosh\left(\lambda_{m,1}b\right) + C_{m,2}\sinh\left(\lambda_{m,1}b\right) & \\
+C_{m,3}\cosh\left(\lambda_{m,3}b\right) + C_{m,4}\sinh\left(\lambda_{m,3}b\right) &= -\frac{P_m}{D_{11}}\left(\frac{a}{m\pi}\right)^4, \\
\lambda_{m,1}C_{m,2} + \lambda_{m,3}C_{m,4} &= 0, \\
\lambda_{m,1}C_{m,1}\sinh\left(\lambda_{m,1}b\right) + \lambda_{m,1}C_{m,2}\cosh\left(\lambda_{m,1}b\right) & \\
+\lambda_{m,3}C_{m,3}\sinh\left(\lambda_{m,3}b\right) + \lambda_{m,3}C_{m,4}\cosh\left(\lambda_{m,3}b\right) &= 0.
\end{aligned}
\tag{7.148}
$$

The result for the coefficients reads:

$$
\begin{aligned}
C_{m,1} &= -\frac{a^4 P_m \lambda_{m,3}}{D_{11} m^4 \pi^4 \Delta}\left[-\lambda_{m,1}\cosh\left(\lambda_{m,1}b\right) + \lambda_{m,1}\cosh\left(\lambda_{m,3}b\right)\right. \\
&\quad \left. + \lambda_{m,1}\cosh\left(\lambda_{m,1}b\right)\cosh\left(\lambda_{m,3}b\right) - \lambda_{m,3}\sinh\left(\lambda_{m,1}b\right)\sinh\left(\lambda_{m,3}b\right) - \lambda_{m,1}\right], \\
C_{m,2} &= \frac{\lambda_{m,3} P_m a^4}{D_{11} m^4 \pi^4 \Delta}\left[-\lambda_{m,1}\sinh\left(\lambda_{m,1}b\right) + \lambda_{m,3}\sinh\left(\lambda_{m,3}b\right)\right. \\
&\quad \left. - \lambda_{m,3}\cosh\left(\lambda_{m,1}b\right)\sinh\left(\lambda_{m,3}b\right) + \lambda_{m,1}\cosh\left(\lambda_{m,3}b\right)\sinh\left(\lambda_{m,1}b\right)\right], \\
C_{m,3} &= \frac{P_m a^4 \lambda_{m,1}}{D_{11} m^4 \pi^4 \Delta}\left[-\lambda_{m,3}\cosh\left(\lambda_{m,1}b\right) + \lambda_{m,3}\cosh\left(\lambda_{m,3}b\right)\right. \\
&\quad \left. - \lambda_{m,3}\cosh\left(\lambda_{m,1}b\right)\cosh\left(\lambda_{m,3}b\right) + \lambda_{m,1}\sinh\left(\lambda_{m,1}b\right)\sinh\left(\lambda_{m,3}b\right) + \lambda_{m,3}\right], \\
C_{m,4} &= -\frac{\lambda_{m,1} P_m a^4}{D_{11} m^4 \pi^4 \Delta}\left[-\lambda_{m,1}\sinh\left(\lambda_{m,1}b\right) + \lambda_{m,3}\sinh\left(\lambda_{m,3}b\right)\right. \\
&\quad \left. - \lambda_{m,3}\cosh\left(\lambda_{m,1}b\right)\sinh\left(\lambda_{m,3}b\right) + \lambda_{m,1}\cosh\left(\lambda_{m,3}b\right)\sinh\left(\lambda_{m,1}b\right)\right],
\end{aligned}
\tag{7.149}
$$

where Δ is the determinant of the coefficient matrix:

$$
\Delta = -\left(\lambda_{m,1}^2 + \lambda_{m,3}^2\right)\sinh\left(\lambda_{m,1}b\right)\sinh\left(\lambda_{m,3}b\right) + 2\lambda_1\lambda_{m,3}\cosh\left(\lambda_{m,1}b\right)\cosh\left(\lambda_3 b\right) - 1.
\tag{7.150}
$$

7.8.3 Isotropic Plates

If the special case of an isotropic plate is given, then it is convenient to use an alternative representation of the Lévy-type solution. The starting point is the plate equation (7.54), which we give again here for the sake of clarity:

$$
\frac{\partial^4 w_0}{\partial x^4} + 2\frac{\partial^4 w_0}{\partial x^2 \partial y^2} + \frac{\partial^4 w_0}{\partial y^4} = \frac{p}{D}.
\tag{7.151}
$$

The expansion of the external load $p(x, y)$ in the form of a Fourier series is analogous to what has been shown so far. The plate deflection w_0 is represented in the form of a series expansion:

$$w_0 = \sum_{m=1}^{m=\infty} \sin\left(\frac{m\pi x}{a}\right) W_m(y). \tag{7.152}$$

Substituting into the plate equation (7.151) yields:

$$\sum_{m=1}^{m=\infty} \left(\frac{d^4 W_m(y)}{dy^4} - 2\frac{m^2\pi^2}{a^2}\frac{d^2 W_m(y)}{dy^2} + \frac{m^4\pi^4}{a^4} W_m(y) - \frac{1}{D} P_m(y) \right) \times \sin\left(\frac{m\pi x}{a}\right) = 0. \tag{7.153}$$

This equation must be satisfied for any values for x, so again it is required that the bracket term must vanish:

$$\frac{d^4 W_m(y)}{dy^4} - 2\frac{m^2\pi^2}{a^2}\frac{d^2 W_m(y)}{dy^2} + \frac{m^4\pi^4}{a^4} W_m(y) = \frac{1}{D} P_m(y). \tag{7.154}$$

For the homogeneous solution of this ordinary differential equation, we again use an approach of the form

$$W_{m,h} = Ce^{\lambda_m y}, \tag{7.155}$$

which after substitution into (7.154) leads to a fourth order polynomial for the unknown exponents λ_m:

$$\lambda_m^4 - 2\frac{m^2\pi^2}{a^2}\lambda_m^2 + \frac{m^4\pi^4}{a^4} = 0. \tag{7.156}$$

The four solutions result as:

$$\lambda_{m,1} = \lambda_{m,2} = -\lambda_{m,3} = -\lambda_{m,4} = \frac{m\pi}{a}. \tag{7.157}$$

The homogeneous solution of the differential equation (7.154) can be written as:

$$W_{m,h} = \left(C_{m,1} + C_{m,2}y\right)\cosh\left(\frac{m\pi y}{a}\right) + \left(C_{m,3} + C_{m,4}y\right)\sinh\left(\frac{m\pi y}{a}\right). \tag{7.158}$$

The total solution for W_m follows again from the addition of the homogeneous and the particular solution: $W_m = W_{m,h} + W_{m,p}$. The total solution for the plate deflection $w_0(x, y)$ is thus:

$$\begin{aligned} w_0 &= \sum_{m=1}^{m=\infty} \sin\left(\frac{m\pi x}{a}\right)\left[\left(C_{m,1} + C_{m,2}y\right)\cosh\left(\frac{m\pi y}{a}\right)\right. \\ &\quad \left. + \left(C_{m,3} + C_{m,4}y\right)\sinh\left(\frac{m\pi y}{a}\right) + W_{m,p}\right]. \end{aligned} \tag{7.159}$$

The constants $C_{m,1}$, $C_{m,2}$, $C_{m,3}$, $C_{m,4}$ are again determined from the boundary conditions at the edges $y = 0$ and $y = b$.

For illustration, consider an isotropic plate under a uniform load $p(x, y) = p_0$. The differential equation (7.154) is then:

$$\frac{\mathrm{d}^4 W_m(y)}{\mathrm{d}y^4} - 2\frac{m^2\pi^2}{a^2}\frac{\mathrm{d}^2 W_m(y)}{\mathrm{d}y^2} + \frac{m^4\pi^4}{a^4}W_m(y) = \frac{1}{D}\frac{4p_0}{m\pi}. \tag{7.160}$$

Let the plate be simply supported at the edges $x = 0$ and $x = a$, and let the plate be rigidly clamped at the edges $y = 0$ and $y = b$. Hence, at $y = 0$ and $y = b$:

$$w_0(y=0) = 0, \quad w_0(y=b) = 0, \quad \left.\frac{\partial w_0}{\partial y}\right|_{y=0} = 0, \quad \left.\frac{\partial w_0}{\partial y}\right|_{y=b} = 0. \tag{7.161}$$

The particular solution for the inhomogeneous differential equation (7.160) results from a right-hand side type approach:

$$W_{m,p} = \frac{4p_0a^4}{Dm^5\pi^5}. \tag{7.162}$$

The solution for the plate deflection w_0 is then obtained as follows:

$$\begin{aligned} w_0 = & \sum_{m=1}^{m=\infty} \sin\left(\frac{m\pi x}{a}\right)\left[\left(C_{m,1} + C_{m,2}y\right)\cosh\left(\frac{m\pi y}{a}\right)\right. \\ & \left. + \left(C_{m,3} + C_{m,4}y\right)\sinh\left(\frac{m\pi y}{a}\right) + \frac{4p_0a^4}{Dm^5\pi^5}\right]. \end{aligned} \tag{7.163}$$

The constants $C_{m,1}$, $C_{m,2}$, $C_{m,3}$, $C_{m,4}$ are determined from the boundary conditions at thc cdges $y = 0$ and $y = b$. The condition $w_0(y = 0) = 0$ results in

$$C_{m,1} = -\frac{4p_0a^4}{Dm^5\pi^5}. \tag{7.164}$$

From the remaining boundary conditions at $y = 0$ and $y = b$, we obtain the following three equations for the three remaining constants $C_{m,2}$, $C_{m,3}$, $C_{m,4}$:

$$\begin{aligned} & C_{m,2}b\cosh\left(\frac{m\pi b}{a}\right) + C_{m,3}\sinh\left(\frac{m\pi b}{a}\right) + C_{m,4}b\sinh\left(\frac{m\pi b}{a}\right) \\ & = \frac{4p_0a^4}{Dm^5\pi^5}\left[\cosh\left(\frac{m\pi b}{a}\right) - 1\right], \\ & C_{m,2} + C_{m,3}\frac{m\pi}{a} = 0, \\ & C_{m,2}\left[\cosh\left(\frac{m\pi b}{a}\right) + \frac{m\pi b}{a}\sinh\left(\frac{m\pi b}{a}\right)\right] + C_{m,3}\frac{m\pi}{a}\cosh\left(\frac{m\pi b}{a}\right) \\ & +C_{m,4}\left[\sinh\left(\frac{m\pi b}{a}\right) + \frac{m\pi b}{a}\cosh\left(\frac{m\pi b}{a}\right)\right] \\ & = \frac{4p_0a^3}{Dm^4\pi^4}\sinh\left(\frac{m\pi b}{a}\right). \end{aligned} \tag{7.165}$$

The solution for this system of equations is:

$$
\begin{aligned}
C_{m,2} &= \frac{4p_0a^2}{Dm^4\pi^4\Delta}\left[a\sinh\left(\frac{m\pi b}{a}\right) - m\pi b\right]\left[\cosh\left(\frac{m\pi b}{a}\right) - 1\right], \\
C_{m,3} &= -\frac{4p_0a^3}{Dm^5\pi^5\Delta}\left[a\sinh\left(\frac{m\pi b}{a}\right) - m\pi b\right]\left[\cosh\left(\frac{m\pi b}{a}\right) - 1\right], \\
C_{m,4} &= \frac{4a^2p_0}{Dm^4\pi^4\Delta}\left[a + m\pi b\sinh\left(\frac{m\pi b}{a}\right) - a\cosh^2\left(\frac{m\pi b}{a}\right)\right],
\end{aligned}
\tag{7.166}
$$

where Δ is the determinant of the coefficient matrix of the system of equations (7.165):

$$
\Delta = \frac{m^2\pi^2b^2}{a^2} + 1 - \cosh^2\left(\frac{m\pi b}{a}\right). \tag{7.167}
$$

7.9 Energetic Consideration of Plate Bending

In many cases, static plate bending problems can be solved very advantageously with the help of energetic methods. For this purpose, the corresponding energy formulations are to be given. The starting point of the considerations is the principle of the minimum of the elastic potential (cf. Chap. 2), where we first want to show how the plate equation and the associated boundary conditions can be derived from this minimum principle. Furthermore, the principle of virtual displacements will be applied, where we also deal with plates with arbitrary boundary shapes.

7.9.1 Principle of the Minimum of the Total Elastic Potential

Consider a rectangular plate with dimensions a and b in the x- and y-directions and thickness h. Let the plate be orthotropic and loaded by an arbitrarily distributed surface load p (see Fig. 7.14, top). Assume that there is no elastic foundation. For this case, we first want to find an expression for the total elastic potential Π of the plate, which results from the addition of the inner potential Π_i and the potential Π_a of the applied loads:

$$
\Pi = \Pi_i + \Pi_a. \tag{7.168}
$$

The inner potential Π_i is in the general three-dimensional case in the presence of all stress components:

$$
\Pi_i = \frac{1}{2}\int_V \left(\sigma_{xx}\varepsilon_{xx} + \sigma_{yy}\varepsilon_{yy} + \sigma_{zz}\varepsilon_{zz} + \tau_{yz}\gamma_{yz} + \tau_{xz}\gamma_{xz} + \tau_{xy}\gamma_{xy}\right) \mathrm{d}V. \tag{7.169}
$$

Since in the framework of Kirchhoff's plate theory we assume shear-rigid plates in the plane stress state with respect to the thickness direction, only the following expression remains at present:

$$\Pi_i = \frac{1}{2}\int_V \left(\sigma_{xx}\varepsilon_{xx} + \sigma_{yy}\varepsilon_{yy} + \tau_{xy}\gamma_{xy}\right) \mathrm{d}V. \tag{7.170}$$

We now use the strains according to (7.8)

$$\varepsilon_{xx} = -z\frac{\partial^2 w_0}{\partial x^2}, \quad \varepsilon_{yy} = -z\frac{\partial^2 w_0}{\partial y^2}, \quad \gamma_{xy} = -2z\frac{\partial^2 w_0}{\partial x \partial y}, \tag{7.171}$$

insert them into (7.170) and split the volume integral into integrals with respect to x, y, z. We obtain:

$$\Pi_i = \frac{1}{2}\int_0^a \int_0^b \int_{-\frac{h}{2}}^{+\frac{h}{2}} \left[-\sigma_{xx} z\frac{\partial^2 w_0}{\partial x^2} - \sigma_{yy} z\frac{\partial^2 w_0}{\partial y^2} - 2\tau_{xy} z\frac{\partial^2 w_0}{\partial x \partial y}\right] \mathrm{d}z\mathrm{d}y\mathrm{d}x. \tag{7.172}$$

With the definitions of moment flows (7.16) according to

$$\begin{pmatrix} M_{xx}^0 \\ M_{yy}^0 \\ M_{xy}^0 \end{pmatrix} = \int_{-\frac{h}{2}}^{\frac{h}{2}} \begin{pmatrix} \sigma_{xx} \\ \sigma_{yy} \\ \tau_{xy} \end{pmatrix} z\mathrm{d}z \tag{7.173}$$

the integration prescribed in (7.172) concerning z can be easily performed, and we obtain in a vector-matrix notation:

$$\Pi_i = \frac{1}{2}\int_0^b \int_0^a \begin{pmatrix} M_{xx}^0 \\ M_{yy}^0 \\ M_{xy}^0 \end{pmatrix}^T \begin{pmatrix} -\frac{\partial^2 w_0}{\partial x^2} \\ -\frac{\partial^2 w_0}{\partial y^2} \\ -2\frac{\partial^2 w_0}{\partial x \partial y} \end{pmatrix} \mathrm{d}x\mathrm{d}y. \tag{7.174}$$

Using the constitutive law (7.18) and the stiffness matrix $\underline{D}$ (7.20) we obtain:

$$\Pi_i = \frac{1}{2}\int_0^b \int_0^a \begin{pmatrix} -\frac{\partial^2 w_0}{\partial x^2} \\ -\frac{\partial^2 w_0}{\partial y^2} \\ -2\frac{\partial^2 w_0}{\partial x \partial y} \end{pmatrix}^T \underline{\underline{D}} \begin{pmatrix} -\frac{\partial^2 w_0}{\partial x^2} \\ -\frac{\partial^2 w_0}{\partial y^2} \\ -2\frac{\partial^2 w_0}{\partial x \partial y} \end{pmatrix} \mathrm{d}x\mathrm{d}y. \tag{7.175}$$

Using the constitutive law (7.18) with plate stiffnesses D_{11}, D_{22}, D_{12} and D_{66} we obtain:

$$\begin{aligned} \Pi_i = \frac{1}{2}\int_0^b \int_0^a \Bigg[& D_{11}\left(\frac{\partial^2 w_0}{\partial x^2}\right)^2 + D_{22}\left(\frac{\partial^2 w_0}{\partial y^2}\right)^2 + 2D_{12}\frac{\partial^2 w_0}{\partial x^2}\frac{\partial^2 w_0}{\partial y^2} \\ & + 4D_{66}\left(\frac{\partial^2 w_0}{\partial x \partial y}\right)^2 \Bigg] \mathrm{d}x\mathrm{d}y. \end{aligned} \tag{7.176}$$

The potential of the surface load p is currently as follows:

$$\Pi_a = -\int_0^b \int_0^a p w_0 \mathrm{d}x \mathrm{d}y. \tag{7.177}$$

The principle of the minimum of the total elastic potential of the plate under consideration requires the disappearance of the first variation $\delta\Pi$:

$$\delta\Pi = \delta\Pi_i + \delta\Pi_a = 0. \tag{7.178}$$

The first variation of the inner potential Π_i is then obtained as:

$$\begin{aligned}
\delta\Pi_i &= \int_0^b \int_0^a \begin{pmatrix} -\frac{\partial^2 w_0}{\partial x^2} \\ -\frac{\partial^2 w_0}{\partial y^2} \\ -2\frac{\partial^2 w_0}{\partial x \partial y} \end{pmatrix}^T \underline{\underline{D}} \begin{pmatrix} -\delta\frac{\partial^2 w_0}{\partial x^2} \\ -\delta\frac{\partial^2 w_0}{\partial y^2} \\ -2\delta\frac{\partial^2 w_0}{\partial x \partial y} \end{pmatrix} \mathrm{d}x\mathrm{d}y \\
&= \int_0^b \int_0^a \begin{pmatrix} M_{xx}^0 \\ M_{yy}^0 \\ M_{xy}^0 \end{pmatrix}^T \begin{pmatrix} -\delta\frac{\partial^2 w_0}{\partial x^2} \\ -\delta\frac{\partial^2 w_0}{\partial y^2} \\ -2\delta\frac{\partial^2 w_0}{\partial x \partial y} \end{pmatrix} \mathrm{d}x\mathrm{d}y \\
&= -\int_0^b \int_0^a \left(M_{xx}^0 \frac{\partial^2 \delta w_0}{\partial x^2} + M_{yy}^0 \frac{\partial^2 \delta w_0}{\partial y^2} + 2M_{xy}^0 \frac{\partial^2 \delta w_0}{\partial x \partial y} \right) \mathrm{d}x\mathrm{d}y.
\end{aligned} \tag{7.179}$$

The first variation of the external potential Π_a is presently:

$$\delta\Pi_a = -\int_0^b \int_0^a p \delta w_0 \mathrm{d}x\mathrm{d}y. \tag{7.180}$$

We perform a stepwise partial integration of the individual terms of the inner potential Π_i, whereas a partial integration of the external potential Π_a is not necessary, no derivatives of the displacements w appear there. In summary, after sorting according to the variations, we obtain:

$$\begin{aligned}
&-\int_0^a \int_0^b \left[\left(\frac{\partial^2 M_{xx}^0}{\partial x^2} + 2\frac{\partial^2 M_{xy}^0}{\partial x \partial y} + \frac{\partial^2 M_{yy}^0}{\partial y^2} + p \right) \delta w_0 \right] \mathrm{d}y\mathrm{d}x \\
&+\int_0^a \left\{ \left[-M_{yy}^0 \delta\left(\frac{\partial w_0}{\partial y}\right) + \left(\frac{\partial M_{yy}^0}{\partial y} + 2\frac{\partial M_{xy}^0}{\partial x} \right) \delta w_0 \right]\Bigg|_0^b \right\} \mathrm{d}x
\end{aligned}$$

$$+\int_0^b \left\{ \left[-M^0_{xx}\delta\left(\frac{\partial w_0}{\partial x}\right) + \left(\frac{\partial M^0_{xx}}{\partial x} + 2\frac{\partial M^0_{xy}}{\partial y}\right)\delta w_0 \right]\Bigg|_0^a \right\} \mathrm{d}y$$

$$-\left(2M^0_{xy}\delta w_0\right)\Big|_0^a\Big|_0^b = 0. \tag{7.181}$$

Since the variations δ of the plate deflection w_0 and its derivatives are completely arbitrary (taking into account the given geometrical boundary conditions of the system) and furthermore are independent of each other, (7.181) can be fulfilled only if $\delta w_0 = 0$, $\delta\frac{\partial w_0}{\partial x} = 0$, $\delta\frac{\partial w_0}{\partial y} = 0$ hold, or if the associated terms become zero. The latter variant then yields the equilibrium conditions of the plate as well as all associated boundary conditions.

The first integral expression in (7.181) yields the condensed plate equilibrium already derived with (7.52):

$$\frac{\partial^2 M^0_{xx}}{\partial x^2} + 2\frac{\partial^2 M^0_{xy}}{\partial x \partial y} + \frac{\partial^2 M^0_{yy}}{\partial y^2} + p = 0. \tag{7.182}$$

Substituting the constitutive law then leads to the already known plate equation.

The terms remaining in (7.181) describe the boundary conditions of the plate under consideration. At the edges at $x = 0$ and $x = a$ for $0 \leq y \leq b$ we obtain:

$$\begin{aligned} M^0_{xx} = 0 \quad \text{or} \quad \frac{\partial w_0}{\partial x} &= \widehat{\frac{\partial w_0}{\partial x}}, \\ \frac{\partial M^0_{xx}}{\partial x} + 2\frac{\partial M^0_{xy}}{\partial y} = 0 \quad \text{or} \quad w_0 &= \widehat{w}_0. \end{aligned}$$

At $y = 0$ and $y = b$ for $0 \leq x \leq a$ we have:

$$\begin{aligned} M^0_{yy} = 0 \quad \text{or} \quad \frac{\partial w_0}{\partial y} &= \widehat{\frac{\partial w_0}{\partial y}}, \\ \frac{\partial M^0_{yy}}{\partial y} + 2\frac{\partial M^0_{xy}}{\partial x} = 0 \quad \text{or} \quad w_0 &= \widehat{w}_0. \end{aligned}$$

At the plate edges at $x = 0, y = 0$, at $x = a, y = 0$, at $x = 0, y = b$ and at $x = a, y = b$ holds:

$$M^0_{xy} = 0 \quad \text{or} \quad \widehat{w}_0 = 0. \tag{7.183}$$

Thus, the energetic consideration of the given plate problem automatically yields those boundary conditions which have already been obtained before in a purely intuitive way.

7.9.2 Principle of Virtual Displacements

In the following we want to show that identical equations and boundary conditions result when the principle of virtual displacements $\delta W_i = \delta W_a$ is used. We consider again the rectangular orthotropic plate of length a and width b under surface load p in z-direction. The virtual inner work δW_i is:

$$\delta W_i = \int_V \left(\sigma_{xx}\delta\varepsilon_{xx} + \sigma_{yy}\delta\varepsilon_{yy} + \tau_{xy}\delta\gamma_{xy}\right) \mathrm{d}V = \int_0^a \int_0^b \int_{-\frac{h}{2}}^{+\frac{h}{2}} \left(\sigma_{xx}\delta\varepsilon_{xx} + \sigma_{yy}\delta\varepsilon_{yy} + \tau_{xy}\delta\gamma_{xy}\right) \mathrm{d}z\mathrm{d}y\mathrm{d}x. \quad (7.184)$$

With

$$\delta\varepsilon_{xx} = -z\frac{\partial^2 \delta w_0}{\partial x^2}, \quad \delta\varepsilon_{yy} = -z\frac{\partial^2 \delta w_0}{\partial y^2}, \quad \delta\gamma_{xy} = -2z\frac{\partial^2 \delta w_0}{\partial x \partial y} \quad (7.185)$$

follows after integration over the plate thickness h:

$$\delta W_i = \int_0^a \int_0^b \left(-M_{xx}^0 \frac{\partial^2 \delta w_0}{\partial x^2} - M_{yy}^0 \frac{\partial^2 \delta w_0}{\partial y^2} - 2M_{xy}^0 \frac{\partial^2 \delta w_0}{\partial x \partial y}\right) \mathrm{d}y\mathrm{d}x. \quad (7.186)$$

The virtual external work δW_a follows presently to:

$$\delta W_a = \int_0^a \int_0^b p\delta w_0 \mathrm{d}y\mathrm{d}x. \quad (7.187)$$

Application of the principle of virtual displacements $\delta W_i = \delta W_a$ and performing partial integrations ultimately yields:

$$\begin{aligned}
&-\int_0^a \int_0^b \left[\left(\frac{\partial^2 M_{xx}^0}{\partial x^2} + 2\frac{\partial^2 M_{xy}^0}{\partial x \partial y} + \frac{\partial^2 M_{yy}^0}{\partial y^2} + p\right)\delta w_0\right] \mathrm{d}y\mathrm{d}x \\
&+\int_0^a \left\{\left[-M_{yy}^0 \delta\left(\frac{\partial w_0}{\partial y}\right) + \left(\frac{\partial M_{yy}^0}{\partial y} + 2\frac{\partial M_{xy}^0}{\partial x}\right)\delta w_0\right]\Bigg|_0^b\right\} \mathrm{d}x \\
&+\int_0^b \left\{\left[-M_{xx}^0 \delta\left(\frac{\partial w_0}{\partial x}\right) + \left(\frac{\partial M_{xx}^0}{\partial x} + 2\frac{\partial M_{xy}^0}{\partial y}\right)\delta w_0\right]\Bigg|_0^a\right\} \mathrm{d}y \\
&-\left(2M_{xy}^0 \delta w_0\right)\Big|_0^a\Big|_0^b = 0.
\end{aligned} \quad (7.188)$$

One can easily convince oneself that this expression agrees with (7.181). A renewed discussion is therefore omitted at this point.

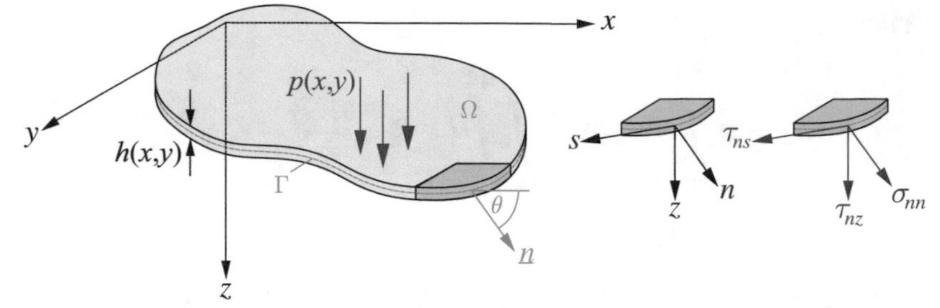

Fig. 7.30 Plate with arbitrary boundary

7.9.3 Plate with Arbitrary Boundary

We now extend the considerations to a plate with arbitrary boundary, as shown in Fig. 7.30. The plate has the area Ω and is bounded by the edge Γ. The edge is characterized by the normal vector $\underline{n}$, which is oriented under the angle θ to the global axis x. The circumferential coordinate is denoted by s. The normal vector $\underline{n} = n_x \underline{e}_x + n_y \underline{e}_y$ is a unit vector, so $|\underline{n}| = 1$ holds. The quantities n_x and n_y are the components of the vector $\underline{n}$, and thus $n_x = \cos\theta$ and $n_y = \sin\theta$ holds. At the edge of the plate, the stress components σ_{nn}, τ_{ns} and τ_{nz} are defined, and let the edge displacement quantities be u_n and u_s. Let the boundary Γ be divided into two parts Γ_u and Γ_σ, on which boundary displacements and boundary stresses are given, respectively: $\Gamma = \Gamma_u + \Gamma_\sigma$. On the boundary part Γ_σ the edge stresses σ_{nn}, τ_{ns}, τ_{nz} are given by the values $\hat{\sigma}_{nn}$, $\hat{\tau}_{ns}$ and $\hat{\tau}_{nz}$. Let the plate be loaded by an arbitrary surface load p in z-direction.

The virtual internal work δW_i is obtained for the present case:

$$\delta W_i = \int_\Omega \left(-M_{xx}^0 \frac{\partial^2 \delta w_0}{\partial x^2} - M_{yy}^0 \frac{\partial^2 \delta w_0}{\partial y^2} - 2M_{xy}^0 \frac{\partial^2 \delta w_0}{\partial x \partial y} \right) \mathrm{d}\Omega. \tag{7.189}$$

For the virtual external work δW_a we obtain:

$$\delta W_a = \int_\Omega p \delta w_0 \mathrm{d}\Omega + \int_{\Gamma_\sigma} \int_{-\frac{h}{2}}^{+\frac{h}{2}} \left(\hat{\sigma}_{nn} \delta u_n + \hat{\tau}_{ns} \delta u_s + \hat{\tau}_{nz} \delta w_0 \right) \mathrm{d}z \mathrm{d}s. \tag{7.190}$$

With

$$\delta u_n = -z \frac{\partial \delta w_0}{\partial n}, \quad \delta u_s = -z \frac{\partial \delta w_0}{\partial s} \tag{7.191}$$

we obtain

$$\delta W_a = \int\limits_{\Omega} p\delta w_0 \mathrm{d}\Omega + \int\limits_{\Gamma_\sigma}\int\limits_{-\frac{h}{2}}^{+\frac{h}{2}} \left(-\hat{\sigma}_{nn} z\frac{\partial \delta w_0}{\partial n} - \hat{\tau}_{ns} z\frac{\partial \delta w_0}{\partial s} + \hat{\tau}_{nz}\delta w_0\right) \mathrm{d}z\mathrm{d}s. \tag{7.192}$$

We introduce at this point the following definitions for the boundary force and moment quantities:

$$\hat{M}^0_{nn} = \int\limits_{-\frac{h}{2}}^{+\frac{h}{2}} \hat{\sigma}_{nn} z\mathrm{d}z, \quad \hat{M}^0_{ns} = \int\limits_{-\frac{h}{2}}^{+\frac{h}{2}} \hat{\tau}_{ns} z\mathrm{d}z, \quad \hat{Q}_n = \int\limits_{-\frac{h}{2}}^{+\frac{h}{2}} \hat{\tau}_{nz}\mathrm{d}z. \tag{7.193}$$

Thus the virtual external work δW_a passes into the following form:

$$\delta W_a = \int\limits_{\Omega} p\delta w_0 \mathrm{d}\Omega + \int\limits_{\Gamma_\sigma} \left(-\hat{M}^0_{nn}\frac{\partial \delta w_0}{\partial n} - \hat{M}^0_{ns}\frac{\partial \delta w_0}{\partial s} + \hat{Q}_n\delta w_0\right) \mathrm{d}s. \tag{7.194}$$

The principle of virtual displacements $\delta W_i = \delta W_a$ then gives the following expression:

$$\int\limits_{\Omega} \left(-M^0_{xx}\frac{\partial^2 \delta w_0}{\partial x^2} - M^0_{yy}\frac{\partial^2 \delta w_0}{\partial y^2} - 2M^0_{xy}\frac{\partial^2 \delta w_0}{\partial x\partial y}\right)\mathrm{d}\Omega$$
$$-\int\limits_{\Omega} p\delta w_0 \mathrm{d}\Omega - \int\limits_{\Gamma_\sigma} \left(-\hat{M}^0_{nn}\frac{\partial \delta w_0}{\partial n} - \hat{M}^0_{ns}\frac{\partial \delta w_0}{\partial s} + \hat{Q}_n\delta w_0\right) \mathrm{d}s = 0. \tag{7.195}$$

At this point, those terms which contain partial derivatives of the virtual displacements are partially integrated. It results using the Gaussian integral theorem:

$$-\int\limits_{\Omega} M^0_{xx}\frac{\partial^2 \delta w_0}{\partial x^2}\mathrm{d}\Omega = -\oint_{\Gamma} M^0_{xx}\frac{\partial \delta w_0}{\partial x} n_x \mathrm{d}s + \oint_{\Gamma} \frac{\partial M^0_{xx}}{\partial x}\delta w_0 n_x \mathrm{d}s - \int\limits_{\Omega} \frac{\partial^2 M^0_{xx}}{\partial x^2}\delta w_0 \mathrm{d}\Omega,$$

$$-\int\limits_{\Omega} M^0_{yy}\frac{\partial^2 \delta w_0}{\partial y^2}\mathrm{d}\Omega = -\oint_{\Gamma} M^0_{yy}\frac{\partial \delta w_0}{\partial y} n_y \mathrm{d}s + \oint_{\Gamma} \frac{\partial M^0_{yy}}{\partial y}\delta w_0 n_y \mathrm{d}s - \int\limits_{\Omega} \frac{\partial^2 M^0_{yy}}{\partial y^2}\delta w_0 \mathrm{d}\Omega,$$

$$
\begin{aligned}
-\int_{\Omega} M^0_{xy}\frac{\partial^2\delta w_0}{\partial x\partial y}\mathrm{d}\Omega &= -\oint_{\Gamma} M^0_{xy} n_x \frac{\partial\delta w_0}{\partial y}\mathrm{d}s \\
&\quad + \oint_{\Gamma}\frac{\partial M^0_{xy}}{\partial x} n_y \delta w_0 \mathrm{d}s - \int_{\Omega}\frac{\partial^2 M^0_{xy}}{\partial x\partial y}\delta w_0 d\Omega, \\
-\int_{\Omega} M^0_{xy}\frac{\partial^2\delta w_0}{\partial x\partial y}\mathrm{d}\Omega &= -\oint_{\Gamma} M^0_{xy} n_y \frac{\partial\delta w_0}{\partial x}\mathrm{d}s \\
&\quad + \oint_{\Gamma}\frac{\partial M^0_{xy}}{\partial y} n_x \delta w_0 \mathrm{d}s - \int_{\Omega}\frac{\partial^2 M^0_{xy}}{\partial x\partial y}\delta w_0 d\Omega. \qquad (7.196)
\end{aligned}
$$

This ultimately leads to the following expression:

$$
\begin{aligned}
&-\int_{\Omega}\left(\frac{\partial^2 M^0_{xx}}{\partial x^2}+\frac{\partial^2 M^0_{yy}}{\partial y^2}+2\frac{\partial^2 M^0_{xy}}{\partial x\partial y}+p\right)\delta w_0 \mathrm{d}\Omega \\
&+\int_{\Gamma_\sigma}\Bigg[\left(\frac{\partial M^0_{xx}}{\partial x}n_x+\frac{\partial M^0_{yy}}{\partial y}n_y+\frac{\partial M^0_{xy}}{\partial x}n_y+\frac{\partial M^0_{xy}}{\partial y}n_x\right)\delta w_0 \\
&-\left(M^0_{xx}n_x+M^0_{xy}n_y\right)\frac{\partial\delta w_0}{\partial x}-\left(M^0_{yy}n_y+M^0_{xy}n_x\right)\frac{\partial\delta w_0}{\partial y}\Bigg]\mathrm{d}s \\
&-\int_{\Gamma_\sigma}\left(-\hat{M}^0_{nn}\frac{\partial\delta w_0}{\partial n}-\hat{M}^0_{ns}\frac{\partial\delta w_0}{\partial s}+\hat{Q}_n\delta w_0\right)\mathrm{d}s = 0. \qquad (7.197)
\end{aligned}
$$

From the integral term of the first line in (7.197) the condensed plate equilibrium (7.52) can be immediately deduced:

$$
\frac{\partial^2 M^0_{xx}}{\partial x^2}+\frac{\partial^2 M^0_{yy}}{\partial y^2}+2\frac{\partial^2 M^0_{xy}}{\partial x\partial y}+p=0. \qquad (7.198)
$$

For further discussion of the boundary expressions in (7.197), the following transformation rules are used to relate all boundary terms uniformly to the curved boundary with the boundary normal vector $\underline{n}$. The following relation holds between the coordinate directions x, y, z and n, s, z:

$$
\begin{aligned}
\begin{pmatrix} x \\ y \\ z \end{pmatrix} &= \begin{bmatrix} \cos\theta & -\sin\theta & 0 \\ \sin\theta & \cos\theta & 0 \\ 0 & 0 & 1 \end{bmatrix}\begin{pmatrix} n \\ s \\ z \end{pmatrix} \\
&= \begin{bmatrix} n_x & -n_y & 0 \\ n_y & n_x & 0 \\ 0 & 0 & 1 \end{bmatrix}\begin{pmatrix} n \\ s \\ z \end{pmatrix}. \qquad (7.199)
\end{aligned}
$$

The angular rotations can be transformed quite analogously:

$$\begin{pmatrix} \frac{\partial w_0}{\partial x} \\ \frac{\partial w_0}{\partial y} \end{pmatrix} = \begin{bmatrix} n_x & -n_y \\ n_y & n_x \end{bmatrix} \begin{pmatrix} \frac{\partial w_0}{\partial n} \\ \frac{\partial w_0}{\partial s} \end{pmatrix}. \tag{7.200}$$

Similarly, a transformation of the following form can be used for the stresses (see Chap. 1):

$$\begin{pmatrix} \sigma_{nn} \\ \tau_{ns} \end{pmatrix} = \begin{bmatrix} n_x^2 & n_y^2 & 2n_x n_y \\ -n_x n_y & n_x n_y & n_x^2 - n_y^2 \end{bmatrix} \begin{pmatrix} \sigma_{xx} \\ \sigma_{yy} \\ \tau_{xy} \end{pmatrix}. \tag{7.201}$$

This results in the following transformation relations for the moment flows of the plate:

$$\begin{pmatrix} M_{nn}^0 \\ M_{ns}^0 \end{pmatrix} = \begin{bmatrix} n_x^2 & n_y^2 & 2n_x n_y \\ -n_x n_y & n_x n_y & n_x^2 - n_y^2 \end{bmatrix} \begin{pmatrix} M_{xx}^0 \\ M_{yy}^0 \\ M_{xy}^0 \end{pmatrix}. \tag{7.202}$$

Then:

$$\begin{aligned} &\left(M_{xx}^0 n_x + M_{xy}^0 n_y\right) \frac{\partial \delta w_0}{\partial x} - \left(M_{xy}^0 n_x + M_{yy}^0 n_y\right) \frac{\partial \delta w_0}{\partial y} \\ &= M_{nn}^0 \frac{\partial \delta w_0}{\partial n} + M_{ns}^0 \frac{\partial \delta w_0}{\partial s}. \end{aligned} \tag{7.203}$$

The boundary term in (7.197) can then be written as:

$$\begin{aligned} \int_{\Gamma_\sigma} \Bigg[&\left(\frac{\partial M_{xx}^0}{\partial x} n_x + \frac{\partial M_{xy}^0}{\partial y} n_x + \frac{\partial M_{yy}^0}{\partial y} n_y + \frac{\partial M_{xy}^0}{\partial x} n_y - \hat{Q}_n \right) \delta w_0 \\ &- \left(M_{nn}^0 - \hat{M}_{nn}^0\right) \frac{\partial \delta w_0}{\partial n} - \left(M_{ns}^0 - \hat{M}_{ns}^0\right) \frac{\partial \delta w_0}{\partial s} \Bigg] \mathrm{d}s = 0. \end{aligned} \tag{7.204}$$

The following term is now partially integrated:

$$- \int_{\Gamma_\sigma} M_{ns}^0 \frac{\partial \delta w_0}{\partial s} \mathrm{d}s = -M_{ns}^0 \delta w_0 \big|_\Gamma + \int_{\Gamma_\sigma} \frac{\partial M_{ns}^0}{\partial s} \delta w_0 \mathrm{d}s. \tag{7.205}$$

The term $-M_{ns}^0 \delta w_0 \big|_\Gamma$ becomes zero if there is a closed boundary curve. The remaining term is added to the shear force flow. The equivalent shear force flow is then:

$$V_n = \frac{\partial M_{xx}^0}{\partial x} n_x + \frac{\partial M_{xy}^0}{\partial y} n_x + \frac{\partial M_{yy}^0}{\partial y} n_y + \frac{\partial M_{xy}^0}{\partial x} n_y + \frac{\partial M_{ns}^0}{\partial s}. \tag{7.206}$$

The boundary integral then transforms into the following form:

$$\int_{\Gamma_\sigma} \left[\left(V_n - \hat{V}_n \right) \delta w_0 - \left(M^0_{nn} - \hat{M}^0_{nn} \right) \frac{\partial \delta w_0}{\partial n} \right] \mathrm{d}s = 0, \tag{7.207}$$

wherein:

$$\hat{V}_n = \hat{Q}_n + \frac{\partial \hat{M}^0_{ns}}{\partial s}. \tag{7.208}$$

7.10 Plate on Elastic Foundation

In the following, we will consider the case where the plate under consideration is not only supported at its edges, but is continuously elastically bedded over its entire surface. The elastic foundation is such that the pressure p_b exerted by the foundation on the plate is proportional to the plate deflection w_0. Accordingly, the following applies:

$$p_b = k w_0. \tag{7.209}$$

This elastic foundation can be thought of as a continuous planar arrangement of displacement springs (Fig. 7.31). Here, the quantity k is the elastic stiffness of the foundation or the so-called bedding number. In this case, one also speaks of the so-called Winkler[4] foundation. Combining the load p as well as the bedding pressure p_b into a total load $p - p_b$, the equation of the plate on elastic foundation follows as:

$$D_{11} \frac{\partial^4 w_0}{\partial x^4} + 2\left(D_{12} + 2D_{66}\right) \frac{\partial^4 w_0}{\partial x^2 \partial y^2} + D_{22} \frac{\partial^4 w_0}{\partial y^4} = p - p_b, \tag{7.210}$$

or

$$D_{11} \frac{\partial^4 w_0}{\partial x^4} + 2\left(D_{12} + 2D_{66}\right) \frac{\partial^4 w_0}{\partial x^2 \partial y^2} + D_{22} \frac{\partial^4 w_0}{\partial y^4} + k w_0 = p. \tag{7.211}$$

As an application we consider again the Navier plate, which is loaded by the arbitrarily distributed surface load p and at the same time is subject to an elastic foundation with the bedding number k. Both the plate deflection w_0 and the plate load p are developed in the form of Fourier series as shown before:

$$w_0 = \sum_{m=1}^{m=\infty} \sum_{n=1}^{n=\infty} W_{mn} \sin\left(\frac{m\pi x}{a}\right) \sin\left(\frac{n\pi y}{b}\right),$$

[4] Emil Ernst Oskar Winkler, 1835–1888, German civil engineer.

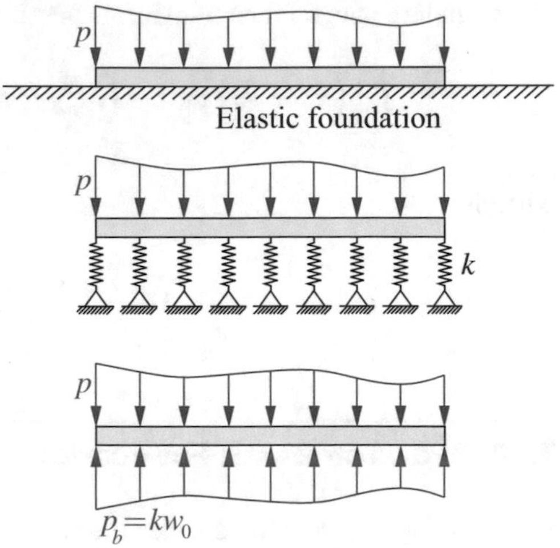

Fig. 7.31 Plate on elastic foundation

$$p = \sum_{m=1}^{m=\infty} \sum_{n=1}^{n=\infty} P_{mn} \sin\left(\frac{m\pi x}{a}\right) \sin\left(\frac{n\pi y}{b}\right). \tag{7.212}$$

Inserting into the plate equation (7.211), after a short calculation, gives the following expression for the Fourier coefficients W_{mn}:

$$W_{mn} = \frac{P_{mn}}{D_{11}\left(\frac{m\pi}{a}\right)^4 + 2\left(D_{12} + 2D_{66}\right)\left(\frac{m\pi}{a}\right)^2\left(\frac{n\pi}{b}\right)^2 + D_{22}\left(\frac{n\pi}{b}\right)^4 + k}. \tag{7.213}$$

Obviously, the elastic foundation makes the Fourier coefficients W_{mn} smaller than in the unbedded case, so that the larger the bedding coefficient k, the smaller the overall deflection of the plate becomes.

The analysis can be applied to a plate strip on an elastic foundation (see also the elaborations in Sect. 7.6). The differential equation describing the problem of the elastically bedded plate strip is:

$$D_{11}\frac{\mathrm{d}^4 w_0}{\mathrm{d}x^4} + kw_0 = p. \tag{7.214}$$

It has the following solution:

$$\begin{aligned} w_0 = & \left[C_1 \sinh\left(\frac{2\beta x}{a}\right) + C_2 \cosh\left(\frac{2\beta x}{a}\right)\right] \sin\left(\frac{2\beta x}{a}\right) \\ & + \left[C_3 \sinh\left(\frac{2\beta x}{a}\right) + C_4 \cosh\left(\frac{2\beta x}{a}\right)\right] \cos\left(\frac{2\beta x}{a}\right) + \frac{p}{k}, \end{aligned} \tag{7.215}$$

wherein $\beta^4 = \frac{ka^4}{64D_{11}}$. The constants C_1, ..., C_4 are adjusted to given boundary conditions. For illustration, consider the plate strip shown in Fig. 7.32, which is simply

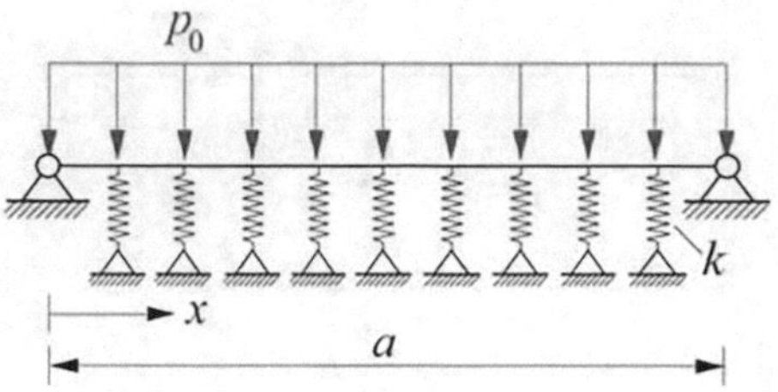

Fig. 7.32 Simply supported plate strip on elastic foundation

supported on both sides (length a) and subjected to a uniformly distributed load $p = p_0$.

To evaluate the boundary conditions to be applied here, the second derivative of the bending line $w(x)$ must be determined. This results in

$$\begin{aligned}\frac{\mathrm{d}w_0}{\mathrm{d}x} &= (C_1 - C_4)\,\lambda\cosh(\lambda x)\sin(\lambda x) + (C_2 + C_3)\,\lambda\cosh(\lambda x)\cos(\lambda x)\\ &\quad + (C_2 - C_3)\,\lambda\sinh(\lambda x)\sin(\lambda x) + (C_1 + C_4)\,\lambda\sinh(\lambda x)\cos(\lambda x),\\ \frac{\mathrm{d}^2w_0}{\mathrm{d}x^2} &= 2C_1\lambda^2\cosh(\lambda x)\cos(\lambda x) + 2C_2\lambda^2\sinh(\lambda x)\cos(\lambda x)\\ &\quad -2C_3\lambda^2\cosh(\lambda x)\sin(\lambda x) - 2C_4\lambda^2\sinh(\lambda x)\sin(\lambda x), \qquad (7.216)\end{aligned}$$

wherein

$$\lambda = \frac{2\beta}{a}. \qquad (7.217)$$

Evaluation of the two boundary conditions

$$w_0(w = 0) = 0, \quad \left.\frac{\mathrm{d}^2w_0}{\mathrm{d}x^2}\right|_{x=0} = 0 \qquad (7.218)$$

gives $C_1 = 0$ and $C_4 = -\frac{p_0}{k}$. The two remaining boundary conditions

$$w_0(w = a) = 0, \quad \left.\frac{\mathrm{d}^2w_0}{\mathrm{d}x^2}\right|_{x=a} = 0 \qquad (7.219)$$

on the other hand, taking into account the two constants already determined, give the following expressions:

$$\begin{aligned}C_2\cosh(\lambda a)\sin(\lambda a) + C_3\sinh(\lambda a)\cos(\lambda a) &= \frac{p_0}{k}\left[\cosh(\lambda a)\cos(\lambda a) - 1\right],\\ C_2\sinh(\lambda a)\cos(\lambda a) - C_3\cosh(\lambda a)\sin(\lambda a) &= -\frac{p_0}{k}\sinh(\lambda a)\sin(\lambda a). \qquad (7.220)\end{aligned}$$

From this, the two remaining constants C_2 and C_3 can be determined. A presentation of the result is omitted here.

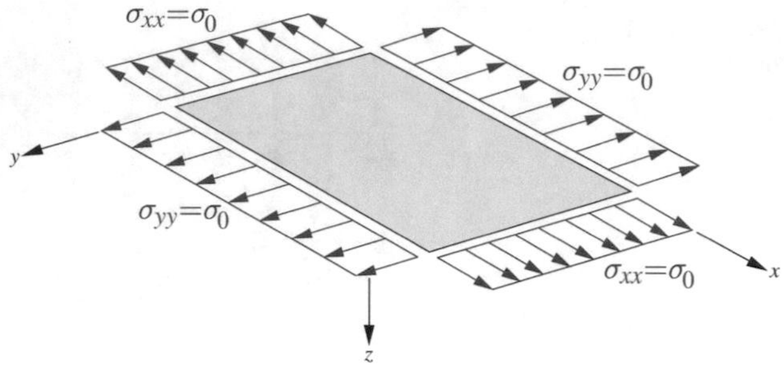

Fig. 7.33 Membrane

7.11 The Membrane

A special case of a plate-like structure is the so-called membrane (Fig. 7.33, see also Szilard 2004; Gross et al. 2014). A membrane is a prestressed thin structure (constant thickness h), which has no bending stiffness and therefore cannot transmit any moment loads. In order for the membrane to be load-bearing at all, it must be supported at its edges in such a way that force flows can form in its plane due to prestressing. We assume here that this prestress is in the form of the plane stress component σ_0, which is assumed to be present at each edge point.

To derive the membrane differential equation, we consider a deflected section element of edge lengths $\mathrm{d}x$, $\mathrm{d}y$ under a surface load p, restricting ourselves here to the case of small deformations. We can assume for small deformations that the plane force flows at each point of the membrane are identical with the value $\sigma_{xx}h = \sigma_{yy}h = \sigma_0 h$. A plane shear stress is not present in this case. The corresponding free body diagram is shown in Fig. 7.34. Equilibrium of forces in z-direction then yields:

$$
\begin{aligned}
& \left(\sigma_0 + \frac{\partial \sigma_0}{\partial x}\mathrm{d}x\right)\left(\frac{\partial w_0}{\partial x} + \frac{\partial^2 w_0}{\partial x^2}\mathrm{d}x\right)\mathrm{d}yh - \sigma_0 \frac{\partial w_0}{\partial x}\mathrm{d}yh \\
+ & \left(\sigma_0 + \frac{\partial \sigma_0}{\partial y}\mathrm{d}y\right)\left(\frac{\partial w_0}{\partial y} + \frac{\partial^2 w_0}{\partial y^2}\mathrm{d}x\right)\mathrm{d}xh - \sigma_0 \frac{\partial w_0}{\partial y}\mathrm{d}xh \\
+ & \; p\mathrm{d}x\mathrm{d}y = 0.
\end{aligned}
\tag{7.221}
$$

Neglecting terms that are small of higher order eventually yields the following inhomogeneous Poisson's equation[5]:

$$
\frac{\partial^2 w_0}{\partial x^2} + \frac{\partial^2 w_0}{\partial y^2} = -\frac{p}{\sigma_0 h}. \tag{7.222}
$$

[5] Siméon Denis Poisson, 1781–1840, French physicist and mathematician.

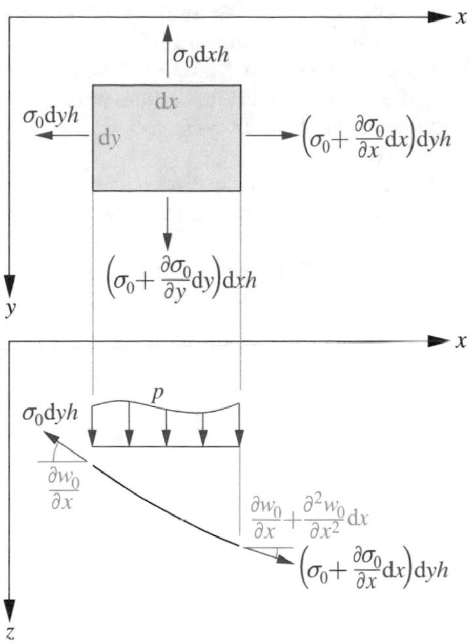

Fig. 7.34 Free body diagram

From the solutions of this differential equation, the ones that satisfy the given boundary conditions have to be selected.

References

Altenbach, H., Altenbach, J., Naumenko, K.: Ebene Flächentragwerke, 2nd edn. Springer, Berlin (2016)

Ambartsumyan, S.A.: Theory of Anisotropic Plates. Technomic Publishing, Stamford (1970)

Becker, W., Gross, D.: Mechanik elastischer Körper und Strukturen. Springer, Berlin (2002)

Czerwenka, G., Schnell, W.: Einführung in die Rechenmethoden des Leichtbaus, vol. 2. Bibliographisches Institut, Mannheim (1970)

Eschenauer, H., Schnell, W.: Elastizitätstheorie, 3rd edn. BI Wissenschaftsverlag, Mannheim (1993)

Eschenauer, H., Olhoff, N., Schnell, W.: Applied Structural Mechanics. Springer, Berlin (1997)

Girkmann, K.: Flächentragwerke, 6th edn. Springer, Wien (1974)

Göldner, H., Altenbach, J., Eschke, K., Garz, K.F., Sähn, S.: Lehrbuch Höhere Festigkeitslehre Band 1. Physik, Weinheim (1979)

Gross, D., Hauger, W., Wriggers, P.: Technische Mechanik 4, 9th edn. Springer, Berlin (2014)

Hake, E., Meskouris, K.: Statik der Flächentragwerke, 2nd edn. Springer, Berlin (2007)

Lekhnitskii, S.G.: Anisotropic Plates. Gordon and Breach, New York (1968)

Mansfield, E.H.: The Bending and Stretching of Plates. Cambridge University Press, Cambridge (1989)

Marguerre, K., Woernle, H.T.: Elastische Platten. Bibliographisches Institut, Mannheim (1975)

Mittelstedt, C., Becker, W.: Strukturmechanik ebener Laminate. Studienbereich Mechanik Technische Universität, Darmstadt (2016)
Reddy, J.N.: Theory and Analysis of Elastic Plates and Shells. CRC Press, Boca Raton (2006)
Szilard, R.: Theories and Applications of Plate Analysis. Wiley, Hoboken (2004)
Timoshenko, S., Woinowsky-Krieger, S.: Theory of Plates and Shells, 2nd edn. McGraw-Hill, New York (1964)
Turner, C.E.: Introduction to Plate and Shell Theory. Longmans Green and Co. Ltd., London (1965)
Ugural, A.C.: Stresses in Plates and Shells. McGraw Hill, New York (1981)
Vinson, J.R.: The Behaviour of Plates and Shells. Wiley, New York (1974)
Wang, C.T.: Applied Elasticity. McGraw-Hill, New York (1953)
Wiedemann, J.: Leichtbau, Elemente und Konstruktion, 3rd edn. Springer, Berlin (2006)

Approximation Methods for the Kirchhoff Plate

8

This chapter is devoted to the application of classical approximation methods to plate bending problems. Specifically, we discuss both the Ritz method, which is a very universal method of approximation, and the Galerkin method. Further elaborations on the methods mentioned here can be found, for example, in Mittelstedt (2021), Reddy (2017), and Szilard (2004).

8.1 The Ritz Method

The Ritz method is suitable for the approximate solution of static bending problems of elastic plates, as we will show below for the case of linear elasticity. It is based on the principle of the minimum of the total elastic potential Π:

$$\Pi = \Pi_i + \Pi_a = \text{Min.} \tag{8.1}$$

We will now derive the Ritz method for rectangular orthotropic linear-elastic plates of dimensions a and b under arbitrary surface load $p(x, y)$ and then illustrate it with some examples.

The inner potential is given with the expression

$$\Pi_i = \frac{1}{2}\int_0^b\int_0^a \left[D_{11}\left(\frac{\partial^2 w_0}{\partial x^2}\right)^2 + D_{22}\left(\frac{\partial^2 w_0}{\partial y^2}\right)^2 + 2D_{12}\frac{\partial^2 w_0}{\partial x^2}\frac{\partial^2 w_0}{\partial y^2} \right.$$
$$\left. + 4D_{66}\left(\frac{\partial^2 w_0}{\partial x \partial y}\right)^2 \right] \mathrm{d}x\mathrm{d}y, \tag{8.2}$$

C. Mittelstedt, *Theory of Plates and Shells*,
https://doi.org/10.1007/978-3-662-66805-4_8

and the external potential as a consequence of the surface load $p(x, y)$ is:

$$\Pi_a = -\int_0^b \int_0^a p w_0 \mathrm{d}x\mathrm{d}y. \tag{8.3}$$

At this point, we do not specify any boundary conditions, but we want to derive the Ritz formulation in a general way.

We use a displacement approach of the form

$$w_0(x, y) = \sum_{m=1}^{m=\infty} \sum_{n=1}^{n=\infty} W_{mn} w_{1m}(x) w_{2n}(y). \tag{8.4}$$

This approach consists of the functions $w_{1m}(x)$ and $w_{2n}(y)$, which have to fulfill the underlying kinematic boundary conditions identically. The dynamic boundary conditions do not have to be satisfied explicitly, but are better and better satisfied in an average sense with increasing degrees of series expansion by using the energy principle. Usually, one will always strive to keep the computational cost as low as possible, so that the series expansion in (8.4) is terminated at a certain degree $m = M$ and $n = N$, thus:

$$w_0(x, y) = \sum_{m=1}^{m=M} \sum_{n=1}^{n=N} W_{mn} w_{1m}(x) w_{2n}(y). \tag{8.5}$$

Substitution of (8.5) into the inner potential (8.2) yields:

$$\begin{aligned}
\Pi_i = \sum_{m=1}^{m=M} \sum_{n=1}^{n=N} \sum_{p=1}^{p=M} \sum_{q=1}^{q=N} W_{mn} W_{pq} \\
\times \Bigg[\frac{1}{2} D_{11} \int_0^a \frac{\mathrm{d}^2 w_{1m}}{\mathrm{d}x^2} \frac{\mathrm{d}^2 w_{1p}}{\mathrm{d}x^2} \mathrm{d}x \int_0^b w_{2n} w_{2q} \mathrm{d}y \\
+ \frac{1}{2} D_{22} \int_0^a w_{1m} w_{1p} \mathrm{d}x \int_0^b \frac{\mathrm{d}^2 w_{2n}}{\mathrm{d}y^2} \frac{\mathrm{d}^2 w_{2q}}{\mathrm{d}y^2} \mathrm{d}y \\
+ D_{12} \int_0^a \frac{\mathrm{d}^2 w_{1m}}{\mathrm{d}x^2} w_{1p} \mathrm{d}x \int_0^b w_{2n} \frac{\mathrm{d}^2 w_{2q}}{\mathrm{d}y^2} \mathrm{d}y \\
+ 2 D_{66} \int_0^a \frac{\mathrm{d}w_{1m}}{\mathrm{d}x} \frac{\mathrm{d}w_{1p}}{\mathrm{d}x} \mathrm{d}x \int_0^b \frac{\mathrm{d}w_{2n}}{\mathrm{d}y} \frac{\mathrm{d}w_{2q}}{\mathrm{d}y} \mathrm{d}y \Bigg].
\end{aligned} \tag{8.6}$$

We introduce the following abbreviations at this point:

$$\lambda_{x,ij}^{mp} = \int_0^a \frac{\mathrm{d}^i w_{1m}}{\mathrm{d}x^i} \frac{\mathrm{d}^j w_{1p}}{\mathrm{d}x^j} \mathrm{d}x, \quad \lambda_{y,ij}^{nq} = \int_0^b \frac{\mathrm{d}^i w_{2n}}{\mathrm{d}y^i} \frac{\mathrm{d}^j w_{2q}}{\mathrm{d}y^j} \mathrm{d}y. \tag{8.7}$$

With this we can write for (8.6):

$$\Pi_i = \sum_{m=1}^{m=M} \sum_{n=1}^{n=N} \sum_{p=1}^{p=M} \sum_{q=1}^{q=N} W_{mn} W_{pq} \left[\frac{1}{2} D_{11} \lambda_{x,22}^{mp} \lambda_{y,00}^{nq} + \frac{1}{2} D_{22} \lambda_{x,00}^{mp} \lambda_{y,22}^{nq} \right.$$
$$\left. + D_{12} \lambda_{x,20}^{mp} \lambda_{y,02}^{nq} + 2 D_{66} \lambda_{x,11}^{mp} \lambda_{y,11}^{nq} \right]. \tag{8.8}$$

Using the displacement approach (8.5), we also obtain the following expression for the potential Π_a (8.3) of the given surface load $p(x, y)$:

$$\Pi_a = - \int_0^b \int_0^a p \sum_{m=1}^{m=M} \sum_{n=1}^{n=N} W_{mn} w_{1m} w_{2n} \mathrm{d}x \mathrm{d}y. \tag{8.9}$$

With the abbreviation

$$\Omega_{mn} = \int_0^b \int_0^a p w_{1m} w_{2n} \mathrm{d}x \mathrm{d}y \tag{8.10}$$

we obtain:

$$\Pi_a = - \sum_{m=1}^{m=M} \sum_{n=1}^{n=N} W_{mn} \Omega_{mn}. \tag{8.11}$$

In the special case of a constant load $p(x, y) = \text{const.} = p_0$ one obtains:

$$\Pi_a = -p_0 \sum_{m=1}^{m=M} \sum_{n=1}^{n=N} W_{mn} \int_0^a w_{1m} \mathrm{d}x \int_0^b w_{2n} \mathrm{d}y. \tag{8.12}$$

With the abbreviations

$$\omega_x^m = \int_0^a w_{1m} \mathrm{d}x, \quad \omega_y^n = \int_0^b w_{2n} \mathrm{d}y \tag{8.13}$$

this leads to:

$$\Pi_a = -p_0 \sum_{m=1}^{m=M} \sum_{n=1}^{n=N} W_{mn} \omega_x^m \omega_y^n. \tag{8.14}$$

The total elastic potential Π assumes a stationary value (here a minimum) exactly when the first variation of the potential Π vanishes:

$$\delta\Pi = 0. \tag{8.15}$$

By using the approach (8.5), however, the functional form of the plate deflection w_0 is already completely fixed, so that the only quantities of the potential which are still amenable to variation are the coefficients W_{mn}. Thus $\Pi = \Pi\,(W_{mn})$ holds. The requirement (8.15) thus passes into the Ritz equations, which are here as follows:

$$\frac{\partial\Pi}{\partial W_{11}} = 0, \quad \frac{\partial\Pi}{\partial W_{12}} = 0, \quad \frac{\partial\Pi}{\partial W_{13}} = 0, \ldots \tag{8.16}$$

or

$$\frac{\partial\Pi}{\partial W_{mn}} = 0. \tag{8.17}$$

If we implement the requirement (8.17) for all coefficients W_{mn}, this leads to a linear system of equations of the form

$$\underline{\underline{K}}\,\underline{W} = \underline{P} \tag{8.18}$$

with the dimension $M \times N$, from which the constants W_{mn} can be determined. The stiffness matrix $\underline{\underline{K}}$ occurring here takes the following form:

$$\underline{\underline{K}} = \begin{bmatrix} \underline{\underline{K}}^{11} & \underline{\underline{K}}^{12} & \underline{\underline{K}}^{13} & \ldots & \underline{\underline{K}}^{1M} \\ \underline{\underline{K}}^{21} & \underline{\underline{K}}^{22} & \underline{\underline{K}}^{23} & \ldots & \underline{\underline{K}}^{2M} \\ \underline{\underline{K}}^{31} & \underline{\underline{K}}^{32} & \underline{\underline{K}}^{33} & \ldots & \underline{\underline{K}}^{3M} \\ \vdots & \vdots & \vdots & \ddots & \vdots \\ \underline{\underline{K}}^{M1} & \underline{\underline{K}}^{M2} & \underline{\underline{K}}^{M3} & \ldots & \underline{\underline{K}}^{MM} \end{bmatrix}. \tag{8.19}$$

The components K_{nq}^{mp} ($m, p = 1, 2, 3, \ldots, M$ and $n, q = 1, 2, 3, \ldots, N$) of the submatrices $\underline{\underline{K}}^{mp}$ with dimension $N \times N$ can be given as follows:

$$\begin{aligned} K_{nq}^{mp} &= \frac{1}{2} D_{11} \left(\lambda_{x,22}^{mp} \lambda_{y,00}^{nq} + \lambda_{x,22}^{pm} \lambda_{y,00}^{qn} \right) + \frac{1}{2} D_{22} \left(\lambda_{x,00}^{mp} \lambda_{y,22}^{nq} + \lambda_{x,00}^{pm} \lambda_{y,22}^{qn} \right) \\ &+ D_{12} \left(\lambda_{x,20}^{mp} \lambda_{y,02}^{nq} + \lambda_{x,20}^{pm} \lambda_{y,02}^{qn} \right) + 2 D_{66} \left(\lambda_{x,11}^{mp} \lambda_{y,11}^{nq} + \lambda_{x,11}^{pm} \lambda_{y,11}^{qn} \right). \end{aligned} \tag{8.20}$$

The vector $\underline{W} = (W_{11}, W_{12}, W_{13}, \ldots, W_{MN})^T$ contains the approach constants. The vector $\underline{P} = \left(\underline{P}^1, \underline{P}^2, \underline{P}^3, \ldots, \underline{P}^M\right)^T$ is composed of the subvectors $\underline{P}^m$ whose components P_n^m are obtained as follows:

$$P_n^m = -\Omega_{mn}. \tag{8.21}$$

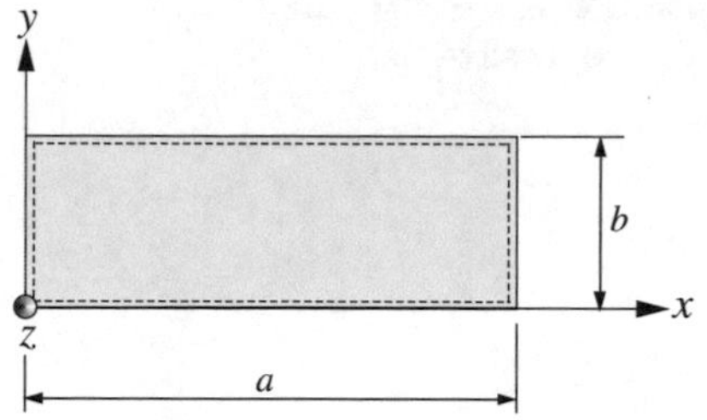

Fig. 8.1 Simply supported rectangular plate

In the case of a constantly distributed surface load $p = p_0$ we obtain:

$$P_n^m = -p_0 \omega_x^m \omega_y^n. \tag{8.22}$$

The Ritz method finds universal application for a variety of static plate bending problems. In the following, we will discuss a selection of approach functions for different types of boundary conditions.

If the plate is rectangular and simply supported at all edges (Fig. 8.1), the following functions can be used in (8.5):

$$w_{1m}(x) = \sin\left(\frac{m\pi x}{a}\right), \quad w_{2n}(y) = \sin\left(\frac{n\pi y}{b}\right). \tag{8.23}$$

It can be shown that in this case for an orthotropic plate under an arbitrary surface load $p(x, y)$ the Navier solution will result from the Ritz method, since the approach (8.5) then contains the exact solution (7.105) of the Navier solution.

If, on the other hand, a rectangular plate is present which is clamped at all four edges (Fig. 8.2), then several different types of approach functions can be used:

$$\begin{aligned} w_{1m}(x) &= \left(\frac{x}{a}\right)^{m+1}\left(1-\frac{x}{a}\right)^2, \\ w_{2n}(y) &= \left(\frac{y}{b}\right)^{n+1}\left(1-\frac{y}{b}\right)^2, \end{aligned} \tag{8.24}$$

or

$$\begin{aligned} w_{1m}(x) &= \sin(\lambda_m x) - \sinh(\lambda_m x) + \alpha_m\Big[\cosh(\lambda_m x) - \cos(\lambda_m x)\Big], \\ w_{2n}(y) &= \sin(\lambda_n y) - \sinh(\lambda_n y) + \alpha_n\Big[\cosh(\lambda_n y) - \cos(\lambda_n y)\Big], \end{aligned} \tag{8.25}$$

where the quantities λ_m and λ_n are the roots of the characteristic equations

$$\begin{aligned} \cos(\lambda_m a)\cosh(\lambda_m a) - 1 &= 0, \\ \cos(\lambda_n b)\cosh(\lambda_n b) - 1 &= 0. \end{aligned} \tag{8.26}$$

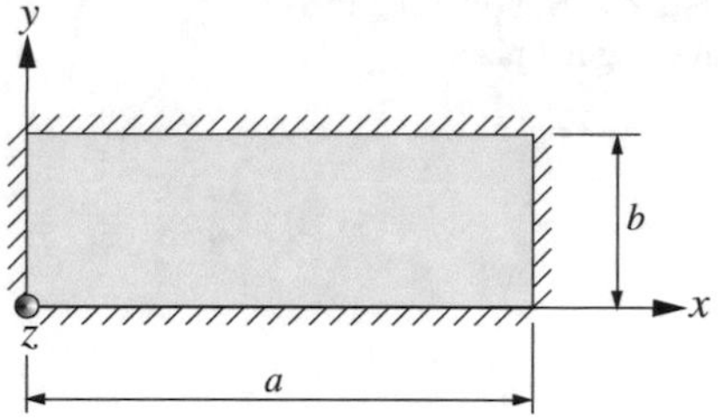

Fig. 8.2 Rectangular plate clamped at all edges

The quantities α_m and α_n are defined as follows:

$$\alpha_m = \frac{\sinh(\lambda_m a) - \sin(\lambda_m a)}{\cosh(\lambda_m a) - \cos(\lambda_m a)} = \frac{\cosh(\lambda_m a) - \cos(\lambda_m a)}{\sinh(\lambda_m a) + \sin(\lambda_m a)},$$
$$\alpha_n = \frac{\sinh(\lambda_n b) - \sin(\lambda_n b)}{\cosh(\lambda_n b) - \cos(\lambda_n b)} = \frac{\cosh(\lambda_n b) - \cos(\lambda_n b)}{\sinh(\lambda_n b) + \sin(\lambda_n b)}. \tag{8.27}$$

The following approaches can also be used:

$$w_{1m}(x) = \cos\left(\frac{(m-2)\pi x}{a}\right) - \cos\left(\frac{m\pi x}{a}\right),$$
$$w_{2n}(y) = \cos\left(\frac{(n-2)\pi y}{b}\right) - \cos\left(\frac{n\pi y}{b}\right), \tag{8.28}$$

where then the summation is to be carried out as follows:

$$w_0(x, y) = \sum_{m=2}^{m=M+1} \sum_{n=2}^{n=N+1} W_{mn} w_{1m}(x) w_{2n}(y). \tag{8.29}$$

If the plate is rectangular and clamped at $x = 0$ and $x = a$ (Fig. 8.3), but simply supported at $y = 0$ and $y = b$, then the following functions can be considered:

$$w_{1m}(x) = \sin(\lambda_m x) - \sinh(\lambda_m x) + \alpha_m \Big[\cosh(\lambda_m x) - \cos(\lambda_m x)\Big],$$
$$w_{2n}(y) = \sin\left(\frac{n\pi y}{b}\right). \tag{8.30}$$

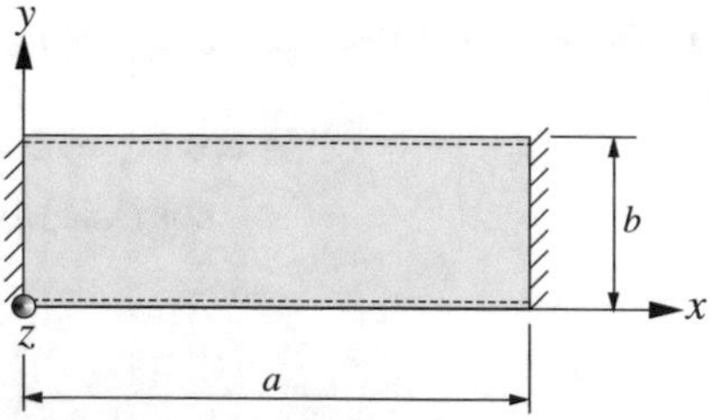

Fig. 8.3 Rectangular plate with two simply supported and two clamped edges

The quantities λ_m are the roots of the characteristic equation

$$\cos(\lambda_m a)\cosh(\lambda_m a) - 1 = 0, \tag{8.31}$$

and the quantities α_m are defined as follows:

$$\alpha_m = \frac{\sinh(\lambda_m a) - \sin(\lambda_m a)}{\cosh(\lambda_m a) - \cos(\lambda_m a)}. \tag{8.32}$$

A rectangular plate with a clamped edge at $x = 0$, a free edge at $x = a$, and two simply supported edges at $y = 0$ and $y = b$ (Fig. 8.4) can be treated using the following approach functions:

$$w_{1m}(x) = \sin(\lambda_m x) - \sinh(\lambda_m x) + \alpha_m\left[\cosh(\lambda_m x) - \cos(\lambda_m x)\right],$$
$$w_{2n}(y) = \sin\left(\frac{n\pi y}{b}\right). \tag{8.33}$$

The quantities λ_m are the roots of the characteristic equation

$$\cos(\lambda_m a)\cosh(\lambda_m a) + 1 = 0, \tag{8.34}$$

and the quantities α_m are defined as follows:

$$\alpha_m = \frac{\sinh(\lambda_m a) + \sin(\lambda_m a)}{\cosh(\lambda_m a) + \cos(\lambda_m a)}. \tag{8.35}$$

A rectangular plate with free edges at $x = 0$ and $x = a$ and simple supports at $y = 0$ and $y = b$ (Fig. 8.5) can be treated with the following approaches:

$$w_{1m}(x) = \sin(\lambda_m x) + \sinh(\lambda_m x) - \alpha_m\left[\cosh(\lambda_m x) + \cos(\lambda_m x)\right],$$
$$w_{2n}(y) = \sin\left(\frac{n\pi y}{b}\right). \tag{8.36}$$

The quantities λ_m are given by the characteristic equation

$$\cos(\lambda_m a)\cosh(\lambda_m a) - 1 = 0, \tag{8.37}$$

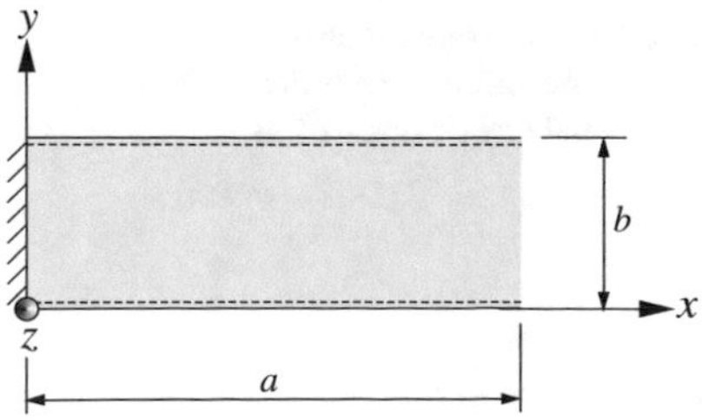

Fig. 8.4 Rectangular plate with one clamped edge, one free edge and two simply supported edges

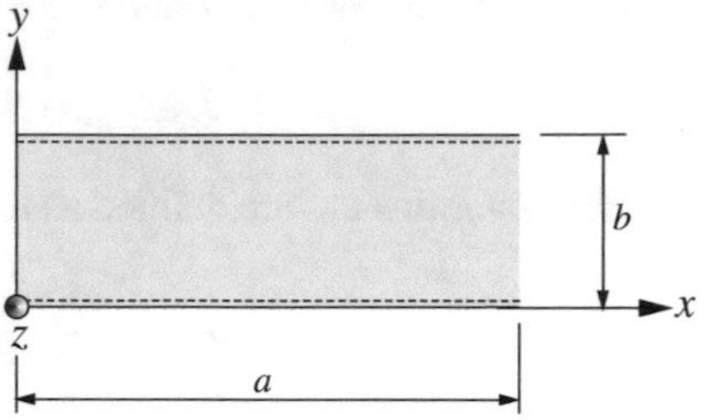

Fig. 8.5 Rectangular plate with two free edges and two simply supported edges

the quantities α_m are defined as follows:

$$\alpha_m = \frac{\sinh(\lambda_m a) - \sin(\lambda_m a)}{\cosh(\lambda_m a) - \cos(\lambda_m a)}. \tag{8.38}$$

If there is a simple support at $x = 0$, $y = 0$ and $y = b$ and the rectangular plate is clamped at $x = a$ (Fig. 8.6), then the following approach functions can be used:

$$w_{1m}(x) = \sinh(\lambda_m a)\sin(\lambda_m x) + \sin(\lambda_m a)\sinh(\lambda_m x),$$
$$w_{2n}(y) = \sin\left(\frac{n\pi y}{b}\right). \tag{8.39}$$

The quantities λ_m follow from the characteristic equation

$$\tan(\lambda_m a) - \tanh(\lambda_m a) = 0. \tag{8.40}$$

A rectangular plate with simple supports at $x = 0$, $y = 0$ and $y = b$ and a free edge at $x = a$ (Fig. 8.7) can be treated using the following approaches:

$$w_{1m}(x) = \sinh(\lambda_m a)\sin(\lambda_m x) - \sin(\lambda_m a)\sinh(\lambda_m x),$$
$$w_{2n}(y) = \sin\left(\frac{n\pi y}{b}\right). \tag{8.41}$$

The quantities λ_m are given by the following characteristic equation

$$\tan(\lambda_m a) - \tanh(\lambda_m a) = 0. \tag{8.42}$$

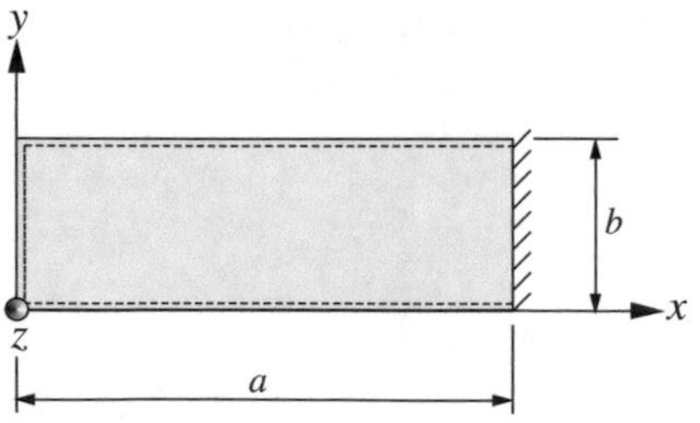

Fig. 8.6 Rectangular plate with three simply supported edges and one clamped edge

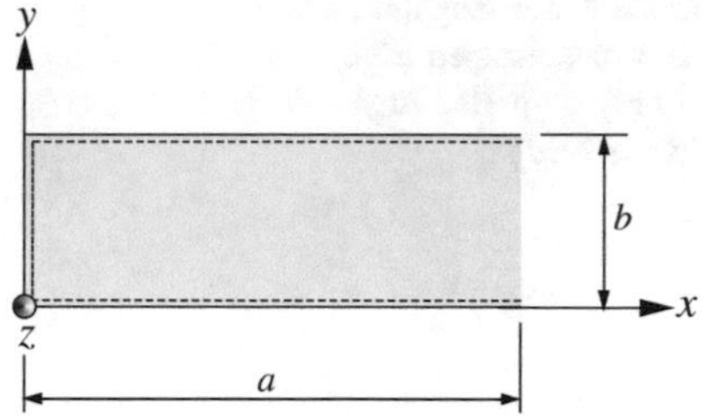

Fig. 8.7 Rectangular plate with three simply supported edges and one free edge

A rectangular plate with a clamped edge at $x = 0$ and free edges at $x = a$, $y = 0$, and $y = b$ (Fig. 8.8) can be analyzed using the following approach functions:

$$w_{1m}(x) = \sin(\lambda_m x) - \sinh(\lambda_m x) + \alpha_m \Big[\cosh(\lambda_m x) - \cos(\lambda_m x)\Big],$$
$$w_{2n}(y) = \sin(\lambda_n y) + \sinh(\lambda_n y) - \alpha_n \Big[\cosh(\lambda_n y) + \cos(\lambda_n y)\Big]. \tag{8.43}$$

The quantities λ_m and λ_n are obtained as the roots of the characteristic equations

$$\cos(\lambda_m a)\cosh(\lambda_m a) + 1 = 0,$$
$$\cos(\lambda_n b)\cosh(\lambda_n b) - 1 = 0. \tag{8.44}$$

The quantities α_m and α_n are defined as follows:

$$\alpha_m = \frac{\sinh(\lambda_m a) + \sin(\lambda_m a)}{\cosh(\lambda_m a) + \cos(\lambda_m a)},$$
$$\alpha_n = \frac{\sinh(\lambda_n b) - \sin(\lambda_n b)}{\cosh(\lambda_n b) - \cos(\lambda_n b)}. \tag{8.45}$$

If there is a clamped edge at $x = 0$ and a simple support at $y = 0$ and the edges of the rectangular plate at $x = a$ and $y = b$ are free of any support (Fig. 8.9), then the following approaches can be used:

$$w_{1m}(x) = \sin(\lambda_m x) - \sinh(\lambda_m x) + \alpha_m \Big[\cosh(\lambda_m x) - \cos(\lambda_m x)\Big],$$
$$w_{2n}(y) = \sinh(\lambda_n b)\sin(\lambda_n y) - \sin(\lambda_n b)\sinh(\lambda_n y). \tag{8.46}$$

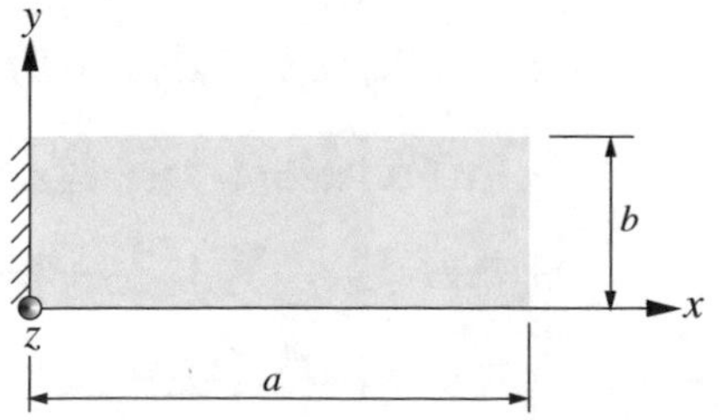

Fig. 8.8 Rectangular plate with three free edges and one clamped edge

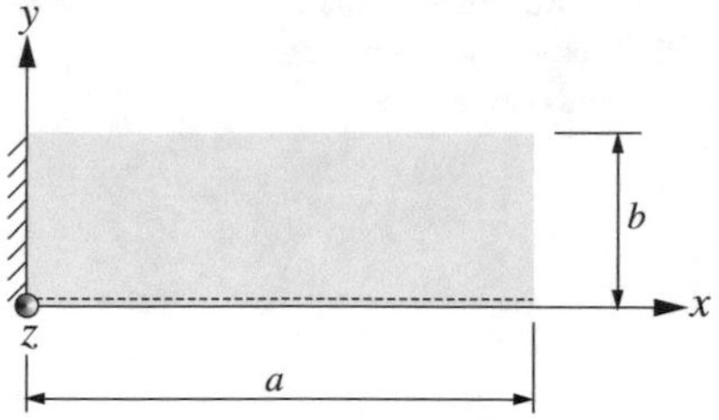

Fig. 8.9 Rectangular plate with one clamped edge, one simply supported edge and two free edges

The quantities λ_m and λ_n are obtained as the roots of the following characteristic equations:

$$\cos(\lambda_m a)\cosh(\lambda_m a) + 1 = 0,$$
$$\tan(\lambda_n b) - \tanh(\lambda_n b) = 0. \tag{8.47}$$

The quantity α_m is given by the following formula

$$\alpha_m = \frac{\sinh(\lambda_m a) + \sin(\lambda_m a)}{\cosh(\lambda_m a) + \cos(\lambda_m a)}. \tag{8.48}$$

We illustrate the application of the Ritz method with a rectangular orthotropic plate (length a, width b) which is simply supported at all four edges. The plate is under a constant surface load $p(x, y) = p_0$. The deflection w_0 is to be found using the Ritz method. An approach of the form

$$w_0(x, y) = \sum_{m=1}^{m=2}\sum_{n=1}^{n=2} W_{mn} w_{1m}(x) w_{2n}(y) \tag{8.49}$$

is employed.

First, the inner potential Π_i of the plate is considered, which can currently be written as follows after performing the summations and sorting the resulting expression:

$$\begin{aligned}
\Pi_i = {} & W_{11}^2 \left(\frac{1}{2} D_{11} \lambda_{x,22}^{11} \lambda_{y,00}^{11} + \frac{1}{2} D_{22} \lambda_{x,00}^{11} \lambda_{y,22}^{11} + D_{12} \lambda_{x,20}^{11} \lambda_{y,02}^{11} + 2 D_{66} \lambda_{x,11}^{11} \lambda_{y,11}^{11} \right) \\
& + W_{11} W_{12} \left(\frac{1}{2} D_{11} \lambda_{x,22}^{11} \left(\lambda_{y,00}^{12} + \lambda_{y,00}^{21} \right) + \frac{1}{2} D_{22} \lambda_{x,00}^{11} \left(\lambda_{y,22}^{12} + \lambda_{y,22}^{21} \right) \right. \\
& \left. + D_{12} \lambda_{x,20}^{11} \left(\lambda_{y,02}^{12} + \lambda_{y,02}^{21} \right) + 2 D_{66} \lambda_{x,11}^{11} \left(\lambda_{y,11}^{12} + \lambda_{y,11}^{21} \right) \right) \\
& + W_{11} W_{21} \left(\frac{1}{2} D_{11} \left(\lambda_{x,22}^{12} + \lambda_{x,22}^{21} \right) \lambda_{y,00}^{11} + \frac{1}{2} D_{22} \left(\lambda_{x,00}^{12} + \lambda_{x,00}^{21} \right) \lambda_{y,22}^{11} \right. \\
& \left. + D_{12} \left(\lambda_{x,20}^{12} + \lambda_{x,20}^{21} \right) \lambda_{y,02}^{11} + 2 D_{66} \left(\lambda_{x,11}^{12} + \lambda_{x,11}^{21} \right) \lambda_{y,11}^{11} \right) \\
& + W_{11} W_{22} \left(\frac{1}{2} D_{11} \left(\lambda_{x,22}^{12} \lambda_{y,00}^{12} + \lambda_{x,22}^{21} \lambda_{y,00}^{21} \right) + \frac{1}{2} D_{22} \left(\lambda_{x,00}^{12} \lambda_{y,22}^{12} + \lambda_{x,00}^{21} \lambda_{y,22}^{21} \right) \right.
\end{aligned}$$

$$
\begin{aligned}
&+ D_{12}\left(\lambda^{12}_{x,20}\lambda^{12}_{y,02} + \lambda^{21}_{x,20}\lambda^{21}_{y,02}\right) + 2D_{66}\left(\lambda^{12}_{x,11}\lambda^{12}_{y,11} + \lambda^{21}_{x,11}\lambda^{21}_{y,11}\right)\Big)\\
&+ W_{12}^2\left(\frac{1}{2}D_{11}\lambda^{11}_{x,22}\lambda^{22}_{y,00} + \frac{1}{2}D_{22}\lambda^{11}_{x,00}\lambda^{22}_{y,22} + D_{12}\lambda^{11}_{x,20}\lambda^{22}_{y,02} + 2D_{66}\lambda^{11}_{x,11}\lambda^{22}_{y,11}\right)\\
&+ W_{12}W_{21}\Big(\frac{1}{2}D_{11}\left(\lambda^{12}_{x,22}\lambda^{21}_{y,00} + \lambda^{21}_{x,22}\lambda^{12}_{y,00}\right) + \frac{1}{2}D_{22}\left(\lambda^{12}_{x,00}\lambda^{21}_{y,22} + \lambda^{21}_{x,00}\lambda^{12}_{y,22}\right)\\
&+ D_{12}\left(\lambda^{12}_{x,20}\lambda^{21}_{y,02} + \lambda^{21}_{x,20}\lambda^{12}_{y,02}\right) + 2D_{66}\left(\lambda^{12}_{x,11}\lambda^{21}_{y,11} + \lambda^{21}_{x,11}\lambda^{12}_{y,11}\right)\Big)\\
&+ W_{12}W_{22}\Big(\frac{1}{2}D_{11}\left(\lambda^{12}_{x,22} + \lambda^{21}_{x,22}\right)\lambda^{22}_{y,00} + \frac{1}{2}D_{22}\left(\lambda^{12}_{x,00} + \lambda^{21}_{x,00}\right)\lambda^{22}_{y,22}\\
&+ D_{12}\left(\lambda^{12}_{x,20} + \lambda^{21}_{x,20}\right)\lambda^{22}_{y,02} + 2D_{66}\left(\lambda^{12}_{x,11} + \lambda^{21}_{x,11}\right)\lambda^{22}_{y,11}\Big)\\
&+ W_{21}^2\left(\frac{1}{2}D_{11}\lambda^{22}_{x,22}\lambda^{11}_{y,00} + \frac{1}{2}D_{22}\lambda^{22}_{x,00}\lambda^{11}_{y,22} + D_{12}\lambda^{22}_{x,20}\lambda^{11}_{y,02} + 2D_{66}\lambda^{22}_{x,11}\lambda^{11}_{y,11}\right)\\
&+ W_{21}W_{22}\Big(\frac{1}{2}D_{11}\lambda^{22}_{x,22}\left(\lambda^{12}_{y,00} + \lambda^{21}_{y,00}\right) + \frac{1}{2}D_{22}\lambda^{22}_{x,00}\left(\lambda^{12}_{y,22} + \lambda^{21}_{y,22}\right)\\
&+ D_{12}\lambda^{22}_{x,20}\left(\lambda^{12}_{y,02} + \lambda^{21}_{y,02}\right) + 2D_{66}\lambda^{22}_{x,11}\left(\lambda^{12}_{y,11} + \lambda^{21}_{y,11}\right)\Big)\\
&+ W_{22}^2\left(\frac{1}{2}D_{11}\lambda^{22}_{x,22}\lambda^{22}_{y,00} + \frac{1}{2}D_{22}\lambda^{22}_{x,00}\lambda^{22}_{y,22} + D_{12}\lambda^{22}_{x,20}\lambda^{22}_{y,02} + 2D_{66}\lambda^{22}_{x,11}\lambda^{22}_{y,11}\right).
\end{aligned}
\tag{8.50}
$$

The external potential Π_a is obtained accordingly as:

$$
\Pi_a = -p_0\left(W_{11}\omega^1_x\omega^1_y + W_{12}\omega^1_x\omega^2_y + W_{21}\omega^2_x\omega^1_y + W_{22}\omega^2_x\omega^2_y\right). \tag{8.51}
$$

At this point, the Ritz equations can be evaluated, which currently read:

$$
\frac{\partial \Pi}{\partial W_{11}} = 0, \quad \frac{\partial \Pi}{\partial W_{12}} = 0, \quad \frac{\partial \Pi}{\partial W_{21}} = 0, \quad \frac{\partial \Pi}{\partial W_{22}} = 0. \tag{8.52}
$$

This then ultimately leads to a linear system of equations consisting of four equations for the four unknown coefficients W_{11}, W_{12}, W_{21}, W_{22}:

$$
\begin{aligned}
\frac{\partial \Pi}{\partial W_{11}} &= 2W_{11}\left(\frac{1}{2}D_{11}\lambda^{11}_{x,22}\lambda^{11}_{y,00} + \frac{1}{2}D_{22}\lambda^{11}_{x,00}\lambda^{11}_{y,22} + D_{12}\lambda^{11}_{x,20}\lambda^{11}_{y,02} + 2D_{66}\lambda^{11}_{x,11}\lambda^{11}_{y,11}\right)\\
&+ W_{12}\Big(\frac{1}{2}D_{11}\lambda^{11}_{x,22}\left(\lambda^{12}_{y,00} + \lambda^{21}_{y,00}\right) + \frac{1}{2}D_{22}\lambda^{11}_{x,00}\left(\lambda^{12}_{y,22} + \lambda^{21}_{y,22}\right)\\
&+ D_{12}\lambda^{11}_{x,20}\left(\lambda^{12}_{y,02} + \lambda^{21}_{y,02}\right) + 2D_{66}\lambda^{11}_{x,11}\left(\lambda^{12}_{y,11} + \lambda^{21}_{y,11}\right)\Big)\\
&+ W_{21}\Big(\frac{1}{2}D_{11}\left(\lambda^{12}_{x,22} + \lambda^{21}_{x,22}\right)\lambda^{11}_{y,00} + \frac{1}{2}D_{22}\left(\lambda^{12}_{x,00} + \lambda^{21}_{x,00}\right)\lambda^{11}_{y,22}\\
&+ D_{12}\left(\lambda^{12}_{x,20} + \lambda^{21}_{x,20}\right)\lambda^{11}_{y,02} + 2D_{66}\left(\lambda^{12}_{x,11} + \lambda^{21}_{x,11}\right)\lambda^{11}_{y,11}\Big)\\
&+ W_{22}\Big(\frac{1}{2}D_{11}\left(\lambda^{12}_{x,22}\lambda^{12}_{y,00} + \lambda^{21}_{x,22}\lambda^{21}_{y,00}\right) + \frac{1}{2}D_{22}\left(\lambda^{12}_{x,00}\lambda^{12}_{y,22} + \lambda^{21}_{x,00}\lambda^{21}_{y,22}\right)\\
&+ D_{12}\left(\lambda^{12}_{x,20}\lambda^{12}_{y,02} + \lambda^{21}_{x,20}\lambda^{21}_{y,02}\right) + 2D_{66}\left(\lambda^{12}_{x,11}\lambda^{12}_{y,11} + \lambda^{21}_{x,11}\lambda^{21}_{y,11}\right)\Big)\\
&- p_0\omega^1_x\omega^1_y = 0,
\end{aligned}
$$

$$
\begin{aligned}
\frac{\partial \Pi}{\partial W_{12}} &= W_{11}\left(\frac{1}{2}D_{11}\lambda_{x,22}^{11}\left(\lambda_{y,00}^{12}+\lambda_{y,00}^{21}\right)+\frac{1}{2}D_{22}\lambda_{x,00}^{11}\left(\lambda_{y,22}^{12}+\lambda_{y,22}^{21}\right)\right.\\
&\quad\left.+\ D_{12}\lambda_{x,20}^{11}\left(\lambda_{y,02}^{12}+\lambda_{y,02}^{21}\right)+2D_{66}\lambda_{x,11}^{11}\left(\lambda_{y,11}^{12}+\lambda_{y,11}^{21}\right)\right)\\
&\quad+2W_{12}\left(\frac{1}{2}D_{11}\lambda_{x,22}^{11}\lambda_{y,00}^{22}+\frac{1}{2}D_{22}\lambda_{x,00}^{11}\lambda_{y,22}^{22}+D_{12}\lambda_{x,20}^{11}\lambda_{y,02}^{22}+2D_{66}\lambda_{x,11}^{11}\lambda_{y,11}^{22}\right)\\
&\quad+W_{21}\left(\frac{1}{2}D_{11}\left(\lambda_{x,22}^{12}\lambda_{y,00}^{21}+\lambda_{x,22}^{21}\lambda_{y,00}^{12}\right)+\frac{1}{2}D_{22}\left(\lambda_{x,00}^{12}\lambda_{y,22}^{21}+\lambda_{x,00}^{21}\lambda_{y,22}^{12}\right)\right.\\
&\quad\left.+\ D_{12}\left(\lambda_{x,20}^{12}\lambda_{y,02}^{21}+\lambda_{x,20}^{21}\lambda_{y,02}^{12}\right)+2D_{66}\left(\lambda_{x,11}^{12}\lambda_{y,11}^{21}+\lambda_{x,11}^{21}\lambda_{y,11}^{12}\right)\right)\\
&\quad+W_{22}\left(\frac{1}{2}D_{11}\left(\lambda_{x,22}^{12}+\lambda_{x,22}^{21}\right)\lambda_{y,00}^{22}+\frac{1}{2}D_{22}\left(\lambda_{x,00}^{12}+\lambda_{x,00}^{21}\right)\lambda_{y,22}^{22}\right.\\
&\quad\left.+\ D_{12}\left(\lambda_{x,20}^{12}+\lambda_{x,20}^{21}\right)\lambda_{y,02}^{22}+2D_{66}\left(\lambda_{x,11}^{12}+\lambda_{x,11}^{21}\right)\lambda_{y,11}^{22}\right)\\
&\quad-\ p_0\omega_x^1\omega_y^2=0,
\end{aligned}
$$

$$
\begin{aligned}
\frac{\partial \Pi}{\partial W_{21}} &= W_{11}\left(\frac{1}{2}D_{11}\left(\lambda_{x,22}^{12}+\lambda_{x,22}^{21}\right)\lambda_{y,00}^{11}+\frac{1}{2}D_{22}\left(\lambda_{x,00}^{12}+\lambda_{x,00}^{21}\right)\lambda_{y,22}^{11}\right.\\
&\quad\left.+\ D_{12}\left(\lambda_{x,20}^{12}+\lambda_{x,20}^{21}\right)\lambda_{y,02}^{11}+2D_{66}\left(\lambda_{x,11}^{12}+\lambda_{x,11}^{21}\right)\lambda_{y,11}^{11}\right)\\
&\quad+W_{12}\left(\frac{1}{2}D_{11}\left(\lambda_{x,22}^{12}\lambda_{y,00}^{21}+\lambda_{x,22}^{21}\lambda_{y,00}^{12}\right)+\frac{1}{2}D_{22}\left(\lambda_{x,00}^{12}\lambda_{y,22}^{21}+\lambda_{x,00}^{21}\lambda_{y,22}^{12}\right)\right.\\
&\quad\left.+\ D_{12}\left(\lambda_{x,20}^{12}\lambda_{y,02}^{21}+\lambda_{x,20}^{21}\lambda_{y,02}^{12}\right)+2D_{66}\left(\lambda_{x,11}^{12}\lambda_{y,11}^{21}+\lambda_{x,11}^{21}\lambda_{y,11}^{12}\right)\right)\\
&\quad+2W_{21}\left(\frac{1}{2}D_{11}\lambda_{x,22}^{22}\lambda_{y,00}^{11}+\frac{1}{2}D_{22}\lambda_{x,00}^{22}\lambda_{y,22}^{11}+D_{12}\lambda_{x,20}^{22}\lambda_{y,02}^{11}+2D_{66}\lambda_{x,11}^{22}\lambda_{y,11}^{11}\right)\\
&\quad+W_{22}\left(\frac{1}{2}D_{11}\lambda_{x,22}^{22}\left(\lambda_{y,00}^{12}+\lambda_{y,00}^{21}\right)+\frac{1}{2}D_{22}\lambda_{x,00}^{22}\left(\lambda_{y,22}^{12}+\lambda_{y,22}^{21}\right)\right.\\
&\quad\left.+\ D_{12}\lambda_{x,20}^{22}\left(\lambda_{y,02}^{12}+\lambda_{y,02}^{21}\right)+2D_{66}\lambda_{x,11}^{22}\left(\lambda_{y,11}^{12}+\lambda_{y,11}^{21}\right)\right)\\
&\quad-\ p_0\omega_x^2\omega_y^1=0,
\end{aligned}
$$

$$
\begin{aligned}
\frac{\partial \Pi}{\partial W_{22}} &= W_{11}\left(\frac{1}{2}D_{11}\left(\lambda_{x,22}^{12}\lambda_{y,00}^{12}+\lambda_{x,22}^{21}\lambda_{y,00}^{21}\right)+\frac{1}{2}D_{22}\left(\lambda_{x,00}^{12}\lambda_{y,22}^{12}+\lambda_{x,00}^{21}\lambda_{y,22}^{21}\right)\right.\\
&\quad\left.+\ D_{12}\left(\lambda_{x,20}^{12}\lambda_{y,02}^{12}+\lambda_{x,20}^{21}\lambda_{y,02}^{21}\right)+2D_{66}\left(\lambda_{x,11}^{12}\lambda_{y,11}^{12}+\lambda_{x,11}^{21}\lambda_{y,11}^{21}\right)\right)\\
&\quad+W_{12}\left(\frac{1}{2}D_{11}\left(\lambda_{x,22}^{12}+\lambda_{x,22}^{21}\right)\lambda_{y,00}^{22}+\frac{1}{2}D_{22}\left(\lambda_{x,00}^{12}+\lambda_{x,00}^{21}\right)\lambda_{y,22}^{22}\right.\\
&\quad\left.+\ D_{12}\left(\lambda_{x,20}^{12}+\lambda_{x,20}^{21}\right)\lambda_{y,02}^{22}+2D_{66}\left(\lambda_{x,11}^{12}+\lambda_{x,11}^{21}\right)\lambda_{y,11}^{22}\right)\\
&\quad+W_{21}\left(\frac{1}{2}D_{11}\lambda_{x,22}^{22}\left(\lambda_{y,00}^{12}+\lambda_{y,00}^{21}\right)+\frac{1}{2}D_{22}\lambda_{x,00}^{22}\left(\lambda_{y,22}^{12}+\lambda_{y,22}^{21}\right)\right.\\
&\quad\left.+\ D_{12}\lambda_{x,20}^{22}\left(\lambda_{y,02}^{12}+\lambda_{y,02}^{21}\right)+2D_{66}\lambda_{x,11}^{22}\left(\lambda_{y,11}^{12}+\lambda_{y,11}^{21}\right)\right)\\
&\quad+2W_{22}\left(\frac{1}{2}D_{11}\lambda_{x,22}^{22}\lambda_{y,00}^{22}+\frac{1}{2}D_{22}\lambda_{x,00}^{22}\lambda_{y,22}^{22}+D_{12}\lambda_{x,20}^{22}\lambda_{y,02}^{22}+2D_{66}\lambda_{x,11}^{22}\lambda_{y,11}^{22}\right)
\end{aligned}
$$

$$- p_0\omega_x^2\omega_y^2 = 0. \tag{8.53}$$

This can be written in a more compact vector-matrix notation as:

$$\begin{bmatrix} \underline{\underline{K}}^{11} & \underline{\underline{K}}^{12} \\ \underline{\underline{K}}^{21} & \underline{\underline{K}}^{22} \end{bmatrix} \begin{pmatrix} \underline{W}^1 \\ \underline{W}^2 \end{pmatrix} = \begin{pmatrix} \underline{P}^1 \\ \underline{P}^2 \end{pmatrix}, \tag{8.54}$$

where the submatrices and vectors are assigned as follows:

$$\underline{\underline{K}}^{11} = \begin{bmatrix} K_{11}^{11} & K_{12}^{11} \\ K_{21}^{11} & K_{22}^{11} \end{bmatrix}, \quad \underline{\underline{K}}^{12} = \begin{bmatrix} K_{11}^{12} & K_{12}^{12} \\ K_{21}^{12} & K_{22}^{12} \end{bmatrix},$$
$$\underline{\underline{K}}^{21} = \begin{bmatrix} K_{11}^{21} & K_{12}^{21} \\ K_{21}^{21} & K_{22}^{21} \end{bmatrix}, \quad \underline{\underline{K}}^{22} = \begin{bmatrix} K_{11}^{22} & K_{12}^{22} \\ K_{21}^{22} & K_{22}^{22} \end{bmatrix},$$
$$\underline{W}^1 = \begin{pmatrix} W_{11} \\ W_{12} \end{pmatrix}, \quad \underline{W}^2 = \begin{pmatrix} W_{21} \\ W_{22} \end{pmatrix}, \quad \underline{P}^1 = \begin{pmatrix} P_1^1 \\ P_2^1 \end{pmatrix}, \quad \underline{P}^2 = \begin{pmatrix} P_1^2 \\ P_2^2 \end{pmatrix}. \tag{8.55}$$

The respective entries are:

$$\begin{aligned}
K_{11}^{11} &= D_{11}\lambda_{x,22}^{11}\lambda_{y,00}^{11} + D_{22}\lambda_{x,00}^{11}\lambda_{y,22}^{11} + 2D_{12}\lambda_{x,20}^{11}\lambda_{y,02}^{11} + 4D_{66}\lambda_{x,11}^{11}\lambda_{y,11}^{11}, \\
K_{12}^{11} &= \frac{1}{2}D_{11}\lambda_{x,22}^{11}\left(\lambda_{y,00}^{12} + \lambda_{y,00}^{21}\right) + \frac{1}{2}D_{22}\lambda_{x,00}^{11}\left(\lambda_{y,22}^{12} + \lambda_{y,22}^{21}\right) \\
&\quad + D_{12}\lambda_{x,20}^{11}\left(\lambda_{y,02}^{12} + \lambda_{y,02}^{21}\right) + 2D_{66}\lambda_{x,11}^{11}\left(\lambda_{y,11}^{12} + \lambda_{y,11}^{21}\right), \\
K_{21}^{11} &= \frac{1}{2}D_{11}\lambda_{x,22}^{11}\left(\lambda_{y,00}^{12} + \lambda_{y,00}^{21}\right) + \frac{1}{2}D_{22}\lambda_{x,00}^{11}\left(\lambda_{y,22}^{12} + \lambda_{y,22}^{21}\right) \\
&\quad + D_{12}\lambda_{x,20}^{11}\left(\lambda_{y,02}^{12} + \lambda_{y,02}^{21}\right) + 2D_{66}\lambda_{x,11}^{11}\left(\lambda_{y,11}^{12} + \lambda_{y,11}^{21}\right), \\
K_{22}^{11} &= D_{11}\lambda_{x,22}^{11}\lambda_{y,00}^{22} + D_{22}\lambda_{x,00}^{11}\lambda_{y,22}^{22} + 2D_{12}\lambda_{x,20}^{11}\lambda_{y,02}^{22} + 4D_{66}\lambda_{x,11}^{11}\lambda_{y,11}^{22}, \\
K_{11}^{12} &= \frac{1}{2}D_{11}\left(\lambda_{x,22}^{12} + \lambda_{x,22}^{21}\right)\lambda_{y,00}^{11} + \frac{1}{2}D_{22}\left(\lambda_{x,00}^{12} + \lambda_{x,00}^{21}\right)\lambda_{y,22}^{11} \\
&\quad + D_{12}\left(\lambda_{x,20}^{12} + \lambda_{x,20}^{21}\right)\lambda_{y,02}^{11} + 2D_{66}\left(\lambda_{x,11}^{12} + \lambda_{x,11}^{21}\right)\lambda_{y,11}^{11}, \\
K_{12}^{12} &= \frac{1}{2}D_{11}\left(\lambda_{x,22}^{12}\lambda_{y,00}^{12} + \lambda_{x,22}^{21}\lambda_{y,00}^{21}\right) + \frac{1}{2}D_{22}\left(\lambda_{x,00}^{12}\lambda_{y,22}^{12} + \lambda_{x,00}^{21}\lambda_{y,22}^{21}\right) \\
&\quad + D_{12}\left(\lambda_{x,20}^{12}\lambda_{y,02}^{12} + \lambda_{x,20}^{21}\lambda_{y,02}^{21}\right) + 2D_{66}\left(\lambda_{x,11}^{12}\lambda_{y,11}^{12} + \lambda_{x,11}^{21}\lambda_{y,11}^{21}\right), \\
K_{21}^{12} &= \frac{1}{2}D_{11}\left(\lambda_{x,22}^{12}\lambda_{y,00}^{21} + \lambda_{x,22}^{21}\lambda_{y,00}^{12}\right) + \frac{1}{2}D_{22}\left(\lambda_{x,00}^{12}\lambda_{y,22}^{21} + \lambda_{x,00}^{21}\lambda_{y,22}^{12}\right) \\
&\quad + D_{12}\left(\lambda_{x,20}^{12}\lambda_{y,02}^{21} + \lambda_{x,20}^{21}\lambda_{y,02}^{12}\right) + 2D_{66}\left(\lambda_{x,11}^{12}\lambda_{y,11}^{21} + \lambda_{x,11}^{21}\lambda_{y,11}^{12}\right), \\
K_{22}^{12} &= \frac{1}{2}D_{11}\left(\lambda_{x,22}^{12} + \lambda_{x,22}^{21}\right)\lambda_{y,00}^{22} + \frac{1}{2}D_{22}\left(\lambda_{x,00}^{12} + \lambda_{x,00}^{21}\right)\lambda_{y,22}^{22} \\
&\quad + D_{12}\left(\lambda_{x,20}^{12} + \lambda_{x,20}^{21}\right)\lambda_{y,02}^{22} + 2D_{66}\left(\lambda_{x,11}^{12} + \lambda_{x,11}^{21}\right)\lambda_{y,11}^{22},
\end{aligned}$$

$$
\begin{aligned}
K_{11}^{21} &= \frac{1}{2} D_{11} \left(\lambda_{x,22}^{12} + \lambda_{x,22}^{21}\right) \lambda_{y,00}^{11} + \frac{1}{2} D_{22} \left(\lambda_{x,00}^{12} + \lambda_{x,00}^{21}\right) \lambda_{y,22}^{11} \\
&\quad + D_{12} \left(\lambda_{x,20}^{12} + \lambda_{x,20}^{21}\right) \lambda_{y,02}^{11} + 2 D_{66} \left(\lambda_{x,11}^{12} + \lambda_{x,11}^{21}\right) \lambda_{y,11}^{11}, \\
K_{12}^{21} &= \frac{1}{2} D_{11} \left(\lambda_{x,22}^{12} \lambda_{y,00}^{21} + \lambda_{x,22}^{21} \lambda_{y,00}^{12}\right) + \frac{1}{2} D_{22} \left(\lambda_{x,00}^{12} \lambda_{y,22}^{21} + \lambda_{x,00}^{21} \lambda_{y,22}^{12}\right) \\
&\quad + D_{12} \left(\lambda_{x,20}^{12} \lambda_{y,02}^{21} + \lambda_{x,20}^{21} \lambda_{y,02}^{12}\right) + 2 D_{66} \left(\lambda_{x,11}^{12} \lambda_{y,11}^{21} + \lambda_{x,11}^{21} \lambda_{y,11}^{12}\right), \\
K_{21}^{21} &= \frac{1}{2} D_{11} \left(\lambda_{x,22}^{12} \lambda_{y,00}^{12} + \lambda_{x,22}^{21} \lambda_{y,00}^{21}\right) + \frac{1}{2} D_{22} \left(\lambda_{x,00}^{12} \lambda_{y,22}^{12} + \lambda_{x,00}^{21} \lambda_{y,22}^{21}\right) \\
&\quad + D_{12} \left(\lambda_{x,20}^{12} \lambda_{y,02}^{12} + \lambda_{x,20}^{21} \lambda_{y,02}^{21}\right) + 2 D_{66} \left(\lambda_{x,11}^{12} \lambda_{y,11}^{12} + \lambda_{x,11}^{21} \lambda_{y,11}^{21}\right), \\
K_{22}^{21} &= \frac{1}{2} D_{11} \left(\lambda_{x,22}^{12} + \lambda_{x,22}^{21}\right) \lambda_{y,00}^{22} + \frac{1}{2} D_{22} \left(\lambda_{x,00}^{12} + \lambda_{x,00}^{21}\right) \lambda_{y,22}^{22} \\
&\quad + D_{12} \left(\lambda_{x,20}^{12} + \lambda_{x,20}^{21}\right) \lambda_{y,02}^{22} + 2 D_{66} \left(\lambda_{x,11}^{12} + \lambda_{x,11}^{21}\right) \lambda_{y,11}^{22}, \\
K_{11}^{22} &= D_{11} \lambda_{x,22}^{22} \lambda_{y,00}^{11} + D_{22} \lambda_{x,00}^{22} \lambda_{y,22}^{11} + 2 D_{12} \lambda_{x,20}^{22} \lambda_{y,02}^{11} + 4 D_{66} \lambda_{x,11}^{22} \lambda_{y,11}^{11}, \\
K_{12}^{22} &= \frac{1}{2} D_{11} \lambda_{x,22}^{22} \left(\lambda_{y,00}^{12} + \lambda_{y,00}^{21}\right) + \frac{1}{2} D_{22} \lambda_{x,00}^{22} \left(\lambda_{y,22}^{12} + \lambda_{y,22}^{21}\right) \\
&\quad + D_{12} \lambda_{x,20}^{22} \left(\lambda_{y,02}^{12} + \lambda_{y,02}^{21}\right) + 2 D_{66} \lambda_{x,11}^{22} \left(\lambda_{y,11}^{12} + \lambda_{y,11}^{21}\right), \\
K_{21}^{22} &= \frac{1}{2} D_{11} \lambda_{x,22}^{22} \left(\lambda_{y,00}^{12} + \lambda_{y,00}^{21}\right) + \frac{1}{2} D_{22} \lambda_{x,00}^{22} \left(\lambda_{y,22}^{12} + \lambda_{y,22}^{21}\right) \\
&\quad + D_{12} \lambda_{x,20}^{22} \left(\lambda_{y,02}^{12} + \lambda_{y,02}^{21}\right) + 2 D_{66} \lambda_{x,11}^{22} \left(\lambda_{y,11}^{12} + \lambda_{y,11}^{21}\right), \\
K_{22}^{22} &= D_{11} \lambda_{x,22}^{22} \lambda_{y,00}^{22} + D_{22} \lambda_{x,00}^{22} \lambda_{y,22}^{22} + 2 D_{12} \lambda_{x,20}^{22} \lambda_{y,02}^{22} + 4 D_{66} \lambda_{x,11}^{22} \lambda_{y,11}^{22}, \\
P_1^1 &= p_0 \omega_x^1 \omega_y^1, \quad P_2^1 = p_0 \omega_x^1 \omega_y^2, \quad P_1^2 = p_0 \omega_x^2 \omega_y^1, \quad P_2^2 = p_0 \omega_x^2 \omega_y^2. \qquad (8.56)
\end{aligned}
$$

The system of equations can then be easily solved using any method for solving systems of linear equations. Then the coefficients W_{11}, W_{12}, W_{21}, W_{22} as well as the approximation solution for the deflection w_0 of the plate are known. From this, approximate force and moments flows as well as stresses in the plate can be calculated.

As an approach for the deflection w_0 a series expansion of the following form shall be used:

$$
w_0(x, y) = \sum_{m=1}^{m=2} \sum_{n=1}^{n=2} W_{mn} \sin\left(\frac{m\pi x}{a}\right) \sin\left(\frac{n\pi y}{b}\right). \qquad (8.57)
$$

For the evaluation of the integrals $\lambda_{x,ij}^{mp}$ and $\lambda_{y,ij}^{nq}$ the following holds:

$$
\lambda_{x,00}^{mp} = \int_0^a w_{1m} w_{1p} \mathrm{d}x = \int_0^a \sin\left(\frac{m\pi x}{a}\right) \sin\left(\frac{p\pi x}{a}\right) \mathrm{d}x,
$$

$$\lambda_{x,01}^{mp} = \int_0^a w_{1m} \frac{\mathrm{d}w_{1p}}{\mathrm{d}x} \mathrm{d}x = \frac{p\pi}{a} \int_0^a \sin\left(\frac{m\pi x}{a}\right) \cos\left(\frac{p\pi x}{a}\right) \mathrm{d}x,$$

$$\lambda_{x,02}^{mp} = \int_0^a w_{1m} \frac{\mathrm{d}^2 w_{1p}}{\mathrm{d}x^2} \mathrm{d}x = -\frac{p^2\pi^2}{a^2} \int_0^a \sin\left(\frac{m\pi x}{a}\right) \sin\left(\frac{p\pi x}{a}\right) \mathrm{d}x,$$

$$\lambda_{x,10}^{mp} = \int_0^a \frac{\mathrm{d}w_{1m}}{\mathrm{d}x} w_{1p} \mathrm{d}x = \frac{m\pi}{a} \int_0^a \cos\left(\frac{m\pi x}{a}\right) \sin\left(\frac{p\pi x}{a}\right) \mathrm{d}x,$$

$$\lambda_{x,11}^{mp} = \int_0^a \frac{\mathrm{d}w_{1m}}{\mathrm{d}x} \frac{\mathrm{d}w_{1p}}{\mathrm{d}x} \mathrm{d}x = \frac{m\pi}{a} \frac{p\pi}{a} \int_0^a \cos\left(\frac{m\pi x}{a}\right) \cos\left(\frac{p\pi x}{a}\right) \mathrm{d}x,$$

$$\lambda_{x,12}^{mp} = \int_0^a \frac{\mathrm{d}w_{1m}}{\mathrm{d}x} \frac{\mathrm{d}^2 w_{1p}}{\mathrm{d}x^2} \mathrm{d}x = -\frac{m\pi}{a} \frac{p^2\pi^2}{a^2} \int_0^a \cos\left(\frac{m\pi x}{a}\right) \sin\left(\frac{p\pi x}{a}\right) \mathrm{d}x,$$

$$\lambda_{x,20}^{mp} = \int_0^a \frac{\mathrm{d}^2 w_{1m}}{\mathrm{d}x^2} w_{1p} \mathrm{d}x = -\frac{m^2\pi^2}{a^2} \int_0^a \sin\left(\frac{m\pi x}{a}\right) \sin\left(\frac{p\pi x}{a}\right) \mathrm{d}x,$$

$$\lambda_{x,21}^{mp} = \int_0^a \frac{\mathrm{d}^2 w_{1m}}{\mathrm{d}x^2} \frac{\mathrm{d}w_{1p}}{\mathrm{d}x} \mathrm{d}x = -\frac{m^2\pi^2}{a^2} \frac{p\pi}{a} \int_0^a \sin\left(\frac{m\pi x}{a}\right) \cos\left(\frac{p\pi x}{a}\right) \mathrm{d}x,$$

$$\lambda_{x,22}^{mp} = \int_0^a \frac{\mathrm{d}^2 w_{1m}}{\mathrm{d}x^2} \frac{\mathrm{d}^2 w_{1p}}{\mathrm{d}x^2} \mathrm{d}x = \frac{m^2\pi^2}{a^2} \frac{p^2\pi^2}{a^2} \int_0^a \sin\left(\frac{m\pi x}{a}\right) \sin\left(\frac{p\pi x}{a}\right) \mathrm{d}x,$$

$$\lambda_{y,00}^{nq} = \int_0^b w_{2n} w_{2q} \mathrm{d}y = \int_0^b \sin\left(\frac{n\pi y}{b}\right) \sin\left(\frac{q\pi y}{b}\right) \mathrm{d}y,$$

$$\lambda_{y,01}^{nq} = \int_0^b w_{2n} \frac{\mathrm{d}w_{2q}}{\mathrm{d}y} \mathrm{d}y = \frac{q\pi}{b} \int_0^b \sin\left(\frac{n\pi y}{b}\right) \cos\left(\frac{q\pi y}{b}\right) \mathrm{d}y,$$

$$\lambda_{y,02}^{nq} = \int_0^b w_{2n} \frac{\mathrm{d}^2 w_{2q}}{\mathrm{d}y^2} \mathrm{d}y = -\frac{q^2\pi^2}{b^2} \int_0^b \sin\left(\frac{n\pi y}{b}\right) \sin\left(\frac{q\pi y}{b}\right) \mathrm{d}y,$$

$$\lambda_{y,10}^{nq} = \int_0^b \frac{\mathrm{d}w_{2n}}{\mathrm{d}y} w_{2q} \mathrm{d}y = \frac{n\pi}{b} \int_0^b \cos\left(\frac{n\pi y}{b}\right) \sin\left(\frac{q\pi y}{b}\right) \mathrm{d}y,$$

$$
\begin{aligned}
\lambda_{y,11}^{nq} &= \int\limits_0^b \frac{\mathrm{d}w_{2n}}{\mathrm{d}y}\frac{\mathrm{d}w_{2q}}{\mathrm{d}y}\mathrm{d}y = \frac{n\pi}{b}\frac{q\pi}{b}\int\limits_0^b \cos\left(\frac{n\pi y}{b}\right)\cos\left(\frac{q\pi y}{b}\right)\mathrm{d}y,\\
\lambda_{y,12}^{nq} &= \int\limits_0^b \frac{\mathrm{d}w_{2n}}{\mathrm{d}y}\frac{\mathrm{d}^2w_{2q}}{\mathrm{d}y^2}\mathrm{d}y = -\frac{n\pi}{b}\frac{q^2\pi^2}{b^2}\int\limits_0^b \cos\left(\frac{n\pi y}{b}\right)\sin\left(\frac{q\pi y}{b}\right)\mathrm{d}y,\\
\lambda_{y,20}^{nq} &= \int\limits_0^b \frac{\mathrm{d}^2w_{2n}}{\mathrm{d}y^2}w_{2q}\mathrm{d}y = -\frac{n^2\pi^2}{b^2}\int\limits_0^b \sin\left(\frac{n\pi y}{b}\right)\sin\left(\frac{q\pi y}{b}\right)\mathrm{d}y,\\
\lambda_{y,21}^{nq} &= \int\limits_0^b \frac{\mathrm{d}^2w_{2n}}{\mathrm{d}y^2}\frac{\mathrm{d}w_{2q}}{\mathrm{d}y}\mathrm{d}y = -\frac{n^2\pi^2}{b^2}\frac{q\pi}{b}\int\limits_0^b \sin\left(\frac{n\pi y}{b}\right)\cos\left(\frac{q\pi y}{b}\right)\mathrm{d}y,\\
\lambda_{y,22}^{nq} &= \int\limits_0^b \frac{\mathrm{d}^2w_{2n}}{\mathrm{d}y^2}\frac{\mathrm{d}^2w_{2q}}{\mathrm{d}y^2}\mathrm{d}y = \frac{n^2\pi^2}{b^2}\frac{q^2\pi^2}{b^2}\int\limits_0^b \sin\left(\frac{n\pi y}{b}\right)\sin\left(\frac{q\pi y}{b}\right)\mathrm{d}y.
\end{aligned}
\tag{8.58}
$$

The integrals can be solved exactly, and the result is obtained as:

$$
\begin{aligned}
\lambda_{x,00}^{mp} &= \begin{cases} 0 \text{ if } m \neq p \\ \frac{a}{2} \text{ if } m = p \end{cases},\\
\lambda_{x,01}^{mp} &= \begin{cases} 0 \text{ if } m+p \text{ even} \\ \frac{2mp}{m^2-p^2} \text{ if } m+p \text{ odd} \end{cases},\\
\lambda_{x,02}^{mp} &= \begin{cases} 0 \text{ if } m \neq p \\ -\frac{p^2\pi^2}{2a} \text{ if } m = p \end{cases},\\
\lambda_{x,10}^{mp} &= \begin{cases} 0 \text{ if } m+p \text{ even} \\ \frac{2mp}{p^2-m^2} \text{ if } m+p \text{ odd} \end{cases},\\
\lambda_{x,11}^{mp} &= \begin{cases} 0 \text{ if } m \neq p \\ \frac{mp\pi^2}{2a} \text{ if } m = p \end{cases},\\
\lambda_{x,12}^{mp} &= \begin{cases} 0 \text{ if } m+p \text{ even} \\ -\frac{2mp^3\pi^2}{a^2(p^2-m^2)} \text{ if } m+p \text{ odd} \end{cases},\\
\lambda_{x,20}^{mp} &= \begin{cases} 0 \text{ if } m \neq p \\ -\frac{m^2\pi^2}{2a} \text{ if } m = p \end{cases},\\
\lambda_{x,21}^{mp} &= \begin{cases} 0 \text{ if } m+p \text{ even} \\ -\frac{2m^3p\pi^2}{a^2(m^2-p^2)} \text{ if } m+p \text{ odd} \end{cases},\\
\lambda_{x,22}^{mp} &= \begin{cases} 0 \text{ if } m \neq p \\ \frac{m^2p^2\pi^4}{2a^3} \text{ if } m = p \end{cases}.
\end{aligned}
\tag{8.59}
$$

The integrals λ_{00}^{nq}, λ_{01}^{nq}, λ_{02}^{nq}, λ_{10}^{nq}, λ_{11}^{nq}, λ_{12}^{nq}, λ_{20}^{nq}, λ_{21}^{nq} and λ_{22}^{nq} can be obtained directly from this by replacing a with b. Furthermore, the running indices m and p must be replaced by n and q. For reasons of brevity this is not shown here.

The integrals ω_x^m and ω_y^n currently result in:

$$\omega_x^m = \int_0^a \sin\left(\frac{m\pi x}{a}\right) \mathrm{d}x = \begin{cases} 0 \text{ if } m \text{ even} \\ \frac{2a}{m\pi} \text{ if } m \text{ odd} \end{cases},$$

$$\omega_y^n = \int_0^b \sin\left(\frac{n\pi y}{b}\right) \mathrm{d}y = \begin{cases} 0 \text{ if } n \text{ even} \\ \frac{2b}{n\pi} \text{ if } n \text{ odd} \end{cases}. \tag{8.60}$$

It can be shown that the Ritz formulation chosen for this concrete example will lead exactly to the exact solution in the form of Navier's solution. The reason is that the chosen form of approach corresponds exactly to the exact solution for the plate deflection. The proof of this fact is left to the reader at this point.

8.2 The Galerkin Method

The Galerkin method also assumes an approach of the form (8.5), i.e. the plate deflection is approximated as

$$\tilde{w}_0(x, y) = \sum_{m=1}^{m=M} \sum_{n=1}^{n=N} W_{mn} w_{1m}(x) w_{2n}(y). \tag{8.61}$$

Here, however, it is now required that the approach (8.61) satisfies all underlying boundary conditions.

An error minimization of the following form is performed to determine the constants W_{mn}:

$$\int_0^a \int_0^b \left[D_{11} \frac{\partial^4 \tilde{w}_0}{\partial x^4} + 2(D_{12} + 2D_{66}) \frac{\partial^4 \tilde{w}_0}{\partial x^2 \partial y^2} + D_{22} \frac{\partial^4 \tilde{w}_0}{\partial y^4} - p \right] w_{1p} w_{2q} \mathrm{d}y \mathrm{d}x = 0, \tag{8.62}$$

with $p = 1, 2, 3, ..., M, q = 1, 2, 3, ..., N$. Substituting the approach (8.61) into (8.62), we obtain:

$$\begin{aligned} &\sum_{m=1}^{m=M} \sum_{n=1}^{n=N} W_{mn} \left[D_{11} \int_0^a \frac{\mathrm{d}^4 w_{1m}}{\mathrm{d}x^4} w_{1p} \mathrm{d}x \int_0^b w_{2n} w_{2q} \mathrm{d}y \right. \\ &\quad + 2(D_{12} + 2D_{66}) \int_0^a \frac{\mathrm{d}^2 w_{1m}}{\mathrm{d}x^2} w_{1p} \mathrm{d}x \int_0^b \frac{\mathrm{d}^2 w_{2n}}{\mathrm{d}y^2} w_{2q} \\ &\quad \left. + D_{22} \int_0^a w_{1m} w_{1p} \mathrm{d}x \int_0^b \frac{\mathrm{d}^4 w_{2n}}{\mathrm{d}y^2} w_{2q} \mathrm{d}y \right] - \int_0^a \int_0^b p w_{1p} w_{2q} \mathrm{d}y \mathrm{d}x = 0. \end{aligned} \tag{8.63}$$

With the abbreviations

$$\lambda_{x,ij}^{mp} = \int_0^a \frac{\mathrm{d}^i w_{1m}}{\mathrm{d}x^i} \frac{\mathrm{d}^j w_{1p}}{\mathrm{d}x^j} \mathrm{d}x, \quad \lambda_{y,ij}^{nq} = \int_0^b \frac{\mathrm{d}^i w_{2n}}{\mathrm{d}y^i} \frac{\mathrm{d}^j w_{2q}}{\mathrm{d}y^j} \mathrm{d}y \tag{8.64}$$

and

$$\Omega_{pq} = \int_0^a \int_0^b p w_{1p} w_{2q} \mathrm{d}y \mathrm{d}x \tag{8.65}$$

we obtain for (8.63):

$$\sum_{m=1}^{m=M} \sum_{n=1}^{n=N} W_{mn} \left[D_{11} \lambda_{x,40}^{mp} \lambda_{y,00}^{nq} + 2\left(D_{12} + 2D_{66}\right) \lambda_{x,20}^{mp} \lambda_{y,20}^{nq} + D_{22} \lambda_{x,00}^{mp} \lambda_{y,40}^{nq} \right] = \Omega_{pq}. \tag{8.66}$$

This is a linear system of equations for the determination of the constants W_{mn}. Once these are available, the approach (8.61) is completely established, and the state quantities of the plate can be determined.

As an example, consider a rectangular orthotropic plate loaded by a constant surface load $p = p_0$. The plate has the dimensions a and b with respect to the $x-$ and $y-$ directions, respectively, and is clamped at all four edges. The following boundary conditions apply. At the two edges $x = 0$ and $x = a$ with $0 \leq y \leq b$ holds:

$$w_0, \quad \frac{\partial w_0}{\partial x} = 0. \tag{8.67}$$

On the other hand, at the two edges $y = 0$ and $y = b$ with $0 \leq x \leq a$ we have:

$$w_0, \quad \frac{\partial w_0}{\partial y} = 0. \tag{8.68}$$

For the given plate situation, an approach of the form

$$\tilde{w}_0 = \sum_{m=2}^{m=M+1} \sum_{n=2}^{n=N+1} W_{mn} w_{1m}(x) w_{2n}(y), \tag{8.69}$$

with

$$w_{1m}(x) = \cos\left(\frac{(m-2)\pi x}{a}\right) - \cos\left(\frac{m\pi x}{a}\right),$$
$$w_{2n}(y) = \cos\left(\frac{(n-2)\pi y}{b}\right) - \cos\left(\frac{m\pi y}{b}\right) \tag{8.70}$$

is suitable. We perform the calculation of the plate deflection for this example using a single-term approach, i.e., for:

$$\tilde{w}_0 = W w_1(x) w_2(y) = W \left[1 - \cos\left(\frac{2\pi x}{a}\right)\right]\left[1 - \cos\left(\frac{2\pi y}{b}\right)\right]. \quad (8.71)$$

From (8.62) we then obtain the following condition:

$$\int_0^a \int_0^b \left[D_{11}\frac{\partial^4 \tilde{w}_0}{\partial x^4} + 2\left(D_{12} + 2D_{66}\right)\frac{\partial^4 \tilde{w}_0}{\partial x^2 \partial y^2} + D_{22}\frac{\partial^4 \tilde{w}_0}{\partial y^4} - p_0\right] w_1 w_2 \mathrm{d}y\mathrm{d}x = 0. \quad (8.72)$$

The partial derivatives necessary herein follow as:

$$\begin{aligned}
\frac{\partial^4 \tilde{w}_0}{\partial x^4} &= -W\left(\frac{2\pi}{a}\right)^4 \cos\left(\frac{2\pi x}{a}\right)\left[1 - \cos\left(\frac{2\pi y}{b}\right)\right], \\
\frac{\partial^4 \tilde{w}_0}{\partial y^4} &= -W\left(\frac{2\pi}{b}\right)^4 \left[1 - \cos\left(\frac{2\pi x}{a}\right)\right]\cos\left(\frac{2\pi y}{b}\right), \\
\frac{\partial^4 \tilde{w}_0}{\partial x^2 \partial y^2} &= W\left(\frac{2\pi}{a}\right)^2\left(\frac{2\pi}{b}\right)^2 \cos\left(\frac{2\pi x}{a}\right)\cos\left(\frac{2\pi y}{b}\right).
\end{aligned} \quad (8.73)$$

With the integrals resulting from (8.72)

$$\begin{aligned}
&\int_0^a \cos\left(\frac{2\pi x}{a}\right)\mathrm{d}x = 0, \quad \int_0^a \cos^2\left(\frac{2\pi x}{a}\right)\mathrm{d}x = \frac{a}{2}, \\
&\int_0^b \cos\left(\frac{2\pi y}{b}\right)\mathrm{d}y = 0, \quad \int_0^b \cos^2\left(\frac{2\pi y}{b}\right)\mathrm{d}y = \frac{b}{2}
\end{aligned} \quad (8.74)$$

the following condition for the determination of the constant W results:

$$4\pi^4\left[3D_{11}\frac{b}{a^3} + 3D_{22}\frac{a}{b^3} + 2\left(D_{12} + 2D_{66}\right)\frac{1}{ab}\right] W = p_0 ab. \quad (8.75)$$

Solving for W results in:

$$W = \frac{p_0 a^4 b^4}{4\pi^4\left[3D_{11}b^4 + 3D_{22}a^4 + 2\left(D_{12} + 2D_{66}\right)a^2 b^2\right]}. \quad (8.76)$$

It can be seen that in this specific case the result is identical to the result obtained by the Ritz method.

Like the Ritz method, the Galerkin method is a very efficient method for the analysis of plate bending problems. Due to the requirement that the approach functions have to fulfill all boundary conditions, the determination of suitable functions may prove to be difficult. Therefore, further explanations of the Galerkin method will not be given at this point.

8.3 The Finite Element Method

In addition to the classical methods presented for the approximate solution of plate bending problems, the finite element method (FEM) has proven itself in many ways for the analysis of even the most complex plate problems. The use of this discretizing method is always justified when exact-analytical methods are not applicable and also the mentioned classical approximation methods reach their limits. The FEM for plate problems will not be discussed here, the interested reader is referred to the works of Altenbach et al. (1996, 2004), Bathe (1996), Crisfield (1991, 1997), Reddy (2004, 2005, 2014), Wriggers (2001), Zienkiewicz et al. (2013a,b), among others.

References

Altenbach, H., Altenbach, J., Rikards, R.: Einführung in die Mechanik der Laminat- und Sandwichtragwerke. Deutscher Verlag für Grundstoffindustrie Stuttgart, Germany (1996)

Altenbach, H., Altenbach, J., Kissing, W.: Mechanics of composite structural elements. Springer, Berlin (2004)

Bathe, K.J.: Finite element procedures. Prentice Hall, New Jersey (1996)

Crisfield, M.A.: Non-linear finite element analysis of solids and structures, volume 1: Essentials. Wiley, Chichester (1991)

Crisfield, M.A.: Non-linear finite element analysis of solids and structures, volume 2: advanced topics. Wiley, Chichester (1997)

Mittelstedt, C.: Rechenmethoden des Leichtbaus. Springer, Heidelberg (2021)

Reddy, J.N.: Mechanics of laminated composite plates and shells, 2nd edn. CRC Press, Boca Raton (2004)

Reddy, J.N.: An introduction to the finite element method, 3rd edn. Mcgraw Hill, New York (2005)

Reddy, J.N.: An introduction to nonlinear finite element analysis with applications to heat transfer, fluid mechanics, and solid mechanics, 2nd edn. Oxford University Press, New York (2014)

Reddy, J.N.: Energy principles and variational methods in applied mechanics, 3rd edn. Wiley, New York (2017)

Szilard, R.: Theories and applications of plate analysis. Wiley Hoboken, New Jersey (2004)

Wriggers, P.: Nichtlineare Finite-Element-Methoden. Springer, Berlin (2001)

Zienkiewicz, O.C., Taylor, R.L., Zhu, J.Z.: The finite element method: its basis and fundamentals, 7th edn. Butterworth Heinemann, Oxford (2013)

Zienkiewicz, O.C., Taylor, R.L.: The finite element method for solid and structural mechanics, 7th edn. Butterworth Heinemann, Oxford (2013)

9 Kirchhoff Plate Theory in Polar Coordinates

In this chapter we consider plate problems in the framework of Kirchhoff's plate theory which can be advantageously described by polar coordinates. Among these are naturally circular plates and circular ring plates. At the beginning of this chapter all necessary basic equations of the Kirchhoff plate theory are provided in polar coordinates. Hereafter, the case of bending of circular and annular plates is discussed where both the structural situation and the loading are rotationally symmetric, and some exemplary solutions are presented. The chapter concludes with a discussion of the case of asymmetric bending. The interested reader can find further elaborations on this subject in Altenbach et al. (2016), Girkmann (1974), Hake and Meskouris (2007), Reddy (2006), Szilard (2004), Timoshenko and Woinowsky-Krieger (1964), Ugural (1981), among others.

9.1 Transition to Polar Coordinates

The following relationship between the Cartesian coordinates x, y used so far and the polar coordinates r, φ holds (Fig. 9.1):

$$x = r\cos\varphi, \quad y = r\sin\varphi, \quad r^2 = x^2 + y^2, \quad \tan\varphi = \frac{y}{x}. \tag{9.1}$$

The following relationships can be derived from this:

$$\begin{gathered}\frac{\partial r}{\partial x} = \frac{x}{r} = \cos\varphi, \quad \frac{\partial r}{\partial y} = \frac{y}{r} = \sin\varphi, \\ \frac{\partial \varphi}{\partial x} = -\frac{y}{r^2} = -\frac{\sin\varphi}{r}, \quad \frac{\partial \varphi}{\partial y} = \frac{x}{r^2} = \frac{\cos\varphi}{r}.\end{gathered} \tag{9.2}$$

C. Mittelstedt, *Theory of Plates and Shells*,
https://doi.org/10.1007/978-3-662-66805-4_9

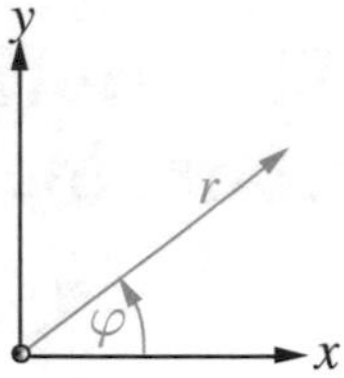

Fig. 9.1 Polar coordinates

The following expressions for the derivatives of the plate deflection w_0 follow:

$$\frac{\partial w_0}{\partial x} = \frac{\partial w_0}{\partial r}\frac{\partial r}{\partial x} + \frac{\partial w_0}{\partial \varphi}\frac{\partial \varphi}{\partial x} = \frac{\partial w_0}{\partial r}\cos\varphi - \frac{1}{r}\frac{\partial w_0}{\partial \varphi}\sin\varphi,$$
$$\frac{\partial w_0}{\partial y} = \frac{\partial w_0}{\partial r}\frac{\partial r}{\partial y} + \frac{\partial w_0}{\partial \varphi}\frac{\partial \varphi}{\partial y} = \frac{\partial w_0}{\partial r}\sin\varphi + \frac{1}{r}\frac{\partial w_0}{\partial y}\cos\varphi. \tag{9.3}$$

We can be proceed analogously for higher derivatives. For the second partial derivative $\frac{\partial^2 w_0}{\partial x^2}$ follows:

$$\begin{aligned}\frac{\partial^2 w_0}{\partial x^2} = \frac{\partial}{\partial x}\left(\frac{\partial w_0}{\partial x}\right) &= \cos\varphi\frac{\partial}{\partial r}\left(\frac{\partial w_0}{\partial x}\right) - \frac{1}{r}\sin\varphi\frac{\partial}{\partial \varphi}\left(\frac{\partial w_0}{\partial x}\right)\\ &= \cos\varphi\frac{\partial}{\partial r}\left(\frac{\partial w_0}{\partial r}\cos\varphi - \frac{1}{r}\frac{\partial w_0}{\partial \varphi}\sin\varphi\right)\\ &\quad - \frac{1}{r}\sin\varphi\frac{\partial}{\partial \varphi}\left(\frac{\partial w_0}{\partial r}\cos\varphi - \frac{1}{r}\frac{\partial w_0}{\partial \varphi}\sin\varphi\right)\\ &= \frac{\partial^2 w_0}{\partial r^2}\cos^2\varphi + 2\left(\frac{1}{r}\frac{\partial w_0}{\partial \varphi} - \frac{\partial^2 w_0}{\partial r\partial \varphi}\right)\frac{\sin\varphi\cos\varphi}{r}\\ &\quad + \left(\frac{\partial w_0}{\partial r} + \frac{1}{r}\frac{\partial^2 w_0}{\partial \varphi^2}\right)\frac{\sin^2\varphi}{r}.\end{aligned} \tag{9.4}$$

For the two derivatives $\frac{\partial^2 w_0}{\partial y^2}$ and $\frac{\partial^2 w_0}{\partial x\partial y}$ results:

$$\begin{aligned}\frac{\partial^2 w_0}{\partial y^2} &= \frac{\partial^2 w_0}{\partial r^2}\sin^2\varphi + 2\left(-\frac{1}{r}\frac{\partial w_0}{\partial \varphi} + \frac{\partial^2 w_0}{\partial r\partial \varphi}\right)\frac{\sin\varphi\cos\varphi}{r}\\ &\quad + \left(\frac{\partial w_0}{\partial r} + \frac{1}{r}\frac{\partial^2 w_0}{\partial \varphi^2}\right)\frac{\cos^2\varphi}{r},\\ \frac{\partial^2 w_0}{\partial x\partial y} &= \left(\frac{\partial^2 w_0}{\partial r^2} - \frac{1}{r}\frac{\partial w_0}{\partial r} - \frac{1}{r^2}\frac{\partial^2 w_0}{\partial \varphi^2}\right)\sin\varphi\cos\varphi\\ &\quad + \left(\frac{\partial^2 w_0}{\partial r\partial \varphi} - \frac{1}{r}\frac{\partial w_0}{\partial \varphi}\right)\frac{\cos^2\varphi}{r}.\end{aligned} \tag{9.5}$$

The Laplace operator Δ can thus be concluded as:

$$\Delta = \frac{\partial^2}{\partial x^2} + \frac{\partial^2}{\partial y^2} = \frac{\partial^2}{\partial r^2} + \frac{1}{r}\frac{\partial}{\partial r} + \frac{1}{r^2}\frac{\partial^2}{\partial \varphi^2}. \tag{9.6}$$

The plate equation (7.54) for the isotropic Kirchhoff plate in Cartesian coordinates

$$\frac{\partial^4 w_0}{\partial x^4} + 2\frac{\partial^4 w_0}{\partial x^2 \partial y^2} + \frac{\partial^4 w_0}{\partial y^4} = \frac{p}{D}. \tag{9.7}$$

then transfers into the following form:

$$\Delta\Delta w_0 = \left(\frac{\partial^2}{\partial r^2} + \frac{1}{r}\frac{\partial}{\partial r} + \frac{1}{r^2}\frac{\partial^2}{\partial \varphi^2}\right)\left(\frac{\partial^2 w_0}{\partial r^2} + \frac{1}{r}\frac{\partial w_0}{\partial r} + \frac{1}{r^2}\frac{\partial^2 w_0}{\partial \varphi^2}\right) = \frac{p}{D}. \tag{9.8}$$

9.2 Basic Equations

The starting point of the considerations is the displacement field in the framework of Kirchhoff's assumptions, which at present can be written as follows:

$$\begin{aligned} u = u_r &= -z\frac{\partial w_0}{\partial r}, \\ v = u_\varphi &= -z\frac{1}{r}\frac{\partial w_0}{\partial \varphi}, \\ w &= w_0. \end{aligned} \tag{9.9}$$

From the kinematic equations (1.209), the strain components can then be determined using (9.9):

$$\begin{aligned} \varepsilon_{rr} &= \frac{\partial u}{\partial r} = -z\frac{\partial^2 w_0}{\partial r^2}, \\ \varepsilon_{\varphi\varphi} &= \frac{1}{r}\frac{\partial v}{\partial \varphi} + \frac{u}{r} = -\frac{z}{r}\left(\frac{1}{r}\frac{\partial^2 w_0}{\partial \varphi^2} + \frac{\partial w_0}{\partial r}\right), \\ \gamma_{r\varphi} &= \frac{\partial v}{\partial r} + \frac{1}{r}\frac{\partial u}{\partial \varphi} - \frac{v}{r} = \frac{2z}{r}\left(\frac{1}{r}\frac{\partial w_0}{\partial \varphi} - \frac{\partial^2 w_0}{\partial r \partial \varphi}\right). \end{aligned} \tag{9.10}$$

In polar coordinates, the two shear force flows Q_r and Q_φ and the moment flows M^0_{rr}, $M^0_{\varphi\varphi}$ and $M^0_{r\varphi}$ are present (Fig. 9.2), which can be procured from the already known constitutive relations (7.28) with the following reasoning. Let the x−axis be the radial coordinate r at the position $\varphi = 0$. Thus the force flows Q_r, Q_φ and the moment flows M^0_{rr}, $M^0_{\varphi\varphi}$, $M^0_{r\varphi}$ at this point have identical values as the force and moment flows Q_x, Q_y, M^0_{xx}, M^0_{yy}, M^0_{xy} in Cartesian coordinates. Thus, setting $\varphi = 0$ and using the known constitutive relations (7.28) with (9.4) and (9.5), we obtain for

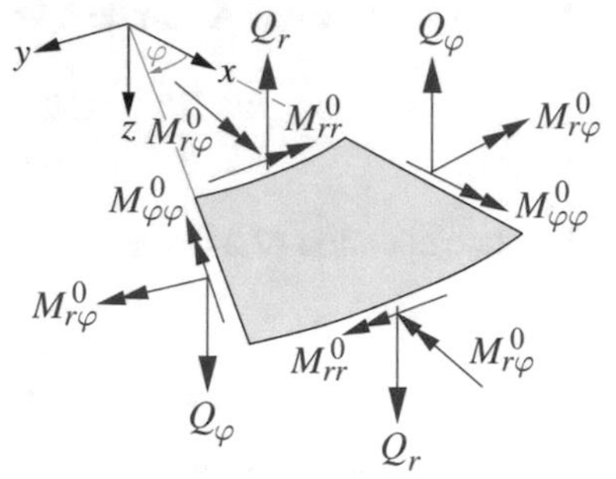

Fig. 9.2 Infinitesimal element, force and moment quantities

the moment flows M^0_{rr}, $M^0_{\varphi\varphi}$, $M^0_{r\varphi}$:

$$
\begin{aligned}
M^0_{rr} &= -D\left(\frac{\partial^2 w_0}{\partial x^2} + \nu\frac{\partial^2 w_0}{\partial y^2}\right) \\
&= -D\left[\frac{\partial^2 w_0}{\partial r^2} + \nu\left(\frac{1}{r}\frac{\partial w_0}{\partial r} + \frac{1}{r^2}\frac{\partial^2 w_0}{\partial \varphi^2}\right)\right], \\
M^0_{\varphi\varphi} &= -D\left(\nu\frac{\partial^2 w_0}{\partial x^2} + \frac{\partial^2 w_0}{\partial y^2}\right) \\
&= -D\left(\frac{1}{r}\frac{\partial w_0}{\partial r} + \frac{1}{r^2}\frac{\partial^2 w_0}{\partial \varphi^2} + \nu\frac{\partial^2 w_0}{\partial r^2}\right), \\
M^0_{r\varphi} &= -D(1-\nu)\frac{\partial^2 w_0}{\partial x \partial y} = -D\left(1-\nu\right)\left(\frac{1}{r}\frac{\partial^2 w_0}{\partial r \partial \varphi} - \frac{1}{r^2}\frac{\partial w_0}{\partial \varphi}\right).
\end{aligned}
\tag{9.11}
$$

For the shear force flows Q_r and Q_φ results analogously with (7.50) and (7.51):

$$
\begin{aligned}
Q_r &= \frac{\partial M^0_{xx}}{\partial x} + \frac{\partial M^0_{xy}}{\partial y} = -D\frac{\partial}{\partial r}\left(\Delta w_0\right), \\
Q_\varphi &= \frac{\partial M^0_{xy}}{\partial x} + \frac{\partial M^0_{yy}}{\partial y} = -D\frac{1}{r}\frac{\partial}{\partial \varphi}\left(\Delta w_0\right).
\end{aligned}
\tag{9.12}
$$

The Kirchhoff equivalent transverse shear force flows for $r = const.$ for any angle φ are:

$$
\begin{aligned}
\bar{Q}_r &= Q_r + \frac{1}{r}\frac{\partial M^0_{r\varphi}}{\partial \varphi} = -D\left[\frac{\partial}{\partial r}\left(\Delta w_0\right) + \frac{1-\nu}{r}\frac{\partial}{\partial \varphi}\left(\frac{1}{r}\frac{\partial^2 w_0}{\partial r \partial \varphi} - \frac{1}{r^2}\frac{\partial w_0}{\partial \varphi}\right)\right], \\
\bar{Q}_\varphi &= Q_\varphi + \frac{\partial M^0_{r\varphi}}{\partial r} = -D\left[\frac{1}{r}\frac{\partial}{\partial \varphi}\left(\Delta w_0\right) + \left(1-\nu\right)\frac{\partial}{\partial r}\left(\frac{1}{r}\frac{\partial^2 w_0}{\partial r \partial \varphi} - \frac{1}{r^2}\frac{\partial w_0}{\partial \varphi}\right)\right].
\end{aligned}
\tag{9.13}
$$

The moment flows can also be obtained directly from the constitutive law (1.211) for the plane stress components σ_{rr},$\sigma_{\varphi\varphi}$, $\tau_{r\varphi}$ with the strains (9.10). If the considered plate is a polar-orthotropic structure, where the material principal directions coincide with the polar coordinates r, φ, then the moment flows result from the constitutive law

$$\begin{pmatrix} \sigma_{rr} \\ \sigma_{\varphi\varphi} \\ \tau_{r\varphi} \end{pmatrix} = \begin{bmatrix} Q_{11} & Q_{12} & 0 \\ Q_{12} & Q_{22} & 0 \\ 0 & 0 & Q_{66} \end{bmatrix} \begin{pmatrix} \varepsilon_{rr} \\ \varepsilon_{\varphi\varphi} \\ \gamma_{r\varphi} \end{pmatrix} \tag{9.14}$$

with the strains (9.10) by integration as follows:

$$\begin{aligned} M^0_{rr} &= \int_{-\frac{h}{2}}^{+\frac{h}{2}} \sigma_{rr} z \mathrm{d}z = -D_{11}\frac{\partial^2 w_0}{\partial r^2} - D_{12}\frac{1}{r}\left(\frac{\partial w_0}{\partial r} + \frac{1}{r}\frac{\partial^2 w_0}{\partial \varphi^2}\right), \\ M^0_{\varphi\varphi} &= \int_{-\frac{h}{2}}^{+\frac{h}{2}} \sigma_{\varphi\varphi} z \mathrm{d}z = -D_{12}\frac{\partial^2 w_0}{\partial r^2} - D_{22}\frac{1}{r}\left(\frac{\partial w_0}{\partial r} + \frac{1}{r}\frac{\partial^2 w_0}{\partial \varphi^2}\right), \\ M^0_{r\varphi} &= \int_{-\frac{h}{2}}^{+\frac{h}{2}} \tau_{r\varphi} z \mathrm{d}z = -D_{66}\frac{2}{r}\left(\frac{\partial^2 w_0}{\partial r \partial \varphi} - \frac{1}{r}\frac{\partial w_0}{\partial \varphi}\right). \end{aligned} \tag{9.15}$$

The quantities D_{ij} are the plate stiffnesses:

$$D_{ij} = \frac{Q_{ij} h^3}{12}. \tag{9.16}$$

We now take into account the free body diagram of Fig. 9.3 and consider the local equilibrium. The sum of forces in $z-$ direction gives:

$$\begin{aligned} &- Q_r r \mathrm{d}\varphi + \left(Q_r + \frac{\partial Q_r}{\partial r}\mathrm{d}r\right)(r + \mathrm{d}r)\mathrm{d}\varphi \\ &- Q_\varphi \mathrm{d}r + \left(Q_\varphi + \frac{\partial Q_\varphi}{\partial \varphi}\mathrm{d}\varphi\right)\mathrm{d}r + p\mathrm{d}r\left(r + \frac{\mathrm{d}r}{2}\right)\mathrm{d}\varphi = 0, \end{aligned} \tag{9.17}$$

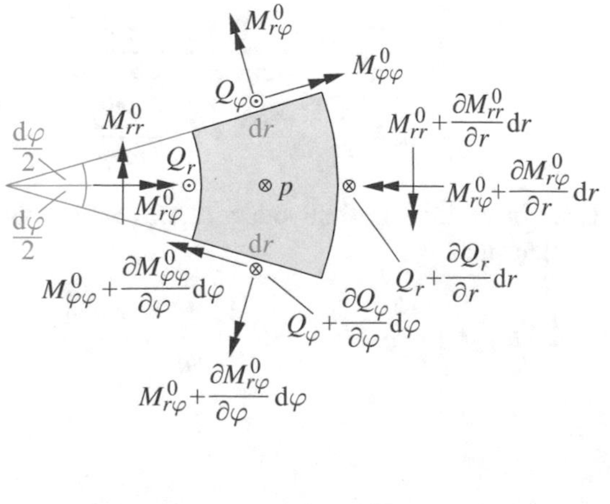

Fig. 9.3 Infinitesimal plate element

or after neglecting terms of higher order:

$$\frac{\partial Q_r}{\partial r} + \frac{1}{r}\frac{\partial Q_\varphi}{\partial \varphi} + \frac{Q_r}{r} + p = 0. \tag{9.18}$$

The moment sum with respect to the $\varphi-$direction with reference point in the center of the infinitesimal element of Fig. 9.3 yields:

$$\begin{aligned}
&- M_{rr}^0 r \mathrm{d}\varphi + \left(M_{rr}^0 + \frac{\partial M_{rr}^0}{\partial r}\mathrm{d}r \right)(r + \mathrm{d}r)\mathrm{d}\varphi \\
&- M_{\varphi\varphi}^0 \mathrm{d}r \sin\frac{\mathrm{d}\varphi}{2} - \left(M_{\varphi\varphi}^0 + \frac{\partial M_{\varphi\varphi}^0}{\partial \varphi}\mathrm{d}\varphi \right) \mathrm{d}r \sin\frac{\mathrm{d}\varphi}{2} \\
&- M_{r\varphi}^0 \mathrm{d}r \cos\frac{\mathrm{d}\varphi}{2} + \left(M_{r\varphi}^0 + \frac{\partial M_{r\varphi}^0}{\partial \varphi}\mathrm{d}\varphi \right) \mathrm{d}r \cos\frac{\mathrm{d}\varphi}{2} \\
&- Q_r r \mathrm{d}\varphi \frac{\mathrm{d}r}{2} - \left(Q_r + \frac{\partial Q_r}{\partial r}\mathrm{d}r \right)(r + \mathrm{d}r)\mathrm{d}\varphi \frac{\mathrm{d}r}{2} = 0.
\end{aligned} \tag{9.19}$$

Neglecting terms of higher order assuming small angles $\cos\frac{\mathrm{d}\varphi}{2} \simeq 1$, $\sin\frac{\mathrm{d}\varphi}{2} \simeq \frac{\mathrm{d}\varphi}{2}$ finally results in:

$$\frac{\partial M_{rr}^0}{\partial r} + \frac{M_{rr}^0 - M_{\varphi\varphi}^0}{r} + \frac{1}{r}\frac{\partial M_{r\varphi}^0}{\partial \varphi} - Q_r = 0. \tag{9.20}$$

An analogous result stems from the moment sum with respect to the radial direction around the center of the infinitesimal element:

$$\frac{\partial M_{r\varphi}^0}{\partial r} + \frac{1}{r}\frac{\partial M_{\varphi\varphi}^0}{\partial \varphi} + \frac{2M_{r\varphi}^0}{r} - Q_\varphi = 0. \tag{9.21}$$

Substituting (9.20) and (9.21) into (9.18) gives the following form of the plate equation:

$$\begin{aligned}
&\frac{\partial^2 M_{rr}^0}{\partial r^2} + \frac{2}{r}\frac{\partial M_{rr}^0}{\partial r} - \frac{1}{r}\frac{\partial M_{\varphi\varphi}^0}{\partial r} \\
&+ \frac{1}{r^2}\frac{\partial^2 M_{\varphi\varphi}^0}{\partial \varphi^2} + \frac{2}{r}\frac{\partial^2 M_{r\varphi}^0}{\partial r \partial \varphi} + \frac{2}{r^2}\frac{\partial M_{r\varphi}^0}{\partial \varphi} + p = 0.
\end{aligned} \tag{9.22}$$

If we insert the constitutive law (9.11), we get the plate equation already given by (9.8) again:

$$\Delta\Delta w_0 = \left(\frac{\partial^2}{\partial r^2} + \frac{1}{r}\frac{\partial}{\partial r} + \frac{1}{r^2}\frac{\partial^2}{\partial \varphi^2} \right)\left(\frac{\partial^2 w_0}{\partial r^2} + \frac{1}{r}\frac{\partial w_0}{\partial r} + \frac{1}{r^2}\frac{\partial^2 w_0}{\partial \varphi^2} \right) = \frac{p}{D}. \tag{9.23}$$

Its solution is composed of a homogeneous solution $w_{0,h}$ and a load-dependent particular solution $w_{0,p}$. The homogeneous form of the plate equation

$$\Delta\Delta w_0 = \left(\frac{\partial^2}{\partial r^2} + \frac{1}{r}\frac{\partial}{\partial r} + \frac{1}{r^2}\frac{\partial^2}{\partial \varphi^2}\right)\left(\frac{\partial^2 w_0}{\partial r^2} + \frac{1}{r}\frac{\partial w_0}{\partial r} + \frac{1}{r^2}\frac{\partial^2 w_0}{\partial \varphi^2}\right) = 0 \quad (9.24)$$

has the following general solution:

$$w_{0,h} = f_0 + \sum_{n=1}^{\infty} f_n(r)\cos n\varphi + \sum_{n=1}^{\infty} g_n(r)\sin n\varphi. \quad (9.25)$$

In this series form the functions f_n and g_n are exclusively functions of the radial coordinate r. Substituting this approach into the plate equation (9.24), we obtain the following solutions for $f_n(r)$ and $g_n(r)$:

$$\begin{aligned}
f_0 &= A_0 + B_0 r^2 + C_0 \ln r + D_0 r^2 \ln r,\\
f_1 &= A_1 r + B_1 r^3 + C_1 r^{-1} + D_1 r \ln r,\\
f_n &= A_n r^n + B_n r^{-n} + C_n r^{n+2} + D_n r^{-n+2},\\
g_1 &= E_1 r + F_1 r^3 + G_1 r^{-1} + H_1 r \ln r,\\
g_n &= E_n r^n + F_n r^{-n} + G_n r^{n+2} + H_n r^{-n+2},
\end{aligned} \quad (9.26)$$

with the constants A, B, C, D, E, F, G, H.

The complete formulation of a given plate problem can only be performed if boundary conditions are specified. At an edge with $r = r_0 = const.$ the following conditions may occur:

$$\begin{aligned}
&w_0 = \widehat{w}_0 \quad \text{oderor} \quad r_0\bar{Q}_r = r_0\widehat{\bar{Q}}_r,\\
&\frac{\partial w_0}{\partial r} = \widehat{\frac{\partial w_0}{\partial r}} \quad \text{or} \quad r_0 M_{rr}^0 = r_0\widehat{M}_{rr}^0.
\end{aligned} \quad (9.27)$$

Herein, prescribed quantities were marked with the symbolism $\widehat{(...)}$.

At an edge $\varphi = \varphi_0 = const.$ the following conditions may exist:

$$\begin{aligned}
&w_0 = \widehat{w}_0 \quad \text{or} \quad \bar{Q}_\varphi = \widehat{\bar{Q}}_\varphi,\\
&\frac{1}{r}\frac{\partial w_0}{\partial \varphi} = \widehat{\frac{1}{r}\frac{\partial w_0}{\partial \varphi}} \quad \text{or} \quad M_{\varphi\varphi}^0 = \widehat{M}_{\varphi\varphi}^0.
\end{aligned} \quad (9.28)$$

9.3 Rotationally Symmetric Bending of Circular Plates

9.3.1 Basic Equations

In this section, we consider the bending of plates with circular planforms, where both the load and the boundary conditions are rotationally symmetric. As a consequence, all state variables depend only on the radial coordinate r, and the displacement $v = u_\varphi$ also vanishes. The strain field is obtained with $\gamma_{r\varphi} = 0$ as:

$$\varepsilon_{rr} = -z\frac{\mathrm{d}^2 w_0}{\mathrm{d}r^2}, \quad \varepsilon_{\varphi\varphi} = -\frac{z}{r}\frac{\mathrm{d}w_0}{\mathrm{d}r}. \tag{9.29}$$

The stress field is with $\tau_{r\varphi} = 0$:

$$\begin{aligned}\sigma_{rr} &= -Q_{11}z\frac{\mathrm{d}^2 w_0}{\mathrm{d}r^2} - Q_{12}\frac{z}{r}\frac{\mathrm{d}w_0}{\mathrm{d}r},\\ \sigma_{\varphi\varphi} &= -Q_{12}z\frac{\mathrm{d}^2 w_0}{\mathrm{d}r^2} - Q_{22}\frac{z}{r}\frac{\mathrm{d}w_0}{\mathrm{d}r}.\end{aligned} \tag{9.30}$$

The nonvanishing moment flows M^0_{rr} and $M^0_{\varphi\varphi}$ are obtained as:

$$\begin{aligned}M^0_{rr} &= -D_{11}\frac{\mathrm{d}^2 w_0}{\mathrm{d}r^2} - D_{12}\frac{1}{r}\frac{\mathrm{d}w_0}{\mathrm{d}r},\\ M^0_{\varphi\varphi} &= -D_{12}\frac{\mathrm{d}^2 w_0}{\mathrm{d}r^2} - D_{22}\frac{1}{r}\frac{\mathrm{d}w_0}{\mathrm{d}r}.\end{aligned} \tag{9.31}$$

The remaining transverse shear force flow Q_r can be given as follows:

$$Q_r = \frac{\mathrm{d}M^0_{rr}}{\mathrm{d}r} + \frac{M^0_{rr} - M^0_{\varphi\varphi}}{r} = -D_{11}\frac{1}{r}\frac{\mathrm{d}}{\mathrm{d}r}\left(r\frac{\mathrm{d}^2 w_0}{\mathrm{d}r^2}\right) + D_{22}\frac{1}{r^2}\frac{\mathrm{d}w_0}{\mathrm{d}r}. \tag{9.32}$$

The plate equation can be formulated as:

$$-\frac{1}{r}\frac{\mathrm{d}}{\mathrm{d}r}\left(rQ_r\right) = p. \tag{9.33}$$

Inserting the shear force flow Q_r yields:

$$\frac{1}{r}\frac{\mathrm{d}}{\mathrm{d}r}\left[D_{11}\frac{\mathrm{d}}{\mathrm{d}r}\left(r\frac{\mathrm{d}^2 w_0}{\mathrm{d}r^2}\right) - D_{22}\frac{1}{r}\frac{\mathrm{d}w_0}{\mathrm{d}r}\right] = p. \tag{9.34}$$

With

$$\frac{\mathrm{d}}{\mathrm{d}r}\left(r\frac{\mathrm{d}^2 w_0}{\mathrm{d}r^2}\right) = r\frac{\mathrm{d}}{\mathrm{d}r}\left[\frac{1}{r}\frac{\mathrm{d}}{\mathrm{d}r}\left(r\frac{\mathrm{d}w_0}{\mathrm{d}r}\right)\right] + \frac{1}{r}\frac{\mathrm{d}w_0}{\mathrm{d}r} \tag{9.35}$$

it follows:

$$D_{11}\frac{1}{r}\frac{\mathrm{d}}{\mathrm{d}r}\left\{r\frac{\mathrm{d}}{\mathrm{d}r}\left[\frac{1}{r}\frac{\mathrm{d}}{\mathrm{d}r}\left(r\frac{\mathrm{d}w_0}{\mathrm{d}r}\right)\right]+\left(\frac{D_{11}-D_{22}}{D_{11}}\right)\frac{1}{r}\frac{\mathrm{d}w_0}{\mathrm{d}r}\right\}=p. \tag{9.36}$$

In the case of an isotropic plate, the stresses and force and moment quantities are as follows:

$$\begin{aligned}
\sigma_{rr} &= -\frac{Ez}{1-\nu^2}\left(\frac{\mathrm{d}^2w_0}{\mathrm{d}r^2}+\frac{\nu}{r}\frac{\mathrm{d}w_0}{\mathrm{d}r}\right),\\
\sigma_{\varphi\varphi} &= -\frac{Ez}{1-\nu^2}\left(\nu\frac{\mathrm{d}^2w_0}{\mathrm{d}r^2}+\frac{1}{r}\frac{\mathrm{d}w_0}{\mathrm{d}r}\right),\\
M^0_{rr} &= -D\left(\frac{\mathrm{d}^2w_0}{\mathrm{d}r^2}+\frac{\nu}{r}\frac{\mathrm{d}w_0}{\mathrm{d}r}\right),\\
M^0_{\varphi\varphi} &= -D\left(\nu\frac{\mathrm{d}^2w_0}{\mathrm{d}r^2}+\frac{1}{r}\frac{\mathrm{d}w_0}{\mathrm{d}r}\right),\\
Q_r &= -D\frac{1}{r}\frac{\mathrm{d}}{\mathrm{d}r}\left(r\frac{\mathrm{d}^2w_0}{\mathrm{d}r^2}\right)+D\frac{1}{r^2}\frac{\mathrm{d}w_0}{\mathrm{d}r}=-D\frac{\mathrm{d}}{\mathrm{d}r}\left[\frac{1}{r}\frac{\mathrm{d}}{\mathrm{d}r}\left(r\frac{\mathrm{d}w_0}{\mathrm{d}r}\right)\right].
\end{aligned} \tag{9.37}$$

The plate equation (9.36) takes on the following form:

$$\frac{1}{r}\frac{\mathrm{d}}{\mathrm{d}r}\left\{r\frac{\mathrm{d}}{\mathrm{d}r}\left[\frac{1}{r}\frac{\mathrm{d}}{\mathrm{d}r}\left(r\frac{\mathrm{d}w_0}{\mathrm{d}r}\right)\right]\right\}=\frac{p}{D}. \tag{9.38}$$

This equation can be solved by successive integration for a given load p, and it follows:

$$w_0=\int\frac{1}{r}\int r\int\frac{1}{r}\int\frac{rp}{D}\mathrm{d}r\mathrm{d}r\mathrm{d}r\mathrm{d}r. \tag{9.39}$$

For the special case of constant load $p = p_0 = const.$ the complete solution can be given as:

$$w_0=A_0+B_0r^2+C_0\ln r+D_0r^2\ln r+\frac{p_0r^4}{64D}, \tag{9.40}$$

i.e., the function f_0 in (9.26) represents the homogeneous solution of the plate equation for a rotationally symmetric plate situation.

9.3.2 Plates Under Constant Surface Load

In this section, we solve some elementary examples of isotropic circular plates under rotationally symmetric loads and boundary conditions. The first plate situation to be considered is the simply supported circular plate loaded by a constant surface load

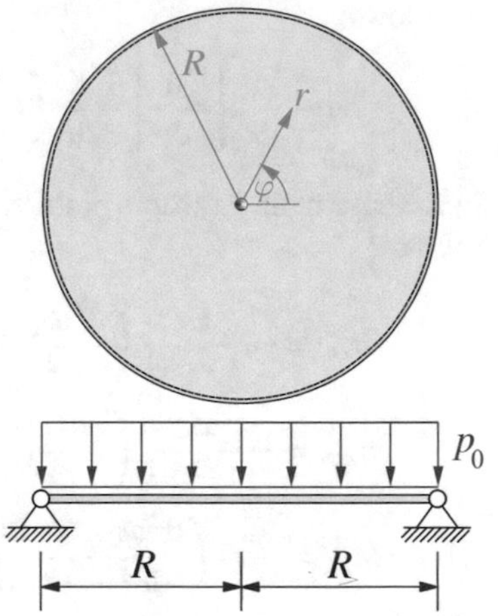

Fig. 9.4 Simply supported circular plate under constant surface load p_0

$p = p_0 = const.$ (Fig. 9.4). The plate has the radius R and the constant plate stiffness D.

For such a plate situation, the two constants C_0 and D_0 are set to zero, since otherwise the solution (9.40) for $r \to \infty$ would describe infinitely large deflections. Thus, it remains:

$$w_0 = A_0 + B_0 r^2 + \frac{p_0 r^4}{64D}. \tag{9.41}$$

The boundary conditions to be applied here are:

$$w_0(r = R) = 0, \quad M_{rr}^0(r = R) = 0. \tag{9.42}$$

From the second condition, the constant B_0 can be obtained immediately with (9.41) as:

$$B_0 = -\frac{p_0 R^2}{32D}\frac{3+\nu}{1+\nu}. \tag{9.43}$$

From the first condition in (9.42) follows.

$$A_0 + B_0 R^2 + \frac{p_0 R^4}{64D} = 0, \tag{9.44}$$

which with (9.43) leads to the following expression for the constant A_0:

$$A_0 = \frac{p_0 R^4}{64D}\frac{5+\nu}{1+\nu}. \tag{9.45}$$

The plate deflection can thus be specified as:

$$w_0 = \frac{p_0 R^4}{64D}\left(\frac{r^4}{R^4} - \frac{2r^2}{R^2}\frac{3+\nu}{1+\nu} + \frac{5+\nu}{1+\nu}\right). \tag{9.46}$$

The maximum deflection is obtained at the center of the plate for $r = 0$:

$$w_{0,\max} = \frac{p_0 R^4}{64D}\frac{5+\nu}{1+\nu}. \tag{9.47}$$

The moment flows M^0_{rr} and $M^0_{\varphi\varphi}$ are given by (9.37):

$$\begin{aligned} M^0_{rr} &= \frac{p_0}{16}(3+\nu)\left(R^2 - r^2\right), \\ M^0_{\varphi\varphi} &= \frac{p_0}{16}\left[(3+\nu)R^2 - (1+3\nu)r^2\right]. \end{aligned} \tag{9.48}$$

For the two normal stresses σ_{rr} and $\sigma_{\varphi\varphi}$ the following expressions are obtained:

$$\begin{aligned} \sigma_{rr} &= \frac{3p_0 z}{4h^3}(3+\nu)\left(R^2 - r^2\right), \\ \sigma_{\varphi\varphi} &= \frac{3p_0 z}{4h^3}\left[(3+\nu)R^2 - (1+3\nu)r^2\right]. \end{aligned} \tag{9.49}$$

The maximum stresses σ_{rr} and $\sigma_{\varphi\varphi}$ in this plate occur at $r = 0$ with the value

$$\sigma_{rr,\max} = \sigma_{\varphi\varphi,\max} = \frac{3(3+\nu)p_0}{8}\left(\frac{R}{h}\right)^2. \tag{9.50}$$

As another example, consider the clamped circular plate under uniform surface load p_0 with constant bending stiffness D (Fig. 9.5).

Also for this case, the two constants C_0 and D_0 are set to zero again, so that we again start from the solution (9.41). The underlying boundary conditions are as follows for the given plate situation:

$$w_0(r = R) = 0, \quad \left.\frac{\mathrm{d}w_0}{\mathrm{d}r}\right|_{r=R} = 0. \tag{9.51}$$

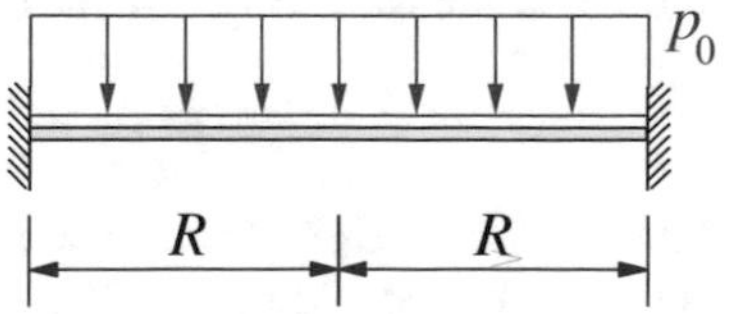

Fig. 9.5 Clamped circular plate under constant surface load p_0

The two remaining constants A_0 and B_0 follow as:

$$A_0 = \frac{p_0 R^4}{64D}, \quad B_0 = -\frac{p_0 R^2}{32D}. \tag{9.52}$$

Thus, the plate deflection w_0 can be given as:

$$w_0 = \frac{p_0}{64D}\left(R^2 - r^2\right)^2. \tag{9.53}$$

Again, the maximum deflection occurs at the center of the plate $r = 0$ and is given by the value

$$w_{0,\max} = \frac{p_0 R^4}{64D}. \tag{9.54}$$

The two moment flows M_{rr}^0 and $M_{\varphi\varphi}^0$ result as:

$$\begin{aligned} M_{rr}^0 &= \frac{p_0}{16}\left[(1+\nu) R^2 - (3+\nu) r^2\right], \\ M_{\varphi\varphi}^0 &= \frac{p_0}{16}\left[(1+\nu) R^2 - (1+3\nu) r^2\right]. \end{aligned} \tag{9.55}$$

The normal stresses σ_{rr} and $\sigma_{\varphi\varphi}$ follow as:

$$\begin{aligned} \sigma_{rr} &= \frac{3p_0 z}{4h^3}\left[(1+\nu) R^2 - (3+\nu) r^2\right], \\ \sigma_{\varphi\varphi} &= \frac{3p_0 z}{4h^3}\left[(1+\nu) R^2 - (1+3\nu) r^2\right]. \end{aligned} \tag{9.56}$$

The maximum normal stress σ_{rr} occurs at the plate edge $r = R$ with the value

$$\sigma_{rr,\max} = -\frac{3p_0}{4}\left(\frac{R}{h}\right)^2. \tag{9.57}$$

9.3.3 Plates Under Centric Point Force

We now consider a rotationally symmetrical isotropic circular plate (constant plate stiffness D), which is loaded by a centric single point force P. Therefore $p = 0$ is valid, and the following form of the displacement solution is valid (9.40):

$$w_0 = A_0 + B_0 r^2 + C_0 \ln r + D_0 r^2 \ln r. \tag{9.58}$$

Again, the term $C_0 \ln r$ is neglected. The term $D_0 r^2 \ln r$, on the other hand, is retained in the further consideration. We thus have:

$$w_0 = A_0 + B_0 r^2 + D_0 r^2 \ln r. \tag{9.59}$$

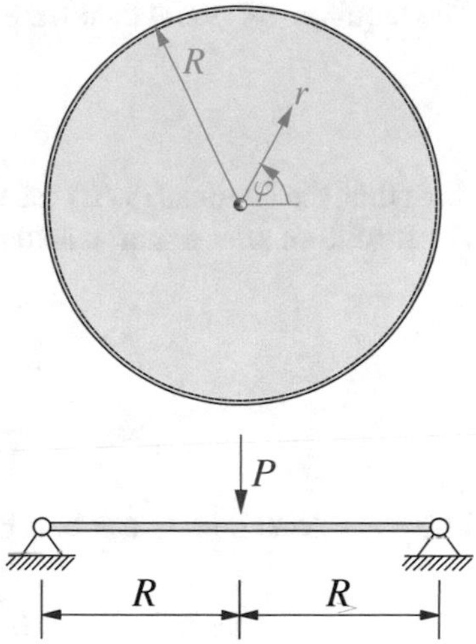

Fig. 9.6 Simply supported circular plate under centric point force P

As a first example, consider the simply supported circular plate under the centric point force P (Fig. 9.6). The boundary conditions to be considered here are:

$$w_0(r = R) = 0, \quad M^0_{rr}(r = R) = 0. \tag{9.60}$$

Additionally, it is required that at any circular section through the plate for any r, the shear force flow Q_r is in equilibrium with the applied load P, i.e.:

$$Q_r = -\frac{P}{2\pi r}. \tag{9.61}$$

The two conditions (9.60) give rise to the following two equations:

$$\begin{aligned} A_0 + B_0 R^2 + D_0 R^2 \ln R &= 0, \\ 2B_0\,(1+\nu) + D_0\,[2 \ln R\,(1+\nu) + 3 + \nu] &= 0. \end{aligned} \tag{9.62}$$

To evaluate the condition (9.61), we consider the expression (9.37) for the transverse shear force flow Q_r:

$$Q_r = -D \frac{\mathrm{d}}{\mathrm{d}r}\left[\frac{1}{r}\frac{\mathrm{d}}{\mathrm{d}r}\left(r\frac{\mathrm{d}w_0}{\mathrm{d}r}\right)\right]. \tag{9.63}$$

Inserting (9.59) yields after a short calculation:

$$Q_r = -\frac{4D}{r} D_0. \tag{9.64}$$

The requirement (9.61) then leads to the following constant D_0:

$$D_0 = \frac{P}{8\pi D}. \tag{9.65}$$

With this, the relations (9.62) can then be solved for the remaining constants A_0 and B_0. It follows after a short calculation:

$$\begin{aligned} A_0 &= \frac{PR^2}{16\pi D}\frac{3+\nu}{1+\nu}, \\ B_0 &= -\frac{P}{16\pi D}\left(2\ln R + \frac{3+\nu}{1+\nu}\right). \end{aligned} \tag{9.66}$$

The plate deflection w_0 can thus be written as:

$$w_0 = \frac{P}{16\pi D}\left[2r^2 \ln\left(\frac{r}{R}\right) + \left(R^2 - r^2\right)\frac{3+\nu}{1+\nu}\right]. \tag{9.67}$$

The maximum value of the deflection w_0 at the point $r = 0$ follows as:

$$w_{0,\max} = \frac{PR^2}{16\pi D}\frac{3+\nu}{1+\nu}. \tag{9.68}$$

For the two normal stresses, the following expressions are obtained:

$$\begin{aligned} \sigma_{rr} &= \frac{3Pz}{\pi h^3}(1+\nu)\ln\left(\frac{R}{r}\right), \\ \sigma_{\varphi\varphi} &= \frac{3Pz}{\pi h^3}\left[(1+\nu)\ln\left(\frac{R}{r}\right) + 1 - \nu\right]. \end{aligned} \tag{9.69}$$

It follows that for $r \to 0$ there is a stress singularity and the normal stresses σ_{rr} and $\sigma_{\varphi\varphi}$ grow without bounds.

We further consider the clamped circular plate under the centric point force P (Fig. 9.7) and again assume the solution (9.59). The boundary conditions to be considered here are:

$$w_0(r = R) = 0, \quad \left.\frac{\mathrm{d}w_0}{\mathrm{d}r}\right|_{r=R} = 0. \tag{9.70}$$

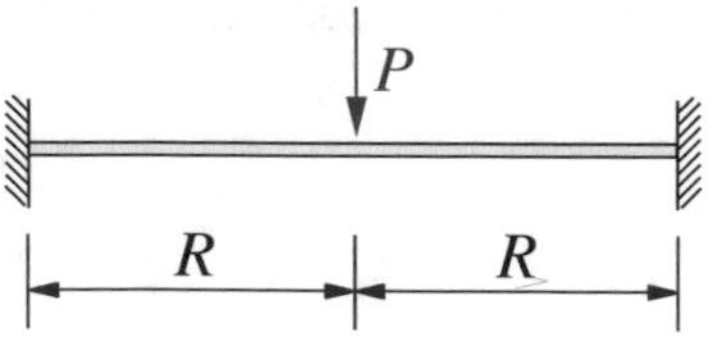

Fig. 9.7 Clamped circular plate under centric single point force P

Moreover, the condition (9.61) is raised again. Evaluating these demands leads to the constants A_0, B_0, D_0:

$$A_0 = \frac{PR^2}{16\pi D}, \quad B_0 = -\frac{P}{16\pi D}\left(2\ln R + 1\right), \quad D_0 = \frac{P}{8\pi D}. \tag{9.71}$$

The deflection w_0 then follows for the clamped plate as:

$$w_0 = \frac{P}{16\pi D}\left[2r^2 \ln\left(\frac{r}{R}\right) + R^2 - r^2\right]. \tag{9.72}$$

The maximum value in the center of the plate is:

$$w_{0,\max} = \frac{PR^2}{16\pi D}. \tag{9.73}$$

The two normal stresses σ_{rr} and $\sigma_{\varphi\varphi}$ follow to:

$$\begin{aligned} \sigma_{rr} &= \frac{3Pz}{\pi h^3}\left[(1+\nu)\ln\left(\frac{R}{r}\right) - 1\right], \\ \sigma_{\varphi\varphi} &= \frac{3Pz}{\pi h^3}\left[(1+\nu)\ln\left(\frac{R}{r}\right) - \nu\right]. \end{aligned} \tag{9.74}$$

Again, a stress singularity for $r \to 0$ occurs.

9.3.4 Plate Under Edge Moments

In this section, we consider a simply supported circular plate (radius R, constant bending stiffness D) under the edge moment M_0 (Fig. 9.8). The boundary conditions to be considered here are:

$$w_0(r = R) = 0, \quad M_{rr}^0(r = R) = M_0. \tag{9.75}$$

Using the approach (9.41) with $p_0 = 0$, the constants A_0 and B_0 are obtained as:

$$A_0 = \frac{M_0 R^2}{2D(1+\nu)}, \quad B_0 = -\frac{M_0}{2D(1+\nu)}. \tag{9.76}$$

This gives the plate deflection as:

$$w_0 = \frac{M_0}{2D(1+\nu)}\left(R^2 - r^2\right). \tag{9.77}$$

The moment flows M_{rr}^0 and $M_{\varphi\varphi}^0$ follow as:

$$M_{rr}^0 = M_{\varphi\varphi}^0 = M_0. \tag{9.78}$$

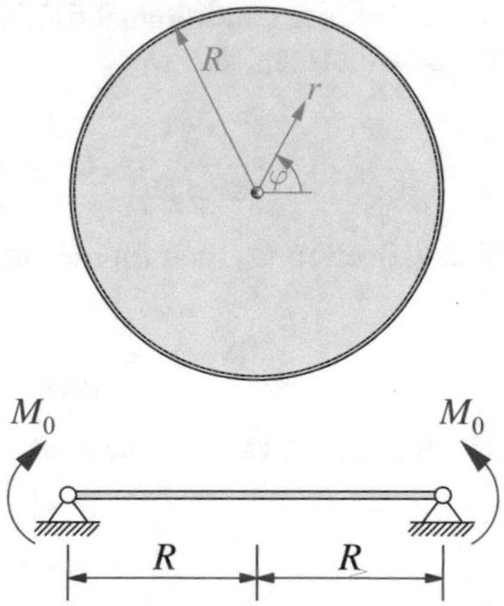

Fig. 9.8 Simply supported circular plate under edge moment M_0

9.3.5 Plate Under Partial Load

Consider a simply supported circular plate of constant bending stiffness D, which is loaded by a constant surface load p_0 on a partial area of radius R_1 (Fig. 9.9). This example has been solved e.g. by Girkmann (1974). The deflection is to be formulated separately for the two subareas $0 \leq r \leq R_i$ and $R_i \leq r \leq R$. For the interior region $0 \leq r \leq R_i$ the expression (9.41) for w_0 is valid:

$$w_0 = A_0 + B_0 r^2 + \frac{p_0 r^4}{64D}. \tag{9.79}$$

For the ring-shaped region $R_i \leq r \leq R$, on the other hand, (9.40) with $p_0 = 0$ is to be employed:

$$\bar{w}_0 = \bar{A}_0 + \bar{B}_0 r^2 + \bar{C}_0 \ln r + \bar{D}_0 r^2 \ln r. \tag{9.80}$$

Here, to distinguish the solutions for the two sub-areas, an overbar has been provided.

Besides the boundary conditions

$$\bar{w}_0(r = R) = 0, \quad \bar{M}^0_{rr}(r = R) = 0 \tag{9.81}$$

at the simply supported edge, suitable continuity conditions have to be formulated at the position $r = R_i$. These are as follows for the given example:

$$w_0(r = R_i) = \bar{w}_0(r = R_i), \quad \left.\frac{\mathrm{d}w_0}{\mathrm{d}r}\right|_{r=R_i} = \left.\frac{\mathrm{d}\bar{w}_0}{\mathrm{d}r}\right|_{r=R_i},$$

$$M^0_{rr}(r = R_i) = \bar{M}^0_{rr}(r = R_i), \quad Q_r(r = R_i) = \bar{Q}_r(r = R_i). \tag{9.82}$$

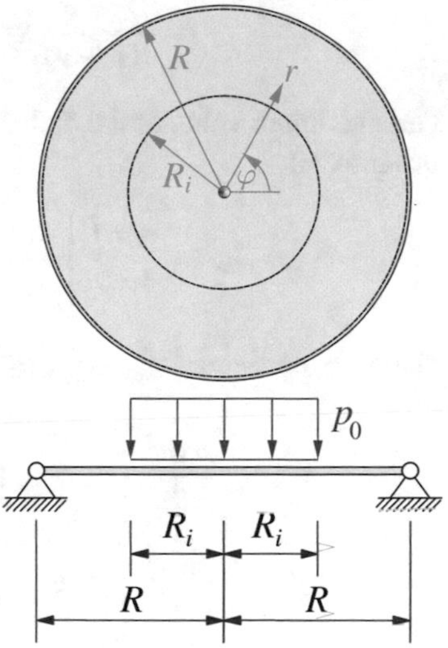

Fig. 9.9 Simply supported circular plate under partial load p_0

From the boundary and continuity conditions, the constants $A_0, B_0, \bar{A}_0, \bar{B}_0, \bar{C}_0, \bar{D}_0$ can be determined as follows:

$$A_0 = \frac{p_0 R_i^2}{64(1+\nu)D}\left[4(3+\nu)R^2 - (7+3\nu)R_i^2 - 4(1+\nu)R_i^2 \ln\frac{R}{R_i}\right],$$

$$B_0 = -\frac{p_0 R_i^2}{32(1+\nu)D}\left[4 - (1-\nu)\frac{R_i^2}{R^2} + 4(1+\nu)\ln\frac{R}{R_i}\right],$$

$$\bar{A}_0 = \frac{p_0 R_i^2}{32(1+\nu)D}\left[2(3+\nu)R^2 - (1-\nu)R_i^2 - 2(1+\nu)R_i^2 \ln R\right],$$

$$\bar{B}_0 = -\frac{p_0 R_i^2}{32(1+\nu)D}\left[2(3+\nu) + 4(1+\nu)\ln R - (1-\nu)\frac{R_i^2}{R^2}\right],$$

$$\bar{C}_0 = \frac{p_0 R_i^4}{16D}, \quad \bar{D}_0 = \frac{p_0 R_i^2}{8D}. \tag{9.83}$$

Thus, the two deflection functions w_0 and $\bar{w}_0$ can be formulated as:

$$w_0 = \frac{p_0 R_i^2}{16D}\left\{\frac{r^4}{4R_i^2} + r^2\left[\frac{(1-\nu)R_i^2 - 4R^2}{2(1+\nu)R^2} + 2\ln\frac{R_i}{R}\right] + \frac{4(3+\nu)R^2 - (7+3\nu)R_i^2}{4(1+\nu)} + R_i^2 \ln\frac{R_i}{R}\right\},$$

$$\bar{w}_0 = \frac{p_0 R_i^2}{16D}\left[\frac{3+\nu}{1+\nu}R^2\left(1 - \frac{r^2}{R^2}\right) - 2r^2 \ln\frac{R}{r}\right.$$

$$- \frac{1-\nu}{2(1+\nu)} R_i^2 \left(1 - \frac{r^2}{R^2}\right) - R_i^2 \ln \frac{R}{r}\Bigg]. \tag{9.84}$$

The maximum value of the deflections occurs in the center of the plate at $r = 0$ and amounts to:

$$w_{0,\max} = \frac{p_0 R_i^2}{16D} \left[\frac{4(3+\nu)R^2 - (7+3\nu)R_i^2}{4(1+\nu)} + R_i^2 \ln \frac{R_i}{R} \right]. \tag{9.85}$$

The moment flows M_{rr}^0, $M_{\varphi\varphi}^0$, $\bar{M}_{rr}^0$, $\bar{M}_{\varphi\varphi}^0$ follow as:

$$\begin{aligned}
M_{rr}^0 &= \frac{p_0 R_i^2}{4} \left[(1+\nu) \ln \frac{R}{R_i} + 1 - \frac{1-\nu}{4} \frac{R_i^2}{R^2} - \frac{3+\nu}{4} \frac{r^2}{R_i^2} \right], \\
M_{\varphi\varphi}^0 &= \frac{p_0 R_i^2}{4} \left[(1+\nu) \ln \frac{R}{R_i} + 1 - \frac{1-\nu}{4} \frac{R_i^2}{R^2} - \frac{3\nu+1}{4} \frac{r^2}{R_i^2} \right], \\
\bar{M}_{rr}^0 &= \frac{p_0 R_i^2}{4} \left[(1+\nu) \ln \frac{R}{R_i} + \frac{1-\nu}{4} \frac{R_i^2}{r^2} \left(1 - \frac{r^2}{R^2}\right) \right], \\
\bar{M}_{\varphi\varphi}^0 &= \frac{p_0 R_i^2}{4} \left[(1+\nu) \ln \frac{R}{R_i} + 1 - \nu - \frac{1-\nu}{4} \frac{R_i^2}{r^2} \left(1 + \frac{r^2}{R^2}\right) \right].
\end{aligned} \tag{9.86}$$

9.3.6 Circular Ring Plates

Another important case of circularly bounded plates are circular ring plates (Fig. 9.10). We consider the example of a simply supported circular ring plate (inner radius R_i, outer radius R_a), which is loaded by the boundary moment flows M_1 and M_2 as shown. The starting point for the considerations is the transverse shear force equation in (9.37):

$$Q_r = -D \frac{\mathrm{d}}{\mathrm{d}r} \left[\frac{1}{r} \frac{\mathrm{d}}{\mathrm{d}r} \left(r \frac{\mathrm{d}w_0}{\mathrm{d}r} \right) \right]. \tag{9.87}$$

For the given case, there is no shear force flow Q_r in the plate at all, so that:

$$\frac{\mathrm{d}}{\mathrm{d}r} \left[\frac{1}{r} \frac{\mathrm{d}}{\mathrm{d}r} \left(r \frac{\mathrm{d}w_0}{\mathrm{d}r} \right) \right] = 0. \tag{9.88}$$

The deflection w_0 can be determined from this by triple integration:

$$\begin{aligned}
\frac{1}{r} \frac{\mathrm{d}}{\mathrm{d}r} \left(r \frac{\mathrm{d}w_0}{\mathrm{d}r} \right) &= C_1, \\
\frac{\mathrm{d}}{\mathrm{d}r} \left(r \frac{\mathrm{d}w_0}{\mathrm{d}r} \right) &= C_1 r,
\end{aligned}$$

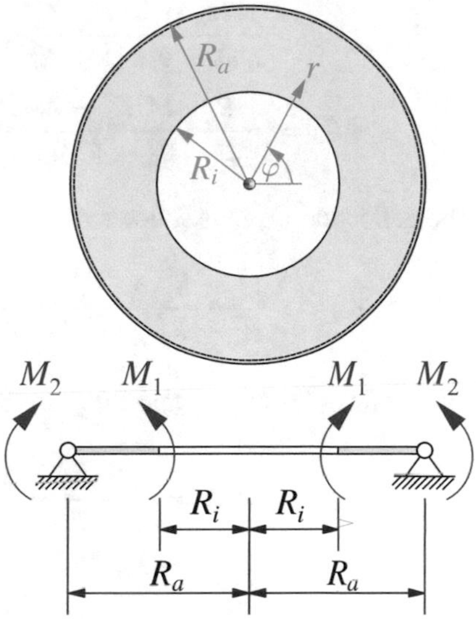

Fig. 9.10 Simply supported circular ring plate under edge moments

$$r\frac{\mathrm{d}w_0}{\mathrm{d}r} = \frac{1}{2}C_1 r^2 + C_2,$$
$$\frac{\mathrm{d}w_0}{\mathrm{d}r} = \frac{1}{2}C_1 r + \frac{C_2}{r},$$
$$w_0 = \frac{1}{4}C_1 r^2 + C_2 \ln r + C_3. \tag{9.89}$$

Thus the moment flow M_{rr}^0

$$M_{rr}^0 = -D\left(\frac{\mathrm{d}^2 w_0}{\mathrm{d}r^2} + \frac{\nu}{r}\frac{\mathrm{d}w_0}{\mathrm{d}r}\right) \tag{9.90}$$

can be determined as:

$$M_{rr}^0 = -D\left[\frac{1}{2}C_1(1+\nu) + \frac{C_2}{r^2}(-1+\nu)\right]. \tag{9.91}$$

The constants C_1, C_2 and C_3 are determined from the following boundary conditions:

$$w_0(r = R_a) = 0, \quad M_{rr}^0(r = R_a) = M_2, \quad M_{rr}^0(r = R_i) = M_1. \tag{9.92}$$

For the constants C_1, C_2 and C_3 the following expressions result after a short calculation:

$$C_1 = \frac{2}{D}\frac{M_1 R_i^2 - M_2 R_a^2}{(R_a^2 - R_i^2)(1+\nu)},$$

$$C_2 = \frac{(M_1 - M_2)R_i^2 R_a^2}{D(1-\nu)(R_a^2 - R_i^2)},$$

$$C_3 = -\frac{R_a^2}{2D}\frac{M_1 R_i^2 - M_2 R_a^2}{(R_a^2 - R_i^2)(1+\nu)} - \frac{(M_1 - M_2)R_i^2 R_a^2}{D(1-\nu)(R_a^2 - R_i^2)} \ln R_a. \tag{9.93}$$

The deflection w_0 thus follows as:

$$w_0 = \frac{1}{2}\frac{r^2 - R_a^2}{R_a^2 - R_i^2}\frac{M_1 R_i^2 - M_2 R_a^2}{D(1+\nu)} + \frac{R_i^2 R_a^2}{R_a^2 - R_i^2}\frac{M_1 - M_2}{D(1-\nu)} \ln \frac{r}{R_a}. \tag{9.94}$$

The moment flows M_{rr}^0 and $M_{\varphi\varphi}^0$ can be given as:

$$M_{rr}^0 = -\frac{M_1 R_i^2 - M_2 R_a^2}{R_a^2 - R_i^2} + \frac{R_i^2 R_a^2}{r^2}\frac{M_1 - M_2}{R_a^2 - R_i^2},$$

$$M_{\varphi\varphi}^0 = -\frac{M_1 R_i^2 - M_2 R_a^2}{R_a^2 - R_i^2} - \frac{R_i^2 R_a^2}{r^2}\frac{M_1 - M_2}{R_a^2 - R_i^2}. \tag{9.95}$$

We also consider the simply supported circular ring plate under the edge shear force flow Q_1 (Fig. 9.11). For this case, the following relationship holds between the shear force flow Q_r and the boundary force flow Q_1:

$$Q_r = -Q_1 \frac{R_i}{r}. \tag{9.96}$$

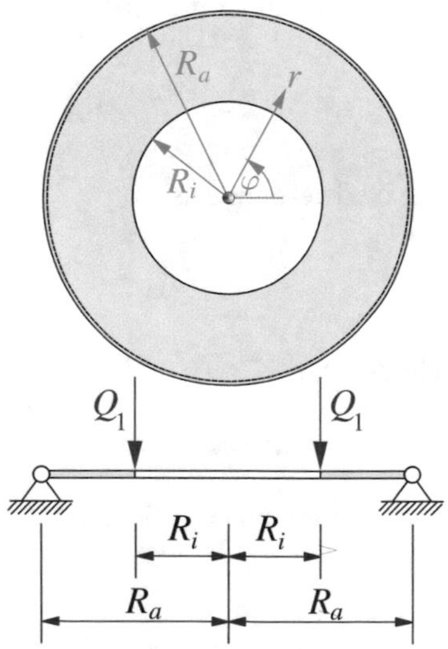

Fig. 9.11 Simply supported circular ring plate under edge forces

It follows accordingly with (9.87):

$$\frac{\mathrm{d}}{\mathrm{d}r}\left[\frac{1}{r}\frac{\mathrm{d}}{\mathrm{d}r}\left(r\frac{\mathrm{d}w_0}{\mathrm{d}r}\right)\right] = \frac{Q_1 R_i}{Dr}. \tag{9.97}$$

The plate deflection w_0 can then be obtained again by integration and is given as:

$$w_0 = \frac{Q_1 R_i r^2}{4D}(\ln r - 1) + \frac{C_1 r^2}{4} + C_2 \ln r + C_3. \tag{9.98}$$

The constants C_1, C_2 and C_3 follow from the boundary conditions

$$w_0(r = R_a) = 0, \quad M^0_{rr}(r = R_a) = 0, \quad M^0_{rr}(r = R_i) = 0, \tag{9.99}$$

and the following expression for w_0 can be given for this example:

$$w_0 = \frac{Q_1 R_a^2 R_i}{4D}\left\{\left(1 - \frac{r^2}{R_a^2}\right)\left[\frac{3+\nu}{2(1+\nu)} - \frac{R_i^2}{R_a^2 - R_i^2}\ln\frac{R_i}{R_a}\right] + \frac{r^2}{R_a^2}\ln\frac{r}{R_a} + \frac{2R_i^2}{R_a^2 - R_i^2}\frac{1+\nu}{1-\nu}\ln\frac{R_i}{R_a}\ln\frac{r}{R_a}\right\}. \tag{9.100}$$

9.4 Asymmetric Bending of Circular Plates

We now consider the case where the plate geometry and the boundary conditions of a circular plate are still rotationally symmetric, but the given load has no rotational symmetry. In this case the plate equation (9.23) is valid, and the homogeneous solution (9.25) with (9.26) can be used to describe the plate deflection.

As an example, consider the circular plate of Fig. 9.12, which is simply supported at its edge and subjected to the load p, where p can be split into the constant part p_0 and the varying part p_1 (Fig. 9.12, bottom). Let the load be described by the following expression:

$$p = p_0 + p_1 \frac{r}{R}\cos\varphi. \tag{9.101}$$

We first consider the particular solution $w_{0,p}$ for the constant load component p_0. Since this is a rotationally symmetric load component, the particular solution can be taken directly from (9.40), i.e.:

$$w_0 = \int \frac{1}{r}\int r \int \frac{1}{r}\int \frac{r p_0}{D}\mathrm{d}r\mathrm{d}r\mathrm{d}r\mathrm{d}r = \frac{p_0 r^4}{64D}. \tag{9.102}$$

Fig. 9.12 Simply supported circular plate under variable asymmetric load

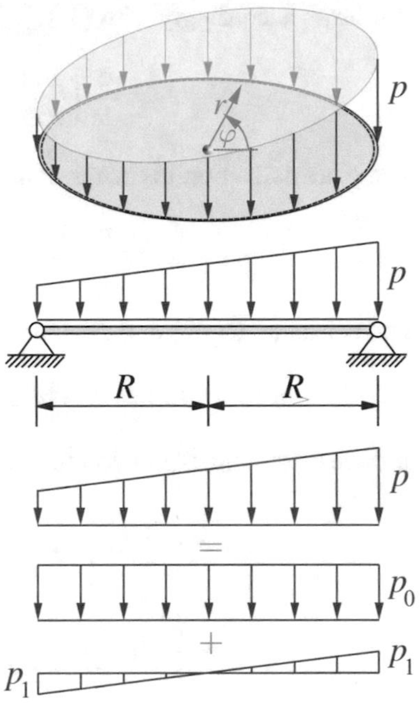

For the variable load component, on the other hand, an approach of the form

$$w_{0,p} = C\frac{p_1 r^5}{R}\cos\varphi \tag{9.103}$$

is used, which after insertion into the plate equation (9.23) gives the following constant C:

$$C = \frac{1}{192D}. \tag{9.104}$$

Thus, the total particular solution $w_{0,p}$ is:

$$w_{0,p} = \frac{p_0 r^4}{64D} + \frac{p_1 r^5}{192DR}\cos\varphi. \tag{9.105}$$

The homogeneous solution is given in the form

$$w_{0,h} = \sum_{n=0}^{\infty} f_n \cos n\varphi, \tag{9.106}$$

where we will use only the first two series elements in the following. In order that this solution leads to finite displacements at $r = 0$, the constants C_0, D_0, C_1, D_1 are set to zero. Thus, for the homogeneous solution $w_{0,h}$ follows:

$$w_{0,h} = A_0 + B_0 r^2 + \left(A_1 r + B_1 r^3\right) \cos\varphi, \tag{9.107}$$

and thus the full solution for the plate deflection can be given as:

$$w_0 = A_0 + B_0 r^2 + \frac{p_0 r^4}{64D} + \left(A_1 r + B_1 r^3 + \frac{p_1 r^5}{192DR}\right) \cos\varphi. \tag{9.108}$$

For the simply supported plate, the following boundary conditions apply at the boundary $r = R$:

$$w_0(r = R) = 0, \quad M_{rr}^0(r = R) = 0, \tag{9.109}$$

where M_{rr}^0 is calculated according to (9.11). Substituting the deflection w_0 (9.108) into the boundary conditions (9.109) gives the following constants A_0, B_0, A_1, B_1, considering that these requirements must be fulfilled for arbitrary angles φ:

$$\begin{aligned} A_0 &= \frac{5+\nu}{1+\nu}\frac{p_0 R^4}{64D}, \quad B_0 = -\frac{3+\nu}{1+\nu}\frac{p_0 R^2}{32D}, \\ A_1 &= \frac{7+\nu}{3+\nu}\frac{p_1 R^3}{192D}, \quad B_1 = -\frac{5+\nu}{3+\nu}\frac{p_1 R}{96D}. \end{aligned} \tag{9.110}$$

The deflection of the plate can then be expressed as follows:

$$\begin{aligned} w_0 = &\frac{p_0 R^4}{64D}\left(1 - \frac{r^2}{R^2}\right)\left(\frac{5+\nu}{1+\nu} - \frac{r^2}{R^2}\right) \\ &+ \frac{p_1 R^4}{192(3+\nu)D}\frac{r}{R}\left(1 - \frac{r^2}{R^2}\right)\left[7 + \nu - (3+\nu)\frac{r^2}{R^2}\right] \cos\varphi. \end{aligned} \tag{9.111}$$

The moment flows M_{rr}^0, $M_{\varphi\varphi}^0$, $M_{r\varphi}^0$ follow as:

$$\begin{aligned} M_{rr}^0 &= \frac{(3+\nu)p_0 R^2}{16}\left(1 - \frac{r^2}{R^2}\right) + \frac{(5+\nu)p_1 R^2}{48}\left(\frac{r}{R} - \frac{r^3}{R^3}\right)\cos\varphi, \\ M_{\varphi\varphi}^0 &= \frac{(3+\nu)p_0 R^2}{16}\left(1 - \frac{1+3\nu}{3+\nu}\frac{r^2}{R^2}\right) \\ &\quad + \frac{(1+3\nu)p_1 R^2}{48}\left(\frac{5+\nu}{3+\nu}\frac{r}{R} - \frac{1+5\nu}{1+3\nu}\frac{r^2}{R^2}\right)\cos\varphi, \\ M_{r\varphi}^0 &= -\frac{(1-\nu)p_1 R^2}{48}\left(\frac{5+\nu}{3+\nu} - \frac{r^2}{R^2}\right)\sin\varphi. \end{aligned} \tag{9.112}$$

If, on the other hand, a clamped plate is present, then the following boundary conditions apply for all angles φ:

$$w_0(r = R) = 0, \quad \left.\frac{\mathrm{d}w_0}{\mathrm{d}r}\right|_{r=R} = 0. \tag{9.113}$$

Evaluation using (9.108) yields the following expressions for the constants A_0, B_0, A_1, B_1:

$$A_0 = \frac{p_0 R^4}{64D}, \quad B_0 = -\frac{p_0 R^2}{32D}, \quad A_1 = \frac{p_1 R^3}{192D}, \quad B_1 = -\frac{p_1 R}{96D}. \tag{9.114}$$

The plate deflection can thus be specified as:

$$w_0 = \left(\frac{p_0 R^4}{64D} + \frac{p_1 R^4}{192D}\frac{r}{R}\cos\varphi\right)\left(1 - \frac{r^2}{R^2}\right)^2. \tag{9.115}$$

From this, the moment flows M^0_{rr}, $M^0_{\varphi\varphi}$, $M^0_{r\varphi}$ can be determined as follows:

$$\begin{aligned}
M^0_{rr} &= \frac{p_0 R^2}{16}\left[1 + \nu - (3 + \nu)\frac{r^2}{R^2}\right] \\
&\quad - \frac{p_1 R^2}{48}\left[(5 + \nu)\frac{r^3}{R^3} - (3 + \nu)\frac{r}{R}\right]\cos\varphi, \\
M^0_{\varphi\varphi} &= \frac{p_0 R^2}{16}\left[1 + \nu - (1 + 3\nu)\frac{r^2}{R^2}\right] \\
&\quad - \frac{p_1 R^2}{48}\left[(1 + 5\nu)\frac{r^3}{R^3} - (1 + 3\nu)\frac{r}{R}\right]\cos\varphi, \\
M^0_{r\varphi} &= -\frac{(1 - \nu) p_1 R^2}{48}\left(1 - \frac{r^2}{R^2}\right)\sin\varphi.
\end{aligned} \tag{9.116}$$

Further solutions for circular plates under non-rotationally symmetric loading can be found, for example, in Girkmann (1974) or Timoshenko and Woinowsky-Krieger (1964).

9.5 Strain Energy

For the application of energy-based approximation methods like the Ritz method or the finite element method, the strain energy / internal potential Π_i must be provided. For the plate with base surface Ω and constant thickness h, Π_i is given as:

$$\Pi_i = \frac{1}{2}\int_\Omega \int_{-\frac{h}{2}}^{+\frac{h}{2}} \left(\sigma_{rr}\varepsilon_{rr} + \sigma_{\varphi\varphi}\varepsilon_{\varphi\varphi} + \tau_{r\varphi}\gamma_{r\varphi}\right) \mathrm{d}z r \mathrm{d}r \mathrm{d}\varphi. \tag{9.117}$$

With the strains ε_{rr}, $\varepsilon_{\varphi\varphi}$, and $\gamma_{r\varphi}$ according to (9.10), integration over the plate thickness h yields:

$$\Pi_i = \frac{1}{2}\int_\Omega \left[-M_{rr}^0 \frac{\partial^2 w_0}{\partial r^2} - M_{\varphi\varphi}^0 \frac{1}{r}\left(\frac{1}{r}\frac{\partial^2 w_0}{\partial \varphi^2} + \frac{\partial w_0}{\partial r}\right) \right. \\ \left. + \frac{2}{r} M_{r\varphi}^0 \left(\frac{1}{r}\frac{\partial w_0}{\partial \varphi} - \frac{\partial^2 w_0}{\partial r \partial \varphi}\right)\right] r\mathrm{d}r\mathrm{d}\varphi. \tag{9.118}$$

Inserting the constitutive law (9.15), one obtains the strain energy Π_i for the polar-orthotropic plate as:

$$\Pi_i = \frac{1}{2}\int_\Omega \left[D_{11}\left(\frac{\partial^2 w_0}{\partial r^2}\right)^2 + 2D_{12}\frac{1}{r}\left(\frac{\partial w_0}{\partial r} + \frac{1}{r}\frac{\partial^2 w_0}{\partial \varphi^2}\right)\frac{\partial^2 w_0}{\partial r^2} \right. \\ \left. + D_{22}\frac{1}{r^2}\left(\frac{\partial w_0}{\partial r} + \frac{1}{r}\frac{\partial^2 w_0}{\partial \varphi^2}\right)^2 + 4D_{66}\frac{1}{r}\left(\frac{1}{r}\frac{\partial w_0}{\partial \varphi} - \frac{\partial^2 w_0}{\partial r \partial \varphi}\right)\right] r\mathrm{d}r\mathrm{d}\varphi. \tag{9.119}$$

For the case of an isotropic plate with $D_{11} = D_{22} = D$, $D_{12} = \nu D$, $D_{66} = \frac{1-\nu}{2} D$ we obtain:

$$\Pi_i = \frac{D}{2}\int_\Omega \left[\left(\frac{\partial^2 w_0}{\partial r^2}\right)^2 + \frac{2\nu}{r}\left(\frac{\partial w_0}{\partial r} + \frac{1}{r}\frac{\partial^2 w_0}{\partial \varphi^2}\right)\frac{\partial^2 w_0}{\partial r^2} \right. \\ \left. + \frac{1}{r^2}\left(\frac{\partial w_0}{\partial r} + \frac{1}{r}\frac{\partial^2 w_0}{\partial \varphi^2}\right)^2 + \frac{2(1-\nu)}{r}\left(\frac{1}{r}\frac{\partial w_0}{\partial \varphi} - \frac{\partial^2 w_0}{\partial r \partial \varphi}\right)\right] r\mathrm{d}r\mathrm{d}\varphi. \tag{9.120}$$

References

Altenbach, H., Altenbach, J., Naumenko, K.: Ebene Flächentragwerke, 2nd edn. Springer, Berlin (2016)
Girkmann, K.: Flächentragwerke, 6th edn. Springer, Wien (1974)
Hake, E., Meskouris, K.: Statik der Flächentragwerke, 2nd edn. Springer, Berlin (2007)
Reddy, J.N.: Theory and analysis of elastic plates and shells. CRC Press, Boca Raton (2006)
Szilard, R.: Theories and applications of plate analysis. Wiley Hoboken, USA (2004)
Timoshenko, S., Woinowsky-Krieger, S.: Theory of plates and shells, 2nd edn. McGraw-Hill, New York (1964)
Ugural, A.C.: Stresses in plates and shells. McGraw Hill, New York (1981)

10 Higher-order Plate Theories

The Kirchhoff plate theory discussed in the previous chapters has proven itself for many technical applications and is widely used in many technical fields of application. It does have some inconsistencies, but these are mostly negligible when considering sufficiently thin plate structures. The main assumptions of Kirchhoff's plate theory are the hypothesis that the cross-sections remain plane also in the deformed state, as well as the normal hypothesis and the assumption of a plane stress state with respect to the thickness direction of the plate. These hypotheses are reasonable and plausible for sufficiently thin plate structures, but produce increasingly poor quality results when dealing with thicker plates and/or plate materials with low transverse shear stiffness. The resulting errors are always on the unsafe side, i.e. one will underestimate, among other things, deflections of the plate as well as overestimate buckling loads, which is of course not acceptable in practical applications. Another problem is the calculation of the transverse shear stresses τ_{xz} and τ_{yz}, for which no constitutive relation is available within the framework of Kirchhoff plate theory and which must be determined from a post-calculation.

In order to circumvent these inconsistencies of the Kirchhoff plate theory, considerable research efforts have been made in the past decades to develop improved plate theories which are based on less stringent assumptions and lead to an improved quality of the computational results. This field of knowledge, which is the subject of this chapter, therefore includes the so-called higher-order plate theories. In this chapter we will discuss two such higher-order plate theories, again for the case of the orthotropic plate. One is the so-called First-Order Shear Deformation Theory (often abbreviated as FSDT), which is often also referred to as the so-called Reissner-Mindlin plate theory. It is in large parts identical with Kirchhoff's plate theory, the main difference being that the normal hypothesis is discarded. FSDT thus represents the equivalent of Timoshenko's beam theory in the treatment of plate structures. Furthermore, we will consider the so-called Reddy's Third-Order Shear Deformation Theory (abbreviated as TSDT), which, in addition to the normal hypothesis, also drops the hypothesis that cross-sections remain plane in the deformed state. In the framework of TSDT, cross-sectional warpings are explicitly allowed. For both higher-order plate theories we will present not only exact solutions but also

C. Mittelstedt, *Theory of Plates and Shells*,
https://doi.org/10.1007/978-3-662-66805-4_10

approximation methods. The interested reader can find further information on the topic discussed here e.g. in Altenbach et al. (2016), Mittelstedt and Becker (2016), and Reddy (2004, 2006).

10.1 First-Order Shear Deformation Theory

10.1.1 Kinematics and Constitutive Equations

The First-Order Shear Deformation Theory is based on the following assumptions:

- It is assumed that the normal stress σ_{zz} disappears at each point of the plate.
- The hypothesis that the cross-sections remain plane is still assumed to be valid, but the normal hypothesis is discarded. A straight line which was normal to the plate middle plane before deformation is also a straight line in the deformed state, but it is no longer necessarily normal to the plate middle plane.
- The thickness h of the plate does not change during the deformation process.
- We assume linear elasticity and small deformations of the plate.

The essential difference between Kirchhoff's plate theory and the First-Order Shear Deformation Theory thus obviously consists in the different treatment of the normal hypothesis. The kinematics according to the First-Order Shear Deformation Theory is shown in Fig. 10.1. By dropping the normal hypothesis, no connection between the bending angle ψ_x and the slope $\frac{\partial w_0}{\partial x}$ of the deflection w_0 can be established within the framework of the First-Order Shear Deformation Theory. Thus, the bending angle ψ_x is to be treated as an independent degree of freedom, which is also true for the bending angle ψ_y that also occurs but is not depicted in Fig. 10.1 due to the chosen perspective.

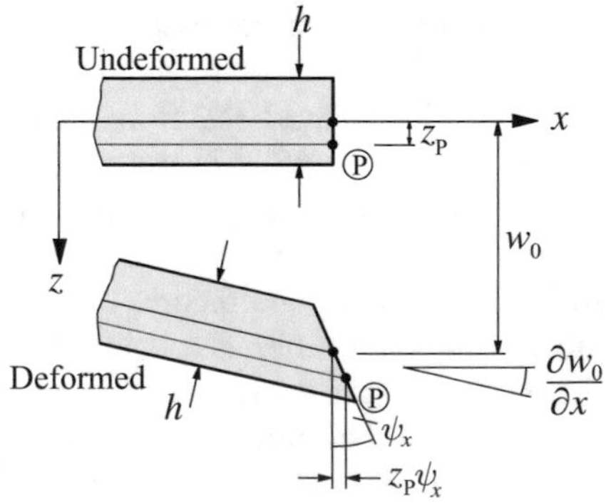

Fig. 10.1 Undeformed and deformed plate segment according to First-Order Shear Deformation Theory

Thus, within the framework of the First-Order Shear Deformation Theory, there are three independent degrees of freedom, namely the deflection w_0 of the plate midplane on the one hand, and the two bending angles ψ_x and ψ_y on the other hand.

From Fig. 10.1 the displacement field in the framework of First-Order Shear Deformation Theory can be deduced as:

$$
\begin{aligned}
u(x, y, z) &= z\psi_x(x, y), \\
v(x, y, z) &= z\psi_y(x, y), \\
w(x, y) &= w_0(x, y).
\end{aligned} \tag{10.1}
$$

This displacement field can be transformed into the one according to Kirchhoff's plate theory if $\psi_x = -\frac{\partial w_0}{\partial x}$ and $\psi_y = -\frac{\partial w_0}{\partial y}$ is used.

From the displacement field (10.1), the strain field in the framework of First-Order Shear Deformation Theory can be derived as follows:

$$
\begin{aligned}
\varepsilon_{xx} &= \frac{\partial u}{\partial x} = z\frac{\partial \psi_x}{\partial x}, \\
\varepsilon_{yy} &= \frac{\partial v}{\partial y} = z\frac{\partial \psi_y}{\partial y}, \\
\varepsilon_{zz} &= \frac{\partial w}{\partial z} = 0, \\
\gamma_{xy} &= \frac{\partial u}{\partial y} + \frac{\partial v}{\partial x} = z\left(\frac{\partial \psi_x}{\partial y} + \frac{\partial \psi_y}{\partial x}\right), \\
\gamma_{xz} &= \frac{\partial u}{\partial z} + \frac{\partial w}{\partial x} = \frac{\partial w_0}{\partial x} + \psi_x, \\
\gamma_{yz} &= \frac{\partial v}{\partial z} + \frac{\partial w}{\partial y} = \frac{\partial w_0}{\partial y} + \psi_y.
\end{aligned} \tag{10.2}
$$

In contrast to Kirchhoff's plate theory, not only the three plane strain components $\varepsilon_{xx}, \varepsilon_{yy}$ and γ_{xy} appear, but also the two transverse shear strains γ_{xz} and γ_{yz}. This is a direct consequence of dropping the normal hypothesis, since shear strains (expressed by the two bending angles ψ_x and ψ_y) are now allowed in the framework of First-Order Shear Deformation Theory. The two transverse shear strains γ_{xz} and γ_{yz} are constant over the plate thickness h and therefore not functions of the thickness coordinate z, which can be explained from the assumption that the cross-sections of the plate remain plane also in the deformed state. This entails some peculiarities, which will be discussed later. As a result of the assumption of invariability of the plate thickness h even in the deformed state, the transverse strain ε_{zz} also becomes zero in the framework of First-Order Shear Deformation Theory.

Once the displacement field and the strain field have been determined, the plane stresses σ_{xx}, σ_{yy} and τ_{xy} in the plate can be determined from Hooke's law:

$$\begin{pmatrix} \sigma_{xx} \\ \sigma_{yy} \\ \tau_{xy} \end{pmatrix} = \begin{bmatrix} Q_{11} & Q_{12} & 0 \\ Q_{12} & Q_{22} & 0 \\ 0 & 0 & Q_{66} \end{bmatrix} \begin{pmatrix} \varepsilon_{xx} \\ \varepsilon_{yy} \\ \gamma_{xy} \end{pmatrix}$$

$$= \begin{bmatrix} Q_{11} & Q_{12} & 0 \\ Q_{12} & Q_{22} & 0 \\ 0 & 0 & Q_{66} \end{bmatrix} \begin{pmatrix} z\frac{\partial \psi_x}{\partial x} \\ z\frac{\partial \psi_y}{\partial y} \\ z\left(\frac{\partial \psi_x}{\partial y} + \frac{\partial \psi_y}{\partial x}\right) \end{pmatrix}. \tag{10.3}$$

As expected, the stresses are distributed linearly over the plate thickness.

Since the transverse shear strains γ_{xz} and γ_{yz} do not vanish within the framework of First-Order Shear Deformation Theory, the transverse shear stresses τ_{xz} and τ_{yz} can now also be determined from a constitutive relation. At this point, a calculation from equilibrium relations in the context of a post-calculation is no longer necessary. The generalized Hooke's law (1.113) yields for τ_{xz} and τ_{yz}:

$$\begin{pmatrix} \tau_{yz} \\ \tau_{xz} \end{pmatrix} = \begin{bmatrix} C_{44} & 0 \\ 0 & C_{55} \end{bmatrix} \begin{pmatrix} \gamma_{yz} \\ \gamma_{xz} \end{pmatrix}, \tag{10.4}$$

with the transverse shear stiffnesses C_{44} and C_{55} of the orthotropic plate material. The resulting shear stresses τ_{xz} and τ_{yz} are constant over the plate thickness, which is a direct result of the kinematic assumptions (discarding the normal hypothesis, but maintaining the hypothesis that the cross-sections remain plane) of the First-Order Shear Deformation Theory. The transverse shear force flows Q_x and Q_y are then obtained by integrating the shear stresses τ_{xz} and τ_{yz} over the plate thickness:

$$\begin{pmatrix} Q_y \\ Q_x \end{pmatrix} = \int_{-\frac{h}{2}}^{+\frac{h}{2}} \begin{pmatrix} \tau_{yz} \\ \tau_{xz} \end{pmatrix} \mathrm{d}z = \int_{-\frac{h}{2}}^{+\frac{h}{2}} \begin{bmatrix} C_{44} & 0 \\ 0 & C_{55} \end{bmatrix} \begin{pmatrix} \gamma_{yz} \\ \gamma_{xz} \end{pmatrix} \mathrm{d}z. \tag{10.5}$$

The result of the integration is:

$$\begin{pmatrix} Q_y \\ Q_x \end{pmatrix} = \begin{bmatrix} A_{44} & 0 \\ 0 & A_{55} \end{bmatrix} \begin{pmatrix} \gamma_{yz} \\ \gamma_{xz} \end{pmatrix}, \tag{10.6}$$

where A_{44} and A_{55} represent the transverse shear stiffnesses of the plate:

$$A_{44} = \int_{-\frac{h}{2}}^{+\frac{h}{2}} C_{44}\mathrm{d}z = C_{44}h, \quad A_{55} = \int_{-\frac{h}{2}}^{+\frac{h}{2}} C_{55}\mathrm{d}z = C_{55}h. \tag{10.7}$$

Within the framework of First-Order Shear Deformation Theory, constant shear stresses τ_{xz} and τ_{yz} are obtained over z. Again, this result is in contrast to the actual parabolic shear stresses, which must become zero at the plate surfaces at $z = \pm\frac{h}{2}$ if there are no applied tangential loads. As a consequence, the transverse shear stiffnesses A_{44}, A_{55} will be higher than would result, for example, from experimental tests. Therefore, within the framework of First-Order Shear Deformation Theory, a so-called shear correction factor K is introduced to modify the transverse shear stiffnesses. For the transverse shear stresses τ_{xz} and τ_{yz} we then obtain:

$$\begin{pmatrix} \tau_{yz} \\ \tau_{xz} \end{pmatrix} = K \begin{bmatrix} C_{44} & 0 \\ 0 & C_{55} \end{bmatrix} \begin{pmatrix} \gamma_{yz} \\ \gamma_{xz} \end{pmatrix}. \tag{10.8}$$

The constitutive equation (10.6) then reads:

$$\begin{pmatrix} Q_y \\ Q_x \end{pmatrix} = K \begin{bmatrix} A_{44} & 0 \\ 0 & A_{55} \end{bmatrix} \begin{pmatrix} \gamma_{yz} \\ \gamma_{xz} \end{pmatrix}. \tag{10.9}$$

The introduction of the shear correction factor K does not eliminate the fundamentally incorrect constant distribution of the transverse shear stresses τ_{xz} and τ_{yz} over the plate thickness, but the computational reduction of the transverse shear stiffnesses attempts to counteract the effects of this contradiction of First-Order Shear Deformation Theory.

Finally, the constitutive law for a plate in the framework of First-Order Shear Deformation Theory is summarized as follows:

$$\begin{pmatrix} M^0_{xx} \\ M^0_{yy} \\ M^0_{xy} \\ Q_y \\ Q_x \end{pmatrix} = \begin{bmatrix} D_{11} & D_{12} & 0 & 0 & 0 \\ D_{12} & D_{22} & 0 & 0 & 0 \\ 0 & 0 & D_{66} & 0 & 0 \\ 0 & 0 & 0 & KA_{44} & 0 \\ 0 & 0 & 0 & 0 & KA_{55} \end{bmatrix} \begin{pmatrix} \frac{\partial \psi_x}{\partial x} \\ \frac{\partial \psi_y}{\partial y} \\ \frac{\partial \psi_x}{\partial y} + \frac{\partial \psi_y}{\partial x} \\ \frac{\partial w_0}{\partial y} + \psi_y \\ \frac{\partial w_0}{\partial x} + \psi_x \end{pmatrix}. \tag{10.10}$$

Besides the plate stiffnesses $D_{11}, D_{22}, D_{12}, D_{66}$ and moment flows M^0_{xx}, M^0_{yy} already known from Kirchhoff's plate theory, M^0_{xy}, the transverse shear force flows Q_x, Q_y and their associated transverse shear stiffnesses A_{44}, A_{55}, multiplied by the shear correction factor K, also appear here. In symbolic form, (10.10) reads:

$$\begin{pmatrix} \underline{M}^0 \\ \underline{Q} \end{pmatrix} = \begin{bmatrix} \underline{\underline{D}} & \underline{\underline{0}} \\ \underline{\underline{0}} & \underline{\underline{A}}_S \end{bmatrix} \begin{pmatrix} \underline{\kappa}^0 \\ \underline{\gamma}_S \end{pmatrix}. \tag{10.11}$$

The First-Order Shear Deformation Theory now provides for the first time a possibility to determine the transverse shear stresses τ_{xz} and τ_{yz} from a constitutive relation, but leads to constant stresses over z and thus contradictory results. The introduction of the shear correction factor does not eliminate this shortcoming, but it does reduce its effects when determining deflections, buckling loads or natural frequencies. If one is interested in more accurate results for the transverse shear stresses, then the local three-dimensional equilibrium can be used to determine the transverse shear stresses

τ_{yz} and τ_{xz} by integrating the plane stresses σ_{xx}, σ_{yy}, τ_{xy} present with (10.3):

$$\tau_{xz} = -\int_{-h/2}^{z} \left(\frac{\partial \sigma_{xx}}{\partial x} + \frac{\partial \tau_{xy}}{\partial y} \right) \mathrm{d}z,$$
$$\tau_{yz} = -\int_{-h/2}^{z} \left(\frac{\partial \tau_{xy}}{\partial x} + \frac{\partial \sigma_{yy}}{\partial y} \right) \mathrm{d}z. \tag{10.12}$$

10.1.2 Determination of the Shear Correction Factor K

The shear correction factor K is determined by equating the shear strain energies due to the quadratic shear stress distribution and according to First-Order Shear Deformation Theory. The shear strain energy reads in its general form:

$$\Pi_i = \frac{1}{2} \int\int\int_V \left(\tau_{xz}\gamma_{xz} + \tau_{yz}\gamma_{yz} \right) \mathrm{d}V. \tag{10.13}$$

We first consider the shear strain energy due to the quadratic shear stress distribution. The transverse shear stresses τ_{xz} and τ_{yz} due to the transverse shear forces Q_x and Q_y are found to be parabolic over the plate cross section and are:

$$\tau_{xz} = \frac{3Q_x}{2h}\left(1 - \frac{4z^2}{h^2}\right), \quad \tau_{yz} = \frac{3Q_y}{2h}\left(1 - \frac{4z^2}{h^2}\right). \tag{10.14}$$

Using the generalized Hooke's law $\tau_{yz} = C_{44}\gamma_{yz}$ and $\tau_{xz} = C_{55}\gamma_{xz}$ and splitting the volume integral into an integral over the cross-section A and an integral in the thickness direction yields:

$$\Pi_i = \frac{1}{2} \int\int_A \int_{-\frac{h}{2}}^{+\frac{h}{2}} \left(\frac{\tau_{xz}^2}{C_{55}} + \frac{\tau_{yz}^2}{C_{44}} \right) \mathrm{d}z\mathrm{d}A. \tag{10.15}$$

Inserting (10.14) yields:

$$\Pi_i = \frac{9}{8h^2} \int\int_A \left(\frac{Q_x^2}{C_{55}} + \frac{Q_y^2}{C_{44}} \right) \mathrm{d}A \int_{-\frac{h}{2}}^{+\frac{h}{2}} \left(1 - \frac{4z^2}{h^2} \right)^2 \mathrm{d}z$$
$$= \frac{9}{15h} \int\int_A \left(\frac{Q_x^2}{C_{55}} + \frac{Q_y^2}{C_{44}} \right) \mathrm{d}A. \tag{10.16}$$

The shear strain energy according to the present First-Order Shear Deformation Theory provides with the constitutive law $Q_x = KA_{55}\gamma_{xz}$ and $Q_y = KA_{44}\gamma_{yz}$:

$$\Pi_i = \frac{1}{2K} \int\int_A \int_{-\frac{h}{2}}^{+\frac{h}{2}} \left(\tau_{xz}\frac{Q_x}{A_{55}} + \tau_{yz}\frac{Q_y}{A_{44}} \right) \mathrm{d}z\mathrm{d}A. \tag{10.17}$$

Performing the integration with respect to z with $A_{44} = C_{44}h$ and $A_{55} = C_{55}h$ yields:

$$\Pi_i = \frac{1}{2hK} \int\int_A \left(\frac{Q_x^2}{C_{55}} + \frac{Q_y^2}{C_{44}} \right) \mathrm{d}A. \tag{10.18}$$

Equating the two expressions (10.16) and (10.18) leads to:

$$\frac{9}{15h} \int\int_A \left(\frac{Q_x^2}{C_{55}} + \frac{Q_y^2}{C_{44}} \right) \mathrm{d}A = \frac{1}{2hK} \int\int_A \left(\frac{Q_x^2}{C_{55}} + \frac{Q_y^2}{C_{44}} \right) \mathrm{d}A. \tag{10.19}$$

Solving for K then gives the classical result for the homogeneous orthotropic plate:

$$K = \frac{5}{6}. \tag{10.20}$$

10.1.3 Equilibrium and Boundary Conditions

The equilibrium conditions which we have already determined in the framework of Kirchhoff's plate theory can also be used in the framework of First-Order Shear Deformation Theory. They read (see Chap. 7):

$$\begin{aligned} \frac{\partial M_{xx}^0}{\partial x} + \frac{\partial M_{xy}^0}{\partial y} &= Q_x, \\ \frac{\partial M_{xy}^0}{\partial x} + \frac{\partial M_{yy}^0}{\partial y} &= Q_y, \\ \frac{\partial Q_x}{\partial x} + \frac{\partial Q_y}{\partial y} + p &= 0, \end{aligned} \tag{10.21}$$

where $p = p(x, y)$ is the load on the plate in the $z-$direction. It should be noted, however, that the use of the condensed plate equilibrium is not suitable here. Substitution of the constitutive relations (10.10) then yields the corresponding displacement differential equations describing the static behavior of the orthotropic plate in the context of First-Order Shear Deformation Theory:

$$\begin{aligned} D_{11}\frac{\partial^2 \psi_x}{\partial x^2} + D_{66}\frac{\partial^2 \psi_x}{\partial y^2} + (D_{12} + D_{66})\frac{\partial^2 \psi_y}{\partial x \partial y} &= K A_{55}\left(\psi_x + \frac{\partial w_0}{\partial x}\right), \\ (D_{12} + D_{66})\frac{\partial^2 \psi_x}{\partial x \partial y} + D_{66}\frac{\partial^2 \psi_y}{\partial x^2} + D_{22}\frac{\partial^2 \psi_y}{\partial y^2} &= K A_{44}\left(\psi_y + \frac{\partial w_0}{\partial y}\right), \\ K\left[A_{55}\left(\frac{\partial \psi_x}{\partial x} + \frac{\partial^2 w_0}{\partial x^2}\right) + A_{44}\left(\frac{\partial \psi_y}{\partial y} + \frac{\partial^2 w_0}{\partial y^2}\right)\right] + p = 0. \end{aligned} \tag{10.22}$$

These are three coupled partial differential equations in the three degrees of freedom according to the First-Order Shear Deformation Theory, i.e., the displacement w_0 of the plate midplane and the two bending angles ψ_x and ψ_y.

The equilibrium and boundary conditions can also be determined from an energetic consideration. For this purpose, we use the principle of virtual displacements $\delta W_i = \delta W_a$, considering a rectangular plate with dimensions a and b in the $x-$direction and the $y-$direction, respectively, and using the following expression for the virtual internal work δW_i:

$$\delta W_i = \int_0^b \int_0^a \left[M_{xx}^0 \frac{\partial \delta\psi_x}{\partial x} + M_{yy}^0 \frac{\partial \delta\psi_y}{\partial y} + M_{xy}^0 \left(\frac{\partial \delta\psi_x}{\partial y} + \frac{\partial \delta\psi_y}{\partial x} \right) \right.$$
$$\left. + Q_y \left(\frac{\partial \delta w_0}{\partial y} + \delta\psi_y \right) + Q_x \left(\frac{\partial \delta w_0}{\partial x} + \delta\psi_x \right) \right] \mathrm{d}x\mathrm{d}y. \tag{10.23}$$

We perform partial integration for those terms in which derivatives of virtual displacement quantities appear as follows:

$$\int_0^b \int_0^a M_{xx}^0 \frac{\partial \delta\psi_x}{\partial x} \mathrm{d}x\mathrm{d}y = \int_0^b M_{xx}^0 \delta\psi_x \Big|_0^a \mathrm{d}y - \int_0^b \int_0^a \frac{\partial M_{xx}^0}{\partial x} \delta\psi_x \mathrm{d}x\mathrm{d}y,$$
$$\int_0^b \int_0^a M_{yy}^0 \frac{\partial \delta\psi_y}{\partial y} \mathrm{d}x\mathrm{d}y = \int_0^a M_{yy}^0 \delta\psi_y \Big|_0^b \mathrm{d}x - \int_0^b \int_0^a \frac{\partial M_{yy}^0}{\partial y} \delta\psi_y \mathrm{d}x\mathrm{d}y,$$
$$\int_0^b \int_0^a M_{xy}^0 \frac{\partial \delta\psi_x}{\partial y} \mathrm{d}x\mathrm{d}y = \int_0^a M_{xy}^0 \delta\psi_x \Big|_0^b \mathrm{d}x - \int_0^b \int_0^a \frac{\partial M_{xy}^0}{\partial y} \delta\psi_x \mathrm{d}x\mathrm{d}y,$$
$$\int_0^b \int_0^a M_{xy}^0 \frac{\partial \delta\psi_y}{\partial x} \mathrm{d}x\mathrm{d}y = \int_0^b M_{xy}^0 \delta\psi_y \Big|_0^a \mathrm{d}y - \int_0^b \int_0^a \frac{\partial M_{xy}^0}{\partial x} \delta\psi_y \mathrm{d}x\mathrm{d}y,$$
$$\int_0^b \int_0^a Q_y \frac{\partial \delta w_0}{\partial y} \mathrm{d}x\mathrm{d}y = \int_0^a Q_y \delta w_0 \Big|_0^b \mathrm{d}x - \int_0^b \int_0^a \frac{\partial Q_y^0}{\partial y} \delta w_0 \mathrm{d}x\mathrm{d}y,$$
$$\int_0^b \int_0^a Q_x \frac{\partial \delta w_0}{\partial x} \mathrm{d}x\mathrm{d}y = \int_0^b Q_x \delta w_0 \Big|_0^a \mathrm{d}y - \int_0^b \int_0^a \frac{\partial Q_x}{\partial x} \delta w_0 \mathrm{d}x\mathrm{d}y. \tag{10.24}$$

The virtual external work is currently as follows:

$$\delta W_a = \int_0^b \int_0^a p \delta w_0 \mathrm{d}x\mathrm{d}y. \tag{10.25}$$

The principle of virtual displacements $\delta W_i = \delta W_a$ then results in:

$$\int_0^b \int_0^a \left[- \left(\frac{\partial Q_x}{\partial x} + \frac{\partial Q_y}{\partial y} + p \right) \delta w_0 - \left(\frac{\partial M_{xx}^0}{\partial x} + \frac{\partial M_{xy}^0}{\partial y} - Q_x \right) \delta\psi_x \right.$$
$$\left. - \left(\frac{\partial M_{xy}^0}{\partial x} + \frac{\partial M_{yy}^0}{\partial y} - Q_y \right) \delta\psi_y \right] \mathrm{d}x\mathrm{d}y$$

$$+\int_0^b \left(Q_x\delta w_0 + M_{xx}^0\delta\psi_x + M_{xy}^0\delta\psi_y\right)\Big|_0^a \mathrm{d}y$$
$$+\int_0^a \left(Q_y\delta w_0 + M_{yy}^0\delta\psi_y + M_{xy}^0\delta\psi_x\right)\Big|_0^b \mathrm{d}x = 0. \tag{10.26}$$

Obviously, the bracket terms of the first two lines contain the three equilibrium conditions (10.21), whereas the remaining terms describe the corresponding potential boundary conditions of the given static plate bending problem in the framework of First-Order Shear Deformation Theory. Accordingly, at the two edges $x = 0$ and $x = a$ for $0 \leq y \leq b$ one of the following quantities is predefined:

$$\begin{aligned} w_0 &\quad \text{or} \quad Q_x, \\ \psi_x &\quad \text{or} \quad M_{xx}^0, \\ \psi_y &\quad \text{or} \quad M_{xy}^0. \end{aligned} \tag{10.27}$$

On the other hand, at the two edges at $y = 0$ and $y = b$ for $0 \leq x \leq a$, the following edge quantities can be given:

$$\begin{aligned} w_0 &\quad \text{or} \quad Q_y, \\ \psi_x &\quad \text{or} \quad M_{xy}^0, \\ \psi_y &\quad \text{or} \quad M_{yy}^0. \end{aligned} \tag{10.28}$$

If we use more generally the indices n and s for the directions normal and tangential to the considered boundary, then the following boundary quantities can be given:

$$\begin{aligned} w_0 &\quad \text{or} \quad Q_n, \\ \psi_n &\quad \text{or} \quad M_{nn}^0, \\ \psi_s &\quad \text{or} \quad M_{ns}^0. \end{aligned} \tag{10.29}$$

At a simply supported plate edge, the moment flows M_{nn}^0 and M_{ns}^0 disappear in addition to the deflection w_0:

$$M_{nn}^0 = M_{ns}^0 = w_0 = 0. \tag{10.30}$$

If there is a clamped plate edge, all displacement boundary terms disappear:

$$\psi_n = \psi_s = w_0 = 0. \tag{10.31}$$

At a free plate edge, on the other hand, all force quantities of the plate must become zero:

$$M_{nn}^0 = M_{ns}^0 = Q_n = 0. \tag{10.32}$$

10.1.4 Strain Energy

For the application of energy-based approximative methods, statements are to be made about the energy stored in the plate under consideration, so that we want to derive the strain energy or the internal potential Π_i of a plate according to First-Order Shear Deformation Theory. It holds:

$$\Pi_i = \frac{1}{2}\int\int\int_V \left(\sigma_{xx}\varepsilon_{xx} + \sigma_{yy}\varepsilon_{yy} + \tau_{xy}\gamma_{xy} + \tau_{xz}\gamma_{xz} + \tau_{yz}\gamma_{yz}\right) \mathrm{d}V. \quad (10.33)$$

Performing the integration with respect to the thickness direction z yields:

$$\begin{aligned}\Pi_i &= \frac{1}{2}\int_0^b\int_0^a \left[M_{xx}^0 \frac{\partial \psi_x}{\partial x} + M_{yy}^0 \frac{\partial \psi_y}{\partial y} + M_{xy}^0 \left(\frac{\partial \psi_x}{\partial y} + \frac{\partial \psi_y}{\partial x}\right)\right. \\ &+ \left. Q_y \left(\frac{\partial w_0}{\partial y} + \psi_y\right) + Q_x \left(\frac{\partial w_0}{\partial x} + \psi_x\right)\right] \mathrm{d}x\mathrm{d}y. \end{aligned} \quad (10.34)$$

Inserting the constitutive law (10.10) delivers:

$$\begin{aligned}\Pi_i &= \frac{1}{2}\int_0^b\int_0^a \left[D_{11}\left(\frac{\partial \psi_x}{\partial x}\right)^2 + 2D_{12}\frac{\partial \psi_x}{\partial x}\frac{\partial \psi_y}{\partial y} + D_{22}\left(\frac{\partial \psi_y}{\partial y}\right)^2 + D_{66}\left(\frac{\partial \psi_x}{\partial y} + \frac{\partial \psi_y}{\partial x}\right)^2\right. \\ &+ \left. KA_{44}\left(\frac{\partial w_0}{\partial y} + \psi_y\right)^2 + KA_{55}\left(\frac{\partial w_0}{\partial x} + \psi_x\right)^2\right] \mathrm{d}x\mathrm{d}y. \end{aligned} \quad (10.35)$$

10.1.5 Bending of Plate Strips

For a number of specific situations, as for the Kirchhoff plate, exact solutions for static boundary value problems can be given in the framework of the First-Order Shear Deformation Theory. We begin the elaborations for the case of a plate strip, i.e., for plate situations as shown in Fig. 7.20 and 7.21. Both moment flows M_{yy}^0 and M_{xy}^0 as well as the transverse shear force flow Q_y and the bending angle ψ_y are set to zero, i.e. $M_{yy}^0 = M_{xy}^0 = 0$, $Q_y = 0$, $\psi_y = 0$. Furthermore, the deflection w_0 as well as the bending angle ψ_x are only functions of the longitudinal coordinate x: $w_0 = w_0(x)$, $\psi_x = \psi_x(x)$. This is also assumed for the load $p = p(x)$.

The constitutive law (10.10) can be inverted as follows:

$$\begin{pmatrix} \frac{\partial \psi_x}{\partial x} \\ \frac{\partial \psi_y}{\partial y} \\ \frac{\partial \psi_x}{\partial y} + \frac{\partial \psi_y}{\partial x} \\ \frac{\partial w_0}{\partial y} + \psi_y \\ \frac{\partial w_0}{\partial x} + \psi_x \end{pmatrix} = \begin{bmatrix} \bar{D}_{11} & \bar{D}_{12} & 0 & 0 & 0 \\ \bar{D}_{12} & \bar{D}_{22} & 0 & 0 & 0 \\ 0 & 0 & \bar{D}_{66} & 0 & 0 \\ 0 & 0 & 0 & \frac{1}{K}\bar{A}_{44} & 0 \\ 0 & 0 & 0 & 0 & \frac{1}{K}\bar{A}_{55} \end{bmatrix} \begin{pmatrix} M_{xx}^0 \\ M_{yy}^0 \\ M_{xy}^0 \\ Q_y \\ Q_x \end{pmatrix}. \quad (10.36)$$

Here, the quantities marked with an overbar are the inverse plate stiffnesses and the inverse transverse shear stiffnesses, respectively. For $\bar{D}_{11}$ and $\bar{D}_{12}$ we obtain:

$$\bar{D}_{11} = \frac{D_{22}}{D_{11}D_{22} - D_{12}^2}, \quad \bar{D}_{12} = -\frac{D_{12}}{D_{11}D_{22} - D_{12}^2}. \tag{10.37}$$

The inverted transverse shear stiffnesses $\bar{A}_{44}$ and $\bar{A}_{55}$ read:

$$\bar{A}_{44} = \frac{1}{A_{44}}, \quad \bar{A}_{55} = \frac{1}{A_{55}}. \tag{10.38}$$

With the assumptions made, the reduced displacement field of the plate strip can be written as:

$$u(x, z) = z\psi_x(x), \quad w(x, y) = w_0(x). \tag{10.39}$$

The resulting reduced strain field is then:

$$\varepsilon_{xx} = z\frac{\mathrm{d}\psi_x}{\mathrm{d}x}, \quad \gamma_{xz} = \frac{\mathrm{d}w_0}{\mathrm{d}x} + \psi_x. \tag{10.40}$$

From the inverted representation of the constitutive law we obtain with $M_{yy}^0 = M_{xy}^0 = 0$:

$$\begin{aligned} \frac{\mathrm{d}\psi_x}{\mathrm{d}x} &= \bar{D}_{11}M_{xx}^0, \\ \frac{\mathrm{d}w_0}{\mathrm{d}x} + \psi_x &= \frac{1}{K}\bar{A}_{55}Q_x. \end{aligned} \tag{10.41}$$

If we relate the force and moment quantities to the width b of the plate strip, i.e. we use $M(x) = M_{xx}^0 b$ and $Q(x) = Q_x b$, we get:

$$\begin{aligned} EI\frac{\mathrm{d}\psi_x}{\mathrm{d}x} &= M, \\ KGA\left(\frac{\mathrm{d}w_0}{\mathrm{d}x} + \psi_x\right) &= Q. \end{aligned} \tag{10.42}$$

The quantity $E = \frac{12}{\bar{D}_{11}h^3}$ is the elastic modulus with respect to bending about the $y-$axis. The geometric quantity $I = I_{yy}$ is the corresponding second-order moment of inertia. The material parameter G is the shear modulus of the plate, defined as $G = \frac{1}{\bar{A}_{55}h}$. The quantity A is the cross-sectional area of the plate strip, i.e. $A = bh$.

With the equilibrium condition $\frac{\mathrm{d}Q}{\mathrm{d}x} = -\bar{p}$ (where $\bar{p}$ is the line load of the plate strip), it follows from the second equation in (10.42) and $\bar{p} = pb$:

$$KGA\left(\frac{\mathrm{d}^2w_0}{\mathrm{d}x^2} + \frac{\mathrm{d}\psi_x}{\mathrm{d}x}\right) = -\bar{p}. \tag{10.43}$$

With $\frac{\mathrm{d}M}{\mathrm{d}x} = Q$ one also obtains:

$$EI\frac{\mathrm{d}^2\psi_x}{\mathrm{d}x^2} = Q, \tag{10.44}$$

or

$$EI\frac{\mathrm{d}^2\psi_x}{\mathrm{d}x^2} - KGA\left(\frac{\mathrm{d}w_0}{\mathrm{d}x} + \psi_x\right) = 0. \tag{10.45}$$

Integration of (10.43) with respect to x provides:

$$KGA\left(\frac{\mathrm{d}w_0}{\mathrm{d}x} + \psi_x\right) = -\int \bar{p}\mathrm{d}x + C_1, \tag{10.46}$$

where C_1 represents an integration constant. Substituting this expression into (10.45) yields:

$$EI\frac{\mathrm{d}^2\psi_x}{\mathrm{d}x^2} = -\int \bar{p}\mathrm{d}x + C_1. \tag{10.47}$$

We perform a twofold integration of this expression and obtain the bending angle ψ_x as follows:

$$\begin{aligned} EI\frac{\mathrm{d}\psi_x}{\mathrm{d}x} &= -\int\int \bar{p}\mathrm{d}x\mathrm{d}x + C_1 x + C_2, \\ EI\psi_x &= -\int\int\int \bar{p}\mathrm{d}x\mathrm{d}x\mathrm{d}x + \frac{1}{2}C_1 x^2 + C_2 x + C_3, \end{aligned} \tag{10.48}$$

where C_2 and C_3 are additional integration constants.

Eq. (10.45) can be solved for $\frac{\mathrm{d}w_0}{\mathrm{d}x}$ as:

$$\frac{\mathrm{d}w_0}{\mathrm{d}x} = \frac{1}{KGA}\left(-\int \bar{p}\mathrm{d}x + C_1\right) - \psi_x. \tag{10.49}$$

Substituting ψ_x according to (10.48) then gives:

$$\begin{aligned} \frac{\mathrm{d}w_0}{\mathrm{d}x} &= \frac{1}{KGA}\left(-\int \bar{p}\mathrm{d}x + C_1\right) \\ &\quad - \frac{1}{EI}\left(-\int\int\int \bar{p}\mathrm{d}x\mathrm{d}x\mathrm{d}x + \frac{1}{2}C_1 x^2 + C_2 x + C_3\right). \end{aligned} \tag{10.50}$$

Integration with respect to x then finally provides the deflection of the plate strip under consideration:

$$\begin{aligned} w_0 &= \frac{1}{KGA}\left(-\int\int \bar{p}\mathrm{d}x\mathrm{d}x + C_1 x\right) \\ &\quad - \frac{1}{EI}\left(-\int\int\int\int \bar{p}\mathrm{d}x\mathrm{d}x\mathrm{d}x\mathrm{d}x + \frac{1}{6}C_1 x^3 + \frac{1}{2}C_2 x^2 + C_3 x + C_4\right). \end{aligned} \tag{10.51}$$

The integration constants C_1, C_2, C_3, C_4 are adapted to given boundary conditions for the plate strip under consideration. With the deflection w_0 present, all other state variables of the plate strip can then be determined. If the case of a plate strip exists for which the shear stiffness KGA becomes large, then the part of the transverse shear strain in (10.51) is negligible, and the present formulation passes into Kirchhoff's plate theory, which is included in First-Order Shear Deformation Theory as a special case.

10.1.6 Navier Solution

As for the Kirchhoff plate, a Navier solution can be derived for the simply supported rectangular orthotropic plate within the framework of First-Order Shear Deformation Theory. The plate has the dimensions a and b and is under the arbitrary surface load $p = p(x, y)$. The load $p(x, y)$ is developed in the form of a Fourier series:

$$p = \sum_{m=1}^{m=\infty} \sum_{n=1}^{n=\infty} P_{mn} \sin\left(\frac{m\pi x}{a}\right) \sin\left(\frac{n\pi y}{b}\right), \tag{10.52}$$

where the coefficients P_{mn} can be determined as:

$$P_{mn} = \frac{4}{ab} \int_0^b \int_0^a p(x, y) \sin\left(\frac{m\pi x}{a}\right) \sin\left(\frac{n\pi y}{b}\right) \mathrm{d}x\mathrm{d}y. \tag{10.53}$$

The boundary conditions are given as follows:

$$\begin{aligned}
w_0 (x = 0) &= 0, \quad w_0 (x = a) = 0,\\
w_0 (y = 0) &= 0, \quad w_0 (y = b) = 0,\\
\psi_x (y = 0) &= 0, \quad \psi_x (y = b) = 0,\\
\psi_y (x = 0) &= 0, \quad \psi_y (x = a) = 0,\\
M_{xx}^0 (x = 0) &= 0, \quad M_{xx}^0 (x = a) = 0,\\
M_{yy}^0 (y = 0) &= 0, \quad M_{yy}^0 (y = b) = 0.
\end{aligned} \tag{10.54}$$

The following series expansions are used for w_0, ψ_x, ψ_y, which satisfy all boundary conditions identically:

$$\begin{aligned}
w_0 &= \sum_{m=1}^{m=\infty} \sum_{n=1}^{n=\infty} W_{mn} \sin\left(\frac{m\pi x}{a}\right) \sin\left(\frac{n\pi y}{b}\right),\\
\psi_x &= \sum_{m=1}^{m=\infty} \sum_{n=1}^{n=\infty} X_{mn} \cos\left(\frac{m\pi x}{a}\right) \sin\left(\frac{n\pi y}{b}\right),\\
\psi_y &= \sum_{m=1}^{m=\infty} \sum_{n=1}^{n=\infty} Y_{mn} \sin\left(\frac{m\pi x}{a}\right) \cos\left(\frac{n\pi y}{b}\right).
\end{aligned} \tag{10.55}$$

The approach coefficients W_{mn}, X_{mn}, Y_{mn} can be determined by substituting (10.55) into the differential equation system (10.22), resulting in the following linear equation system:

$$\begin{bmatrix} \lambda_{11} & \lambda_{12} & \lambda_{13} \\ \lambda_{12} & \lambda_{22} & \lambda_{23} \\ \lambda_{13} & \lambda_{23} & \lambda_{33} \end{bmatrix} \begin{pmatrix} W_{mn} \\ X_{mn} \\ Y_{mn} \end{pmatrix} = \begin{pmatrix} P_{mn} \\ 0 \\ 0 \end{pmatrix}. \tag{10.56}$$

The following abbreviations have been used herein:

$$\lambda_{11} = K\left[A_{55}\left(\frac{m\pi}{a}\right)^2 + A_{44}\left(\frac{n\pi}{b}\right)^2\right], \quad \lambda_{12} = KA_{55}\frac{m\pi}{a}, \quad \lambda_{13} = KA_{44}\frac{n\pi}{b},$$
$$\lambda_{22} = D_{11}\left(\frac{m\pi}{a}\right)^2 + D_{66}\left(\frac{n\pi}{b}\right)^2 + KA_{55}, \quad \lambda_{23} = (D_{12} + D_{66})\frac{m\pi}{a}\frac{n\pi}{b},$$
$$\lambda_{33} = D_{66}\left(\frac{m\pi}{a}\right)^2 + D_{22}\left(\frac{n\pi}{b}\right)^2 + KA_{44}. \tag{10.57}$$

This system of equations can be solved for any pair of values $m, n = 1, 2, 3, \ldots$ and yields the coefficients W_{mn}, X_{mn}, Y_{mn}. Thus, the quantities w_0, ψ_x, ψ_y are completely available and all further state variables of the plate can be determined.

10.1.7 Lévy-type solutions

It is possible to derive Lévy-type solutions for static plate problems within the framework of First-Order Shear Deformation Theory as well. A prerequisite for the existence of a Lévy-type solution is that two opposing plate edges, e.g. the two edges at $x = 0$ and $x = a$, are simply supported. The two remaining edges at $y = 0$ and $y = b$ can be subject to any kind of boundary conditions. However, we will not elaborate on this type of solution method for plates in the context of First-Order Shear Deformation Theory here, the reader is referred to e.g. Reddy (2004, 2006).

10.1.8 The Ritz Method

Also for plates in the framework of First-Order Shear Deformation Theory, the Ritz method can be used advantageously to solve static boundary value problems. We consider a rectangular plate with dimensions a and b with respect to the $x-$ and $y-$directions and assume an arbitrary surface load $p = p(x, y)$. In the framework of the Ritz method, we make assumptions for the displacement quantities w_0, ψ_x, ψ_y as follows:

$$w_0 = \sum_{m=1}^{m=M}\sum_{n=1}^{n=N} W_{mn} w_{1m}(x)\, w_{2n}(y),$$

$$\psi_x = \sum_{m=1}^{m=K} \sum_{n=1}^{n=L} X_{mn} \psi_{x1m}(x) \, \psi_{x2n}(y),$$

$$\psi_y = \sum_{m=1}^{m=I} \sum_{n=1}^{n=J} Y_{mn} \psi_{y1m}(x) \, \psi_{y2n}(y). \tag{10.58}$$

The approach functions are required to exactly fulfill all kinematic boundary conditions.

We use the principle of the minimum of the total elastic potential Π of the plate under consideration. The inner potential Π_i has already been given by the expression (10.35). The external potential Π_a in the case of a surface load $p = p(x, y)$ is given as:

$$\Pi_a = -\int_0^b \int_0^a p(x, y) w(x, y) \mathrm{d}x \mathrm{d}y. \tag{10.59}$$

In the equilibrium state, the total elastic potential assumes a minimum:

$$\Pi = \Pi_i + \Pi_a = \text{Minimum}. \tag{10.60}$$

Using the approach functions (10.58), the potential Π can be expressed in the form

$$\Pi = \Pi\left(W_{mn}, X_{mn}, Y_{mn}\right). \tag{10.61}$$

The minimum requirement $\delta\Pi = 0$ then passes into the Ritz equations:

$$\frac{\partial \Pi}{\partial W_{mn}} = 0, \quad \frac{\partial \Pi}{\partial X_{mn}} = 0, \quad \frac{\partial \Pi}{\partial Y_{mn}} = 0. \tag{10.62}$$

This results in a system of linear equations for the unknown coefficients W_{mn}, X_{mn}, Y_{mn} for a geometrically linear problem. Thus, the approximate quantities w_0, ψ_x, ψ_y are known, and all other approximate state variables of the plate can be determined.

We substitute (10.58) into (10.35) and after some transformations we obtain the following expression for the inner potential Π_i:

$$\begin{aligned}
\Pi_i = {} & \frac{1}{2} D_{11} \sum_{m=1}^{m=K} \sum_{n=1}^{n=L} \sum_{p=1}^{p=K} \sum_{q=1}^{q=L} X_{mn} X_{pq} \int_0^a \frac{\mathrm{d}\psi_{x1m}}{\mathrm{d}x} \frac{\mathrm{d}\psi_{x1p}}{\mathrm{d}x} \mathrm{d}x \int_0^b \psi_{x2n} \psi_{x2q} \mathrm{d}y \\
& + \frac{1}{2} D_{22} \sum_{m=1}^{m=I} \sum_{n=1}^{n=J} \sum_{p=1}^{p=I} \sum_{q=1}^{q=J} Y_{mn} Y_{pq} \int_0^a \psi_{y1m} \psi_{y1p} \mathrm{d}x \int_0^b \frac{\mathrm{d}\psi_{y2n}}{\mathrm{d}y} \frac{\mathrm{d}\psi_{y2q}}{\mathrm{d}y} \mathrm{d}y \\
& + D_{12} \sum_{m=1}^{m=K} \sum_{n=1}^{n=L} \sum_{p=1}^{p=I} \sum_{q=1}^{q=J} X_{mn} Y_{pq} \int_0^a \frac{\mathrm{d}\psi_{x1m}}{\mathrm{d}x} \psi_{y1p} \mathrm{d}x \int_0^b \frac{\mathrm{d}\psi_{y2q}}{\mathrm{d}y} \psi_{x2n} \mathrm{d}y \\
& + \frac{1}{2} D_{66} \left[\sum_{m=1}^{m=K} \sum_{n=1}^{n=L} \sum_{p=1}^{p=K} \sum_{q=1}^{q=L} X_{mn} X_{pq} \int_0^a \psi_{x1m} \psi_{x1p} \mathrm{d}x \int_0^b \frac{\mathrm{d}\psi_{x2n}}{\mathrm{d}y} \frac{\mathrm{d}\psi_{x2q}}{\mathrm{d}y} \mathrm{d}y \right.
\end{aligned}$$

$$+2\sum_{m=1}^{m=K}\sum_{n=1}^{n=L}\sum_{p=1}^{p=I}\sum_{q=1}^{q=J}X_{mn}Y_{pq}\int_0^a \psi_{x1m}\frac{\mathrm{d}\psi_{y1p}}{\mathrm{d}x}\mathrm{d}x\int_0^b \frac{\mathrm{d}\psi_{x2n}}{\mathrm{d}y}\psi_{y2q}\mathrm{d}y$$

$$\left.+\sum_{m=1}^{m=I}\sum_{n=1}^{n=J}\sum_{p=1}^{p=I}\sum_{q=1}^{q=J}Y_{mn}Y_{pq}\int_0^a \frac{\mathrm{d}\psi_{y1m}}{\mathrm{d}x}\frac{\mathrm{d}\psi_{y1p}}{\mathrm{d}x}\mathrm{d}x\int_0^b \psi_{y2n}\psi_{y2q}\mathrm{d}y\right]$$

$$+\frac{1}{2}KA_{44}\left[\sum_{m=1}^{m=M}\sum_{n=1}^{n=N}\sum_{p=1}^{p=M}\sum_{q=1}^{q=N}W_{mn}W_{pq}\int_0^a w_{1m}w_{1p}\mathrm{d}x\int_0^b \frac{\mathrm{d}w_{2n}}{\mathrm{d}y}\frac{\mathrm{d}w_{2q}}{\mathrm{d}y}\mathrm{d}y\right.$$

$$+2\sum_{m=1}^{m=M}\sum_{n=1}^{n=N}\sum_{p=1}^{p=I}\sum_{q=1}^{q=J}W_{mn}Y_{pq}\int_0^a w_{1m}\psi_{y1p}\mathrm{d}x\int_0^b \frac{\mathrm{d}w_{2n}}{\mathrm{d}y}\psi_{y2q}\mathrm{d}y$$

$$\left.+\sum_{m=1}^{m=I}\sum_{n=1}^{n=J}\sum_{p=1}^{p=I}\sum_{q=1}^{q=J}Y_{mn}Y_{pq}\int_0^a \psi_{y1m}\psi_{y1p}\mathrm{d}x\int_0^b \psi_{y2n}\psi_{y2q}\mathrm{d}y\right]$$

$$+\frac{1}{2}KA_{55}\left[\sum_{m=1}^{m=M}\sum_{n=1}^{n=N}\sum_{p=1}^{p=M}\sum_{q=1}^{q=N}W_{mn}W_{pq}\int_0^a \frac{\mathrm{d}w_{1m}}{\mathrm{d}x}\frac{\mathrm{d}w_{1p}}{\mathrm{d}x}\mathrm{d}x\int_0^b w_{2n}w_{2q}\mathrm{d}y\right.$$

$$+2\sum_{m=1}^{m=M}\sum_{n=1}^{n=N}\sum_{p=1}^{p=K}\sum_{q=1}^{q=L}W_{mn}X_{pq}\int_0^a \frac{\mathrm{d}w_{1m}}{\mathrm{d}x}\psi_{x1p}\mathrm{d}x\int_0^b w_{2n}\psi_{x2q}\mathrm{d}y$$

$$\left.+\sum_{m=1}^{m=K}\sum_{n=1}^{n=L}\sum_{p=1}^{p=K}\sum_{q=1}^{q=L}X_{mn}X_{pq}\int_0^a \psi_{x1m}\psi_{x1p}\mathrm{d}x\int_0^b \psi_{x2m}\psi_{x2p}\mathrm{d}y\right]. \tag{10.63}$$

At this point, we define the resultants of the approach functions as:

$$\Omega^1_{ij}=\int_0^a \frac{\mathrm{d}^i\psi_{x1m}}{\mathrm{d}x^i}\frac{\mathrm{d}^j\psi_{x1p}}{\mathrm{d}x^j}\mathrm{d}x,\quad \Omega^2_{ij}=\int_0^a \frac{\mathrm{d}^i\psi_{x1m}}{\mathrm{d}x^i}\frac{\mathrm{d}^j\psi_{y1p}}{\mathrm{d}x^j}\mathrm{d}x,$$

$$\Omega^3_{ij}=\int_0^a \frac{\mathrm{d}^i\psi_{y1m}}{\mathrm{d}x^i}\frac{\mathrm{d}^j\psi_{y1p}}{\mathrm{d}x^j}\mathrm{d}x,\quad \Omega^4_{ij}=\int_0^a \frac{\mathrm{d}^i w_{1m}}{\mathrm{d}x^i}\frac{\mathrm{d}^j w_{1p}}{\mathrm{d}x^j}\mathrm{d}x,$$

$$\Omega^5_{ij}=\int_0^a \frac{\mathrm{d}^i w_{1m}}{\mathrm{d}x^i}\frac{\mathrm{d}^j\psi_{y1p}}{\mathrm{d}x^j}\mathrm{d}x,\quad \Omega^6_{ij}=\int_0^a \frac{\mathrm{d}^i w_{1m}}{\mathrm{d}x^i}\frac{\mathrm{d}^j\psi_{x1p}}{\mathrm{d}x^j}\mathrm{d}x,$$

$$\Delta^1_{ij}=\int_0^b \frac{\mathrm{d}^i\psi_{x2n}}{\mathrm{d}y^i}\frac{\mathrm{d}^j\psi_{x2q}}{\mathrm{d}y^j}\mathrm{d}y,\quad \Delta^2_{ij}=\int_0^b \frac{\mathrm{d}^i\psi_{x2n}}{\mathrm{d}y^i}\frac{\mathrm{d}^j\psi_{y2q}}{\mathrm{d}y^j}\mathrm{d}y,$$

$$\Delta^3_{ij}=\int_0^b \frac{\mathrm{d}^i\psi_{y2n}}{\mathrm{d}y^i}\frac{\mathrm{d}^j\psi_{y2q}}{\mathrm{d}y^j}\mathrm{d}y,\quad \Delta^4_{ij}=\int_0^b \frac{\mathrm{d}^i w_{2n}}{\mathrm{d}y^i}\frac{\mathrm{d}^j w_{2q}}{\mathrm{d}y^j}\mathrm{d}y,$$

$$\Delta^5_{ij}=\int_0^b \frac{\mathrm{d}^i w_{2n}}{\mathrm{d}y^i}\frac{\mathrm{d}^j\psi_{y2q}}{\mathrm{d}y^j}\mathrm{d}y,\quad \Delta^6_{ij}=\int_0^b \frac{\mathrm{d}^i w_{2n}}{\mathrm{d}y^i}\frac{\mathrm{d}^j\psi_{x2q}}{\mathrm{d}y^j}\mathrm{d}y. \tag{10.64}$$

With the abbreviations

$$
\begin{aligned}
\Gamma_i^{klkl} &= \frac{1}{2}\left(D_{11}\Omega_{11}^1\Delta_{00}^1 + D_{66}\Omega_{00}^1\Delta_{11}^1 + KA_{55}\Omega_{00}^1\Delta_{00}^1\right),\\
\Gamma_i^{ijij} &= \frac{1}{2}\left(D_{22}\Omega_{00}^3\Delta_{11}^3 + D_{66}\Omega_{11}^3\Delta_{00}^3 + KA_{44}\Omega_{00}^3\Delta_{00}^3\right),\\
\Gamma_i^{klij} &= D_{12}\Omega_{10}^2\Delta_{01}^2 + D_{66}\Omega_{01}^2\Delta_{10}^2,\\
\Gamma_i^{mnmn} &= \frac{K}{2}\left(A_{44}\Omega_{00}^4\Delta_{11}^4 + A_{55}\Omega_{11}^4\Delta_{00}^4\right),\\
\Gamma_i^{mnij} &= KA_{44}\Omega_{00}^5\Delta_{10}^5,\\
\Gamma_i^{mnkl} &= KA_{55}\Omega_{10}^6\Delta_{00}^6
\end{aligned}
\tag{10.65}
$$

the following representation for the inner potential Π_i results:

$$
\begin{aligned}
\Pi_i &= \sum_{m=1}^{m=K}\sum_{n=1}^{n=L}\sum_{p=1}^{p=K}\sum_{q=1}^{q=L} X_{mn}X_{pq}\Gamma_i^{klkl} + \sum_{m=1}^{m=I}\sum_{n=1}^{n=J}\sum_{p=1}^{p=I}\sum_{q=1}^{q=J} Y_{mn}Y_{pq}\Gamma_i^{ijij}\\
&+ \sum_{m=1}^{m=K}\sum_{n=1}^{n=L}\sum_{p=1}^{p=I}\sum_{q=1}^{q=J} X_{mn}Y_{pq}\Gamma_i^{klij} + \sum_{m=1}^{m=M}\sum_{n=1}^{n=N}\sum_{p=1}^{p=M}\sum_{q=1}^{q=N} W_{mn}W_{pq}\Gamma_i^{mnmn}\\
&+ \sum_{m=1}^{m=M}\sum_{n=1}^{n=N}\sum_{p=1}^{p=I}\sum_{q=1}^{q=J} W_{mn}Y_{pq}\Gamma_i^{mnij} + \sum_{m=1}^{m=M}\sum_{n=1}^{n=N}\sum_{p=1}^{p=K}\sum_{q=1}^{q=L} W_{mn}X_{pq}\Gamma_i^{mnkl}.
\end{aligned}
\tag{10.66}
$$

For the external potential Π_a we refer to the explanations in Chapter 8.

We also consider the case where the plate under consideration is elastically restrained at the two edges at $y = 0$ and $y = b$ by elastic torsional springs with stiffnesses k_l and k_r, respectively. The corresponding energy Π_s stored in the springs is:

$$
\Pi_s = \frac{1}{2}\int_0^a \left[k_l\left(\psi_y(y=0)\right)^2 + k_r\left(\psi_y(y=b)\right)^2\right]\mathrm{d}x. \tag{10.67}
$$

Substituting the approach functions (10.58) for ψ_y yields:

$$
\begin{aligned}
\Pi_s &= \frac{k_l}{2}\sum_{m=1}^{m=I}\sum_{n=1}^{n=J}\sum_{p=1}^{p=I}\sum_{q=1}^{q=J} Y_{mn}Y_{pq}\int_0^a \psi_{y1m}\psi_{y1p}\mathrm{d}x\left(\psi_{y2n}(y=0)\psi_{y2q}(y=0)\right)\\
&+ \frac{k_r}{2}\sum_{m=1}^{m=I}\sum_{n=1}^{n=J}\sum_{p=1}^{p=I}\sum_{q=1}^{q=J} Y_{mn}Y_{pq}\int_0^a \psi_{y1m}\psi_{y1p}\mathrm{d}x\left(\psi_{y2n}(y=b)\psi_{y2q}(y=b)\right).
\end{aligned}
\tag{10.68}
$$

With the abbreviations

$$
\begin{aligned}
\Phi_{11}^{nq} &= \psi_{y2n}(y=0)\psi_{y2q}(y=0),\\
\Theta_{11}^{nq} &= \psi_{y2n}(y=b)\psi_{y2q}(y=b)
\end{aligned}
\tag{10.69}
$$

and

$$\Gamma_s^{ijij} = \frac{1}{2}\Omega_{00}^3 \left(k_l \Phi_{11}^{nq} + k_r \Theta_{11}^{nq}\right) \tag{10.70}$$

we obtain:

$$\Pi_s = \sum_{m=1}^{m=I} \sum_{n=1}^{n=J} \sum_{p=1}^{p=I} \sum_{q=1}^{q=J} Y_{mn} Y_{pq} \Gamma_s^{ijij}. \tag{10.71}$$

The total elastic potential Π of the plate under consideration is then the sum of Π_i, Π_a and Π_s. The Ritz equations (10.62) then yield a linear system of equations for the approach coefficients, and with the displacement quantities thus approximated, all other state quantities of the plate can be determined.

10.2 Third-Order Shear Deformation Theory According to Reddy

A further higher-order shear deformation theory, which further relaxes the assumptions of the already discussed plate theories, but can be applied with reasonable effort, is the so-called Third-Order Shear Deformation Theory according to Reddy[1]. It is based on the fundamental idea to develop the plane displacements u and v of the plate up to a certain degree in the form of a polynomial with respect to the thickness direction z and thus to admit the cross-sectional warpings that were excluded so far. This circumvents the already mentioned contradictions of Kirchhoff's plate theory and First-Order Shear Deformation Theory. Specifically, it can be shown that for a cubic distribution of the displacements u and v with respect to z (which explains the notion of Third-Order Shear Deformation Theory), parabolic distributions of the transverse shear stresses τ_{yz} and τ_{xz} can be realized. Thus, in the context of such a plate theory of higher order, the use of a shear correction factor can be omitted.

10.2.1 Kinematics

Within the framework of Reddy's Third-Order Shear Deformation Theory, both the normal hypothesis and the hypothesis that the cross-sections remain plane in the deformed state are discarded. Thus, warpings of the plate cross-section, as they are to be expected for shear compliant plates, are explicitly allowed. Starting point of the considerations is the following displacement field:

$$\begin{aligned} u(x, y, z) &= z\psi_x(x, y) + z^2\theta_x(x, y) + z^3\lambda_x(x, y), \\ v(x, y, z) &= z\psi_y(x, y) + z^2\theta_y(x, y) + z^3\lambda_y(x, y), \\ w(x, y) &= w_0(x, y). \end{aligned} \tag{10.72}$$

[1] Junuthula N. Reddy, born 1945, US-American scientist.

Here $\psi_x, \psi_y, \theta_x, \theta_y, \lambda_x, \lambda_y$ are functions still to be determined, by which the warping of the plate cross-section is represented. Thus, a total of seven unknown displacement quantities have to be determined within the framework of Third-Order Shear Deformation Theory. Here, the two bending angles ψ_x and ψ_y indicate the inclination of the cross-sectional warping at the point $z = 0$:

$$\psi_x = \left.\frac{\partial u}{\partial z}\right|_{z=0}, \quad \psi_y = \left.\frac{\partial v}{\partial z}\right|_{z=0}. \tag{10.73}$$

The quantities $\theta_x, \theta_y, \lambda_x, \lambda_y$ can be interpreted as follows:

$$2\theta_x = \left.\frac{\partial^2 u}{\partial z^2}\right|_{z=0}, \quad 2\theta_y = \left.\frac{\partial^2 v}{\partial z^2}\right|_{z=0}, \quad 6\lambda_x = \left.\frac{\partial^3 u}{\partial z^3}\right|_{z=0}, \quad 6\lambda_y = \left.\frac{\partial^3 v}{\partial z^3}\right|_{z=0}. \tag{10.74}$$

At this point, it is required that the transverse shear stresses τ_{yz} and τ_{xz} on the free plate surfaces at $z = -\frac{h}{2}$ and $z = +\frac{h}{2}$ vanish:

$$\tau_{xz}\left(z = -\frac{h}{2}\right) = \tau_{xz}\left(z = +\frac{h}{2}\right) = \tau_{yz}\left(z = -\frac{h}{2}\right) = \tau_{yz}\left(z = +\frac{h}{2}\right) = 0. \tag{10.75}$$

From the generalized Hooke's law it follows:

$$\begin{aligned} C_{55}\gamma_{xz}\left(z = -\frac{h}{2}\right) &= 0, \quad C_{55}\gamma_{xz}\left(z = +\frac{h}{2}\right) = 0, \\ C_{44}\gamma_{yz}\left(z = -\frac{h}{2}\right) &= 0, \quad C_{44}\gamma_{yz}\left(z = +\frac{h}{2}\right) = 0. \end{aligned} \tag{10.76}$$

Obviously, the requirement for vanishing shear stresses τ_{yz} and τ_{xz} can be fulfilled only if the associated shear strains γ_{yz} and γ_{xz} at the plate surfaces also vanish. With the shear strains according to

$$\begin{aligned} \gamma_{yz} &= \frac{\partial v}{\partial z} + \frac{\partial w}{\partial y} = \frac{\partial w_0}{\partial y} + \psi_y + 2z\theta_y + 3z^2\lambda_y, \\ \gamma_{xz} &= \frac{\partial u}{\partial z} + \frac{\partial w}{\partial x} = \frac{\partial w_0}{\partial x} + \psi_x + 2z\theta_x + 3z^2\lambda_x \end{aligned} \tag{10.77}$$

four conditions can be derived:

$$\begin{aligned} &\frac{\partial w_0}{\partial y} + \psi_y - h\theta_y + \frac{3h^2}{4}\lambda_y = 0, \\ &\frac{\partial w_0}{\partial y} + \psi_y + h\theta_y + \frac{3h^2}{4}\lambda_y = 0, \\ &\frac{\partial w_0}{\partial x} + \psi_x - h\theta_x + \frac{3h^2}{4}\lambda_x = 0, \end{aligned}$$

$$\frac{\partial w_0}{\partial x} + \psi_x + h\theta_x + \frac{3h^2}{4}\lambda_x = 0. \tag{10.78}$$

We determine the quantities $\theta_x, \theta_y, \lambda_x, \lambda_y$ from this as follows:

$$\theta_x = 0, \quad \theta_y = 0, \quad \lambda_x = -\frac{4}{3h^2}\left(\psi_x + \frac{\partial w_0}{\partial x}\right),$$
$$\lambda_y = -\frac{4}{3h^2}\left(\psi_y + \frac{\partial w_0}{\partial y}\right). \tag{10.79}$$

The displacement field (10.72) then transforms into the following form:

$$u(x, y, z) = z\psi_x(x, y) - \frac{4z^3}{3h^2}\left(\psi_x(x, y) + \frac{\partial w_0(x, y)}{\partial x}\right),$$
$$v(x, y, z) = z\psi_y(x, y) - \frac{4z^3}{3h^2}\left(\psi_y(x, y) + \frac{\partial w_0(x, y)}{\partial y}\right),$$
$$w(x, y) = w_0(x, y). \tag{10.80}$$

Thus, also in the framework of Reddy's Third-Order Shear Deformation Theory, as in the First-Order Shear Deformation Theory, only three degrees of freedom occur (Fig. 10.2).

10.2.2 Strains and Constitutive Equations

With the help of the kinematic relations (1.60), the components of the infinitesimal strain tensor can be determined as follows:

$$\varepsilon_{xx} = \frac{\partial u}{\partial x} = z\frac{\partial \psi_x}{\partial x} - \frac{4z^3}{3h^2}\left(\frac{\partial \psi_x}{\partial x} + \frac{\partial^2 w_0}{\partial x^2}\right),$$

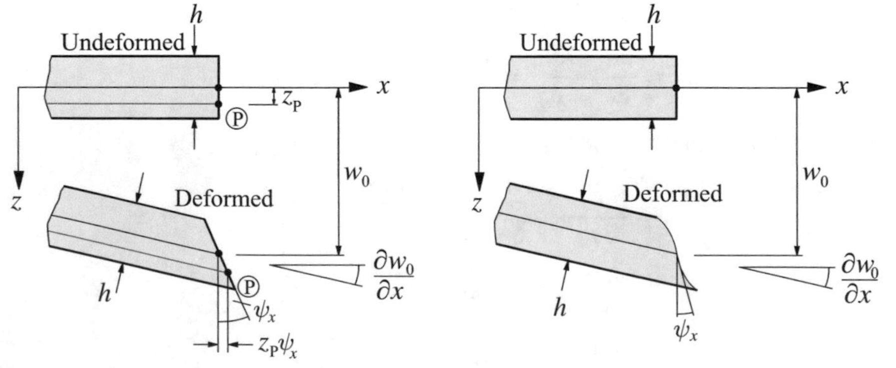

Fig. 10.2 Undeformed and deformed plate segment according to First-Order Shear Deformation Theory (left) and according to Third-Order Shear Deformation Theory (right)

$$
\begin{aligned}
\varepsilon_{yy} &= \frac{\partial v}{\partial y} = z\frac{\partial \psi_y}{\partial y} - \frac{4z^3}{3h^2}\left(\frac{\partial \psi_y}{\partial y} + \frac{\partial^2 w_0}{\partial y^2}\right), \\
\varepsilon_{zz} &= \frac{\partial w}{\partial z} = 0, \\
\gamma_{yz} &= \frac{\partial v}{\partial z} + \frac{\partial w}{\partial y} = \psi_y - \frac{4z^2}{h^2}\left(\psi_y + \frac{\partial w_0}{\partial y}\right) + \frac{\partial w_0}{\partial y}, \\
\gamma_{xz} &= \frac{\partial u}{\partial z} + \frac{\partial w}{\partial x} = \psi_x - \frac{4z^2}{h^2}\left(\psi_x + \frac{\partial w_0}{\partial x}\right) + \frac{\partial w_0}{\partial x}, \\
\gamma_{xy} &= \frac{\partial u}{\partial y} + \frac{\partial v}{\partial x} = z\left(\frac{\partial \psi_x}{\partial y} + \frac{\partial \psi_y}{\partial x}\right) - \frac{4z^3}{3h^2}\left(\frac{\partial \psi_x}{\partial y} + \frac{\partial \psi_y}{\partial x} + 2\frac{\partial^2 w_0}{\partial x \partial y}\right),
\end{aligned}
\tag{10.81}
$$

which we can specify in a vector notation as:

$$
\begin{aligned}
\underline{\varepsilon}^{(1)} &= \begin{pmatrix} \varepsilon_{xx}^{(1)} \\ \varepsilon_{yy}^{(1)} \\ \gamma_{xy}^{(1)} \end{pmatrix} = \begin{pmatrix} \frac{\partial \psi_x}{\partial x} \\ \frac{\partial \psi_y}{\partial y} \\ \frac{\partial \psi_x}{\partial y} + \frac{\partial \psi_y}{\partial x} \end{pmatrix}, \\
\underline{\varepsilon}^{(3)} &= \begin{pmatrix} \varepsilon_{xx}^{(3)} \\ \varepsilon_{yy}^{(3)} \\ \gamma_{xy}^{(3)} \end{pmatrix} = \begin{pmatrix} -\frac{4}{3h^2}\left(\frac{\partial \psi_x}{\partial x} + \frac{\partial^2 w_0}{\partial x^2}\right) \\ -\frac{4}{3h^2}\left(\frac{\partial \psi_y}{\partial y} + \frac{\partial^2 w_0}{\partial y^2}\right) \\ -\frac{4}{3h^2}\left(\frac{\partial \psi_x}{\partial y} + \frac{\partial \psi_y}{\partial x} + 2\frac{\partial^2 w_0}{\partial x \partial y}\right) \end{pmatrix}, \\
\underline{\gamma}^{(0)} &= \begin{pmatrix} \gamma_{yz}^{(0)} \\ \gamma_{xz}^{(0)} \end{pmatrix} = \begin{pmatrix} \psi_y + \frac{\partial w_0}{\partial y} \\ \psi_x + \frac{\partial w_0}{\partial x} \end{pmatrix}, \\
\underline{\gamma}^{(2)} &= \begin{pmatrix} \gamma_{yz}^{(2)} \\ \gamma_{xz}^{(2)} \end{pmatrix} = \begin{pmatrix} -\frac{4}{h^2}\left(\psi_y + \frac{\partial w_0}{\partial y}\right) \\ -\frac{4}{h^2}\left(\psi_x + \frac{\partial w_0}{\partial x}\right) \end{pmatrix}.
\end{aligned}
\tag{10.82}
$$

We can then write the strain field as:

$$
\begin{aligned}
\underline{\varepsilon} &= z\underline{\varepsilon}^{(1)} + z^3\underline{\varepsilon}^{(3)}, \\
\underline{\gamma} &= \underline{\gamma}^{(0)} + z^2\underline{\gamma}^{(2)}.
\end{aligned}
\tag{10.83}
$$

The following already known force and moment quantities are applied.

$$
\begin{aligned}
\underline{M}^0 &= \begin{pmatrix} M_{xx}^0 \\ M_{yy}^0 \\ M_{xy}^0 \end{pmatrix} = \int_{-\frac{h}{2}}^{+\frac{h}{2}} \begin{pmatrix} \sigma_{xx} \\ \sigma_{yy} \\ \tau_{xy} \end{pmatrix} z\mathrm{d}z, \\
\underline{Q} &= \begin{pmatrix} Q_y \\ Q_x \end{pmatrix} = \int_{-\frac{h}{2}}^{+\frac{h}{2}} \begin{pmatrix} \tau_{yz} \\ \tau_{xz} \end{pmatrix} \mathrm{d}z.
\end{aligned}
\tag{10.84}
$$

Furthermore, the following force and moment quantities newly added within the framework of Third-Order Shear Deformation Theory occur:

$$\underline{P} = \begin{pmatrix} P_{xx} \\ P_{yy} \\ P_{xy} \end{pmatrix} = \int_{-\frac{h}{2}}^{+\frac{h}{2}} \begin{pmatrix} \sigma_{xx} \\ \sigma_{yy} \\ \tau_{xy} \end{pmatrix} z^3 \mathrm{d}z,$$

$$\underline{R} = \begin{pmatrix} R_y \\ R_x \end{pmatrix} = \int_{-\frac{h}{2}}^{+\frac{h}{2}} \begin{pmatrix} \tau_{yz} \\ \tau_{xz} \end{pmatrix} z^2 \mathrm{d}z. \tag{10.85}$$

The constitutive law for a plate in the framework of Third-Order Shear Deformation Theory according to Reddy then reads as follows:

$$\begin{pmatrix} M^0_{xx} \\ M^0_{yy} \\ M^0_{xy} \\ P_{xx} \\ P_{yy} \\ P_{xy} \\ Q_y \\ Q_x \\ R_y \\ R_y \end{pmatrix} = \begin{bmatrix} \underline{\underline{D}} & \underline{\underline{F}} & \underline{\underline{0}} & \underline{\underline{0}} \\ \underline{\underline{F}} & \underline{\underline{H}} & \underline{\underline{0}} & \underline{\underline{0}} \\ \underline{\underline{0}} & \underline{\underline{0}} & \underline{\underline{A}}_S & \underline{\underline{D}}_S \\ \underline{\underline{0}} & \underline{\underline{0}} & \underline{\underline{D}}_S & \underline{\underline{F}}_S \end{bmatrix} \begin{pmatrix} \frac{\partial \psi_x}{\partial x} \\ \frac{\partial \psi_y}{\partial y} \\ \frac{\partial \psi_x}{\partial y} + \frac{\partial \psi_y}{\partial x} \\ -\frac{4}{3h^2}\left(\frac{\partial \psi_x}{\partial x} + \frac{\partial^2 w_0}{\partial x^2}\right) \\ -\frac{4}{3h^2}\left(\frac{\partial \psi_y}{\partial y} + \frac{\partial^2 w_0}{\partial y^2}\right) \\ -\frac{4}{3h^2}\left(\frac{\partial \psi_x}{\partial y} + \frac{\partial \psi_y}{\partial x} + 2\frac{\partial^2 w_0}{\partial x \partial y}\right) \\ \psi_y + \frac{\partial w_0}{\partial y} \\ \psi_x + \frac{\partial w_0}{\partial x} \\ -\frac{4}{h^2}\left(\psi_y + \frac{\partial w_0}{\partial y}\right) \\ -\frac{4}{h^2}\left(\psi_x + \frac{\partial w_0}{\partial x}\right) \end{pmatrix}. \tag{10.86}$$

In symbolic form we obtain:

$$\begin{pmatrix} \underline{M}^0 \\ \underline{P} \\ \underline{Q} \\ \underline{R} \end{pmatrix} = \begin{bmatrix} \underline{\underline{D}} & \underline{\underline{F}} & \underline{\underline{0}} & \underline{\underline{0}} \\ \underline{\underline{F}} & \underline{\underline{H}} & \underline{\underline{0}} & \underline{\underline{0}} \\ \underline{\underline{0}} & \underline{\underline{0}} & \underline{\underline{A}}_S & \underline{\underline{D}}_S \\ \underline{\underline{0}} & \underline{\underline{0}} & \underline{\underline{D}}_S & \underline{\underline{F}}_S \end{bmatrix} \begin{pmatrix} \underline{\varepsilon}^{(1)} \\ \underline{\varepsilon}^{(3)} \\ \underline{\gamma}^{(0)} \\ \underline{\gamma}^{(3)} \end{pmatrix}. \tag{10.87}$$

The stiffness matrix is composed as follows:

$$\begin{bmatrix} \underline{\underline{D}} & \underline{\underline{F}} & \underline{\underline{0}} & \underline{\underline{0}} \\ \underline{\underline{F}} & \underline{\underline{H}} & \underline{\underline{0}} & \underline{\underline{0}} \\ \underline{\underline{0}} & \underline{\underline{0}} & \underline{\underline{A}}_S & \underline{\underline{D}}_S \\ \underline{\underline{0}} & \underline{\underline{0}} & \underline{\underline{D}}_S & \underline{\underline{F}}_S \end{bmatrix}$$

$$= \begin{bmatrix} D_{11} & D_{12} & 0 & F_{11} & F_{12} & 0 & 0 & 0 & 0 & 0 \\ D_{12} & D_{22} & 0 & F_{12} & F_{22} & 0 & 0 & 0 & 0 & 0 \\ 0 & 0 & D_{66} & 0 & 0 & F_{66} & 0 & 0 & 0 & 0 \\ F_{11} & F_{12} & 0 & H_{11} & H_{12} & 0 & 0 & 0 & 0 & 0 \\ F_{12} & F_{22} & 0 & H_{12} & H_{22} & 0 & 0 & 0 & 0 & 0 \\ 0 & 0 & F_{66} & 0 & 0 & H_{66} & 0 & 0 & 0 & 0 \\ 0 & 0 & 0 & 0 & 0 & 0 & A_{44} & 0 & D_{44} & 0 \\ 0 & 0 & 0 & 0 & 0 & 0 & 0 & A_{55} & 0 & D_{55} \\ 0 & 0 & 0 & 0 & 0 & 0 & D_{44} & 0 & F_{44} & 0 \\ 0 & 0 & 0 & 0 & 0 & 0 & 0 & D_{55} & 0 & F_{55} \end{bmatrix}. \tag{10.88}$$

In this form of the constitutive law some quantities appear which are already known from the plate theories discussed before. These are, besides the moment and force flows $\underline{M}^0$ and $\underline{Q}$, the matrix of plate stiffnesses $\underline{\underline{D}}$ and the matrix $\underline{\underline{A}}_S$ of the transverse shear stiffnesses, as they also appear in the framework of First-Order Shear Deformation Theory. New here are the submatrices $\underline{\underline{F}}$, $\underline{\underline{H}}$ as well as $\underline{\underline{D}}_S$ and $\underline{\underline{F}}_S$, which can be interpreted as transverse shear stiffnesses. On the other hand, the newly added force quantities $\underline{P}$ and $\underline{R}$ can be interpreted as higher-order warping moments.

The new submatrices added here are defined as:

$$\underline{\underline{F}} = \begin{bmatrix} F_{11} & F_{12} & 0 \\ F_{12} & F_{22} & 0 \\ 0 & 0 & F_{66} \end{bmatrix} = \int_{-\frac{h}{2}}^{\frac{h}{2}} \begin{bmatrix} Q_{11} & Q_{12} & 0 \\ Q_{12} & Q_{22} & 0 \\ 0 & 0 & Q_{66} \end{bmatrix} z^4 \mathrm{d}z,$$

$$\underline{\underline{H}} = \begin{bmatrix} H_{11} & H_{12} & 0 \\ H_{12} & H_{22} & 0 \\ 0 & 0 & H_{66} \end{bmatrix} = \int_{-\frac{h}{2}}^{\frac{h}{2}} \begin{bmatrix} Q_{11} & Q_{12} & 0 \\ Q_{12} & Q_{22} & 0 \\ 0 & 0 & Q_{66} \end{bmatrix} z^6 \mathrm{d}z, \tag{10.89}$$

and

$$\underline{\underline{D}}_S = \begin{bmatrix} D_{44} & 0 \\ 0 & D_{55} \end{bmatrix} = \int_{-\frac{h}{2}}^{\frac{h}{2}} \begin{bmatrix} C_{44} & 0 \\ 0 & C_{55} \end{bmatrix} z^2 \mathrm{d}z,$$

$$\underline{\underline{F}}_S = \begin{bmatrix} F_{44} & 0 \\ 0 & F_{55} \end{bmatrix} = \int_{-\frac{h}{2}}^{\frac{h}{2}} \begin{bmatrix} C_{44} & 0 \\ 0 & C_{55} \end{bmatrix} z^4 \mathrm{d}z. \tag{10.90}$$

With constant elastic plate properties, this gives for F_{ij} and H_{ij} $(i, j = 1, 2, 6)$:

$$F_{ij} = \frac{Q_{ij} h^5}{80}, \quad H_{ij} = \frac{Q_{ij} h^7}{448}, \tag{10.91}$$

where $F_{16} = F_{26} = 0$ and $H_{16} = H_{26} = 0$. For D_{ij} and F_{ij} with $i, j = 4, 5$ follows:

$$D_{ij} = \frac{C_{ij} h^3}{12}, \quad F_{ij} = \frac{C_{ij} h^5}{80}, \tag{10.92}$$

where $D_{45} = 0$ and $F_{45} = 0$.

The new stiffness quantities which appear here contain powers of z of higher degrees. It may therefore be assumed that these terms will play only a minor role for sufficiently thin plates and will only come into play for plate situations where a thin-walled structure can no longer be assumed and where the transverse shear strains will play an important role in the structural response.

10.2.3 Equilibrium Conditions

The equilibrium conditions describing a plate in the framework of Reddy's Third-Order Shear Deformation Theory will now be derived from the principle of virtual displacements $\delta W_i = \delta W_a$. For the inner virtual work δw_i we obtain:

$$\begin{aligned}\delta W_i = \int_0^b \int_0^a \int_{-\frac{h}{2}}^{\frac{h}{2}} \Bigg\{ & \sigma_{xx} \left[z\delta\left(\frac{\partial \psi_x}{\partial x}\right) - \frac{4}{3h^2} z^3 \delta\left(\frac{\partial \psi_x}{\partial x} + \frac{\partial^2 w_0}{\partial x^2}\right)\right] \\ & + \sigma_{yy}\left[z\delta\left(\frac{\partial \psi_y}{\partial y}\right) - \frac{4}{3h^2} z^3 \delta\left(\frac{\partial \psi_y}{\partial y} + \frac{\partial^2 w_0}{\partial y^2}\right)\right] \\ & + \tau_{xy}\left[z\delta\left(\frac{\partial \psi_x}{\partial y} + \frac{\partial \psi_y}{\partial x}\right) - \frac{4}{3h^2} z^3 \delta\left(\frac{\partial \psi_x}{\partial y} + \frac{\partial \psi_y}{\partial x} + 2\frac{\partial^2 w_0}{\partial x \partial y}\right)\right] \\ & + \tau_{xz}\left[\delta\left(\psi_x + \frac{\partial w_0}{\partial x}\right) - \frac{4}{h^2} z^2 \delta\left(\psi_x + \frac{\partial w_0}{\partial x}\right)\right] \\ & + \tau_{yz}\left[\delta\left(\psi_y + \frac{\partial w_0}{\partial y}\right) - \frac{4}{h^2} z^2 \delta\left(\psi_y + \frac{\partial w_0}{\partial y}\right)\right]\Bigg\} \mathrm{d}z\mathrm{d}x\mathrm{d}y. \end{aligned} \quad (10.93)$$

Integration with respect to z yields:

$$\begin{aligned}\delta W_i = \int_0^b \int_0^a \Bigg[& M_{xx}^0 \delta \frac{\partial \psi_x}{\partial x} - \frac{4}{3h^2} P_{xx} \delta\left(\frac{\partial \psi_x}{\partial x} + \frac{\partial^2 w_0}{\partial x^2}\right) \\ & + M_{yy}^0 \delta\left(\frac{\partial \psi_y}{\partial y}\right) - \frac{4}{3h^2} P_{yy} \delta\left(\frac{\partial \psi_y}{\partial y} + \frac{\partial^2 w_0}{\partial y^2}\right) \\ & + M_{xy}^0 \delta\left(\frac{\partial \psi_x}{\partial y} + \frac{\partial \psi_y}{\partial x}\right) - \frac{4}{3h^2} P_{xy} \delta\left(\frac{\partial \psi_x}{\partial y} + \frac{\partial \psi_y}{\partial x} + 2\frac{\partial^2 w_0}{\partial x \partial y}\right) \\ & + Q_x \delta\left(\psi_x + \frac{\partial w_0}{\partial x}\right) - \frac{4}{h^2} R_x \delta\left(\psi_x + \frac{\partial w_0}{\partial x}\right) \\ & + Q_y \delta\left(\psi_y + \frac{\partial w_0}{\partial y}\right) - \frac{4}{h^2} R_y \delta\left(\psi_y + \frac{\partial w_0}{\partial y}\right)\Bigg] \mathrm{d}x\mathrm{d}y. \end{aligned} \quad (10.94)$$

The virtual work in the case of a surface load is $p = p(x, y)$:

$$\delta W_a = \int_0^b \int_0^a p \delta w_0 \mathrm{d}x\mathrm{d}y. \quad (10.95)$$

We now perform partial integrations of those terms where partial derivatives of the virtual quantities δw_0, $\delta\psi_x$, $\delta\psi_y$ appear:

$$\begin{aligned}
&\int_0^b\int_0^a M_{xx}^0\frac{\partial\delta\psi_x}{\partial x}\mathrm{d}x\mathrm{d}y = \int_0^b M_{xx}^0\delta\psi_x\Big|_0^a\mathrm{d}y - \int_0^b\int_0^a\frac{\partial M_{xx}^0}{\partial x}\delta\psi_x\mathrm{d}x\mathrm{d}y,\\
&-\frac{4}{3h^2}\int_0^b\int_0^a P_{xx}\frac{\partial\delta\psi_x}{\partial x}\mathrm{d}x\mathrm{d}y = -\frac{4}{3h^2}\int_0^b P_{xx}\delta\psi_x\Big|_0^a\mathrm{d}y + \frac{4}{3h^2}\int_0^b\int_0^a\frac{\partial P_{xx}}{\partial x}\delta\psi_x\mathrm{d}x\mathrm{d}y,\\
&-\frac{4}{3h^2}\int_0^b\int_0^a P_{xx}\frac{\partial^2\delta w_0}{\partial x^2}\mathrm{d}x\mathrm{d}y = -\frac{4}{3h^2}\int_0^b P_{xx}\frac{\partial\delta w_0}{\partial x}\Big|_0^a\mathrm{d}y\\
&\qquad + \frac{4}{3h^2}\int_0^b\frac{\partial P_{xx}}{\partial x}\delta w_0\Big|_0^a\mathrm{d}y - \frac{4}{3h^2}\int_0^b\int_0^a\frac{\partial^2 P_{xx}}{\partial x^2}\delta w_0\mathrm{d}x\mathrm{d}y,\\
&\int_0^b\int_0^a M_{yy}^0\frac{\partial\delta\psi_y}{\partial y}\mathrm{d}x\mathrm{d}y = \int_0^a M_{yy}^0\delta\psi_y\Big|_0^b\mathrm{d}x - \int_0^b\int_0^a\frac{\partial M_{yy}^0}{\partial y}\delta\psi_y\mathrm{d}x\mathrm{d}y,\\
&-\frac{4}{3h^2}\int_0^b\int_0^a P_{yy}\frac{\partial\delta\psi_y}{\partial y}\mathrm{d}x\mathrm{d}y = -\frac{4}{3h^2}\int_0^a P_{yy}\delta\psi_y\Big|_0^b\mathrm{d}x + \frac{4}{3h^2}\int_0^b\int_0^a\frac{\partial P_{yy}}{\partial y}\delta\psi_y\mathrm{d}x\mathrm{d}y,\\
&-\frac{4}{3h^2}\int_0^b\int_0^a P_{yy}\frac{\partial^2\delta w_0}{\partial y^2}\mathrm{d}x\mathrm{d}y = -\frac{4}{3h^2}\int_0^a P_{yy}\frac{\partial\delta w_0}{\partial y}\Big|_0^b\mathrm{d}x\\
&\qquad + \frac{4}{3h^2}\int_0^a\frac{\partial P_{yy}}{\partial y}\delta w_0\Big|_0^b\mathrm{d}x - \frac{4}{3h^2}\int_0^b\int_0^a\frac{\partial^2 P_{yy}}{\partial y^2}\delta w_0\mathrm{d}x\mathrm{d}y,\\
&\int_0^b\int_0^a M_{xy}^0\frac{\partial\delta\psi_x}{\partial y}\mathrm{d}x\mathrm{d}y = \int_0^a M_{xy}^0\delta\psi_x\Big|_0^b\mathrm{d}x - \int_0^b\int_0^a\frac{\partial M_{xy}^0}{\partial y}\delta\psi_x\mathrm{d}x\mathrm{d}y,\\
&\int_0^b\int_0^a M_{xy}^0\frac{\partial\delta\psi_y}{\partial x}\mathrm{d}x\mathrm{d}y = \int_0^b M_{xy}^0\delta\psi_y\Big|_0^a\mathrm{d}y - \int_0^b\int_0^a\frac{\partial M_{xy}^0}{\partial x}\delta\psi_y\mathrm{d}x\mathrm{d}y,\\
&-\frac{4}{3h^2}\int_0^b\int_0^a P_{xy}\frac{\partial\delta\psi_x}{\partial y}\mathrm{d}x\mathrm{d}y = -\frac{4}{3h^2}\int_0^a P_{xy}\delta\psi_x\Big|_0^b\mathrm{d}x + \frac{4}{3h^2}\int_0^b\int_0^a\frac{\partial P_{xy}}{\partial y}\delta\psi_x\mathrm{d}x\mathrm{d}y,\\
&-\frac{4}{3h^2}\int_0^b\int_0^a P_{xy}\frac{\partial\delta\psi_y}{\partial x}\mathrm{d}x\mathrm{d}y = -\frac{4}{3h^2}\int_0^b P_{xy}\delta\psi_y\Big|_0^a\mathrm{d}y + \frac{4}{3h^2}\int_0^b\int_0^a\frac{\partial P_{xy}}{\partial x}\delta\psi_y\mathrm{d}x\mathrm{d}y,\\
&-\frac{8}{3h^2}\int_0^b\int_0^a P_{xy}\frac{\partial^2\delta w_0}{\partial x\partial y}\mathrm{d}x\mathrm{d}y = -\frac{8}{3h^2}P_{xy}\delta w_0\Big|_0^a\Big|_0^b + \frac{8}{3h^2}\int_0^b\frac{\partial P_{xy}}{\partial y}\delta w_0\Big|_0^a\mathrm{d}y\\
&\qquad + \frac{8}{3h^2}\int_0^a\frac{\partial P_{xy}}{\partial x}\delta w_0\Big|_0^b\mathrm{d}x - \frac{8}{3h^2}\int_0^b\int_0^a\frac{\partial^2 P_{xy}}{\partial x\partial y}\delta w_0\mathrm{d}x\mathrm{d}y,\\
&-\frac{4}{h^2}\int_0^b\int_0^a R_x\frac{\partial\delta w_0}{\partial x}\mathrm{d}x\mathrm{d}y = -\frac{4}{h^2}\int_0^b R_x\delta w_0\Big|_0^a\mathrm{d}y + \frac{4}{h^2}\int_0^b\int_0^a\frac{\partial R_x}{\partial x}\delta w_0\mathrm{d}x\mathrm{d}y,\\
&\int_0^b\int_0^a Q_x\frac{\partial\delta w_0}{\partial x}\mathrm{d}x\mathrm{d}y = \int_0^b Q_x\delta w_0\Big|_0^a\mathrm{d}y - \int_0^b\int_0^a\frac{\partial Q_x}{\partial x}\delta w_0\mathrm{d}x\mathrm{d}y,\\
&-\frac{4}{h^2}\int_0^b\int_0^a R_y\frac{\partial\delta w_0}{\partial y}\mathrm{d}x\mathrm{d}y = -\frac{4}{h^2}\int_0^a R_y\delta w_0\Big|_0^b\mathrm{d}x + \frac{4}{h^2}\int_0^b\int_0^a\frac{\partial R_y}{\partial y}\delta w_0\mathrm{d}x\mathrm{d}y,\\
&\int_0^b\int_0^a Q_y\frac{\partial\delta w_0}{\partial y}\mathrm{d}x\mathrm{d}y = \int_0^a Q_y\delta w_0\Big|_0^b\mathrm{d}x - \int_0^b\int_0^a\frac{\partial Q_y}{\partial y}\delta w_0\mathrm{d}x\mathrm{d}y.
\end{aligned} \tag{10.96}$$

The principle of virtual displacements $\delta W_i = \delta W_a$ then leads after some transformations to the following expression:

$$\begin{aligned}
&\int_0^b \int_0^a \Bigg[\delta w_0 \left(-\frac{4}{3h^2}\frac{\partial^2 P_{xx}}{\partial x^2} - \frac{4}{3h^2}\frac{\partial^2 P_{yy}}{\partial y^2} - \frac{8}{3h^2}\frac{\partial^2 P_{xy}}{\partial x \partial y} \right.\\
&\left. + \frac{4}{h^2}\frac{\partial R_x}{\partial x} + \frac{4}{h^2}\frac{\partial R_y}{\partial y} - \frac{\partial Q_x}{\partial x} - \frac{\partial Q_y}{\partial y} - p \right)\\
&+ \delta\psi_x \left(-\frac{\partial M^0_{xx}}{\partial x} - \frac{\partial M^0_{xy}}{\partial y} + \frac{4}{3h^2}\frac{\partial P_{xx}}{\partial x} + \frac{4}{3h^2}\frac{\partial P_{xy}}{\partial y} + Q_x - \frac{4}{h^2}R_x \right)\\
&+ \delta\psi_y \left(-\frac{\partial M^0_{xy}}{\partial x} - \frac{\partial M^0_{yy}}{\partial y} + \frac{4}{3h^2}\frac{\partial P_{xy}}{\partial x} + \frac{4}{3h^2}\frac{\partial P_{yy}}{\partial y} + Q_y - \frac{4}{h^2}R_y \right) \Bigg] \mathrm{d}x\mathrm{d}y\\
&+ \int_0^a \delta w_0 \left(\frac{4}{3h^2}\frac{\partial P_{yy}}{\partial y} + \frac{8}{3h^2}\frac{\partial P_{xy}}{\partial x} - \frac{4}{h^2}R_y + Q_y \right)\Bigg|_0^b \mathrm{d}x\\
&+ \int_0^b \delta w_0 \left(\frac{4}{3h^2}\frac{\partial P_{xx}}{\partial x} + \frac{8}{3h^2}\frac{\partial P_{xy}}{\partial y} - \frac{4}{h^2}R_x + Q_x \right)\Bigg|_0^a \mathrm{d}y\\
&+ \int_0^a \delta\psi_x \left(M^0_{xy} - \frac{4}{3h^2}P_{xy} \right)\Bigg|_0^b \mathrm{d}x + \int_0^b \delta\psi_y \left(M^0_{xx} - \frac{4}{3h^2}P_{xx} \right)\Bigg|_0^a \mathrm{d}y\\
&+ \int_0^a \delta\psi_y \left(M^0_{yy} - \frac{4}{3h^2}P_{yy} \right)\Bigg|_0^b \mathrm{d}x + \int_0^b \delta\psi_y \left(M^0_{xy} - \frac{4}{3h^2}P_{xy} \right)\Bigg|_0^a \mathrm{d}y\\
&- \frac{4}{3h^2}\int_0^b P_{xx}\frac{\partial \delta w_0}{\partial x}\Bigg|_0^a \mathrm{d}y - \frac{4}{3h^2}\int_0^a P_{yy}\frac{\partial \delta w_0}{\partial y}\Bigg|_0^b \mathrm{d}x - \frac{8}{3h^2}P_{xy}\delta w_0\Bigg|_0^a\Bigg|_0^b = 0.
\end{aligned} \tag{10.97}$$

From the bracket terms of the first integral expression, the equilibrium conditions of a plate in the context of Third-Order Shear Deformation Theory can be read as follows:

$$\begin{aligned}
\frac{\partial Q_x}{\partial x} + \frac{\partial Q_y}{\partial y} - \frac{4}{h^2}\left(\frac{\partial R_x}{\partial x} + \frac{\partial R_y}{\partial y}\right) + \frac{4}{3h^2}\left(\frac{\partial^2 P_{xx}}{\partial x^2} + 2\frac{\partial^2 P_{xy}}{\partial x \partial y} + \frac{\partial^2 P_{yy}}{\partial y^2}\right) + p &= 0,\\
\frac{\partial M^0_{xx}}{\partial x} + \frac{\partial M^0_{xy}}{\partial y} - \frac{4}{3h^2}\left(\frac{\partial P_{xx}}{\partial x} + \frac{\partial P_{xy}}{\partial y}\right) - Q_x + \frac{4R_x}{h^2} &= 0,\\
\frac{\partial M^0_{xy}}{\partial x} + \frac{\partial M^0_{yy}}{\partial y} - \frac{4}{3h^2}\left(\frac{\partial P_{xy}}{\partial x} + \frac{\partial P_{yy}}{\partial y}\right) - Q_y + \frac{4R_y}{h^2} &= 0.
\end{aligned} \tag{10.98}$$

Insertion of the constitutive law according to the Third-Order Shear Deformation Theory then leads to a set of differential equations in the displacement quantities. A presentation is omitted here.

The remaining terms in (10.97) represent the corresponding boundary conditions of the given static plate bending problem.

10.2.4 Navier Solution

In this section we want to derive the Navier solution for a rectangular plate (length a, width b) under an arbitrary surface load $p = p(x, y)$. Let the plate be simply supported at all edges. The surface load $p = p(x, y)$ is developed in the form of a Fourier series:

$$p(x, y) = \sum_{m=1}^{m=\infty} \sum_{n=1}^{n=\infty} P_{mn} \sin\left(\frac{m\pi x}{a}\right) \sin\left(\frac{n\pi y}{b}\right), \tag{10.99}$$

wherein:

$$P_{mn} = \frac{4}{ab} \int_0^b \int_0^a p(x, y) \sin\left(\frac{m\pi x}{a}\right) \sin\left(\frac{n\pi y}{b}\right) \mathrm{d}x \mathrm{d}y. \tag{10.100}$$

The boundary conditions are:

$$\begin{aligned}
&\psi_x\,(y=0) = 0,\ \psi_x\,(y=b) = 0,\ \psi_y\,(x=0) = 0,\ \psi_y\,(x=a) = 0,\\
&w_0\,(x=0) = 0,\ w_0\,(x=a) = 0,\ w_0\,(y=0) = 0,\ w_0\,(y=b) = 0,\\
&\left(M_{xx}^0 - \frac{4P_{xx}}{3h^2}\right)(x=0) = 0,\ \left(M_{xx}^0 - \frac{4P_{xx}}{3h^2}\right)(x=a) = 0,\\
&\left(M_{yy}^0 - \frac{4P_{yy}}{3h^2}\right)(y=0) = 0,\ \left(M_{yy}^0 - \frac{4P_{yy}}{3h^2}\right)(y=b) = 0.
\end{aligned} \tag{10.101}$$

The following series expansion for w_0, ψ_x, ψ_y satisfies the given boundary conditions identically:

$$\begin{aligned}
w_0 &= \sum_{m=1}^{m=\infty} \sum_{n=1}^{n=\infty} W_{mn} \sin\left(\frac{m\pi x}{a}\right) \sin\left(\frac{n\pi y}{b}\right),\\
\psi_x &= \sum_{m=1}^{m=\infty} \sum_{n=1}^{n=\infty} X_{mn} \cos\left(\frac{m\pi x}{a}\right) \sin\left(\frac{n\pi y}{b}\right),\\
\psi_y &= \sum_{m=1}^{m=\infty} \sum_{n=1}^{n=\infty} Y_{mn} \sin\left(\frac{m\pi x}{a}\right) \cos\left(\frac{n\pi y}{b}\right).
\end{aligned} \tag{10.102}$$

Substituting into the equilibrium conditions (10.98) leads to a linear system of equations for the unknown coefficients W_{mn}, X_{mn}, W_{mn}:

$$\begin{bmatrix} \lambda_{11} & \lambda_{12} & \lambda_{13} \\ \lambda_{12} & \lambda_{22} & \lambda_{23} \\ \lambda_{13} & \lambda_{23} & \lambda_{33} \end{bmatrix} \begin{pmatrix} W_{mn} \\ X_{mn} \\ Y_{mn} \end{pmatrix} = \begin{pmatrix} P_{mn} \\ 0 \\ 0 \end{pmatrix}, \tag{10.103}$$

with:

$$\lambda_{11} = \left(A_{55} - \frac{8D_{55}}{3h^2} + \frac{16F_{55}}{9h^4}\right)\left(\frac{m\pi}{a}\right)^2 + \left(A_{44} - \frac{8D_{44}}{3h^2} + \frac{16F_{44}}{9h^4}\right)\left(\frac{n\pi}{b}\right)^2$$

$$
\begin{aligned}
&+\frac{16}{9h^4}\left[H_{11}\left(\frac{m\pi}{a}\right)^4+2\left(H_{12}+2H_{66}\right)\left(\frac{m\pi}{a}\right)^2\left(\frac{n\pi}{b}\right)^2+H_{22}\left(\frac{n\pi}{b}\right)^4\right],\\
\lambda_{12}=&\left(A_{55}-\frac{8D_{55}}{3h^2}+\frac{16F_{55}}{9h^4}\right)\frac{m\pi}{a}-\frac{4}{3h^2}\left[\left(F_{11}-\frac{4H_{11}}{3h^2}\right)\left(\frac{m\pi}{a}\right)^3\right.\\
&\left.+\left(F_{12}-\frac{4H_{12}}{3h^2}+2\left\{F_{66}-\frac{4H_{66}}{3h^2}\right\}\right)\frac{m\pi}{a}\left(\frac{n\pi}{b}\right)^2\right],\\
\lambda_{13}=&\left(A_{44}-\frac{8D_{44}}{3h^2}+\frac{16F_{44}}{9h^4}\right)\frac{n\pi}{b}-\frac{4}{3h^2}\left[\left(F_{22}-\frac{4H_{22}}{3h^2}\right)\left(\frac{n\pi}{b}\right)^3\right.\\
&\left.+\left(F_{12}-\frac{4H_{12}}{3h^2}+2\left\{F_{66}-\frac{4H_{66}}{3h^2}\right\}\right)\left(\frac{m\pi}{a}\right)^2\frac{n\pi}{b}\right],\\
\lambda_{22}=&A_{55}-\frac{8D_{55}}{3h^2}+\frac{16F_{55}}{9h^4}+\left(D_{11}-\frac{8F_{11}}{3h^2}+\frac{16H_{11}}{9h^4}\right)\left(\frac{m\pi}{a}\right)^2\\
&+\left(D_{66}-\frac{8F_{66}}{3h^2}+\frac{16H_{66}}{9h^4}\right)\left(\frac{n\pi}{b}\right)^2,\\
\lambda_{23}=&\left[\left(D_{12}-\frac{8F_{12}}{3h^2}+\frac{16H_{12}}{9h^4}\right)+\left(D_{66}-\frac{8F_{66}}{3h^2}+\frac{16H_{66}}{9h^4}\right)\right]\frac{m\pi}{a}\frac{n\pi}{b},\\
\lambda_{33}=&A_{44}-\frac{8D_{44}}{3h^2}+\frac{16F_{44}}{9h^4}+\left(D_{66}-\frac{8F_{66}}{3h^2}+\frac{16H_{66}}{9h^4}\right)\left(\frac{m\pi}{a}\right)^2\\
&+\left(D_{22}-\frac{8F_{22}}{3h^2}+\frac{16H_{22}}{9h^4}\right)\left(\frac{n\pi}{b}\right)^2.
\end{aligned}
\tag{10.104}
$$

Once the coefficients W_{mn}, X_{mn}, Y_{mn} are determined, the kinematic equations and the material law can be used to determine the strains and stresses in the plate, respectively. For the plane stress components we obtain:

$$
\begin{aligned}
\begin{pmatrix}\sigma_{xx}\\ \sigma_{yy}\\ \tau_{xy}\end{pmatrix}&=\begin{bmatrix}Q_{11}&Q_{12}&0\\ Q_{12}&Q_{22}&0\\ 0&0&Q_{66}\end{bmatrix}\left[z\begin{pmatrix}\frac{\partial\psi_x}{\partial x}\\ \frac{\partial\psi_y}{\partial y}\\ \frac{\partial\psi_x}{\partial y}+\frac{\partial\psi_y}{\partial x}\end{pmatrix}-\frac{4}{3h^2}z^3\begin{pmatrix}\frac{\partial\psi_x}{\partial x}+\frac{\partial^2w_0}{\partial x^2}\\ \frac{\partial\psi_y}{\partial y}+\frac{\partial^2w_0}{\partial y^2}\\ \frac{\partial\psi_x}{\partial y}+\frac{\partial\psi_y}{\partial x}+2\frac{\partial^2w_0}{\partial x\partial y}\end{pmatrix}\right]\\
&=\sum_{m=1}^{\infty}\sum_{n=1}^{\infty}\begin{pmatrix}F_1(z)\sin\left(\frac{m\pi x}{a}\right)\sin\left(\frac{n\pi y}{b}\right)\\ F_2(z)\sin\left(\frac{m\pi x}{a}\right)\sin\left(\frac{n\pi y}{b}\right)\\ F_3(z)\cos\left(\frac{m\pi x}{a}\right)\cos\left(\frac{n\pi y}{b}\right)\end{pmatrix},
\end{aligned}
\tag{10.105}
$$

with:

$$
\begin{aligned}
F_1(z)&=\frac{m\pi}{a}\left[-zX_{mn}+\frac{4}{3h^2}z^3\left(X_{mn}+\frac{m\pi}{a}W_{mn}\right)\right],\\
F_2(z)&=\frac{n\pi}{b}\left[-zY_{mn}+\frac{4}{3h^2}z^3\left(Y_{mn}+\frac{n\pi}{b}W_{mn}\right)\right],\\
F_3(z)&=z\left(\frac{n\pi}{b}X_{mn}+\frac{m\pi}{a}Y_{mn}\right)\\
&\quad-\frac{4}{3h^2}z^3\left(\frac{n\pi}{b}X_{mn}+\frac{m\pi}{a}Y_{mn}+2\frac{m\pi}{a}\frac{n\pi}{b}W_{mn}\right).
\end{aligned}
\tag{10.106}
$$

The transverse shear stresses τ_{xz} and τ_{yz} are obtained from the constitutive law as follows:

$$\begin{aligned}\begin{pmatrix}\tau_{yz}\\ \tau_{xz}\end{pmatrix} &= \sum_{m=1}^{\infty}\sum_{n=1}^{\infty}\begin{bmatrix}C_{44} & 0\\ 0 & C_{55}\end{bmatrix}\begin{pmatrix}\psi_y + \frac{\partial w_0}{\partial y} - \frac{4}{h^2}z^2\left(\psi_y + \frac{\partial w_0}{\partial y}\right)\\ \psi_x + \frac{\partial w_0}{\partial x} - \frac{4}{h^2}z^2\left(\psi_x + \frac{\partial w_0}{\partial x}\right)\end{pmatrix}\\ &= \left(1-\frac{4}{h^2}z^2\right)\sum_{m=1}^{\infty}\sum_{n=1}^{\infty}\begin{bmatrix}C_{44} & 0\\ 0 & C_{55}\end{bmatrix}\\ &\quad \begin{pmatrix}\left(Y_{mn} + \frac{n\pi}{b}W_{mn}\right)\sin\left(\frac{m\pi x}{a}\right)\cos\left(\frac{n\pi y}{b}\right)\\ \left(X_{mn} + \frac{m\pi}{a}W_{mn}\right)\cos\left(\frac{m\pi x}{a}\right)\sin\left(\frac{n\pi y}{b}\right)\end{pmatrix}.\end{aligned} \tag{10.107}$$

Obviously, the transverse shear stresses τ_{xz}, τ_{yz} result as quadratic functions of the thickness coordinate z, but they can also be obtained from the integration of the equilibrium conditions (1.36) as follows ($z_0 = -\frac{h}{2}$):

$$\begin{aligned}\tau_{xz} &= \sum_{m=1}^{\infty}\sum_{n=1}^{\infty}\left[\frac{1}{2}\left(z^2 - z_0^2\right)G_{mn1} + \frac{1}{3h^2}\left(z^4 - z_0^4\right)G_{mn2}\right]\cos\left(\frac{m\pi x}{a}\right)\sin\left(\frac{n\pi y}{b}\right),\\ \tau_{yz} &= \sum_{m=1}^{\infty}\sum_{n=1}^{\infty}\left[\frac{1}{2}\left(z^2 - z_0^2\right)G_{mn3} + \frac{1}{3h^2}\left(z^4 - z_0^4\right)G_{mn4}\right]\sin\left(\frac{m\pi x}{a}\right)\cos\left(\frac{n\pi y}{b}\right),\end{aligned} \tag{10.108}$$

with:

$$\begin{aligned}G_{mn1} &= \left[\left(\frac{m\pi}{a}\right)^2 Q_{11} + \left(\frac{n\pi}{b}\right)^2 Q_{66}\right]X_{mn} + \frac{m\pi}{a}\frac{n\pi}{b}\left(Q_{12} + Q_{66}\right)Y_{mn},\\ G_{mn2} &= -\left[\left(\frac{m\pi}{a}\right)^3 Q_{11} + \left(\frac{m\pi}{a}\right)\left(\frac{n\pi}{b}\right)^2\left(Q_{12} + 2Q_{66}\right)\right]W_{mn} - G_{mn1},\\ G_{mn3} &= \frac{m\pi}{a}\frac{n\pi}{b}\left(Q_{12} + Q_{66}\right)X_{mn} + \left[\left(\frac{m\pi}{a}\right)^2 Q_{66} + \left(\frac{n\pi}{b}\right)^2 Q_{22}\right]Y_{mn},\\ G_{mn4} &= -\left[\left(\frac{n\pi}{b}\right)^3 Q_{22} + \left(\frac{m\pi}{a}\right)^2\left(\frac{n\pi}{b}\right)\left(Q_{12} + 2Q_{66}\right)\right]W_{mn} - G_{mn3}.\end{aligned} \tag{10.109}$$

10.2.5 The Ritz Method

In this section we present the Ritz method for rectangular orthotropic plates in the framework of Third-Order Shear Deformation Theory and start from the principle of the minimum of the total elastic potential Π. The inner potential reads:

$$\Pi_i = \frac{1}{2}\int_{-\frac{h}{2}}^{\frac{h}{2}}\int_0^b\int_0^a\left(\sigma_{xx}\varepsilon_{xx} + \sigma_{yy}\varepsilon_{yy} + \tau_{xy}\gamma_{xy} + \tau_{xz}\gamma_{xz} + \tau_{yz}\gamma_{yz}\right)\mathrm{d}x\mathrm{d}y\mathrm{d}z. \tag{10.110}$$

Performing the integration with respect to z using the kinematic equations (10.81) yields:

$$\Pi_i = \frac{1}{2}\int_0^b\int_0^a\left[M_{xx}^0\frac{\partial\psi_x}{\partial x} + M_{yy}^0\frac{\partial\psi_y}{\partial y} + M_{xy}^0\left(\frac{\partial\psi_x}{\partial y} + \frac{\partial\psi_y}{\partial x}\right)\right.$$

$$
\begin{aligned}
& -\frac{4}{3h^2} P_{xx}\left(\frac{\partial \psi_x}{\partial x}+\frac{\partial^2 w_0}{\partial x^2}\right)-\frac{4}{3h^2} P_{yy}\left(\frac{\partial \psi_y}{\partial y}+\frac{\partial^2 w_0}{\partial y^2}\right) \\
& -\frac{4}{3h^2} P_{xy}\left(\frac{\partial \psi_x}{\partial y}+\frac{\partial \psi_y}{\partial x}+\frac{\partial^2 w_0}{\partial x \partial y}\right) \\
& +Q_x\left(\psi_x+\frac{\partial w_0}{\partial x}\right)+Q_y\left(\psi_y+\frac{\partial w_0}{\partial y}\right) \\
& \left.-\frac{4}{h^2} R_y\left(\psi_y+\frac{\partial w_0}{\partial y}\right)-\frac{4}{h^2} R_x\left(\psi_x+\frac{\partial w_0}{\partial x}\right)\right] \mathrm{d}x \mathrm{d}y . \qquad (10.111)
\end{aligned}
$$

Using the constitutive equations (10.86), we get:

$$
\begin{aligned}
\Pi_i= & \frac{1}{2} \int_0^b \int_0^a\left\{D_{11}\left(\frac{\partial \psi_x}{\partial x}\right)^2+2 D_{12} \frac{\partial \psi_x}{\partial x} \frac{\partial \psi_y}{\partial y}+D_{22}\left(\frac{\partial \psi_y}{\partial y}\right)^2\right. \\
& +D_{66}\left(\frac{\partial \psi_x}{\partial y}+\frac{\partial \psi_y}{\partial x}\right)^2 \\
& -\frac{8}{3h^2} F_{11}\left(\frac{\partial \psi_x}{\partial x}+\frac{\partial^2 w_0}{\partial x^2}\right) \frac{\partial \psi_x}{\partial x}-\frac{8}{3h^2} F_{22}\left(\frac{\partial \psi_y}{\partial y}+\frac{\partial^2 w_0}{\partial y^2}\right) \frac{\partial \psi_y}{\partial y} \\
& -\frac{8}{3h^2} F_{12}\left(\frac{\partial^2 w_0}{\partial x^2} \frac{\partial \psi_y}{\partial y}+2 \frac{\partial \psi_x}{\partial y} \frac{\partial \psi_y}{\partial y}+\frac{\partial^2 w_0}{\partial y^2} \frac{\partial \psi_x}{\partial x}\right) \\
& -\frac{8}{3h^2} F_{66}\left[\left(\frac{\partial \psi_x}{\partial y}+\frac{\partial \psi_y}{\partial x}\right)^2+2\left(\frac{\partial \psi_x}{\partial y}+\frac{\partial \psi_y}{\partial x}\right) \frac{\partial^2 w_0}{\partial x \partial y}\right] \\
& +\frac{16}{9h^4} H_{11}\left(\frac{\partial \psi_x}{\partial x}+\frac{\partial^2 w_0}{\partial x^2}\right)^2+\frac{16}{9h^4} H_{22}\left(\frac{\partial \psi_y}{\partial y}+\frac{\partial^2 w_0}{\partial y^2}\right)^2 \\
& +\frac{32}{9h^4} H_{12}\left(\frac{\partial \psi_x}{\partial x} \frac{\partial \psi_y}{\partial y}+\frac{\partial \psi_x}{\partial x} \frac{\partial^2 w_0}{\partial y^2}+\frac{\partial \psi_y}{\partial y} \frac{\partial^2 w_0}{\partial x^2}+\frac{\partial^2 w_0}{\partial x^2} \frac{\partial^2 w_0}{\partial y^2}\right) \\
& +\frac{16}{9h^4} H_{66}\left[\left(\frac{\partial \psi_x}{\partial y}+\frac{\partial \psi_y}{\partial x}\right)^2+4\left(\frac{\partial \psi_x}{\partial y}+\frac{\partial \psi_y}{\partial x}+\frac{\partial^2 w_0}{\partial x \partial y}\right) \frac{\partial^2 w_0}{\partial x \partial y}\right] \\
& +A_{44}\left(\psi_y+\frac{\partial w_0}{\partial y}\right)^2+A_{55}\left(\psi_x+\frac{\partial w_0}{\partial x}\right)^2 \\
& -\frac{8}{h^2} D_{44}\left(\psi_y+\frac{\partial w_0}{\partial y}\right)^2-\frac{8}{h^2} D_{55}\left(\psi_x+\frac{\partial w_0}{\partial x}\right)^2 \\
& \left.+\frac{16}{h^2} F_{44}\left(\psi_y+\frac{\partial w_0}{\partial y}\right)^2+\frac{16}{h^2} F_{55}\left(\psi_x+\frac{\partial w_0}{\partial x}\right)^2\right\} \mathrm{d}x \mathrm{d}y . \qquad (10.112)
\end{aligned}
$$

For the quantities w_0, ψ_x, and ψ_y, we now use an approach of the following form:

$$
\begin{aligned}
w_0 & =\sum_{m=1}^{m=M} \sum_{n=1}^{n=N} W_{mn} w_{1m}(x) w_{2n}(y), \\
\psi_x & =\sum_{m=1}^{m=K} \sum_{n=1}^{n=L} X_{mn} \psi_{x1m}(x) \psi_{x2n}(y), \\
\psi_y & =\sum_{m=1}^{m=I} \sum_{n=1}^{n=J} Y_{mn} \psi_{y1m}(x) \psi_{y2n}(y) . \qquad (10.113)
\end{aligned}
$$

We define the resultants of the approach functions $w_{1m}, w_{2n}, \psi_{x1m}, \psi_{x2n}, \psi_{y1m}$, and ψ_{y2n} as

$$
\begin{aligned}
\Omega_{ij}^1 &= \int_0^a \frac{d^i\psi_{x1m}}{dx^i}\frac{d^j\psi_{x1p}}{dx^j}dx, \quad \Omega_{ij}^2 = \int_0^a \frac{d^i\psi_{x1m}}{dx^i}\frac{d^j\psi_{y1p}}{dx^j}dx,\\
\Omega_{ij}^3 &= \int_0^a \frac{d^i\psi_{y1m}}{dx^i}\frac{d^j\psi_{y1p}}{dx^j}dx, \quad \Omega_{ij}^4 = \int_0^a \frac{d^i w_{1m}}{dx^i}\frac{d^j w_{1p}}{dx^j}dx,\\
\Omega_{ij}^5 &= \int_0^a \frac{d^i w_{1m}}{dx^i}\frac{d^j\psi_{y1p}}{dx^j}dx, \quad \Omega_{ij}^6 = \int_0^a \frac{d^i w_{1m}}{dx^i}\frac{d^j\psi_{x1p}}{dx^j}dx,\\
\Delta_{ij}^1 &= \int_0^b \frac{d^i\psi_{x2n}}{dy^i}\frac{d^j\psi_{x2q}}{dy^j}dy, \quad \Delta_{ij}^2 = \int_0^b \frac{d^i\psi_{x2n}}{dy^i}\frac{d^j\psi_{y2q}}{dy^j}dy,\\
\Delta_{ij}^3 &= \int_0^b \frac{d^i\psi_{y2n}}{dy^i}\frac{d^j\psi_{y2q}}{dy^j}dy, \quad \Delta_{ij}^4 = \int_0^b \frac{d^i w_{2n}}{dy^i}\frac{d^j w_{2q}}{dy^j}dy,\\
\Delta_{ij}^5 &= \int_0^b \frac{d^i w_{2n}}{dy^i}\frac{d^j\psi_{y2q}}{dy^j}dy, \quad \Delta_{ij}^6 = \int_0^b \frac{d^i w_{2n}}{dy^i}\frac{d^j\psi_{x2q}}{dy^j}dy
\end{aligned}
\tag{10.114}
$$

as well as the abbreviations

$$
\begin{aligned}
\Gamma_i^{klkl} &= \Omega_{11}^1\Delta_{00}^1\left(\frac{1}{2}D_{11} - \frac{4}{3h^2}F_{11} + \frac{8}{9h^4}H_{11}\right)\\
&+ \Omega_{00}^1\Delta_{11}^1\left(\frac{1}{2}D_{66} - \frac{4}{3h^2}F_{66} + \frac{8}{9h^4}H_{66}\right)\\
&+ \Omega_{00}^1\Delta_{00}^1\left(\frac{1}{2}A_{55} + \frac{8}{h^4}F_{55} - \frac{4}{h^2}D_{55}\right),\\
\Gamma_i^{ijij} &= \Omega_{00}^3\Delta_{11}^3\left(\frac{1}{2}D_{22} - \frac{4}{3h^2}F_{22} + \frac{8}{9h^4}H_{22}\right)\\
&+ \Omega_{11}^3\Delta_{00}^3\left(\frac{1}{2}D_{66} - \frac{4}{3h^2}F_{66} + \frac{8}{9h^4}H_{66}\right)\\
&+ \Omega_{00}^3\Delta_{00}^3\left(\frac{1}{2}A_{44} + \frac{8}{h^4}F_{44} - \frac{4}{h^2}D_{44}\right),\\
\Gamma_i^{klij} &= \Omega_{10}^2\Delta_{01}^2\left(D_{12} - \frac{8}{3h^2}F_{12} + \frac{16}{9h^4}H_{12}\right)\\
&+ \Omega_{01}^2\Delta_{10}^2\left(D_{66} - \frac{8}{3h^2}F_{66} + \frac{16}{9h^4}H_{66}\right),\\
\Gamma_i^{mnmn} &= \Omega_{00}^4\Delta_{11}^4\left(\frac{1}{2}A_{44} - \frac{4}{h^2}D_{44} + \frac{8}{h^2}F_{44}\right)\\
&+ \Omega_{11}^4\Delta_{00}^4\left(\frac{1}{2}A_{55} - \frac{4}{h^2}D_{55} + \frac{8}{h^2}F_{55}\right)\\
&+ \frac{8}{9h^4}\left(\Omega_{22}^4\Delta_{00}^4H_{11} + 2\Omega_{20}^4\Delta_{02}^4H_{12} + \Omega_{00}^4\Delta_{22}^4H_{22}\right)\\
&+ \frac{32}{9h^2}\Omega_{11}^4\Delta_{11}^4H_{66},\\
\Gamma_i^{mnij} &= \Omega_{00}^5\Delta_{21}^5\left(-\frac{4}{3h^2}F_{22} + \frac{16}{9h^4}H_{22}\right)
\end{aligned}
$$

$$
\begin{aligned}
&+\Omega_{11}^{5}\Delta_{10}^{5}\left(-\frac{8}{3h^2}F_{66}+\frac{32}{9h^4}H_{66}\right)\\
&+\Omega_{20}^{5}\Delta_{01}^{5}\left(-\frac{4}{3h^2}F_{12}+\frac{16}{9h^4}H_{12}\right)\\
&+\Omega_{00}^{5}\Delta_{10}^{5}\left(A_{44}-\frac{8}{h^2}D_{44}+\frac{16}{h^4}F_{44}\right),\\
\Gamma_i^{mnkl} &= \Omega_{21}^{6}\Delta_{00}^{6}\left(-\frac{4}{3h^2}F_{11}+\frac{16}{9h^4}H_{11}\right)\\
&+\Omega_{10}^{6}\Delta_{11}^{6}\left(-\frac{8}{3h^2}F_{66}+\frac{32}{9h^4}H_{66}\right)\\
&+\Omega_{01}^{6}\Delta_{20}^{6}\left(-\frac{4}{3h^2}F_{12}+\frac{16}{9h^4}H_{12}\right)\\
&+\Omega_{10}^{6}\Delta_{00}^{6}\left(A_{55}-\frac{8}{h^2}D_{55}+\frac{16}{h^4}F_{55}\right). \qquad (10.115)
\end{aligned}
$$

This gives the following representation for the inner potential Π_i:

$$
\begin{aligned}
\Pi_i = &\sum_{m=1}^{m=K}\sum_{n=1}^{n=L}\sum_{p=1}^{p=K}\sum_{q=1}^{q=L} X_{mn}X_{pq}\Gamma_i^{klkl} + \sum_{m=1}^{m=I}\sum_{n=1}^{n=J}\sum_{p=1}^{p=I}\sum_{q=1}^{q=J} Y_{mn}Y_{pq}\Gamma_i^{ijij}\\
&+\sum_{m=1}^{m=K}\sum_{n=1}^{n=L}\sum_{p=1}^{p=I}\sum_{q=1}^{q=J} X_{mn}Y_{pq}\Gamma_i^{klij} + \sum_{m=1}^{m=M}\sum_{n=1}^{n=N}\sum_{p=1}^{p=M}\sum_{q=1}^{q=N} W_{mn}W_{pq}\Gamma_i^{mnmn}\\
&+\sum_{m=1}^{m=M}\sum_{n=1}^{n=N}\sum_{p=1}^{p=I}\sum_{q=1}^{q=J} W_{mn}Y_{pq}\Gamma_i^{mnij}\\
&+\sum_{m=1}^{m=M}\sum_{n=1}^{n=N}\sum_{p=1}^{p=K}\sum_{q=1}^{q=L} W_{mn}X_{pq}\Gamma_i^{mnkl}. \qquad (10.116)
\end{aligned}
$$

References

Altenbach, H., Altenbach, J., Naumenko, K.: Ebene Flächentragwerke, 2nd edn. Springer, Berlin et al., Germany (2016)

Mittelstedt, C., Becker, W.: Strukturmechanik ebener Laminate. Studienbereich Mechanik Technische Universität Darmstadt, Germany (2016)

Reddy, J.N.: Mechanics of laminated composite plates and shells, 2nd edn. CRC Press Boca Raton et al, USA (2004)

Reddy, J.N.: Theory and analysis of elastic plates and shells. CRC Press Boca Raton et al, USA (2006)

Plate Buckling 11

Plates are thin-walled structures. In this respect, attention must be paid to their buckling behavior. Accordingly, this chapter is devoted to the analytical treatment of the buckling behavior of plates within the framework of the Kirchhoff plate theory. First, all necessary basic equations describing the buckling of orthotropic plates are provided. Then, the Navier solution is discussed, i.e. the exact analytical determination of the buckling load of rectangular simply supported orthotropic plates under uniaxial and biaxial loading. The chapter concludes with the energetic treatment of plate buckling problems and describes as a simple method the so-called Rayleigh quotient and as a more universally applicable method the Ritz method. In the works of Girkmann (1974), Mittelstedt and Becker (2016), Petersen (1982), Reddy (2004, 2006), Szilard (2004), Wiedemann (2006) further information on this subject can be found.

11.1 Basic Equations

We consider an orthotropic plate under boundary loads N_{xx}^0, N_{yy}^0, N_{xy}^0 (Fig. 11.1, top left). In this figure, the boundary loads are plotted in such a way that N_{xx}^0 and N_{yy}^0 are counted as positive when they act as compressive forces. Similarly, N_{xy}^0 is also counted as positive contrary to the usual sign convention.

Now consider a plate structure under such inplane loads. As long as the given combination of loads is below a certain critical limit, a plane stress state is present, the so-called initial or pre-buckling stress state. This state must satisfy the inplane force equilibrium conditions:

C. Mittelstedt, *Theory of Plates and Shells*,
https://doi.org/10.1007/978-3-662-66805-4_11

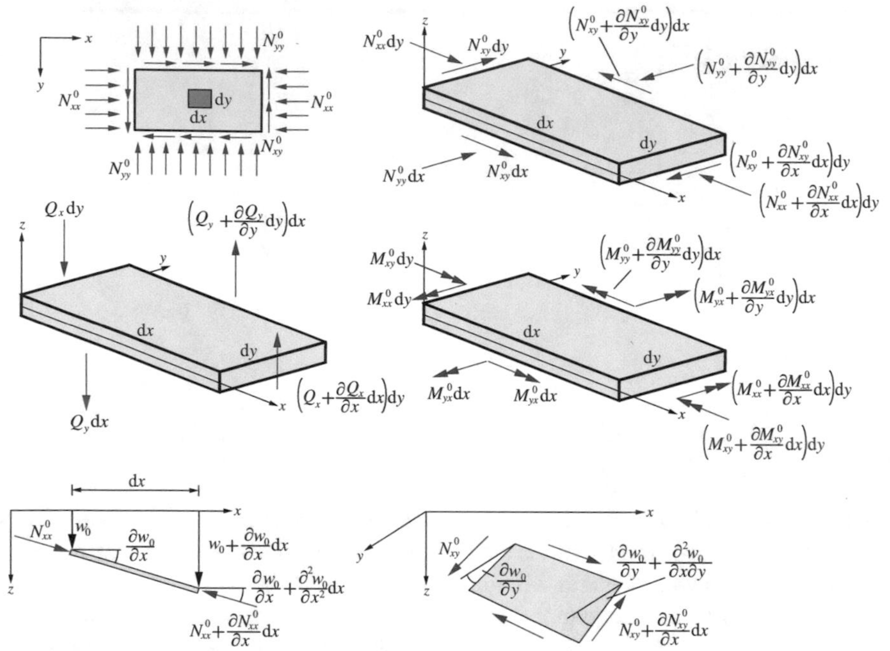

Fig. 11.1 Plate under combined load (top left), infinitesimal element (top right and center), section in the xz-plane (bottom left), contribution from the shear force flow N_{xy}^0 (bottom right)

$$\frac{\partial N_{xx}^0}{\partial x} + \frac{\partial N_{xy}^0}{\partial y} = 0,$$
$$\frac{\partial N_{xy}^0}{\partial x} + \frac{\partial N_{yy}^0}{\partial y} = 0. \tag{11.1}$$

In all further elaborations, we assume that the internal forces in the plate structure under consideration correspond to the respective boundary loads applied at each point. In order to obtain a statement about the critical load or load combination at which buckling occurs, it is necessary to consider the equilibrium on the deformed system, assuming infinitesimal deflections. Due to the smallness of the deflections, it is assumed that the applied loads and thus also the pre-buckling stress state do not change due to the onset of buckling.

We first consider the forces in the $z-$direction arising at the infinitesimally deflected element and examine the section in the $xz-$plane as shown in Fig. 11.1, bottom left. The resulting force in the $z-$direction follows as:

$$V_1 = N_{xx}^0 \mathrm{d}y \frac{\partial w_0}{\partial x} - \left(N_{xx}^0 + \frac{\partial N_{xx}^0}{\partial x} \mathrm{d}x \right) \left(\frac{\partial w_0}{\partial x} + \frac{\partial^2 w_0}{\partial x^2} \mathrm{d}x \right) \mathrm{d}y. \tag{11.2}$$

Neglecting terms of higher order, the following expression for V_1 results:

$$V_1 = -\left(N_{xx}^0 \frac{\partial^2 w_0}{\partial x^2} + \frac{\partial N_{xx}^0}{\partial x} \frac{\partial w_0}{\partial x} \right) \mathrm{d}x\mathrm{d}y. \quad (11.3)$$

Obviously, V_1 is the force component obtained from the force flow N_{xx}^0 due to the deflection of the infinitesimal element. Quite analogously, one can determine the force component V_2 due to N_{yy}^0 at a cut in the $yz-$plane:

$$V_2 = -\left(N_{yy}^0 \frac{\partial^2 w_0}{\partial y^2} + \frac{\partial N_{yy}^0}{\partial y} \frac{\partial w_0}{\partial y} \right) \mathrm{d}x\mathrm{d}y. \quad (11.4)$$

The force component V_3 due to the shear force N_{xy}^0 can be obtained by looking at the free body diagram of Fig. 11.1, bottom right:

$$V_3 = N_{xy}^0 \mathrm{d}y \frac{\partial w_0}{\partial y} - \left(N_{xy}^0 + \frac{\partial N_{xy}^0}{\partial x} \mathrm{d}x \right) \mathrm{d}y \left(\frac{\partial w_0}{\partial y} + \frac{\partial^2 w_0}{\partial x \partial y} \mathrm{d}x \right). \quad (11.5)$$

If we neglect terms of higher order, we obtain the following result:

$$V_3 = -\left(N_{xy}^0 \frac{\partial^2 w_0}{\partial x \partial y} + \frac{\partial N_{xy}^0}{\partial x} \frac{\partial w_0}{\partial y} \right) \mathrm{d}x\mathrm{d}y. \quad (11.6)$$

Quite analogously, one obtains a contribution from $N_{yx}^0 = N_{xy}^0$:

$$V_4 = -\left(N_{xy}^0 \frac{\partial^2 w_0}{\partial x \partial y} + \frac{\partial N_{xy}^0}{\partial y} \frac{\partial w_0}{\partial x} \right) \mathrm{d}x\mathrm{d}y. \quad (11.7)$$

The total resulting force component V in $z-$ direction due to the observation at the deflected infninitesimal element is then obtained by summing up the partial forces V_1, V_2, V_3, V_4. We obtain:

$$\begin{aligned} V = & -N_{xx}^0 \frac{\partial^2 w_0}{\partial x^2} \mathrm{d}x\mathrm{d}y - 2N_{xy}^0 \frac{\partial^2 w_0}{\partial x \partial y} \mathrm{d}x\mathrm{d}y - N_{yy}^0 \frac{\partial^2 w_0}{\partial y^2} \mathrm{d}x\mathrm{d}y \\ & - \left(\frac{\partial N_{xx}^0}{\partial x} + \frac{\partial N_{xy}^0}{\partial y} \right) \frac{\partial w_0}{\partial x} \mathrm{d}x\mathrm{d}y - \left(\frac{\partial N_{xy}^0}{\partial x} + \frac{\partial N_{yy}^0}{\partial y} \right) \frac{\partial w_0}{\partial y} \mathrm{d}x\mathrm{d}y. \end{aligned} \quad (11.8)$$

Obviously, the bracket terms in the second line of (11.8) correspond exactly to the equilibrium conditions (11.1), so they must vanish. Accordingly, the following force V remains in the $z-$direction:

$$V = -\left(N_{xx}^0 \frac{\partial^2 w_0}{\partial x^2} + 2N_{xy}^0 \frac{\partial^2 w_0}{\partial x \partial y} + N_{yy}^0 \frac{\partial^2 w_0}{\partial y^2} \right) \mathrm{d}x\mathrm{d}y. \quad (11.9)$$

The force V can be interpreted as the resultant of a load in $z-$direction acting on the considered section element.

We now form the sum of transverse forces at the section element of Fig. 11.1 and obtain:

$$\frac{\partial Q_x}{\partial x}+\frac{\partial Q_y}{\partial y}-N_{xx}^0\frac{\partial^2 w_0}{\partial x^2}-2N_{xy}^0\frac{\partial^2 w_0}{\partial x\partial y}-N_{yy}^0\frac{\partial^2 w_0}{\partial y^2}=0. \tag{11.10}$$

The two moment equilibria on the deformed section element are obtained as:

$$\begin{aligned}\frac{\partial M_{xx}^0}{\partial x}+\frac{\partial M_{xy}^0}{\partial y}&=Q_x,\\ \frac{\partial M_{xy}^0}{\partial x}+\frac{\partial M_{yy}^0}{\partial y}&=Q_y.\end{aligned} \tag{11.11}$$

If in (11.10) the transverse forces are replaced by (11.11)s, then the following equation is obtained:

$$\frac{\partial^2 M_{xx}^0}{\partial x^2}+2\frac{\partial^2 M_{xy}^0}{\partial x\partial y}+\frac{\partial^2 M_{yy}^0}{\partial y^2}-N_{xx}^0\frac{\partial^2 w_0}{\partial x^2}-2N_{xy}^0\frac{\partial^2 w_0}{\partial x\partial y}-N_{yy}^0\frac{\partial^2 w_0}{\partial y^2}=0. \tag{11.12}$$

Thus, the plane equilibrium conditions (11.1) and the condensed plate equilibrium (11.12) are those equations by which the equilibrium at the infinitesimally deflected infinitesimal element is described. We can express in this the force and moment quantities by the constitutive law as follows:

$$\begin{pmatrix}N_{xx}^0\\N_{yy}^0\\N_{xy}^0\\M_{xx}^0\\M_{yy}^0\\M_{xy}^0\end{pmatrix}=\begin{bmatrix}A_{11}&A_{12}&0&0&0&0\\A_{12}&A_{22}&0&0&0&0\\0&0&A_{66}&0&0&0\\0&0&0&D_{11}&D_{12}&0\\0&0&0&D_{12}&D_{22}&0\\0&0&0&0&0&D_{66}\end{bmatrix}\begin{pmatrix}\frac{\partial u_0}{\partial x}\\\frac{\partial v_0}{\partial y}\\\frac{\partial u_0}{\partial y}+\frac{\partial v_0}{\partial x}\\-\frac{\partial^2 w_0}{\partial x^2}\\-\frac{\partial^2 w_0}{\partial y^2}\\-2\frac{\partial^2 w_0}{\partial x\partial y}\end{pmatrix}. \tag{11.13}$$

From this one obtains the differential equations in the framework of second-order theory as follows:

$$\begin{aligned}A_{11}\frac{\partial^2 u_0}{\partial x^2}+A_{66}\frac{\partial^2 u_0}{\partial y^2}+(A_{12}+A_{66})\frac{\partial^2 v_0}{\partial x\partial y}&=0,\\(A_{12}+A_{66})\frac{\partial^2 u_0}{\partial x\partial y}+A_{66}\frac{\partial^2 v_0}{\partial x^2}+A_{22}\frac{\partial^2 v_0}{\partial y^2}&=0,\\D_{11}\frac{\partial^4 w_0}{\partial x^4}+2\left(D_{12}+2D_{66}\right)\frac{\partial^4 w_0}{\partial x^2\partial y^2}+D_{22}\frac{\partial^4 w_0}{\partial y^4}&\end{aligned}$$

$$+N_{xx}^0\frac{\partial^2 w_0}{\partial x^2}+2N_{xy}^0\frac{\partial^2 w_0}{\partial x\partial y}+N_{yy}^0\frac{\partial^2 w_0}{\partial y^2}=0. \tag{11.14}$$

The displacements u_0, v_0, w_0 are the displacements of the plate middle plane which occur at the onset of buckling. One speaks also of the so-called buckling mode. It can be seen that the first two differential equations in (11.14) contain exclusively the two inplane displacements u_0 and v_0, whereas in the third equation in (11.14) only the displacement w_0 occurs. Consequently, the first two equations in (11.14) describe the disk behavior of the plane surface structure under a given boundary load, while the third equation describes the buckling behavior. This equation will be used in the following to derive analytical solutions of plate buckling problems.

11.2 Navier Solution

For a few load cases in connection with defined boundary conditions, exact-analytical solutions for plate buckling problems can be derived. One such class of solutions is the so-called Navier solution, which we will discuss in the following for the case of uniaxial compressive load. Consider a rectangular orthotropic plate of length a and width b (Fig. 11.2). It is loaded by the uniaxial compressive edge load N_{xx}^0. In addition, let the plate under consideration be simply supported at all four edges, so that the following boundary conditions apply:

$$\begin{gathered}
w_0\,(x=0)=0,\quad w_0\,(x=a)=0,\quad w_0\,(y=0)=0,\quad w_0\,(y=b)=0,\\
M_{xx}^0\,(x=0)=-D_{11}\left.\frac{\partial^2 w_0}{\partial x^2}\right|_{x=0}=0,\quad M_{xx}^0\,(x=a)=-D_{11}\left.\frac{\partial^2 w_0}{\partial x^2}\right|_{x=a}=0,\\
M_{yy}^0\,(y=0)=-D_{22}\left.\frac{\partial^2 w_0}{\partial y^2}\right|_{y=0}=0,\quad M_{yy}^0\,(y=b)=-D_{22}\left.\frac{\partial^2 w_0}{\partial y^2}\right|_{y=b}=0.
\end{gathered} \tag{11.15}$$

We have already made use of the fact that the curvatures

$$\left.\frac{\partial^2 w_0}{\partial y^2}\right|_{x=0,a},\quad \left.\frac{\partial^2 w_0}{\partial x^2}\right|_{y=0,b} \tag{11.16}$$

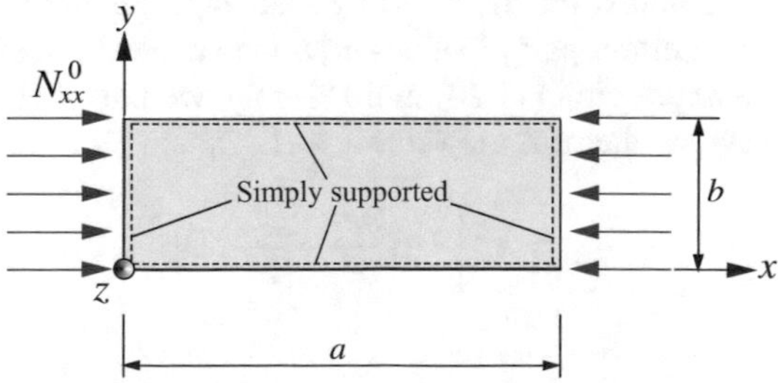

Fig. 11.2 Simply supported plate under uniaxial compressive load

are zero.

The following approach for the buckling mode w_0 satisfies all boundary conditions (11.15) identically:

$$w_0 = W_{mn} \sin\left(\frac{m\pi x}{a}\right) \sin\left(\frac{n\pi y}{b}\right). \tag{11.17}$$

Here m and n are unknown positive integer half-wave numbers of the used sine functions. Inserting the approach (11.17) into the buckling differential equation (11.14) yields:

$$\left[D_{11}\frac{m^4\pi^4}{a^4} + 2\left(D_{12} + 2D_{66}\right)\frac{m^2\pi^2}{a^2}\frac{n^2\pi^2}{b^2} + D_{22}\frac{n^4\pi^4}{b^4} - N_{xx}^0\frac{m^2\pi^2}{a^2}\right] \times W_{mn} \sin\left(\frac{m\pi x}{a}\right) \sin\left(\frac{n\pi x}{b}\right) = 0. \tag{11.18}$$

To avoid the trivial solution $W_{mn} = 0$, the expression in the square brackets in (11.18) is set to zero, i.e.:

$$D_{11}\frac{m^4\pi^4}{a^4} + 2\left(D_{12} + 2D_{66}\right)\frac{m^2\pi^2}{a^2}\frac{n^2\pi^2}{b^2} + D_{22}\frac{n^4\pi^4}{b^4} - N_{xx}^0\frac{m^2\pi^2}{a^2} = 0. \tag{11.19}$$

One can solve this expression for the boundary load N_{xx}^0 as follows:

$$N_{xx}^0 = \pi^2\left[D_{11}\frac{m^2}{a^2} + 2\left(D_{12} + 2D_{66}\right)\frac{n^2}{b^2} + D_{22}\frac{a^2}{m^2}\frac{n^4}{b^4}\right]. \tag{11.20}$$

This result can be interpreted as follows. The boundary load N_{xx}^0, which can be calculated by (11.20), is the load at which the formerly plane surface structure changes into a deflected configuration, i.e. an infinitesimally deflected equilibrium position exists. Thus, the load N_{xx}^0 represents the buckling load.

The buckling load N_{xx}^0 according to (11.20) depends not only on the plate dimensions a, b and the plate stiffnesses D_{11}, D_{22}, D_{12}, D_{66}, but also on the two wave numbers m and n. For each pair of values m, n, there is then a buckling load associated with a buckling mode described by (11.17). The plate thus buckles in the form of two sinusoidal functions, with m half-waves in the $x-$direction and n half-waves in the $y-$direction. Accordingly, the expression (11.20) describes an infinite number of solutions for the buckling load N_{xx}^0. Technically relevant is only the case that for a certain pair of values m, n the smallest buckling load N_{xx}^0 results. Looking at the expression (11.20) more closely, we find that the only relevant solution for the wavenumber n is the value $n = 1$. Thus:

$$N_{xx}^0 = \pi^2\left[D_{11}\frac{m^2}{a^2} + \frac{2}{b^2}\left(D_{12} + 2D_{66}\right) + D_{22}\frac{a^2}{b^4 m^2}\right]. \tag{11.21}$$

The wave number m, on the other hand, cannot be determined a priori without further considerations. It is useful to require that the buckling load N_{xx}^0 takes an extreme value with respect to m:

$$\frac{\partial N_{xx}^0}{\partial m} = 0. \tag{11.22}$$

The resulting expression can be solved for the wavenumber m:

$$m = \frac{a}{b}\sqrt[4]{\frac{D_{22}}{D_{11}}}. \tag{11.23}$$

In general, the expression (11.23) leads to non-integer numerical values, so it is convenient to round up the resulting value for m once to the next higher integer and once to the next lower integer. Both of these integer values for m are substituted into the buckling load (11.21), and the smaller resulting load N_{xx}^0 is the buckling load we are looking for.

It is important to note here that the coefficient W_{mn} of the approach (11.17), as usual in the context of a consideration in the framework of second-order theory, remains undetermined. Consequently, within the framework of the second-order theory it is only possible to determine the onset of buckling and the associated buckling load and the shape of the buckling mode, but it is not possible to make statements beyond this load and thus also for the amplitude W_{mn}. For this, a geometrically nonlinear analysis would be necessary which is not addressed in this chapter.

An important limiting case is the infinite plate with $a \to \infty$. It is shown that the buckling load N_{xx}^0 approaches an asymptotic limit as the length a increases. This limit is obtained by substituting the expression (11.23) for the wavenumber m into (11.21). One obtains:

$$N_{xx}^0 = \frac{2\pi^2}{b^2}\left(\sqrt{D_{11}D_{22}} + D_{12} + 2D_{66}\right). \tag{11.24}$$

Figure 11.3 shows the buckling load N_{xx}^0 for a symmetrical cross-ply laminate with the layup $[0°/90°/0°/90°]_S$. The width of the laminate was set as $b = 200\,\text{mm}$ and the length a was varied. The underlying elastic properties of the single layers were set with the numerical values $E_{11} = 135000\,\text{MPa}$, $E_{22} = 10000\,\text{MPa}$, $G_{12} = 5000\,\text{MPa}$, $\nu_{12} = 0.27$, the single layer thickness was chosen as $h_k = 0.125\,\text{mm}$.

In Fig. 11.3 the results for the buckling load N_{xx}^0 according to (11.21) for different wavenumbers m are plotted as dashed lines. Each of these curves exhibits a local minimum, with the location of the minimum shifting further and further to the right as the wavenumber m increases. Furthermore, it can be seen that each of the curves N_{xx}^0 tends to infinity with increasing distance from the local minimum. Furthermore, the individual curves have various intersections with each other. Technically relevant from all these possible results for the buckling load N_{xx}^0 for a given plate length a is only the smallest possible value, in Fig. 11.3 indicated by the solid line. This lower envelope is also called buckling curve, or garland curve because of its specific appearance. The results of Fig. 11.3 can be interpreted as follows. For very short

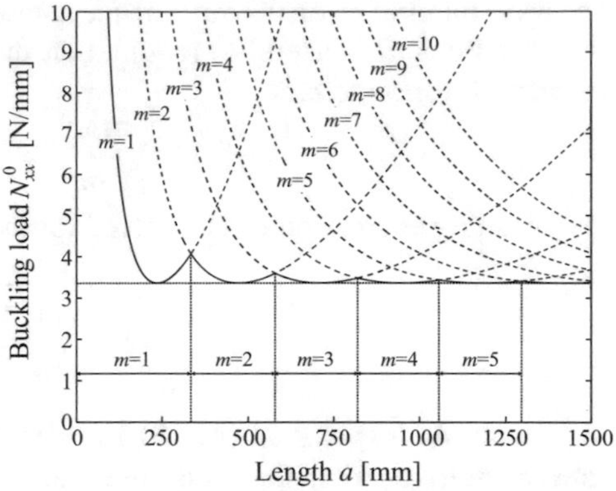

Fig. 11.3 Buckling curve for a cross-ply laminate $[0°/90°/0°/90°]_S$ under uniaxial compressive load

plates, the value of the buckling load tends to infinity. If the plate length a increases, then the plate buckles with $m = 1$ half-wave. The garland curve reaches a local minimum at about $a \approx 240$mm. At a plate length of about $a \approx 330$ mm the two curves for $m = 1$ and $m = 2$ intersect, and a local maximum for the garland curve is obtained at this point. From this value for the plate length a the plate buckles into $m = 2$ half-waves, and in this context one also speaks of a mode change from $m = 1$ to $m = 2$ half-waves having taken place. The next mode change to $m = 3$ half-waves occurs at about $a \approx 580$ mm. For higher plate lengths, further mode changes and local minima and maxima can be observed. Thus, for the given plate buckling problem, clear intervals can be identified within which exactly one half-wavenumber m is technically relevant. Furthermore, it is obvious that the height of the local maxima decreases with increasing plate length and wavenumber. Thus, the garland curve flattens more and more and approaches an asymptotic limit which corresponds exactly to the value (11.24).

Figure 11.4 shows the buckling modes for different lengths a for the considered symmetrical cross-ply laminate. The increase in wavenumber m with increasing length a can be clearly seen here.

11.2.1 Biaxial Load

We again consider a Navier plate, but now under a biaxial load (Fig. 11.5). The given edge loads consist of the axial load N_{xx}^0 and the edge load N_{yy}^0 acting perpendicular to it, which can be represented as a multiple of the axial load N_{xx}^0: $N_{yy}^0 = \psi N_{xx}^0$. The buckling differential equation then takes the following form:

$$D_{11}\frac{\partial^4 w_0}{\partial x^4} + 2\left(D_{12} + 2D_{66}\right)\frac{\partial^4 w_0}{\partial x^2 \partial y^2} + D_{22}\frac{\partial^4 w_0}{\partial y^4} + N_{xx}^0\frac{\partial^2 w_0}{\partial x^2} + \psi N_{xx}^0\frac{\partial^2 w_0}{\partial y^2} = 0. \tag{11.25}$$

At this point, we again use an approach of the form (11.17), which, after insertion into the buckling differential equation (11.25), leads after brief transformation to the

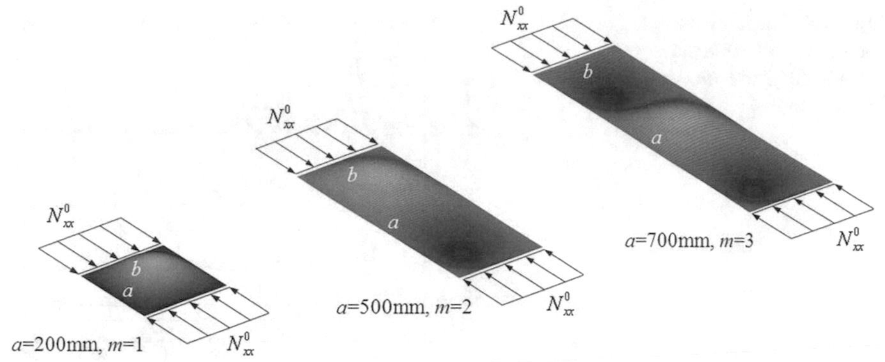

Fig. 11.4 Buckling modes for a cross-ply laminate $[0°/90°/0°/90°]_S$ under uniaxial compressive load

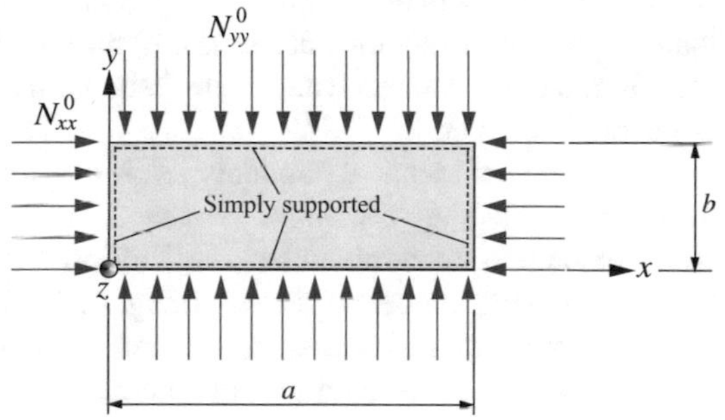

Fig. 11.5 Plate under biaxial load

following expression for the buckling load N_{xx}^0:

$$N_{xx}^0 = \pi^2 \frac{D_{11}\dfrac{m^2}{a^2} + 2\left(D_{12} + 2D_{66}\right)\dfrac{n^2}{b^2} + D_{22}\dfrac{a^2}{m^2}\dfrac{n^4}{b^4}}{1 + \psi\dfrac{a^2}{m^2}\dfrac{n^2}{b^2}}. \tag{11.26}$$

Again, the buckling load N_{xx}^0 depends on the two wavenumbers m and n. In contrast to the plate under uniaxial compressive load, however, the wavenumber n cannot be set a priori to $n = 1$, but rather the relevant pair of values m, n must be determined by trial and error.

An interesting case occurs when one of the two boundary loads N_{xx}^0 or N_{yy}^0 is a tensile load. For example, if N_{yy}^0 is a tensile load, then the factor ψ is negative. From the solution (11.26) it follows that this has a stabilizing effect on the result, the buckling load increases. A tensile load therefore stabilizes the plate. This is also shown in Fig. 11.6 for selected buckling curves of a symmetric cross-ply laminate

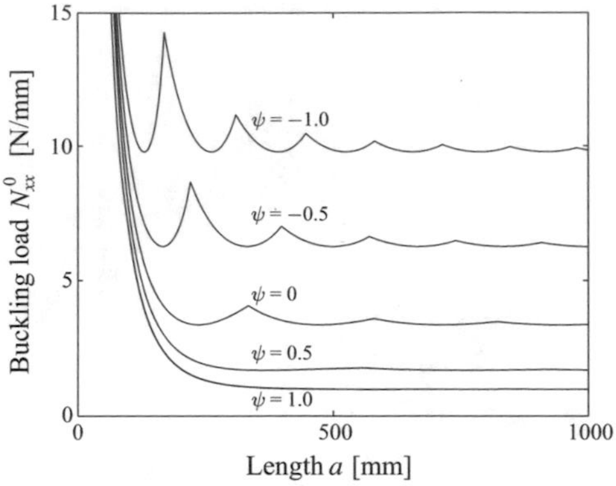

Fig. 11.6 Buckling curves for a cross-ply laminate $[0°/90°/0°/90°]_S$ under biaxial loading

$[0°/90°/0°/90°]_S$ under biaxial loading. Obviously, the presence of an edge tensile load N_{yy}^0 results in a significant increase of the buckling load. Furthermore, it becomes clear that more pronounced and more frequent mode changes occur when the factor ψ becomes smaller.

For the case of identical boundary loads N_{xx}^0 and N_{yy}^0, the two relevant half-wave numbers m and n are always $m = n = 1$, independent of the length of the plate, which explains the continuous shape of the corresponding buckling curve in Fig. 11.6. Selected buckling modes are shown in Fig. 11.7 for a symmetrical cross-ply laminate with $a = 500$ mm. From this, an increase in the longitudinal half-wave number m can be seen as the edge load N_{yy}^0 increases.

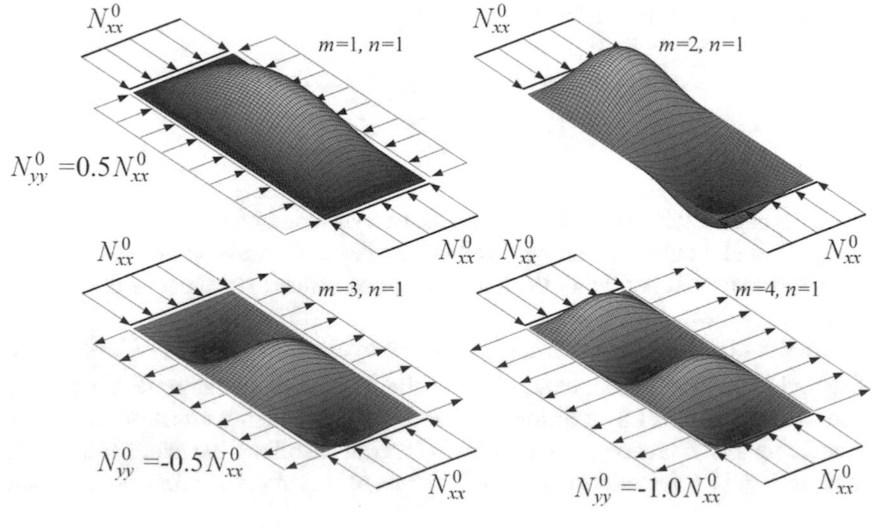

Fig. 11.7 Buckling modes for a cross-ply laminate $[0°/90°/0°/90°]_S$ under biaxial loading

11.3 Energy Methods for the Solution of Plate Buckling Problems

11.3.1 Introduction

Plate buckling problems can be advantageously treated with energy-based methods. To be mentioned in this context are the so-called Rayleigh quotient as well as the Ritz method, both of which are discussed in this section. Discretizing methods like the Finite Element Method have become an integral part of everyday engineering work, but are not treated in this section. We first consider the total elastic potential of a plate in the buckled state, which is composed of the internal potential Π_i and the potential Π_a of the applied forces:

$$\Pi = \Pi_i + \Pi_a. \tag{11.27}$$

We consider a rectangular plate of length a, width b. The inner potential Π_i of the plate can be written as:

$$\begin{aligned}\Pi_i = &\frac{1}{2}\int_0^a\int_0^b \left[N_{xx}^0\frac{\partial u_0}{\partial x} + N_{yy}^0\frac{\partial v_0}{\partial y} + N_{xy}^0\left(\frac{\partial u_0}{\partial y} + \frac{\partial v_0}{\partial x}\right)\right]\mathrm{d}y\mathrm{d}x \\ &- \frac{1}{2}\int_0^a\int_0^b \left[M_{xx}^0\frac{\partial^2 w_0}{\partial x^2} + M_{yy}^0\frac{\partial^2 w_0}{\partial y^2} + 2M_{xy}^0\frac{\partial^2 w_0}{\partial x\partial y}\right]\mathrm{d}y\mathrm{d}x.\end{aligned} \tag{11.28}$$

The potential of the edge loads, i.e. the normal force flows $\hat{N}_{xx}^0$ and $\hat{N}_{yy}^0$ as well as the shear force flow $\hat{N}_{xy}^0$, can be derived from the onset of buckling. For this purpose, we consider the boundary load $\hat{N}_{xx}^0$ as an example (Fig. 11.8). The potential Π_a of the varying boundary load $\hat{N}_{xx}^0(y)$ over y follows as:

$$\Pi_a = -W_a = -\int_0^b \hat{N}_{xx}^0(y)\, u_0(y)\, \mathrm{d}y. \tag{11.29}$$

Herein, $u_0(y)$ is the displacement of the plate midplane at the edge $x = a$. The integration prescribed in (11.29) implies that all infinitesimal resulting forces $\hat{N}_{xx}^0(y)\,\mathrm{d}y$ from $y = 0$ to $y = b$ are to be summed up. To provide an expression for the boundary displacement $u_0(y)$, we consider the infinitesimal section element with length $\mathrm{d}x$ (Fig. 11.8, left) and the deformed state at the onset of buckling shown in Fig. 11.8, right. It follows for $\mathrm{d}u_0$:

$$\mathrm{d}u_0 = (1 - \cos(\varphi))\,\mathrm{d}x. \tag{11.30}$$

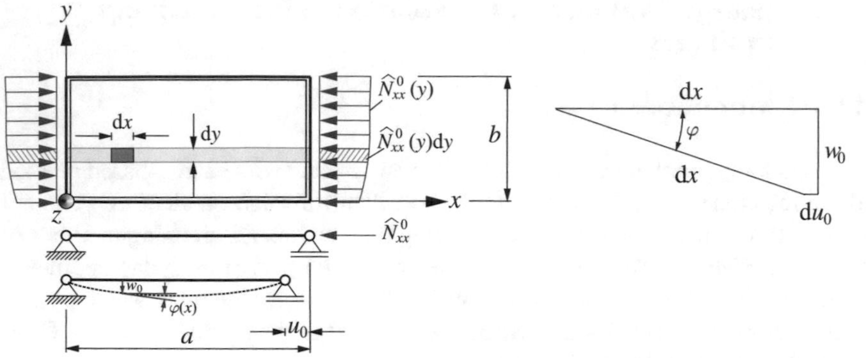

Fig. 11.8 Derivation of the potential of the boundary load $\hat{N}^0_{xx}$

The cosine term is developed into a Taylor series according to $\cos(\varphi) = 1 - \frac{1}{2}\varphi^2$, so we can also write:

$$du_0 = \frac{1}{2}\varphi^2 dx = \frac{1}{2}\left(\frac{\partial w_0}{\partial x}\right)^2 dx. \tag{11.31}$$

Integrating over the longitudinal coordinate x, we obtain the boundary displacement $u_0(y)$ as:

$$u_0(y) = \frac{1}{2}\int_0^a \left(\frac{\partial w_0}{\partial x}\right)^2 dx. \tag{11.32}$$

The potential Π_a of the edge load $\hat{N}^0_{xx}(y)$ then follows as:

$$\Pi_a = -\frac{1}{2}\int_0^b \hat{N}^0_{xx}(y)\left[\int_0^a \left(\frac{\partial w_0}{\partial x}\right)^2 dx\right] dy. \tag{11.33}$$

For a constant boundary load $\hat{N}^0_{xx}$ one obtains:

$$\Pi_a = -\frac{1}{2}\hat{N}^0_{xx}\int_0^a\int_0^b \left(\frac{\partial w_0}{\partial x}\right)^2 dy dx. \tag{11.34}$$

One can proceed quite analogously for the edge loads $\hat{N}^0_{yy}$ and $\hat{N}^0_{xy}$. The external potential in the presence of all three boundary loads $\hat{N}^0_{xx}$, $\hat{N}^0_{yy}$, $\hat{N}^0_{xy}$ is:

$$\Pi_a = -\frac{1}{2}\int_0^a\int_0^b \left[\hat{N}_{xx}^0\left(\frac{\partial w_0}{\partial x}\right)^2 + \hat{N}_{yy}^0\left(\frac{\partial w_0}{\partial y}\right)^2 + 2\hat{N}_{xy}^0\frac{\partial w_0}{\partial x}\frac{\partial w_0}{\partial y}\right]\mathrm{d}y\mathrm{d}x. \tag{11.35}$$

The total elastic potential of the plate $\Pi = \Pi_i + \Pi_a$ in the buckled state follows after inserting the constitutive law

$$\begin{pmatrix} M_{xx}^0 \\ M_{yy}^0 \\ M_{xy}^0 \end{pmatrix} = \begin{bmatrix} D_{11} & D_{12} & 0 \\ D_{12} & D_{22} & 0 \\ 0 & 0 & D_{66} \end{bmatrix} \begin{pmatrix} -\frac{\partial^2 w_0}{\partial x^2} \\ -\frac{\partial^2 w_0}{\partial y^2} \\ -2\frac{\partial^2 w_0}{\partial x \partial y} \end{pmatrix} \tag{11.36}$$

as:

$$\begin{aligned}\Pi = &\int_0^a\int_0^b \left[\frac{1}{2}D_{11}\left(\frac{\partial^2 w_0}{\partial x^2}\right)^2 + \frac{1}{2}D_{22}\left(\frac{\partial^2 w_0}{\partial y^2}\right)^2 + D_{12}\frac{\partial^2 w_0}{\partial x^2}\frac{\partial^2 w_0}{\partial y^2} + 2D_{66}\left(\frac{\partial^2 w_0}{\partial x \partial y}\right)^2\right]\mathrm{d}y\mathrm{d}x \\ &- \frac{1}{2}\int_0^a\int_0^b \left[N_{xx}^0\left(\frac{\partial w_0}{\partial x}\right)^2 + N_{yy}^0\left(\frac{\partial w_0}{\partial y}\right)^2 + 2N_{xy}^0\frac{\partial w_0}{\partial x}\frac{\partial w_0}{\partial y}\right]\mathrm{d}y\mathrm{d}x\end{aligned} \tag{11.37}$$

Here, the overbars identifying the edge loads have been omitted. At least for the buckling problems discussed in this chapter, there is no confusion between edge loads and internal forces - it is assumed that the stress state throughout the laminate is identical to the respective edge loads. In all the following explanations, N_{xx}^0, N_{yy}^0, N_{xy}^0 always refer to the given edge loads and the associated buckling loads, respectively.

11.3.2 The Rayleigh Quotient

The Rayleigh quotient uses the potential formulation (11.37) and assumes a separation approach for the buckling shape as follows:

$$w_0 = W w_1(x)\, w_2(y). \tag{11.38}$$

The approach functions w_1 and w_2 are required to fulfill all static and geometric boundary conditions identically. The coefficient W is an indeterminate approach coefficient. In the following, we only consider plates under a constant longitudinal compressive load N_{xx}^0. Using the approach (11.38), the total elastic potential (11.37) of the buckled plate is evaluated, and we obtain:

$$\Pi = \left[\frac{1}{2}D_{11}\int_0^a\left(\frac{\mathrm{d}^2 w_1}{\mathrm{d}x^2}\right)^2\mathrm{d}x\int_0^b w_2^2\mathrm{d}y + \frac{1}{2}D_{22}\int_0^a w_1^2\mathrm{d}x\int_0^b\left(\frac{\mathrm{d}^2 w_2}{\mathrm{d}y^2}\right)^2\mathrm{d}y\right.$$

$$+ D_{12} \int_0^a w_1 \frac{d^2 w_1}{dx^2} dx \int_0^b w_2 \frac{d^2 w_2}{dy^2} dy + 2D_{66} \int_0^a \left(\frac{dw_1}{dx} \right)^2 dx \int_0^b \left(\frac{dw_2}{dy} \right)^2 dy$$

$$\left. - \frac{1}{2} N_{xx}^0 \int_0^a \left(\frac{dw_1}{dx} \right)^2 dx \int_0^b w_2^2 dy \right] W^2. \tag{11.39}$$

For convenience, the following abbreviations are introduced for the integrals of the shape functions:

$$\lambda_{x,ij} = \int_0^a \frac{d^i w_1}{dx^i} \frac{d^j w_1}{dx^j} dx, \quad \lambda_{y,ij} = \int_0^b \frac{d^i w_2}{dy^i} \frac{d^j w_2}{dy^j} dy. \tag{11.40}$$

Thus:

$$\Pi = \left[\frac{1}{2} D_{11} \lambda_{x,22} \lambda_{y,00} + \frac{1}{2} D_{22} \lambda_{x,00} \lambda_{y,22} + D_{12} \lambda_{x,02} \lambda_{y,02} \right.$$
$$\left. + 2D_{66} \lambda_{x,11} \lambda_{y,11} - \frac{1}{2} N_{xx}^0 \lambda_{x,11} \lambda_{y,00} \right] W^2. \tag{11.41}$$

In the implementation of the Rayleigh quotient, one will usually strive to use simple shape functions w_1 and w_2, which should represent the buckling mode as closely as possible. It can therefore be assumed that the integral expressions occurring in (11.41) can be solved in a closed-form analytical manner.

The buckling condition $\delta \Pi = 0$ reduces to the Ritz equation using the approach (11.38):

$$\frac{\partial \Pi}{\partial W} = 0. \tag{11.42}$$

It follows from (11.41):

$$\left[D_{11} \lambda_{x,22} \lambda_{y,00} + D_{22} \lambda_{x,00} \lambda_{y,22} + 2D_{12} \lambda_{x,02} \lambda_{y,02} + 4D_{66} \lambda_{x,11} \lambda_{y,11} - N_{xx}^0 \lambda_{x,11} \lambda_{y,00} \right] W = 0. \tag{11.43}$$

The relevant solution results from setting the term in parentheses in (11.43) to zero:

$$D_{11} \lambda_{x,22} \lambda_{y,00} + D_{22} \lambda_{x,00} \lambda_{y,22} + 2D_{12} \lambda_{x,02} \lambda_{y,02}$$
$$+ 4D_{66} \lambda_{x,11} \lambda_{y,11} - N_{xx}^0 \lambda_{x,11} \lambda_{y,00} = 0. \tag{11.44}$$

This expression can be solved for the buckling load N_{xx}^0, and the following closed-form analytical solution is obtained:

$$N_{xx}^0 = \frac{D_{11} \lambda_{x,22} \lambda_{y,00} + D_{22} \lambda_{x,00} \lambda_{y,22} + 2D_{12} \lambda_{x,02} \lambda_{y,02} + 4D_{66} \lambda_{x,11} \lambda_{y,11}}{\lambda_{x,11} \lambda_{y,00}}. \tag{11.45}$$

This is the so-called Rayleigh quotient for the approximate determination of the buckling load of an orthotropic plate under uniaxial compression. It is important to note here that the buckling load is approximated 'from above' with this expression, so the buckling load is always overestimated. The error committed here is therefore always on the non-conservative side.

The Rayleigh quotient is illustrated below using two simple examples.

Simply supported plate

As a first example, consider the simply supported orthotropic plate (see Fig. 11.2) under uniaxial loading N_{xx}^0. The boundary conditions are:

$$\begin{aligned} &w_0\,(x=0)=0,\quad w_0\,(x=a)=0,\quad w_0\,(y=0)=0,\quad w_0\,(y=b)=0,\\ &M_{xx}^0\,(x=0)=-D_{11}\left.\frac{\partial^2 w_0}{\partial x^2}\right|_{x=0}=0,\quad M_{xx}^0\,(x=a)=-D_{11}\left.\frac{\partial^2 w_0}{\partial x^2}\right|_{x=a}=0,\\ &M_{yy}^0\,(y=0)=-D_{22}\left.\frac{\partial^2 w_0}{\partial y^2}\right|_{y=0}=0,\quad M_{yy}^0\,(y=b)=-D_{22}\left.\frac{\partial^2 w_0}{\partial y^2}\right|_{y=b}=0. \end{aligned} \tag{11.46}$$

An approach to the buckling mode w_0 can be found in the following form:

$$w_0=W\sin\left(\frac{m\pi x}{a}\right)\sin\left(\frac{n\pi y}{b}\right). \tag{11.47}$$

This approach not only satisfies the geometric boundary conditions in (11.46), but satisfies all boundary conditions identically. The integrals (11.40) are obtained with the approach (11.47) as follows:

$$\lambda_{x,00}=\int_0^a w_1^2\mathrm{d}x=\int_0^a\sin^2\left(\frac{m\pi x}{a}\right)\mathrm{d}x=\frac{a}{2},$$

$$\lambda_{x,11}=\int_0^a\left(\frac{\mathrm{d}w_1}{\mathrm{d}x}\right)^2\mathrm{d}x=\frac{m^2\pi^2}{a^2}\int_0^a\cos^2\left(\frac{m\pi x}{a}\right)\mathrm{d}x=\frac{m^2\pi^2}{2a},$$

$$\lambda_{x,02}=\int_0^a w_1\frac{\mathrm{d}^2w_1}{\mathrm{d}x^2}\mathrm{d}x=-\frac{m^2\pi^2}{a^2}\int_0^a\sin^2\left(\frac{m\pi x}{a}\right)\mathrm{d}x=-\frac{m^2\pi^2}{2a},$$

$$\lambda_{x,22}=\int_0^a\frac{\mathrm{d}^2w_1}{\mathrm{d}x^2}\frac{\mathrm{d}^2w_1}{\mathrm{d}x^2}\mathrm{d}x=\frac{m^4\pi^4}{a^4}\int_0^a\sin^2\left(\frac{m\pi x}{a}\right)\mathrm{d}x=\frac{m^4\pi^4}{2a^3},$$

$$\lambda_{y,00}=\int_0^b w_2^2\mathrm{d}y=\int_0^b\sin^2\left(\frac{n\pi y}{b}\right)\mathrm{d}y=\frac{b}{2},$$

$$\lambda_{y,11}=\int_0^b\frac{\mathrm{d}w_2}{\mathrm{d}y}\frac{\mathrm{d}w_2}{\mathrm{d}y}\mathrm{d}y=\frac{n^2\pi^2}{b^2}\int_0^b\cos^2\left(\frac{n\pi y}{b}\right)\mathrm{d}y=\frac{n^2\pi^2}{2b},$$

$$\lambda_{y,02} = \int_0^b w_2 \frac{\mathrm{d}^2 w_2}{\mathrm{d}y^2}\,\mathrm{d}y = -\frac{n^2\pi^2}{b^2}\int_0^b \sin^2\left(\frac{n\pi y}{b}\right)\mathrm{d}y = -\frac{n^2\pi^2}{2b},$$

$$\lambda_{y,22} = \int_0^b \frac{\mathrm{d}^2 w_2}{\mathrm{d}y^2}\frac{\mathrm{d}^2 w_2}{\mathrm{d}y^2}\,\mathrm{d}y = \frac{n^4\pi^4}{b^4}\int_0^b \sin^2\left(\frac{n\pi y}{b}\right)\mathrm{d}y = \frac{n^4\pi^4}{2b^3}. \quad (11.48)$$

The buckling load for the considered simply supported plate can thus be given by (11.45) as:

$$N_{xx}^0 = \pi^2\left[D_{11}\frac{m^2}{a^2} + D_{22}\frac{a^2n^4}{b^4m^2} + \frac{2n^2}{b^2}\left(D_{12}+2D_{66}\right)\right]. \quad (11.49)$$

Technically relevant in this case is only the value $n = 1$, so that:

$$N_{xx}^0 = \pi^2\left[D_{11}\frac{m^2}{a^2} + D_{22}\frac{a^2}{b^4m^2} + \frac{2}{b^2}\left(D_{12}+2D_{66}\right)\right]. \quad (11.50)$$

Apparently, this expression is identical to the exact solution (11.21) for the buckling load of the Navier plate. This can be justified by the fact that the approach (11.47) corresponds exactly to the exact solution for the buckling shape of the rectangular plate under uniaxial compressive load N_{xx}^0. It should be noted that also when the Rayleigh quotient is used, the approach constant W remains undetermined.

Plate with clamped unloaded edges

We now consider a plate that is still simply supported at the loaded edges but is clamped at the unloaded edges (Fig. 11.9). The boundary conditions are as follows for this plate situation:

$$w_0\left(x=0\right)=0,\quad w_0\left(x=a\right)=0,\quad w_0\left(y=0\right)=0,\quad w_0\left(y=b\right)=0,$$

$$M_{xx}^0\left(x=0\right) = -D_{11}\left.\frac{\partial^2 w_0}{\partial x^2}\right|_{x=0} = 0,\quad M_{xx}^0\left(x=a\right) = -D_{11}\left.\frac{\partial^2 w_0}{\partial x^2}\right|_{x=a} = 0,$$

$$\left.\frac{\partial w_0}{\partial y}\right|_{y=0} = 0,\quad \left.\frac{\partial w_0}{\partial y}\right|_{y=b} = 0. \quad (11.51)$$

A suitable approach for the buckling mode w_0, which satisfies both the static and the dynamic boundary conditions, is as follows:

$$w_0 = W\sin\left(\frac{m\pi x}{a}\right)\left\{\frac{1}{2}\left[1-\cos\left(\frac{2\pi y}{b}\right)\right]\right\}. \quad (11.52)$$

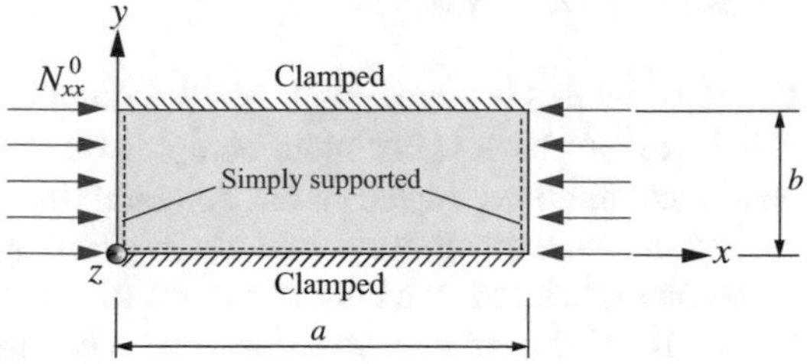

Fig. 11.9 Plate under uniaxial compressive load with simply supported loaded edges and clamped unloaded edges

The integrals (11.40) can be taken from (11.48) with respect to the shape function w_1. The remaining expressions are as follows:

$$\lambda_{y,00} = \int_0^b w_2^2 \mathrm{d}y = \frac{1}{4}\int_0^b \left[1-\cos\left(\frac{2\pi y}{b}\right)\right]^2 \mathrm{d}y = \frac{3}{8}b,$$

$$\lambda_{y,11} = \int_0^b \left(\frac{\mathrm{d}w_2}{\mathrm{d}y}\right)^2 \mathrm{d}y = \frac{\pi^2}{b^2}\int_0^b \sin^2\left(\frac{2\pi y}{b}\right)\mathrm{d}y = \frac{\pi^2}{2b},$$

$$\lambda_{y,02} = \int_0^b w_2 \frac{\mathrm{d}^2 w_2}{\mathrm{d}y^2}\mathrm{d}y = \frac{\pi^2}{b^2}\int_0^b \left[1-\cos\left(\frac{2\pi y}{b}\right)\right]\cos\left(\frac{2\pi y}{b}\right)\mathrm{d}y = -\frac{\pi^2}{2b},$$

$$\lambda_{y,22} = \int_0^b \frac{\mathrm{d}^2 w_2}{\mathrm{d}y^2}\frac{\mathrm{d}^2 w_2}{\mathrm{d}y^2}\mathrm{d}y = \frac{4\pi^4}{b^4}\int_0^b \cos^2\left(\frac{2\pi y}{b}\right)\mathrm{d}y = \frac{2\pi^4}{b^3}. \qquad (11.53)$$

Thus, the buckling load for the given plate is:

$$N_{xx}^0 = \pi^2\left[D_{11}\frac{m^2}{a^2} + \frac{16}{3}D_{22}\frac{a^2}{b^4 m^2} + \frac{8}{3b^2}\left(D_{12}+2D_{66}\right)\right]. \qquad (11.54)$$

The relevant wavenumber m follows again from the extreme value statement $\frac{\partial N_{xx}^0}{\partial m} = 0$ and results as:

$$m = \frac{2}{\sqrt{3}}\frac{a}{b}\sqrt[4]{3\frac{D_{22}}{D_{11}}}. \qquad (11.55)$$

The asymptotic limit that results for sufficiently long plates is currently:

$$N_{xx}^0 = \frac{8\pi^2}{3b^2}\left(\sqrt{3D_{11}D_{22}} + D_{12} + 2D_{66}\right). \qquad (11.56)$$

11.3.3 The Ritz Method

The Ritz method is a universally applicable method for the determination of buckling loads of plates under more complex boundary and load conditions than the plate configurations treated so far. Although the Ritz method is comparatively old, it continues to enjoy great popularity in practical applications because of its computational efficiency. We consider the Ritz method for the configuration shown in Fig. 11.10. Consider a rectangular plate with dimensions a and b, which is subjected to a linearly varying normal force flow $N^0_{xx}(y) = \frac{N^0_{xx}}{b}\left[(\psi - 1)\, y + b\right]$ with boundary values $N^0_{xxl} = N^0_{xx}(y = 0) = N^0_{xx}$ and $N^0_{xxr} = N^0_{xx}(y = b) = \psi N^0_{xx}$ (where ψ is any real number and indicates the multiple of the boundary value $N^0_{xxl} = N^0_{xx}(y = 0) = N^0_{xx}$). Moreover, the inplane force flows N^0_{yy} and N^0_{xy}, assumed to be constant, are present. Let the plate be simply supported at all edges, but note that the following formulation of the Ritz method is valid for arbitrary boundary conditions and arbitrary shape functions. The specialization to simply supported boundary conditions will be shown later. In addition to the simple supports, let the plate under consideration be elastically restrained at the two edges $y = 0$ and $y = b$, where the two restraint stiffnesses k_l and k_r are given.

The total elastic potential Π of the considered elastically restrained plate in the buckled state is composed of the inner potential / strain energy Π_i, the energy Π_s stored in the springs / elastic restraints, and the potential of the applied loads Π_a:

$$\Pi = \Pi_i + \Pi_a + \Pi_s, \tag{11.57}$$

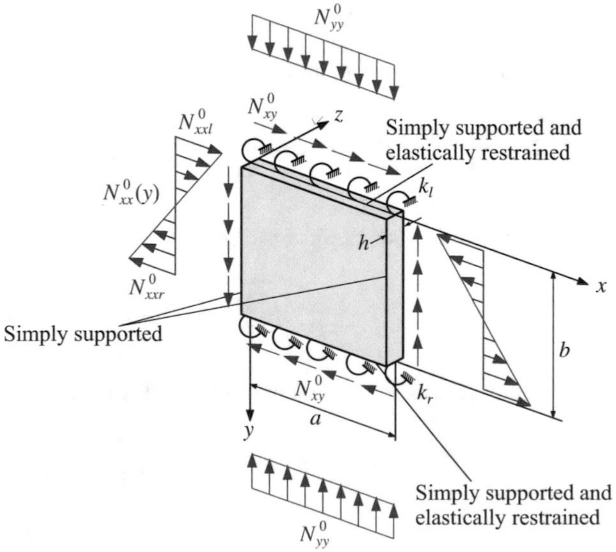

Fig. 11.10 Elastically restrained plate under combined load

with:

$$\Pi_i = \frac{1}{2}\int_0^a\int_0^b \left[D_{11}\left(\frac{\partial^2 w_0}{\partial x^2}\right)^2 + D_{22}\left(\frac{\partial^2 w_0}{\partial y^2}\right)^2\right] \mathrm{d}y\mathrm{d}x$$
$$+\int_0^a\int_0^b \left[D_{12}\frac{\partial^2 w_0}{\partial x^2}\frac{\partial^2 w_0}{\partial y^2} + 2D_{66}\left(\frac{\partial^2 w_0}{\partial x \partial y}\right)^2\right] \mathrm{d}y\mathrm{d}x,$$
$$\Pi_a = -\frac{1}{2}\int_0^a\int_0^b \left[\Lambda N_{xx}^0\left(\frac{\partial w_0}{\partial x}\right)^2 + \Lambda N_{yy}^0\left(\frac{\partial w_0}{\partial y}\right)^2 + 2\Lambda N_{xy}^0\frac{\partial w_0}{\partial x}\frac{\partial w_0}{\partial y}\right] \mathrm{d}y\mathrm{d}x,$$
$$\Pi_s = \frac{1}{2}\int_0^a \left[k_l\left(\left.\frac{\partial w_0}{\partial y}\right|_{y=0}\right)^2 + k_r\left(\left.\frac{\partial w_0}{\partial y}\right|_{y=b}\right)^2\right] \mathrm{d}x. \tag{11.58}$$

The load combination of N_{xx}^0, N_{yy}^0 and N_{xy}^0 occurring here is multiplied by the load multiplier Λ. This load multiplier is ultimately the target of the calculation and indicates by how much the given load combination can be increased before buckling occurs.

The following approach is used for the buckling shape w_0:

$$w_0 = \sum_{m=1}^{m=M}\sum_{n=1}^{n=N} W_{mn} w_{1m}(x)\, w_{2n}(y). \tag{11.59}$$

Herein, the quantities W_{mn} are indeterminate approach coefficients. The shape functions w_{1m} and w_{2n} (of which there are M and N, respectively) must satisfy the given geometric boundary conditions. The special case $M = 1$, $N = 1$ then leads to the already discussed Rayleigh quotient as a special case of the Ritz method.

Substituting the approach (11.59) into the potential formulation (11.58), we obtain:

$$\Pi_i = \frac{D_{11}}{2}\sum_{m=1}^{m=M}\sum_{n=1}^{n=N}\sum_{p=1}^{p=M}\sum_{q=1}^{q=N} W_{mn}W_{pq}\int_0^a \frac{\mathrm{d}^2 w_{1m}}{\mathrm{d}x^2}\frac{\mathrm{d}^2 w_{1p}}{\mathrm{d}x^2}\mathrm{d}x \int_0^b w_{2n}w_{2q}\mathrm{d}y$$
$$+\frac{D_{22}}{2}\sum_{m=1}^{m=M}\sum_{n=1}^{n=N}\sum_{p=1}^{p=M}\sum_{q=1}^{q=N} W_{mn}W_{pq}\int_0^a w_{1m}w_{1p}\mathrm{d}x \int_0^b \frac{\mathrm{d}^2 w_{2n}}{\mathrm{d}y^2}\frac{\mathrm{d}^2 w_{2q}}{\mathrm{d}y^2}\mathrm{d}y$$
$$+D_{12}\sum_{m=1}^{m=M}\sum_{n=1}^{n=N}\sum_{p=1}^{p=M}\sum_{q=1}^{q=N} W_{mn}W_{pq}\int_0^a \frac{\mathrm{d}^2 w_{1m}}{\mathrm{d}x^2} w_{1p}\mathrm{d}x \int_0^b w_{2n}\frac{\mathrm{d}^2 w_{2q}}{\mathrm{d}y^2}\mathrm{d}y,$$
$$+2D_{66}\sum_{m=1}^{m=M}\sum_{n=1}^{n=N}\sum_{p=1}^{p=M}\sum_{q=1}^{q=N} W_{mn}W_{pq}\int_0^a \frac{\mathrm{d}w_{1m}}{\mathrm{d}x}\frac{\mathrm{d}w_{1p}}{\mathrm{d}x}\mathrm{d}x \int_0^b \frac{\mathrm{d}w_{2n}}{\mathrm{d}y}\frac{\mathrm{d}w_{2q}}{\mathrm{d}y}\mathrm{d}y,$$

$$
\begin{aligned}
\Pi_a = &-\Lambda \frac{N_{xx}^0}{2b}(\psi-1)\sum_{m=1}^{m=M}\sum_{n=1}^{n=N}\sum_{p=1}^{p=M}\sum_{q=1}^{q=N} W_{mn}W_{pq}\int_0^a \frac{\mathrm{d}w_{1m}}{\mathrm{d}x}\frac{\mathrm{d}w_{1p}}{\mathrm{d}x}\mathrm{d}x\int_0^b w_{2n}w_{2q}\,y\,\mathrm{d}y\\
&-\Lambda \frac{N_{xx}^0}{2}\sum_{m=1}^{m=M}\sum_{n=1}^{n=N}\sum_{p=1}^{p=M}\sum_{q=1}^{q=N} W_{mn}W_{pq}\int_0^a \frac{\mathrm{d}w_{1m}}{\mathrm{d}x}\frac{\mathrm{d}w_{1p}}{\mathrm{d}x}\mathrm{d}x\int_0^b w_{2n}w_{2q}\mathrm{d}y\\
&-\frac{\Lambda N_{yy}^0}{2}\sum_{m=1}^{m=M}\sum_{n=1}^{n=N}\sum_{p=1}^{p=M}\sum_{q=1}^{q=N} W_{mn}W_{pq}\int_0^a w_{1m}w_{1p}\mathrm{d}x\int_0^b \frac{\mathrm{d}w_{2n}}{\mathrm{d}y}\frac{\mathrm{d}w_{2q}}{\mathrm{d}y}\mathrm{d}y\\
&-\Lambda N_{xy}^0\sum_{m=1}^{m=M}\sum_{n=1}^{n=N}\sum_{p=1}^{p=M}\sum_{q=1}^{q=N} W_{mn}W_{pq}\int_0^a \frac{\mathrm{d}w_{1m}}{\mathrm{d}x}w_{1p}\mathrm{d}x\int_0^a w_{2n}\frac{\mathrm{d}w_{2q}}{\mathrm{d}y}\mathrm{d}y,\\
\Pi_s = &\frac{k_l}{2}\sum_{m=1}^{m=M}\sum_{n=1}^{n=N}\sum_{p=1}^{p=M}\sum_{q=1}^{q=N} W_{mn}W_{pq}\int_0^a w_{1m}w_{1p}\mathrm{d}x\left.\left(\frac{\mathrm{d}w_{2n}}{\mathrm{d}y}\frac{\mathrm{d}w_{2q}}{\mathrm{d}y}\right)\right|_{y=0}\\
&+\frac{k_r}{2}\sum_{m=1}^{m=M}\sum_{n=1}^{n=N}\sum_{p=1}^{p=M}\sum_{q=1}^{q=N} W_{mn}W_{pq}\int_0^a w_{1m}w_{1p}\mathrm{d}x\left.\left(\frac{\mathrm{d}w_{2n}}{\mathrm{d}y}\frac{\mathrm{d}w_{2q}}{\mathrm{d}y}\right)\right|_{y=b}.
\end{aligned}
\tag{11.60}
$$

.

This can be written in a summarized form as

$$
\begin{aligned}
\Pi_i &= \sum_{m=1}^{m=M}\sum_{n=1}^{n=N}\sum_{p=1}^{p=M}\sum_{q=1}^{q=N} W_{mn}W_{pq}\Gamma_i^{mnpq},\\
\Pi_a &= -\Lambda\sum_{m=1}^{m=M}\sum_{n=1}^{n=N}\sum_{p=1}^{p=M}\sum_{q=1}^{q=N} W_{mn}W_{pq}\Gamma_a^{mnpq},\\
\Pi_s &= \sum_{m=1}^{m=M}\sum_{n=1}^{n=N}\sum_{p=1}^{p=M}\sum_{q=1}^{q=N} W_{mn}W_{pq}\Gamma_s^{mnpq},
\end{aligned}
\tag{11.61}
$$

with the following abbreviations:

$$
\begin{aligned}
\Gamma_i^{mnpq} &= \frac{1}{2}\left(D_{11}R_{22}^{mp}S_{00}^{nq} + D_{22}R_{00}^{mp}S_{22}^{nq}\right) + D_{12}R_{20}^{mp}S_{02}^{nq} + 2D_{66}R_{11}^{mp}S_{11}^{nq},\\
\Gamma_a^{mnpq} &= \frac{1}{2}N_{xx}^0R_{11}^{mp}\left[S_{00}^{nq} + (\psi-1)\,T_{00}^{nq}\right] + \frac{1}{2}N_{yy}^0R_{00}^{mp}S_{11}^{nq} + N_{xy}^0R_{10}^{mp}S_{01}^{nq},\\
\Gamma_s^{mnpq} &= \frac{1}{2}R_{00}^{mp}\left(k_lU_{11}^{nq} + k_rV_{11}^{nq}\right).
\end{aligned}
\tag{11.62}
$$

The following abbreviations have been introduced for convenience:

$$
R_{ij}^{mp} = \int_0^a \frac{\mathrm{d}^i w_{1m}}{\mathrm{d}x^i}\frac{\mathrm{d}^j w_{1p}}{\mathrm{d}x^j}\mathrm{d}x,\quad S_{ij}^{nq} = \int_0^b \frac{\mathrm{d}^i w_{2n}}{\mathrm{d}y^i}\frac{\mathrm{d}^j w_{2q}}{\mathrm{d}y^j}\mathrm{d}y,
$$

$$T_{00}^{nq} = \frac{1}{b} \int_0^b w_{2n} w_{2q} y \mathrm{d}y,$$

$$U_{11}^{nq} = \left.\frac{\mathrm{d}w_{2n}}{\mathrm{d}y}\right|_{y=0} \left.\frac{\mathrm{d}w_{2q}}{\mathrm{d}y}\right|_{y=0}, \quad V_{11}^{nq} = \left.\frac{\mathrm{d}w_{2n}}{\mathrm{d}y}\right|_{y=b} \left.\frac{\mathrm{d}w_{2q}}{\mathrm{d}y}\right|_{y=b}. \tag{11.63}$$

The buckling condition $\delta\Pi = 0$ passes into the Ritz equations

$$\frac{\partial \Pi}{\partial W_{mn}} = 0. \tag{11.64}$$

Performing the differentiations required here according to the constants W_{mn} then leads to the following linear eigenvalue problem:

$$\left[\underline{\underline{K}}_i + \underline{\underline{K}}_s - \Lambda \underline{\underline{K}}_a\right] \underline{W} = \underline{0}. \tag{11.65}$$

Therein, the load multiplier Λ is the eigenvalue we are looking for, and the eigenvector $\underline{W}$ contains the associated constants W_{mn}. The matrices $\underline{\underline{K}}_i$ and $\underline{\underline{K}}_s$ are the stiffness matrix of the plate under consideration and the stiffness matrix of the elastic edge restraints, respectively. The matrix $\underline{\underline{K}}_a$ is called the geometric stiffness matrix. These matrices can be represented as follows:

$$\underline{\underline{K}}_i = \begin{bmatrix} \underline{\underline{K}}_i^{11} & \underline{\underline{K}}_i^{12} & \underline{\underline{K}}_i^{13} & \dots & \underline{\underline{K}}_i^{1M} \\ \underline{\underline{K}}_i^{21} & \underline{\underline{K}}_i^{22} & \underline{\underline{K}}_i^{23} & \dots & \underline{\underline{K}}_i^{2M} \\ \underline{\underline{K}}_i^{31} & \underline{\underline{K}}_i^{32} & \underline{\underline{K}}_i^{33} & \dots & \underline{\underline{K}}_i^{3M} \\ \vdots & \vdots & \vdots & \ddots & \vdots \\ \underline{\underline{K}}_i^{M1} & \underline{\underline{K}}_i^{M2} & \underline{\underline{K}}_i^{M3} & \dots & \underline{\underline{K}}_i^{MM} \end{bmatrix} = \left[\underline{\underline{K}}_i^{\gamma\delta}\right],$$

$$\underline{\underline{K}}_s = \begin{bmatrix} \underline{\underline{K}}_s^{11} & \underline{\underline{K}}_s^{12} & \underline{\underline{K}}_s^{13} & \dots & \underline{\underline{K}}_s^{1M} \\ \underline{\underline{K}}_s^{21} & \underline{\underline{K}}_s^{22} & \underline{\underline{K}}_s^{23} & \dots & \underline{\underline{K}}_s^{2M} \\ \underline{\underline{K}}_s^{31} & \underline{\underline{K}}_s^{32} & \underline{\underline{K}}_s^{33} & \dots & \underline{\underline{K}}_s^{3M} \\ \vdots & \vdots & \vdots & \ddots & \vdots \\ \underline{\underline{K}}_s^{M1} & \underline{\underline{K}}_s^{M2} & \underline{\underline{K}}_s^{M3} & \dots & \underline{\underline{K}}_s^{MM} \end{bmatrix} = \left[\underline{\underline{K}}_s^{\gamma\delta}\right],$$

$$\underline{\underline{K}}_a = \begin{bmatrix} \underline{\underline{K}}_a^{11} & \underline{\underline{K}}_a^{12} & \underline{\underline{K}}_a^{13} & \dots & \underline{\underline{K}}_a^{1M} \\ \underline{\underline{K}}_a^{21} & \underline{\underline{K}}_a^{22} & \underline{\underline{K}}_a^{23} & \dots & \underline{\underline{K}}_a^{2M} \\ \underline{\underline{K}}_a^{31} & \underline{\underline{K}}_a^{32} & \underline{\underline{K}}_a^{33} & \dots & \underline{\underline{K}}_a^{3M} \\ \vdots & \vdots & \vdots & \ddots & \vdots \\ \underline{\underline{K}}_a^{M1} & \underline{\underline{K}}_a^{M2} & \underline{\underline{K}}_a^{M3} & \dots & \underline{\underline{K}}_a^{MM} \end{bmatrix} = \left[\underline{\underline{K}}_a^{\gamma\delta}\right]. \tag{11.66}$$

The components $K^{\gamma\delta}_{\varrho\omega,i}$, $K^{\gamma\delta}_{\varrho\omega,s}$, $K^{\gamma\delta}_{\varrho\omega,a}$ of the submatrices $\underline{\underline{K}}^{\gamma\delta}_i = \left[K^{\gamma\delta}_{\varrho\omega,i}\right]$, $\underline{\underline{K}}^{\gamma\delta}_s = \left[K^{\gamma\delta}_{\varrho\omega,s}\right]$, $\underline{\underline{K}}^{\gamma\delta}_a = \left[K^{\gamma\delta}_{\varrho\omega,a}\right]$ with N rows and N columns can be represented as:

$$K^{\gamma\delta}_{\varrho\omega,i} = \Gamma_i^{\varrho\gamma\omega\delta} + \Gamma_i^{\omega\delta\varrho\gamma}, \quad K^{\gamma\delta}_{\varrho\omega,s} = \Gamma_s^{\varrho\gamma\omega\delta} + \Gamma_s^{\omega\delta\varrho\gamma}, \quad K^{\gamma\delta}_{\varrho\omega,a} = \Gamma_a^{\varrho\gamma\omega\delta} + \Gamma_a^{\omega\delta\varrho\gamma}. \tag{11.67}$$

The eigenvalue problem for the determination of the load multiplier Λ can in principle be solved quickly and efficiently with any available numerical software.

The representation of the Ritz method has validity for arbitrary shape functions $w_{1m}(x)$ and $w_{2n}(y)$. If all the edges of the laminate are simply supported, as discussed earlier, then an approach of the following type can be used:

$$w_0 = \sum_{m=1}^{m=M}\sum_{n=1}^{n=N} W_{mn} w_{1m}(x)\, w_{2n}(y) = \sum_{m=1}^{m=M}\sum_{n=1}^{n=N} W_{mn} \sin\left(\frac{m\pi x}{a}\right)\sin\left(\frac{n\pi y}{b}\right). \tag{11.68}$$

This satisfies all geometrical boundary conditions identically. The resultants of the shape functions (11.63) can then be given as follows:

$$R_{00}^{mp} = \int_0^a w_{1m} w_{1p}\mathrm{d}x = \int_0^a \sin\left(\frac{m\pi x}{a}\right)\sin\left(\frac{p\pi x}{a}\right)\mathrm{d}x$$
$$= \begin{cases} 0 & \text{if } m \neq p \\ \frac{a}{2} & \text{if } m = p \end{cases},$$

$$R_{10}^{mp} = \int_0^a \frac{\mathrm{d}w_{1m}}{\mathrm{d}x} w_{1p}\mathrm{d}x = \frac{m\pi}{a}\int_0^a \cos\left(\frac{m\pi x}{a}\right)\sin\left(\frac{p\pi x}{a}\right)\mathrm{d}x$$
$$= \begin{cases} 0 & \text{if } m + p = \text{even} \\ \dfrac{2mp}{p^2 - m^2} & \text{if } m + p = \text{odd} \end{cases},$$

$$R_{11}^{mp} = \int_0^a \frac{\mathrm{d}w_{1m}}{\mathrm{d}x}\frac{\mathrm{d}w_{1p}}{\mathrm{d}x}\mathrm{d}x = \frac{m\pi}{a}\frac{p\pi}{a}\int_0^a \cos\left(\frac{m\pi x}{a}\right)\cos\left(\frac{p\pi x}{a}\right)\mathrm{d}x$$
$$= \begin{cases} 0 & \text{if } m \neq p \\ \dfrac{mp\pi^2}{2a} & \text{if } m = p \end{cases},$$

$$R_{20}^{mp} = \int_0^a \frac{\mathrm{d}^2 w_{1m}}{\mathrm{d}x^2} w_{1p}\mathrm{d}x = -\left(\frac{m\pi}{a}\right)^2\int_0^a \sin\left(\frac{m\pi x}{a}\right)\sin\left(\frac{p\pi x}{a}\right)\mathrm{d}x$$
$$= \begin{cases} 0 & \text{if } m \neq p \\ -\dfrac{m^2\pi^2}{2a} & \text{if } m = p \end{cases},$$

$$
\begin{aligned}
R_{22}^{mp} &= \int\limits_0^a \frac{\mathrm{d}^2 w_{1m}}{\mathrm{d}x^2}\frac{\mathrm{d}^2 w_{1p}}{\mathrm{d}x^2}\mathrm{d}x = \left(\frac{m\pi}{a}\right)^2\left(\frac{p\pi}{a}\right)^2 \int\limits_0^a \sin\left(\frac{m\pi x}{a}\right)\sin\left(\frac{p\pi x}{a}\right)\mathrm{d}x \\
&= \begin{cases} 0 & \text{if } m \neq p \\ \dfrac{m^2 p^2 \pi^4}{2a^3} & \text{if } m = p \end{cases}, \\
T_{00}^{nq} &= \frac{1}{b}\int\limits_0^b w_{2n} w_{2q}\, y \mathrm{d}y = \frac{1}{b}\int\limits_0^b \sin\left(\frac{n\pi y}{b}\right)\sin\left(\frac{q\pi y}{b}\right) y \mathrm{d}y \\
&= \begin{cases} \dfrac{b}{4} & \text{if } n = q \\ \dfrac{2bnq}{\pi^2\left(n^2 - q^2\right)^2}\left[(-1)^{n+q} - 1\right] & \text{if } n \neq q \end{cases}, \\
U_{11}^{nq} &= \left.\frac{\mathrm{d}w_{2n}}{\mathrm{d}y}\right|_{y=0} \left.\frac{\mathrm{d}w_{2q}}{\mathrm{d}y}\right|_{y=0} = nq\left(\frac{\pi}{b}\right)^2, \\
V_{11}^{nq} &= \left.\frac{\mathrm{d}w_{2n}}{\mathrm{d}y}\right|_{y=b} \left.\frac{\mathrm{d}w_{2q}}{\mathrm{d}y}\right|_{y=b} = nq\left(\frac{\pi}{b}\right)^2 \cos(n\pi)\cos(q\pi) \\
&= \begin{cases} nq\left(\frac{\pi}{b}\right)^2 & \text{if } n + q = \text{even} \\ -nq\left(\frac{\pi}{b}\right)^2 & \text{if } n + q = odd \end{cases},
\end{aligned}
\tag{11.69}
$$

The resultants

$$
S_{ij}^{nq} = \int\limits_0^b \frac{\mathrm{d}^i w_{2n}}{\mathrm{d}y^i}\frac{\mathrm{d}^j w_{2q}}{\mathrm{d}y^j}\mathrm{d}y \tag{11.70}
$$

can be determined from the already given resultants R_{ij}^{mp} by replacing the length a by the width b in the corresponding expressions and exchanging the indices m and p with the indices n and q.

References

Girkmann, K.: Flächentragwerke, 6th edn. Springer, Wien (1974)

Mittelstedt, C., Becker, W.: Strukturmechanik ebener Laminate. Studienbereich Mechanik Technische Universität Darmstadt, Germany (2016)

Petersen, C.: Statik und Stabilität der Baukonstruktionen, 2nd edn. Vieweg Braunschweig, Wiesbaden (1982)

Reddy, J.N.: Mechanics of laminated composite plates and shells, 2nd edn. CRC Press, Boca Raton (2004)

Reddy, J.N.: Theory and analysis of elastic plates and shells. CRC Press, Boca Raton (2006)

Szilard, R.: Theories and applications of plate analysis. Wiley Hoboken (2004)

Wiedemann, J.: Leichtbau, Elemente und Konstruktion, 3rd edn. Springer, Berlin (2006)

12 Geometrically Nonlinear Analysis

Plates are thin-walled structures, so in addition to the already discussed plate bending in the context of a geometrically linear analysis, the geometrically nonlinear behavior is also of quite fundamental importance. For this reason, this chapter considers the bending behavior of linear elastic plates in the framework of v. Kármán plate theory (Chap. 1, Sect. 1.3.3), both based on the kinematic assumptions of Kirchhoff plate theory and in the framework of First-Order Shear Deformation Theory. The governing equations and boundary conditions can be obtained particularly advantageously in an energetic way, and a part of these considerations is devoted to the energetic derivation and the discussion of the boundary terms arising in the framework of the assumptions of Kirchhoff plate theory. After this, static bending problems are considered and various solution methods for such problems are discussed. This chapter concludes by providing the governing equations in the framework of First-Order Shear Deformation Theory. Further information on the subject of this chapter can be found, for example, in Altenbach et al. (2016), Chia (1980), Reddy (2003, 2006), Sathyamoorthy (1997), or Szilard (2004).

12.1 Kirchhoff Plate Theory

12.1.1 Energetic Consideration

Starting point of the considerations is the displacement field (7.7) of Kirchhoff's plate theory, which has already been derived in Chap. 7 and is given here again for the sake of better readability:

C. Mittelstedt, *Theory of Plates and Shells*,
https://doi.org/10.1007/978-3-662-66805-4_12

$$
\begin{aligned}
u(x, y, z) &= u_0(x, y) - z\frac{\partial w_0(x, y)}{\partial x}, \\
v(x, y, z) &= v_0(x, y) - z\frac{\partial w_0(x, y)}{\partial y}, \\
w(x, y) &= w_0(x, y).
\end{aligned} \tag{12.1}
$$

At this point, the plane displacements u_0 and v_0 of the plate midplane have been taken into account, which, as will be shown in the further course of this chapter, play an important role in the treatment of plate problems with large deflections. The formulation (12.1) thus formally corresponds to the superposition of a disk action with a plate action.

In the following, we consider the case where the given plane thin-walled structure assumes a deformation state characterized by moderate rotations of the surface normals at small strains. For such a case, it is convenient to refer to the v. Kármán strains (1.59), which are also written here again for the sake of a better readability:

$$
\begin{aligned}
\varepsilon_{xx} &= \frac{\partial u_0}{\partial x} + \frac{1}{2}\left(\frac{\partial w_0}{\partial x}\right)^2 - z\frac{\partial^2 w_0}{\partial x^2}, \\
\varepsilon_{yy} &= \frac{\partial v_0}{\partial y} + \frac{1}{2}\left(\frac{\partial w_0}{\partial y}\right)^2 - z\frac{\partial^2 w_0}{\partial y^2}, \\
\gamma_{xy} &= \frac{\partial u_0}{\partial y} + \frac{\partial v_0}{\partial x} + \frac{\partial w_0}{\partial x}\frac{\partial w_0}{\partial y} - 2z\frac{\partial^2 w_0}{\partial x \partial y}.
\end{aligned} \tag{12.2}
$$

Let the remaining strain quantities be zero, i.e. $\gamma_{xz} = \gamma_{yz} = \varepsilon_{zz} = 0$.

A more compact notation of the relations (12.2) can be given as follows:

$$
\begin{pmatrix} \varepsilon_{xx} \\ \varepsilon_{yy} \\ \gamma_{xy} \end{pmatrix} = \begin{pmatrix} \varepsilon_{xx}^0 \\ \varepsilon_{yy}^0 \\ \gamma_{xy}^0 \end{pmatrix} + z \begin{pmatrix} \kappa_{xx}^0 \\ \kappa_{yy}^0 \\ \kappa_{xy}^0 \end{pmatrix}, \tag{12.3}
$$

wherein

$$
\begin{aligned}
\varepsilon_{xx}^0 &= \frac{\partial u_0}{\partial x} + \frac{1}{2}\left(\frac{\partial w_0}{\partial x}\right)^2, \\
\varepsilon_{yy}^0 &= \frac{\partial v_0}{\partial y} + \frac{1}{2}\left(\frac{\partial w_0}{\partial y}\right)^2, \\
\gamma_{xy}^0 &= \frac{\partial u_0}{\partial y} + \frac{\partial v_0}{\partial x} + \frac{\partial w_0}{\partial x}\frac{\partial w_0}{\partial y}, \\
\kappa_{xx}^0 &= -\frac{\partial^2 w_0}{\partial x^2}, \quad \kappa_{yy}^0 = -\frac{\partial^2 w_0}{\partial y^2}, \quad \kappa_{xy}^0 = -2\frac{\partial^2 w_0}{\partial x \partial y}.
\end{aligned} \tag{12.4}
$$

To derive the governing differential equations as well as the associated boundary conditions, the principle of virtual displacements $\delta W_i = \delta W_a$ (Chap. 2) is used. For

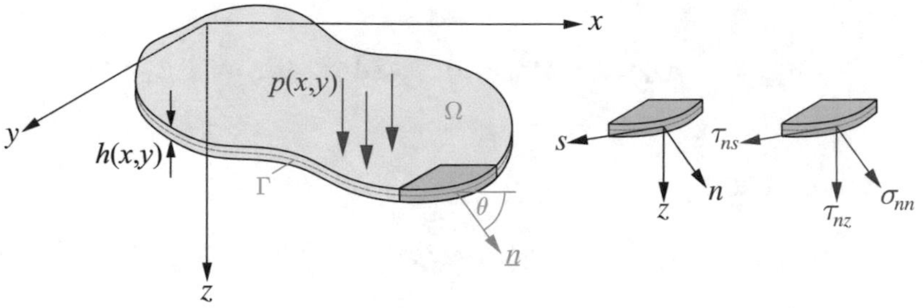

Fig. 12.1 Considered plate

this purpose, we consider the plate situation of Fig. 12.1. The plate has the area Ω and is bounded by the edge Γ, on which the normal direction is marked by the coordinate n. The circumferential direction is s. The normal vector $\underline{n}$ of the boundary is oriented by the angle θ with respect to the x-axis. Let the normal vector $\underline{n} = n_x \underline{e}_x + n_y \underline{e}_y$ be a unit vector, i.e. $|\underline{n}| = 1$. Thus, for the components n_x and n_y of the vector $\underline{n}$, $n_x = \cos\theta$ and $n_y = \sin\theta$ holds. Let the stress components σ_{nn}, τ_{ns} and τ_{nz} be defined at the boundary. Let the corresponding plane boundary displacement quantities be u_n and u_s. The boundary Γ is divided into two parts Γ_u and Γ_σ, on which boundary displacements and boundary stresses are given: $\Gamma = \Gamma_u + \Gamma_\sigma$. On the edge piece Γ_σ the edge stresses σ_{nn}, τ_{ns} and τ_{nz} with the values $\hat{\sigma}_{nn}$, $\hat{\tau}_{ns}$ and $\hat{\tau}_{nz}$ are prescribed. Let a surface load $p(x, y)$ be given as the load on the plate. In addition, let the plate be resting on an elastic foundation with the foundation coefficient k.

We first consider the virtual inner work δW_i done by the internal stresses of the plate at an infinitesimal virtual admissible strain. If the stresses σ_{xx}, σ_{yy}, τ_{xy} are present, then δW_i reads:

$$\begin{aligned}\delta W_i &= \int_\Omega \int_{-\frac{h}{2}}^{+\frac{h}{2}} \left[\sigma_{xx}\delta\varepsilon_{xx} + \sigma_{yy}\delta\varepsilon_{yy} + \tau_{xy}\delta\gamma_{xy}\right] \mathrm{d}z\mathrm{d}\Omega \\ &= \int_\Omega \int_{-\frac{h}{2}}^{+\frac{h}{2}} \left[\sigma_{xx}\left(\delta\varepsilon_{xx}^0 + z\delta\kappa_{xx}^0\right) + \sigma_{yy}\left(\delta\varepsilon_{yy}^0 + z\delta\kappa_{yy}^0\right) + \tau_{xy}\left(\delta\gamma_{xy}^0 + z\delta\kappa_{xy}^0\right)\right] \mathrm{d}z\mathrm{d}\Omega.\end{aligned} \tag{12.5}$$

Introduction of the force and moment flows

$$N_{xx}^0 = \int_{-\frac{h}{2}}^{+\frac{h}{2}} \sigma_{xx}\mathrm{d}z, \quad N_{yy}^0 = \int_{-\frac{h}{2}}^{+\frac{h}{2}} \sigma_{yy}\mathrm{d}z, \quad N_{xy}^0 = \int_{-\frac{h}{2}}^{+\frac{h}{2}} \tau_{xy}\mathrm{d}z,$$

$$M_{xx}^0 = \int_{-\frac{h}{2}}^{+\frac{h}{2}} \sigma_{xx} z \mathrm{d}z, \quad M_{yy}^0 = \int_{-\frac{h}{2}}^{+\frac{h}{2}} \sigma_{yy} z \mathrm{d}z, \quad M_{xy}^0 = \int_{-\frac{h}{2}}^{+\frac{h}{2}} \tau_{xy} z \mathrm{d}z \tag{12.6}$$

then gives for (12.5):

$$\delta W_i = \int_{\Omega} \left(N_{xx}^0 \delta\varepsilon_{xx}^0 + N_{yy}^0 \delta\varepsilon_{yy}^0 + N_{xy}^0 \delta\gamma_{xy}^0 + M_{xx}^0 \delta\kappa_{xx}^0 + M_{yy}^0 \delta\kappa_{yy}^0 + M_{xy}^0 \delta\kappa_{xy}^0 \right) \mathrm{d}\Omega. \tag{12.7}$$

External virtual work is performed on the one hand by the surface load p and by the reaction forces of the elastic foundation (bedding number k), and on the other hand by the boundary stresses $\hat{\sigma}_{nn}$, $\hat{\tau}_{ns}$, $\hat{\tau}_{nz}$ given on the boundary piece Γ_σ. Hence:

$$\delta W_a = \int_{\Omega} (p\delta w_0 - k w_0 \delta w_0) \mathrm{d}\Omega + \int_{\Gamma_\sigma} \int_{-\frac{h}{2}}^{+\frac{h}{2}} \left(\hat{\sigma}_{nn} \delta u_n + \hat{\tau}_{ns} \delta u_s + \hat{\tau}_{nz} \delta w_0 \right) \mathrm{d}z \mathrm{d}s. \tag{12.8}$$

The virtual boundary displacements δu_n and δu_s can be divided analogously to (12.1) into a disk part and a bending part, i.e.:

$$\delta u_n = \delta u_{0n} - z \frac{\partial \delta w_0}{\partial n}, \quad \delta u_s = \delta u_{0s} - z \frac{\partial \delta w_0}{\partial s}. \tag{12.9}$$

With the force and moment flows resulting from the given boundary stresses $\hat{\sigma}_{nn}$, $\hat{\tau}_{ns}$, $\hat{\tau}_{nz}$

$$\hat{N}_{nn}^0 = \int_{-\frac{h}{2}}^{+\frac{h}{2}} \hat{\sigma}_{nn} \mathrm{d}z, \quad \hat{N}_{ns}^0 = \int_{-\frac{h}{2}}^{+\frac{h}{2}} \hat{\tau}_{ns} \mathrm{d}z,$$

$$\hat{M}_{nn}^0 = \int_{-\frac{h}{2}}^{+\frac{h}{2}} \hat{\sigma}_{nn} z \mathrm{d}z, \quad \hat{M}_{ns}^0 = \int_{-\frac{h}{2}}^{+\frac{h}{2}} \hat{\tau}_{ns} z \mathrm{d}z, \quad \hat{Q}_n = \int_{-\frac{h}{2}}^{+\frac{h}{2}} \hat{\tau}_{nz} \mathrm{d}z \tag{12.10}$$

Eq. (12.8) results in:

$$\delta W_a = \int_{\Omega} (q - k w_0) \delta w_0 \mathrm{d}\Omega + \int_{\Gamma_\sigma} \left(\hat{N}_{nn}^0 \delta u_{0n} - \hat{M}_{nn}^0 \frac{\partial \delta w_0}{\partial n} + \hat{N}_{ns}^0 \delta u_{0s} - \hat{M}_{ns}^0 \frac{\partial \delta w_0}{\partial s} + \hat{Q}_n \delta w_0 \right) \mathrm{d}s. \tag{12.11}$$

The principle of virtual displacements $\delta W_i = \delta W_a$ with the virtual strain quantities

$$
\begin{aligned}
\delta\varepsilon_{xx}^0 &= \frac{\partial \delta u_0}{\partial x} + \frac{\partial w_0}{\partial x}\frac{\partial \delta w_0}{\partial x}, \\
\delta\varepsilon_{yy}^0 &= \frac{\partial \delta v_0}{\partial y} + \frac{\partial w_0}{\partial y}\frac{\partial \delta w_0}{\partial y}, \\
\delta\gamma_{xy}^0 &= \frac{\partial \delta u_0}{\partial y} + \frac{\partial \delta v_0}{\partial x} + \frac{\partial w_0}{\partial y}\frac{\partial \delta w_0}{\partial x} + \frac{\partial w_0}{\partial x}\frac{\partial \delta w_0}{\partial y}, \\
\delta\kappa_{xx}^0 &= -\frac{\partial^2 \delta w_0}{\partial x^2}, \quad \delta\kappa_{yy}^0 = -\frac{\partial^2 \delta w_0}{\partial y^2}, \quad \delta\kappa_{xy}^0 = -2\frac{\partial^2 \delta w_0}{\partial x \partial y}
\end{aligned} \tag{12.12}
$$

then results in:

$$
\begin{aligned}
&\int_\Omega \Bigg[N_{xx}^0 \left(\frac{\partial \delta u_0}{\partial x} + \frac{\partial w_0}{\partial x}\frac{\partial \delta w_0}{\partial x} \right) + N_{yy}^0 \left(\frac{\partial \delta v_0}{\partial y} + \frac{\partial w_0}{\partial y}\frac{\partial \delta w_0}{\partial y} \right) \\
&+ N_{xy}^0 \left(\frac{\partial \delta u_0}{\partial y} + \frac{\partial \delta v_0}{\partial x} + \frac{\partial w_0}{\partial y}\frac{\partial \delta w_0}{\partial x} + \frac{\partial w_0}{\partial x}\frac{\partial \delta w_0}{\partial y} \right) \\
&- M_{xx}^0 \frac{\partial^2 \delta w_0}{\partial x^2} - M_{yy}^0 \frac{\partial^2 \delta w_0}{\partial y^2} - 2M_{xy}^0 \frac{\partial^2 \delta w_0}{\partial x \partial y} + k w_0 \delta w_0 - p \delta w_0 \Bigg] \mathrm{d}\Omega \\
&- \int_{\Gamma_\sigma} \left(\hat{N}_{nn}^0 \delta u_{0n} - \hat{M}_{nn}^0 \frac{\partial \delta w_0}{\partial n} + \hat{N}_{ns}^0 \delta u_{0s} - \hat{M}_{ns}^0 \frac{\partial \delta w_0}{\partial s} + \hat{Q}_n \delta w_0 \right) \mathrm{d}s = 0.
\end{aligned} \tag{12.13}
$$

Those terms, which contain partial derivatives of the virtual displacement quantities, are now partially integrated considering the Gaussian integral theorem. In detail, this results in:

$$
\int_\Omega N_{xx}^0 \frac{\partial \delta u_0}{\partial x} \mathrm{d}\Omega = \oint_\Gamma N_{xx}^0 n_x \delta u_0 \mathrm{d}s \quad \int_\Omega \frac{\partial N_{xx}^0}{\partial x} \delta u_0 \mathrm{d}\Omega,
$$

$$
\int_\Omega N_{xx}^0 \frac{\partial w_0}{\partial x}\frac{\partial \delta w_0}{\partial x} \mathrm{d}\Omega = \oint_\Gamma N_{xx}^0 \frac{\partial w_0}{\partial x} n_x \delta w_0 \mathrm{d}s - \int_\Omega \frac{\partial}{\partial x}\left(N_{xx}^0 \frac{\partial w_0}{\partial x} \right) \delta w_0 \mathrm{d}\Omega,
$$

$$
\int_\Omega N_{yy}^0 \frac{\partial \delta v_0}{\partial y} \mathrm{d}\Omega = \oint_\Gamma N_{yy}^0 n_y \delta v_0 \mathrm{d}s - \int_\Omega \frac{\partial N_{yy}^0}{\partial y} \delta v_0 \mathrm{d}\Omega,
$$

$$
\int_\Omega N_{yy}^0 \frac{\partial w_0}{\partial y}\frac{\partial \delta w_0}{\partial y} \mathrm{d}\Omega = \oint_\Gamma N_{yy}^0 \frac{\partial w_0}{\partial y} n_y \delta w_0 \mathrm{d}s - \int_\Omega \frac{\partial}{\partial y}\left(N_{yy}^0 \frac{\partial w_0}{\partial y} \right) \delta w_0 \mathrm{d}\Omega,
$$

$$
\int_\Omega N_{xy}^0 \frac{\partial \delta u_0}{\partial y} \mathrm{d}\Omega = \oint_\Gamma N_{xy}^0 n_y \delta u_0 \mathrm{d}s - \int_\Omega \frac{\partial N_{xy}^0}{\partial y} \delta u_0 \mathrm{d}\Omega,
$$

$$
\int_\Omega N_{xy}^0 \frac{\partial \delta v_0}{\partial x} \mathrm{d}\Omega = \oint_\Gamma N_{xy}^0 n_x \delta v_0 \mathrm{d}s - \int_\Omega \frac{\partial N_{xy}^0}{\partial y} \delta v_0 \mathrm{d}\Omega,
$$

$$
\int_\Omega N_{xy}^0 \frac{\partial \delta w_0}{\partial x}\frac{\partial w_0}{\partial y} \mathrm{d}\Omega = \oint_\Gamma N_{xy}^0 \frac{\partial w_0}{\partial y} n_x \delta w_0 \mathrm{d}s - \int_\Omega \frac{\partial}{\partial x}\left(N_{xy}^0 \frac{\partial w_0}{\partial y} \right) \delta w_0 \mathrm{d}\Omega,
$$

$$\int_{\Omega} N_{xy}^0 \frac{\partial w_0}{\partial x}\frac{\partial \delta w_0}{\partial y}\mathrm{d}\Omega = \oint_{\Gamma} N_{xy}^0 \frac{\partial w_0}{\partial x} n_y \delta w_0 \mathrm{d}s - \int_{\Omega} \frac{\partial}{\partial y}\left(N_{xy}^0 \frac{\partial w_0}{\partial x}\right)\delta w_0 \mathrm{d}\Omega,$$

$$-\int_{\Omega} M_{xx}^0 \frac{\partial^2 \delta w_0}{\partial x^2}\mathrm{d}\Omega = -\oint_{\Gamma} M_{xx}^0 n_x \frac{\partial \delta w_0}{\partial x}\mathrm{d}s + \oint_{\Gamma} \frac{\partial M_{xx}^0}{\partial x} n_x \delta w_0 \mathrm{d}s - \int_{\Omega} \frac{\partial^2 M_{xx}^0}{\partial x^2}\delta w_0 \mathrm{d}\Omega,$$

$$-\int_{\Omega} M_{yy}^0 \frac{\partial^2 \delta w_0}{\partial y^2}\mathrm{d}\Omega = -\oint_{\Gamma} M_{yy}^0 n_y \frac{\partial \delta w_0}{\partial y}\mathrm{d}s + \oint_{\Gamma} \frac{\partial M_{yy}^0}{\partial y} n_y \delta w_0 \mathrm{d}s - \int_{\Omega} \frac{\partial^2 M_{yy}^0}{\partial y^2}\delta w_0 \mathrm{d}\Omega,$$

$$-\int_{\Omega} M_{xy}^0 \frac{\partial^2 \delta w_0}{\partial x \partial y}\mathrm{d}\Omega = -\oint_{\Gamma} M_{xy}^0 n_x \frac{\partial \delta w_0}{\partial y}\mathrm{d}s + \oint_{\Gamma} \frac{\partial M_{xy}^0}{\partial x} n_y \delta w_0 \mathrm{d}s - \int_{\Omega} \frac{\partial^2 M_{xy}^0}{\partial x \partial y}\delta w_0 \mathrm{d}\Omega,$$

$$-\int_{\Omega} M_{xy}^0 \frac{\partial^2 \delta w_0}{\partial x \partial y}\mathrm{d}\Omega = -\oint_{\Gamma} M_{xy}^0 n_y \frac{\partial \delta w_0}{\partial x}\mathrm{d}s + \oint_{\Gamma} \frac{\partial M_{xy}^0}{\partial y} n_x \delta w_0 \mathrm{d}s - \int_{\Omega} \frac{\partial^2 M_{xy}^0}{\partial x \partial y}\delta w_0 \mathrm{d}\Omega. \tag{12.14}$$

Merging terms of the same kind, taking into account that virtual displacements vanish on the boundary piece Γ_u, leads to the following expression:

$$\begin{aligned}
&\int_{\Omega} \Bigg\{ -\left(\frac{\partial N_{xx}^0}{\partial x} + \frac{\partial N_{xy}^0}{\partial y}\right)\delta u_0 - \left(\frac{\partial N_{xy}^0}{\partial x} + \frac{\partial N_{yy}^0}{\partial y}\right)\delta v_0 \\
&\quad - \left[\frac{\partial^2 M_{xx}^0}{\partial x^2} + 2\frac{\partial^2 M_{xy}^0}{\partial x \partial y} + \frac{\partial^2 M_{yy}^0}{\partial y^2} + \frac{\partial}{\partial x}\left(N_{xx}^0 \frac{\partial w_0}{\partial x} + N_{xy}^0 \frac{\partial w_0}{\partial y}\right)\right. \\
&\quad \left. + \frac{\partial}{\partial y}\left(N_{xy}^0 \frac{\partial w_0}{\partial x} + N_{yy}^0 \frac{\partial w_0}{\partial y}\right) - k w_0 + p\right]\delta w_0 \Bigg\}\mathrm{d}\Omega \\
&\quad + \int_{\Gamma_\sigma} \Bigg\{ \left(N_{xx}^0 n_x + N_{xy}^0 n_y\right)\delta u_0 + \left(N_{xy}^0 n_x + N_{yy}^0 n_y\right)\delta v_0 \\
&\quad + \left[\frac{\partial M_{xx}^0}{\partial x} n_x + \frac{\partial M_{xy}^0}{\partial y} n_x + \frac{\partial M_{yy}^0}{\partial y} n_y + \frac{\partial M_{xy}^0}{\partial x} n_y \right. \\
&\quad \left. + \left(N_{xx}^0 \frac{\partial w_0}{\partial x} + N_{xy}^0 \frac{\partial w_0}{\partial y}\right) n_x + \left(N_{xy}^0 \frac{\partial w_0}{\partial x} + N_{yy}^0 \frac{\partial w_0}{\partial y}\right) n_y\right]\delta w_0 \\
&\quad - \left(M_{xx}^0 n_x + M_{xy}^0 n_y\right)\frac{\partial \delta w_0}{\partial x} - \left(M_{xy}^0 n_x + M_{yy}^0 n_y\right)\frac{\partial \delta w_0}{\partial y}\Bigg\}\mathrm{d}s \\
&\quad - \int_{\Gamma_\sigma} \left(\hat{N}_{nn}^0 \delta u_{0n} - \hat{M}_{nn}^0 \frac{\partial \delta w_0}{\partial n} + \hat{N}_{ns}^0 \delta u_{0s} - \hat{M}_{ns}^0 \frac{\partial \delta w_0}{\partial s} + \hat{Q}_n \delta w_0\right)\mathrm{d}s = 0.
\end{aligned} \tag{12.15}$$

From the area integrals, the governing differential equations of the given nonlinear plate problem can be read:

$$\frac{\partial N_{xx}^0}{\partial x} + \frac{\partial N_{xy}^0}{\partial y} = 0,$$

$$\frac{\partial N_{xy}^0}{\partial x} + \frac{\partial N_{yy}^0}{\partial y} = 0,$$

$$\frac{\partial^2 M_{xx}^0}{\partial x^2} + 2\frac{\partial^2 M_{xy}^0}{\partial x \partial y} + \frac{\partial^2 M_{yy}^0}{\partial y^2} + \frac{\partial}{\partial x}\left(N_{xx}^0 \frac{\partial w_0}{\partial x} + N_{xy}^0 \frac{\partial w_0}{\partial y}\right)$$
$$+\frac{\partial}{\partial y}\left(N_{xy}^0 \frac{\partial w_0}{\partial x} + N_{yy}^0 \frac{\partial w_0}{\partial y}\right) - k w_0 + p = 0. \quad (12.16)$$

In the last differential equation, the following transformations can be made:

$$\frac{\partial}{\partial x}\left(N_{xx}^0 \frac{\partial w_0}{\partial x} + N_{xy}^0 \frac{\partial w_0}{\partial y}\right) + \frac{\partial}{\partial y}\left(N_{xy}^0 \frac{\partial w_0}{\partial x} + N_{yy}^0 \frac{\partial w_0}{\partial y}\right)$$
$$= \left(\frac{\partial N_{xx}^0}{\partial x} + \frac{\partial N_{xy}^0}{\partial y}\right)\frac{\partial w_0}{\partial x} + \left(\frac{\partial N_{xy}^0}{\partial x} + \frac{\partial N_{yy}^0}{\partial y}\right)\frac{\partial w_0}{\partial y}$$
$$+N_{xx}^0 \frac{\partial^2 w_0}{\partial x^2} + 2N_{xy}^0 \frac{\partial^2 w_0}{\partial x \partial y} + N_{yy}^0 \frac{\partial^2 w_0}{\partial y^2}. \quad (12.17)$$

Herein, the two new terms in parentheses correspond exactly to the first two equilibrium conditions in (12.16), so they can be set to zero. Therefore, it remains for the third equation in (12.16):

$$\frac{\partial^2 M_{xx}^0}{\partial x^2} + 2\frac{\partial^2 M_{xy}^0}{\partial x \partial y} + \frac{\partial^2 M_{yy}^0}{\partial y^2} + N_{xx}^0 \frac{\partial^2 w_0}{\partial x^2} + 2N_{xy}^0 \frac{\partial^2 w_0}{\partial x \partial y} + N_{yy}^0 \frac{\partial^2 w_0}{\partial y^2} - k w_0 + p = 0. \quad (12.18)$$

It is shown that the equilibrium conditions in the plate plane leads exactly to the equilibrium conditions already known from disk structures. Furthermore, the third equation represents the condensed plate equilibrium condition, but here again the downward components from the plane force flows are added (cf. Chap. 11).

12.1.2 Th. V. Kármán equations

In the next step, we represent the force flows N_{xx}^0, N_{yy}^0, N_{xy}^0 and the moment flows M_{xx}^0, M_{yy}^0, M_{xy}^0 in the displacements. The constitutive law is currently:

$$\begin{pmatrix} N_{xx}^0 \\ N_{yy}^0 \\ N_{xy}^0 \end{pmatrix} = \begin{bmatrix} A_{11} & A_{12} & 0 \\ A_{12} & A_{22} & 0 \\ 0 & 0 & A_{66} \end{bmatrix} \begin{pmatrix} \frac{\partial u_0}{\partial x} + \frac{1}{2}\left(\frac{\partial w_0}{\partial x}\right)^2 \\ \frac{\partial v_0}{\partial y} + \frac{1}{2}\left(\frac{\partial w_0}{\partial y}\right)^2 \\ \frac{\partial u_0}{\partial y} + \frac{\partial v_0}{\partial x} + \frac{\partial w_0}{\partial x}\frac{\partial w_0}{\partial y} \end{pmatrix},$$

$$\begin{pmatrix} M_{xx}^0 \\ M_{yy}^0 \\ M_{xy}^0 \end{pmatrix} = -\begin{bmatrix} D_{11} & D_{12} & 0 \\ D_{12} & D_{22} & 0 \\ 0 & 0 & D_{66} \end{bmatrix} \begin{pmatrix} \frac{\partial^2 w_0}{\partial x^2} \\ \frac{\partial^2 w_0}{\partial y^2} \\ 2\frac{\partial^2 w_0}{\partial x \partial y} \end{pmatrix}. \tag{12.19}$$

For plates with constant membrane stiffnesses A_{ij} and plate stiffnesses D_{ij}, substituting in (12.16) and (12.18) gives:

$$\begin{aligned}
&A_{11}\left(\frac{\partial^2 u_0}{\partial x^2} + \frac{\partial w_0}{\partial x}\frac{\partial^2 w_0}{\partial x^2}\right) + A_{12}\left(\frac{\partial^2 v_0}{\partial x \partial y} + \frac{\partial w_0}{\partial y}\frac{\partial^2 w_0}{\partial x \partial y}\right) \\
&\quad + A_{66}\left(\frac{\partial^2 u_0}{\partial y^2} + \frac{\partial^2 v_0}{\partial x \partial y} + \frac{\partial^2 w_0}{\partial x \partial y}\frac{\partial w_0}{\partial y} + \frac{\partial w_0}{\partial x}\frac{\partial^2 w_0}{\partial y^2}\right) = 0, \\
&A_{22}\left(\frac{\partial^2 v_0}{\partial y^2} + \frac{\partial w_0}{\partial y}\frac{\partial^2 w_0}{\partial y^2}\right) + A_{12}\left(\frac{\partial^2 u_0}{\partial x \partial y} + \frac{\partial w_0}{\partial x}\frac{\partial^2 w_0}{\partial x \partial y}\right) \\
&\quad + A_{66}\left(\frac{\partial^2 u_0}{\partial x \partial y} + \frac{\partial^2 v_0}{\partial x^2} + \frac{\partial^2 w_0}{\partial x^2}\frac{\partial w_0}{\partial y} + \frac{\partial w_0}{\partial x}\frac{\partial^2 w_0}{\partial x \partial y}\right) = 0, \\
&-D_{11}\frac{\partial^4 w_0}{\partial x^4} - 2\left(D_{12} + 2D_{66}\right)\frac{\partial^4 w_0}{\partial x^2 \partial y^2} - D_{22}\frac{\partial^4 w_0}{\partial y^4} - k w_0 + p \\
&\quad + N_{xx}^0 \frac{\partial^2 w_0}{\partial x^2} + 2N_{xy}^0 \frac{\partial^2 w_0}{\partial x \partial y} + N_{yy}^0 \frac{\partial^2 w_0}{\partial y^2} = 0.
\end{aligned} \tag{12.20}$$

The special case of an isotropic plate with the membrane stiffnesses $A_{11} = A_{22} = A = \frac{Eh}{1-\nu^2}$, $A_{12} = \nu A$, $2A_{66} = (1-\nu)A$ and the plate stiffnesses $D_{11} = D_{22} = D = \frac{Eh^3}{12(1-\nu^2)}$, $D_{12} = \nu D$, $2D_{66} = (1-\nu)D$ yields:

$$\begin{aligned}
&A\left(\frac{\partial^2 u_0}{\partial x^2} + \frac{\partial w_0}{\partial x}\frac{\partial^2 w_0}{\partial x^2}\right) + \nu A\left(\frac{\partial^2 v_0}{\partial x \partial y} + \frac{\partial w_0}{\partial y}\frac{\partial^2 w_0}{\partial x \partial y}\right) \\
&+ \frac{1-\nu}{2}A\left(\frac{\partial^2 u_0}{\partial y^2} + \frac{\partial^2 v_0}{\partial x \partial y} + \frac{\partial^2 w_0}{\partial x \partial y}\frac{\partial w_0}{\partial y} + \frac{\partial w_0}{\partial x}\frac{\partial^2 w_0}{\partial y^2}\right) = 0, \\
&A\left(\frac{\partial^2 v_0}{\partial y^2} + \frac{\partial w_0}{\partial y}\frac{\partial^2 w_0}{\partial y^2}\right) + \nu A\left(\frac{\partial^2 u_0}{\partial x \partial y} + \frac{\partial w_0}{\partial x}\frac{\partial^2 w_0}{\partial x \partial y}\right) \\
&+ \frac{1-\nu}{2}A\left(\frac{\partial^2 u_0}{\partial x \partial y} + \frac{\partial^2 v_0}{\partial x^2} + \frac{\partial^2 w_0}{\partial x^2}\frac{\partial w_0}{\partial y} + \frac{\partial w_0}{\partial x}\frac{\partial^2 w_0}{\partial x \partial y}\right) = 0, \\
&-D\frac{\partial^4 w_0}{\partial x^4} - 2D\frac{\partial^4 w_0}{\partial x^2 \partial y^2} - D\frac{\partial^4 w_0}{\partial y^4} - k w_0 + p \\
&+ N_{xx}^0 \frac{\partial^2 w_0}{\partial x^2} + 2N_{xy}^0 \frac{\partial^2 w_0}{\partial x \partial y} + N_{yy}^0 \frac{\partial^2 w_0}{\partial y^2} = 0.
\end{aligned} \tag{12.21}$$

When using the Airy stress function

$$N_{xx}^0 = \frac{\partial^2 F}{\partial y^2}, \quad N_{yy}^0 = \frac{\partial^2 F}{\partial x^2}, \quad N_{xy}^0 = -\frac{\partial^2 F}{\partial x \partial y}, \tag{12.22}$$

the system of equations (12.20) can be represented in a particularly convenient mixed form. The first two conditions in (12.20) are identically satisfied due to the form of

(12.22), regardless of the choice of Airy's stress function. We consider the inverted material law

$$\begin{pmatrix} \varepsilon^0_{xx} \\ \varepsilon^0_{yy} \\ \gamma^0_{xy} \end{pmatrix} = \begin{bmatrix} \alpha_{11} & \alpha_{12} & 0 \\ \alpha_{12} & \alpha_{22} & 0 \\ 0 & 0 & \alpha_{66} \end{bmatrix} \begin{pmatrix} N^0_{xx} \\ N^0_{yy} \\ N^0_{xy} \end{pmatrix} = \begin{bmatrix} \alpha_{11} & \alpha_{12} & 0 \\ \alpha_{12} & \alpha_{22} & 0 \\ 0 & 0 & \alpha_{66} \end{bmatrix} \begin{pmatrix} \frac{\partial^2 F}{\partial y^2} \\ \frac{\partial^2 F}{\partial x^2} \\ -\frac{\partial^2 F}{\partial x \partial y} \end{pmatrix}, \tag{12.23}$$

where the quantities α_{11}, α_{22}, α_{12} and α_{66} are the inverted disk stiffnesses:

$$\alpha_{11} = \frac{A_{22}}{A_{11}A_{22} - A_{12}^2}, \quad \alpha_{22} = \frac{A_{11}}{A_{11}A_{22} - A_{12}^2}, \quad \alpha_{12} = -\frac{A_{12}}{A_{11}A_{22} - A_{12}^2}, \quad \alpha_{66} = \frac{1}{A_{66}}. \tag{12.24}$$

We also consider the compatibility condition (3.13):

$$\frac{\partial^2 \varepsilon^0_{xx}}{\partial y^2} + \frac{\partial^2 \varepsilon^0_{yy}}{\partial x^2} - \frac{\partial^2 \gamma^0_{xy}}{\partial x \partial y} = 0. \tag{12.25}$$

Inserting the membrane strains ε^0_{xx}, ε^0_{yy}, γ^0_{xy} according to (12.4) gives the following expression for (12.25):

$$\frac{\partial^2 \varepsilon^0_{xx}}{\partial y^2} + \frac{\partial^2 \varepsilon^0_{yy}}{\partial x^2} - \frac{\partial^2 \gamma^0_{xy}}{\partial x \partial y} = \left(\frac{\partial^2 w_0}{\partial x \partial y}\right)^2 - \frac{\partial^2 w_0}{\partial x^2}\frac{\partial^2 w_0}{\partial y^2}. \tag{12.26}$$

Moreover, from (12.23) it follows:

$$\begin{aligned} \frac{\partial^2 \varepsilon^0_{xx}}{\partial y^2} &= \alpha_{11}\frac{\partial^4 F}{\partial y^4} + \alpha_{12}\frac{\partial^4 F}{\partial x^2 \partial y^2}, \\ \frac{\partial^2 \varepsilon^0_{yy}}{\partial x^2} &= \alpha_{12}\frac{\partial^4 F}{\partial x^2 \partial y^2} + \alpha_{22}\frac{\partial^4 F}{\partial x^4}, \\ -\frac{\partial^2 \gamma^0_{xy}}{\partial x \partial y} &= \alpha_{66}\frac{\partial^4 F}{\partial x^2 \partial y^2}, \end{aligned} \tag{12.27}$$

which when inserted into (12.26) leads to the following expression:

$$\alpha_{11}\frac{\partial^4 F}{\partial y^4} + (2\alpha_{12} + \alpha_{66})\frac{\partial^4 F}{\partial x^2 \partial y^2} + \alpha_{22}\frac{\partial^4 F}{\partial x^4} = \left(\frac{\partial^2 w_0}{\partial x \partial y}\right)^2 - \frac{\partial^2 w_0}{\partial x^2}\frac{\partial^2 w_0}{\partial y^2}. \tag{12.28}$$

For the case of the isotropic plate follows:

$$\frac{1}{Eh}\Delta\Delta F = \left(\frac{\partial^2 w_0}{\partial x \partial y}\right)^2 - \frac{\partial^2 w_0}{\partial x^2}\frac{\partial^2 w_0}{\partial y^2}. \tag{12.29}$$

The second equation follows from (12.18) by expressing there the membrane force flows N^0_{xx}, N^0_{yy}, N^0_{xy} according to (12.22) and also applying the constitutive law (12.19). The final result is:

$$D_{11}\frac{\partial^4 w_0}{\partial x^4} + 2\left(D_{12} + 2D_{66}\right)\frac{\partial^4 w_0}{\partial x^2 \partial y^2} + D_{22}\frac{\partial^4 w_0}{\partial y^4}$$

$$= \frac{\partial^2 F}{\partial y^2}\frac{\partial^2 w_0}{\partial x^2} + \frac{\partial^2 F}{\partial x^2}\frac{\partial^2 w_0}{\partial y^2} - 2\frac{\partial^2 F}{\partial x \partial y}\frac{\partial^2 w_0}{\partial x \partial y} - k w_0 + p. \tag{12.30}$$

In the case of the isotropic plate, the result is:

$$D\Delta\Delta w_0 = \frac{\partial^2 F}{\partial y^2}\frac{\partial^2 w_0}{\partial x^2} + \frac{\partial^2 F}{\partial x^2}\frac{\partial^2 w_0}{\partial y^2} - 2\frac{\partial^2 F}{\partial x \partial y}\frac{\partial^2 w_0}{\partial x \partial y} - k w_0 + p. \tag{12.31}$$

12.1.3 Discussion of the Boundary Terms

The boundary terms in (12.15) require further discussion. We have:

$$\begin{aligned}
&\int_{\Gamma_\sigma} \Bigg\{ \left(N^0_{xx} n_x + N^0_{xy} n_y\right) \delta u_0 + \left(N^0_{xy} n_x + N^0_{yy} n_y\right) \delta v_0 \\
&+ \Bigg[\frac{\partial M^0_{xx}}{\partial x} n_x + \frac{\partial M^0_{xy}}{\partial y} n_x + \frac{\partial M^0_{yy}}{\partial y} n_y + \frac{\partial M^0_{xy}}{\partial x} n_y \\
&+ \left(N^0_{xx} \frac{\partial w_0}{\partial x} + N^0_{xy} \frac{\partial w_0}{\partial y} \right) n_x + \left(N^0_{xy} \frac{\partial w_0}{\partial x} + N^0_{yy} \frac{\partial w_0}{\partial y} \right) n_y \Bigg] \delta w_0 \\
&- \left(M^0_{xx} n_x + M^0_{xy} n_y\right) \frac{\partial \delta w_0}{\partial x} - \left(M^0_{xy} n_x + M^0_{yy} n_y\right) \frac{\partial \delta w_0}{\partial y} \Bigg\} \mathrm{d}s \\
&- \int_{\Gamma_\sigma} \left(\hat{N}^0_{nn} \delta u_{0n} - \hat{M}^0_{nn} \frac{\partial \delta w_0}{\partial n} + \hat{N}^0_{ns} \delta u_{0s} - \hat{M}^0_{ns} \frac{\partial \delta w_0}{\partial s} + \hat{Q}_n \delta w_0 \right) \mathrm{d}s = 0.
\end{aligned} \tag{12.32}$$

The goal is to rewrite the boundary terms uniformly with respect to the curved boundary with the boundary normal $\underline{n}$. The following relation between the coordinate directions x, y, z and n, s, z holds:

$$\begin{pmatrix} x \\ y \\ z \end{pmatrix} = \begin{bmatrix} \cos\theta & -\sin\theta & 0 \\ \sin\theta & \cos\theta & 0 \\ 0 & 0 & 1 \end{bmatrix} \begin{pmatrix} n \\ s \\ z \end{pmatrix} = \begin{bmatrix} n_x & -n_y & 0 \\ n_y & n_x & 0 \\ 0 & 0 & 1 \end{bmatrix} \begin{pmatrix} n \\ s \\ z \end{pmatrix}. \tag{12.33}$$

The displacements and the angular rotations can be transformed analogously:

$$\begin{pmatrix} u_0 \\ v_0 \\ w_0 \end{pmatrix} = \begin{bmatrix} n_x & -n_y & 0 \\ n_y & n_x & 0 \\ 0 & 0 & 1 \end{bmatrix} \begin{pmatrix} u_{0n} \\ u_{0s} \\ w_0 \end{pmatrix}, \quad \begin{pmatrix} \frac{\partial w_0}{\partial x} \\ \frac{\partial w_0}{\partial y} \end{pmatrix} = \begin{bmatrix} n_x & -n_y \\ n_y & n_x \end{bmatrix} \begin{pmatrix} \frac{\partial w_0}{\partial n} \\ \frac{\partial w_0}{\partial s} \end{pmatrix}. \tag{12.34}$$

For the transformation of stresses, according to Chap. 1 we obtain:

$$\begin{pmatrix} \sigma_{nn} \\ \tau_{ns} \end{pmatrix} = \begin{bmatrix} n_x^2 & n_y^2 & 2n_x n_y \\ -n_x n_y & n_x n_y & n_x^2 - n_y^2 \end{bmatrix} \begin{pmatrix} \sigma_{xx} \\ \sigma_{yy} \\ \tau_{xy} \end{pmatrix}. \tag{12.35}$$

From this, the following transformation relations for the force and moment flows can be concluded:

$$\begin{pmatrix} N_{nn}^0 \\ N_{ns}^0 \end{pmatrix} = \begin{bmatrix} n_x^2 & n_y^2 & 2n_x n_y \\ -n_x n_y & n_x n_y & n_x^2 - n_y^2 \end{bmatrix} \begin{pmatrix} N_{xx}^0 \\ N_{yy}^0 \\ N_{xy}^0 \end{pmatrix},$$

$$\begin{pmatrix} M_{nn}^0 \\ M_{ns}^0 \end{pmatrix} = \begin{bmatrix} n_x^2 & n_y^2 & 2n_x n_y \\ -n_x n_y & n_x n_y & n_x^2 - n_y^2 \end{bmatrix} \begin{pmatrix} M_{xx}^0 \\ M_{yy}^0 \\ M_{xy}^0 \end{pmatrix}. \qquad (12.36)$$

It then follows after a short calculation:

$$\left(N_{xx}^0 n_x + N_{xy}^0 n_y\right) \delta u_0 + \left(N_{xy}^0 n_x + N_{yy}^0 n_y\right) \delta v_0 = N_{nn}^0 \delta u_{0n} + N_{ns}^0 \delta u_{0s}. \qquad (12.37)$$

Analogously we have:

$$\left(M_{xx}^0 n_x + M_{xy}^0 n_y\right) \frac{\partial \delta w_0}{\partial x} - \left(M_{xy}^0 n_x + M_{yy}^0 n_y\right) \frac{\partial \delta w_0}{\partial y} = M_{nn}^0 \frac{\partial \delta w_0}{\partial n} + M_{ns}^0 \frac{\partial \delta w_0}{\partial s}. (12.38)$$

For the boundary term (12.32) then follows:

$$\begin{aligned} \int_{\Gamma_\sigma} \Bigg\{ & \left(N_{nn}^0 - \hat{N}_{nn}^0\right) \delta u_{0n} + \left(N_{ns}^0 - \hat{N}_{ns}^0\right) \delta u_{0s} \\ & + \Bigg[\frac{\partial M_{xx}^0}{\partial x} n_x + \frac{\partial M_{xy}^0}{\partial y} n_x + \frac{\partial M_{yy}^0}{\partial y} n_y + \frac{\partial M_{xy}^0}{\partial x} n_y \\ & + \left(N_{xx}^0 \frac{\partial w_0}{\partial x} + N_{xy}^0 \frac{\partial w_0}{\partial y} \right) n_x + \left(N_{xy}^0 \frac{\partial w_0}{\partial x} + N_{yy}^0 \frac{\partial w_0}{\partial y} \right) n_y - \hat{Q}_n \Bigg] \delta w_0 \\ & - \left(M_{nn}^0 - \hat{M}_{nn}^0\right) \frac{\partial \delta w_0}{\partial n} - \left(M_{ns}^0 - \hat{M}_{ns}^0\right) \frac{\partial \delta w_0}{\partial s} \Bigg\} \mathrm{d}s = 0. \end{aligned} \qquad (12.39)$$

It can be seen at this point that there are a total of ten boundary conditions, whereas the underlying theory only allows for eight boundary conditions. This contradiction is met by partially integrating the following term:

$$-\int_{\Gamma_\sigma} M_{ns}^0 \frac{\partial \delta w_0}{\partial s} \mathrm{d}s = -M_{ns}^0 \delta w_0 \big|_\Gamma + \int_{\Gamma_\sigma} \frac{\partial M_{ns}^0}{\partial s} \delta w_0 \mathrm{d}s. \qquad (12.40)$$

The term $-M_{ns}^0 \delta w_0 \big|_\Gamma$ vanishes, provided it is a closed boundary curve. The remaining term is added to the shear force flow, and the so-called Kirchhoff equivalent shear force reads:

$$\begin{aligned} \bar{Q}_n = & \frac{\partial M_{xx}^0}{\partial x} n_x + \frac{\partial M_{xy}^0}{\partial y} n_x + \frac{\partial M_{yy}^0}{\partial y} n_y + \frac{\partial M_{xy}^0}{\partial x} n_y + \frac{\partial M_{ns}^0}{\partial s} \\ & + \left(N_{xx}^0 \frac{\partial w_0}{\partial x} + N_{xy}^0 \frac{\partial w_0}{\partial y} \right) n_x + \left(N_{xy}^0 \frac{\partial w_0}{\partial x} + N_{yy}^0 \frac{\partial w_0}{\partial y} \right) n_y. \end{aligned} \qquad (12.41)$$

The boundary integral (12.39) then passes into the following form:

$$\int_{\Gamma_\sigma} \Big[\left(N_{nn}^0 - \hat{N}_{nn}^0 \right) \delta u_{0n} + \left(N_{ns}^0 - \hat{N}_{ns}^0 \right) \delta u_{0s} + \left(\bar{Q}_n - \hat{\bar{Q}}_n \right) \delta w_0 - \left(M_{nn}^0 - \hat{M}_{nn}^0 \right) \frac{\partial \delta w_0}{\partial n} \Big] \mathrm{d}s, \tag{12.42}$$

wherein:

$$\hat{\bar{Q}}_n = \hat{Q}_n + \frac{\partial \hat{M}_{ns}^0}{\partial s}. \tag{12.43}$$

Thus, the quantities u_{0n}, u_{0s}, w_0, $\frac{\partial w_0}{\partial n}$ are the prime variables of the given plate problem, whereas the force quantities N_{nn}^0, N_{ns}^0, V_n, M_{nn}^0 are the secondary variables.

12.1.4 Inner and External Potential

For the application of energy-based approximation methods (cf. Chap. 2) it is necessary to consider the energy contributions stored in the system under consideration. The inner potential Π_i is currently:

$$\begin{aligned}
\Pi_i &= \frac{1}{2} \int_V \left(\sigma_{xx} \varepsilon_{xx} + \sigma_{yy} \varepsilon_{yy} + \tau_{xy} \gamma_{xy} \right) \mathrm{d}V \\
&= \frac{1}{2} \int_\Omega \int_{-\frac{h}{2}}^{+\frac{h}{2}} \left[\sigma_{xx} \left(\varepsilon_{xx}^0 + z\kappa_{xx}^0 \right) + \sigma_{yy} \left(\varepsilon_{yy}^0 + z\kappa_{yy}^0 \right) + \tau_{xy} \left(\gamma_{xy}^0 + z\kappa_{xy}^0 \right) \right] \mathrm{d}z\mathrm{d}\Omega \\
&= \frac{1}{2} \int_\Omega \left(N_{xx}^0 \varepsilon_{xx}^0 + M_{xx}^0 \kappa_{xx}^0 + N_{yy}^0 \varepsilon_{yy}^0 + M_{yy}^0 \kappa_{yy}^0 + N_{xy}^0 \gamma_{xy}^0 + M_{xy}^0 \kappa_{xy}^0 \right) \mathrm{d}\Omega \\
&= \frac{1}{2} \int_\Omega \Bigg\{ N_{xx}^0 \left[\frac{\partial u_0}{\partial x} + \frac{1}{2} \left(\frac{\partial w_0}{\partial x} \right)^2 \right] - M_{xx}^0 \frac{\partial^2 w_0}{\partial x^2} \\
&\quad + N_{yy}^0 \left[\frac{\partial v_0}{\partial y} + \frac{1}{2} \left(\frac{\partial w_0}{\partial y} \right)^2 \right] - M_{yy}^0 \frac{\partial^2 w_0}{\partial y^2} \\
&\quad + N_{xy}^0 \left(\frac{\partial u_0}{\partial y} + \frac{\partial v_0}{\partial x} + \frac{\partial w_0}{\partial x} \frac{\partial w_0}{\partial y} \right) - 2M_{xy}^0 \frac{\partial^2 w_0}{\partial x \partial y} \Bigg\} \mathrm{d}\Omega.
\end{aligned} \tag{12.44}$$

Inserting the constitutive law (12.19) and neglecting terms of the order

$$\left(\frac{\partial w_0}{\partial x} \right)^4, \quad \left(\frac{\partial w_0}{\partial y} \right)^4, \quad \left(\frac{\partial w_0}{\partial x} \right)^2 \left(\frac{\partial w_0}{\partial y} \right)^2 \tag{12.45}$$

results after a short transformation:

$$\Pi_i = \frac{1}{2} \int_\Omega \Bigg\{ A_{11} \left[\left(\frac{\partial u_0}{\partial x} \right)^2 + \frac{\partial u_0}{\partial x} \left(\frac{\partial w_0}{\partial x} \right)^2 \right] + A_{22} \left[\left(\frac{\partial v_0}{\partial y} \right)^2 + \frac{\partial v_0}{\partial y} \left(\frac{\partial w_0}{\partial y} \right)^2 \right]$$

$$
\begin{aligned}
&+2A_{12}\left[\frac{\partial u_0}{\partial x}\frac{\partial v_0}{\partial y}+\frac{1}{2}\frac{\partial u_0}{\partial x}\left(\frac{\partial w_0}{\partial y}\right)^2+\frac{1}{2}\frac{\partial v_0}{\partial y}\left(\frac{\partial w_0}{\partial x}\right)^2\right]\\
&+A_{66}\left[\left(\frac{\partial u_0}{\partial y}\right)^2+\left(\frac{\partial v_0}{\partial x}\right)^2+2\frac{\partial u_0}{\partial y}\frac{\partial v_0}{\partial x}+2\frac{\partial u_0}{\partial y}\frac{\partial w_0}{\partial x}\frac{\partial w_0}{\partial y}+2\frac{\partial v_0}{\partial x}\frac{\partial w_0}{\partial x}\frac{\partial w_0}{\partial y}\right]\\
&+D_{11}\left(\frac{\partial^2 w_0}{\partial x^2}\right)^2+D_{22}\left(\frac{\partial^2 w_0}{\partial y^2}\right)^2+2D_{12}\frac{\partial^2 w_0}{\partial x^2}\frac{\partial^2 w_0}{\partial y^2}+4D_{66}\left(\frac{\partial^2 w_0}{\partial x\partial y}\right)^2\Bigg\}\,\mathrm{d}\Omega.
\end{aligned}
\tag{12.46}
$$

In the case of the isotropic plate, this expression changes to:

$$
\begin{aligned}
\Pi_i=\frac{A}{2}\int_\Omega&\Bigg\{\left(\frac{\partial u_0}{\partial x}\right)^2+\frac{\partial u_0}{\partial x}\left(\frac{\partial w_0}{\partial x}\right)^2+\left(\frac{\partial v_0}{\partial y}\right)^2+\frac{\partial v_0}{\partial y}\left(\frac{\partial w_0}{\partial y}\right)^2\\
&+2\nu\left[\frac{\partial u_0}{\partial x}\frac{\partial v_0}{\partial y}+\frac{1}{2}\frac{\partial u_0}{\partial x}\left(\frac{\partial w_0}{\partial y}\right)^2+\frac{1}{2}\frac{\partial v_0}{\partial y}\left(\frac{\partial w_0}{\partial x}\right)^2\right]\\
&+\frac{1-\nu}{2}\left[\left(\frac{\partial u_0}{\partial y}\right)^2+\left(\frac{\partial v_0}{\partial x}\right)^2+2\frac{\partial u_0}{\partial y}\frac{\partial v_0}{\partial x}+2\frac{\partial u_0}{\partial y}\frac{\partial w_0}{\partial x}\frac{\partial w_0}{\partial y}+2\frac{\partial v_0}{\partial x}\frac{\partial w_0}{\partial x}\frac{\partial w_0}{\partial y}\right]\Bigg\}\,\mathrm{d}\Omega\\
&+\frac{D}{2}\int_\Omega\left[\left(\frac{\partial^2 w_0}{\partial x^2}\right)^2+\left(\frac{\partial^2 w_0}{\partial y^2}\right)^2+2\nu\frac{\partial^2 w_0}{\partial x^2}\frac{\partial^2 w_0}{\partial y^2}+2(1-\nu)\left(\frac{\partial^2 w_0}{\partial x\partial y}\right)^2\right]\mathrm{d}\Omega.
\end{aligned}
\tag{12.47}
$$

The external potential is composed of parts of the surface load p, the elastic foundation (bedding number k), the edge moments $\hat{M}^0_{nn}$ and the edge transverse forces $\hat{\hat{Q}}_n$ as well as the edge membrane forces $\hat{N}^0_{nn}$, $\hat{N}^0_{ns}$ as follows:

$$
\begin{aligned}
\Pi_a=\frac{1}{2}\int_\Omega\left(kw_0^2-2pw_0\right)\mathrm{d}\Omega+\oint_{\Gamma_\sigma}\left(\hat{M}^0_{nn}\frac{\partial w_0}{\partial n}-\hat{\hat{Q}}_n w_o\right)\mathrm{d}s\\
-\oint_{\Gamma_\sigma}\left(\hat{N}^0_{nn}u_{0n}+\hat{N}^0_{ns}u_{0s}\right)\mathrm{d}s.
\end{aligned}
\tag{12.48}
$$

In some cases it may be useful to use a mixed energy formulation. Such a mixed formulation is obtained starting from the virtual internal work δW_i:

$$
\begin{aligned}
\delta W_i=&\int_\Omega\Big(N^0_{xx}\delta\varepsilon^0_{xx}+N^0_{yy}\delta\varepsilon^0_{yy}+N^0_{xy}\delta\gamma^0_{xy}\\
&+M^0_{xx}\delta\kappa^0_{xx}+M^0_{yy}\delta\kappa^0_{yy}+M^0_{xy}\delta\kappa^0_{xy}\Big)\,\mathrm{d}\Omega\\
=&\,\delta W_{i,m}+\delta W_{i,b}.
\end{aligned}
\tag{12.49}
$$

We now perform a partial integration of $\delta W_{i,m}$ and obtain:

$$
\delta W_{i,m}=\int_\Omega\left(N^0_{xx}\delta\varepsilon^0_{xx}+N^0_{yy}\delta\varepsilon^0_{yy}+N^0_{xy}\delta\gamma^0_{xy}\right)\mathrm{d}\Omega
$$

$$
\begin{aligned}
&= \delta \int_\Omega \left(N_{xx}^0 \varepsilon_{xx}^0 + N_{yy}^0 \varepsilon_{yy}^0 + N_{xy}^0 \gamma_{xy}^0 \right) \mathrm{d}\Omega \\
&\quad - \int_\Omega \left(\delta N_{xx}^0 \delta \varepsilon_{xx}^0 + \delta N_{yy}^0 \delta \varepsilon_{yy}^0 + \delta N_{xy}^0 \delta \gamma_{xy}^0 \right) \mathrm{d}\Omega \\
&= \delta W_{i,m1} - \delta W_{i,m2}.
\end{aligned} \tag{12.50}
$$

In the expression $\delta W_{i,m1}$ the kinematic relations (12.4) are now used. It follows:

$$
\begin{aligned}
\delta W_{i,m1} = \delta \int_\Omega \Bigg[& N_{xx}^0 \frac{\partial u_0}{\partial x} + N_{yy}^0 \frac{\partial v_0}{\partial y} + N_{xy}^0 \left(\frac{\partial u_0}{\partial y} + \frac{\partial v_0}{\partial x} \right) \\
& + \frac{1}{2} N_{xx}^0 \left(\frac{\partial w_0}{\partial x} \right)^2 + \frac{1}{2} N_{yy}^0 \left(\frac{\partial w_0}{\partial y} \right)^2 + N_{xy}^0 \frac{\partial w_0}{\partial x} \frac{\partial w_0}{\partial y} \Bigg] \mathrm{d}\Omega.
\end{aligned} \tag{12.51}
$$

The first line of this expression is now partially integrated:

$$
\begin{aligned}
\delta W_{i,m1} &= \delta \Bigg[\int_\Gamma N_{xx}^0 u_0 n_x \mathrm{d}s + \int_\Gamma N_{yy}^0 v_0 n_y \mathrm{d}s \\
&\quad + \int_\Gamma N_{xy}^0 u_0 n_y \mathrm{d}s + \int_\Gamma N_{xy}^0 v_0 n_x \mathrm{d}s \Bigg] \\
&\quad - \delta \int_\Omega \left[\frac{\partial N_{xx}^0}{\partial x} u_0 + \frac{\partial N_{yy}^0}{\partial y} v_0 + \frac{\partial N_{xy}^0}{\partial y} u_0 + \frac{\partial N_{xy}^0}{\partial x} v_0 \right] \mathrm{d}\Omega \\
&\quad + \delta \left[\frac{1}{2} N_{xx}^0 \left(\frac{\partial w_0}{\partial x} \right)^2 + \frac{1}{2} N_{yy}^0 \left(\frac{\partial w_0}{\partial y} \right)^2 + N_{xy}^0 \frac{\partial w_0}{\partial x} \frac{\partial w_0}{\partial y} \right] \mathrm{d}\Omega \\
&= \delta W_{i,m1}^\Gamma + \delta W_{i,m1}^\Omega.
\end{aligned} \tag{12.52}
$$

With $n_x = \cos\theta$ and $n_y = \sin\theta$ we obtain for $\delta W_{i,m1}^\Gamma$:

$$
\delta W_{i,m1}^\Gamma = \delta \int_\Gamma \left[\left(N_{xx}^0 \cos\theta + N_{xy}^0 \sin\theta \right) u_0 + \left(N_{yy}^0 \sin\theta + N_{xy}^0 \cos\theta \right) v_0 \right] \mathrm{d}s. \tag{12.53}
$$

If one adds the virtual work of the boundary forces here, then it follows:

$$
\begin{aligned}
\delta W_{i,m1}^\Gamma - \delta W_{a,m} = \delta \int_\Gamma \Big[& \left(N_{xx}^0 \cos\theta + N_{xy}^0 \sin\theta - \hat{N}_{xx}^0 \right) u_0 \\
& + \left(N_{yy}^0 \sin\theta + N_{xy}^0 \cos\theta - \hat{N}_{yy}^0 \right) v_0 \Big] \mathrm{d}s.
\end{aligned} \tag{12.54}
$$

On the edge piece on which edge membrane forces are prescribed this integral disappears. If, on the other hand, edge displacements are prescribed, then the result is:

$$\delta W^{\Gamma}_{i,m1} - \delta W_{a,m} = \delta \int_{\Gamma} \left[\left(N^0_{xx} \cos\theta + N^0_{xy} \sin\theta \right) \hat{u}_0 + \left(N^0_{yy} \sin\theta + N^0_{xy} \cos\theta \right) \hat{v}_0 \right] \mathrm{d}s, \tag{12.55}$$

or

$$\delta W^{\Gamma}_{i,m1} - \delta W_{a,m} = \delta \int_{\Gamma} \left[N^0_{nn} \hat{u}_{0n} + N^0_{ns} \hat{u}_{0s} \right] \mathrm{d}s. \tag{12.56}$$

At this point, the Airy stress function according to

$$N^0_{xx} = \frac{\partial^2 F}{\partial y^2}, \quad N^0_{yy} = \frac{\partial^2 F}{\partial x^2}, \quad N^0_{xy} = -\frac{\partial^2 F}{\partial x \partial y} \tag{12.57}$$

is introduced. Then it follows for the expression $\delta W_{i,m1} - \delta W_{a,m}$:

$$\begin{aligned} \delta W_{i,m1} - \delta W_{a,m} = \delta \int_{\Gamma} \left(\frac{\partial^2 F}{\partial s^2} \hat{u}_{0n} - \frac{\partial^2 F}{\partial n \partial s} \hat{u}_{0s} \right) \mathrm{d}s \\ + \delta \int_{\Omega} \left[\frac{1}{2} \frac{\partial^2 F}{\partial y^2} \left(\frac{\partial w_0}{\partial x} \right)^2 + \frac{1}{2} \frac{\partial^2 F}{\partial x^2} \left(\frac{\partial w_0}{\partial y} \right)^2 - \frac{\partial^2 F}{\partial x \partial y} \frac{\partial w_0}{\partial x} \frac{\partial w_0}{\partial y} \right] \mathrm{d}\Omega. \end{aligned} \tag{12.58}$$

We now also consider the expression $\delta W_{i,m2}$ and replace the strains of the mid-plane by the moment and force quantities according to (12.23). It results after a short transformation:

$$\delta W_{i,m2} = \delta \int_{\Omega} \left[\frac{1}{2} \alpha_{11} \left(\frac{\partial^2 F}{\partial y^2} \right)^2 + \alpha_{12} \frac{\partial^2 F}{\partial x^2} \frac{\partial^2 F}{\partial y^2} + \frac{1}{2} \alpha_{22} \left(\frac{\partial^2 F}{\partial x^2} \right)^2 + \frac{1}{2} \alpha_{66} \left(\frac{\partial^2 F}{\partial x \partial y} \right)^2 \right] \mathrm{d}\Omega. \tag{12.59}$$

Finally, we consider the term $\delta W_{i,b}$ which, after inserting the constitutive law (12.19), transitions to the following form:

$$\delta W_{i,b} = \delta \int_{\Omega} \left[\frac{1}{2} D_{11} \left(\frac{\partial^2 w_0}{\partial x^2} \right)^2 + D_{12} \frac{\partial^2 w_0}{\partial x^2} \frac{\partial^2 w_0}{\partial y^2} + \frac{1}{2} D_{22} \left(\frac{\partial^2 w_0}{\partial y^2} \right)^2 + 2 D_{66} \left(\frac{\partial^2 w_0}{\partial x \partial y} \right)^2 \right] \mathrm{d}\Omega. \tag{12.60}$$

From the principle of virtual displacements $\delta W_i = \delta W_a$, the mixed potential formulation $\Pi\,(w_0, F)$ can be derived, taking into account the edge loads and the load p:

$$
\begin{aligned}
\Pi = &-\int\limits_{\Omega} \left[\frac{1}{2}\alpha_{11} \left(\frac{\partial^2 F}{\partial y^2} \right)^2 + \alpha_{12} \frac{\partial^2 F}{\partial x^2} \frac{\partial^2 F}{\partial y^2} + \frac{1}{2}\alpha_{22} \left(\frac{\partial^2 F}{\partial x^2} \right)^2 + \frac{1}{2}\alpha_{66} \left(\frac{\partial^2 F}{\partial x \partial y} \right)^2 \right] \mathrm{d}\Omega \\
&+ \int\limits_{\Omega} \left[\frac{1}{2} D_{11} \left(\frac{\partial^2 w_0}{\partial x^2} \right)^2 + D_{12} \frac{\partial^2 w_0}{\partial x^2} \frac{\partial^2 w_0}{\partial y^2} + \frac{1}{2} D_{22} \left(\frac{\partial^2 w_0}{\partial y^2} \right)^2 + 2 D_{66} \left(\frac{\partial^2 w_0}{\partial x \partial y} \right)^2 \right] \mathrm{d}\Omega \\
&+ \int\limits_{\Omega} \left[\frac{1}{2} \frac{\partial^2 F}{\partial y^2} \left(\frac{\partial w_0}{\partial x} \right)^2 + \frac{1}{2} \frac{\partial^2 F}{\partial x^2} \left(\frac{\partial w_0}{\partial y} \right)^2 - \frac{\partial^2 F}{\partial x \partial y} \frac{\partial w_0}{\partial x} \frac{\partial w_0}{\partial y} \right] \mathrm{d}\Omega \\
&+ \int\limits_{\Gamma} \left(\frac{\partial^2 F}{\partial s^2} \hat{u}_{0n} - \frac{\partial^2 F}{\partial n \partial s} \hat{u}_{0s} \right) - \int\limits_{\Omega} p w_0 \mathrm{d}\Omega \\
&+ \int\limits_{\Gamma} \left(\hat{M}^0_{nn} \frac{\partial w_0}{\partial n} - \hat{\hat{Q}}_n w_0 \right) \mathrm{d}s.
\end{aligned}
\tag{12.61}
$$

12.1.5 Special Cases

From the equations derived so far, some special cases can be derived (see e.g. Altenbach et al. (2016)).

The first case is the consideration of a plane thin-walled structure in the framework of a geometrically linear theory. In this case, all nonlinear terms are omitted and the problem is described by the following form of the compatibility equation (12.28):

$$
\alpha_{11} \frac{\partial^4 F}{\partial y^4} + (2\alpha_{12} + \alpha_{66}) \frac{\partial^4 F}{\partial x^2 \partial y^2} + \alpha_{22} \frac{\partial^4 F}{\partial x^4} = 0, \tag{12.62}
$$

respectively for the isotropic disk:

$$
\Delta\Delta F = 0. \tag{12.63}
$$

The neglect of membrane forces in the condensed plate equilibrium leads to:

$$
D_{11} \frac{\partial^4 w_0}{\partial x^4} + 2\,(D_{12} + 2D_{66}) \frac{\partial^4 w_0}{\partial x^2 \partial y^2} + D_{22} \frac{\partial^4 w_0}{\partial y^4} + k w_0 = p. \tag{12.64}
$$

For the isotropic plate follows:

$$
D\Delta\Delta w_0 + k w_0 = p. \tag{12.65}
$$

The second special case is the consideration of the plate in the framework of the second-order theory, i.e. with coupled disk and plate behavior. The plate is stressed by the surface load p and by membrane forces. The assumption made here is that deflections are small and linearized strains apply accordingly. This gives the following form of the disk equation:

$$\alpha_{11}\frac{\partial^4 F}{\partial y^4} + (2\alpha_{12} + \alpha_{66})\frac{\partial^4 F}{\partial x^2 \partial y^2} + \alpha_{22}\frac{\partial^4 F}{\partial x^4} = 0, \tag{12.66}$$

as well as for the isotropic disk:

$$\Delta\Delta F = 0. \tag{12.67}$$

The plate equation, on the other hand, reads:

$$\begin{aligned} &D_{11}\frac{\partial^4 w_0}{\partial x^4} + 2\left(D_{12} + 2D_{66}\right)\frac{\partial^4 w_0}{\partial x^2 \partial y^2} + D_{22}\frac{\partial^4 w_0}{\partial y^4} \\ &= \frac{\partial^2 F}{\partial y^2}\frac{\partial^2 w_0}{\partial x^2} + \frac{\partial^2 F}{\partial x^2}\frac{\partial^2 w_0}{\partial y^2} - 2\frac{\partial^2 F}{\partial x \partial y}\frac{\partial^2 w_0}{\partial x \partial y} - k w_0 + p, \end{aligned} \tag{12.68}$$

or for an isotropic plate:

$$D\Delta\Delta w_0 = \frac{\partial^2 F}{\partial y^2}\frac{\partial^2 w_0}{\partial x^2} + \frac{\partial^2 F}{\partial x^2}\frac{\partial^2 w_0}{\partial y^2} - 2\frac{\partial^2 F}{\partial x \partial y}\frac{\partial^2 w_0}{\partial x \partial y} - k w_0 + p. \tag{12.69}$$

The third special case is formed by the membrane equations according to Föppl[1], which result from neglecting the bending stiffness. They are:

$$\begin{aligned} &\frac{1}{Eh}\Delta\Delta F = \left(\frac{\partial^2 w_0}{\partial x \partial y}\right)^2 - \frac{\partial^2 w_0}{\partial x^2}\frac{\partial^2 w_0}{\partial y^2}, \\ &-\frac{\partial^2 F}{\partial y^2}\frac{\partial^2 w_0}{\partial x^2} - \frac{\partial^2 F}{\partial x^2}\frac{\partial^2 w_0}{\partial y^2} + 2\frac{\partial^2 F}{\partial x \partial y}\frac{\partial^2 w_0}{\partial x \partial y} = p. \end{aligned} \tag{12.70}$$

The fourth and last special case discussed here is the linear buckling analysis. Here, the linear disk equation of the form

$$\alpha_{11}\frac{\partial^4 F}{\partial y^4} + (2\alpha_{12} + \alpha_{66})\frac{\partial^4 F}{\partial x^2 \partial y^2} + \alpha_{22}\frac{\partial^4 F}{\partial x^4} = 0, \tag{12.71}$$

or

$$\Delta\Delta F = 0 \tag{12.72}$$

and the plate equation in the following form with $k = 0$ and $p = 0$:

$$D_{11}\frac{\partial^4 w_0}{\partial x^4} + 2\left(D_{12} + 2D_{66}\right)\frac{\partial^4 w_0}{\partial x^2 \partial y^2} + D_{22}\frac{\partial^4 w_0}{\partial y^4}$$

[1] August Otto Föppl, 1854–1924, German engineer.

$$+N_{xx}\frac{\partial^2 w_0}{\partial x^2}+N_{yy}\frac{\partial^2 w_0}{\partial y^2}+2N_{xy}\frac{\partial^2 w_0}{\partial x\partial y}=0 \tag{12.73}$$

describe the problem at hand. In the case of the isotropic plate follows:

$$D\Delta\Delta w_0+N_{xx}\frac{\partial^2 w_0}{\partial x^2}+N_{yy}\frac{\partial^2 w_0}{\partial y^2}+2N_{xy}\frac{\partial^2 w_0}{\partial x\partial y}=0. \tag{12.74}$$

It can be seen that these equations have already been derived in Chap. 11 by equilibrium of the infinitesimally deflected plate element. It should be noted that in the context of buckling analyses, compressive forces are usually assumed to be positive.

12.2 Bending of Plates with Large Deflections

In the following, we consider the static bending of a rectangular orthotropic plate (length a, width b, thickness h) loaded by the surface load p in the z-direction. There is no elastic foundation. The governing differential equations are then:

$$\begin{aligned}
&\alpha_{11}\frac{\partial^4 F}{\partial y^4}+(2\alpha_{12}+\alpha_{66})\frac{\partial^4 F}{\partial x^2\partial y^2}+\alpha_{22}\frac{\partial^4 F}{\partial x^4}=\left(\frac{\partial^2 w_0}{\partial x\partial y}\right)^2-\frac{\partial^2 w_0}{\partial x^2}\frac{\partial^2 w_0}{\partial y^2},\\
&D_{11}\frac{\partial^4 w_0}{\partial x^4}+2\left(D_{12}+2D_{66}\right)\frac{\partial^4 w_0}{\partial x^2\partial y^2}+D_{22}\frac{\partial^4 w_0}{\partial y^4}\\
&=\frac{\partial^2 F}{\partial y^2}\frac{\partial^2 w_0}{\partial x^2}+\frac{\partial^2 F}{\partial x^2}\frac{\partial^2 w_0}{\partial y^2}-2\frac{\partial^2 F}{\partial x\partial y}\frac{\partial^2 w_0}{\partial x\partial y}+p.
\end{aligned} \tag{12.75}$$

There are several ways to solve such a static bending problem, which are briefly described below.

12.2.1 Solution by Series Expansion

A first way is to develop both the plate deflection w_0 and the Airy stress function F in series as follows:

$$\begin{aligned}
F&=\sum_{m=1}^{\infty}\sum_{n=1}^{\infty}F_{mn}X_m(x)Y_n(y),\\
w_0&=\sum_{m=1}^{\infty}\sum_{n=1}^{\infty}W_{mn}X_m(x)Y_n(y).
\end{aligned} \tag{12.76}$$

The functions $X_m(x)$ and $Y_n(y)$ are required to satisfy all underlying boundary conditions. Putting these approaches (12.76) into the two governing differential equations (12.75), we obtain the following expressions with the summation upper bounds M

and N:

$$\sum_{m=1}^{M}\sum_{n=1}^{N} F_{mn}\left[\alpha_{11} X_m \frac{d^4 Y_n}{dy^4} + (2\alpha_{12}+\alpha_{66})\frac{d^2 X_m}{dx^2}\frac{d^2 Y_n}{dy^2} + \alpha_{22}\frac{d^4 X_m}{dx^4} y_n\right]$$
$$= \sum_{m=1}^{M}\sum_{n=1}^{N}\sum_{p=1}^{M}\sum_{q=1}^{N} W_{mn} W_{pq}\left[\frac{dX_m}{dx}\frac{dX_p}{dx}\frac{dY_n}{dy}\frac{dY_q}{dy} - \frac{d^2 X_m}{dx^2} X_p Y_n \frac{d^2 Y_q}{dy^2}\right],$$
$$\sum_{m=1}^{M}\sum_{n=1}^{N} W_{mn}\left[D_{11}\frac{d^4 X_m}{dx^4} Y_n + 2(D_{12}+2D_{66})\frac{d^2 X_m}{dx^2}\frac{d^2 Y_n}{dy^2} + D_{22} X_m \frac{d^4 Y_n}{dy^4}\right]$$
$$= p + \sum_{m=1}^{M}\sum_{n=1}^{N}\sum_{p=1}^{M}\sum_{q=1}^{N} F_{mn} W_{pq}\left[X_m \frac{d^2 X_p}{dx^2}\frac{d^2 Y_n}{dy^2} Y_q + \frac{d^2 X_m}{dx^2} X_p Y_n \frac{d^2 Y_q}{dy^2} - 2\frac{dX_m}{dx}\frac{dX_p}{dx}\frac{dY_n}{dy}\frac{dY_q}{dy}\right]. \tag{12.77}$$

Multiplying these two expressions by X_i and Y_j and integrating the resulting terms over the area of the plate Ω, we obtain with the abbreviations

$$\Omega_{ri} = \int_0^a \frac{d^r X_m}{dx^r} X_i dx, \quad \Omega_{rsi} = \int_0^a \frac{d^r X_m}{dx^r}\frac{d^s X_p}{dx^s} X_i dx,$$
$$\Delta_{rj} = \int_0^b \frac{d^r Y_n}{dy^r} Y_j dy, \quad \Delta_{rsj} = \int_0^b \frac{d^r Y_n}{dy^r}\frac{d^s Y_q}{dy^s} Y_j dy,$$
$$P_{ij} = \int_0^a \int_0^b p X_i Y_j dy dx \tag{12.78}$$

the following nonlinear system of algebraic equations for the constants W_{mn} and F_{mn}:

$$\sum_{m=1}^{M}\sum_{n=1}^{N} F_{mn}\left[\alpha_{11}\Omega_{0i}\Delta_{4j} + (2\alpha_{12}+\alpha_{66})\Omega_{2i}\Delta_{2j} + \alpha_{22}\Omega_{4i}\Delta_{0j}\right]$$
$$= \sum_{m=1}^{M}\sum_{n=1}^{N}\sum_{p=1}^{M}\sum_{q=1}^{N} W_{mn} W_{pq}\left[\Omega_{11i}\Delta_{11j} - \Omega_{20i}\Delta_{02j}\right],$$
$$\sum_{m=1}^{M}\sum_{n=1}^{N} W_{mn}\left[D_{11}\Omega_{4i}\Delta_{0j} + 2(D_{12}+2D_{66})\Omega_{2i}\Delta_{2j} + D_{22}\Omega_{0i}\Delta_{4j}\right]$$
$$= P_{ij} + \sum_{m=1}^{M}\sum_{n=1}^{N}\sum_{p=1}^{M}\sum_{q=1}^{N} F_{mn} W_{pq}\left[\Omega_{02i}\Delta_{20j} + \Omega_{20i}\Delta_{02j} - 2\Omega_{11i}\Delta_{11j}\right]. \tag{12.79}$$

12.2.2 The Galerkin Method

A suitable approximation method for the solution of nonlinear plate bending problems is the Galerkin method, in which suitable shape functions for all occurring displacement components and an error minimization are used. For the displacement components, approaches of the form

$$u_0 = \sum_{m=1}^{M_u} \sum_{n=1}^{N_u} U_{mn} u_{1m}(x) u_{2n}(y),$$
$$v_0 = \sum_{m=1}^{M_v} \sum_{n=1}^{N_v} V_{mn} v_{1m}(x) v_{2n}(y),$$
$$w_0 = \sum_{m=1}^{M_w} \sum_{n=1}^{N_w} W_{mn} w_{1m}(x) w_{2n}(y) \tag{12.80}$$

are used. The functions u_{1m}, u_{2n}, v_{1m}, v_{2n}, w_{1m}, w_{2n} are required to satisfy all underlying boundary conditions of the given plate problem. From the variational formulation $\delta\Pi = 0$, we can then derive the following conditions:

$$\int_0^a \int_0^b L_1 (u_0, v_0, w_0)\, u_{1r} u_{2s} \mathrm{d}y\mathrm{d}x = 0,$$
$$\int_0^a \int_0^b L_2 (u_0, v_0, w_0)\, v_{1r} v_{2s} \mathrm{d}y\mathrm{d}x = 0,$$
$$\int_0^a \int_0^b L_3 (u_0, v_0, w_0)\, w_{1r} w_{2s} \mathrm{d}y\mathrm{d}x = 0, \tag{12.81}$$

where the operators $L_1 (u_0, v_0, w_0)$, $L_2 (u_0, v_0, w_0)$, $L_3 (u_0, v_0, w_0)$ are the left-hand sides of the differential equations (12.20) with (12.19) and r and s can take values between 1 and M respectively N.

12.2.3 The Ritz Method

The Ritz method proves to be much easier to use in contrast to the Galerkin method. Here, too, an approach of the form (12.80) is assumed, although in this case the shape functions are only required to satisfy the given geometric boundary conditions, which makes it easier to formulate suitable approaches. The total potential $\Pi = \Pi_i + \Pi_a$ then takes the form $\Pi (U_{mn}, V_{mn}, W_{mn})$, and the requirement $\delta\Pi = 0$ leads to the Ritz equations:

$$\frac{\partial \Pi}{\partial U_{mn}} = 0, \quad \frac{\partial \Pi}{\partial V_{mn}} = 0, \quad \frac{\partial \Pi}{\partial W_{mn}} = 0. \tag{12.82}$$

This ultimately leads to a nonlinear system of equations from which the constants U_{mn}, V_{mn}, W_{mn} can be determined numerically-iteratively.

The Ritz method is illustrated below using the example of a rectangular plate (length a, width b) under constant surface load $p = p_0$. A simple approach of the form

$$u_0 = U u_1(x) u_2(y), \quad v_0 = V v_1(x) v_2(y), \quad w_0 = W w_1(x) w_2(y) \tag{12.83}$$

is chosen. The total elastic potential (12.46) with (12.48) then takes the following form:

$$\begin{aligned}
\Pi = \frac{1}{2} \Bigg[& A_{11} U^2 \int_0^a \frac{du_1}{dx}\frac{du_1}{dx} dx \int_0^b u_2 u_2 dy + A_{11} U W^2 \int_0^a \frac{du_1}{dx}\frac{dw_1}{dx}\frac{dw_1}{dx} dx \int_0^b u_2 w_2 w_2 dy \\
& + A_{22} V^2 \int_0^a v_1 v_1 dx \int_0^b \frac{dv_2}{dy}\frac{dv_2}{dy} dy + A_{22} V W^2 \int_0^a v_1 w_1 w_1 dx \int_0^b \frac{dv_2}{dy}\frac{dw_2}{dy}\frac{dw_2}{dy} dy \\
& + 2A_{12} U V \int_0^a \frac{du_1}{dx} v_1 dx \int_0^b u_2 \frac{dv_2}{dy} dy + A_{12} U W^2 \int_0^a \frac{du_1}{dx} w_1 w_1 dx \int_0^b u_2 \frac{dw_2}{dy}\frac{dw_2}{dy} dy \\
& + A_{12} V W^2 \int_0^a v_1 \frac{dw_1}{dx}\frac{dw_1}{dx} dx \int_0^b \frac{dv_2}{dy} w_2 w_2 dy + A_{66} U^2 \int_0^a u_1 u_1 dx \int_0^b \frac{du_2}{dy}\frac{du_2}{dy} dy \\
& + A_{66} V^2 \int_0^a \frac{dv_1}{dx}\frac{dv_1}{dx} dx \int_0^b v_2 v_2 dy + 2A_{66} U V \int_0^a u_1 \frac{dv_1}{dx} dx \int_0^b \frac{du_2}{dy} v_2 dy \\
& + 2A_{66} U W^2 \int_0^a u_1 \frac{dw_1}{dx} w_1 dx \int_0^b \frac{du_2}{dy} w_2 \frac{dw_2}{dy} dy \\
& + 2A_{66} V W^2 \int_0^a \frac{dv_1}{dx}\frac{dw_1}{dx} w_1 dx \int_0^b v_2 w_2 \frac{dw_2}{dy} dy \\
& + D_{11} W^2 \int_0^a \frac{d^2 w_1}{dx^2}\frac{d^2 w_1}{dx^2} dx \int_0^b w_2 w_2 dy + D_{22} W^2 \int_0^a w_1 w_1 dx \int_0^b \frac{d^2 w_2}{dy^2}\frac{d^2 w_2}{dy^2} dy \\
& + 2D_{12} W^2 \int_0^a \frac{d^2 w_1}{dx^2} w_1 dx \int_0^b w_2 \frac{d^2 w_2}{dy^2} dy + 4D_{66} W^2 \int_0^a \frac{dw_1}{dx}\frac{dw_1}{dx} dx \int_0^b \frac{dw_2}{dy}\frac{dw_2}{dy} dy \Bigg] \\
& - p_0 W \int_0^a w_1 dx \int_0^b w_2 dy.
\end{aligned} \tag{12.84}$$

With the abbreviations

$$I_{uu}^{ij} = \int_0^a \frac{d^i u_1}{dx^i}\frac{d^j u_1}{dx^j} dx, \quad J_{uu}^{ij} = \int_0^b \frac{d^i u_2}{dy^i}\frac{d^j u_2}{dy^j} dy,$$

$$I_{vv}^{ij} = \int_0^a \frac{\mathrm{d}^i v_1}{\mathrm{d}x^i}\frac{\mathrm{d}^j v_1}{\mathrm{d}x^j}\mathrm{d}x, \quad J_{vv}^{ij} = \int_0^b \frac{\mathrm{d}^i v_2}{\mathrm{d}y^i}\frac{\mathrm{d}^j v_2}{\mathrm{d}y^j}\mathrm{d}y,$$

$$I_{ww}^{ij} = \int_0^a \frac{\mathrm{d}^i w_1}{\mathrm{d}x^i}\frac{\mathrm{d}^j w_1}{\mathrm{d}x^j}\mathrm{d}x, \quad J_{uu}^{ij} = \int_0^b \frac{\mathrm{d}^i w_2}{\mathrm{d}y^i}\frac{\mathrm{d}^j w_2}{\mathrm{d}y^j}\mathrm{d}y,$$

$$I_{uv}^{ij} = \int_0^a \frac{\mathrm{d}^i u_1}{\mathrm{d}x^i}\frac{\mathrm{d}^j v_1}{\mathrm{d}x^j}\mathrm{d}x, \quad J_{uv}^{ij} = \int_0^b \frac{\mathrm{d}^i u_2}{\mathrm{d}y^i}\frac{\mathrm{d}^j v_2}{\mathrm{d}y^j}\mathrm{d}y,$$

$$K_{uww}^{ijk} = \int_0^a \frac{\mathrm{d}^i u_1}{\mathrm{d}x^i}\frac{\mathrm{d}^j w_1}{\mathrm{d}x^j}\frac{\mathrm{d}^k w_1}{\mathrm{d}x^k}\mathrm{d}x, \quad L_{uww}^{ijk} = \int_0^a \frac{\mathrm{d}^i u_2}{\mathrm{d}y^i}\frac{\mathrm{d}^j w_2}{\mathrm{d}y^j}\frac{\mathrm{d}^k w_2}{\mathrm{d}y^k}\mathrm{d}y,$$

$$K_{vww}^{ijk} = \int_0^a \frac{\mathrm{d}^i v_1}{\mathrm{d}x^i}\frac{\mathrm{d}^j w_1}{\mathrm{d}x^j}\frac{\mathrm{d}^k w_1}{\mathrm{d}x^k}\mathrm{d}x, \quad L_{vww}^{ijk} = \int_0^a \frac{\mathrm{d}^i v_2}{\mathrm{d}y^i}\frac{\mathrm{d}^j w_2}{\mathrm{d}y^j}\frac{\mathrm{d}^k w_2}{\mathrm{d}y^k}\mathrm{d}y,$$

$$P_1 = \int_0^a w_1 \mathrm{d}x, \quad P_2 = \int_0^b w_2 \mathrm{d}y \tag{12.85}$$

and

$$\begin{aligned}
\Omega_1 &= \frac{1}{2}A_{11}I_{uu}^{11}J_{uu}^{00} + \frac{1}{2}A_{66}I_{uu}^{00}J_{uu}^{11},\\
\Omega_2 &= \frac{1}{2}A_{22}I_{vv}^{00}J_{vv}^{11} + \frac{1}{2}A_{66}I_{vv}^{11}J_{vv}^{00},\\
\Omega_3 &= \frac{1}{2}D_{11}I_{ww}^{22}J_{ww}^{00} + \frac{1}{2}D_{22}I_{ww}^{00}J_{ww}^{22} + D_{12}I_{ww}^{20}J_{ww}^{20} + 2D_{66}I_{ww}^{11}J_{ww}^{11},\\
\Lambda_1 &= A_{12}I_{uv}^{10}J_{uv}^{01} + A_{66}I_{uv}^{01}J_{uv}^{10},\\
\Lambda_2 &= \frac{1}{2}A_{11}K_{uww}^{111}L_{uww}^{000} + \frac{1}{2}A_{12}K_{uww}^{100}L_{uww}^{011} + A_{66}K_{uww}^{010}L_{uww}^{110},\\
\Lambda_3 &= \frac{1}{2}A_{22}K_{vww}^{000}L_{vww}^{111} + \frac{1}{2}A_{12}K_{vww}^{011}L_{vww}^{100} + A_{66}K_{vww}^{110}L_{vww}^{010}
\end{aligned} \tag{12.86}$$

the following expression for the total elastic potential $\Pi = \Pi_i + \Pi_a = \Pi(U, V, W)$ is obtained:

$$\Pi = \Omega_1 U^2 + \Omega_2 V^2 + \Omega_3 W^2 + \Lambda_1 UV + \Lambda_2 UW^2 + \Lambda_3 VW^2 - p_0 W P_1 P_2. \tag{12.87}$$

Evaluating the Ritz equations

$$\frac{\partial \Pi}{\partial U} = 0, \quad \frac{\partial \Pi}{\partial V} = 0, \quad \frac{\partial \Pi}{\partial W} = 0 \tag{12.88}$$

then leads to the following nonlinear equation system for the determination of the constants U, V, W:

$$\frac{\partial\Pi}{\partial U} = 2\Omega_1 U + \Lambda_1 V + \Lambda_2 W^2 = 0,$$
$$\frac{\partial\Pi}{\partial V} = 2\Omega_2 V + \Lambda_1 U + \Lambda_3 W^2 = 0,$$
$$\frac{\partial\Pi}{\partial W} = 2\Omega_3 W + 2\Lambda_2 U W + 2\Lambda_3 V W - p_0 P_1 P_2 = 0. \tag{12.89}$$

In many cases, one will apply the Ritz method using series expansions. Let the following approach be given:

$$u_0 = \sum_{m=1}^{M_u}\sum_{n=1}^{N_u} U_{mn} u_{1m}(x) u_{2n}(y),$$
$$v_0 = \sum_{m=1}^{M_v}\sum_{n=1}^{N_v} V_{mn} v_{1m}(x) v_{2n}(y),$$
$$w_0 = \sum_{m=1}^{M_w}\sum_{n=1}^{N_w} W_{mn} w_{1m}(x) w_{2n}(y). \tag{12.90}$$

Substituting this approach into the total elastic potential (12.46) with (12.48), we obtain $p = p_0$ for the special case of a constant surface load:

$$\Pi = \sum_{m=1}^{M_u}\sum_{n=1}^{N_u}\sum_{p=1}^{M_u}\sum_{q=1}^{N_u}\left(\frac{1}{2}A_{11}U_{mn}U_{pq}\int_0^a \frac{du_{1m}}{dx}\frac{du_{1p}}{dx}dx\int_0^b u_{2n}u_{2q}dy\right.$$
$$\left. + \frac{1}{2}A_{66}U_{mn}U_{pq}\int_0^a u_{1m}u_{1p}dx\int_0^b \frac{du_{2n}}{dy}\frac{du_{2q}}{dy}dy\right)$$
$$+\sum_{m=1}^{M_v}\sum_{n=1}^{N_v}\sum_{p=1}^{M_v}\sum_{q=1}^{N_v}\left(\frac{1}{2}A_{22}V_{mn}V_{pq}\int_0^a v_{1m}v_{1p}dx\int_0^b \frac{dv_{2n}}{dy}\frac{dv_{2q}}{dy}dy\right.$$
$$+\frac{1}{2}A_{66}V_{mn}V_{pq}\int_0^a \frac{dv_{1m}}{dx}\frac{v_{1p}}{dx}dx\int_0^b v_{2n}v_{2q}dy$$
$$+\sum_{m=1}^{M_u}\sum_{n=1}^{N_u}\sum_{p=1}^{M_v}\sum_{q=1}^{N_v}\left(A_{12}U_{mn}V_{pq}\int_0^a \frac{du_{1m}}{dx}v_{1p}dx\int_0^b u_{2n}\frac{dv_{2q}}{dy}dy\right.$$
$$\left. + A_{66}U_{mn}V_{pq}\int_0^a u_{1m}\frac{dv_{1p}}{dx}dx\int_0^b \frac{du_{2n}}{dy}v_{2q}dy\right)$$
$$+\frac{1}{2}\sum_{m=1}^{M_w}\sum_{n=1}^{N_w}\sum_{p=1}^{M_w}\sum_{q=1}^{N_w} W_{mn}W_{pq}\left(D_{11}\int_0^a \frac{d^2w_{1m}}{dx^2}\frac{d^2w_{1p}}{dx^2}dx\int_0^b w_{2n}w_{2q}dy\right.$$
$$+D_{22}\int_0^a w_{1m}w_{1p}dx\int_0^b \frac{d^2w_{2n}}{dy^2}\frac{d^2w_{2q}}{dy^2}dy + 2D_{12}\int_0^a \frac{d^2w_{1m}}{dx^2}w_{1p}dx\int_0^b w_{2n}\frac{d^2w_{2q}}{dy^2}dy$$

$$
+4D_{66}\int_0^a \frac{\mathrm{d}w_{1m}}{\mathrm{d}x}\frac{\mathrm{d}w_{1p}}{\mathrm{d}x}\mathrm{d}x\int_0^b \frac{\mathrm{d}w_{2n}}{\mathrm{d}y}\frac{\mathrm{d}w_{2q}}{\mathrm{d}y}\mathrm{d}y
$$

$$
+\sum_{m=1}^{M_u}\sum_{n=1}^{N_u}\sum_{p=1}^{M_w}\sum_{q=1}^{N_w}\sum_{r=1}^{M_w}\sum_{s=1}^{N_w} U_{mn}W_{pq}W_{rs}\left(\frac{1}{2}A_{11}\int_0^a \frac{\mathrm{d}u_{1m}}{\mathrm{d}x}\frac{\mathrm{d}w_{1p}}{\mathrm{d}x}\frac{\mathrm{d}w_{1r}}{\mathrm{d}x}\mathrm{d}x\int_0^b u_{2n}w_{2q}w_{2s}\mathrm{d}y\right.
$$

$$
+\frac{1}{2}A_{12}\int_0^a \frac{\mathrm{d}u_{1m}}{\mathrm{d}x}w_{1p}w_{1r}\mathrm{d}x\int_0^b u_{2n}\frac{\mathrm{d}w_{2q}}{\mathrm{d}y}\frac{\mathrm{d}w_{1s}}{\mathrm{d}y}\mathrm{d}y
$$

$$
\left.+A_{66}\int_0^a u_{1m}\frac{\mathrm{d}w_{1p}}{\mathrm{d}x}w_{1r}\mathrm{d}x\int_0^b \frac{\mathrm{d}u_{2n}}{\mathrm{d}y}w_{2q}\frac{\mathrm{d}w_{2s}}{\mathrm{d}y}\mathrm{d}y\right)
$$

$$
+\sum_{m=1}^{M_v}\sum_{n=1}^{N_v}\sum_{p=1}^{M_w}\sum_{q=1}^{N_w}\sum_{r=1}^{M_w}\sum_{s=1}^{N_w} V_{mn}W_{pq}W_{rs}\left(\frac{1}{2}A_{22}\int_0^a v_{1m}w_{1p}w_{1r}\mathrm{d}x\int_0^b \frac{\mathrm{d}v_{2n}}{\mathrm{d}y}\frac{\mathrm{d}w_{2q}}{\mathrm{d}y}\frac{\mathrm{d}w_{2s}}{\mathrm{d}y}\mathrm{d}y\right.
$$

$$
+\frac{1}{2}A_{12}\int_0^a v_{1m}\frac{\mathrm{d}w_{1p}}{\mathrm{d}x}\frac{\mathrm{d}w_{1r}}{\mathrm{d}x}\mathrm{d}x\int_0^b \frac{\mathrm{d}v_{2n}}{\mathrm{d}y}w_{2q}w_{2s}\mathrm{d}y
$$

$$
\left.+A_{66}\int_0^a \frac{\mathrm{d}v_{1m}}{\mathrm{d}x}\frac{\mathrm{d}w_{1p}}{\mathrm{d}x}w_{1r}\mathrm{d}x\int_0^b v_{2n}w_{2q}\frac{\mathrm{d}w_{2s}}{\mathrm{d}y}\mathrm{d}y\right)
$$

$$
-p_0\sum_{m=1}^{M_w}\sum_{n=1}^{N_w} W_{mn}\int_0^a w_{1m}\mathrm{d}x\int_0^b w_{2n}\mathrm{d}y. \tag{12.91}
$$

Again, it is convenient to define the following resultants of the approach functions:

$$
I_{uu}^{ij,mp}=\int_0^a \frac{\mathrm{d}^i u_{1m}}{\mathrm{d}x^i}\frac{\mathrm{d}^j u_{1p}}{\mathrm{d}x^j}\mathrm{d}x,\quad J_{uu}^{ij,nq}=\int_0^b \frac{\mathrm{d}^i u_{2n}}{\mathrm{d}y^i}\frac{\mathrm{d}^j u_{2q}}{\mathrm{d}y^j}\mathrm{d}y,
$$

$$
I_{vv}^{ij,mp}=\int_0^a \frac{\mathrm{d}^i v_{1m}}{\mathrm{d}x^i}\frac{\mathrm{d}^j v_{1p}}{\mathrm{d}x^j}\mathrm{d}x,\quad J_{vv}^{ij,nq}=\int_0^b \frac{\mathrm{d}^i v_{2n}}{\mathrm{d}y^i}\frac{\mathrm{d}^j v_{2q}}{\mathrm{d}y^j}\mathrm{d}y,
$$

$$
I_{ww}^{ij,mp}=\int_0^a \frac{\mathrm{d}^i w_{1m}}{\mathrm{d}x^i}\frac{\mathrm{d}^j w_{1p}}{\mathrm{d}x^j}\mathrm{d}x,\quad J_{ww}^{ij,nq}=\int_0^b \frac{\mathrm{d}^i w_{2n}}{\mathrm{d}y^i}\frac{\mathrm{d}^j w_{2q}}{\mathrm{d}y^j}\mathrm{d}y,
$$

$$
I_{uv}^{ij,mp}=\int_0^a \frac{\mathrm{d}^i u_{1m}}{\mathrm{d}x^i}\frac{\mathrm{d}^j v_{1p}}{\mathrm{d}x^j}\mathrm{d}x,\quad J_{uv}^{ij,nq}=\int_0^b \frac{\mathrm{d}^i u_{2n}}{\mathrm{d}y^i}\frac{\mathrm{d}^j v_{2q}}{\mathrm{d}y^j}\mathrm{d}y,
$$

$$
K_{uww}^{ijk,mpr}=\int_0^a \frac{\mathrm{d}^i u_{1m}}{\mathrm{d}x^i}\frac{\mathrm{d}^j w_{1p}}{\mathrm{d}x^j}\frac{\mathrm{d}^k w_{1r}}{\mathrm{d}x^k}\mathrm{d}x,\quad L_{uww}^{ijk,nqs}=\int_0^b \frac{\mathrm{d}^i u_{2n}}{\mathrm{d}y^i}\frac{\mathrm{d}^j w_{2q}}{\mathrm{d}y^j}\frac{\mathrm{d}^k w_{2s}}{\mathrm{d}y^k}\mathrm{d}y,
$$

$$
K_{vww}^{ijk,mpr}=\int_0^a \frac{\mathrm{d}^i v_{1m}}{\mathrm{d}x^i}\frac{\mathrm{d}^j w_{1p}}{\mathrm{d}x^j}\frac{\mathrm{d}^k w_{1r}}{\mathrm{d}x^k}\mathrm{d}x,\quad L_{vww}^{ijk,nqs}=\int_0^b \frac{\mathrm{d}^i v_{2n}}{\mathrm{d}y^i}\frac{\mathrm{d}^j w_{2q}}{\mathrm{d}y^j}\frac{\mathrm{d}^k w_{2s}}{\mathrm{d}y^k}\mathrm{d}y,
$$

$$P_{1m} = \int\limits_0^a w_{1m} \mathrm{d}x, \quad P_{2n} = \int\limits_0^b w_{2n} \mathrm{d}y. \tag{12.92}$$

With the abbreviations

$$\begin{aligned}
\Omega_1^{mnpq} &= A_{11} I_{uu}^{11,mp} J_{uu}^{00,nq} + A_{66} I_{uu}^{00,mp} J_{uu}^{11,nq}, \\
\Omega_2^{mnpq} &= A_{22} I_{vv}^{00,mp} J_{vv}^{11,nq} + A_{66} I_{vv}^{11,mp} J_{vv}^{00,nq}, \\
\Omega_3^{mnpq} &= A_{12} I_{uv}^{10,mp} J_{uv}^{01,nq} + A_{66} I_{uv}^{01,mp} J_{uv}^{10,nq}, \\
\Omega_4^{mnpq} &= D_{11} I_{ww}^{22,mp} J_{ww}^{00,nq} + D_{22} I_{ww}^{00,mp} J_{ww}^{22,nq} \\
&\quad +2D_{12} I_{ww}^{20,mp} J_{ww}^{02,nq} + 4D_{66} I_{ww}^{11,mp} J_{ww}^{11,nq}, \\
\Omega_5^{mnpqrs} &= \frac{1}{2} A_{11} K_{uww}^{111,mpr} L_{uww}^{000,nqs} + \frac{1}{2} A_{12} K_{uww}^{100,mpr} L_{uww}^{011,nqs} \\
&\quad +A_{66} K_{uww}^{010,mpr} L_{uww}^{101,nqs}, \\
\Omega_6^{mnpqrs} &= \frac{1}{2} A_{22} K_{vww}^{000,mpr} L_{vww}^{111,nqs} + \frac{1}{2} A_{12} K_{vww}^{011,mpr} L_{vww}^{100,nqs} \\
&\quad +A_{66} K_{vww}^{110,mpr} L_{vww}^{001,nqs}
\end{aligned} \tag{12.93}$$

the total elastic potential Π (12.91) can be written as:

$$\begin{aligned}
\Pi = &\frac{1}{2} \sum_{m=1}^{M_u} \sum_{n=1}^{N_u} \sum_{p=1}^{M_u} \sum_{q=1}^{N_u} U_{mn} U_{pq} \Omega_1^{mnpq} + \frac{1}{2} \sum_{m=1}^{M_v} \sum_{n=1}^{N_v} \sum_{p=1}^{M_v} \sum_{q=1}^{N_v} V_{mn} V_{pq} \Omega_2^{mnpq} \\
&+ \sum_{m=1}^{M_u} \sum_{n=1}^{N_u} \sum_{p=1}^{M_v} \sum_{q=1}^{N_v} U_{mn} V_{pq} \Omega_3^{mnpq} + \frac{1}{2} \sum_{m=1}^{M_w} \sum_{n=1}^{N_w} \sum_{p=1}^{M_w} \sum_{q=1}^{N_w} W_{mn} W_{pq} \Omega_4^{mnpq} \\
&+ \sum_{m=1}^{M_u} \sum_{n=1}^{N_u} \sum_{p=1}^{M_w} \sum_{q=1}^{N_w} \sum_{r=1}^{M_w} \sum_{s=1}^{N_w} U_{mn} W_{pq} W_{rs} \Omega_5^{mnpqrs} \\
&+ \sum_{m=1}^{M_v} \sum_{n=1}^{N_v} \sum_{p=1}^{M_w} \sum_{q=1}^{N_w} \sum_{r=1}^{M_w} \sum_{s=1}^{N_w} V_{mn} W_{pq} W_{rs} \Omega_6^{mnpqrs} \\
&- p_0 \sum_{m=1}^{M_w} \sum_{n=1}^{N_w} W_{mn} P_{1m} P_{2n}.
\end{aligned} \tag{12.94}$$

Evaluating the Ritz equations

$$\frac{\partial \Pi}{\partial U_{mn}} = 0, \quad \frac{\partial \Pi}{\partial V_{mn}} = 0, \quad \frac{\partial \Pi}{\partial W_{mn}} = 0 \tag{12.95}$$

then leads to the following nonlinear system of equations for finding the constants U_{mn}, V_{mn}, W_{mn}:

$$\frac{\partial \Pi}{\partial U_{mn}} = \frac{1}{2} \sum_{p=1}^{M_u} \sum_{q=1}^{N_u} U_{pq} \left(\Omega_1^{mnpq} + \Omega_1^{pqmn} \right) + \sum_{p=1}^{M_v} \sum_{q=1}^{N_v} V_{pq} \Omega_3^{mnpq}$$

$$+\sum_{p=1}^{M_w}\sum_{q=1}^{N_w}\sum_{r=1}^{M_w}\sum_{s=1}^{N_w} W_{pq} W_{rs} \Omega_5^{mnpqrs} = 0 \quad (m = 1, 2, ..., M_u; n = 1, 2, ..., N_u)$$

$$\frac{\partial \Pi}{\partial V_{mn}} = \frac{1}{2}\sum_{p=1}^{M_v}\sum_{q=1}^{N_v} V_{pq} \left(\Omega_2^{mnpq} + \Omega_2^{pqmn}\right) + \sum_{p=1}^{M_u}\sum_{q=1}^{N_u} U_{pq} \Omega_3^{pqmn}$$
$$+\sum_{p=1}^{M_w}\sum_{q=1}^{N_w}\sum_{r=1}^{M_w}\sum_{s=1}^{N_w} W_{pq} W_{rs} \Omega_6^{mnpqrs} = 0 \quad (m = 1, 2, ..., M_v; n = 1, 2, ..., N_v)$$

$$\frac{\partial \Pi}{\partial W_{mn}} = \frac{1}{2}\sum_{p=1}^{M_w}\sum_{q=1}^{N_w} W_{pq} \left(\Omega_4^{mnpq} + \Omega_4^{pqmn}\right) + \sum_{p=1}^{M_u}\sum_{q=1}^{N_u}\sum_{r=1}^{M_w}\sum_{s=1}^{N_w} U_{pq} W_{rs} \left(\Omega_5^{pqrsmn} + \Omega_5^{pqmnrs}\right)$$
$$+\sum_{p=1}^{M_v}\sum_{q=1}^{N_v}\sum_{r=1}^{M_w}\sum_{s=1}^{N_w} V_{pq} W_{rs} \left(\Omega_6^{pqrsmn} + \Omega_6^{pqmnrs}\right) - p_0 P_{1m} P_{2n} = 0$$
$$(m = 1, 2, ..., M_w; n = 1, 2, ..., N_w) \qquad (12.96)$$

Alternatively, the Ritz method can be formulated in a mixed form by using approaches for the plate deflection w_0 and for the Airy stress function F:

$$w_0 = \sum_{m=1}^{M_w}\sum_{n=1}^{N_w} W_{mn} w_{1m}(x) w_{2n}(y),$$
$$F = \sum_{m=1}^{M_F}\sum_{n=1}^{N_F} F_{mn} f_{1m}(x) f_{2n}(y). \qquad (12.97)$$

Here, the mixed potential formulation (12.61) is used.

12.3 First-Order Shear Deformation Theory

In this section we want to derive the governing equations in the framework of First-Order Shear Deformation Theory for geometrically nonlinear analysis (see also Reddy (2003, 2006)). We assume the displacement field of the following form (see Chap. 10, Eq. (10.1)):

$$u(x, y, z) = u_0(x, y) + z\psi_x(x, y),$$
$$v(x, y, z) = v_0(x, y) + z\psi_y(x, y),$$
$$w(x, y) = w_0(x, y),$$

where we have also added the two plane displacements u_0 and v_0 of the plate midplane at this point.

The strains in the framework of v.-Kármán plate theory are obtained as:

$$\begin{aligned}
\varepsilon_{xx} &= \frac{\partial u_0}{\partial x} + \frac{1}{2}\left(\frac{\partial w_0}{\partial x}\right)^2 + z\frac{\partial \psi_x}{\partial x},\\
\varepsilon_{yy} &= \frac{\partial v_0}{\partial y} + \frac{1}{2}\left(\frac{\partial w_0}{\partial y}\right)^2 + z\frac{\partial \psi_y}{\partial y},\\
\varepsilon_{zz} &= 0,\\
\gamma_{xy} &= \frac{\partial u_0}{\partial y} + \frac{\partial v_0}{\partial x} + \frac{\partial w_0}{\partial x}\frac{\partial w_0}{\partial y} + z\left(\frac{\partial \psi_x}{\partial y} + \frac{\partial \psi_y}{\partial x}\right),\\
\gamma_{xz} &= \frac{\partial w_0}{\partial x} + \psi_x,\\
\gamma_{yz} &= \frac{\partial w_0}{\partial y} + \psi_y.
\end{aligned} \tag{12.98}$$

The following vector notation proves to be useful:

$$\underline{\varepsilon} = \underline{\varepsilon}^0 + z\underline{\varepsilon}^1 = \begin{pmatrix} \varepsilon_{xx}^0 \\ \varepsilon_{yy}^0 \\ \gamma_{yz}^0 \\ \gamma_{xz}^0 \\ \gamma_{xy}^0 \end{pmatrix} + z\begin{pmatrix} \varepsilon_{xx}^1 \\ \varepsilon_{yy}^1 \\ 0 \\ 0 \\ \gamma_{xy}^1 \end{pmatrix}, \tag{12.99}$$

wherein:

$$\begin{aligned}
\varepsilon_{xx}^0 &= \frac{\partial u_0}{\partial x} + \frac{1}{2}\left(\frac{\partial w_0}{\partial x}\right)^2,\\
\varepsilon_{yy}^0 &= \frac{\partial v_0}{\partial y} + \frac{1}{2}\left(\frac{\partial w_0}{\partial y}\right)^2,\\
\gamma_{yz}^0 &= \frac{\partial w_0}{\partial y} + \psi_y,\\
\gamma_{xz}^0 &= \frac{\partial w_0}{\partial x} + \psi_x,\\
\gamma_{xy}^0 &= \frac{\partial u_0}{\partial y} + \frac{\partial v_0}{\partial x} + \frac{\partial w_0}{\partial x}\frac{\partial w_0}{\partial y},\\
\varepsilon_{xx}^1 &= \frac{\partial \psi_x}{\partial x}, \quad \varepsilon_{yy}^1 = \frac{\partial \psi_y}{\partial y}, \quad \gamma_{xy}^1 = \frac{\partial \psi_x}{\partial y} + \frac{\partial \psi_y}{\partial x}.
\end{aligned} \tag{12.100}$$

The principle of virtual displacements $\delta W_i = \delta W_a$ can then be written in the following form. For the virtual inner work δW_i follows:

$$\delta W_i = \int_{\Omega}\int_{-\frac{h}{2}}^{+\frac{h}{2}} \left[\sigma_{xx}\left(\delta\varepsilon_{xx}^0 + z\delta\varepsilon_{xx}^1\right) + \sigma_{yy}\left(\delta\varepsilon_{yy}^0 + z\delta\varepsilon_{yy}^1\right)\right.$$

$$+ \tau_{xy}\left(\delta\gamma_{xy}^{0} + z\delta\gamma_{xy}^{1}\right) + \tau_{xz}\delta\gamma_{xz}^{0} + \tau_{yz}\delta\gamma_{yz}^{0}\Big] \mathrm{d}z\mathrm{d}\Omega$$

$$= \int_{\Omega}\Big[N_{xx}^{0}\delta\varepsilon_{xx}^{0} + M_{xx}^{0}\delta\varepsilon_{xx}^{1} + N_{yy}^{0}\delta\varepsilon_{yy}^{0} + M_{yy}^{0}\delta\varepsilon_{yy}^{1} + N_{xy}^{0}\delta\gamma_{xy}^{0} + M_{xy}^{0}\delta\gamma_{xy}^{1} +$$

$$+ Q_x\delta\gamma_{xz}^{0} + Q_y\delta\gamma_{yz}^{0}\Big] \mathrm{d}\Omega. \tag{12.101}$$

The virtual external work follows as:

$$\delta W_a = \int_{\Gamma_\sigma}\int_{-\frac{h}{2}}^{+\frac{h}{2}}\left[\hat{\sigma}_{nn}\left(\delta u_n + z\delta\psi_n\right) + \hat{\tau}_{ns}\left(\delta u_s + z\delta\psi_s\right) + \hat{\tau}_{nz}\delta w_0\right]\mathrm{d}z\mathrm{d}s$$

$$+ \int_{\Omega}\left(p - kw_0\right)\delta w_0 \mathrm{d}\Omega$$

$$= \int_{\Gamma_\sigma}\left[\hat{N}_{nn}^{0}\delta u_{0n} + \hat{M}_{nn}^{0}\delta\psi_n + \hat{N}_{ns}^{0}\delta u_{0s} + \hat{M}_{ns}^{0}\delta\psi_s + \hat{Q}_n\delta w_0\right]\mathrm{d}s$$

$$+ \int_{\Omega}\left(p - kw_0\right)\delta w_0 \mathrm{d}\Omega. \tag{12.102}$$

Application of the principle of virtual displacements $\delta W_i = \delta W_a$ leads to the following expression after inserting the virtual strains and performing partial integrations (see Reddy (2006)):

$$-\int_{\Omega}\Bigg\{\left(\frac{\partial N_{xx}^{0}}{\partial x} + \frac{\partial N_{xy}^{0}}{\partial y}\right)\delta u_0 + \left(\frac{\partial N_{xy}^{0}}{\partial x} + \frac{\partial N_{yy}^{0}}{\partial y}\right)\delta v_0$$

$$+ \left(\frac{\partial M_{xx}^{0}}{\partial x} + \frac{\partial M_{xy}^{0}}{\partial y} - Q_x\right)\delta\psi_x + \left(\frac{\partial M_{xy}^{0}}{\partial x} + \frac{\partial M_{yy}^{0}}{\partial y} - Q_y\right)\delta\psi_y$$

$$+ \left[\frac{\partial Q_x}{\partial x} + \frac{\partial Q_y}{\partial y} - kw_0 + p + \frac{\partial}{\partial x}\left(N_{xx}^{0}\frac{\partial w_0}{\partial x} + N_{xy}^{0}\frac{\partial w_0}{\partial y}\right)\right.$$

$$\left. + \frac{\partial}{\partial y}\left(N_{xy}^{0}\frac{\partial w_0}{\partial x} + N_{yy}^{0}\frac{\partial w_0}{\partial y}\right)\right]\delta w_0\Bigg\}\mathrm{d}\Omega$$

$$+ \int_{\Gamma}\Big[\left(N_{nn}^{0} - \hat{N}_{nn}^{0}\right)\delta u_{0n} + \left(N_{ns}^{0} - \hat{N}_{ns}^{0}\right)\delta u_{0s} + \left(Q_n - \hat{Q}_n\right)\delta w_0$$

$$+ \left(M_{nn}^{0} - \hat{M}_{nn}^{0}\right)\delta\psi_n + \left(M_{ns}^{0} - \hat{M}_{ns}^{0}\right)\delta\psi_s\Big]\mathrm{d}s = 0. \tag{12.103}$$

From this, the governing differential equations can be read as:

$$
\begin{aligned}
\frac{\partial N_{xx}^0}{\partial x} + \frac{\partial N_{xy}^0}{\partial y} &= 0, \\
\frac{\partial N_{xy}^0}{\partial x} + \frac{\partial N_{yy}^0}{\partial y} &= 0, \\
\frac{\partial M_{xx}^0}{\partial x} + \frac{\partial M_{xy}^0}{\partial y} - Q_x &= 0, \\
\frac{\partial M_{xy}^0}{\partial x} + \frac{\partial M_{yy}^0}{\partial y} - Q_y &= 0, \\
\frac{\partial Q_x}{\partial x} + \frac{\partial Q_y}{\partial y} - k w_0 + p + \frac{\partial}{\partial x}\left(N_{xx}^0 \frac{\partial w_0}{\partial x} + N_{xy}^0 \frac{\partial w_0}{\partial y}\right) + \frac{\partial}{\partial y}\left(N_{xy}^0 \frac{\partial w_0}{\partial x} + N_{yy}^0 \frac{\partial w_0}{\partial y}\right) &= 0.
\end{aligned}
\tag{12.104}
$$

Inserting the constitutive relations to be applied here yields:

$$
\begin{aligned}
& A_{11}\left(\frac{\partial^2 u_0}{\partial x^2} + \frac{\partial w_0}{\partial x}\frac{\partial^2 w_0}{\partial x^2}\right) + A_{12}\left(\frac{\partial^2 v_0}{\partial x \partial y} + \frac{\partial w_0}{\partial y}\frac{\partial^2 w_0}{\partial x \partial y}\right) \\
& \quad + A_{66}\left(\frac{\partial^2 u_0}{\partial y^2} + \frac{\partial^2 v_0}{\partial x \partial y} + \frac{\partial^2 w_0}{\partial x \partial y}\frac{\partial w_0}{\partial y} + \frac{\partial w_0}{\partial x}\frac{\partial^2 w_0}{\partial y^2}\right) = 0, \\
& A_{22}\left(\frac{\partial^2 v_0}{\partial y^2} + \frac{\partial w_0}{\partial y}\frac{\partial^2 w_0}{\partial y^2}\right) + A_{12}\left(\frac{\partial^2 u_0}{\partial x \partial y} + \frac{\partial w_0}{\partial x}\frac{\partial^2 w_0}{\partial x \partial y}\right) \\
& \quad + A_{66}\left(\frac{\partial^2 u_0}{\partial x \partial y} + \frac{\partial^2 v_0}{\partial x^2} + \frac{\partial^2 w_0}{\partial x^2}\frac{\partial w_0}{\partial y} + \frac{\partial w_0}{\partial x}\frac{\partial^2 w_0}{\partial x \partial y}\right) = 0, \\
& K A_{44}\left(\frac{\partial^2 w_0}{\partial y^2} + \frac{\partial \psi_y}{\partial y}\right) + K A_{55}\left(\frac{\partial^2 w_0}{\partial x^2} + \frac{\partial \psi_x}{\partial x}\right) - k w_0 + p \\
& \quad + \frac{\partial}{\partial x}\left(N_{xx}^0 \frac{\partial w_0}{\partial x} + N_{xy}^0 \frac{\partial w_0}{\partial y}\right) + \frac{\partial}{\partial y}\left(N_{xy}^0 \frac{\partial w_0}{\partial x} + N_{yy}^0 \frac{\partial w_0}{\partial y}\right) = 0, \\
& D_{11}\frac{\partial^2 \psi_x}{\partial x^2} + D_{12}\frac{\partial^2 \psi_y}{\partial x \partial y} + D_{66}\left(\frac{\partial^2 \psi_x}{\partial y^2} + \frac{\partial^2 \psi_y}{\partial x \partial y}\right) - K A_{55}\left(\frac{\partial w_0}{\partial x} + \psi_x\right) = 0, \\
& D_{22}\frac{\partial^2 \psi_y}{\partial y^2} + D_{12}\frac{\partial^2 \psi_x}{\partial x \partial y} + D_{66}\left(\frac{\partial^2 \psi_y}{\partial x^2} + \frac{\partial^2 \psi_x}{\partial x \partial y}\right) - K A_{44}\left(\frac{\partial w_0}{\partial y} + \psi_y\right) = 0.
\end{aligned}
\tag{12.105}
$$

In the third equation, a transformation of the form (12.17) can then be applied again, which remains without representation at this point.

References

Altenbach, H., Altenbach, J., Naumenko, K.: Ebene Flächentragwerke, 2nd edn. Springer, Berlin (2016)

Chia, C.Y.: Nonlinear Analysis of Plates. McGraw-Hill, New York (1980)

Reddy, J.N.: Mechanics of Laminated Composite Plates and Shells, 2nd edn. CRC Press, Boca Raton (2003)

Reddy, J.N.: Theory and Analysis of Elastic Plates and Shells. CRC Press, Boca Raton (2006)

Sathyamoorthy, M.: Nonlinear Analysis of Structures. CRC Press, Boca Raton (1997)

Szilard, R.: Theories and Applications of Plate Analysis. Wiley, Hoboken (2004)

13 Laminated Plates

13.1 Introduction

Laminates are thin-walled structures in the form of disks, plates or shells consisting of any number of individual layers, each of which can have different properties (thickness, material properties, principal material directions, etc.). A very common application in lightweight construction is laminates whose layers consist of fiber-reinforced plastics. Such a fibrous composite material is a composite of at least two different components (constituents) that are combined by means of a specific manufacturing process to form a new material, i.e. the composite material. In the case of fiber-reinforced plastics, these are typically high-strength and high-stiffness fibers (e.g. carbon fibers) inserted into a matrix, e.g. an epoxy matrix. The resulting new material is characterized in its mechanical properties by the properties of the individual constituents and will have significantly more advantageous properties over the individual constituents. In lightweight construction, such composite materials are widely used, especially in aerospace engineering, but also in automotive engineering and generally in any application where the weight of a structure is an important factor. Carbon fiber-reinforced plastics, for example, exhibit particularly high stiffness and strength compared with their own weight.

Fiber reinforced plastics usually occur in layered form, as shown in Fig. 13.1. Such layered structures, which are composed of a number of so-called individual layers of arbitrary properties, are called laminates. The individual layers can generally be arranged in any order with any fiber direction. In many cases, all layers of a laminate will have identical thicknesses and also be made of identical materials, but lightweight construction applications also know cases where individual layers of different materials are combined in a laminate. However, it is already clear at this point that engineers designing and planning a laminate construction have a great deal of freedom and flexibility to design the laminate in such a way that the desired mechanical properties can be achieved through a clever choice of material, the type of reinforcement, the arrangement and alignment of the individual layers in the laminate, and so on.

C. Mittelstedt, *Theory of Plates and Shells*,
https://doi.org/10.1007/978-3-662-66805-4_13

This chapter addresses the so-called Classical Laminated Plate Theory as one of the most important and widely used laminate theories in lightweight design practice. Further literature on the topic discussed in this chapter is available, for example, with the works of Altenbach et al. (2018), Ambartsumyan (1970), Ashton and Whitney (1970), Jones (1975), Lekhnitskii (1968), Mittelstedt and Becker (2016), or Reddy (2004).

13.2 Classical Laminated Plate Theory

13.2.1 Introduction

This chapter is devoted to the presentation and discussion of one of the simplest and at the same time most used laminate theories, namely the so-called Classical Laminated Plate Theory (CLPT). We will consider laminates of orthotropic single layers (cf. Fig. 13.1) reinforced by unidirectional fibers for which we will assume in the following that the effective layer properties E_{11}, E_{22}, ν_{12} and G_{12} with respect to the principal material directions are known. The fiber direction of the layer k or its principal material direction is determined by specifying the fiber angle θ_k.

The so-called laminate code has proven its worth for the unambiguous designation of the layup/stacking sequence of a laminate. Here, the individual layers are identified by their fiber orientation, and counting starts from the lowest layer in the positive

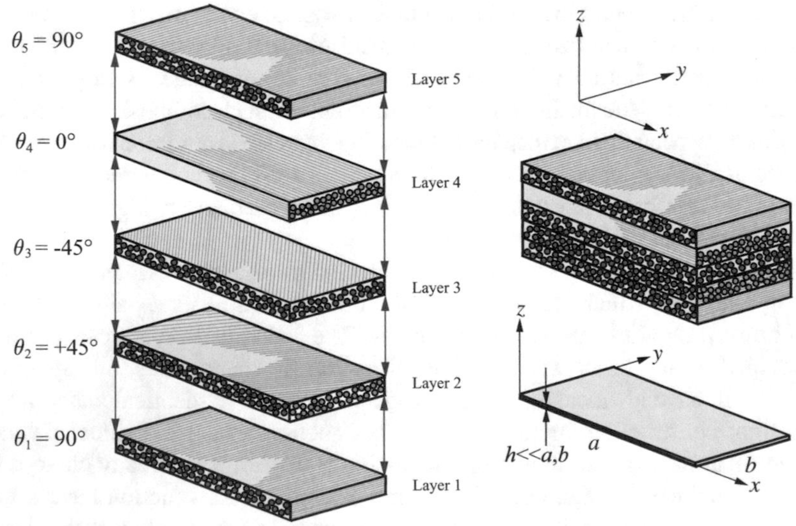

Fig. 13.1 Exemplary composite laminate consisting of five individual layers with different fiber orientations

z-direction. Using the example of Fig. 13.1, the laminate code would thus be

$$\left[90^\circ/45^\circ/-45^\circ/0^\circ/90^\circ\right].$$

It is assumed that all layers have the same material properties and the same thickness. If this is not the case in a given application, this must be noted separately.

Some abbreviations are common when using the laminate code. If there is a symmetrical laminate, where the top half of the laminate can be created by mirroring the bottom half at the laminate midplane, then it is sufficient to specify only one half of the layering and close the square brackets with the subscript S:

$$\left[0^\circ/90^\circ/90^\circ/0^\circ\right] = \left[0^\circ/90^\circ\right]_S.$$

If there are directly successive layers with the same fiber angles but with different signs, they can be marked with the $\pm$- symbol:

$$\left[+45^\circ/-45^\circ\right] = [\pm 45].$$

Combinations of the abbreviations mentioned are commonly used:

$$\left[+45^\circ/-45^\circ/-45^\circ/+45^\circ\right] = [\pm 45]_S.$$

If multiple individual layers of the same orientation occur directly one after the other, they are abbreviated by an index that indicates the number of these individual layers:

$$\left[0^\circ/90^\circ/90^\circ/90^\circ/0^\circ\right] = \left[0^\circ/90^\circ_3/0^\circ\right].$$

Figure 13.2 shows a section through a laminate. We consider laminates of thickness h consisting of N arbitrary layers. The laminate is divided into two halves of equal thickness $\frac{h}{2}$ at each point by the so-called laminate middle plane, and the coordinate origin of the global reference frame x, y, z is located in the midplane. It is spanned by the two coordinates x and y, and consequently z indicates the thickness direction. The position of each single layer k is defined by the coordinates z_{k-1} (lower boundary of the single layer k) and z_k (upper boundary surface of the layer k). We also refer to the contact plane of the two layers k and $k+1$ as interface k. Thus, for example, the interface between layer 1 and layer 2 is located at $z = z_1$.

13.2.2 Assumptions and Kinematics

We consider laminates of thickness h. Classical Laminated Plate Theory assumes thin laminates, although an exact definition of this term is not readily possible. However, it is assumed here that the thickness h of the laminate is significantly smaller than its inplane dimensions, so that in the case of rectangular laminates $h << a, b$ holds. It is now the goal of Classical Laminated Plate Theory to describe the behavior of the entire laminate as a multilayer composite based on the properties of the individual layers. The assumptions to be made here are:

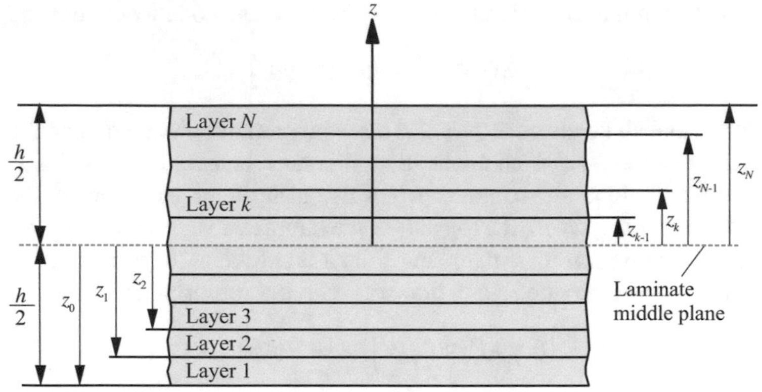

Fig. 13.2 Section through a laminate, nomenclature

(i) We assume a perfect laminate, which means that there is a perfect bond between the individual layers. The laminate is assumed to be free of damage.
(ii) We assume a plane stress state with respect to the thickness direction z throughout the laminate.
(iii) We assume the kinematics of a Kirchhoff plate. This means that also in the case of laminates we want to assume the hypothesis that the cross-sections of the laminate remain plane also in the deformed state and that the normal hypothesis as valid. A straight line which is normal to the laminate middle plane before deformation thus remains a straight line in the deformed state and also normal to the deformed laminate middle plane.
(iv) The thickness h of the laminate does not change during the deformation process.
(v) We assume linear elasticity and small deformations in all further elaborations.

Apparently, the Classical Laminated Plate Theory is a generalization of Kirchhoff's plate theory for the analytical treatment of laminate structures. Assumptions (iii) and (iv) concerning the kinematics of the laminate are shown in Fig. 13.3 for a sectional element. A point located on the laminate middle plane undergoes the longitudinal displacement u_0 and the deflection w_0. The deformed laminate element then has the inclination $\frac{\partial w_0}{\partial x}$ to the laminate middle plane. We introduce the displacements $u_0(x, y)$, $v_0(x, y)$, $w_0(x, y)$ of the laminate midplane, where u_0, v_0, w_0 are exclusive functions of the plane coordinates x, y. As further displacement quantities we introduce the displacements $u(x, y, z)$, $v(x, y, z)$, $w(x, y, z)$, which represent the displacements of a point outside the laminate midplane at any location z and thus must depend on all three coordinates x, y, z. As a further displacement quantity we have the angle ψ_x, which indicates the rotation of the cross-section around the y-axis. Since we assume that the cross-sections remain plane, ψ_x is exclusively a function of the two plane coordinates x and y. Furthermore, and not visible in Fig. 13.3 due

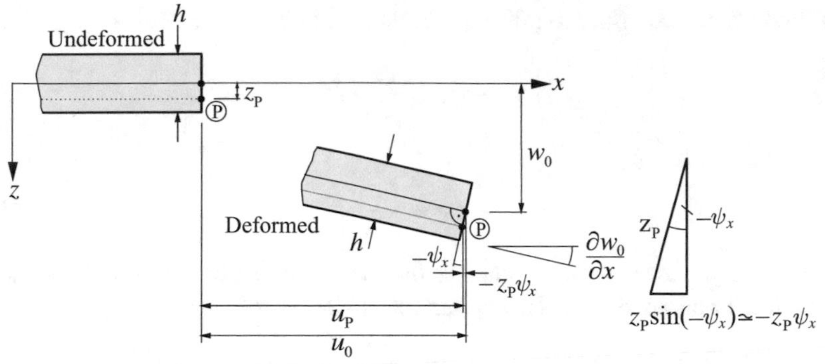

Fig. 13.3 Kinematics according to classical laminated plate theory

to the chosen perspective, there is an angle $\psi_y\,(x, y)$ which describes the rotation of the cross-section around the x-axis.

We now consider the deformations indicated in Fig. 13.3 and establish a relation between the displacement u_0 of the midplane and the displacement u_P of the point P. From Fig. 13.3 it can be deduced:

$$u_P = u_0 - z_P \sin(-\psi_x) = u_0 + z_P \sin\psi_x. \tag{13.1}$$

Since we again assume small deformations also in the framework of Classical Laminated Plate Theory, we can work with the approximation $\sin\psi_x \simeq \psi_x$ such that:

$$u_p = u_0 + z_p\psi_x. \tag{13.2}$$

Of course, this is true for any point on the cross section at any location z, so that we want to drop the index P from now on. Therefore, for the longitudinal displacement u of any point at location z reads:

$$u = u_0 + z\psi_x. \tag{13.3}$$

In the same way we can proceed for the displacement v in y-direction and obtain:

$$v = v_0 + z\psi_y. \tag{13.4}$$

As already demonstrated in the framework of Kirchhoff's plate theory, we assume that the thickness of the laminate h does not change in the deformed state compared to the undeformed configuration. Accordingly, the two deflections w and w_0 may be set equal:

$$w = w_0. \tag{13.5}$$

Finally, we use the normal hypothesis, according to which a straight line which was normal to the laminate middle plane in the undeformed state is also a normal in the deformed state. According to this hypothesis we obtain on the basis of Fig. 13.3:

$$-\psi_x = \frac{\partial w_0}{\partial x}. \tag{13.6}$$

The same holds for ψ_y:

$$-\psi_y = \frac{\partial w_0}{\partial y}. \tag{13.7}$$

With the kinematics now determined in this way, the displacement field according to Classical Laminated Plate Theory can be written as follows:

$$\begin{aligned} u(x, y, z) &= u_0(x, y) + z\psi_x(x, y) = u_0 - z\frac{\partial w_0}{\partial x}, \\ v(x, y, z) &= v_0(x, y) + z\psi_y(x, y) = v_0 - z\frac{\partial w_0}{\partial y}, \\ w(x, y, z) &= w_0(x, y). \end{aligned} \tag{13.8}$$

It turns out that those terms in (13.8) which contain the deflection w_0 are completely identical to the corresponding expressions (7.7) according to Kirchhoff's plate theory. New here are the components from the plane displacements u_0 and v_0 that describe a disk-like behaviour.

13.2.3 Strains and Stresses

With the displacement field of Classical Laminated Plate Theory now available, the strain field can be determined from (1.60):

$$\begin{aligned} \varepsilon_{xx} &= \frac{\partial u}{\partial x} = \frac{\partial u_0}{\partial x} - z\frac{\partial^2 w_0}{\partial x^2}, \\ \varepsilon_{yy} &= \frac{\partial v}{\partial y} = \frac{\partial v_0}{\partial y} - z\frac{\partial^2 w_0}{\partial y^2}, \\ \varepsilon_{zz} &= \frac{\partial w}{\partial z} = 0, \\ \gamma_{xy} &= \frac{\partial u}{\partial y} + \frac{\partial v}{\partial x} = \frac{\partial u_0}{\partial y} + \frac{\partial v_0}{\partial x} - 2z\frac{\partial^2 w_0}{\partial x \partial y}, \\ \gamma_{xz} &= \frac{\partial u}{\partial z} + \frac{\partial w}{\partial x} = 0, \\ \gamma_{yz} &= \frac{\partial v}{\partial z} + \frac{\partial w}{\partial y} = 0. \end{aligned} \tag{13.9}$$

As in Kirchhoff's plate theory, the plane strain components ε_{xx}, ε_{yy}, and γ_{xy} remain in agreement with the aforementioned assumptions of Classical Laminated Plate

Theory. We summarize the remaining strains in a vector notation as follows:

$$\begin{pmatrix} \varepsilon_{xx}(x, y, z) \\ \varepsilon_{yy}(x, y, z) \\ \gamma_{xy}(x, y, z) \end{pmatrix} = \begin{pmatrix} \varepsilon^0_{xx}(x, y) \\ \varepsilon^0_{yy}(x, y) \\ \gamma^0_{xy}(x, y) \end{pmatrix} + z \begin{pmatrix} \kappa^0_{xx}(x, y) \\ \kappa^0_{yy}(x, y) \\ \kappa^0_{xy}(x, y) \end{pmatrix}. \tag{13.10}$$

We refer to the quantities $\varepsilon_{xx}(x, y, z)$, $\varepsilon_{yy}(x, y, z)$ and $\gamma_{xy}(x, y, z)$ as the normal strains and the shear strain at any point in the laminate. They are composed of the normal strains $\varepsilon^0_{xx}(x, y)$, $\varepsilon^0_{yy}(x, y)$ and the shear strain $\gamma^0_{xy}(x, y)$ of the laminate midplane, as well as the curvatures $\kappa^0_{xx}(x, y)$, $\kappa^0_{yy}(x, y)$ and the twist $\kappa^0_{xy}(x, y)$ of the laminate midplane. The superscript '0' here again means that these are quantities related to the laminate midplane. Figure 13.4 represents these elementary deformation states of a laminate. Strains, curvatures and twist of the laminate middle plane are defined as follows:

$$\underline{\varepsilon}^0 = \begin{pmatrix} \varepsilon^0_{xx} \\ \varepsilon^0_{yy} \\ \gamma^0_{xy} \end{pmatrix} = \begin{pmatrix} \frac{\partial u_0}{\partial x} \\ \frac{\partial v_0}{\partial y} \\ \frac{\partial u_0}{\partial y} + \frac{\partial v_0}{\partial x} \end{pmatrix}, \quad \underline{\kappa}^0 = \begin{pmatrix} \kappa^0_{xx} \\ \kappa^0_{yy} \\ \kappa^0_{xy} \end{pmatrix} = - \begin{pmatrix} \frac{\partial^2 w_0}{\partial x^2} \\ \frac{\partial^2 w_0}{\partial y^2} \\ 2\frac{\partial^2 w_0}{\partial x \partial y} \end{pmatrix}. \tag{13.11}$$

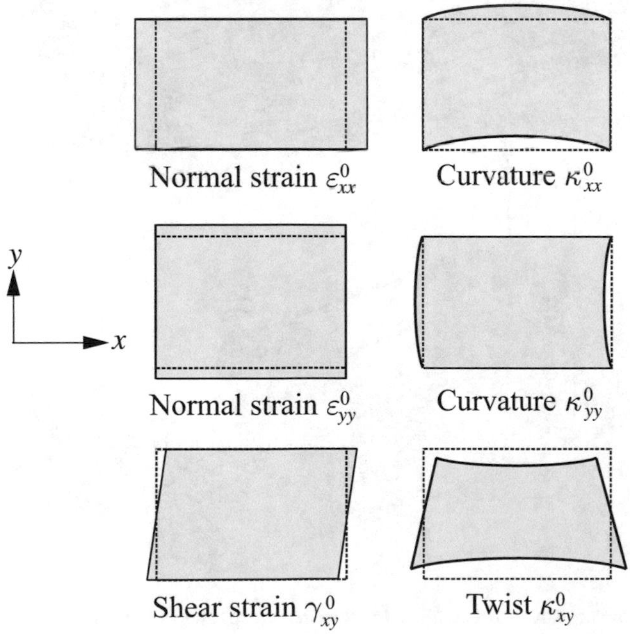

Fig. 13.4 Elementary deformation states of a laminate: strains of the laminate middle plane (top), curvatures and twist of the laminate middle plane (bottom)

Once the strains, curvatures, and twist of the laminate midplane have been determined, Hooke's law can be used to calculate the layerwise stresses of the laminate. For the layer k we obtain from (1.201):

$$\begin{pmatrix} \sigma_{xx} \\ \sigma_{yy} \\ \tau_{xy} \end{pmatrix}_k = \begin{bmatrix} \bar{Q}_{11} & \bar{Q}_{12} & \bar{Q}_{16} \\ \bar{Q}_{12} & \bar{Q}_{22} & \bar{Q}_{26} \\ \bar{Q}_{16} & \bar{Q}_{26} & \bar{Q}_{66} \end{bmatrix}_k \begin{pmatrix} \varepsilon_{xx} \\ \varepsilon_{yy} \\ \gamma_{xy} \end{pmatrix}$$
$$= \begin{bmatrix} \bar{Q}_{11} & \bar{Q}_{12} & \bar{Q}_{16} \\ \bar{Q}_{12} & \bar{Q}_{22} & \bar{Q}_{26} \\ \bar{Q}_{16} & \bar{Q}_{26} & \bar{Q}_{66} \end{bmatrix}_k \left[\begin{pmatrix} \varepsilon^0_{xx} \\ \varepsilon^0_{yy} \\ \gamma^0_{xy} \end{pmatrix} + z \begin{pmatrix} \kappa^0_{xx} \\ \kappa^0_{yy} \\ \kappa^0_{xy} \end{pmatrix} \right], \qquad (13.12)$$

where for layer k $z_{k-1} \leq z \leq z_k$ must hold.

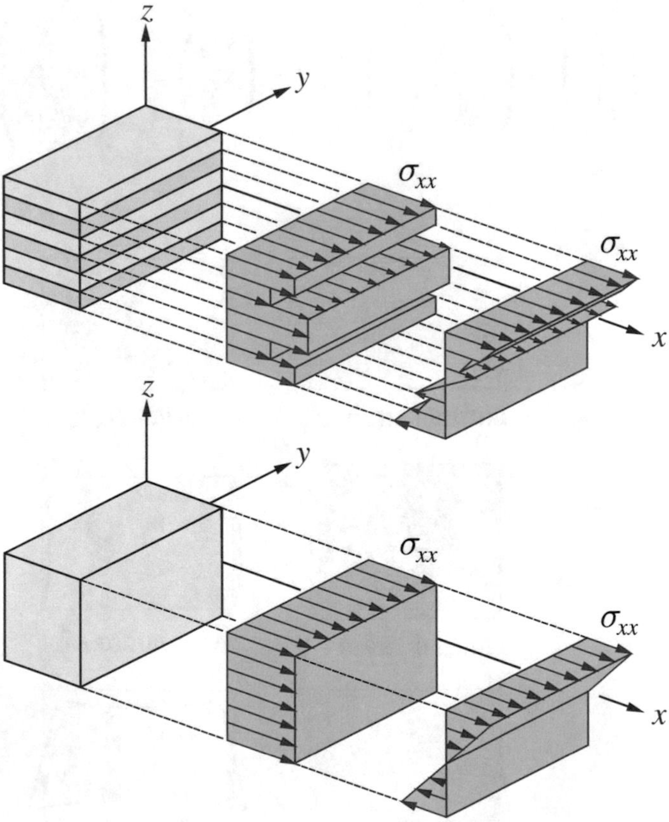

Fig. 13.5 Exemplary stress distributions in a laminate (top) and in a homogeneous plate (bottom)

In a laminate, due to the applied layer materials or due to the orientations of the material principal axes, different elastic properties will be present in the individual layers, so that there will be discontinuities in the stress distributions in the interfaces of adjacent layers, as shown in Fig. 13.5, top, for the normal stress σ_{xx}. Here, the stress σ_{xx} was divided into membrane stresses (stress image on the left) and bending stresses (stress image on the right). As a comparison, in Fig. 13.5, bottom, the characteristic distribution for σ_{xx} in a homogeneous plate is shown. Here, of course, these typical discontinuities in the stress distributions over the thickness do not occur.

13.3 Constitutive Law

We now define the following force and moment quantities of a laminate as follows. In addition to the moment flows M^0_{xx}, M^0_{yy}, M^0_{xy} (Figs. 7.4 and 7.5), which are already known from the Kirchhoff plate, and the shear force flows Q_x, Q_y (cf. Fig. 7.6), the membrane force flows N^0_{xx}, N^0_{yy}, N^0_{xy} (Fig. 3.2) which are already known from disk analysis occur as well. The laminate force flows N^0_{xx}, N^0_{yy}, N^0_{xy} are obtained from the integration of the corresponding stresses over the thickness h of the laminate:

$$\begin{pmatrix} N^0_{xx} \\ N^0_{yy} \\ N^0_{xy} \end{pmatrix} = \int_{-\frac{h}{2}}^{\frac{h}{2}} \begin{pmatrix} \sigma_{xx} \\ \sigma_{yy} \\ \tau_{xy} \end{pmatrix} \mathrm{d}z. \tag{13.13}$$

It should be noted that the integrations required in (13.13) cannot generally be performed continuously in the case of a laminate, since the elastic properties in the individual layers will usually be different. Therefore, the integration over h is decomposed into N partial integrals as follows:

$$\begin{pmatrix} N^0_{xx} \\ N^0_{yy} \\ N^0_{xy} \end{pmatrix} = \sum_{k=1}^{k=N} \int_{z_{k-1}}^{z_k} \begin{pmatrix} \sigma_{xx} \\ \sigma_{yy} \\ \tau_{xy} \end{pmatrix}_k \mathrm{d}z. \tag{13.14}$$

The laminate moment flows are determined quite similarly as the resultants of the corresponding stress components multiplied by the lever arm z:

$$\begin{pmatrix} M^0_{xx} \\ M^0_{yy} \\ M^0_{xy} \end{pmatrix} = \int_{-\frac{h}{2}}^{\frac{h}{2}} \begin{pmatrix} \sigma_{xx} \\ \sigma_{yy} \\ \tau_{xy} \end{pmatrix} z\mathrm{d}z = \sum_{k=1}^{k=N} \int_{z_{k-1}}^{z_k} \begin{pmatrix} \sigma_{xx} \\ \sigma_{yy} \\ \tau_{xy} \end{pmatrix}_k z\mathrm{d}z. \tag{13.15}$$

Substituting Hooke's law yields:

$$\begin{pmatrix} N_{xx}^0 \\ N_{yy}^0 \\ N_{xy}^0 \end{pmatrix} = \sum_{k=1}^{k=N} \int_{z_{k-1}}^{z_k} \begin{bmatrix} \bar{Q}_{11} & \bar{Q}_{12} & \bar{Q}_{16} \\ \bar{Q}_{12} & \bar{Q}_{22} & \bar{Q}_{26} \\ \bar{Q}_{16} & \bar{Q}_{26} & \bar{Q}_{66} \end{bmatrix}_k \left[\begin{pmatrix} \varepsilon_{xx}^0 \\ \varepsilon_{yy}^0 \\ \gamma_{xy}^0 \end{pmatrix} + z \begin{pmatrix} \kappa_{xx}^0 \\ \kappa_{yy}^0 \\ \kappa_{xy}^0 \end{pmatrix} \right] \mathrm{d}z,$$

$$\begin{pmatrix} M_{xx}^0 \\ M_{yy}^0 \\ M_{xy}^0 \end{pmatrix} = \sum_{k=1}^{k=N} \int_{z_{k-1}}^{z_k} \begin{bmatrix} \bar{Q}_{11} & \bar{Q}_{12} & \bar{Q}_{16} \\ \bar{Q}_{12} & \bar{Q}_{22} & \bar{Q}_{26} \\ \bar{Q}_{16} & \bar{Q}_{26} & \bar{Q}_{66} \end{bmatrix}_k \left[\begin{pmatrix} \varepsilon_{xx}^0 \\ \varepsilon_{yy}^0 \\ \gamma_{xy}^0 \end{pmatrix} + z \begin{pmatrix} \kappa_{xx}^0 \\ \kappa_{yy}^0 \\ \kappa_{xy}^0 \end{pmatrix} \right] z\mathrm{d}z. \tag{13.16}$$

Performing the integrations prescribed herein provides:

$$\begin{pmatrix} N_{xx}^0 \\ N_{yy}^0 \\ N_{xy}^0 \end{pmatrix} = \begin{bmatrix} A_{11} & A_{12} & A_{16} \\ A_{12} & A_{22} & A_{26} \\ A_{16} & A_{26} & A_{66} \end{bmatrix} \begin{pmatrix} \varepsilon_{xx}^0 \\ \varepsilon_{yy}^0 \\ \gamma_{xy}^0 \end{pmatrix} + \begin{bmatrix} B_{11} & B_{12} & B_{16} \\ B_{12} & B_{22} & B_{26} \\ B_{16} & B_{26} & B_{66} \end{bmatrix} \begin{pmatrix} \kappa_{xx}^0 \\ \kappa_{yy}^0 \\ \kappa_{xy}^0 \end{pmatrix},$$

$$\begin{pmatrix} M_{xx}^0 \\ M_{yy}^0 \\ M_{xy}^0 \end{pmatrix} = \begin{bmatrix} B_{11} & B_{12} & B_{16} \\ B_{12} & B_{22} & B_{26} \\ B_{16} & B_{26} & B_{66} \end{bmatrix} \begin{pmatrix} \varepsilon_{xx}^0 \\ \varepsilon_{yy}^0 \\ \gamma_{xy}^0 \end{pmatrix} + \begin{bmatrix} D_{11} & D_{12} & D_{16} \\ D_{12} & D_{22} & D_{26} \\ D_{16} & D_{26} & D_{66} \end{bmatrix} \begin{pmatrix} \kappa_{xx}^0 \\ \kappa_{yy}^0 \\ \kappa_{xy}^0 \end{pmatrix}. \tag{13.17}$$

Herein we have used the following abbreviations ($i, j = 1, 2, 6$):

$$A_{ij} = \int_{-\frac{h}{2}}^{\frac{h}{2}} \bar{Q}_{ij}\mathrm{d}z,$$

$$B_{ij} = \int_{-\frac{h}{2}}^{\frac{h}{2}} \bar{Q}_{ij}z\mathrm{d}z,$$

$$D_{ij} = \int_{-\frac{h}{2}}^{\frac{h}{2}} \bar{Q}_{ij}z^2\mathrm{d}z. \tag{13.18}$$

The quantities appearing herein are, on the one hand, the membrane stiffnesses A_{ij}. They establish a relation between the laminate membrane force flows N_{xx}^0, N_{yy}^0, N_{xy}^0 and the strains ε_{xx}^0, ε_{yy}^0, γ_{xy}^0 of the laminate midplane. On the other hand, the plate stiffnesses D_{ij}, which we have already introduced in the treatment of the Kirchhoff plate (cf. Chap. 7), appear at this point. They establish a connection between the laminate moment flows M_{xx}^0, M_{yy}^0, M_{xy}^0 and the curvatures κ_{xx}^0, κ_{yy}^0 as well as the twist κ_{xy}^0 of the laminate midplane. Apparently, then, the Classical Laminated Plate Theory represents a combination of the disk theory and the Kirchhoff plate theory. On the other hand, the terms B_{ij} also appearing in (13.17) represent a special feature in the consideration of laminates and are denoted as coupling stiffnesses. They couple the laminate membrane forces N_{xx}^0, N_{yy}^0, N_{xy}^0 with the curvatures κ_{xx}^0, κ_{yy}^0 and the twist κ_{xy}^0 of the laminate midplane. Moreover, they cause coupling of the laminate moment flows $M^0_{xx}, M^0_{yy}, M^0_{xy}$ with the strains $\varepsilon^0_{xx}, \varepsilon^0_{yy}, \gamma^0_{xy}$ of the laminate midplane. Such coupling terms do not occur in homogeneous surface structures.

If there is a laminate in which all individual layers in the respective interval $z_{k-1} \leq z \leq z_k$ have constant elastic properties, then the integrals in (13.18) can be solved elementarily and decomposed into sums as follows:

$$A_{ij} = \sum_{k=1}^{k=N} \bar{Q}_{ij,k} \left(z_k - z_{k-1}\right),$$

$$B_{ij} = \frac{1}{2} \sum_{k=1}^{k=N} \bar{Q}_{ij,k} \left(z_k^2 - z_{k-1}^2\right),$$

$$D_{ij} = \frac{1}{3} \sum_{k=1}^{k=N} \bar{Q}_{ij,k} \left(z_k^3 - z_{k-1}^3\right). \tag{13.19}$$

We now present (13.17) in a summarized manner as follows:

$$\begin{pmatrix} N^0_{xx} \\ N^0_{yy} \\ N^0_{xy} \\ M^0_{xx} \\ M^0_{yy} \\ M^0_{xy} \end{pmatrix} = \begin{bmatrix} A_{11} & A_{12} & A_{16} & B_{11} & B_{12} & B_{16} \\ A_{12} & A_{22} & A_{26} & B_{12} & B_{22} & B_{26} \\ A_{16} & A_{26} & A_{66} & B_{16} & B_{26} & B_{66} \\ B_{11} & B_{12} & B_{16} & D_{11} & D_{12} & D_{16} \\ B_{12} & B_{22} & B_{26} & D_{12} & D_{22} & D_{26} \\ B_{16} & B_{26} & B_{66} & D_{16} & D_{26} & D_{66} \end{bmatrix} \begin{pmatrix} \varepsilon^0_{xx} \\ \varepsilon^0_{yy} \\ \gamma^0_{xy} \\ \kappa^0_{xx} \\ \kappa^0_{yy} \\ \kappa^0_{xy} \end{pmatrix}. \tag{13.20}$$

The stiffness matrix of the laminate contained herein is also called the laminate stiffness matrix or colloquially the ABD matrix. In symbolic form we obtain:

$$\begin{pmatrix} \underline{N}^0 \\ \underline{M}^0 \end{pmatrix} = \begin{bmatrix} \underline{\underline{A}} & \underline{\underline{B}} \\ \underline{\underline{B}} & \underline{\underline{D}} \end{bmatrix} \begin{pmatrix} \underline{\varepsilon}^0 \\ \underline{\kappa}^0 \end{pmatrix}. \tag{13.21}$$

The submatrices $\underline{\underline{A}}$, $\underline{\underline{B}}$ and $\underline{\underline{D}}$ are usually called membrane quadrant, coupling quadrant and plate quadrant, respectively.

As in the case of disks and plates, laminates are also computationally reduced to their midplane, whose properties are described by characteristic smeared quantities in the form of the stiffnesses A_{ij}, B_{ij} and D_{ij}.

13.4 Coupling Effects

A special feature in the treatment of laminates are the so-called coupling effects. The constitutive law (13.20) reveals that in the presence of a fully occupied laminate stiffness matrix, any force or moment quantity will lead to all strains $\varepsilon^0_{xx}, \varepsilon^0_{yy}, \gamma^0_{xy}$ and both curvatures $\kappa^0_{xx}, \kappa^0_{yy}$ and also the twist κ^0_{xy} of the laminate midplane. This

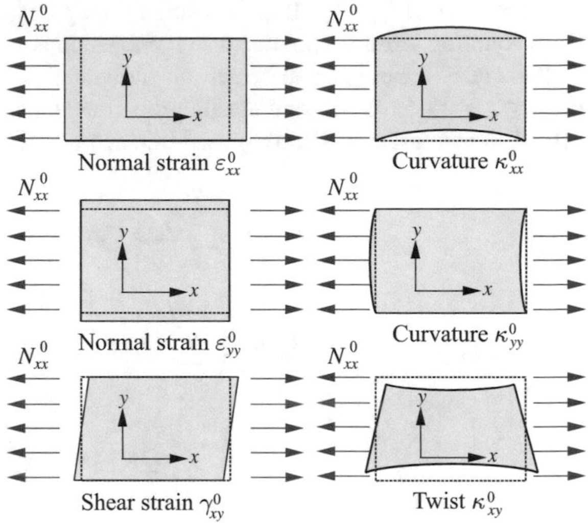

Fig. 13.6 Coupling effects in an arbitrary laminate with fully occupied laminate stiffness matrix; strains as well as curvatures and twist of the laminate midplane caused by the laminate membrane force N_{xx}^0

is shown in Fig. 13.6 using the example of the laminate membrane force N_{xx}^0. Thus, it can be seen that for arbitrary laminates the disk and plate behavior are coupled, expressed by the coupling stiffnesses B_{ij}. Thus, the common distinction of plane structures into disks and plates is not meaningful in the treatment of laminates, since membrane force flows can develop a plate effect and moment flows can result in a membrane effect.

Those entries of the laminate stiffness matrix in the constitutive law (13.20) which entail coupling effects, and which do not occur in either orthotropic disks or plates, are the terms A_{16}, A_{26} in the membrane quadrant, the terms D_{16}, D_{26} in the plate quadrant, and all coupling stiffnesses B_{ij}. The coupling effects associated with these terms are denoted as follows:

A_{16}, A_{26} : Shear coupling,
D_{16}, : Bending-twisting coupling,
B_{ij} : Bending-extension coupling

13.4.1 Shear Coupling

Shear coupling occurs in laminates where the terms A_{16}, A_{26} do not become zero. In this case, the strains ε_{xx}^0, ε_{yy}^0 cause the occurrence of the laminate shear force flow N_{xy}^0. Likewise, a shear strain γ_{xy}^0 evokes the laminate membrane force flows N_{xx}^0 and N_{yy}^0. A simple example of the occurrence of shear coupling is a laminate single layer whose material principal axes x_1, x_2 are rotated by the angle θ with respect to the reference frame x, y (Fig. 13.7). In such a so-called off-axis layer, the normal force flow N_{xx}^0 then provides for an occurrence of the shear strain γ_{xy}^0. Laminates

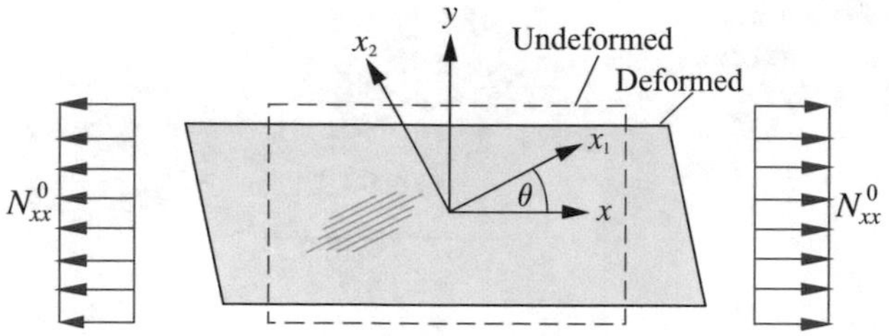

Fig. 13.7 Shear coupling at the example of an off-axis single layer

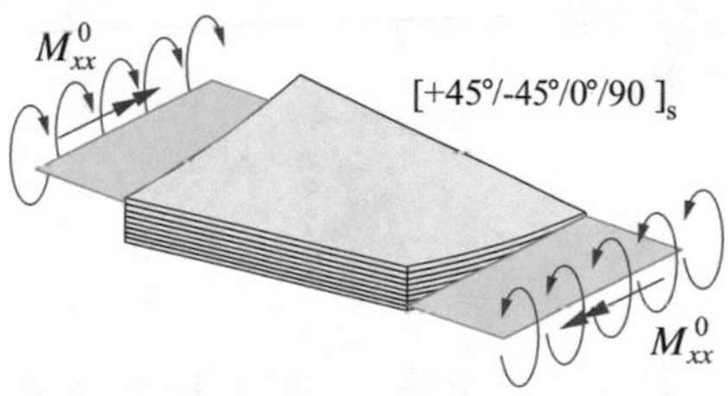

Fig. 13.8 Bending-twisting coupling at the example of a typical aircraft laminate

that are free of shear coupling and in which the two membrane stiffnesses A_{16}, A_{26} vanish are also called balanced laminates.

13.4.2 Bending-Twisting Coupling

The so-called bending-twisting coupling (Fig. 13.8) occurs in laminates where the plate stiffnesses D_{16}, D_{26} do not become zero. In such laminates, the curvatures κ_{xx}^0, κ_{yy}^0 of the laminate midplane provide for the occurrence of the twisting moment flow M_{xy}^0. Conversely, the twist κ_{xy}^0 of the laminate midplane is involved in the occurrence of the two bending moment flows M_{xx}^0 and M_{yy}^0.

13.4.3 Bending-extension Coupling

We speak of the so-called bending-extension coupling when one or more coupling stiffnesses B_{ij} are non-zero. The strains ε_{xx}^0, ε_{yy}^0, γ_{xy}^0 of the laminate midplane are then involved in the generation of the laminate moment flows M_{xx}^0, M_{yy}^0, M_{xy}^0. In the same way, the curvatures κ_{xx}^0, κ_{yy}^0 and the twist κ_{xy}^0 of the laminate midplane provide for an occurrence of the laminate force flows N_{xx}^0, N_{yy}^0, N_{xy}^0. Examples are shown in Figs. 13.9 and 13.10. Using the example of the unsymmetric cross-ply laminate, it can be shown that the coupling stiffnesses B_{11} and B_{22} do not become zero. On the other hand, for the unsymmetric angle-ply laminate, it can be shown that the coupling stiffnesses B_{16} and B_{26} are not zero.

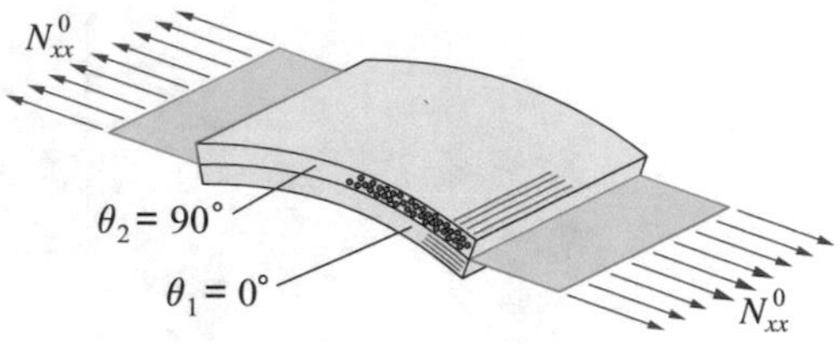

Fig. 13.9 Bending-extension coupling at the example of an unsymmetric cross-ply laminate

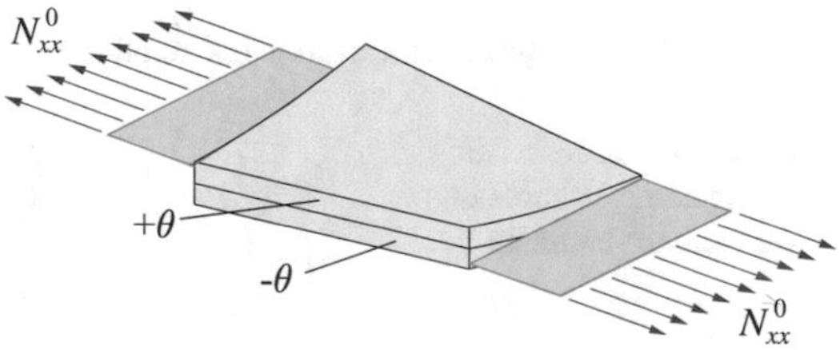

Fig. 13.10 Bending-extension coupling at the example of an unsymmetric angle-ply laminate

Bending-extension coupling only occurs in unsymmetric laminates. Conversely, symmetrically layered laminates are always free of bending-extension coupling. The reason can be found in the governing equation (13.18) for the coupling stiffnesses B_{ij}. The integration required herein is performed over the transformed reduced stiffnesses $\bar{Q}_{ij}$ and the thickness coordinate z. For symmetric laminates, the transformed reduced stiffnesses are a symmetric function of z, whereas the function z itself is point symmetric. Accordingly, the integral in (13.18) will always take the value zero for symmetric laminates due to the symmetric integration limits $z = -\frac{h}{2}$ and $z = \frac{h}{2}$.

13.5 Special Laminates

In many technically relevant cases, laminates occur whose constitutive law is not characterized by a fully occupied stiffness matrix. On the contrary, it can be shown that the constitutive law is significantly simplified in many cases.

13.5.1 Isotropic Single Layer

The special case of a laminate which consists of only one single isotropic layer (i.e. N=1) corresponds to an isotropic disk or an isotropic plate. Thus, both the isotropic disk and the isotropic Kirchhoff plate are included in the Classical Laminated Plate

Theory as special cases. The stiffness matrix simplifies significantly in this case:

$$\begin{bmatrix} \underline{\underline{A}} & \underline{\underline{B}} \\ \underline{\underline{B}} & \underline{\underline{D}} \end{bmatrix} = \begin{bmatrix} A_{11} & A_{12} & 0 & 0 & 0 & 0 \\ A_{12} & A_{22} & 0 & 0 & 0 & 0 \\ 0 & 0 & A_{66} & 0 & 0 & 0 \\ 0 & 0 & 0 & D_{11} & D_{12} & 0 \\ 0 & 0 & 0 & D_{12} & D_{22} & 0 \\ 0 & 0 & 0 & 0 & 0 & D_{66} \end{bmatrix}. \tag{13.22}$$

Thus, an isotropic disk is always free of shear coupling (the terms A_{16}, A_{26} are zero), the isotropic plate is free of bending-twisting coupling (the terms D_{16}, D_{26} do not appear here). Moreover, such a structure is always free of bending-extension coupling, the entire coupling quadrant $\underline{\underline{B}}$ vanishes.

The remaining entries of the stiffness matrix in (13.22) can be written using the engineering constants E, G and ν:

$$\begin{aligned} A_{11} &= Q_{11}h = \frac{Eh}{1-\nu^2}, \quad A_{22} = Q_{22}h = \frac{Eh}{1-\nu^2}, \\ A_{12} &= Q_{12}h = \frac{\nu Eh}{1-\nu^2}, \quad A_{66} = Q_{66}h = Gh, \\ D_{11} &= Q_{11}\frac{h^3}{12} = \frac{Eh^3}{12\left(1-\nu^2\right)}, \quad D_{22} = Q_{22}\frac{h^3}{12} = \frac{Eh^3}{12\left(1-\nu^2\right)}, \\ D_{12} &= Q_{12}\frac{h^3}{12} = \frac{\nu Eh^3}{12\left(1-\nu^2\right)}, \quad D_{66} = Q_{66}\frac{h^3}{12} = \frac{Gh^3}{12}. \end{aligned} \tag{13.23}$$

13.5.2 Orthotropic Single Layer

If a laminate with a single orthotropic single layer is given, then the stiffness matrix is identical to (13.22), but here using the engineering constants E_{11}, E_{22}, ν_{12}, ν_{21}, and G_{12} yields the following expressions for the individual components in (13.22):

$$\begin{aligned} A_{11} &= Q_{11}h = \frac{E_{11}h}{1-\nu_{12}\nu_{21}}, \quad A_{22} = Q_{22}h = \frac{E_{22}h}{1-\nu_{12}\nu_{21}}, \\ A_{12} &= Q_{12}h = \frac{\nu_{12}E_{22}h}{1-\nu_{12}\nu_{21}}, \quad A_{66} = Q_{66}h = G_{12}h, \\ D_{11} &= Q_{11}\frac{h^3}{12} = \frac{E_{11}h^3}{12\left(1-\nu_{12}\nu_{21}\right)}, \quad D_{22} = Q_{22}\frac{h^3}{12} = \frac{E_{22}h^3}{12\left(1-\nu_{12}\nu_{21}\right)}, \\ D_{12} &= Q_{12}\frac{h^3}{12} = \frac{\nu_{12}E_{22}h^3}{12\left(1-\nu_{12}\nu_{21}\right)}, \quad D_{66} = Q_{66}\frac{h^3}{12} = \frac{G_{12}h^3}{12}. \end{aligned} \tag{13.24}$$

13.5.3 Anisotropic Single Layer/Off-axis Layer

We now consider the special case of an anisotropic single layer, for which we can specify the assignment of the stiffness matrix as follows:

$$\begin{bmatrix} \underline{\underline{A}} & \underline{\underline{B}} \\ \underline{\underline{B}} & \underline{\underline{D}} \end{bmatrix} = \begin{bmatrix} A_{11} & A_{12} & A_{16} & 0 & 0 & 0 \\ A_{12} & A_{22} & A_{26} & 0 & 0 & 0 \\ A_{16} & A_{26} & A_{66} & 0 & 0 & 0 \\ 0 & 0 & 0 & D_{11} & D_{12} & D_{16} \\ 0 & 0 & 0 & D_{12} & D_{22} & D_{26} \\ 0 & 0 & 0 & D_{16} & D_{26} & D_{66} \end{bmatrix}. \tag{13.25}$$

Naturally, a single layer does not show any bending-extension coupling, but usually both shear coupling and bending-twisting coupling occur. The terms appearing in the stiffness matrix (13.25) can be written as:

$$A_{ij} = \bar{Q}_{ij} h, \quad D_{ij} = \bar{Q}_{ij} \frac{h^3}{12}. \tag{13.26}$$

13.5.4 Symmetric Laminates

Let us now consider the case of an arbitrary but symmetric laminate. It can be easily shown (see also Sect. 13.4.3 and the related explanations) that a symmetric laminate is always free of bending-extension coupling. However, the symmetry properties of a laminate do not allow any conclusions on the occurrence of shear coupling or bending-twisting coupling. Whether or not the latter two coupling effects occur depends exclusively on the chosen layup of the laminate under consideration. The stiffness matrix of any symmetric laminate can thus be given as:

$$\begin{bmatrix} \underline{\underline{A}} & \underline{\underline{B}} \\ \underline{\underline{B}} & \underline{\underline{D}} \end{bmatrix} = \begin{bmatrix} A_{11} & A_{12} & A_{16} & 0 & 0 & 0 \\ A_{12} & A_{22} & A_{26} & 0 & 0 & 0 \\ A_{16} & A_{26} & A_{66} & 0 & 0 & 0 \\ 0 & 0 & 0 & D_{11} & D_{12} & D_{16} \\ 0 & 0 & 0 & D_{12} & D_{22} & D_{26} \\ 0 & 0 & 0 & D_{16} & D_{26} & D_{66} \end{bmatrix}. \tag{13.27}$$

13.5.5 Cross-ply Laminates

Cross-ply laminates are laminates where exclusively single layers with orientation angles $\theta_k = 0°$ and $\theta_k = 90°$ occur. Such laminates are always balanced, so that the shear coupling terms $A_{16} = A_{26} = 0$ always become zero. In addition, bending-twisting coupling never occurs in cross-ply laminates. Whether or not cross-ply laminates exhibit bending-extension coupling depends exclusively on whether or not they exhibit symmetry properties.

For an unsymmetric cross-ply laminate with the layup [0°/90°], the stiffness matrix can be given as:

$$\begin{bmatrix} \underline{\underline{A}} & \underline{\underline{B}} \\ \underline{\underline{B}} & \underline{\underline{D}} \end{bmatrix} = \begin{bmatrix} A_{11} & A_{12} & 0 & B_{11} & 0 & 0 \\ A_{12} & A_{22} & 0 & 0 & -B_{11} & 0 \\ 0 & 0 & A_{66} & 0 & 0 & 0 \\ B_{11} & 0 & 0 & D_{11} & D_{12} & 0 \\ 0 & -B_{11} & 0 & D_{12} & D_{22} & 0 \\ 0 & 0 & 0 & 0 & 0 & D_{66} \end{bmatrix}. \tag{13.28}$$

For a symmetric cross-ply laminate $[0°/90°]_S$, on the other hand, the bending-extension coupling does not occur, and we obtain:

$$\begin{bmatrix} \underline{\underline{A}} & \underline{\underline{B}} \\ \underline{\underline{B}} & \underline{\underline{D}} \end{bmatrix} = \begin{bmatrix} A_{11} & A_{12} & 0 & 0 & 0 & 0 \\ A_{12} & A_{22} & 0 & 0 & 0 & 0 \\ 0 & 0 & A_{66} & 0 & 0 & 0 \\ 0 & 0 & 0 & D_{11} & D_{12} & 0 \\ 0 & 0 & 0 & D_{12} & D_{22} & 0 \\ 0 & 0 & 0 & 0 & 0 & D_{66} \end{bmatrix}. \tag{13.29}$$

13.5.6 Angle-ply Laminates

Angle-ply laminates are laminates in which for each layer with the angle θ there is a layer with opposite angle $-\theta$ at an arbitrary position in the laminate. Such laminates are always balanced such that $A_{16} = A_{26} = 0$. Whether an angle-ply laminate exhibits bending-twisting coupling and/or bending-extension coupling depends exclusively on the given layup and cannot be determined a priori without further knowledge of the stacking sequence.

If an unsymmetric angle-ply laminate is given, such as the layup $[\pm\theta]$, then it can also be shown that here the bending-twisting coupling always vanishes as well. For unsymmetric angle-ply laminates, the laminate stiffness matrix is obtained as:

$$\begin{bmatrix} \underline{\underline{A}} & \underline{\underline{B}} \\ \underline{\underline{B}} & \underline{\underline{D}} \end{bmatrix} = \begin{bmatrix} A_{11} & A_{12} & 0 & 0 & 0 & B_{16} \\ A_{12} & A_{22} & 0 & 0 & 0 & B_{26} \\ 0 & 0 & A_{66} & B_{16} & B_{26} & 0 \\ 0 & 0 & B_{16} & D_{11} & D_{12} & 0 \\ 0 & 0 & B_{26} & D_{12} & D_{22} & 0 \\ B_{16} & B_{26} & 0 & 0 & 0 & D_{66} \end{bmatrix}. \tag{13.30}$$

Thus, unsymmetric angle-ply laminates are free from shear coupling ($A_{16} = A_{26} = 0$) and bending-twisting coupling ($D_{16} = D_{26} = 0$), but the bending-extension coupling terms B_{16} and B_{26} occur.

If a symmetric angle-ply laminate e.g. of the type $[\pm\theta]_S$ is considered, then this laminate is free of any bending-extension coupling (alls B_{ij} result as zero), and also shear coupling will not occur here naturally ($A_{16} = A_{26} = 0$), but in contrast to the unsymmetric angle-ply laminate bending-extension coupling occurs:

$$\begin{bmatrix} \underline{\underline{A}} & \underline{\underline{B}} \\ \underline{\underline{B}} & \underline{\underline{D}} \end{bmatrix} = \begin{bmatrix} A_{11} & A_{12} & 0 & 0 & 0 & 0 \\ A_{12} & A_{22} & 0 & 0 & 0 & 0 \\ 0 & 0 & A_{66} & 0 & 0 & 0 \\ 0 & 0 & 0 & D_{11} & D_{12} & D_{16} \\ 0 & 0 & 0 & D_{12} & D_{22} & D_{26} \\ 0 & 0 & 0 & D_{16} & D_{26} & D_{66} \end{bmatrix}. \tag{13.31}$$

13.5.7 Quasi-isotropic Laminates

Quasi-isotropic laminates are those laminates in which the membrane quadrant exhibits isotropic properties such that

$$A_{11} = A_{22}. \tag{13.32}$$

Moreover, due to in-plane isotropy, such laminates exhibit the property

$$A_{66} = \frac{1}{2}\left(A_{11} - A_{12}\right). \tag{13.33}$$

If a symmetric quasi-isotropic laminate is given, then the laminate stiffness matrix is obtained as follows:

$$\begin{bmatrix} \underline{\underline{A}} & \underline{\underline{B}} \\ \underline{\underline{B}} & \underline{\underline{D}} \end{bmatrix} = \begin{bmatrix} A_{11} & A_{12} & 0 & 0 & 0 & 0 \\ A_{12} & A_{11} & 0 & 0 & 0 & 0 \\ 0 & 0 & \frac{1}{2}\left(A_{11} - A_{12}\right) & 0 & 0 & 0 \\ 0 & 0 & 0 & D_{11} & D_{12} & D_{16} \\ 0 & 0 & 0 & D_{12} & D_{22} & D_{26} \\ 0 & 0 & 0 & D_{16} & D_{26} & D_{66} \end{bmatrix}. \tag{13.34}$$

Such laminates apparently always have a membrane quadrant corresponding to an isotropic disk. Thus, shear coupling does not occur. However, it must be emphasized at this point that the term quasi-isotropy refers exclusively to the membrane quadrant and does not allow any statements about the plate quadrant. This remains generally anisotropic in its properties, so that in general $D_{11} \neq D_{22}$ holds. Similarly, bending-twisting coupling cannot be excluded a priori, so that in general $D_{16}, D_{26} \neq 0$ must be assumed. Whether a quasi-isotropic laminate also exhibits bending-extension coupling depends exclusively on its symmetry properties. Typical examples of quasi-isotropic laminates are the layups $[-60°/0°/60°]$, $[-60°/0°/60°]_S$, $[0°/\pm 45°/90°]$, $[0°/\pm 45°/90°]_S$,or $[0°/-36°/-72°/36°/72°]$, among others.

13.6 Basic Equations and Boundary Conditions

13.6.1 Equilibrium Conditions

We consider the situation of Fig. 13.11 and formulate the local equilibrium conditions for an infinitesimal element with edge lengths dx and dy. The laminate is assumed to be sufficiently thin and is under the two plane tangential loads p_x and p_y as well as the transverse load p, which can all be functions of the two plane coordinates x and y. The equilibrium of forces in the x- direction results in:

$$\left(N_{xx}^0 + \frac{\partial N_{xx}^0}{\partial x}\mathrm{d}x\right)\mathrm{d}y + \left(N_{yx}^0 + \frac{\partial N_{yx}^0}{\partial y}\mathrm{d}y\right)\mathrm{d}x - N_{xx}^0\mathrm{d}y - N_{yx}^0\mathrm{d}x + p_x\mathrm{d}x\mathrm{d}y = 0, \tag{13.35}$$

or

$$\frac{\partial N_{xx}^0}{\partial x}\mathrm{d}x\mathrm{d}y + \frac{\partial N_{yx}^0}{\partial y}\mathrm{d}x\mathrm{d}y + p_x\mathrm{d}x\mathrm{d}y = 0. \tag{13.36}$$

With $N_{yx}^0 = N_{xy}^0$ we obtain:

$$\frac{\partial N_{xx}^0}{\partial x} + \frac{\partial N_{xy}^0}{\partial y} = -p_x. \tag{13.37}$$

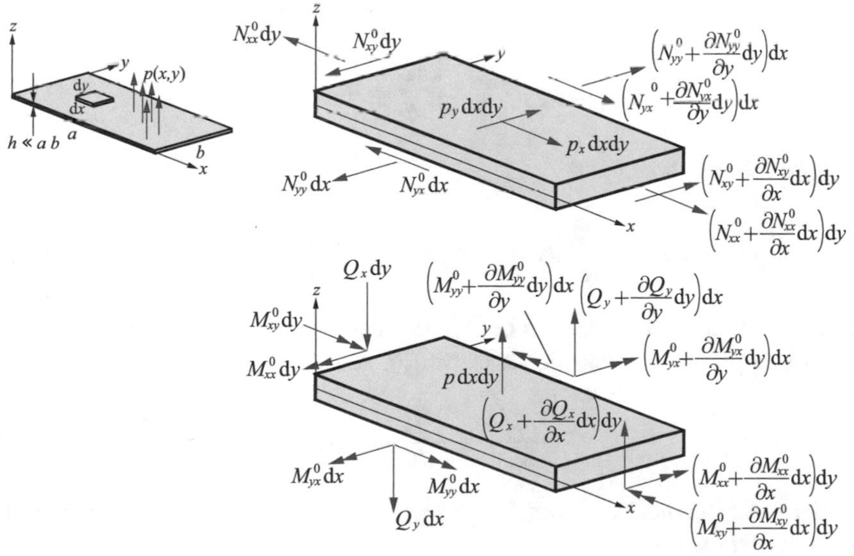

Fig. 13.11 Infinitesimal element of a laminate

The equilibrium of forces in y-direction provides analogously:

$$\frac{\partial N_{xy}^0}{\partial x} + \frac{\partial N_{yy}^0}{\partial y} = -p_y. \tag{13.38}$$

The two equations (13.37) and (13.38) represent the already known disk equilibrium (see Chap. 3).

The force balance in z- direction results in:

$$\left(Q_x + \frac{\partial Q_x}{\partial x}\mathrm{d}x\right)\mathrm{d}y + \left(Q_y + \frac{\partial Q_y}{\partial y}\mathrm{d}y\right)\mathrm{d}x - Q_x\mathrm{d}y - Q_y\mathrm{d}x + p\mathrm{d}x\mathrm{d}y = 0, \tag{13.39}$$

which leads to the following expression:

$$\frac{\partial Q_x}{\partial x}\mathrm{d}x\mathrm{d}y + \frac{\partial Q_y}{\partial y}\mathrm{d}x\mathrm{d}y + p\mathrm{d}x\mathrm{d}y = 0. \tag{13.40}$$

This can be transformed to:

$$\frac{\partial Q_x}{\partial x} + \frac{\partial Q_y}{\partial y} = -p. \tag{13.41}$$

We further consider the moment equilibrium about the y-axis with respect to the center of gravity of the infinitesimal element:

$$\begin{aligned}&\left(M_{xx}^0 + \frac{\partial M_{xx}^0}{\partial x}\mathrm{d}x\right)\mathrm{d}y + \left(M_{yx}^0 + \frac{\partial M_{yx}^0}{\partial y}\mathrm{d}y\right)\mathrm{d}x - M_{xx}^0\mathrm{d}y - M_{yx}^0\mathrm{d}x\\ &-\left(Q_x + \frac{\partial Q_x}{\partial x}\mathrm{d}x\right)\mathrm{d}y\frac{\mathrm{d}x}{2} - Q_x\mathrm{d}y\frac{\mathrm{d}x}{2} = 0.\end{aligned} \tag{13.42}$$

With $M_{yx}^0 = M_{xy}^0$ we obtain:

$$\frac{\partial M_{xx}^0}{\partial x}\mathrm{d}x\mathrm{d}y + \frac{\partial M_{xy}^0}{\partial y}\mathrm{d}x\mathrm{d}y - Q_x\mathrm{d}x\mathrm{d}y - \frac{\partial Q_x}{\partial x}\frac{\mathrm{d}x^2\mathrm{d}y}{2} = 0. \tag{13.43}$$

The last term appearing here can be regarded as small of higher order compared to the other terms and is consequently discarded, so that:

$$\frac{\partial M_{xx}^0}{\partial x} + \frac{\partial M_{xy}^0}{\partial y} = Q_x. \tag{13.44}$$

Similarly, the moment sum can be set up around the x-axis, which leads to the following expression:

$$\frac{\partial M_{xy}^0}{\partial x} + \frac{\partial M_{yy}^0}{\partial y} = Q_y. \tag{13.45}$$

The two equations (13.44) and (13.45) as well as the force equilibrium condition (13.41) together form the plate equilibrium, which is also already known (see Chap. 7) from the Kirchhoff plate theory.

The plate equilibrium can be summarized even further. We differentiate the first moment equilibrium condition (13.44) with respect to x:

$$\frac{\partial^2 M_{xx}^0}{\partial x^2}+\frac{\partial^2 M_{xy}^0}{\partial x \partial y}=\frac{\partial Q_x}{\partial x}. \tag{13.46}$$

Partial differentiation of the second moment sum (13.45) with respect to y yields:

$$\frac{\partial^2 M_{xy}^0}{\partial x \partial y}+\frac{\partial^2 M_{yy}^0}{\partial y^2}=\frac{\partial Q_y}{\partial y}. \tag{13.47}$$

Substituting these two expressions into the transverse force sum (13.41), we obtain the so-called condensed plate equilibrium, which we had already derived in Chap. 7 in the context of Kirchhoff's plate theory:

$$\frac{\partial^2 M_{xx}^0}{\partial x^2}+\frac{\partial^2 M_{yy}^0}{\partial y^2}+2\frac{\partial^2 M_{xy}^0}{\partial x \partial y}=-p. \tag{13.48}$$

13.6.2 Displacement Differential Equations

From the set of equilibrium conditions in the form of equations (13.37), (13.38) and (13.48) the governing diaplacement differential equations can be obtained by means of the constitutive law

$$\begin{pmatrix} N_{xx}^0 \\ N_{yy}^0 \\ N_{xy}^0 \\ M_{xx}^0 \\ M_{yy}^0 \\ M_{xy}^0 \end{pmatrix}=\begin{bmatrix} A_{11} & A_{12} & A_{16} & B_{11} & B_{12} & B_{16} \\ A_{12} & A_{22} & A_{26} & B_{12} & B_{22} & B_{26} \\ A_{16} & A_{26} & A_{66} & B_{16} & B_{26} & B_{66} \\ B_{11} & B_{12} & B_{16} & D_{11} & D_{12} & D_{16} \\ B_{12} & B_{22} & B_{26} & D_{12} & D_{22} & D_{26} \\ B_{16} & B_{26} & B_{66} & D_{16} & D_{26} & D_{66} \end{bmatrix}\begin{pmatrix} \frac{\partial u_0}{\partial x} \\ \frac{\partial v_0}{\partial y} \\ \frac{\partial u_0}{\partial y}+\frac{\partial v_0}{\partial x} \\ -\frac{\partial^2 w_0}{\partial x^2} \\ -\frac{\partial^2 w_0}{\partial y^2} \\ -2\frac{\partial^2 w_0}{\partial x \partial y} \end{pmatrix} \tag{13.49}$$

which describe a given static laminate problem in terms of the displacements u_0, v_0, w_0 of the laminate midplane. They read after substituting (13.49) into (13.37), (13.38) and (13.48):

$$A_{11}\frac{\partial^2 u_0}{\partial x^2}+2A_{16}\frac{\partial^2 u_0}{\partial x \partial y}+A_{66}\frac{\partial^2 u_0}{\partial y^2}+A_{16}\frac{\partial^2 v_0}{\partial x^2}+(A_{12}+A_{66})\frac{\partial^2 v_0}{\partial x \partial y}+A_{26}\frac{\partial^2 v_0}{\partial y^2}$$
$$-B_{11}\frac{\partial^3 w_0}{\partial x^3}-3B_{16}\frac{\partial^3 w_0}{\partial x^2 \partial y}-(B_{12}+2B_{66})\frac{\partial^3 w_0}{\partial x \partial y^2}-B_{26}\frac{\partial^3 w_0}{\partial y^3}+p_x=0,$$

$$A_{16}\frac{\partial^2 u_0}{\partial x^2}+(A_{12}+A_{66})\frac{\partial^2 u_0}{\partial x \partial y}+A_{26}\frac{\partial^2 u_0}{\partial y^2}+A_{66}\frac{\partial^2 v_0}{\partial x^2}+2A_{26}\frac{\partial^2 v_0}{\partial x \partial y}+A_{22}\frac{\partial^2 v_0}{\partial y^2}$$

$$-B_{16}\frac{\partial^3 w_0}{\partial x^3} - (B_{12}+2B_{66})\frac{\partial^3 w_0}{\partial x^2 \partial y} - 3B_{26}\frac{\partial^3 w_0}{\partial x \partial y^2} - B_{22}\frac{\partial^3 w_0}{\partial y^3} + p_y = 0,$$

$$D_{11}\frac{\partial^4 w_0}{\partial x^4} + 4D_{16}\frac{\partial^4 w_0}{\partial x^3 \partial y} + 2\left(D_{12}+2D_{66}\right)\frac{\partial^4 w_0}{\partial x^2 \partial y^2} + 4D_{26}\frac{\partial^4 w_0}{\partial x \partial y^3} + D_{22}\frac{\partial^4 w_0}{\partial y^4}$$
$$-B_{11}\frac{\partial^3 u_0}{\partial x^3} - 3B_{16}\frac{\partial^3 u_0}{\partial x^2 \partial y} - (B_{12}+2B_{66})\frac{\partial^3 u_0}{\partial x \partial y^2} - B_{26}\frac{\partial^3 u_0}{\partial y^3}$$
$$-B_{16}\frac{\partial^3 v_0}{\partial x^3} - (B_{12}+2B_{66})\frac{\partial^3 v_0}{\partial x^2 \partial y} - 3B_{26}\frac{\partial^3 v_0}{\partial x \partial y^2} - B_{22}\frac{\partial^3 v_0}{\partial y^3} - p = 0. \tag{13.50}$$

This is a system of three coupled partial differential equations in the displacement components u_0, v_0, w_0. Obviously, each displacement component appears in each of the three differential equations, so that they have to be solved simultaneously. By means of the material law (13.49) a coupling of the disk and the plate behavior results, which is consequently also reflected in the governing differential equations (13.50).

If there is an arbitrary but symmetric laminate, then the entire coupling quadrant vanishes and the coupling stiffnesses B_{11}, B_{22}, B_{12}, B_{16}, B_{26} and B_{66} become zero. Then the differential equation system (13.50) simplifies quite significantly as follows:

$$A_{11}\frac{\partial^2 u_0}{\partial x^2} + 2A_{16}\frac{\partial^2 u_0}{\partial x \partial y} + A_{66}\frac{\partial^2 u_0}{\partial y^2} + A_{16}\frac{\partial^2 v_0}{\partial x^2} + (A_{12}+A_{66})\frac{\partial^2 v_0}{\partial x \partial y}$$
$$+A_{26}\frac{\partial^2 v_0}{\partial y^2} + p_x = 0,$$

$$A_{16}\frac{\partial^2 u_0}{\partial x^2} + (A_{12}+A_{66})\frac{\partial^2 u_0}{\partial x \partial y} + A_{26}\frac{\partial^2 u_0}{\partial y^2} + A_{66}\frac{\partial^2 v_0}{\partial x^2} + 2A_{26}\frac{\partial^2 v_0}{\partial x \partial y}$$
$$+A_{22}\frac{\partial^2 v_0}{\partial y^2} + p_y = 0,$$

$$D_{11}\frac{\partial^4 w_0}{\partial x^4} + 4D_{16}\frac{\partial^4 w_0}{\partial x^3 \partial y} + 2\left(D_{12}+2D_{66}\right)\frac{\partial^4 w_0}{\partial x^2 \partial y^2} + 4D_{26}\frac{\partial^4 w_0}{\partial x \partial y^3}$$
$$+D_{22}\frac{\partial^4 w_0}{\partial y^4} - p = 0. \tag{13.51}$$

Obviously, the first two differential equations are now decoupled from the third equation. The first two equations in (13.51) contain exclusively the membrane stiffnesses A_{11}, A_{22}, A_{12}, A_{66}, A_{16} and A_{26}, the still unknown displacement functions u_0 and v_0 as well as the tangential loads p_x and p_y and thus describe the membrane behavior of the considered laminate. The third differential equation in (13.51), on the other hand, includes exclusively the plate stiffnesses D_{11}, D_{22}, D_{12}, D_{66}, D_{16}, and D_{26}, the surface load p, and the unknown deflection w_0 of the midplane of the laminate, and thus describes the plate behavior of the symmetric laminate.

13.6.3 Boundary Conditions

A static boundary value problem for a laminate is uniquely described by the differential equations (13.50) or (13.51) and given boundary conditions. We distinguish between the geometric or kinematic boundary conditions on the one hand, and the dynamic boundary conditions on the other hand. If a given laminate problem is described in Cartesian coordinates, then at a boundary with $x = const.$ the following boundary conditions come into question:

$$\begin{aligned} u_0 &= \widehat{u_0} \quad \text{or } N_{xx}^0 = \widehat{N}_{xx}^0, \\ v_0 &= \widehat{v_0} \quad \text{or } N_{xy}^0 = \widehat{N}_{xy}^0, \\ \frac{\partial w_0}{\partial x} &= \widehat{\frac{\partial w_0}{\partial x}} \quad \text{or } M_{xx}^0 = \widehat{M}_{xx}^0, \\ w_0 &= \widehat{w_0} \quad \text{or } \bar{Q}_x = \widehat{\bar{Q}}_x. \end{aligned} \tag{13.52}$$

At an edge with $y = const.$ the following boundary conditions may apply:

$$\begin{aligned} u_0 &= \widehat{u_0} \quad \text{or } N_{xy}^0 = \widehat{N}_{xy}^0, \\ v_0 &= \widehat{v_0} \quad \text{or } N_{yy}^0 = \widehat{N}_{yy}^0, \\ \frac{\partial w_0}{\partial y} &= \widehat{\frac{\partial w_0}{\partial y}} \quad \text{or } M_{yy}^0 = \widehat{M}_{yy}^0, \\ w_0 &= \widehat{w_0} \quad \text{or } \bar{Q}_y = \widehat{\bar{Q}}_y. \end{aligned} \tag{13.53}$$

Herein, $\bar{Q}_x$ and $\bar{Q}_y$ are the Kirchhoff equivalent transverse shear force flows. The notation $\widehat{(...)}$ indicates a given quantity.

With respect to the formulation of boundary conditions for laminates, in contrast to isotropic or orthotropic plates, ambiguities may occur. We want to elaborate on this by means of a rectangular plane laminated structure of length a and width b with simple supports at all edges. If this structure is in the form of an isotropic or orthotropic plate under a load that leads to a transverse deflection of the plate, then simple supports mean that at all four edges both the deflection w_0 and the corresponding bending moment flows M_{xx}^0 and M_{yy}^0 must become zero:

$$\begin{aligned} &w_0(x=0)=0, \quad w_0(x=a)=0, \quad w_0(y=0)=0, \quad w_0(y=b)=0, \\ &M_{xx}^0(x=0)=0, \quad M_{xx}^0(x=a)=0, \quad M_{yy}^0(y=0)=0, \quad M_{yy}^0(y=b)=0. \end{aligned} \tag{13.54}$$

Here, the moment flows can be expressed by partial derivatives of the deflection w_0 by means of the constitutive law of the plate. However, it is important to note that for such homogeneous structures it is sufficient to formulate boundary conditions with respect to the deflection w_0. As long as geometrically linear problems are treated, the plane displacements u_0 and v_0 do not play any role at this point.

If, on the other hand, an arbitrary laminate is given, which in addition to the shear coupling and the bending-twisting coupling can also typically exhibit bending-extension coupling, then for an unambiguous formulation of boundary conditions statements must also be made about the two plane displacements u_0, v_0 and the associated laminate inplane force flows. For an arbitrary laminate, the requirement for a simple support, e.g., at the edge $x = 0$, can have several meanings:

$$\begin{aligned} \text{either } & w_0(x=0)=0, \quad M_{xx}^0(x=0)=0, \quad u_0=\widehat{u}_0, \quad v_0=\widehat{v}_0, \\ \text{or } & w_0(x=0)=0, \quad M_{xx}^0(x=0)=0, \quad N_{xx}^0=\widehat{N}_{xx}^0, \quad v_0=\widehat{v}_0, \\ \text{or } & w_0(x=0)=0, \quad M_{xx}^0(x=0)=0, \quad u_0=\widehat{u}_0, \quad N_{xy}^0=\widehat{N}_{xy}^0, \\ \text{or } & w_0(x=0)=0, \quad M_{xx}^0(x=0)=0, \quad N_{xx}^0=\widehat{N}_{xx}^0, \quad N_{xy}^0=\widehat{N}_{xy}^0. \end{aligned} \tag{13.55}$$

13.7 Navier Solutions

As already discussed in Chap. 7 for isotropic and orthotropic plates, we want to present the so-called Navier solutions also in the case of laminates. We consider a rectangular laminate (length a, width b, thickness h), which is simply supported at all edges. It has already been discussed that the presence of simply supported edges may well involve a certain ambiguity, so that some attention must be paid to the unambiguous formulation of the given boundary conditions. The laminates considered here are under the arbitrary surface load $p(x, y)$, which, as in the case of the plate, is developed as a Fourier series:

$$p(x,y) = \sum_{m=1}^{m=\infty} \sum_{n=1}^{n=\infty} P_{mn} \sin\left(\frac{m\pi x}{a}\right) \sin\left(\frac{n\pi y}{b}\right), \tag{13.56}$$

wherein:

$$P_{mn} = \frac{4}{ab} \int_0^b \int_0^a p(x,y) \sin\left(\frac{m\pi x}{a}\right) \sin\left(\frac{n\pi y}{b}\right) \mathrm{d}x\mathrm{d}y. \tag{13.57}$$

13.7.1 Bending of a Symmetric Cross-Ply Laminate

We consider a symmetric cross-ply laminate where the shear coupling terms $A_{16} = A_{26}$ become zero. Furthermore, there is no bending-twisting coupling ($D_{16} = D_{26} = 0$), and the laminate is free of any bending-extension coupling due to its symmetry properties (all coupling stiffnesses $B_{ij} = 0$). The laminate stiffness matrix to be considered here is then:

$$\begin{bmatrix} \underline{\underline{A}} & \underline{\underline{B}} \\ \underline{\underline{B}} & \underline{\underline{D}} \end{bmatrix} = \begin{bmatrix} A_{11} & A_{12} & 0 & 0 & 0 & 0 \\ A_{12} & A_{22} & 0 & 0 & 0 & 0 \\ 0 & 0 & A_{66} & 0 & 0 & 0 \\ 0 & 0 & 0 & D_{11} & D_{12} & 0 \\ 0 & 0 & 0 & D_{12} & D_{22} & 0 \\ 0 & 0 & 0 & 0 & 0 & D_{66} \end{bmatrix}. \tag{13.58}$$

It can be easily shown that for such a symmetric cross-ply laminate, the disk and the plate behavior are completely decoupled, so that only the third displacement differential equation in (13.51) needs to be considered here, which simplifies with $D_{16} = D_{26} = 0$ as follows:

$$D_{11}\frac{\partial^4 w_0}{\partial x^4} + 2\left(D_{12} + 2D_{66}\right)\frac{\partial^4 w_0}{\partial x^2 \partial y^2} + D_{22}\frac{\partial^4 w_0}{\partial y^4} - p = 0. \tag{13.59}$$

The comparison with the plate equation in Sect. 7.7 shows that this is the same differential equation, so that the solution of the present problem is analogous to the procedure of Sect. 7.7. Therefore, further elaborations can be omitted at this point.

13.7.2 Bending of an Unsymmetric Cross-Ply Laminate $\left[(0°/90°)_N\right]$

A further Navier solution which can be derived for laminates is obtained for the bending of an unsymmetric cross-ply laminate $\left[(0°/90°)_N\right]$. The laminate stiffness matrix in this case is:

$$\begin{bmatrix} \underline{\underline{A}} & \underline{\underline{B}} \\ \underline{\underline{B}} & \underline{\underline{D}} \end{bmatrix} = \begin{bmatrix} A_{11} & A_{12} & 0 & B_{11} & 0 & 0 \\ A_{12} & A_{11} & 0 & 0 & -B_{11} & 0 \\ 0 & 0 & A_{66} & 0 & 0 & 0 \\ B_{11} & 0 & 0 & D_{11} & D_{12} & 0 \\ 0 & -B_{11} & 0 & D_{12} & D_{11} & 0 \\ 0 & 0 & 0 & 0 & 0 & D_{66} \end{bmatrix}. \tag{13.60}$$

The displacement differential equations valid here are with $p_x = p_y = 0$:

$$\begin{aligned} A_{11}\frac{\partial^2 u_0}{\partial x^2} + A_{66}\frac{\partial^2 u_0}{\partial y^2} + \left(A_{12} + A_{66}\right)\frac{\partial^2 v_0}{\partial x \partial y} - B_{11}\frac{\partial^3 w_0}{\partial x^3} &= 0, \\ \left(A_{12} + A_{66}\right)\frac{\partial^2 u_0}{\partial x \partial y} + A_{66}\frac{\partial^2 v_0}{\partial x^2} + A_{11}\frac{\partial^2 v_0}{\partial y^2} + B_{11}\frac{\partial^3 w_0}{\partial y^3} &= 0, \\ D_{11}\left(\frac{\partial^4 w_0}{\partial x^4} + \frac{\partial^4 w_0}{\partial y^4}\right) + 2\left(D_{12} + 2D_{66}\right)\frac{\partial^4 w_0}{\partial x^2 \partial y^2} - B_{11}\left(\frac{\partial^3 u_0}{\partial x^3} - \frac{\partial^3 v_0}{\partial y^3}\right) &= p. \end{aligned} \tag{13.61}$$

Obviously, for the unsymmetric cross-ply laminate considered here, the plate and disk behavior are still coupled.

We want to apply the so-called SS1 boundary conditions, where the abbreviation SS is to be traced back to the term ‘simply supported’. The number 1 indicates that there are different types of simply supported boundary conditions, as we will discuss in the following section. The boundary conditions are for $x = 0$ and $x = a$ with $0 \leq y \leq b$:

$$
\begin{aligned}
w_0 &= 0, \\
v_0 &= 0, \\
M_{xx}^0 &= B_{11}\frac{\partial u_0}{\partial x} - D_{11}\frac{\partial^2 w_0}{\partial x^2} - D_{12}\frac{\partial^2 w_0}{\partial y^2} = 0, \\
N_{xx}^0 &= A_{11}\frac{\partial u_0}{\partial x} + A_{12}\frac{\partial v_0}{\partial y} - B_{11}\frac{\partial^2 w_0}{\partial x^2} = 0.
\end{aligned}
\tag{13.62}
$$

At the edges $y = 0$ and $y = b$ with $0 \leq x \leq a$ holds:

$$
\begin{aligned}
w_0 &= 0, \\
u_0 &= 0, \\
M_{yy}^0 &= -B_{11}\frac{\partial v_0}{\partial y} - D_{12}\frac{\partial^2 w_0}{\partial x^2} - D_{11}\frac{\partial^2 w_0}{\partial y^2} = 0, \\
N_{yy}^0 &= A_{12}\frac{\partial u_0}{\partial x} + A_{11}\frac{\partial v_0}{\partial y} + B_{11}\frac{\partial^2 w_0}{\partial y^2} = 0.
\end{aligned}
\tag{13.63}
$$

A series approach is now used for all displacements u_0, v_0, w_0 as follows:

$$
\begin{aligned}
u_0 &= \sum_{m=1}^{m=\infty}\sum_{n=1}^{n=\infty} U_{mn}\cos\left(\frac{m\pi x}{a}\right)\sin\left(\frac{n\pi y}{b}\right), \\
v_0 &= \sum_{m=1}^{m=\infty}\sum_{n=1}^{n=\infty} V_{mn}\sin\left(\frac{m\pi x}{a}\right)\cos\left(\frac{n\pi y}{b}\right), \\
w_0 &= \sum_{m=1}^{m=\infty}\sum_{n=1}^{n=\infty} W_{mn}\sin\left(\frac{m\pi x}{a}\right)\sin\left(\frac{n\pi y}{b}\right).
\end{aligned}
\tag{13.64}
$$

The surface load p is also developed at this point as a Fourier series as previously demonstrated. Using the approaches (13.64), we then obtain:

$$
\begin{aligned}
&\sum_{m=1}^{m=\infty}\sum_{n=1}^{n=\infty}\left[-\left(A_{11}\frac{m^2\pi^2}{a^2} + A_{66}\frac{n^2\pi^2}{b^2}\right)U_{mn} - (A_{12} + A_{66})\frac{m\pi}{a}\frac{n\pi}{b}V_{mn}\right. \\
&\left. + B_{11}\frac{m^3\pi^3}{a^3}W_{mn}\right]\cos\left(\frac{m\pi x}{a}\right)\sin\left(\frac{n\pi y}{b}\right) = 0, \\
&\sum_{m=1}^{m=\infty}\sum_{n=1}^{n=\infty}\left[-(A_{12} + A_{66})\frac{m\pi}{a}\frac{n\pi}{b}U_{mn} - \left(A_{66}\frac{m^2\pi^2}{a^2} + A_{11}\frac{n^2\pi^2}{b^2}\right)V_{mn}\right. \\
&\left. - B_{11}\frac{n^3\pi^3}{b^3}W_{mn}\right]\sin\left(\frac{m\pi x}{a}\right)\cos\left(\frac{n\pi y}{b}\right) = 0,
\end{aligned}
$$

$$\sum_{m=1}^{m=\infty}\sum_{n=1}^{n=\infty}\left\{B_{11}\frac{m^3\pi^3}{a^3}U_{mn}-B_{11}\frac{n^3\pi^3}{b^3}V_{mn}\right.$$
$$\left.-\left[D_{11}\frac{m^4\pi^4}{a^4}+2\left(D_{12}+2D_{66}\right)\frac{m^2\pi^2}{a^2}\frac{n^2\pi^2}{b^2}+D_{11}\frac{n^4\pi^4}{b^4}\right]W_{mn}\right\}$$
$$\times\sin\left(\frac{m\pi x}{a}\right)\sin\left(\frac{n\pi y}{b}\right)=-p. \tag{13.65}$$

We obtain a meaningful solution if we set the respective bracket terms to zero. We obtain a system of three linear equations for the unknown coefficients U_{mn}, V_{mn}, W_{mn}:

$$-\left(A_{11}\frac{m^2\pi^2}{a^2}+A_{66}\frac{n^2\pi^2}{b^2}\right)U_{mn}-\left(A_{12}+A_{66}\right)\frac{m\pi}{a}\frac{n\pi}{b}V_{mn}+B_{11}\frac{m^3\pi^3}{a^3}W_{mn}=0,$$
$$-\left(A_{12}+A_{66}\right)\frac{m\pi}{a}\frac{n\pi}{b}U_{mn}-\left(A_{66}\frac{m^2\pi^2}{a^2}+A_{11}\frac{n^2\pi^2}{b^2}\right)V_{mn}-B_{11}\frac{n^3\pi^3}{b^3}W_{mn}=0,$$
$$B_{11}\frac{m^3\pi^3}{a^3}U_{mn}-B_{11}\frac{n^3\pi^3}{b^3}V_{mn}$$
$$-\left[D_{11}\frac{m^4\pi^4}{a^4}+2\left(D_{12}+2D_{66}\right)\frac{m^2\pi^2}{a^2}\frac{n^2\pi^2}{b^2}+D_{11}\frac{n^4\pi^4}{b^4}\right]W_{mn}=-P_{mn}. \tag{13.66}$$

This can also be represented in a vector matrix form as:

$$\begin{bmatrix}\lambda_{11}&\lambda_{12}&\lambda_{13}\\\lambda_{12}&\lambda_{22}&\lambda_{23}\\\lambda_{13}&\lambda_{23}&\lambda_{33}\end{bmatrix}\begin{pmatrix}U_{mn}\\V_{mn}\\W_{mn}\end{pmatrix}=\begin{pmatrix}0\\0\\-P_{mn}\end{pmatrix}, \tag{13.67}$$

with:

$$\lambda_{11}=-A_{11}\frac{m^2\pi^2}{a^2}-A_{66}\frac{n^2\pi^2}{b^2},\quad \lambda_{12}=-\left(A_{12}+A_{66}\right)\frac{mn\pi^2}{ab},$$
$$\lambda_{13}=B_{11}\frac{m^3\pi^3}{a^3},\quad \lambda_{22}=-A_{66}\frac{m^2\pi^2}{a^2}-A_{11}\frac{n^2\pi^2}{b^2},\quad \lambda_{23}=-B_{11}\frac{n^3\pi^3}{b^3},$$
$$\lambda_{33}=-\left[D_{11}\left(\frac{m^4\pi^4}{a^4}+\frac{n^4\pi^4}{b^4}\right)+2\left(D_{12}+2D_{66}\right)\frac{m^2n^2\pi^4}{a^2b^2}\right]. \tag{13.68}$$

Thus, the Navier solution for the unsymmetric cross-ply laminate under transverse load p is completely determined with respect to the displacements u_0, v_0 and w_0, and all state quantities and stress components can be computed.

For example, if the special case of a sinusoidal load p of the form

$$p\left(x,y\right)=P_0\sin\left(\frac{\pi x}{a}\right)\sin\left(\frac{\pi y}{b}\right) \tag{13.69}$$

is given, then an approach of the form

$$u_0=U_{11}\cos\left(\frac{\pi x}{a}\right)\sin\left(\frac{\pi y}{b}\right),$$

$$v_0 = V_{11} \sin\left(\frac{\pi x}{a}\right) \cos\left(\frac{\pi y}{b}\right),$$
$$w_0 = W_{11} \sin\left(\frac{\pi x}{a}\right) \sin\left(\frac{\pi y}{b}\right) \tag{13.70}$$

is sufficient. Substituting in the displacement differential equations (13.61) then yields the following linear system of equations for the determination of the coefficients U_{11}, V_{11}, W_{11}:

$$\begin{bmatrix} -A_{11}\frac{\pi^2}{a^2} - A_{66}\frac{\pi^2}{b^2} & -(A_{12}+A_{66})\frac{\pi^2}{ab} & B_{11}\frac{\pi^3}{a^3} \\ -(A_{12}+A_{66})\frac{\pi^2}{ab} & -A_{66}\frac{\pi^2}{a^2} - A_{11}\frac{\pi^2}{b^2} & -B_{11}\frac{\pi^3}{b^3} \\ -B_{11}\frac{\pi^3}{a^3} & B_{11}\frac{\pi^3}{b^3} & D_{11}\left(\frac{\pi^4}{a^4}+\frac{\pi^4}{b^4}\right) + 2(D_{12}+2D_{66})\frac{\pi^4}{a^2b^2} \end{bmatrix}$$
$$\times \begin{pmatrix} U_{11} \\ V_{11} \\ W_{11} \end{pmatrix} = \begin{pmatrix} 0 \\ 0 \\ p_0 \end{pmatrix}. \tag{13.71}$$

All other coefficients U_{mn}, V_{mn}, W_{mn} result in zero for the present case. The system of equations (13.71) can be simply solved for the coefficients U_{11}, V_{11}, W_{11}, which then allows both the strain field and the stress field to be determined and the present boundary value problem to be solved.

13.7.3 Bending of an Unsymmetric Angle-ply Laminate $\left[(\pm\theta)_N\right]$

An analogous solution as in the previous case of the unsymmetric cross-ply laminate also exists for the treatment of an unsymmetric angle-ply laminate of the type $\left[(\pm\theta)_N\right]$. The laminate stiffness matrix can be given as:

$$\begin{bmatrix} \underline{\underline{A}} & \underline{\underline{B}} \\ \underline{\underline{B}} & \underline{\underline{D}} \end{bmatrix} = \begin{bmatrix} A_{11} & A_{12} & 0 & 0 & 0 & B_{16} \\ A_{12} & A_{22} & 0 & 0 & 0 & B_{26} \\ 0 & 0 & A_{66} & B_{16} & B_{26} & 0 \\ 0 & 0 & B_{16} & D_{11} & D_{12} & 0 \\ 0 & 0 & B_{26} & D_{12} & D_{22} & 0 \\ B_{16} & B_{26} & 0 & 0 & 0 & D_{66} \end{bmatrix}, \tag{13.72}$$

and the displacement differential equations are for the present case:

$$A_{11}\frac{\partial^2 u_0}{\partial x^2} + A_{66}\frac{\partial^2 u_0}{\partial y^2} + (A_{12}+A_{66})\frac{\partial^2 v_0}{\partial x \partial y} - 3B_{16}\frac{\partial^3 w_0}{\partial x^2 \partial y} - B_{26}\frac{\partial^3 w_0}{\partial y^3} = 0,$$
$$(A_{12}+A_{66})\frac{\partial^2 u_0}{\partial x \partial y} + A_{66}\frac{\partial^2 v_0}{\partial x^2} + A_{22}\frac{\partial^2 v_0}{\partial y^2} - B_{16}\frac{\partial^3 w_0}{\partial x^3} - 3B_{26}\frac{\partial^3 w_0}{\partial x \partial y^2} = 0,$$
$$D_{11}\frac{\partial^4 w_0}{\partial x^4} + 2(D_{12}+2D_{66})\frac{\partial^4 w_0}{\partial x^2 \partial y^2} + D_{22}\frac{\partial^4 w_0}{\partial y^4}$$
$$-B_{16}\left(3\frac{\partial^3 u_0}{\partial x^2 \partial y} + \frac{\partial^3 v_0}{\partial x^3}\right) - B_{26}\left(3\frac{\partial^3 v_0}{\partial x \partial y^2} + \frac{\partial^3 u_0}{\partial y^3}\right) - p = 0. \tag{13.73}$$

We want to apply the so-called SS2 boundary conditions, which are as follows. At the edges at $x = 0$ and $x = a$ for $0 \leq y \leq b$ we have:

$$\begin{aligned} w_0 &= 0, \\ u_0 &= 0, \\ M^0_{xx} &= B_{16}\left(\frac{\partial u_0}{\partial y} + \frac{\partial v_0}{\partial x}\right) - D_{11}\frac{\partial^2 w_0}{\partial x^2} - D_{12}\frac{\partial^2 w_0}{\partial y^2} = 0, \\ N^0_{xy} &= A_{66}\left(\frac{\partial u_0}{\partial y} + \frac{\partial v_0}{\partial x}\right) - B_{16}\frac{\partial^2 w_0}{\partial x^2} - B_{26}\frac{\partial^2 w_0}{\partial y^2} = 0. \end{aligned} \tag{13.74}$$

For $y = 0$ and $y = b$ for $0 \leq x \leq a$ the following conditions are to be fulfilled:

$$\begin{aligned} w_0 &= 0, \\ v_0 &= 0, \\ M^0_{yy} &= B_{26}\left(\frac{\partial u_0}{\partial y} + \frac{\partial v_0}{\partial x}\right) - D_{12}\frac{\partial^2 w_0}{\partial x^2} - D_{22}\frac{\partial^2 w_0}{\partial y^2} = 0, \\ N^0_{xy} &= A_{66}\left(\frac{\partial u_0}{\partial y} + \frac{\partial v_0}{\partial x}\right) - B_{16}\frac{\partial^2 w_0}{\partial x^2} - B_{26}\frac{\partial^2 w_0}{\partial y^2} = 0. \end{aligned} \tag{13.75}$$

A displacement approach of the type

$$\begin{aligned} u_0 &= \sum_{m=1}^{m=\infty}\sum_{n=1}^{n=\infty} U_{mn} \sin\left(\frac{m\pi x}{a}\right)\cos\left(\frac{n\pi y}{b}\right), \\ v_0 &= \sum_{m=1}^{m=\infty}\sum_{n=1}^{n=\infty} V_{mn} \cos\left(\frac{m\pi x}{a}\right)\sin\left(\frac{n\pi y}{b}\right), \\ w_0 &= \sum_{m=1}^{m=\infty}\sum_{n=1}^{n=\infty} W_{mn} \sin\left(\frac{m\pi x}{a}\right)\sin\left(\frac{n\pi y}{b}\right) \end{aligned} \tag{13.76}$$

satisfies all boundary conditions identically and, after insertion into the displacement differential equations, yields the following system of equations for the determination of the coefficients U_{mn}, V_{mn}, W_{mn}:

$$\begin{bmatrix} \lambda_{11} & \lambda_{12} & \lambda_{13} \\ \lambda_{12} & \lambda_{22} & \lambda_{23} \\ \lambda_{13} & \lambda_{23} & \lambda_{33} \end{bmatrix} \begin{pmatrix} U_{mn} \\ V_{mn} \\ W_{mn} \end{pmatrix} = \begin{pmatrix} 0 \\ 0 \\ P_{mn} \end{pmatrix}, \tag{13.77}$$

with:

$$\lambda_{11} = A_{11}\frac{m^2\pi^2}{a^2} + A_{66}\frac{n^2\pi^2}{b^2}, \quad \lambda_{12} = (A_{12} + A_{66})\frac{m\pi}{a}\frac{n\pi}{b},$$

$$\lambda_{13} = -\left(3B_{16}\frac{m^2\pi^2}{a^2} + B_{26}\frac{n^2\pi^2}{b^2}\right)\frac{n\pi}{b}, \quad \lambda_{22} = A_{66}\frac{m^2\pi^2}{a^2} + A_{22}\frac{n^2\pi^2}{b^2},$$

$$\lambda_{23} = -\left(B_{16}\frac{m^2\pi^2}{a^2} + 3B_{26}\frac{n^2\pi^2}{b^2} \right)\frac{m\pi}{a},$$

$$\lambda_{33} = D_{11}\frac{m^4\pi^4}{a^4} + 2\left(D_{12} + 2D_{66}\right)\frac{m^2\pi^2}{a^2}\frac{n^2\pi^2}{b^2} + D_{22}\frac{n^4\pi^4}{b^4}. \tag{13.78}$$

This system of equations is easily solvable for the constants U_{mn}, V_{mn}, W_{mn}, from which the displacement field of the unsymmetric angle-ply laminate can then be determined. From this, all strain and stress components can then be computed.

References

Altenbach, H., Altenbach, J., Kissing, W.: Mechanics of Composite Structural Elements. Springer Singapore, Singapore (2018)

Ambartsumyan, S.A.: Theory of Anisotropic Plates. Technomic Publishing Co., Inc., Stamford (1970)

Ashton, J.E., Whitney, J.M.: Theory of Laminated Plates. Technomic Publishing Co., Inc., Stamford (1970)

Jones, R.M.: Mechanics of Composite Materials. Scripta Book Co., Washington (1975)

Lekhnitskii, S.G.: Anisotropic Plates. Gordon and Breach, London (1968)

Mittelstedt, C., Becker, W.: Strukturmechanik ebener Laminate. Verlag Studienbereich Mechanik, Technische Universität, Darmstadt (2016)

Reddy, J.N.: Mechanics of Laminated Composite Plates and Shells, 2nd edn. CRC Press, Boca Raton (2004)

Part IV
Shells

Introduction to Shell Structures 14

This chapter introduces the basics of the analysis of shells. It provides a very basic introduction to the consideration of shells as load-bearing structures and clarifies some basic concepts. In addition to some geometric relationships, the basic assumptions of Classical Shell Theory are explained and the stress, displacement, and strain quantities that occur are introduced along with relevant load types.

The treatment of shell problems is the subject of a number of books, of which the reference list contains a selection.

14.1 Introduction

Curved thin-walled structures, known as shells, are of considerable importance in engineering. A shell is a structure arbitrarily curved in space, which can be computationally characterized (analogously to disks and plates) by its midsurface, if it is assumed that the thickness of the shell is small compared to its other dimensions. While, assuming geometric linearity, the corresponding load-bearing effects of disks and plates can be considered separately, this cannot generally be assumed for a shell. Due to the curvature of the structure, membrane and bending effects have to be considered simultaneously, which naturally complicates the analysis.

With respect to their geometry, shells are divided into certain basic types. One has to distinguish between analytically definable shell surfaces on the one hand and free-form surfaces on the other hand. The analytically definable shell surfaces are subdivided into certain classes which we will not discuss in detail here. So-called shells of revolution are shells formed by the rotation of a plane curve about an axis, where the axis lies in the plane of the generatrix. A selection of shells of revolution is shown in Fig. 14.1.

In the following, we will present the essential basics for the analysis of shell structures and discuss them in detail for some selected cases. In all explanations,

C. Mittelstedt, *Theory of Plates and Shells*,
https://doi.org/10.1007/978-3-662-66805-4_14

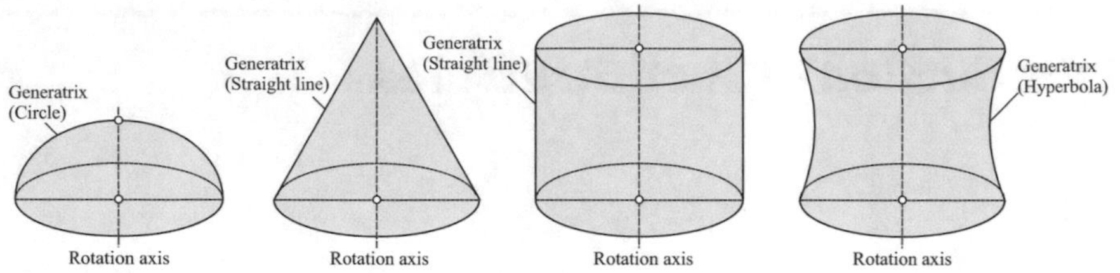

Fig. 14.1 Selected shells of revolution: Spherical shell, conical shell, cylindrical shell, rotational hyperboloid

we will assume isotropic and sufficiently thin shell structures, assuming not only geometric linearity but also material linearity and thus assuming the validity of Hooke's law. We will restrict ourselves to the analysis of shells of revolution, where not only the shell shape but also the boundary conditions are rotationally symmetric. The shell has the thickness h, which does not necessarily have to be constant. A description of the shell then is performed, as is usual for disks and plates, via the computational reduction to the shell midsurface (which bisects the shell at each point) and the thickness h, which in combination with the elastic properties of the shell material is found in effective stiffnesses, here membrane and bending stiffnesses. Let the thickness direction of the shell be denoted as z and point in the direction of the normal to the shell midsurface at any point on the shell.

14.2 Shells of Revolution

In the description of shells of revolution, several geometrical parameters have to be defined as shown in Fig. 14.2. Consider a point P on a surface of revolution (see Fig. 14.2, top). The corresponding normal vector to the surface of revolution is denoted by $\underline{n}$. The surface of revolution is described on the one hand by its longitudinal circle or meridian, which is also the generatrix. On the other hand, there is the so-called circle of latitude, which can be understood as a directrix at the same time. Meridian and directrix are always perpendicular to each other and are used here as coordinates. The position of the point P on the circle of latitude is described by the angle ϑ, i.e. the angle of the meridian plane against a reference meridian plane. The inclination of the normal vector to the rotation axis and thus the position of the point P is defined by the angle φ, i.e. the angle of the normal of the rotation surface with respect to the rotation axis. We denote the corresponding radii of curvature as r_ϑ and r_φ (cf. the angle shown in Fig. 14.2, bottom left, shown in meridian section with ϑ =const.). Since these are the main radii of curvature of the rotational body under consideration, we can also use the terms $r_1 = r_\varphi$ and $r_2 = r_\vartheta$. The following geometrical relation can be deduced:

$$r = r_2 \sin \varphi. \tag{14.1}$$

The meridian section shown in Fig. 14.2, bottom left, also shows the case where we consider a point infinitesimally adjacent to P on the surface of revolution. The corresponding angle is $\mathrm{d}\varphi$. The length $\mathrm{d}s_1$ of the meridian section can then be given by

$$\mathrm{d}s_1 = r_1 \mathrm{d}\varphi. \tag{14.2}$$

Likewise, on the latitudinal section shown in Fig. 14.2, bottom right, with φ =const. the length $\mathrm{d}s_2$ can be read:

$$\mathrm{d}s_2 = r \mathrm{d}\vartheta, \tag{14.3}$$

or with (14.1):

$$\mathrm{d}s_2 = r_2 \sin\varphi \mathrm{d}\vartheta. \tag{14.4}$$

The computation of an infinitesimal surface element $\mathrm{d}A$ of the surface of revolution can be performed using Fig. 14.3, left. We obtain:

$$\mathrm{d}A = \mathrm{d}s_1 \mathrm{d}s_2 = r_1 r_2 \sin\varphi \mathrm{d}\varphi \mathrm{d}\vartheta. \tag{14.5}$$

To uniquely describe the shape of the surface of revolution, the radius r is described as $r = r(\zeta)$ using the vertical coordinate ζ along the axis of rotation. From Fig. 14.3,

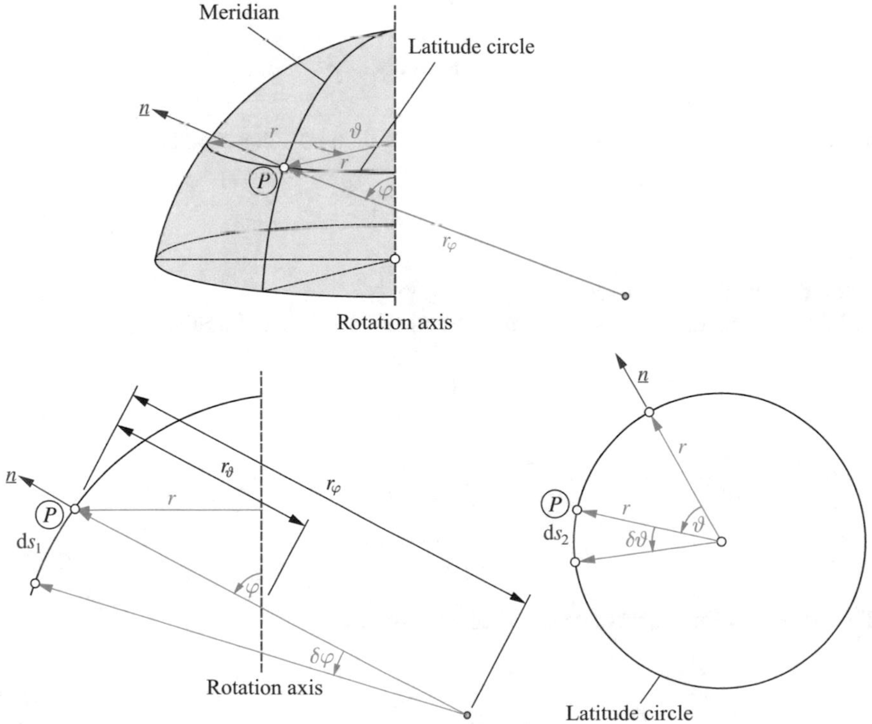

Fig. 14.2 Surface of revolution

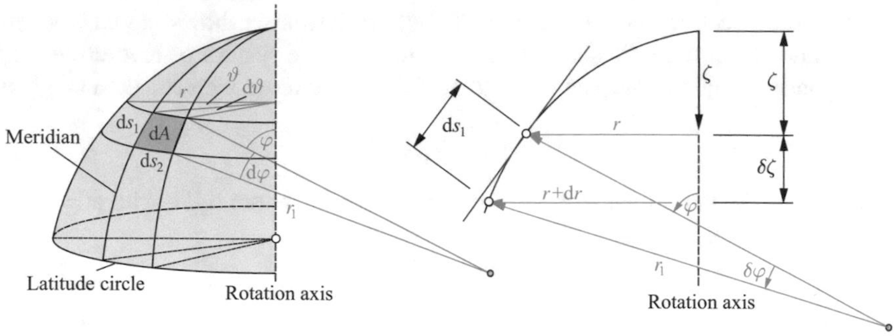

Fig. 14.3 Infinitesimal surface element of a surface of revolution (left), relation between ζ and $r(\zeta)$ (right)

right, we can deduce:

$$\mathrm{d}r = \cos\varphi \mathrm{d}s_1 = r_1 \cos\varphi \mathrm{d}\varphi. \tag{14.6}$$

For $\mathrm{d}\zeta$ results:

$$\mathrm{d}\zeta = r_1 \sin\varphi \mathrm{d}\varphi, \tag{14.7}$$

and thus the change of r over ζ follows as:

$$\frac{\mathrm{d}r}{\mathrm{d}\zeta} = r' = \cot\varphi. \tag{14.8}$$

The surface area A of a body of revolution can then be calculated as:

$$A = 2\pi \int_{\zeta_0}^{\zeta_1} r\sqrt{1+r'^2}\mathrm{d}\zeta, \tag{14.9}$$

when ζ takes values between $\zeta = \zeta_0$ and $\zeta = \zeta_1$.

We now consider the radii of curvature of the meridian. It holds:

$$r_1 = -\frac{\left(1+r'^2\right)^{\frac{3}{2}}}{r''}. \tag{14.10}$$

With $r_2 = \frac{r}{\sin\varphi}$ mit $\sin\varphi = \frac{1}{\sqrt{1+\cot^2\varphi}} = \frac{1}{\sqrt{1+r'^2}}$ one obtains:

$$r_2 = r\sqrt{1+r'^2}. \tag{14.11}$$

The corresponding curvatures can thus be given as:

$$\kappa_1 = \frac{1}{r_1} = -\frac{r''}{\left(1+r'^2\right)^{\frac{3}{2}}}, \quad \kappa_2 = \frac{1}{r_2} = \frac{1}{r\sqrt{1+r'^2}}. \tag{14.12}$$

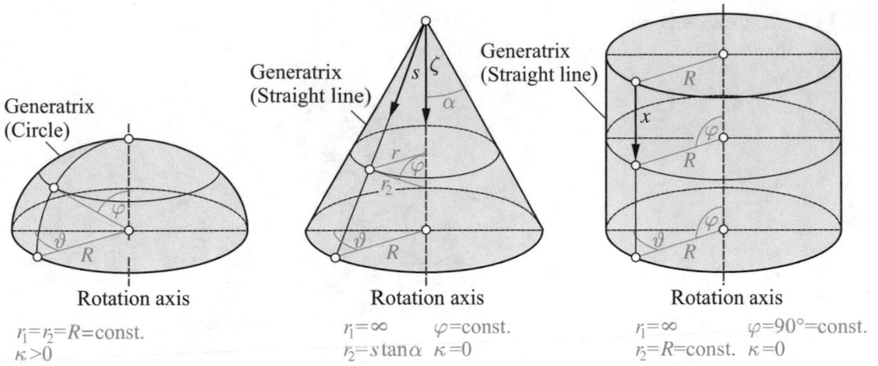

Fig. 14.4 Examples of surfaces of revolution: Spherical shell (left), conical shell (center), cylindrical shell (right)

The Gaussian curvature κ follows from this as:

$$\kappa = \frac{1}{r_1}\frac{1}{r_2} = -\frac{r''}{r\left(1 + r'^2\right)^2}. \tag{14.13}$$

This curvature measure is not only crucial in describing a curved surface, but also in evaluating the static behavior of a shell of revolution. If we assume that $r > 0$ holds, then the sign of κ is determined by the sign of r''. The case where $r'' < 0$ holds leads to $\kappa > 0$. Such a surface of revolution is called elliptic, and the center of curvature M is then on the side of the axis of revolution as seen from the point of the curve. Examples are spherical surfaces (see Fig. 14.4, left) or paraboloidal surfaces. If the case $r'' = 0$ is given, then also the Gaussian curvature κ becomes zero. The center of curvature is then at infinity. Such surfaces are called parabolic, examples are conical shells (Fig. 14.4, center) or cylindrical surfaces (Fig. 14.4, right). The case $r'' > 0$ with $\kappa < 0$, on the other hand, is called hyperbolic. The center of curvature, as seen from the point of the curve, is on the side away from the axis of rotation, and such a surface is called a rotational hyperboloid.

14.3 Load Cases

In this subsection we present typical types of loading relevant for the purposes of this book, based on shells of revolution.

The load on a shell of revolution can be in the form of surface loads, which can act tangentially on the shell surface (p_ϑ, p_φ) or normal to the shell surface (in the form of p_z, if z is the direction of the surface normal). This is shown in Fig. 14.5.

Fig. 14.5 Surface loads p_ϑ, p_φ, p_z

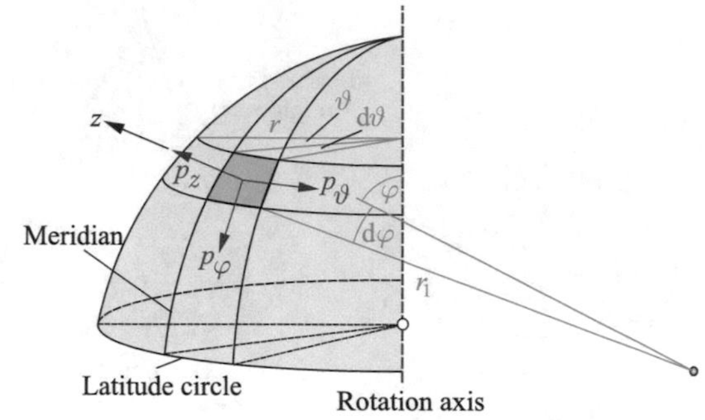

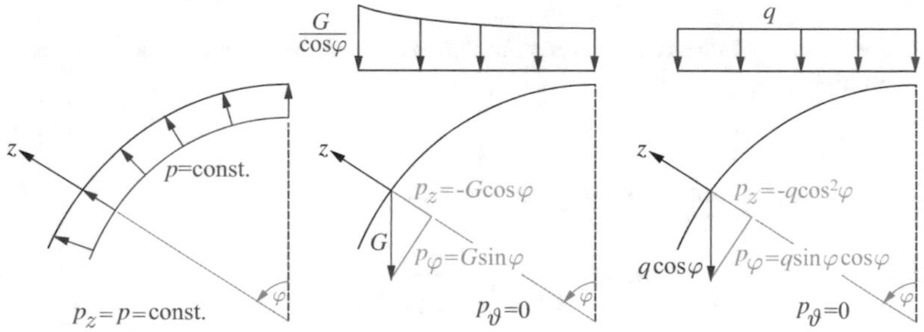

Fig. 14.6 Selected surface loads: constant internal pressure (left), dead weight (center), load uniformly distributed over the floor plan (right)

Some examples for rotationally symmetric surface loads with $p = f(\varphi)$ and $\frac{\partial p}{\partial \vartheta} = 0$ are shown in Fig. 14.6. However, rotationally symmetric loads can also be in the form of force and moment flows and as individual forces and moments (Fig. 14.7). In the case of non-rotationally symmetric loads, a given load can be approximated by development in the form of a Fourier series:

$$\begin{aligned}
p_\varphi &= \sum_{m=1}^{\infty} P_{\varphi,m}(\varphi)\cos m\vartheta + \sum_{n=1}^{\infty} \bar{P}_{\varphi,n}(\varphi)\sin n\vartheta, \\
p_\vartheta &= \sum_{m=1}^{\infty} P_{\vartheta,m}(\varphi)\cos m\vartheta + \sum_{n=1}^{\infty} \bar{P}_{\vartheta,n}(\varphi)\sin n\vartheta, \\
p_z &= \sum_{m=1}^{\infty} P_{z,m}(\varphi)\cos m\vartheta + \sum_{n=1}^{\infty} \bar{P}_{z,n}(\varphi)\sin n\vartheta.
\end{aligned} \tag{14.14}$$

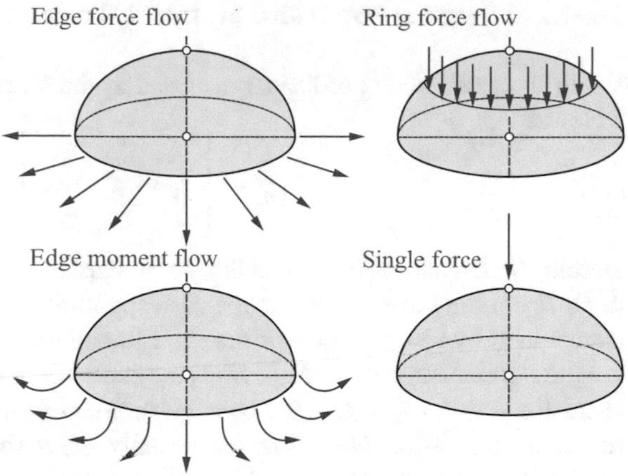

Fig. 14.7 Selected rotational symmetric loads

14.4 Classical Shell Theory

14.4.1 Assumptions

In this section we summarize the basics of the so-called Classical Shell Theory used in the following. It will be seen that this approach to the analysis of the static behavior of shell structures is closely related to the treatment of disks and plates when using Kirchhoff's plate theory. We make the following assumptions:

(i) We assume a geometrically and materially linear theory. Therefore, we can linearize all kinematic relations. Moreover, Hooke's law is assumed to be valid and the considerations are limited to isotropic shells.

(ii) We assume a thin shell structure. Let the shell thickness h be very small compared to the shell surface dimensions. This allows, just as in the case of plates and disks, to reduce the considerations to the shell's midsurface.

(iii) As in the case of the Kirchhoff plate, we assume the hypothesis that the cross-sections of the shell remain plane as well as the normal hypothesis to be valid. A straight line that is normal to the shell midsurface before deformation remains a straight line in the deformed state and, moreover, normal to the deformed midsurface. In addition, we assume that the thickness of the shell does not change due to the deformation.

(iv) We assume a plane stress state with respect to the thickness direction of the shell.

14.4.2 Stresses; Force and Moment Quantities

Using the reference coordinates φ, ϑ and z, the stress tensor in a shell point is as follows:

$$\underline{\underline{\sigma}} = \begin{bmatrix} \sigma_{\varphi\varphi} & \tau_{\varphi\vartheta} & \tau_{\varphi z} \\ \tau_{\varphi\vartheta} & \sigma_{\vartheta\vartheta} & \tau_{\vartheta z} \\ \tau_{\varphi z} & \tau_{\vartheta z} & \sigma_{zz} \end{bmatrix}. \tag{14.15}$$

Herein, we have already applied the fact that associated shear stresses must be identical. From the stress components, corresponding force and moment flows can be determined by integration over the shell thickness h, as shown in Fig. 14.8. These are the normal force flows $N_{\varphi\varphi}$, $N_{\vartheta\vartheta}$, the shear force flows $N_{\varphi\vartheta}$, $N_{\vartheta\varphi}$ the transverse shear force flows Q_φ, Q_ϑ, the bending moment flows $M_{\varphi\varphi}$, $M_{\vartheta\vartheta}$ and the twisting moment flows $M_{\varphi\vartheta}$, $M_{\vartheta\varphi}$. As we can easily show, the shear force flows $N_{\varphi\vartheta}$, $N_{\vartheta\varphi}$ and the twisting moment flows $M_{\varphi\vartheta}$, $M_{\vartheta\varphi}$ are not identical, unlike in the case of the disk and the plate.

To motivate the calculation of force and moment quantities, we consider Fig. 14.9, in which a shell element is shown. The perspective runs tangential to the shell plane in the direction of the angle φ. Let the length of the segment of the shell midsurface shown here be the unit length 1. The surface element dA, also shown here, then is:

$$\mathrm{d}A = \mathrm{d}z\left(1 + \frac{z}{r_2}\right). \tag{14.16}$$

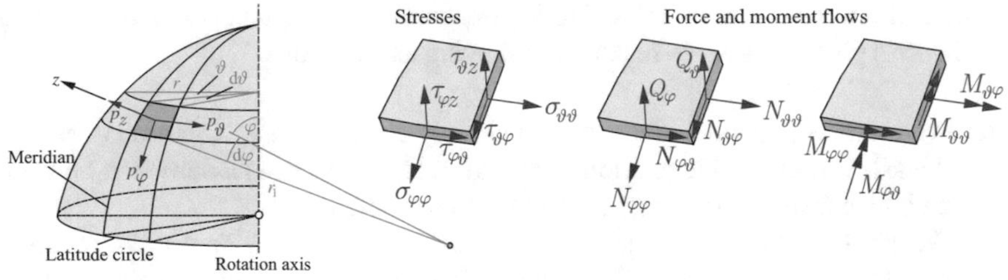

Fig. 14.8 Stress state, force and moment flows of a shell

Fig. 14.9 On the determination of force and moment quantities of the shell

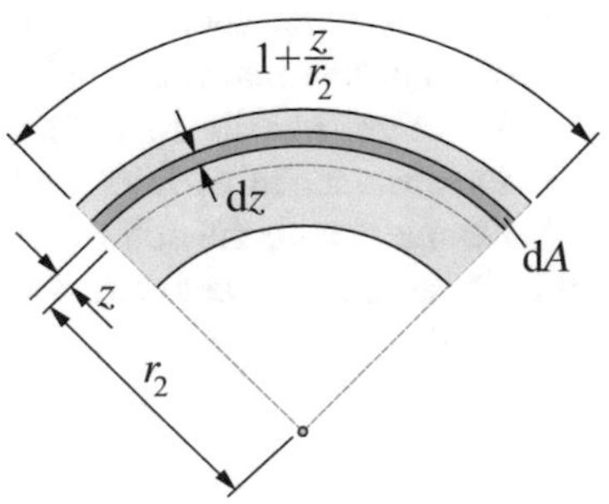

Thus, for example, the normal force flow $N_{\varphi\varphi}$ can be calculated as:

$$N_{\varphi\varphi} = \int_{-\frac{h}{2}}^{\frac{h}{2}} \sigma_{\varphi\varphi} \mathrm{d}A = \int_{-\frac{h}{2}}^{\frac{h}{2}} \sigma_{\varphi\varphi} \left(1 + \frac{z}{r_2}\right) \mathrm{d}z. \tag{14.17}$$

Analogously, for example, the bending moment flow $M_{\varphi\varphi}$ can be determined:

$$M_{\varphi\varphi} = \int_{-\frac{h}{2}}^{\frac{h}{2}} \sigma_{\varphi\varphi} z \mathrm{d}A = \int_{-\frac{h}{2}}^{\frac{h}{2}} \sigma_{\varphi\varphi} z \left(1 + \frac{z}{r_2}\right) \mathrm{d}z. \tag{14.18}$$

In this context, one also speaks of the so-called trapezoidal effect. The force and moment flows of the shell can be determined as follows:

$$\begin{aligned}
N_{\varphi\varphi} &= \int_{-\frac{h}{2}}^{\frac{h}{2}} \sigma_{\varphi\varphi} \left(1 + \frac{z}{r_2}\right) \mathrm{d}z, \quad N_{\vartheta\vartheta} = \int_{-\frac{h}{2}}^{\frac{h}{2}} \sigma_{\vartheta\vartheta} \left(1 + \frac{z}{r_1}\right) \mathrm{d}z, \\
N_{\varphi\vartheta} &= \int_{-\frac{h}{2}}^{\frac{h}{2}} \tau_{\varphi\vartheta} \left(1 + \frac{z}{r_2}\right) \mathrm{d}z, \quad N_{\vartheta\varphi} = \int_{-\frac{h}{2}}^{\frac{h}{2}} \tau_{\vartheta\varphi} \left(1 + \frac{z}{r_1}\right) \mathrm{d}z, \\
M_{\varphi\varphi} &= \int_{-\frac{h}{2}}^{\frac{h}{2}} \sigma_{\varphi\varphi} z \left(1 + \frac{z}{r_2}\right) \mathrm{d}z, \quad M_{\vartheta\vartheta} = \int_{-\frac{h}{2}}^{\frac{h}{2}} \sigma_{\vartheta\vartheta} z \left(1 + \frac{z}{r_1}\right) \mathrm{d}z, \\
M_{\varphi\vartheta} &= \int_{-\frac{h}{2}}^{\frac{h}{2}} \tau_{\varphi\vartheta} z \left(1 + \frac{z}{r_2}\right) \mathrm{d}z, \quad M_{\vartheta\varphi} = \int_{-\frac{h}{2}}^{\frac{h}{2}} \tau_{\vartheta\varphi} z \left(1 + \frac{z}{r_1}\right) \mathrm{d}z, \\
Q_{\varphi} &= \int_{-\frac{h}{2}}^{\frac{h}{2}} \tau_{\varphi z} \left(1 + \frac{z}{r_2}\right) \mathrm{d}z, \quad Q_{\vartheta} = \int_{-\frac{h}{2}}^{\frac{h}{2}} \tau_{\vartheta z} \left(1 + \frac{z}{r_1}\right) \mathrm{d}z.
\end{aligned} \tag{14.19}$$

The result for the shell is that the associated shear force flows $N_{\varphi\vartheta}$ and $N_{\vartheta\varphi}$ as well as the two associated twisting moment flows $M_{\varphi\vartheta}$ and $M_{\vartheta\varphi}$ will not be identical due to the generally different radii r_1 and r_2, despite the equality of the shear stresses $\tau_{\varphi\vartheta}$ and $\tau_{\vartheta\varphi}$ determining them. An exception is, for example, the spherical shell where $r_1 = r_2$ holds. The bending moment flows $M_{\varphi\varphi}$ and $M_{\vartheta\vartheta}$ are defined such that a positive moment flow induces tensile stresses in the range $z > 0$. The twisting moment flow $M_{\varphi\vartheta}$ is positive for shear stresses $\tau_{\varphi\vartheta} = \tau_{\vartheta\varphi}$ for $z > 0$ acting at cuts with positive outer normal in positive coordinate direction. The force flows are positive if they act in positive coordinate direction at a section with positive outer normal.

Figure 14.10 shows the force and moment flows at an infinitesimal element of the shell.

It should be noted that bending and twisting moment flows can occur in a shell even in the presence of constant stresses across the cross-section. This is explained by the curvature of the shell and the definitions (14.19).

In some cases it can be assumed that the recurrent terms $\frac{z}{r_1}$ and $\frac{z}{r_2}$ in (14.19) are negligibly small compared to the value 1. This is denoted as the so-called

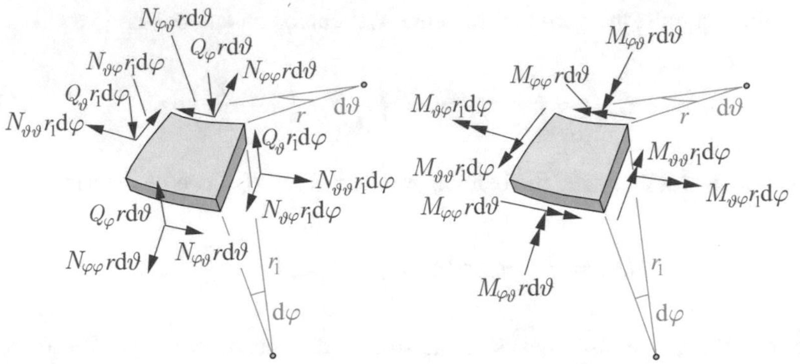

Fig. 14.10 Force and moment flows at an infinitesimal shell element

Kirchhoff-Love approximation. Then, for the shell force and moment flows one obtains:

$$
\begin{aligned}
N_{\varphi\varphi} &= \int_{-\frac{h}{2}}^{\frac{h}{2}} \sigma_{\varphi\varphi} \mathrm{d}z, \quad N_{\vartheta\vartheta} = \int_{-\frac{h}{2}}^{\frac{h}{2}} \sigma_{\vartheta\vartheta} \mathrm{d}z, \\
N_{\varphi\vartheta} &= \int_{-\frac{h}{2}}^{\frac{h}{2}} \tau_{\varphi\vartheta} \mathrm{d}z = N_{\vartheta\varphi}, \quad M_{\varphi\varphi} = \int_{-\frac{h}{2}}^{\frac{h}{2}} \sigma_{\varphi\varphi} z \mathrm{d}z, \\
M_{\vartheta\vartheta} &= \int_{-\frac{h}{2}}^{\frac{h}{2}} \sigma_{\vartheta\vartheta} z \mathrm{d}z, \quad M_{\varphi\vartheta} = \int_{-\frac{h}{2}}^{\frac{h}{2}} \tau_{\varphi\vartheta} z \mathrm{d}z = M_{\vartheta\varphi}, \\
Q_\varphi &= \int_{-\frac{h}{2}}^{\frac{h}{2}} \tau_{\varphi z} \mathrm{d}z, \quad Q_\vartheta = \int_{-\frac{h}{2}}^{\frac{h}{2}} \tau_{\vartheta z} \mathrm{d}z.
\end{aligned}
\tag{14.20}
$$

Obviously, when neglecting the trapezoidal effect the associated shear force flows $N_{\varphi\vartheta}$ and $N_{\vartheta\varphi}$ as well as the two associated twisting moment flows $M_{\varphi\vartheta}$ and $M_{\vartheta\varphi}$ are identical.

If the shell force and moment flows are available, then the corresponding stress distributions can be determined. We will illustrate this by assuming the Kirchhoff-Love approximation $\frac{z}{r_1} << 1$ and $\frac{z}{r_2} << 1$ to be valid. We further assume that the stress distributions are linear across the cross section despite the curvature of the shell. The nonlinearity resulting from the curvature thus will not be considered here. As an example, we consider the normal stress $\sigma_{\varphi\varphi}$, which we divide into the two parts $\sigma_{\varphi\varphi 0}$ and $\sigma_{\varphi\varphi 1}$ (see Fig. 14.11). Then:

$$
\begin{aligned}
N_{\varphi\varphi} &= \int_{-\frac{h}{2}}^{\frac{h}{2}} \sigma_{\varphi\varphi} \mathrm{d}z = \sigma_{\varphi\varphi 0} h, \\
M_{\varphi\varphi} &= \int_{-\frac{h}{2}}^{\frac{h}{2}} \sigma_{\varphi\varphi} z \mathrm{d}z = \sigma_{\varphi\varphi 1} \frac{h^2}{6}.
\end{aligned}
\tag{14.21}
$$

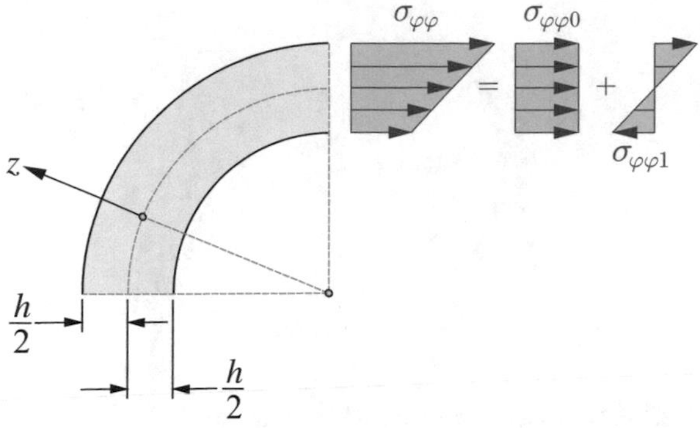

Fig. 14.11 Normal stress $\sigma_{\varphi\varphi}$

The boundary values of the normal stress $\sigma_{\varphi\varphi}$ according to Fig. 14.11 are then:

$$\sigma_{\varphi\varphi}\left(z = \pm\frac{h}{2}\right) = \frac{N_{\varphi\varphi}}{h} \pm \frac{6M_{\varphi\varphi}}{h^2}. \tag{14.22}$$

An analogous expression can be given for the boundary values of the normal stress $\sigma_{\vartheta\vartheta}$:

$$\sigma_{\vartheta\vartheta}\left(z = \pm\frac{h}{2}\right) = \frac{N_{\vartheta\vartheta}}{h} \pm \frac{6M_{\vartheta\vartheta}}{h^2}. \tag{14.23}$$

Similarly, the boundary values of the shear stress $\tau_{\varphi\vartheta}$ can be assumed to be identical to the boundary values of $\tau_{\vartheta\varphi}$ due to the assumption of Kirchhoff-Love approximation $\frac{z}{r_1} << 1$ and $\frac{z}{r_2} << 1$:

$$\tau_{\varphi\vartheta}\left(z = \pm\frac{h}{2}\right) = \tau_{\vartheta\varphi}\left(z = \pm\frac{h}{2}\right) = \frac{N_{\varphi\vartheta}}{h} \pm \frac{6M_{\varphi\vartheta}}{h^2}. \tag{14.24}$$

The transverse shear stresses associated with the transverse shear flows Q_φ and Q_ϑ will be parabolically distributed over the shell thickness similar to those for the Euler-Bernoulli beam or the Kirchhoff plate and can be calculated as follows:

$$\tau_{\varphi z} = \frac{3Q_\varphi}{2h}\left[1 - \left(\frac{2z}{h}\right)^2\right], \quad \tau_{\vartheta z} = \frac{3Q_\vartheta}{2h}\left[1 - \left(\frac{2z}{h}\right)^2\right]. \tag{14.25}$$

14.4.3 Strains and Displacements

The infinitesimal strain tensor $\underline{\underline{\varepsilon}}$ in the reference coordinates employed here reads:

$$\underline{\underline{\varepsilon}} = \begin{bmatrix} \varepsilon_{\varphi\varphi} & \gamma_{\varphi\vartheta} & \gamma_{\varphi z} \\ \gamma_{\vartheta\varphi} & \varepsilon_{\vartheta\vartheta} & \gamma_{\vartheta z} \\ \gamma_{\varphi z} & \gamma_{\vartheta z} & \varepsilon_{zz} \end{bmatrix}. \tag{14.26}$$

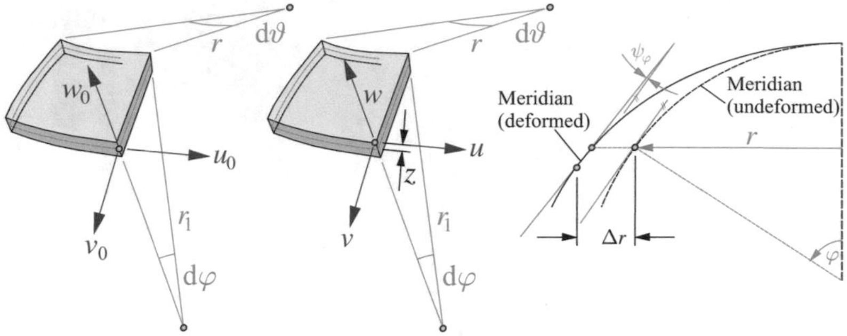

Fig. 14.12 Displacement quantities of the shell

Due to the hypothesis that the cross-sections remain plane and the normal hypothesis, the transverse shear strains $\gamma_{\varphi z}$ and $\gamma_{\vartheta z}$ are omitted. The assumption that the shell thickness h does not change during the deformation process also yields $\varepsilon_{zz} = 0$. Thus, the strain state of a shell in the context of Classical Shell Theory is given by the strains $\varepsilon_{\varphi\varphi}$ (strain in meridian direction) and $\varepsilon_{\vartheta\vartheta}$ (strain in the annular direction) as well as the shear strain $\gamma_{\varphi\vartheta} = \gamma_{\vartheta\varphi}$ (angle change between the coordinate lines φ and ϑ). We also determine the corresponding strain quantities with respect to the shell midsurface $\varepsilon^0_{\varphi\varphi}$, $\varepsilon^0_{\vartheta\vartheta}$, $\gamma^0_{\varphi\vartheta} = \gamma^0_{\vartheta\varphi}$, which do not depend on the thickness coordinate z.

The displacement quantities are the displacements u, v and w for displacements of a shell point outside the shell midsurface, where u is the displacement in the ring direction, v is the displacement in the meridian direction and w is the displacement in the thickness direction z. They are functions of the two angles φ and ϑ and the thickness coordinate z. In addition, there are the displacements u_0, v_0 and w_0 of the shell midsurface, which are not functions of z. The displacement quantities are shown in Fig. 14.12. A displacement quantity derived from this is the rotation angle of the tangent of the meridian, which we will call ψ_φ. It is shown in 14.12, right. The displacement of the considered shell point perpendicular to the rotation axis is called Δr.

References

Altenbach, H., Altenbach, J., Kissing, W.: Mechanics of Composite Structural Elements. Springer, Berlin (2004)

Ambartsumyan, S.A.: Theory of Anisotropic Shells. NASA Technical Translation TT F-118. Washington DC (1964)

Axelrad, E.L.: Schalentheorie. Teubner, Stuttgart (1983)

Axelrad, E.L.: Theory of Flexible Shells. North-Holland, Amsterdam (1987)

Baker, E.H., Kovalevsky, L., Rish, F.L.: Structural Analysis of Shells. McGraw Hill, New York (1972)

Becker, W., Gross, D.: Mechanik elastischer Körper und Strukturen. Springer, Berlin (2002)

Beles, A.A., Soare, M.V.: Berechnung von Schalentragwerken. Bauverlag, Wiesbaden (1972)

Born, J. (1968): Praktische Schalenstatik, Bd. I: Die Rotationsschalen, 2nd edn. Ernst und Sohn, Berlin (1968)

Calladine, C.R.: Theory of Shell Structures. Cambridge University Press, Cambridge (1983)

Czerwenka, G., Schnell, W.: Einführung in die Rechenmethoden des Leichtbaus, vol. 1. Bibliographisches Institut, Mannheim (1970)

Czerwenka, G., Schnell, W.: Einführung in die Rechenmethoden des Leichtbaus, vol. 2. Bibliographisches Institut, Mannheim (1970)

Dym, C.L.: Introduction to the Theory of Shells. Hemisphere Publishing Co., New York (1990)

Eschenauer, H., Schnell, W.: Elastizitätstheorie. Bibliographisches Institut, Mannheim (1993)

Eschenauer, H., Olhoff, N., Schnell, W.: Applied Structural Mechanics. Springer, Berlin (1997)

Flügge, W.: Statik und Dynamik der Schalen, 3rd edn. Springer, Berlin (1981)

Gibson, J.E.: Linear Elastic Theory of Thin Shells. Pergamon, Oxford (1965)

Gibson, J.E.: Thin Shells: Computing and Theory. Pergamon, Oxford (1980)

Girkmann, K.: Flächentragwerke, 6th edn. Springer, Wien (1974)

Göldner, H., Altenbach, J., Eschke, K., Garz, K.F., Sähn, S.: Lehrbuch Höhere Festigkeitslehre, vol. 1. Physik, Weinheim (1979)

Göldner, H., Altenbach, J., Eschke, K., Garz, K.F., Sähn, S.: Lehrbuch Höhere Festigkeitslehre, vol. 2. Physik, Weinheim (1985)

Gould, P.L.: Static Analysis of Shells. DC Heath and Company, Lexington (1977)

Hake, E., Meskouris, K.: Statik der Flächentragwerke, 2nd edn. Springer, Berlin (2007)

Hampe, E.: Statik Rotationssymmetrischer Flächentragwerke Bd. 1: Allgemeine Rotationsschale, Kreis- und Kreisringscheibe, Kreis- und Kreisringplatte, 3rd edn. Verlag für Bauwesen, Berlin (1963)

Hampe, E.: Statik Rotationssymmetrischer Flächentragwerke Bd. 2: Kreiszylinderschale, 3rd edn. Verlag für Bauwesen, Berlin (1964)

Hampe, E.: Statik Rotationssymmetrischer Flächentragwerke Bd. 3: Kegelschale. Kugelschale, 3rd edn. Verlag für Bauwesen, Berlin (1968)

Hampe, E.: Statik Rotationssymmetrischer Flächentragwerke Bd. 4: Zusammengesetzte Flächentragwerke. Zahlentafeln, 3rd edn. Verlag für Bauwesen, Berlin (1971)

Hampe, E.: Statik Rotationssymmetrischer Flächentragwerke Bd. 5: Hyperbelschalen, 3rd edn. Verlag für Bauwesen, Berlin (1973)

Mazurkiewicz, E.M., Nagorski, R.T.: Shells of Revolution. Elsevier, Amsterdam (1991)

Møllmann, H.: Introduction to the Theory of Thin Shells. Wiley, Chichester (1981)

Novozhilov, V.V.: The Theory of Thin Shells. Noordhoff, Groningen (1959)

Pflüger, A.: Elementare Schalenstatik, 4th edn. Springer, Berlin (1967)

Ramm, E., Müller, J.: Schalen. Universität Stuttgart, Institut für Baustatik und Baudynamik, Stuttgart (1995)

Reddy, J.N.: Mechanics of Laminated Composite Plates and Shells, 2nd edn. CRC Press, Boca Raton (2004)

Reddy, J.N.: Theory and Analysis of Elastic Plates and Shells. CRC Press, Boca Raton (2006)

Timoshenko, S., Woinowsky-Krieger, S.: Theory of Plates and Shells, 2nd edn. McGraw Hill, New York (1964)

Turner, C.E.: Introduction to Plate and Shell Theory. Longmans Green and Co. Ltd., London (1965)

Ugural, A.C.: Stresses in Plates and Shells. McGraw Hill, New York (1981)

Vinson, J.R.: The Behaviour of Plates and Shells. Wiley, New York (1974)

Vinson, J.R.: The Behaviour of Thin Walled Structures: Beams, Plates, and Shells. Kluwer, Dordrecht (1989)

Vinson, J.R.: The Behaviour of Shells Composed of Isotropic and Composite Materials. Kluwer, Dordrecht (1993)
Wiedemann, J.: Leichtbau 1: Elemente. Springer, Berlin (2007)
Wlassow, W.S.: Allgemeine Schalentheorie und ihre Anwendung in der Technik. Akademie-Verlag, Berlin (1958)

Membrane Theory of Shells of Revolution

15

In this chapter, the so-called membrane theory is described as a very useful analytical approach to the analysis of shell structures. After describing the limits of applicability of this theory, the equilibrium conditions for shells of revolution are derived and then specialized for the case of rotationally symmetric loading, from which the membrane force flows of the shells can be determined. Basic examples describe the application of the membrane theory to circular cylindrical, spherical and conical shells. The chapter concludes with a consideration of the kinematics of shells of revolution in the context of membrane theory, and both the necessary strain measures and expressions for the resulting displacement quantities are derived. Relevant literature that can be consulted for more in-depth elaborations can be found, e.g., with Becker and Gross (2002), Born (1968), Eschenauer and Schnell (1993), Eschenauer et al. (1997), Flügge (1981), Girkmann (1974), Hake and Meskouris (2007), Pflüger (1967), Ramm and Müller (1995), Reddy (2006), Timoshenko and Woinowsky-Krieger (1964), among others.

15.1 Assumptions

A very useful simplification of shell problems exists if one assumes that a shell structure under consideration carries a given load by a pure membrane action and that no bending action occurs. This is comparable to a disk in which only membrane force flows and plane stresses that are constantly distributed over the thickness occur. This is called a pure membrane stress state, and the corresponding theory is the so-called membrane theory. It is based on the following assumptions:

(i) It is assumed that the plane stress components $\sigma_{\varphi\varphi}, \sigma_{\vartheta\vartheta}, \tau_{\varphi\vartheta} = \tau_{\vartheta\varphi}$ are not only the only stresses acting, but are also constantly distributed over the shell thickness h and thus are not functions of z. Thus, the generally also occurring moment

C. Mittelstedt, *Theory of Plates and Shells*,
https://doi.org/10.1007/978-3-662-66805-4_15

quantities $M_{\varphi\varphi}$, $M_{\vartheta\vartheta}$, $M_{\varphi\vartheta} = M_{\vartheta\varphi}$ are negligibly small. This is accompanied by the disappearance of the transverse shear force flows Q_φ and Q_ϑ.

(ii) The bending and twisting stiffnesses of the shell are considered as negligible compared to the membrane stiffnesses. This, however, assumes that the shell is a thin shell, where all expressions in which the shell thickness h occurs in powers are negligible.

The assumptions made thus are equal to the requirement that the terms $\frac{z}{r_1}$ and $\frac{z}{r_2}$ vanish. This is also called the Kirchhoff-Love approximation. For the remaining force flows, the result is thus:

$$N_{\varphi\varphi} = \sigma_{\varphi\varphi} h, \quad N_{\vartheta\vartheta} = \sigma_{\vartheta\vartheta} h, \quad N_{\varphi\vartheta} = N_{\vartheta\varphi} = \tau_{\varphi\vartheta} h. \tag{15.1}$$

The only remaining equilibrium conditions that can be employed are then the force equilibria with respect to the meridional direction φ, the ring direction ϑ and the thickness direction z. The moment equilibria either yield no result due to the omission of all moment quantities (here the moment balances with respect to φ and ϑ) or are already included in the equality of the plane shear force flows $N_{\varphi\vartheta}$ and $N_{\vartheta\varphi}$ (here the moment equilibrium with respect to the z-axis).

With the three force flows and the three force balances, the considered membrane problem of the shell proves to be internally statically determinate. Thus, a consideration of the deformation state can be omitted if only the determination of the stress state of the shell is of interest and no shell structure is present which is internally statically indeterminate or statically indeterminately supported.

In addition to the mentioned formal conditions, however, there are a number of other points to be considered under which a shell can be regarded as free from bending action, and which must be assumed to be mandatory for the adoption of the membrane theory.

(i) The shell midsurface is required to be continuous with respect to its course, slope and curvature. The situation shown in Fig. 15.1, left, would thus be amenable to analysis by the membrane theory, the one shown in Fig. 15.1, right, is not. The reason is that due to the discontinuous slope of the shell midsurface shown here, local equilibrium with only the membrane force flows can no longer be ensured and a local bending effect occurs at this point.

(ii) The shell thickness h, which can vary over φ and ϑ, must not change abruptly (Fig. 15.2). According to this, the situation shown in Fig. 15.2, left, would be admissible, the shell section shown in Fig. 15.2, right, is not.

(iii) The applied surface loads are required to be continuously distributed and not to exhibit high variability or excessive waviness. Accordingly, the calculation of shells under concentrated loads is also not possible according to the membrane theory. Both single forces and highly variable surface loads cause deflection forces inside the shell, which are associated with a bending action.

(iv) Edge force flows of a shell are to be applied in such a way that they act tangentially to the shell's midsurface. Accordingly, only longitudinal and shear

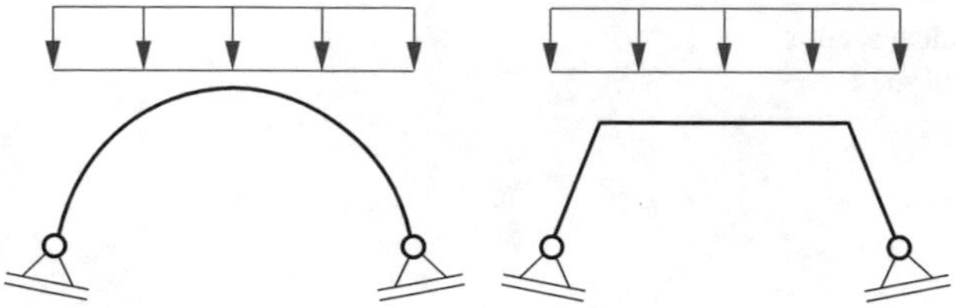

Fig. 15.1 On the shape of the shell midsurface

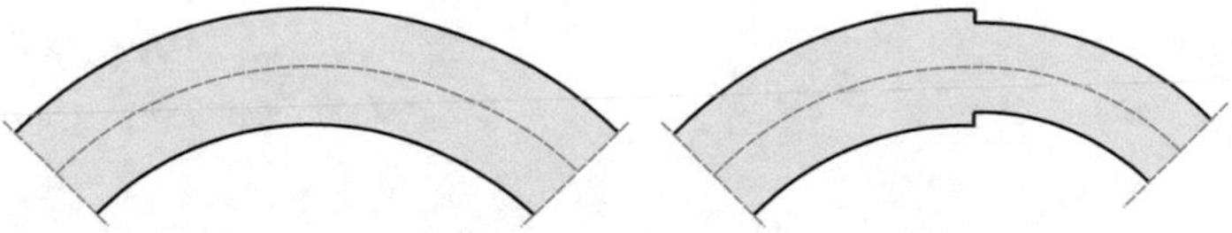

Fig. 15.2 On the distribution of the shell thickness h

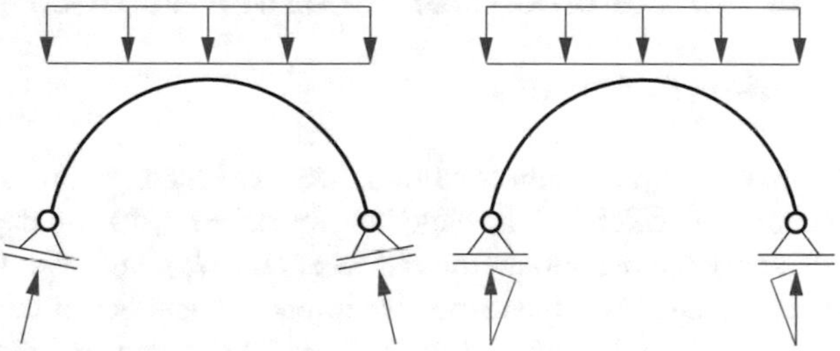

Fig. 15.3 Introduction of edge force flows

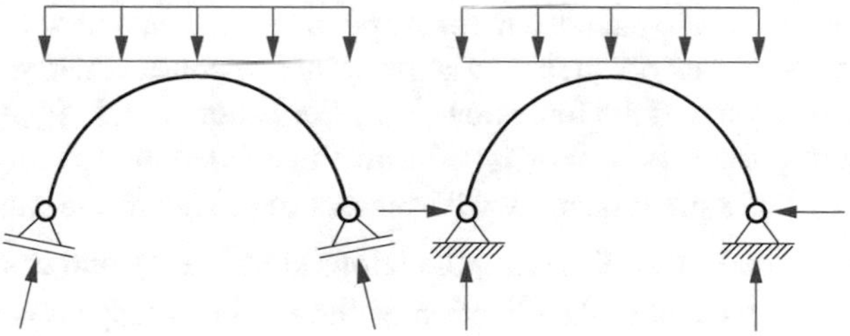

Fig. 15.4 Support conditions and edge displacements

forces are permitted (Fig. 15.3, left), but not transverse forces that would result in a bending action (Fig. 15.3, right). This must be ensured by an appropriate support design.

(v) Deformations that occur at bearings or at the boundaries to other components may only be restricted to the extent that the tangential direction of action of the forces occurring there is not changed (Fig. 15.4, left). Otherwise, a local bending effect will result at such locations (Fig. 15.4, right), which is no longer accessible to a calculation within the framework of the membrane theory.

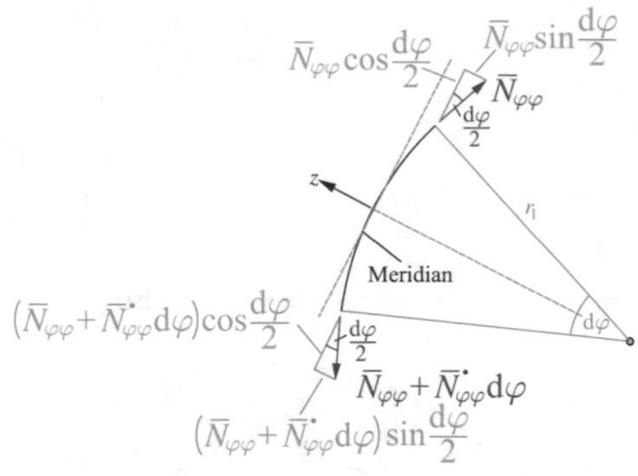

Fig. 15.5 Meridian section, decomposition of the force component $\bar{N}_{\varphi\varphi}$

15.2 Equilibrium Conditions for Shells of Revolution

15.2.1 Equilibrium Conditions

For the derivation of the equilibrium conditions we consider an infinitesimal element of the shell midsurface, which has the edge lengths $ds_1 = r_1 d\varphi$ and $ds_2 = r d\vartheta$. For this purpose we apply the membrane force flows $N_{\varphi\varphi}$, $N_{\vartheta\vartheta}$ and $N_{\varphi\vartheta}$ together with their infinitesimal increments at the respective intersection edges of the element and form the force equilibria in the direction of the meridian tangent and the ring tangent as well as with respect to the thickness direction. In order to derive the equilibrium conditions, we consider in the following various free body images and identify the force components that are included in the respective force balances.

Figure 15.5 shows a meridian section at the point ϑ =const. where we can identify the corresponding parts of the force flow $N_{\varphi\varphi}$. For better readability, we have used here the resulting force $\bar{N}_{\varphi\varphi} = N_{\varphi\varphi} r d\vartheta$ and the abbreviation $(....)^{\cdot}$ for a partial derivative $\frac{\partial}{\partial\varphi}(...)$ of a given quantity with respect to φ. Due to the smallness of the angle $d\varphi$, we can make use of the simplifications $\sin\frac{d\varphi}{2} \simeq \frac{d\varphi}{2}$ and $\cos\frac{d\varphi}{2} \simeq 1$. The resulting force component in the direction of the meridian tangent is then $\bar{N}^{\cdot}_{\varphi\varphi} d\varphi$, and in the normal direction the component $-\bar{N}_{\varphi\varphi} d\varphi$ remains.

In Fig. 15.6 the corresponding force components due to the shear forces $\bar{N}_{\vartheta\varphi} = N_{\vartheta\varphi} r_1 d\varphi$ and $\bar{N}_{\varphi\vartheta} = N_{\varphi\vartheta} r d\vartheta$ are shown. Here again we make use of the small angle approximation and use the abbreviation $(...)'$ for a partial derivative $\frac{\partial}{\partial\vartheta}(...)$ of a given quantity with respect to ϑ. Terms of higher order in the angles $d\varphi$ and $d\vartheta$ are neglected. The resulting force component in the direction of the meridian tangent is $\bar{N}'_{\vartheta\varphi} d\vartheta$, and in the ring direction $\bar{N}_{\vartheta\varphi} d\vartheta \cos\varphi + \bar{N}^{\cdot}_{\varphi\vartheta} d\varphi$.

The decomposition of the force $\bar{N}_{\vartheta\vartheta}$ is given in Fig. 15.7, using an annular section with φ =const. (Fig. 15.7, left) and a meridian section (Fig. 15.7, right). With the simplifications and designations already mentioned, the force component remaining in the meridional direction is $-\bar{N}_{\vartheta\vartheta} \cos\varphi d\vartheta$. In the annular direction the component $\bar{N}'_{\vartheta\vartheta} d\vartheta$ remains, and in the normal direction the component $-\bar{N}_{\vartheta\vartheta} \sin\varphi d\vartheta$ results.

Fig. 15.6 Decomposition of the force components $\bar{N}_{\varphi\vartheta}$ and $\bar{N}_{\vartheta\varphi}$

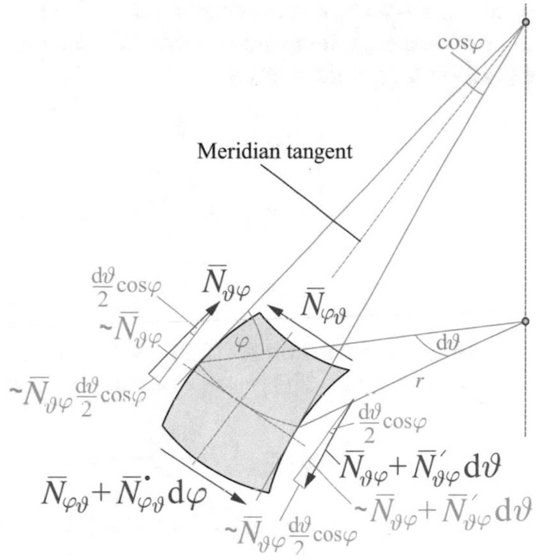

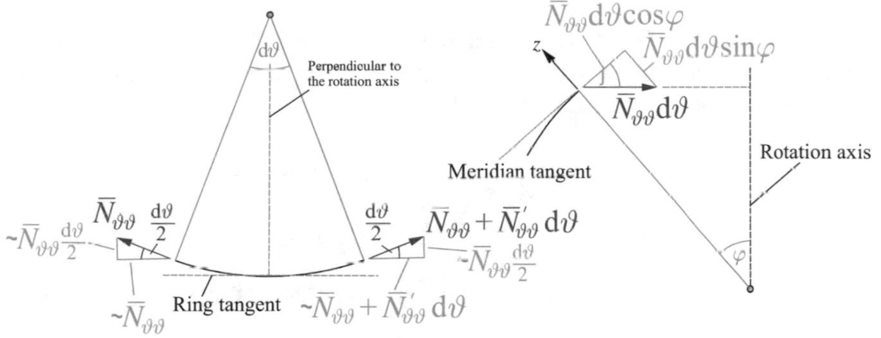

Fig. 15.7 Decomposition of the force component $\bar{N}_{\vartheta\vartheta}$

When balancing the forces in meridian, ring and normal directions, the shares of the surface loads p_φ, p_ϑ and p_z, each multiplied by $dA = r_1 r d\varphi d\vartheta$, must be taken into account. Hence:

$$\frac{\partial}{\partial\varphi}\left(N_{\varphi\varphi}r\right)+\frac{\partial N_{\varphi\vartheta}}{\partial\vartheta}r_1-N_{\vartheta\vartheta}\cos\varphi r_1+p_\varphi r_1 r=0,$$
$$\frac{\partial}{\partial\varphi}\left(N_{\varphi\vartheta}r\right)+\frac{\partial N_{\vartheta\vartheta}}{\partial\vartheta}r_1+N_{\vartheta\varphi}\cos\varphi r_1+p_\vartheta r_1 r=0,$$
$$N_{\varphi\varphi}r_2+N_{\vartheta\vartheta}r_1-p_z r_1 r_2=0. \tag{15.2}$$

These are three coupled differential equations for the three force flows acting in the membrane stress state $N_{\varphi\varphi}$, $N_{\vartheta\vartheta}$, $N_{\varphi\vartheta}=N_{\vartheta\varphi}$.

In such cases where the description by the angle φ is not suitable, it is advisable to provide the meridian arc length s as a parameter. With $\mathrm{d}s = r_1 \mathrm{d}\varphi$ the following equilibrium conditions result:

$$\begin{aligned}
\frac{\partial}{\partial s}(r N_{ss}) + \frac{\partial N_{\vartheta s}}{\partial \vartheta} - N_{\vartheta\vartheta} \cos\varphi + p_s r &= 0,\\
\frac{\partial}{\partial s}(r N_{s\vartheta}) + \frac{\partial N_{\vartheta\vartheta}}{\partial \vartheta} + N_{\vartheta s} \cos\varphi + p_\vartheta r &= 0,\\
\frac{N_{\vartheta\vartheta}}{r_2} + \frac{N_{ss}}{r_1} &= p_z.
\end{aligned} \tag{15.3}$$

The equilibrium conditions (15.2) can be specialized for different particular cases. If a spherical shell with $r_1 = r_2 = R$ and $r = R \sin\varphi$ is considered, then it follows:

$$\begin{aligned}
\frac{\partial}{\partial \varphi}\left(N_{\varphi\varphi} \sin\varphi\right) + \frac{\partial N_{\vartheta\varphi}}{\partial \vartheta} - N_{\vartheta\vartheta} \cos\varphi + p_\varphi R \sin\varphi &= 0,\\
\frac{\partial}{\partial \varphi}\left(N_{\varphi\vartheta} \sin\varphi\right) + \frac{\partial N_{\vartheta\vartheta}}{\partial \vartheta} + N_{\vartheta\varphi} \cos\varphi + p_\vartheta R \sin\varphi &= 0,\\
N_{\varphi\varphi} + N_{\vartheta\vartheta} - p_z R &= 0.
\end{aligned} \tag{15.4}$$

For a conical shell with the opening angle α with the reference axis s tangential to the meridian we obtain with $\mathrm{d}s = r_1 \mathrm{d}\varphi$, $r_1 \to \infty$ and $r_2 = s \tan\alpha$ and with $r = r_2 \sin\varphi = s \sin\alpha$ and $\cos\varphi = \sin\alpha$:

$$\begin{aligned}
\frac{\partial}{\partial s}(N_{ss} s) + \frac{1}{\sin\alpha} \frac{\partial N_{\vartheta s}}{\partial \vartheta} - N_{\vartheta\vartheta} + p_s s &= 0,\\
\frac{\partial}{\partial s}(N_{\vartheta s} s) + \frac{1}{\sin\alpha} \frac{\partial N_{\vartheta\vartheta}}{\partial \vartheta} + N_{\vartheta s} + p_\vartheta s &= 0,\\
N_{\vartheta\vartheta} - p_z s \tan\alpha &= 0.
\end{aligned} \tag{15.5}$$

Circular cylindrical shells result with $r_1 \to \infty$, $r_2 = R$ and $\varphi = 90°$. Using the vertical reference axis x with $r_1 \mathrm{d}\varphi = \mathrm{d}x$ and observing $r = r_2 \sin\varphi = s \sin\alpha$ as well as $\cos\varphi = \sin\alpha$, it follows:

$$\begin{aligned}
\frac{\partial N_{xx}}{\partial x} + \frac{1}{R} \frac{\partial N_{\vartheta x}}{\partial \vartheta} + p_x &= 0,\\
\frac{\partial N_{x\vartheta}}{\partial x} + \frac{1}{R} \frac{\partial N_{\vartheta\vartheta}}{\partial \vartheta} + p_\vartheta &= 0,\\
N_{\vartheta\vartheta} - p_z R &= 0.
\end{aligned} \tag{15.6}$$

15.2.2 Rotational Symmetric Load

In this section we will consider in more detail the case when a shell of revolution under a rotationally symmetric load is considered. In this case, all derivatives with respect to ϑ become zero, and the membrane force flows of the shell are exclusively functions of the angle φ. The equilibrium conditions (15.2) then reduce as follows:

$$\frac{\mathrm{d}}{\mathrm{d}\varphi}\left(N_{\varphi\varphi}r\right) - N_{\vartheta\vartheta}\cos\varphi r_1 + p_\varphi r_1 r = 0,$$
$$\frac{\mathrm{d}}{\mathrm{d}\varphi}\left(N_{\varphi\vartheta}r\right) + N_{\vartheta\varphi}\cos\varphi r_1 + p_\vartheta r_1 r = 0,$$
$$N_{\varphi\varphi}r_2 + N_{\vartheta\vartheta}r_1 - p_z r_1 r_2 = 0. \tag{15.7}$$

It can be seen that the first and third differential equations in (15.7) are coupled and describe the membrane force flows $N_{\varphi\varphi}$ and $N_{\vartheta\vartheta}$ that will occur due to the loads p_φ and p_z. The second differential equation, on the other hand, is decoupled from the other two equations and describes the membrane force flow $N_{\varphi\vartheta} = N_{\vartheta\varphi}$ that will occur due to the load p_ϑ. This can be interpreted as a torsional load case which we will not discuss further at this point. We will focus our elaborations on the solution of the first and third differential equations in (15.7).

The equilibrium conditions can again be formulated concretely for different technically relevant shell types. If there is a spherical shell with radii $r_1 = r_2 = R$, then it follows from (15.7):

$$\frac{\mathrm{d}}{\mathrm{d}\varphi}\left(N_{\varphi\varphi}\sin\varphi\right) - N_{\vartheta\vartheta}\cos\varphi + p_\varphi R\sin\varphi = 0,$$
$$\frac{\mathrm{d}}{\mathrm{d}\varphi}\left(N_{\varphi\vartheta}\sin\varphi\right) + N_{\vartheta\varphi}\cos\varphi + p_\vartheta R\sin\varphi = 0,$$
$$N_{\varphi\varphi} + N_{\vartheta\vartheta} - p_z R = 0. \tag{15.8}$$

The case of a conical shell with the opening angle α with the reference axis s tangential to the meridian can be derived from the equilibrium conditions (15.7) using $\mathrm{d}s = r_1\mathrm{d}\varphi$ and $r_1 \rightarrow \infty$ and $r_2 = s\tan\alpha$. With $r = r_2\sin\varphi = s\sin\alpha$ and $\cos\varphi = \sin\alpha$ we then obtain:

$$\frac{\mathrm{d}}{\mathrm{d}s}\left(N_{ss}s\right) - N_{\vartheta\vartheta} + p_s s = 0,$$
$$\frac{\mathrm{d}}{\mathrm{d}s}\left(N_{\vartheta s}s\right) + N_{\vartheta s} + p_\vartheta s = 0,$$
$$N_{\vartheta\vartheta} - p_z s\tan\alpha = 0. \tag{15.9}$$

For a circular cylindrical shell with $r_1 \rightarrow \infty$, $r_2 = R$ and $\varphi = 90°$ using the vertical reference axis x with $r_1\mathrm{d}\varphi = \mathrm{d}x$ and the relations $r = r_2\sin\varphi = s\sin\alpha$ and $\cos\varphi = \sin\alpha$, we obtain from (15.7):

$$\frac{\mathrm{d}N_{xx}}{\mathrm{d}xs} + p_x = 0,$$
$$\frac{\mathrm{d}N_{x\vartheta}}{\mathrm{d}x} + p_\vartheta = 0,$$
$$N_{\vartheta\vartheta} - p_z R = 0. \tag{15.10}$$

The solution for the differential equation system (15.7) is discussed below for the first and third equation occurring there. We want to exclude here the case of the surface load p_ϑ, so that the following remains:

$$\frac{\mathrm{d}}{\mathrm{d}\varphi}\left(N_{\varphi\varphi}r\right) - N_{\vartheta\vartheta}\cos\varphi r_1 + p_\varphi r_1 r = 0,$$
$$N_{\varphi\varphi}r_2 + N_{\vartheta\vartheta}r_1 - p_z r_1 r_2 = 0. \tag{15.11}$$

From the second equation in (15.11) we obtain:

$$N_{\vartheta\vartheta} = r_2\left(p_z - \frac{N_{\varphi\varphi}}{r_1}\right), \tag{15.12}$$

so that the first equation in (15.11) can be transformed as:

$$\frac{\mathrm{d}}{\mathrm{d}\varphi}\left(N_{\varphi\varphi}r_2\right) + 2N_{\varphi\varphi}r_2\cot\varphi + \left(p_\varphi - p_z\cot\varphi\right)r_1 r_2 = 0. \tag{15.13}$$

This is a first order ordinary linear inhomogeneous differential equation with non-constant coefficients for the membrane force flow $N_{\varphi\varphi}$, multiplied by the radius r_2. We transform by means of

$$\frac{1}{\sin^2\varphi}\frac{\mathrm{d}}{\mathrm{d}\varphi}\left(N_{\varphi\varphi}r_2\sin^2\varphi\right) = \frac{\mathrm{d}}{\mathrm{d}\varphi}\left(N_{\varphi\varphi}r_2\right) + 2N_{\varphi\varphi}r_2\cot\varphi \tag{15.14}$$

and obtain:

$$\frac{\mathrm{d}}{\mathrm{d}\varphi}\left(N_{\varphi\varphi}r_2\sin^2\varphi\right) = \sin^2\varphi\left(p_z\cot\varphi - p_\varphi\right)r_1 r_2. \tag{15.15}$$

This differential equation can be solved by simple integration:

$$N_{\varphi\varphi} = \frac{1}{r_2\sin^2\varphi}\left[\int_\varphi \left(p_z\cot\varphi - p_\varphi\right)r_1 r_2\sin^2\varphi\mathrm{d}\varphi + C_1\right]. \tag{15.16}$$

Thus, the solution for the membrane force flow $N_{\varphi\varphi}$ is found. The quantity C_1 is an integration constant to be determined from given boundary conditions. Substituting the solution thus found for $N_{\varphi\varphi}$ into (15.12) gives the membrane force flow $N_{\vartheta\vartheta}$ as:

$$N_{\vartheta\vartheta} = r_2\left\{p_z - \frac{1}{r_1 r_2\sin^2\varphi}\left[\int_\varphi \left(p_z\cot\varphi - p_\varphi\right)r_1 r_2\sin^2\varphi\mathrm{d}\varphi + C_1\right]\right\}. \tag{15.17}$$

For the special cases of spherical, conical and cylindrical shells discussed earlier, these equations can be reformulated as follows. For the spherical shell we obtain:

$$N_{\varphi\varphi} = \frac{1}{R\sin^2\varphi}\left[R^2\int_\varphi \left(p_z\cot\varphi - p_\varphi\right)\sin^2\varphi\mathrm{d}\varphi + C_1\right],$$
$$N_{\vartheta\vartheta} = p_z R - N_{\varphi\varphi}. \tag{15.18}$$

For the conical shell the following results:

$$N_{ss} = \frac{1}{s}\left[\int_s (p_z \tan\alpha - p_s)\, s\mathrm{d}s + C_1\right],$$
$$N_{\vartheta\vartheta} = p_z s \tan\alpha. \tag{15.19}$$

In the case of the cylindrical shell one obtains:

$$N_{xx} = -\int_x p_x \mathrm{d}x + C_1,$$
$$N_{\vartheta\vartheta} = R p_z. \tag{15.20}$$

15.3 Selected Solutions for Shells of Revolution

15.3.1 Circular Cylindrical Shells

In this section we will consider some selected solutions for shells of revolution in the framework of membrane theory. We start the elaborations with a circular cylindrical shell (height H, radius R, see Fig. 15.8), which we want to consider under three different load cases. Let the shell be open at the point $x = 0$ and free of any support. At its lower end at $x = H$, the shell is closed and annularly supported at the edge such that both the shell walls and the shell bottom are hinged.

Let the first load case be the dead weight (Fig. 15.8, left), represented by the distributed surface load G. Then $p_x = G$ and $p_z = 0$ hold, and it can be immediately concluded from (15.20):

$$N_{xx} = -\int_x G\mathrm{d}x + C_1 = -Gx + C_1, \quad N_{\vartheta\vartheta} = 0. \tag{15.21}$$

The integration constant can be determined from the condition that no membrane force flow N_{xx} can appear at the upper shell edge at $x = 0$. From this, $C_1 = 0$ is

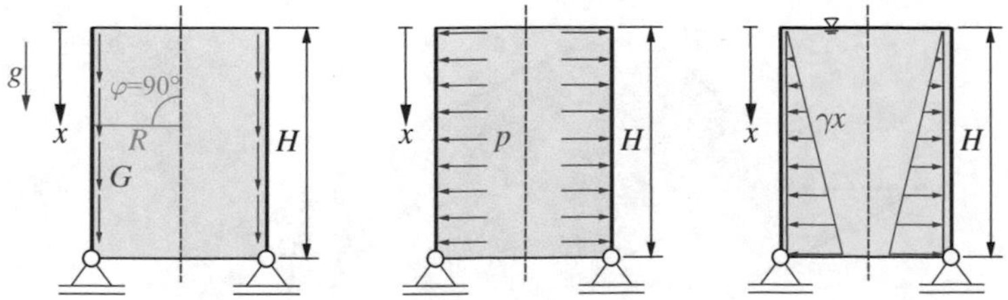

Fig. 15.8 Circular cylindrical shell under different load conditions

immediately obtained, and for N_{xx} remains:

$$N_{xx} = -Gx. \tag{15.22}$$

We further consider the case of constant internal pressure p (Fig. 15.8, middle), where $p_z = p$ and $p_x = 0$. It then follows:

$$\begin{aligned} N_{xx} &= -\int_x p_x \mathrm{d}x + C_1 = C_1, \\ N_{\vartheta\vartheta} &= Rp. \end{aligned} \tag{15.23}$$

We determine the constant C_1 from the condition that no membrane force flow N_{xx} can occur at the upper shell edge at $x = 0$, which immediately leads to $C_1 = 0$ and thus a membrane force $N_{xx} = 0$ vanishing everywhere in the shell.

Finally, we consider the case where the shell is completely filled with a fluid (weight γ). This hydrostatic loading condition corresponds to a linearly distributed internal pressure $p_z(x) = \gamma x$ over x, the load p_x is zero in this case. Using (15.20) we then obtain:

$$N_{xx} = 0, \quad N_{\vartheta\vartheta} = R\gamma x. \tag{15.24}$$

As another example, consider the semicircular cylindrical shell (radius R, length l) of Fig. 15.9, which is loaded by its own weight G (see Fig. 14.6). The shell is supported at its four corners, while the longitudinal edges are free of any support. The shell is free of axial loads at its ends $x = \pm\frac{l}{2}$, so that $N_{xx}\left(x = \pm\frac{l}{2}\right) = 0$ applies. For the self-weight load case, the surface loads to be applied are as follows (Fig. 14.6):

$$p_x = 0, \quad p_\vartheta = G \sin\varphi, \quad p_z = -G\cos\vartheta. \tag{15.25}$$

From the third condition in (15.6), $p_z = -G\cos\vartheta$ can be used to directly determine the membrane force flow $N_{\vartheta\vartheta}$ as:

$$N_{\vartheta\vartheta} = -GR\cos\vartheta. \tag{15.26}$$

Thus, from the two other equilibrium conditions in (15.6), the membrane force flows N_{xx} and $N_{x\varphi}$ can be determined as follows:

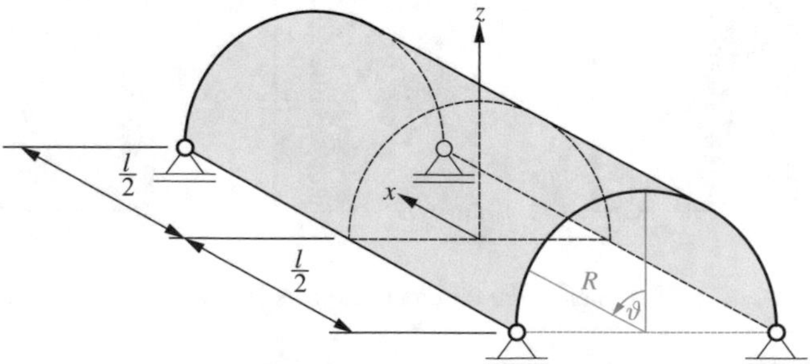

Fig. 15.9 Semicircular cylindrical shell

$$\begin{aligned} N_{x\vartheta} &= -2xG \sin\vartheta + C_1(\vartheta), \\ N_{xx} &= \frac{G}{R}x^2\cos\vartheta - \frac{x}{R}\frac{\mathrm{d}C_1(\vartheta)}{\mathrm{d}\vartheta} + C_2(\vartheta). \end{aligned} \tag{15.27}$$

The function $C_1(\vartheta)$ is zero due to the requirement $N_{x\vartheta}\left(x = -\frac{l}{2}\right) = -N_{x\vartheta}\left(x = \frac{l}{2}\right)$, i.e. $C_1(\vartheta) = 0$. The function $C_2(\vartheta)$, on the other hand, results from the requirement for the disappearance of the membrane force flow N_{xx} at the shell ends $x = \pm\frac{l}{2}$ as:

$$C_2(\vartheta) = -\frac{Gl^2}{4R}\cos\vartheta. \tag{15.28}$$

The membrane force flows for the given shell situation thus follow as:

$$\begin{aligned} N_{xx} &= -\frac{G}{4R}\left(l^2 - 4x^2\right)\cos\vartheta, \\ N_{x\vartheta} &= -2xG\sin\vartheta, \\ N_{\vartheta\vartheta} &= -GR\cos\vartheta. \end{aligned} \tag{15.29}$$

An example of a circular cylindrical shell (radius R, length l) under a non-rotationally symmetric load is given by the horizontal fluid container of Fig. 15.10, which is completely filled with a fluid of density ρ. The circular-cylindrical vessel is freely movable at both ends and closed by vessel seals. The pressure p_z to be applied as a result of the liquid filling varies linearly over z. The pressure p_z can be given projected on the circle as:

$$p_z = \rho g z = \rho g R\left(1 - \cos\vartheta\right). \tag{15.30}$$

From the third equilibrium condition in (15.6), the internal force flow $N_{\vartheta\vartheta}$ can be determined immediately. The two remaining internal force flows N_{xx} and $N_{x\vartheta}$ then follow from the first and second equations in (15.6) by integration. It follows after a short calculation:

$$\begin{aligned} N_{\vartheta\vartheta} &= \rho g R^2\left(1 - \cos\vartheta\right), \\ N_{x\vartheta} &= \left(-\rho g R x + C_1\right)\sin\vartheta, \\ N_{xx} &= C_2 + \left(\frac{\rho g x^2}{2} - C_1\frac{x}{R} + C_3\right)\cos\vartheta. \end{aligned} \tag{15.31}$$

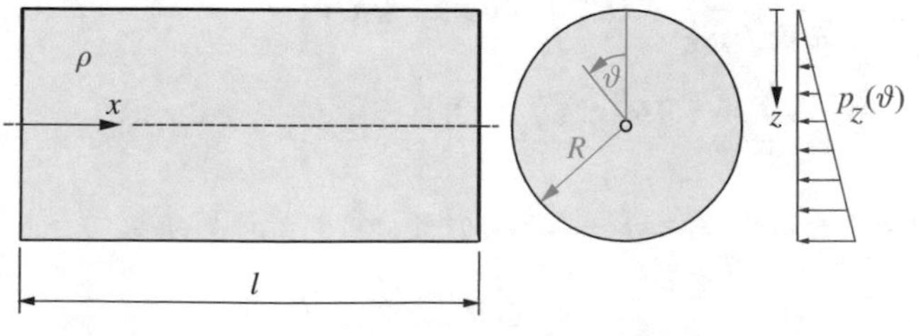

Fig. 15.10 Horizontal liquid tank

The integration constants C_1, C_2, C_3 can be deduced from the boundary conditions. For this purpose, it is required that the internal force flow N_{xx} vanishes at the two ends $x = 0$ and $x = l$. This leads to:

$$C_1 = \frac{\rho g l R}{2}, \quad C_2 = 0, \quad C_3 = 0. \tag{15.32}$$

Thus, the internal force flows are:

$$\begin{aligned} N_{\vartheta\vartheta} &= \rho g R^2 (1 - \cos\vartheta), \\ N_{x\vartheta} &= \rho g R \left(\frac{l}{2} - x\right) \sin\vartheta, \\ N_{xx} &= \frac{\rho g x}{2} (x - l) \cos\vartheta. \end{aligned} \tag{15.33}$$

It should be noted here that, within the framework of a structural design of the fluid tank, edge stiffeners would have to be provided due to the shear force flows $N_{x\vartheta}$ not disappearing at the two shell ends $x = 0$ and $x = l$. The deformations of the stiffeners and the container will generally not be compatible, resulting in a local bending effect that must be taken into account (see Chap. 16).

15.3.2 Spherical Shells

Furthermore, we consider spherical shells under different load cases and boundary conditions. As an introductory case, we treat the case of a hemispherical shell (radius R) under its self-weight (Fig. 15.11, left), which is represented by the surface load G. The surface loads p_φ and p_z are then obtained as:

$$p_\varphi = G \sin\varphi, \quad p_z = -G\cos\varphi. \tag{15.34}$$

We obtain the membrane force flow $N_{\varphi\varphi}$ from (15.18):

$$\begin{aligned} N_{\varphi\varphi} &= \frac{1}{R\sin^2\varphi}\left[R^2 \int_\varphi \left(p_z \cot\varphi - p_\varphi\right) \sin^2\varphi \mathrm{d}\varphi + C_1\right] \\ &= \frac{1}{R\sin^2\varphi}\left[-GR^2 \int_\varphi (\cos\varphi\cot\varphi + \sin\varphi)\sin^2\varphi \mathrm{d}\varphi + C_1\right] \\ &= \frac{1}{R\sin^2\varphi}\left[-GR^2 \int_\varphi \sin\varphi \mathrm{d}\varphi + C_1\right] \\ &= \frac{1}{R\sin^2\varphi}\left(GR^2\cos\varphi + C_1\right). \end{aligned} \tag{15.35}$$

With $\sin^2\varphi = 1 - \cos^2\varphi = (1 - \cos\varphi)(1 + \cos\varphi)$ this yields:

$$N_{\varphi\varphi} = -\frac{GR}{1 + \cos\varphi} \frac{\cos\varphi + \frac{C_1}{GR^2}}{\cos\varphi - 1}. \tag{15.36}$$

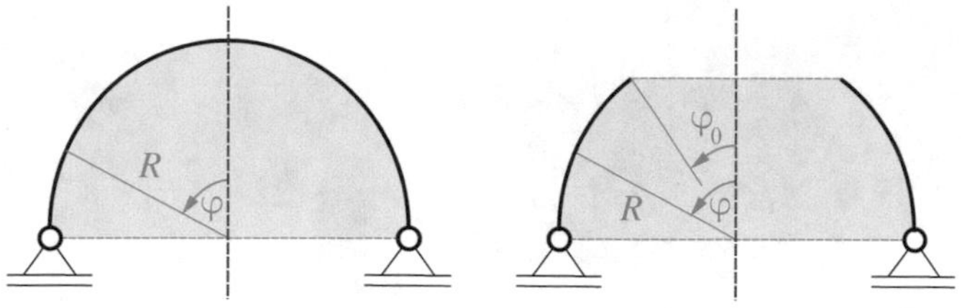

Fig. 15.11 Closed and open spherical shell

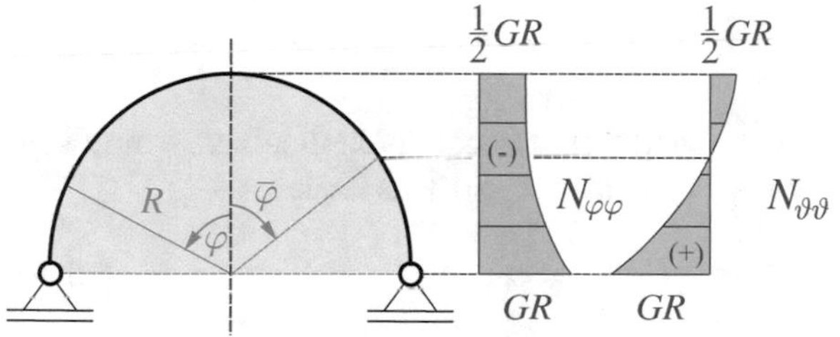

Fig. 15.12 Distribution of membrane force flows $N_{\varphi\varphi}$ and $N_{\vartheta\vartheta}$ for the hemispherical shell under self-weight

It turns out that this solution for $N_{\varphi\varphi}$ tends to infinitely large values for $N_{\varphi\varphi}$ in the case of $\varphi \to 0$. The integration constant C_1 is therefore determined from the requirement that $N_{\varphi\varphi}(\varphi = 0)$ takes finite values. This is achieved when $C_1 = -GR^2$ holds. Hence:

$$N_{\varphi\varphi} = -\frac{GR}{1+\cos\varphi}. \tag{15.37}$$

From (15.18) then also follows the membrane force flow $N_{\vartheta\vartheta}$:

$$N_{\vartheta\vartheta} = p_z R - N_{\varphi\varphi} = GR\left(\frac{1}{1+\cos\varphi} - \cos\varphi\right). \tag{15.38}$$

A qualitative graphical representation of the solution can be found in Fig. 15.12. Obviously, the membrane force flow $N_{\varphi\varphi}$ running in the meridian direction across the entire shell is present as a compressive force flow, whereas $N_{\vartheta\vartheta}$ has a change of sign. The angle $\varphi = \bar{\varphi}$ at which the membrane force flow $N_{\vartheta\vartheta}$ becomes zero can be determined as $\bar{\varphi} = 51.8°$ by setting (15.38) to zero.

The next case is the spherical shell open at the top, with a free edge at $\varphi = \varphi_0$ and loaded by its own weight (Fig. 15.11, right). The solution for the membrane force flow $N_{\varphi\varphi}$ can be taken directly from (15.35).

$$N_{\varphi\varphi} = \frac{1}{R\sin^2\varphi}\left(GR^2\cos\varphi + C_1\right). \tag{15.39}$$

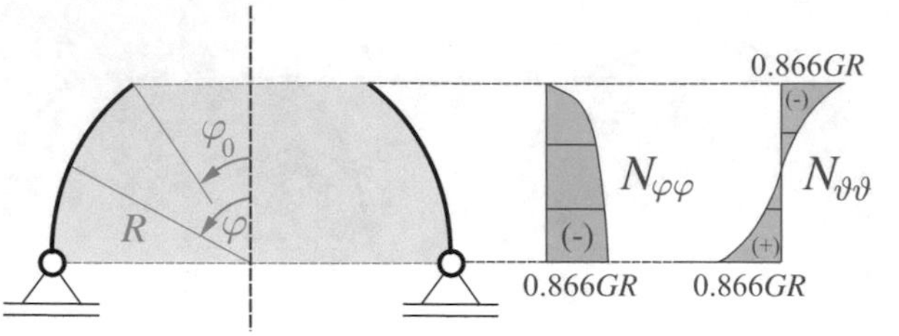

Fig. 15.13 Distribution of the membrane force flows $N_{\varphi\varphi}$ and $N_{\vartheta\vartheta}$ for the open spherical shell under self-weight for $\varphi_0 = 30°$

The integration constant C_1 is determined from the condition that there is no membrane force flow $N_{\varphi\varphi}$ at the free shell edge at $\varphi = \varphi_0$. This leads to:

$$C_1 = -GR^2 \cos\varphi_0, \tag{15.40}$$

and thus:

$$N_{\varphi\varphi} = -\frac{GR}{\sin^2\varphi}\left(\cos\varphi_0 - \cos\varphi\right). \tag{15.41}$$

Analogously we obtain for $N_{\vartheta\vartheta}$:

$$N_{\vartheta\vartheta} = -GR\left(\cos\varphi - \frac{\cos\varphi_0 - \cos\varphi}{\sin^2\varphi}\right). \tag{15.42}$$

The case of the closed shell is included here with $\varphi_0 = 0$ as a special case.

A qualitative graphical representation of the results for the membrane force flows is given in Fig. 15.13 for the case $\varphi_0 = 30°$.

As another case, we consider a closed spherical shell of radius R, which is loaded at every point by the constant internal pressure p (Fig. 15.14). In this case $p_\varphi = 0$ and $p_z = p =$const. From (15.18) then follows:

$$\begin{aligned} N_{\varphi\varphi} &= \frac{1}{R\sin^2\varphi}\left[pR^2\int_\varphi \sin\varphi\cos\varphi \mathrm{d}\varphi + C_1\right] \\ &= \frac{1}{R\sin^2\varphi}\left[\frac{1}{2}pR^2\sin^2\varphi + C_1\right]. \end{aligned} \tag{15.43}$$

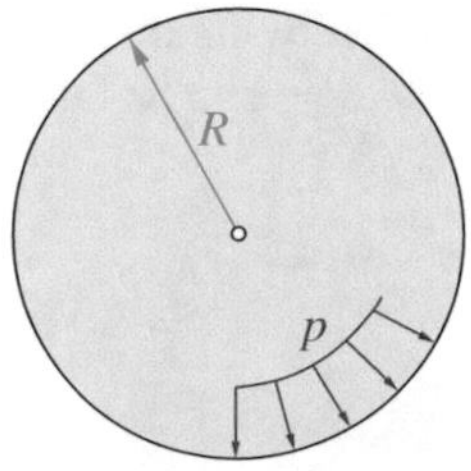

Fig. 15.14 Closed spherical shell under internal pressure p

The integration constant C_1 is procured from the requirement that $N_{\varphi\varphi}$ takes finite values at the vertex $\varphi = 0$, leading to $C_1 = 0$. Thus:

$$N_{\varphi\varphi} = \frac{pR}{2}. \tag{15.44}$$

Analogously, we obtain for $N_{\vartheta\vartheta}$:

$$N_{\vartheta\vartheta} = pR - N_{\varphi\varphi} = \frac{pR}{2}. \tag{15.45}$$

Thus, it is shown that the closed spherical shell under constant internal pressure is under homogeneous membrane force flows with the magnitude $N_{\varphi\varphi} = N_{\vartheta\vartheta} = \frac{pR}{2}$.

Another example is the spherical shell under a load q uniformly distributed over the planform:

$$p_\varphi = q\cos\varphi\sin\varphi, \quad p_z = -q\cos^2\varphi. \tag{15.46}$$

With (15.18) one obtains:

$$\begin{aligned} N_{\varphi\varphi} &= \frac{1}{R\sin^2\varphi}\left[R^2\int_\varphi \left(p_z\cot\varphi - p_\varphi\right)\sin^2\varphi \mathrm{d}\varphi + C_1\right] \\ &= \frac{1}{R\sin^2\varphi}\left[-qR^2\int_\varphi \cos\varphi\sin\varphi \mathrm{d}\varphi + C_1\right] \\ &= \frac{1}{R\sin^2\varphi}\left[-\frac{qR^2}{2}\sin^2\varphi \mathrm{d}\varphi + C_1\right]. \end{aligned} \tag{15.47}$$

The integration constant C_1 must become zero here to ensure finite values of the membrane force flow $N_{\varphi\varphi}$ at the vertex $\varphi = 0$. Therefore, it follows:

$$N_{\varphi\varphi} = -\frac{qR}{2}. \tag{15.48}$$

The membrane force flow $N_{\vartheta\vartheta}$ is then given as:

$$\begin{aligned} N_{\vartheta\vartheta} &= p_z R - N_{\varphi\varphi} \\ &= qR\left(\frac{1}{2} - \cos^2\varphi\right) = -\frac{1}{2}qR\cos 2\varphi. \end{aligned} \tag{15.49}$$

The variation of the membrane force flows $N_{\varphi\varphi}$ and $N_{\vartheta\vartheta}$ is shown in Fig. 15.15. Obviously, $N_{\varphi\varphi}$, independent of the angle φ, is present at every position of the shell with the constant value $-\frac{qR}{2}$, whereas for $N_{\vartheta\vartheta}$ a sign change occurs at $\bar{\varphi} = 45°$.

As a final example of a spherical shell, consider the fluid tank of Fig. 15.16. Given is a spherical shell of radius R supported on a specific circle of latitude at $\varphi = \varphi_0$ by a supported circumferential ring. Let the shell be completely filled with a liquid of weight γ. The fluid pressure can thus be given as:

$$p_z = \gamma R\left(1 - \cos\varphi\right). \tag{15.50}$$

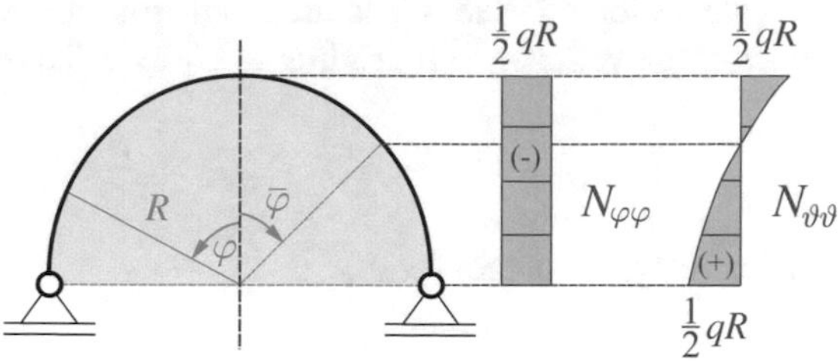

Fig. 15.15 Distribution of the membrane force flows $N_{\varphi\varphi}$ and $N_{\vartheta\vartheta}$ for the spherical shell under a load q uniformly distributed over its planform

Let the surface loads p_φ and p_ϑ be identically zero for this example.

The force flow $N_{\varphi\varphi}$ is obtained by integration from (15.16) as:

$$N_{\varphi\varphi} = \frac{\gamma R^2}{6 \sin^2 \varphi} \left[(2 \cos \varphi - 3) \cos^2 \varphi + 6C_1 \right]. \tag{15.51}$$

At the point $\varphi = 0$ this expression exhibits a singularity, so that it is required that also the bracket expression must become zero at this point. From this, the constant C_1 can be determined as $C_1 = \frac{1}{6}$. Finally, for the two force flows $N_{\varphi\varphi}$ and $N_{\vartheta\vartheta}$ we obtain:

$$\begin{aligned} N_{\varphi\varphi} &= \frac{\gamma R^2}{6} \left(1 - \frac{2 \cos^2 \varphi}{1 + \cos \varphi} \right), \\ N_{\vartheta\vartheta} &= \frac{\gamma R^2}{6} \left(5 - 6 \cos \varphi + \frac{2 \cos^2 \varphi}{1 + \cos \varphi} \right). \end{aligned} \tag{15.52}$$

These expressions are valid for the upper part of the shell between $\varphi = 0$ and $\varphi = \varphi_0$. For the lower range between $\varphi = \varphi_0$ and $\varphi = \pi$, the integration constant C_1 has to be redetermined in such a way that at the lowest shell point the membrane force flows remain finite. From this one obtains $C_1 = \frac{5}{6}$, and the force flows can be written

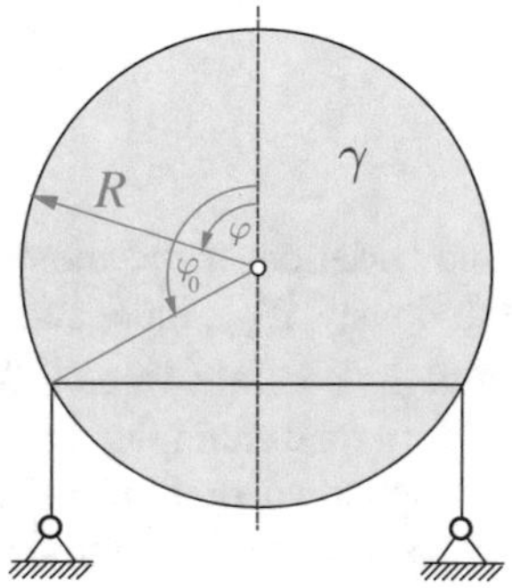

Fig. 15.16 Supported spherical fluid tank

as:

$$N_{\varphi\varphi} = \frac{\gamma R^2}{6}\left(5 + \frac{2\cos^2\varphi}{1-\cos\varphi}\right),$$

$$N_{\vartheta\vartheta} = \frac{\gamma R^2}{6}\left(1 - 6\cos\varphi - \frac{2\cos^2\varphi}{1-\cos\varphi}\right). \qquad (15.53)$$

It can be seen that the exact position of the support ring is not included in the calculations. In addition, there is an abrupt change of the force flows at the level of the support ring, which is associated with a local bending effect (Chap. 16). The difference between the force flows must be transferred by the support ring.

15.3.3 Conical Shells

We consider the example of a conical shell (Fig. 15.17), which is under the boundary ring force flow p (caused by a single force F acting centrically perpendicular to the shell boundary) and the uniformly distributed self-weight G. The conical shell has the opening angle α as shown. Further necessary dimensions are s_0 and R_0 as indicated. Obviously, the given shell situation is rotationally symmetric, so that all derivatives with respect to ϑ as well as the shear force flow $N_{s\vartheta}$ vanish. We start from the set of equations (15.9), which currently can be written as follows with $p_s = G\sin\varphi = G\cos\alpha$ and $p_z = -G\sin\alpha$:

$$\frac{\mathrm{d}}{\mathrm{d}s}(N_{ss}s) - N_{\vartheta\vartheta} + Gs\cos\alpha = 0,$$

$$N_{\vartheta\vartheta} + Gs\sin\alpha\tan\alpha = 0. \qquad (15.54)$$

Note that the second equation in (15.9) becomes zero for the present case and thus is not considered further.

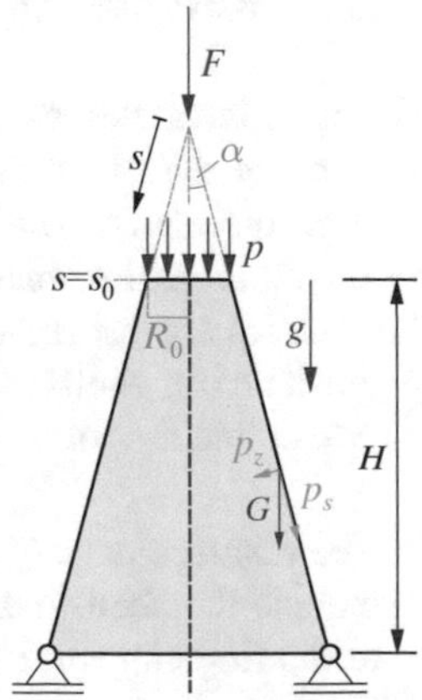

Fig. 15.17 Conical shell under self-weight G and boundary ring force flow p

Transferring the second equation in (15.54) into the first equation, we obtain:

$$\frac{\mathrm{d}}{\mathrm{d}s}(N_{ss}s) + Gs\frac{\sin^2\alpha}{\cos\alpha} + Gs\cos\alpha = 0. \tag{15.55}$$

Integration over s yields:

$$N_{ss}s = -\frac{Gs^2}{2\cos\alpha} + C_1, \tag{15.56}$$

or

$$N_{ss} = -\frac{Gs}{2\cos\alpha} + \frac{C_1}{s}. \tag{15.57}$$

The integration constant C_1 occurring here results from the requirement that at the point $s = s_0$ the force flow N_{ss} must be in equilibrium with the ring force flow p. This results in:

$$C_1 = \frac{Gs_0^2}{2\cos\alpha} - \frac{F}{\pi\sin 2\alpha}. \tag{15.58}$$

The two force flows N_{ss} and $N_{\vartheta\vartheta}$ of the conical shell are thus:

$$\begin{aligned} N_{ss} &= -\frac{G}{2\cos\alpha}\frac{s^2 - s_0^2}{s} - \frac{F}{\pi s\sin 2\alpha}, \\ N_{\vartheta\vartheta} &= -Gs\frac{\sin^2\alpha}{\cos\alpha}. \end{aligned} \tag{15.59}$$

Typically, in practice one will strive to accommodate the horizontal component of the force flow N_{ss} by a stiffener that is continuously applied at the top $s = s_0$.

15.4 Kinematics of Shells of Revolution

If one is interested exclusively in the membrane force flows in the context of the membrane theory of shells, it is sufficient to use equilibrium considerations only. The problem is internally statically determinate. However, if one is also interested in the strain and displacement state of a shell, then considerations with respect to kinematics are necessary. The present section is devoted to this subject, and here exclusively for shells of revolution.

We define the following displacement quantities:

- The displacement quantities u, v and w are the displacements of a shell point outside the shell midsurface $z = 0$. Here u is the displacement in the direction of the ring tangent, v is the displacement in the direction of the meridian tangent, and w is the displacement in the normal direction z. They are functions of the two angles φ and ϑ and the thickness coordinate z.

- The displacement quantities u_0, v_0 and w_0 are the corresponding displacements of a point on the shell midsurface. They depend on φ and ϑ, but not on z.

In addition, strain quantities are required, which we want to relate to the shell midsurface consistently with Classical Shell Theory. These are the strain in the meridional direction $\varepsilon^0_{\varphi\varphi}$, the strain in the annular direction $\varepsilon^0_{\vartheta\vartheta}$, and the shear strain $\gamma^0_{\varphi\vartheta}$ of the shell midsurface respectively the change of the right angle between the coordinate lines. They are, like the displacements u_0, v_0 and w_0, functions of φ and ϑ, but not of z.

We now want to derive relations between the mentioned displacement quantities on the one hand and the strain quantities on the other hand and consider the infinitesimal element $\mathrm{d}A$ of the shell midsurface shown in Fig. 15.18 with the edge lengths $\mathrm{d}s_1$ and $\mathrm{d}s_2$. The vertices of the surface element are denoted as points 1–4 as indicated. Let the corresponding displacements of the vertices be denoted as $u_{0,i}$, $v_{0,i}$ and $w_{0,i}$ $(i = 1, 2, 3, 4)$. We express the displacements of points 2, 3, and 4 by the displacements of point 1 and develop them accordingly as Taylor series which we terminate after the first term in each case. The displacements u_0, v_0, w_0 are present at point 1. Then for the displacements in point 2 we obtain:

$$u_{0,2} = u_0 + \frac{\partial u_0}{\partial \varphi}\mathrm{d}\varphi,$$
$$v_{0,2} = v_0 + \frac{\partial v_0}{\partial \varphi}\mathrm{d}\varphi,$$
$$w_{0,2} = w_0 + \frac{\partial w_0}{\partial \varphi}\mathrm{d}\varphi. \tag{15.60}$$

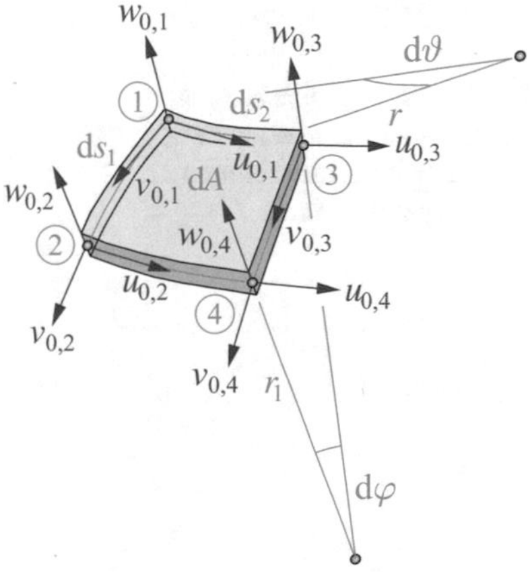

Fig. 15.18 Infinitesimal element of the shell midsurface and displacements of the vertices 1–4

In point 3 applies accordingly:

$$u_{0,3} = u_0 + \frac{\partial u_0}{\partial \vartheta} \mathrm{d}\vartheta,$$
$$v_{0,3} = v_0 + \frac{\partial v_0}{\partial \vartheta} \mathrm{d}\vartheta,$$
$$w_{0,3} = w_0 + \frac{\partial w_0}{\partial \vartheta} \mathrm{d}\vartheta. \tag{15.61}$$

For point 4 results:

$$u_{0,4} = u_0 + \frac{\partial u_0}{\partial \varphi} \mathrm{d}\varphi + \frac{\partial u_0}{\partial \vartheta} \mathrm{d}\vartheta,$$
$$v_{0,4} = v_0 + \frac{\partial v_0}{\partial \varphi} \mathrm{d}\varphi + \frac{\partial v_0}{\partial \vartheta} \mathrm{d}\vartheta,$$
$$w_{0,4} = w_0 + \frac{\partial w_0}{\partial \varphi} \mathrm{d}\varphi + \frac{\partial w_0}{\partial \vartheta} \mathrm{d}\vartheta. \tag{15.62}$$

The strains of the shell midsurface are then defined as:

$$\varepsilon^0_{\varphi\varphi} = \frac{(\mathrm{d}s_1 + \Delta \mathrm{d}s_1) - \mathrm{d}s_1}{\mathrm{d}s_1}, \quad \varepsilon^0_{\vartheta\vartheta} = \frac{(\mathrm{d}s_2 + \Delta \mathrm{d}s_2) - \mathrm{d}s_2}{\mathrm{d}s_2}, \tag{15.63}$$

if $\Delta \mathrm{d}s_1$ and $\Delta \mathrm{d}s_2$ are the changes of the edge lengths of the shell element $\mathrm{d}A$ due to the displacements. The shear strain $\gamma^0_{\varphi\vartheta}$ then corresponds to the change of the right angle between the edges of the element.

To determine the kinematic quantities, we again consider various sections through the shell and determine the respective contributions of the individual displacements to the strain quantities we are looking for, using the relations $\mathrm{d}s_1 = r_1 \mathrm{d}\varphi$ and $\mathrm{d}s_2 = r \mathrm{d}\vartheta$ and neglecting terms of higher order. In Fig. 15.19 a meridian section for ϑ =const. is shown through the two points 1 and 2, both in the undeformed and in the deformed

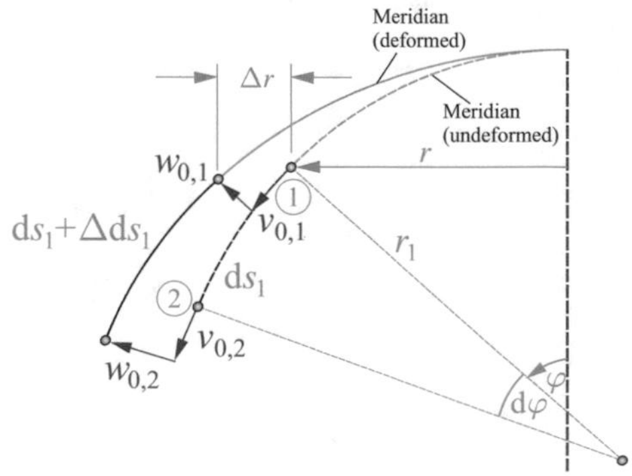

Fig. 15.19 Meridian section of the infinitesimal element of the shell midsurface through the two points 1 and 2

state. The length of the edge section in the deformed state is then obtained as:

$$\mathrm{d}s_1 + \Delta\mathrm{d}s_1 = \left(\mathrm{d}s_1 + \frac{\partial v_0}{\partial\varphi}\mathrm{d}\varphi\right)\left(1 + \frac{\Delta r_1}{r_1}\right), \tag{15.64}$$

which with $\Delta r_1 = w_0$ and neglecting the term $\frac{w}{r_1}\frac{\partial v_0}{\partial\varphi}\mathrm{d}\varphi$ we can also write as:

$$\mathrm{d}s_1 + \Delta\mathrm{d}s_1 = \mathrm{d}s_1\left(1 + \frac{w_0}{r_1}\right) + \frac{\partial v_0}{\partial\varphi}\mathrm{d}\varphi. \tag{15.65}$$

By (15.63) the normal strain $\varepsilon^0_{\varphi\varphi}$ follows as:

$$\begin{aligned}\varepsilon^0_{\varphi\varphi} &= \frac{(\mathrm{d}s_1 + \Delta\mathrm{d}s_1) - \mathrm{d}s_1}{\mathrm{d}s_1}\\ &= \frac{\mathrm{d}s_1\left(1 + \frac{w_0}{r_1}\right) + \frac{\partial v_0}{\partial\varphi}\mathrm{d}\varphi - \mathrm{d}s_1}{\mathrm{d}s_1}\\ &= \frac{1}{r_1}\left(\frac{\partial v_0}{\partial\varphi} + w_0\right),\end{aligned} \tag{15.66}$$

where we have used $\mathrm{d}s_1 = r_1\mathrm{d}\varphi$. The expansion Δr of the radius r can be read as:

$$\Delta r = v_0\cos\varphi + w_0\sin\varphi. \tag{15.67}$$

Figure 15.20 shows an annular section for $\varphi =$ const. through points 1 and 3 with the representation of the undeformed and the deformed annular segment of length $\mathrm{d}s_2$, which in the deformed state has the length $\mathrm{d}s_2 + \delta\mathrm{d}s_2$. The length $\mathrm{d}s_2 + \delta\mathrm{d}s_2$ of the deformed sectional element can be written as:

$$\mathrm{d}s_2 + \Delta\mathrm{d}s_2 = \left(\mathrm{d}s_2 + \frac{\partial u_0}{\partial\vartheta}\mathrm{d}\vartheta\right)\left(1 + \frac{\Delta r}{r}\right), \tag{15.68}$$

which we can also specify as (15.67) by neglecting the term $\frac{\partial u_0}{\partial\vartheta}\mathrm{d}\vartheta\frac{\delta r}{r}$:

$$\mathrm{d}s_2 + \Delta\mathrm{d}s_2 = \mathrm{d}s_2\left(1 + \frac{v_0}{r}\cos\varphi + \frac{w_0}{r}\sin\varphi\right) + \frac{\partial u_0}{\partial\vartheta}\mathrm{d}\vartheta. \tag{15.69}$$

For the normal strain $\varepsilon^0_{\vartheta\vartheta}$ then follows with (15.63):

$$\begin{aligned}\varepsilon^0_{\vartheta\vartheta} &= \frac{(\mathrm{d}s_2 + \Delta\mathrm{d}s_2) - \mathrm{d}s_2}{\mathrm{d}s_2}\\ &= \frac{1}{r}\left(\frac{\partial u_0}{\partial\vartheta} + v_0\cos\varphi + w_0\sin\varphi\right),\end{aligned} \tag{15.70}$$

where we have used $r = \frac{\mathrm{d}s_2}{\mathrm{d}\vartheta}$.

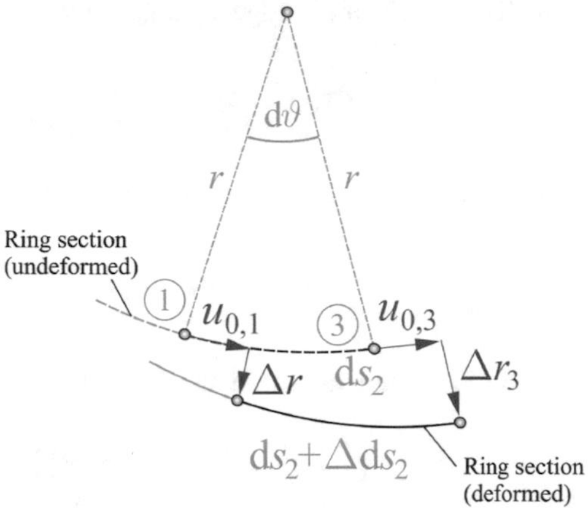

Fig. 15.20 Annular section of the infinitesimal element of the shell midsurface through the two points 1 and 3

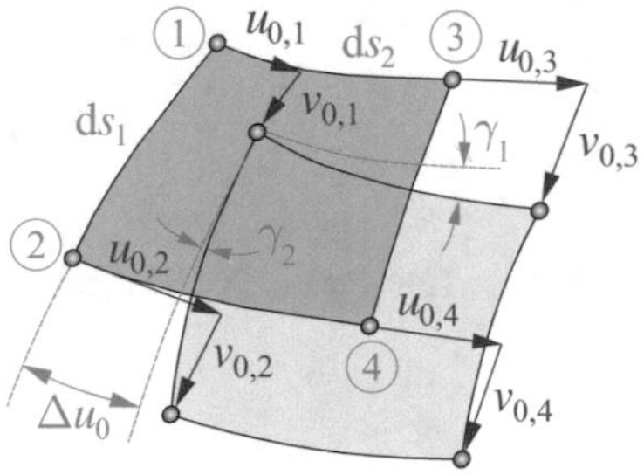

Fig. 15.21 Deformation state of the infinitesimal element of the shell midsurface

In Fig. 15.21 the deformation state of the infinitesimal element of the shell midsurface is shown. The two angles γ_1 and γ_2 can be deduced from this with $\Delta u_0 = u_0\left(1 + \frac{dr}{r}\right)$ as:

$$\gamma_1 = \frac{\frac{\partial v_0}{\partial \vartheta} d\vartheta}{ds_2 + \Delta ds_2} \simeq \frac{\frac{\partial v_0}{\partial \vartheta} d\vartheta}{ds_2},$$

$$\gamma_2 = \frac{u_0 + \frac{\partial u_0}{\partial \varphi} d\varphi - u_0\left(1 + \frac{dr}{r}\right)}{ds_1 + \Delta ds_1} \simeq \frac{\frac{\partial u_0}{\partial \varphi} d\varphi - u_0 \frac{dr}{r}}{ds_1}. \tag{15.71}$$

The shear strain $\gamma_{\varphi\vartheta}$ is given by the sum of the two angles γ_1 and γ_2 and is therefore given with $dr = ds_1 \cos\varphi$ as:

$$\gamma_{\varphi\vartheta} = \gamma_1 + \gamma_2 = \frac{\frac{\partial v_0}{\partial \vartheta} d\vartheta}{ds_2} + \frac{\frac{\partial u_0}{\partial \varphi} d\varphi}{ds_1} - \frac{u_0 \cos\varphi}{r}. \tag{15.72}$$

With $ds_1 = r_1 d\varphi$ and $ds_2 = r d\vartheta$ we obtain:

$$\gamma_{\varphi\vartheta} = \frac{1}{r}\frac{\partial v_0}{\partial \vartheta} + \frac{1}{r_1}\frac{\partial u_0}{\partial \varphi} - \frac{u_0 \cos\varphi}{r}. \tag{15.73}$$

Summarizing, the kinematic equations for the shell of revolution are as follows:

$$
\begin{aligned}
\varepsilon_{\varphi\varphi}^{0} &= \frac{1}{r_1}\left(\frac{\partial v_0}{\partial \varphi} + w_0\right), \\
\varepsilon_{\vartheta\vartheta}^{0} &= \frac{1}{r}\left(\frac{\partial u_0}{\partial \vartheta} + v_0 \cos\varphi + w_0 \sin\varphi\right), \\
\gamma_{\varphi\vartheta} &= \frac{1}{r}\frac{\partial v_0}{\partial \vartheta} + \frac{1}{r_1}\frac{\partial u_0}{\partial \varphi} - \frac{u_0 \cos\varphi}{r}.
\end{aligned} \tag{15.74}
$$

15.5 Constitutive Equations

The constitutive equations establish a relationship between the stresses of the shell and the strains of the shell middle surface:

$$
\begin{aligned}
\varepsilon_{\varphi\varphi}^{0} &= \frac{1}{E}\left(\sigma_{\varphi\varphi} - \nu\sigma_{\vartheta\vartheta}\right), \\
\varepsilon_{\vartheta\vartheta}^{0} &= \frac{1}{E}\left(\sigma_{\vartheta\vartheta} - \nu\sigma_{\varphi\varphi}\right), \\
\gamma_{\varphi\vartheta}^{0} &= \frac{2(1+\nu)}{E}\tau_{\varphi\vartheta}.
\end{aligned} \tag{15.75}
$$

If we use here the relations $N_{\varphi\varphi} = \sigma_{\varphi\varphi}h$, $N_{\vartheta\vartheta} = \sigma_{\vartheta\vartheta}h$, $N_{\varphi\vartheta} = \tau_{\varphi\vartheta}h$, then we obtain:

$$
\begin{aligned}
\varepsilon_{\varphi\varphi}^{0} &= \frac{1}{Eh}\left(N_{\varphi\varphi} - \nu N_{\vartheta\vartheta}\right), \\
\varepsilon_{\vartheta\vartheta}^{0} &= \frac{1}{Eh}\left(N_{\vartheta\vartheta} - \nu N_{\varphi\varphi}\right), \\
\gamma_{\varphi\vartheta}^{0} &= \frac{2(1+\nu)}{Eh}N_{\varphi\vartheta}.
\end{aligned} \tag{15.76}
$$

This relation can be inverted as follows:

$$
\begin{aligned}
N_{\varphi\varphi} &= A\left(\varepsilon_{\varphi\varphi}^{0} + \nu\varepsilon_{\vartheta\vartheta}^{0}\right), \\
N_{\vartheta\vartheta} &= A\left(\nu\varepsilon_{\varphi\varphi}^{0} + \varepsilon_{\vartheta\vartheta}^{0}\right), \\
N_{\varphi\vartheta} &= A\frac{1-\nu}{2}\gamma_{\varphi\vartheta}^{0},
\end{aligned} \tag{15.77}
$$

where A represents the extensional stiffness of the shell:

$$
A = \frac{Eh}{1-\nu^2}. \tag{15.78}
$$

In a similar way orthotropic or anisotropic properties of the shell material can be treated in addition to isotropic material properties which, however, remains without representation at this point.

15.6 Displacement Solutions for Rotationally Symmetric Loads

If the case of rotationally symmetric load is present, the kinematic relations (15.74) simplify as follows. Since all state quantities depend exclusively on φ, all partial derivatives with respect to ϑ vanish, and we obtain:

$$\begin{aligned}\varepsilon_{\varphi\varphi}^{0} &= \frac{1}{r_1}\left(\frac{\mathrm{d}v_0}{\mathrm{d}\varphi} + w_0\right),\\ \varepsilon_{\vartheta\vartheta}^{0} &= \frac{1}{r}\left(v_0\cos\varphi + w_0\sin\varphi\right),\\ \gamma_{\varphi\vartheta} &= \frac{1}{r_1}\frac{\mathrm{d}u_0}{\mathrm{d}\varphi} - \frac{u_0\cos\varphi}{r}.\end{aligned} \tag{15.79}$$

For this case, closed-form analytical solutions for the deformations of the considered shells of revolution for elementary shell shapes can be derived. The relation between the ring displacement u and the shear strain $\gamma_{\varphi\vartheta}$ described by the third equation in (15.79) can be interpreted as a torsion of the shell and will not be considered further. We therefore focus on the first two equations in (15.79).

Using the second equation in (15.79), the displacement w_0 can be eliminated from the first equation, and we obtain:

$$\frac{\mathrm{d}v_0}{\mathrm{d}\varphi} - v_0\cot\varphi = r_1\varepsilon_{\varphi\varphi}^{0} - r_2\varepsilon_{\vartheta\vartheta}^{0}, \tag{15.80}$$

or

$$\frac{\mathrm{d}}{\mathrm{d}\varphi}\left(\frac{v_0}{\sin\varphi}\right) = \frac{1}{\sin\varphi}\left(r_1\varepsilon_{\varphi\varphi}^{0} - r_2\varepsilon_{\vartheta\vartheta}^{0}\right). \tag{15.81}$$

This is an inhomogeneous differential equation to determine the displacement v_0, which can be solved quite easily. The particular solution follows from direct integration:

$$v_{0,p} = \sin\varphi\int_{\varphi}\left(r_1\varepsilon_{\varphi\varphi}^{0} - r_2\varepsilon_{\vartheta\vartheta}^{0}\right)\frac{\mathrm{d}\varphi}{\sin\varphi}. \tag{15.82}$$

The homogeneous solution follows by neglecting the right-hand side of (15.81) as:

$$v_{0,h} = C_2\sin\varphi. \tag{15.83}$$

Obviously, this term describes a pure rigid body translation of the shell.

The total solution for v_0 follows as:

$$v_0 = v_{0,h} + v_{0,p} = \sin\varphi \left[\int_\varphi \left(r_1 \varepsilon_{\varphi\varphi}^0 - r_2 \varepsilon_{\vartheta\vartheta}^0 \right) \frac{\mathrm{d}\varphi}{\sin\varphi} + C_2 \right]. \tag{15.84}$$

From (15.79), the displacement w_0 is then determined as:

$$w_0 = r_2 \varepsilon_{\vartheta\vartheta}^0 - v_0 \cot\varphi. \tag{15.85}$$

If one uses here the constitutive relations (15.76), then it follows:

$$\begin{aligned} v_0 &= \sin\varphi \left[\int_\varphi \frac{1}{Eh} \left(N_{\varphi\varphi} \left(r_1 + \nu r_2 \right) - N_{\vartheta\vartheta} \left(r_2 + \nu r_1 \right) \right) \frac{\mathrm{d}\varphi}{\sin\varphi} + C_2 \right], \\ w_0 &= \frac{r_2}{Eh} \left(N_{\vartheta\vartheta} - \nu N_{\varphi\varphi} \right) - v_0 \cot\varphi. \end{aligned} \tag{15.86}$$

The integration constant C_2 is fitted to given displacement boundary conditions.
The ring expansion Δr is obtained from (15.67) as:

$$\Delta r = v_0 \cos\varphi + w_0 \sin\varphi = \frac{r_2 \sin\varphi}{Eh} \left(N_{\vartheta\vartheta} - \nu N_{\varphi\varphi} \right). \tag{15.87}$$

The rotation ψ_φ of the meridian tangent is composed of two components, one being a component due to the displacement v_0 in the meridian direction, and the other being a component due to the displacement w_0 in the direction of the normal of the shell midsurface. This results in:

$$\psi_\varphi = \frac{1}{r_1} \left(\frac{\partial w_0}{\partial \varphi} - v_0 \right). \tag{15.88}$$

Inserting the already determined displacement solutions and the constitutive relations finally yields:

$$\begin{aligned} \psi_\varphi = & \frac{1}{r_1} \frac{\partial}{\partial \varphi} \left[\frac{r^2}{Eh} \left(N_{\vartheta\vartheta} - \nu N_{\varphi\varphi} \right) \right] \\ & - \frac{\cot\varphi}{Eh} \left[N_{\varphi\varphi} \left(r_1 + \nu r_2 \right) - N_{\vartheta\vartheta} \left(r_2 + \nu r_1 \right) \right]. \end{aligned} \tag{15.89}$$

For the elementary shell geometries already discussed (spherical shell, conical shell, circular cylindrical shell), the displacement solutions (15.86) and the quantities derived from them (15.87) and (15.89) can be written as follows. For the spherical shell (radius R) we have:

$$\begin{aligned} v_0 &= \frac{R \sin\varphi}{Eh} (1 + \nu) \left[\int_\varphi \frac{N_{\varphi\varphi} - N_{\vartheta\vartheta}}{\sin\varphi} \mathrm{d}\varphi + C_2 \right], \\ w_0 &= \frac{R}{Eh} \left(N_{\vartheta\vartheta} - \nu N_{\varphi\varphi} \right) - v_0 \cot\varphi, \end{aligned}$$

$$\Delta r = \frac{R\sin\varphi}{Eh}\left(N_{\vartheta\vartheta} - \nu N_{\varphi\varphi}\right),$$
$$\psi_\varphi = \frac{1}{Eh}\left[\frac{\partial}{\partial\varphi}\left(N_{\vartheta\vartheta} - \nu N_{\varphi\varphi}\right) - (1+\nu)\cot\varphi\left(N_{\varphi\varphi} - N_{\vartheta\vartheta}\right)\right]. \quad (15.90)$$

Using the example of the conical shell (opening angle α, reference axis s) we obtain:

$$v_0 = \frac{1}{Eh}\int_s (N_{ss} - \nu N_{\vartheta\vartheta})\,\mathrm{d}s + C_2,$$
$$w_0 = \tan\alpha\left[\frac{s}{Eh}(N_{\vartheta\vartheta} - \nu N_{ss}) - v_0\right],$$
$$\Delta r = \frac{s\sin\alpha}{Eh}(N_{\vartheta\vartheta} - \nu N_{ss}),$$
$$\psi_s = \frac{\tan\alpha}{Eh}\left[s\frac{\mathrm{d}}{\mathrm{d}s}(N_{\vartheta\vartheta} - \nu N_{ss}) + (1+\nu)(N_{\vartheta\vartheta} - N_{ss})\right]. \quad (15.91)$$

For the circular cylindrical shell (radius R, reference axis x) we obtain:

$$v_0 = \frac{1}{Eh}\int_x (N_x - \nu N_{\vartheta\vartheta})\,\mathrm{d}x + C_2,$$
$$w_0 = \frac{R}{Eh}(N_{\vartheta\vartheta} - \nu N_x),$$
$$\Delta r = \frac{R}{Eh}(N_{\vartheta\vartheta} - \nu N_s) = w_0,$$
$$\psi_x = \frac{R}{Eh}\frac{\mathrm{d}}{\mathrm{d}x}(N_{\vartheta\vartheta} - \nu N_x). \quad (15.92)$$

As an example, consider the spherical shell of Fig. 15.22 (radius R, opening angle φ_0) under a uniformly distributed load p. The membrane force flows $N_{\varphi\varphi}$ and $N_{\vartheta\vartheta}$ in this case are obtained as:

$$N_{\varphi\varphi} = -\frac{Rp}{2}, \quad N_{\vartheta\vartheta} = \frac{Rp}{2}\left(1 - 2\cos^2\varphi\right). \quad (15.93)$$

From this, the normal strains $\varepsilon^0_{\varphi\varphi}$ and $\varepsilon^0_{\vartheta\vartheta}$ can be determined as follows:

$$\varepsilon^0_{\varphi\varphi} = \frac{1}{R}\left(\frac{\mathrm{d}v_0}{\mathrm{d}\varphi} + w_0\right) = \frac{1}{Eh}\left(N_{\varphi\varphi} - \nu N_{\vartheta\vartheta}\right),$$
$$\varepsilon^0_{\vartheta\vartheta} = \frac{1}{R}(v_0\cot\varphi + w_0) = \frac{1}{Eh}\left(N_{\vartheta\vartheta} - \nu N_{\varphi\varphi}\right). \quad (15.94)$$

Subtraction yields:

$$\frac{\mathrm{d}v_0}{\mathrm{d}\varphi} - v_0\cot\varphi = \frac{R}{Eh}(1+\nu)(N_{\varphi\varphi} - N_{\vartheta\vartheta}) = -\frac{(1+\nu)R^2}{Eh}p\sin^2\varphi. \quad (15.95)$$

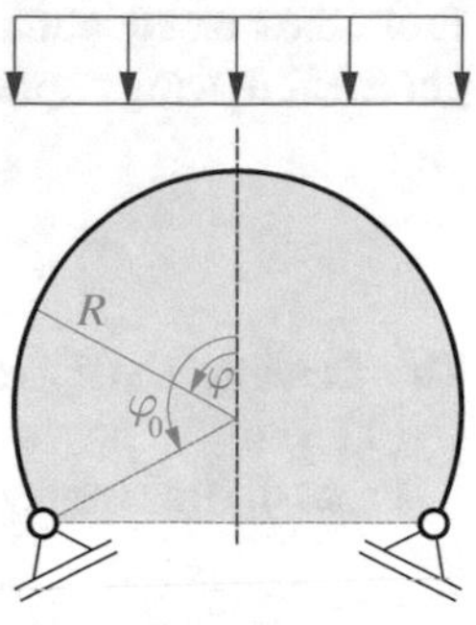

Fig. 15.22 Spherical shell under uniform load p

This is a differential equation for determining the displacement v_0. Its solution is:

$$v_0 = \frac{(1+\nu)R^2 p}{Eh}\left(\sin\varphi\cos\varphi + C\sin\varphi\right). \tag{15.96}$$

The constant C is determined from the condition that at the support the displacement v_0 must vanish: $v_0(\varphi = \varphi_0) = 0$. It follows:

$$C = -\cos\varphi_0. \tag{15.97}$$

Thus it follows from (15.96):

$$v_0 = \frac{(1+\nu)R^2 p\sin\varphi}{Eh}\left(\cos\varphi - \cos\varphi_0\right). \tag{15.98}$$

From this, the displacement w_0 can be determined as:

$$w_0 = \frac{R^2 p}{2Eh}\left[1+\nu - 2\cos^2\varphi - 2(1+\nu)\cos\varphi\left(\cos\varphi - \cos\varphi_0\right)\right]. \tag{15.99}$$

For the special case $\varphi_0 = \frac{\pi}{2}$ it follows:

$$w_0\left(\varphi = \varphi_0\right) = \frac{R^2 p}{2Eh}(1+\nu), \quad w_0\left(\varphi = 0\right) = -\frac{R^2 p}{2Eh}(3+\nu). \tag{15.100}$$

We also consider again the example of the semicircular cylindrical shell of Fig. 15.9 with the membrane force flows according to (15.29). The strain components are for this case:

$$\begin{aligned}
\varepsilon_{xx}^0 &= \frac{\partial v_0}{\partial x} = \frac{1}{Eh}\left(N_{xx} - \nu N_{\vartheta\vartheta}\right),\\
\varepsilon_{\vartheta\vartheta}^0 &= \frac{1}{R}\left(\frac{\partial u_0}{\partial\vartheta} + w_0\right) = \frac{1}{Eh}\left(N_{\vartheta\vartheta} - \nu N_{xx}\right),\\
\gamma_{x\vartheta} &= \frac{1}{R}\frac{\partial v_0}{\partial\vartheta} + \frac{\partial u_0}{\partial x} = \frac{2(1+\nu)}{Eh}N_{x\vartheta}.
\end{aligned} \tag{15.101}$$

From the first equation, after substituting N_{xx} and $N_{\vartheta\vartheta}$ (15.29) and integration, we get the following expression for the displacement v_0:

$$v_0 = \frac{Gx}{ERh}\left(\frac{x^2}{3} - \frac{l^2}{4} + \nu R^2\right)\cos\vartheta + C_3(\vartheta). \tag{15.102}$$

The function $C_3(\vartheta)$ follows to zero due to the symmetry condition $v_0(x = 0)$: $C_3(\vartheta) = 0$.

From the third equation in (15.101), the displacement u_0 after integration follows as:

$$u_0 = \frac{Gx^2}{2ER^2h}\left(\frac{x^2}{6} - \frac{l^2}{4} - (4+3\nu)R^2\right)\sin\vartheta + C_4(\vartheta). \tag{15.103}$$

The function $C_4(\vartheta)$ can be determined from the condition $u_0\left(x = \frac{l}{2}\right) = 0$ as:

$$C_4(\vartheta) = \frac{Gl^2}{8ER^2h}\left(\frac{5l^2}{24} + (4+3\nu)R^2\right)\sin\vartheta. \tag{15.104}$$

Finally, the displacement w_0 has to be determined, which can be obtained directly from the second equation in (15.101). The displacement field of the shell is finally:

$$\begin{aligned} u_0 &= \frac{G}{192EhR^2}(l^2 - 4x^2)\left[5l^2 - 4x^2 + 24(4+3\nu)R^2\right]\sin\vartheta, \\ v_0 &= \frac{Gx}{12ERh}\left(4x^2 - 3l^2 + 12\nu R^2\right)\cos\vartheta, \\ w_0 &= -\frac{G}{192EhR^2}\left((l^2 - 4x^2)\left[5l^2 - 4x^2 + 24(4+\nu)R^2\right] + 192R^4\right)\cos\vartheta. \end{aligned} \tag{15.105}$$

15.7 Energetic Derivation of the Basic Equations

The principle of virtual displacements can be used for an energetic derivation of the equilibrium conditions and boundary conditions for a shell of revolution in the framework of membrane theory. The principle of virtual displacements requires that the inner and the external virtual work are equal, i.e. $\delta W_i = \delta W_a$. We consider a shell of revolution distinguished by the two angles $\varphi = \varphi_0$ and $\varphi = \varphi_1$ at its lower and upper edges, respectively. The virtual internal work is presently given as:

$$\delta W_i = \delta \int_V \left(\sigma_{\varphi\varphi}\varepsilon^0_{\varphi\varphi} + \sigma_{\vartheta\vartheta}\varepsilon^0_{\vartheta\vartheta} + \tau_{\varphi\vartheta}\gamma^0_{\varphi\vartheta}\right) \mathrm{d}V, \tag{15.106}$$

where we can divide the volume integral with the surface element $\mathrm{d}A = r_1 r_2 \sin \varphi \mathrm{d}\varphi \mathrm{d}\vartheta$ into three partial integrals as follows:

$$\delta W_i = \delta \int_{-\frac{h}{2}}^{\frac{h}{2}} \int_0^{2\pi} \int_{\varphi_0}^{\varphi_1} \left(\sigma_{\varphi\varphi} \varepsilon_{\varphi\varphi}^0 + \sigma_{\vartheta\vartheta} \varepsilon_{\vartheta\vartheta}^0 + \tau_{\varphi\vartheta} \gamma_{\varphi\vartheta}^0 \right) r_1 r_2 \sin \varphi \mathrm{d}\varphi \mathrm{d}\vartheta dz. \quad (15.107)$$

The integral with respect to z can be easily solved, and we obtain:

$$\delta W_i = \delta \int_0^{2\pi} \int_{\varphi_0}^{\varphi_1} \left(N_{\varphi\varphi} \varepsilon_{\varphi\varphi}^0 + N_{\vartheta\vartheta} \varepsilon_{\vartheta\vartheta}^0 + N_{\varphi\vartheta} \gamma_{\varphi\vartheta}^0 \right) r_1 r_2 \sin \varphi \mathrm{d}\varphi \mathrm{d}\vartheta, \quad (15.108)$$

or after inserting the kinematic equations (15.74):

$$\begin{aligned} \delta W_i = \delta \int_0^{2\pi} \int_{\varphi_0}^{\varphi_1} & \left[N_{\varphi\varphi} \frac{1}{r_1} \left(\frac{\partial v_0}{\partial \varphi} + w_0 \right) \right. \\ & + N_{\vartheta\vartheta} \frac{1}{r} \left(\frac{\partial u_0}{\partial \vartheta} + v_0 \cos \varphi + w_0 \sin \varphi \right) \\ & \left. + N_{\varphi\vartheta} \left(\frac{1}{r} \frac{\partial v_0}{\partial \vartheta} + \frac{1}{r_1} \frac{\partial u_0}{\partial \varphi} - \frac{u_0 \cos \varphi}{r} \right) \right] r_1 r_2 \sin \varphi \mathrm{d}\varphi \mathrm{d}\vartheta. (15.109) \end{aligned}$$

Variations δu_0, δv_0, δw_0 of the displacements u_0, v_0, w_0 of the shell midsurface and their derivatives appear in this expression. Terms in which derivatives of the variations δu_0, δv_0, δw_0 appear are integrated partially in thc following. It results with $r = r_2 \sin \varphi$:

$$\begin{aligned} \int_0^{2\pi} \int_{\varphi_0}^{\varphi_1} N_{\varphi\varphi} \frac{\partial \delta v_0}{\partial \varphi} r \mathrm{d}\varphi \mathrm{d}\vartheta &= \int_0^{2\pi} N_{\varphi\varphi} r \delta v_0 \mathrm{d}\vartheta \Big|_{\varphi_0}^{\varphi_1} - \int_0^{2\pi} \int_{\varphi_0}^{\varphi_1} \frac{\partial \left(N_{\varphi\varphi} r \right)}{\partial \varphi} \delta v_0 \mathrm{d}\varphi \mathrm{d}\vartheta, \\ \int_0^{2\pi} \int_{\varphi_0}^{\varphi_1} N_{\vartheta\vartheta} \frac{\partial \delta u_0}{\partial \vartheta} r_1 \mathrm{d}\varphi \mathrm{d}\vartheta &= \int_{\varphi_0}^{\varphi_1} N_{\vartheta\vartheta} \delta u_0 r_1 \mathrm{d}\varphi \Big|_0^{2\pi} - \int_0^{2\pi} \int_{\varphi_0}^{\varphi_1} \frac{\partial N_{\vartheta\vartheta}}{\mathrm{d}\vartheta} \delta u_0 r_1 \mathrm{d}\varphi \mathrm{d}\vartheta, \\ \int_0^{2\pi} \int_{\varphi_0}^{\varphi_1} N_{\varphi\vartheta} \frac{\partial \delta v_0}{\partial \vartheta} r_1 \mathrm{d}\varphi \mathrm{d}\vartheta &= \int_{\varphi_0}^{\varphi_1} N_{\varphi\vartheta} \delta v_0 r_1 \mathrm{d}\varphi \Big|_0^{2\pi} - \int_0^{2\pi} \int_{\varphi_0}^{\varphi_1} \frac{\partial N_{\varphi\vartheta}}{\mathrm{d}\vartheta} \delta v_0 r_1 \mathrm{d}\varphi \mathrm{d}\vartheta, \\ \int_0^{2\pi} \int_{\varphi_0}^{\varphi_1} N_{\varphi\vartheta} \frac{\partial \delta u_0}{\partial \varphi} r \mathrm{d}\varphi \mathrm{d}\vartheta &= \int_0^{2\pi} N_{\varphi\vartheta} r \delta u_0 \mathrm{d}\vartheta \Big|_{\varphi_0}^{\varphi_1} - \int_0^{2\pi} \int_{\varphi_0}^{\varphi_1} \frac{\partial \left(N_{\varphi\vartheta} r \right)}{\partial \varphi} \delta u_0 \mathrm{d}\varphi \mathrm{d}\vartheta. \end{aligned} \quad (15.110)$$

The virtual external work δW_a reads in the presence of the surface loads p_φ, p_ϑ and p_z:

$$\delta W_a = \int_0^{2\pi} \int_{\varphi_0}^{\varphi_1} \left(p_\vartheta \delta u_0 + p_\varphi \delta v_0 + p_z \delta w_0 \right) r_1 r \mathrm{d}\varphi \mathrm{d}\vartheta. \quad (15.111)$$

The principle of virtual displacements $\delta W_i = \delta W_a$ can then be written with (15.109), (15.110) and (15.111) as:

$$\begin{aligned}
&\int_0^{2\pi}\int_{\varphi_0}^{\varphi_1}\Bigg[\delta u_0\left(-\frac{\partial N_{\vartheta\vartheta}}{\partial\vartheta}r_1-\frac{\partial\left(N_{\varphi\vartheta}r\right)}{\partial\varphi}-N_{\varphi\vartheta}\cos\varphi r_1-p_\vartheta r_1 r\right)\\
&+\delta v_0\left(-\frac{\partial\left(N_{\varphi\varphi}r\right)}{\partial\varphi}+N_{\vartheta\vartheta}\cos\varphi r_1-\frac{\partial N_{\varphi\vartheta}}{\partial\vartheta}r_1-p_\varphi r_1 r\right)\\
&+\delta w_0\left(N_{\varphi\varphi}r_2\sin\varphi+N_{\vartheta\vartheta}r_1\sin\varphi-p_z r_1 r_2\sin\varphi\right)\Bigg]\mathrm{d}\varphi\mathrm{d}\vartheta\\
&+\int_0^{2\pi}N_{\varphi\varphi}r\delta v_0\mathrm{d}\vartheta\Big|_{\varphi_0}^{\varphi_1}+\int_{\varphi_0}^{\varphi_1}N_{\vartheta\vartheta}\delta u_0 r_1\mathrm{d}\varphi\Big|_0^{2\pi}\\
&+\int_{\varphi_0}^{\varphi_1}N_{\varphi\vartheta}\delta v_0 r_1\mathrm{d}\varphi\Big|_0^{2\pi}+\int_0^{2\pi}N_{\varphi\vartheta}r\delta u_0\mathrm{d}\vartheta\Big|_{\varphi_0}^{\varphi_1}=0.
\end{aligned}\tag{15.112}$$

If the solutions $\delta u_0 = 0$, $\delta v_0 = 0$ and $\delta w_0 = 0$ are not considered further, then the terms in the round brackets of the integral terms give the equilibrium conditions of the considered shell situation:

$$\begin{aligned}
\frac{\partial N_{\vartheta\vartheta}}{\partial\vartheta}r_1+\frac{\partial\left(N_{\varphi\vartheta}r\right)}{\partial\varphi}+N_{\varphi\vartheta}\cos\varphi r_1+p_\vartheta r_1 r&=0,\\
\frac{\partial\left(N_{\varphi\varphi}r\right)}{\partial\varphi}-N_{\vartheta\vartheta}\cos\varphi r_1+\frac{\partial N_{\varphi\vartheta}}{\partial\vartheta}r_1+p_\varphi r_1 r&=0,\\
N_{\varphi\varphi}r_2+N_{\vartheta\vartheta}r_1-p_z r_1 r_2&=0.
\end{aligned}\tag{15.113}$$

It is easy to see that these equations agree with those already derived from elementary equilibrium considerations (15.2). The boundary terms appearing in (15.112) can be interpreted as follows. The terms

$$\begin{aligned}
\int_0^{2\pi}N_{\varphi\varphi}r\delta v_0\mathrm{d}\vartheta\Big|_{\varphi_0}^{\varphi_1}&=0,\\
\int_0^{2\pi}N_{\varphi\vartheta}r\delta u_0\mathrm{d}\vartheta\Big|_{\varphi_0}^{\varphi_1}&=0
\end{aligned}\tag{15.114}$$

are equivalent to the requirement that at the locations $\varphi = \varphi_0$ and $\varphi = \varphi_1$ for any angle ϑ between $\vartheta = 0$ and $\vartheta = 2\pi$ either the membrane force flow $N_{\varphi\varphi}$ resp. $N_{\varphi\vartheta}$ becomes zero, or the variations of the midsurface displacements δv_0 and δu_0 become zero, corresponding to prescribed displacements v_0 and u_0, respectively. The two remaining boundary terms

$$\begin{aligned}
\int_{\varphi_0}^{\varphi_1}N_{\vartheta\vartheta}\delta u_0 r_1\mathrm{d}\varphi\Big|_0^{2\pi}&=0,\\
\int_{\varphi_0}^{\varphi_1}N_{\varphi\vartheta}\delta v_0 r_1\mathrm{d}\varphi\Big|_0^{2\pi}&=0
\end{aligned}\tag{15.115}$$

lead to similar requirements, but vanish due to the equalities $N_{\vartheta\vartheta}(\vartheta = 0) = N_{\vartheta\vartheta}(\vartheta = 2\pi)$ and $N_{\varphi\vartheta}(\vartheta = 0) = N_{\varphi\vartheta}(\vartheta = 2\pi)$.

References

Becker, W., Gross, D.: Mechanik elastischer Körper und Strukturen. Springer, Berlin (2002)
Born, J. (1968): Praktische Schalenstatik, Band I: Die Rotationsschalen, 2nd edn. Ernst und Sohn, Berlin (1968)
Eschenauer, H., Schnell, W.: Elastizitätstheorie. Bibliographisches Institut, Mannheim (1993)
Eschenauer, H., Olhoff, N., Schnell, W.: Applied Structural Mechanics. Springer, Berlin (1997)
Flügge, W.: Statik und Dynamik der Schalen, 3rd edn. Springer, Berlin (1981)
Girkmann, K.: Flächentragwerke, 6th edn. Springer, Wien (1974)
Hake, E., Meskouris, K.: Statik der Flächentragwerke, 2nd edn. Springer, Berlin (2007)
Pflüger, A.: Elementare Schalenstatik, 4th edn. Springer, Berlin (1967)
Ramm, E., Müller, J.: Schalen. Universität Stuttgart, Institut für Baustatik und Baudynamik, Stuttgart (1995)
Reddy, J.N.: Theory and Analysis of Elastic Plates and Shells. CRC Press, Boca Raton (2006)
Timoshenko, S., Woinowsky-Krieger, S.: Theory of Plates and Shells, 2nd edn. McGraw Hill, New York (1964)

16 Bending Theory of Shells of Revolution

This chapter is devoted to the bending theory of shells of revolution. After deriving the equilibrium conditions for arbitrary shells of revolution, the corresponding kinematic equations and the constitutive law in the presence of bending action are discussed. This is followed by the discussion of the so-called container theory of the circular cylindrical shell, i.e. a shell of revolution under rotationally symmetric loading. In addition to providing the so-called container equation, the typical rapidly decaying boundary stress behavior is discussed here. This is followed by the presentation of the force method as a way to solve statically indeterminate shell problems, before the boundary stress problems for spherical shells as well as for arbitrary shells of revolution are discussed. Another section is devoted to the consideration of the circular cylindrical shell under arbitrary loads when the trapezoidal effect is taken into account. The chapter concludes with a consideration of the container theory for laminated circular cylindrical shells. Further information on the contents of this chapter can be found, for example, in Altenbach et al. (2004), Ambartsumyan (1964), Becker and Gross (2002), Eschenauer and Schnell (1993), Eschenauer et al. (1997), Flügge (1981), Göldner et al. (1979, 1985), Ramm and Müller (1995), Reddy (2006), or Timoshenko and Woinowsky-Krieger (1964).

16.1 Basic Equations

16.1.1 Equilibrium Conditions

In addition to the membrane state of stress which is always desirable for thin-walled shell structures, in many technically relevant situations loads on the shell can no longer be transferred exclusively by membrane stresses. Rather, a bending effect is additionally introduced, and the corresponding analysis model is referred to as the so-called bending theory. We want to restrict ourselves here to shells of revolution and

C. Mittelstedt, *Theory of Plates and Shells*,
https://doi.org/10.1007/978-3-662-66805-4_16

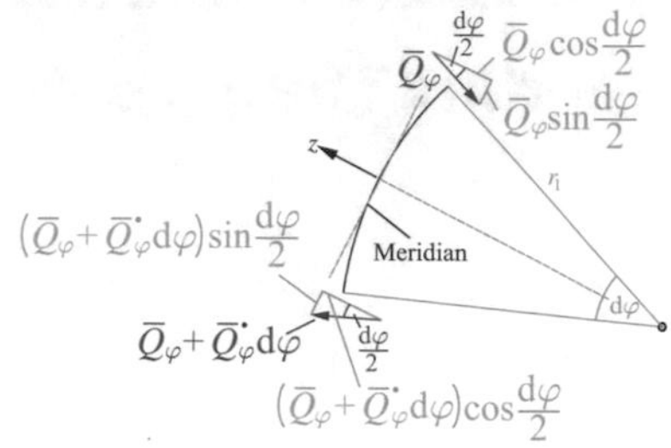

Fig. 16.1 Contribution of the shear force flow to the force equilibrium in the meridional direction

consider shells whose thickness h, assumed to be constant here, is small compared to the other characteristic dimensions r_1, r_2 of the shell. We therefore start from the Kirchhoff-Love approximation and define the force and moment quantities of the shell as already given by (14.20):

$$\begin{aligned} N_{\varphi\varphi} &= \int_{-\frac{h}{2}}^{\frac{h}{2}} \sigma_{\varphi\varphi} dz, \quad N_{\vartheta\vartheta} = \int_{-\frac{h}{2}}^{\frac{h}{2}} \sigma_{\vartheta\vartheta} dz, \\ N_{\varphi\vartheta} &= \int_{-\frac{h}{2}}^{\frac{h}{2}} \tau_{\varphi\vartheta} dz = N_{\vartheta\varphi}, \quad M_{\varphi\varphi} = \int_{-\frac{h}{2}}^{\frac{h}{2}} \sigma_{\varphi\varphi} z dz, \\ M_{\vartheta\vartheta} &= \int_{-\frac{h}{2}}^{\frac{h}{2}} \sigma_{\vartheta\vartheta} z dz, \quad M_{\varphi\vartheta} = \int_{-\frac{h}{2}}^{\frac{h}{2}} \tau_{\varphi\vartheta} z dz = M_{\vartheta\varphi}, \\ Q_{\varphi} &= \int_{-\frac{h}{2}}^{\frac{h}{2}} \tau_{\varphi z} dz, \quad Q_{\vartheta} = \int_{-\frac{h}{2}}^{\frac{h}{2}} \tau_{\vartheta z} dz. \end{aligned} \tag{16.1}$$

Thus, there are a total of eight quantities to be determined. We can provide six equilibrium conditions which we will derive below with the help of the conditions already established in the context of the membrane theory.

We first consider the force equilibrium in the direction of the meridian tangent. As already shown in the context of the membrane theory, this equilibrium condition is composed of contributions of the membrane force flow $N_{\varphi\vartheta}$ in ϑ-direction as well as of $N_{\varphi\varphi}$ in the φ-direction. Added to this is the external load p_φ as well as the fraction of the membrane force flow $N_{\vartheta\vartheta}$ in φ-direction. In addition, in the context of bending theory a fraction from the shear force flow Q_φ, as shown in Fig. 16.1 using the resultants $\bar{Q}_\varphi = Q_\varphi r d\vartheta$ occurs. Decomposition of the two shear force flows occurring at the cutting edges shown here, using the small angle approximation and neglecting higher order terms, indicates that the remaining component of Q_φ acting in the meridional direction has the size $Q_\varphi r d\varphi d\vartheta$. The corresponding equilibrium of forces (15.2) then takes the following form:

$$\frac{\partial}{\partial\varphi}\left(N_{\varphi\varphi} r\right) + \frac{\partial N_{\varphi\vartheta}}{\partial\vartheta} r_1 - N_{\vartheta\vartheta}\cos\varphi r_1 + Q_\varphi r + p_\varphi r_1 r = 0. \tag{16.2}$$

Analogously, the two other force equilibrium relations in (15.2) can be extended to include the shear force flow components, and we obtain:

$$\frac{\partial}{\partial \varphi}\left(N_{\varphi\vartheta} r\right) + \frac{\partial N_{\vartheta\vartheta}}{\partial \vartheta} r_1 + N_{\vartheta\varphi} \cos\varphi r_1 + Q_{\vartheta} r_1 \sin\varphi + p_{\vartheta} r_1 r = 0,$$
$$\frac{\partial}{\partial \varphi}\left(Q_{\varphi} r\right) + \frac{\partial Q_{\vartheta}}{\partial \vartheta} r_1 - N_{\varphi\varphi} r - N_{\vartheta\vartheta} r_1 \sin\varphi + p_z r_1 r = 0. \quad (16.3)$$

With respect to the moment equilibrium conditions, it should be noted that the equilibrium condition concerning the normal axis is no longer applicable due to the equality of $N_{\varphi\vartheta}$ and $N_{\vartheta\varphi}$. Thus, the two moment equilibria with respect to the meridian axis and the ring axis remain. The equilibrium of moments with respect to the direction of the circle of latitude finally results:

$$\frac{\partial}{\partial \varphi}\left(M_{\varphi\varphi} r\right) + \frac{\partial M_{\varphi\vartheta}}{\partial \vartheta} r_1 - M_{\vartheta\vartheta} r_1 \cos\varphi - Q_{\varphi} r_1 r = 0. \quad (16.4)$$

Analogously, we obtain for the moment equilibrium about the meridian tangent:

$$\frac{\partial}{\partial \varphi}\left(M_{\varphi\vartheta} r\right) + \frac{\partial M_{\vartheta\vartheta}}{\partial \vartheta} r_1 + M_{\varphi\vartheta} r_1 \cos\varphi - Q_{\vartheta} r_1 r = 0. \quad (16.5)$$

In summary, the five available equilibrium conditions are thus:

$$\frac{\partial}{\partial \varphi}\left(N_{\varphi\varphi} r\right) + \frac{\partial N_{\varphi\vartheta}}{\partial \vartheta} r_1 - N_{\vartheta\vartheta} \cos\varphi r_1 + Q_{\varphi} r + p_{\varphi} r_1 r = 0,$$
$$\frac{\partial}{\partial \varphi}\left(N_{\varphi\vartheta} r\right) + \frac{\partial N_{\vartheta\vartheta}}{\partial \vartheta} r_1 + N_{\vartheta\varphi} \cos\varphi r_1 + Q_{\vartheta} r_1 \sin\varphi + p_{\vartheta} r_1 r = 0,$$
$$\frac{\partial}{\partial \varphi}\left(Q_{\varphi} r\right) + \frac{\partial Q_{\vartheta}}{\partial \vartheta} r_1 - N_{\varphi\varphi} r - N_{\vartheta\vartheta} r_1 \sin\varphi + p_z r_1 r = 0,$$
$$\frac{\partial}{\partial \varphi}\left(M_{\varphi\varphi} r\right) + \frac{\partial M_{\varphi\vartheta}}{\partial \vartheta} r_1 - M_{\vartheta\vartheta} r_1 \cos\varphi - Q_{\varphi} r_1 r = 0,$$
$$\frac{\partial}{\partial \varphi}\left(M_{\varphi\vartheta} r\right) + \frac{\partial M_{\vartheta\vartheta}}{\partial \vartheta} r_1 + M_{\varphi\vartheta} r_1 \cos\varphi - Q_{\vartheta} r_1 r = 0. \quad (16.6)$$

This is opposed by eight force and moment quantities to be determined, so that a shell problem in the context of bending theory turns out to be statically indeterminate. Thus, we have to use both kinematic and constitutive relations to solve a bending theory problem.

From the equilibrium conditions (16.6) some special cases can be deduced. We first consider the spherical shell with radius R, for which $r_1 = r_2 = R$ and $r = R \sin\varphi$ holds. This gives the following equilibrium conditions:

$$\frac{\partial}{\partial\varphi}(N_{\varphi\varphi}\sin\varphi)+\frac{\partial N_{\varphi\vartheta}}{\partial\vartheta}-N_{\vartheta\vartheta}\cos\varphi+Q_{\varphi}\sin\varphi+p_{\varphi}R\sin\varphi=0,$$
$$\frac{\partial}{\partial\varphi}(N_{\varphi\vartheta}\sin\varphi)+\frac{\partial N_{\vartheta\vartheta}}{\partial\vartheta}+N_{\varphi\vartheta}\cos\varphi+Q_{\vartheta}\sin\varphi+p_{\vartheta}R\sin\varphi=0,$$
$$\frac{\partial}{\partial\varphi}(Q_{\varphi}\sin\varphi)+\frac{\partial Q_{\vartheta}}{\partial\vartheta}-N_{\varphi\varphi}\sin\varphi-N_{\vartheta\vartheta}\sin\varphi+p_{z}R\sin\varphi=0,$$
$$\frac{\partial}{\partial\varphi}(M_{\varphi\varphi}\sin\varphi)+\frac{\partial M_{\varphi\vartheta}}{\partial\vartheta}-M_{\vartheta\vartheta}\cos\varphi-Q_{\varphi}R\sin\varphi=0,$$
$$\frac{\partial}{\partial\varphi}(M_{\varphi\vartheta}\sin\varphi)+\frac{\partial M_{\vartheta\vartheta}}{\partial\vartheta}+M_{\varphi\vartheta}\cos\varphi-Q_{\vartheta}R\sin\varphi=0. \quad (16.7)$$

For the special case of the rotationally symmetrically loaded spherical shell remains with $p_{\vartheta}=0$ and $N_{\varphi\vartheta}=0$, $M_{\varphi\vartheta}=0$, $Q_{\vartheta}=0$:

$$\frac{d}{d\varphi}(N_{\varphi\varphi}\sin\varphi)-N_{\vartheta\vartheta}\cos\varphi+Q_{\varphi}\sin\varphi+p_{\varphi}R\sin\varphi=0,$$
$$\frac{d}{d\varphi}(Q_{\varphi}\sin\varphi)-N_{\varphi\varphi}\sin\varphi-N_{\vartheta\vartheta}\sin\varphi+p_{z}R\sin\varphi=0,$$
$$\frac{d}{d\varphi}(M_{\varphi\varphi}\sin\varphi)-M_{\vartheta\vartheta}\cos\varphi-Q_{\varphi}R\sin\varphi=0. \quad (16.8)$$

The special case of the circular cylinder shell with $r_1\to\infty$, $r_2=r=R$, $\varphi=\frac{\pi}{2}$ and the relation $r_1 d\varphi = dx$ can be inferred from (16.6) as:

$$\frac{\partial N_{xx}}{\partial x}+\frac{1}{R}\frac{\partial N_{x\vartheta}}{\partial\vartheta}+p_x=0,$$
$$\frac{\partial N_{x\vartheta}}{\partial x}+\frac{1}{R}\frac{\partial N_{\vartheta\vartheta}}{\partial\vartheta}+\frac{Q_{\vartheta}}{R}+p_{\vartheta}=0,$$
$$\frac{\partial Q_x}{\partial x}+\frac{1}{R}\frac{\partial Q_{\vartheta}}{\partial\vartheta}-\frac{N_{\vartheta\vartheta}}{R}+p_z=0,$$
$$\frac{\partial M_{xx}}{\partial x}+\frac{1}{R}\frac{\partial M_{x\vartheta}}{\partial\vartheta}-Q_x=0,$$
$$\frac{\partial M_{x\vartheta}}{\partial x}+\frac{1}{R}\frac{\partial M_{\vartheta\vartheta}}{\partial\vartheta}-Q_{\vartheta}=0. \quad (16.9)$$

The special case of the rotationally symmetrically loaded circular cylindrical shell with $p_{\vartheta}=0$ and $N_{x\vartheta}=0$, $M_{x\vartheta}=0$, $Q_{\vartheta}=0$ results as:

$$\frac{dN_{xx}}{dx}+p_x=0,$$
$$\frac{dQ_x}{dx}-\frac{N_{\vartheta\vartheta}}{R}+p_z=0,$$
$$\frac{dM_{xx}}{dx}-Q_x=0. \quad (16.10)$$

The equations of the circular cylindrical shell using a dimensionless reference axis $\xi = \frac{x}{R}$ are obtained from (16.9) as:

$$\begin{aligned}
\frac{\partial N_{\xi\xi}}{\partial \xi} + \frac{\partial N_{\xi\vartheta}}{\partial \vartheta} + R p_{\xi} &= 0, \\
\frac{\partial N_{\xi\vartheta}}{\partial \xi} + \frac{\partial N_{\vartheta\vartheta}}{\partial \vartheta} + Q_{\vartheta} + R p_{\vartheta} &= 0, \\
\frac{\partial Q_{\xi}}{\partial \xi} + \frac{\partial Q_{\vartheta}}{\partial \vartheta} - N_{\vartheta\vartheta} + R p_{z} &= 0, \\
\frac{\partial M_{\xi\xi}}{\partial \xi} + \frac{\partial M_{\xi\vartheta}}{\partial \vartheta} - Q_{\xi} R &= 0, \\
\frac{\partial M_{\xi\vartheta}}{\partial \xi} + \frac{\partial M_{\vartheta\vartheta}}{\partial \vartheta} - Q_{\vartheta} R &= 0,
\end{aligned} \tag{16.11}$$

or for the special case of rotationally symmetric loading:

$$\begin{aligned}
\frac{\mathrm{d}N_{\xi\xi}}{\mathrm{d}\xi} + R p_{\xi} &= 0, \\
\frac{\mathrm{d}Q_{\xi}}{\mathrm{d}\xi} - N_{\vartheta\vartheta} + R p_{z} &= 0, \\
\frac{\mathrm{d}M_{\xi\xi}}{\mathrm{d}\xi} - Q_{\xi} R &= 0.
\end{aligned} \tag{16.12}$$

16.1.2 Kinematic Equations

Besides the displacement quantities u_0, v_0, w_0 already known from the membrane theory as well as the strains $\varepsilon^0_{\varphi\varphi}$, $\varepsilon^0_{\vartheta\vartheta}$, $\gamma^0_{\varphi\vartheta}$, in the context of bending theory the curvatures $\kappa^0_{\varphi\varphi}$ (meridian curvature), $\kappa^0_{\vartheta\vartheta}$ (circumferential curvature) and the twist $\kappa^0_{\varphi\vartheta}$ also occur. For the normal strains $\varepsilon^0_{\varphi\varphi}$, $\varepsilon^0_{\vartheta\vartheta}$ and the shear strain $\gamma^0_{\varphi\vartheta}$ of the shell midsurface, the expressions (15.74) are directly reapplicable.

To derive an expression for the curvature $\kappa^0_{\varphi\varphi}$ of the shell midsurface we introduce the angle χ, which describes the angle of rotation of the shell with respect to the undeformed initial configuration for a section φ = const. This angle is counted as positive when points $z > 0$ are rotated in the direction of increasing values for φ. We again consider Fig. 15.19 and first take a look at the displacement difference between the two normal displacements $w_{0,1}$ and $w_{0,2}$. We denote this difference, which arises over the distance $r_1 \mathrm{d}\varphi$, as w_0. The resulting angle χ_1 can then be determined from the relationship that the displacement difference w_0 corresponds to the angle χ_1 multiplied by the distance $r_1 \mathrm{d}\varphi$. This gives

$$\chi_1 = -\frac{1}{r_1}\frac{\partial w_0}{\partial \varphi}. \tag{16.13}$$

In quite the same way, the fraction χ_2 due to the meridian shift can be determined. If we denote the displacement difference between $v_{0,1}$ and $v_{0,2}$ as v_0, we get:

$$\chi_2 = \frac{v_0}{r_1}. \tag{16.14}$$

The angle χ then follows from adding up the two parts χ_1 and χ_2, so that:

$$\chi = -\frac{1}{r_1}\frac{\partial w_0}{\partial \varphi} + \frac{v_0}{r_1}. \tag{16.15}$$

We then obtain the curvature $\kappa^0_{\varphi\varphi}$ from the change of the angle χ across the meridian, i.e.:

$$\kappa^0_{\varphi\varphi} = \frac{1}{r_1}\frac{\partial \chi}{\partial \varphi} = \frac{1}{r_1}\frac{\partial}{\partial \varphi}\left(\frac{v_0 - \frac{\partial w_0}{\partial \varphi}}{r_1}\right). \tag{16.16}$$

Based on quite similar considerations, one also obtains expressions for the curvature $\kappa^0_{\vartheta\vartheta}$ and the twist $\kappa^0_{\varphi\vartheta}$, which we give below:

$$\begin{aligned}
\kappa^0_{\vartheta\vartheta} &= \frac{1}{r^2}\frac{\partial}{\partial \vartheta}\left(u_0 \sin\varphi - \frac{\partial w_0}{\partial \vartheta}\right) + \frac{\cos\varphi}{r r_1}\left(v_0 - \frac{\partial w_0}{\partial \varphi}\right),\\
\kappa^0_{\varphi\vartheta} &= \frac{1}{r r_1}\frac{\partial v_0}{\partial \vartheta} - \frac{2}{r r_1}\frac{\partial^2 w_0}{\partial \varphi \partial \vartheta} + 2\frac{\cos\varphi}{r^2}\frac{\partial w_0}{\partial \vartheta} - 2\frac{\cos\varphi}{r r_2}u_0 + \frac{1}{r_1 r_2}\frac{\partial u_0}{\partial \varphi} + \frac{\cos\varphi}{r r_1}u_0.
\end{aligned} \tag{16.17}$$

Overall, the following kinematic equations can be given in the context of the bending theory of shells of revolution:

$$\begin{aligned}
\varepsilon^0_{\varphi\varphi} &= \frac{1}{r_1}\left(\frac{\partial v_0}{\partial \varphi} + w_0\right),\\
\varepsilon^0_{\vartheta\vartheta} &= \frac{1}{r}\left(\frac{\partial u_0}{\partial \vartheta} + v_0 \cos\varphi + w_0 \sin\varphi\right),\\
\gamma^0_{\varphi\vartheta} &= \frac{1}{r}\frac{\partial v_0}{\partial \vartheta} + \frac{1}{r_1}\frac{\partial u_0}{\partial \varphi} - \frac{u_0 \cos\varphi}{r},\\
\kappa^0_{\varphi\varphi} &= \frac{1}{r_1}\frac{\partial \chi}{\partial \varphi} = \frac{1}{r_1}\frac{\partial}{\partial \varphi}\left(\frac{v_0 - \frac{\partial w_0}{\partial \varphi}}{r_1}\right),\\
\kappa^0_{\vartheta\vartheta} &= \frac{1}{r^2}\frac{\partial}{\partial \vartheta}\left(u_0 \sin\varphi - \frac{\partial w_0}{\partial \vartheta}\right) + \frac{\cos\varphi}{r r_1}\left(v_0 - \frac{\partial w_0}{\partial \varphi}\right),\\
\kappa^0_{\varphi\vartheta} &= \frac{1}{r r_1}\frac{\partial v_0}{\partial \vartheta} - \frac{2}{r r_1}\frac{\partial^2 w_0}{\partial \varphi \partial \vartheta} + 2\frac{\cos\varphi}{r^2}\frac{\partial w_0}{\partial \vartheta} - 2\frac{\cos\varphi}{r r_2}u_0 + \frac{1}{r_1 r_2}\frac{\partial u_0}{\partial \varphi} + \frac{\cos\varphi}{r r_1}u_0.
\end{aligned} \tag{16.18}$$

For the special case of rotationally symmetric load with $\gamma^0_{\varphi\vartheta} = 0$ and $\kappa^0_{\varphi\vartheta} = 0$ follows:

$$\begin{aligned}
\varepsilon^0_{\varphi\varphi} &= \frac{1}{r_1}\left(\frac{\mathrm{d}v_0}{\mathrm{d}\varphi} + w_0\right), \\
\varepsilon^0_{\vartheta\vartheta} &= \frac{1}{r}\left(v_0 \cos\varphi + w_0 \sin\varphi\right), \\
\kappa^0_{\varphi\varphi} &= \frac{1}{r_1}\frac{\mathrm{d}}{\mathrm{d}\varphi}\left(\frac{v_0 - \frac{\mathrm{d}w_0}{\mathrm{d}\varphi}}{r_1}\right), \\
\kappa^0_{\vartheta\vartheta} &= \frac{\cos\varphi}{r r_1}\left(v_0 - \frac{\mathrm{d}w_0}{\mathrm{d}\varphi}\right).
\end{aligned} \tag{16.19}$$

The kinematic equations (16.18) and (16.19) can be specialized again for certain special forms of shells of revolution. If a spherical shell with radius $r_1 = r_2 = R$ and $r = R\sin\varphi$ is considered, then it follows from (16.18):

$$\begin{aligned}
\varepsilon^0_{\varphi\varphi} &= \frac{1}{R}\left(\frac{\partial v_0}{\partial\varphi} + w_0\right), \\
\varepsilon^0_{\vartheta\vartheta} &= \frac{1}{R}\left(\frac{1}{\sin\varphi}\frac{\partial u_0}{\partial\vartheta} + v_0\cot\varphi + w_0\right), \\
\gamma^0_{\varphi\vartheta} &= \frac{1}{R\sin\varphi}\frac{\partial v_0}{\partial\vartheta} + \frac{1}{R}\frac{\partial u_0}{\partial\varphi} - \frac{u_0}{R}\cot\varphi, \\
\kappa^0_{\varphi\varphi} &= \frac{1}{R^2}\frac{\partial}{\partial\varphi}\left(v_0 - \frac{\partial w_0}{\partial\varphi}\right), \\
\kappa^0_{\vartheta\vartheta} &= \frac{1}{R^2\sin^2\varphi}\frac{\partial}{\partial\vartheta}\left(u_0\sin\varphi - \frac{\partial w_0}{\partial\vartheta}\right) + \frac{\cot\varphi}{R^2}\left(v_0 - \frac{\partial w_0}{\partial\varphi}\right), \\
\kappa^0_{\varphi\vartheta} &= \frac{1}{R^2\sin\varphi}\left(\frac{\partial v_0}{\partial\vartheta} - 2\frac{\partial^2 w_0}{\partial\varphi\partial\vartheta} + 2\cot\varphi\frac{\partial w_0}{\partial\vartheta} - \cos\varphi u_0 + \sin\varphi\frac{\partial u_0}{\partial\varphi}\right).
\end{aligned} \tag{16.20}$$

For the case of rotationally symmetric loading, the following strains and curvatures remain:

$$\begin{aligned}
\varepsilon^0_{\varphi\varphi} &= \frac{1}{R}\left(\frac{\mathrm{d}v_0}{\mathrm{d}\varphi} + w_0\right), \\
\varepsilon^0_{\vartheta\vartheta} &= \frac{1}{R}\left(v_0\cot\varphi + w_0\right), \\
\kappa^0_{\varphi\varphi} &= \frac{1}{R^2}\frac{\mathrm{d}}{\mathrm{d}\varphi}\left(v_0 - \frac{\mathrm{d}w_0}{\mathrm{d}\varphi}\right), \\
\kappa^0_{\vartheta\vartheta} &= \frac{\cot\varphi}{R^2}\left(v_0 - \frac{\mathrm{d}w_0}{\mathrm{d}\varphi}\right).
\end{aligned} \tag{16.21}$$

In the case of a circular cylindrical shell with radius $r = r_2 = R$ and with $r_1 \to \infty$ and $\varphi = \frac{\pi}{2}$ as well as $r_1 \mathrm{d}\varphi = \mathrm{d}x$ results:

$$\begin{aligned}
\varepsilon_{xx}^0 &= \frac{\partial v_0}{\partial x}, \\
\varepsilon_{\vartheta\vartheta}^0 &= \frac{1}{R}\left(\frac{\partial u_0}{\partial \vartheta} + w_0\right), \\
\gamma_{x\vartheta}^0 &= \frac{1}{R}\frac{\partial v_0}{\partial \vartheta} + \frac{\partial u_0}{\partial x}, \\
\kappa_{xx}^0 &= -\frac{\partial^2 w_0}{\partial x^2}, \\
\kappa_{\vartheta\vartheta}^0 &= \frac{1}{R^2}\frac{\partial}{\partial \vartheta}\left(u_0 - \frac{\partial w_0}{\partial \vartheta}\right), \\
\kappa_{x\vartheta}^0 &= \frac{1}{R}\left(\frac{\partial u_0}{\partial x} - 2\frac{\partial^2 w_0}{\partial x \partial \vartheta}\right).
\end{aligned} \tag{16.22}$$

These equations reduce for the case of rotationally symmetric loading as follows:

$$\begin{aligned}
\varepsilon_{xx}^0 &= \frac{\mathrm{d}v_0}{\mathrm{d}x}, \\
\varepsilon_{\vartheta\vartheta}^0 &= \frac{w_0}{R}, \\
\kappa_{xx}^0 &= -\frac{\mathrm{d}^2 w_0}{\mathrm{d}x^2}.
\end{aligned} \tag{16.23}$$

At this point, again, the dimensionless coordinate $\xi = \frac{x}{R}$ can be used. It follows from (16.22):

$$\begin{aligned}
\varepsilon_{\xi\xi}^0 &= \frac{1}{R}\frac{\partial v_0}{\partial \xi}, \\
\varepsilon_{\vartheta\vartheta}^0 &= \frac{1}{R}\left(\frac{\partial u_0}{\partial \vartheta} + w_0\right), \\
\gamma_{\xi\vartheta}^0 &= \frac{1}{R}\left(\frac{\partial v_0}{\partial \vartheta} + \frac{\partial u_0}{\partial \xi}\right), \\
\kappa_{\xi\xi}^0 &= -\frac{1}{R^2}\frac{\partial^2 w_0}{\partial \xi^2}, \\
\kappa_{\vartheta\vartheta}^0 &= \frac{1}{R^2}\frac{\partial}{\partial \vartheta}\left(u_0 - \frac{\partial w_0}{\partial \vartheta}\right), \\
\kappa_{\xi\vartheta}^0 &= \frac{1}{R^2}\left(\frac{\partial u_0}{\partial \xi} - 2\frac{\partial^2 w_0}{\partial \xi \partial \vartheta}\right).
\end{aligned} \tag{16.24}$$

In the case of rotationally symmetric loading, this set of equations reduces to:

$$\begin{aligned}\varepsilon_{\xi\xi}^0 &= \frac{1}{R}\frac{\mathrm{d}v_0}{\mathrm{d}\xi},\\ \varepsilon_{\vartheta\vartheta}^0 &= \frac{w_0}{R},\\ \kappa_{\xi\xi}^0 &= -\frac{1}{R^2}\frac{\mathrm{d}^2 w_0}{\mathrm{d}\xi^2}.\end{aligned} \tag{16.25}$$

16.1.3 Constitutive Equations

In the framework of the bending theory of the shells of revolution, first of all, the relations between the membrane force flows and the strains of the shell midplane have to be taken into account which we present here again for the sake of readability:

$$\begin{aligned}N_{\varphi\varphi} &= A\left(\varepsilon_{\varphi\varphi}^0 + \nu\varepsilon_{\vartheta\vartheta}^0\right),\\ N_{\vartheta\vartheta} &= A\left(\nu\varepsilon_{\varphi\varphi}^0 + \varepsilon_{\vartheta\vartheta}^0\right),\\ N_{\varphi\vartheta} &= A\frac{1-\nu}{2}\gamma_{\varphi\vartheta}^0,\end{aligned} \tag{16.26}$$

with the membrane stiffness of the shell midsurface:

$$A = \frac{Eh}{1-\nu^2}. \tag{16.27}$$

In addition, similar to Kirchhoff's plate theory, the relationships between the moment flows on the one hand and the curvatures and the twist on the other hand have to be considered:

$$\begin{aligned}M_{\varphi\varphi} &= D\left(\kappa_{\varphi\varphi}^0 + \nu\kappa_{\vartheta\vartheta}^0\right),\\ M_{\vartheta\vartheta} &= D\left(\nu\kappa_{\varphi\varphi}^0 + \kappa_{\vartheta\vartheta}^0\right),\\ M_{\varphi\vartheta} &= D\frac{1-\nu}{2}\kappa_{\varphi\vartheta}^0,\end{aligned} \tag{16.28}$$

with the bending stiffness D as:

$$D = \frac{Eh^3}{12\left(1-\nu^2\right)}. \tag{16.29}$$

An arbitrary static problem for a shell of revolution in the framework of bending theory is described by 17 unknown quantities. These are the membrane force flows $N_{\varphi\varphi}$, $N_{\vartheta\vartheta}$, $N_{\varphi\vartheta}$, the transverse shear force flows Q_φ, Q_ϑ, the moment flows $M_{\varphi\varphi}$, $M_{\vartheta\vartheta}$, $M_{\varphi\vartheta}$, the displacements u_0, v_0, w_0 of the shell midsurface, and the six strain quantities $\varepsilon_{\varphi\varphi}^0$, $\varepsilon_{\vartheta\vartheta}^0$, $\gamma_{\varphi\vartheta}$, $\kappa_{\varphi\varphi}^0$, $\kappa_{\vartheta\vartheta}^0$, $\kappa_{\varphi\vartheta}^0$. This is counterbalanced 17 equations, i.e. by the equilibrium conditions (16.6), the kinematic equations (16.18) and the material law (16.26) and (16.28), so that for given boundary conditions an unambiguous solution of a given shell problem is possible. However, the application shows that the closed-form analytical solution of a shell problem in the framework of bending theory is possible only in a few and then often highly idealized cases. Therefore, numerical methods are often used for the analysis of shell problems.

16.1.4 Displacement Differential Equations for the Circular Cylindrical Shell

With the help of the constitutive equations (16.26), (16.28) and the kinematic relations (16.22) the displacement differential equations for the circular cylindrical shell can be obtained from the equilibrium conditions (16.9) after elimination of the transverse forces. They are given by:

$$
\begin{aligned}
&\frac{A}{2R}(1+\nu)\frac{\partial^2 u_0}{\partial x \partial \vartheta} + A\left(\frac{\partial^2}{\partial x^2} + \frac{1-\nu}{2R^2}\frac{\partial^2}{\partial \vartheta^2}\right)v_0 + \frac{\nu A}{R}\frac{\partial w_0}{\partial x} = -p_x,\\
&\left(A + \frac{D}{R^2}\right)\left(\frac{1-\nu}{2}\frac{\partial^2}{\partial x^2} + \frac{1}{R^2}\frac{\partial^2}{\partial \vartheta^2}\right)u_0 + \frac{A}{2R}(1+\nu)\frac{\partial^2 v_0}{\partial x \partial \vartheta}\\
&\quad + \frac{1}{R^2}\left(-\frac{D}{R^2}\frac{\partial^3}{\partial \vartheta^3} - D\frac{\partial^3}{\partial x^2 \partial \vartheta} + A\frac{\partial}{\partial \vartheta}\right)w_0 = -p_\vartheta,\\
&\frac{1}{R}\left(\frac{D}{R}\frac{\partial^3}{\partial x^2 \partial \vartheta} + \frac{D}{R^3}\frac{\partial^3}{\partial \vartheta^3} - A\frac{\partial}{\partial \vartheta}\right)u_0 - \frac{\nu A}{R}\frac{\partial v_0}{\partial x}\\
&\quad + \left(-D\frac{\partial^4}{\partial x^4} - \frac{2D}{R^2}\frac{\partial^4}{\partial x^2 \partial \vartheta^2} - \frac{D}{R^4}\frac{\partial^4}{\partial \vartheta^4} - \frac{A}{R}\right)w_0 = -p_z.
\end{aligned}
\tag{16.30}
$$

16.1.5 Boundary Conditions under Rotationally Symmetric Load

A given static shell problem is not only described by the basic equations presented above, but also by given boundary conditions. In the case of any shell of revolution under rotationally symmetric loading, three boundary force quantities and three boundary displacement quantities are to be considered at each boundary (Fig. 16.2).

Consider an edge at a point φ, which is rigidly clamped. At this boundary the boundary displacements v_0 and w_0 as well as the rotation ψ_φ of the meridian tangent will disappear (Fig. 16.2, left). If, on the other hand, there is a free edge at this point, then the force flows $N_{\varphi\varphi}$, Q_φ, and the moment flow $M_{\varphi\varphi}$ disappear there (Fig. 16.2, right). In a similar way, boundary conditions can be formulated for other types of support.

For the special case of the rotationally symmetrically loaded circular cylindrical shell, identical boundary conditions are obtained, as shown in Fig. 16.3.

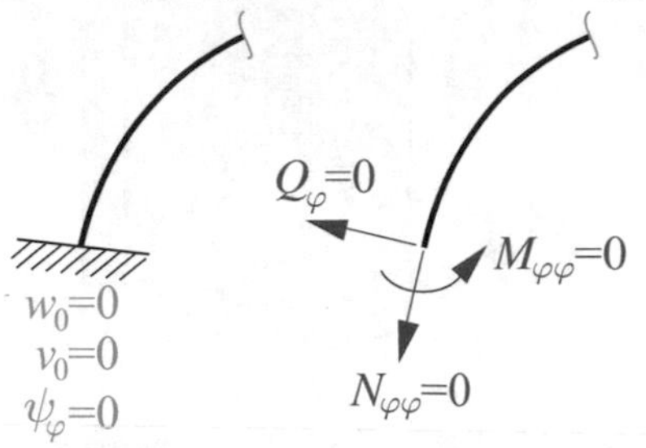

Fig. 16.2 Boundary conditions for an arbitrary shell of revolution under rotationally symmetric loading

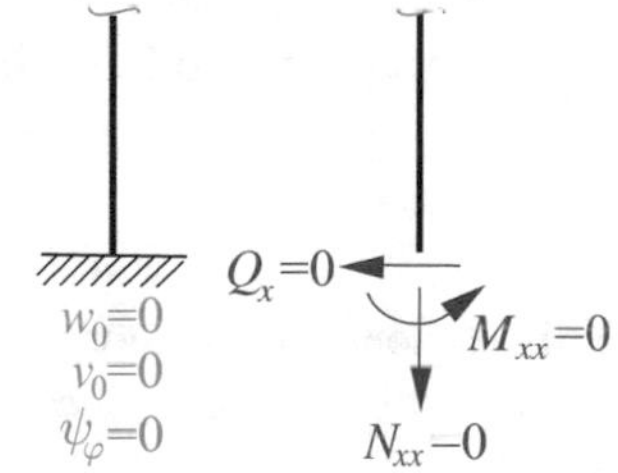

Fig. 16.3 Boundary conditions for the circular cylindrical shell under rotationally symmetric loading

16.2 Container Theory of the Circular Cylindrical Shell

16.2.1 Basic Equations

Shells of revolution under a rotationally symmetric load are often also referred to as containers, and the corresponding bending theory can be properly designated as container theory. In this case, all partial derivatives to ϑ vanish, and it is also assumed that there is no surface load p_ϑ. In this section, we will focus the considerations on the circular cylindrical shell, where with $N_{x\vartheta} = 0$, $M_{x\vartheta} = 0$, $Q_\vartheta = 0$, $u_0 = 0$, $\gamma^0_{x\vartheta} = 0$, $\kappa_{\vartheta\vartheta} = 0$, $\kappa_{x\vartheta} = 0$ the basic equations (16.10) resp. (16.12), (16.23), (16.25) and (16.26), (16.28) can be used.

The equilibrium conditions can be derived particularly easily for the present case from the equilibrium at the infinitesimal sectional element, for which we consider Fig. 16.4. Given here is an element with the edge lengths $\mathrm{d}x$ and $R\mathrm{d}\vartheta$. In addition to the resultants of the surface loads, the force flow and moment flow resultants are indicated here, each marked with a transverse bar. In addition, the fact that the section sizes do not change in the circumferential direction ϑ has already been taken into account. The sum of the forces in x-direction then gives:

$$N_{xx} R\mathrm{d}\vartheta + \frac{\mathrm{d}N_{xx}}{\mathrm{d}x}\mathrm{d}x\, R\mathrm{d}\vartheta - N_{xx} R\mathrm{d}\vartheta + p_x R\mathrm{d}x\mathrm{d}\vartheta = 0, \tag{16.31}$$

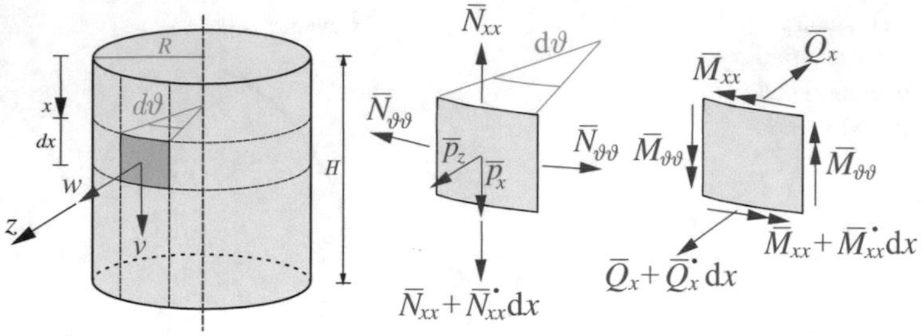

Fig. 16.4 Infinitesimal element of the circular cylindrical shell

which reduces to the following expression:

$$\frac{\mathrm{d}N_{xx}}{\mathrm{d}x} + p_x = 0. \tag{16.32}$$

This obviously corresponds to the first equilibrium condition in (16.10).

We further consider the sum of all forces in z- direction:

$$Q_x R\mathrm{d}\vartheta + \frac{\mathrm{d}Q_x}{\mathrm{d}x}\mathrm{d}x\,R\mathrm{d}\vartheta - Q_x R\mathrm{d}\vartheta - N_{\vartheta\vartheta}\mathrm{d}x\frac{\mathrm{d}\vartheta}{2} - N_{\vartheta\vartheta}\mathrm{d}x\frac{\mathrm{d}\vartheta}{2} + p_z\mathrm{d}x\,R\mathrm{d}\vartheta = 0. \tag{16.33}$$

This results in:

$$R\frac{\mathrm{d}Q_x}{\mathrm{d}x} - N_{\vartheta\vartheta} + Rp_z = 0. \tag{16.34}$$

This corresponds to the second equilibrium condition in (16.10).

Finally, we consider the sum of all moments with respect to the circumferential direction ϑ, with reference point at the location $x + \mathrm{d}x$:

$$M_{xx} R\mathrm{d}\vartheta + \frac{\mathrm{d}M_{xx}}{\mathrm{d}x}\mathrm{d}x\,R\mathrm{d}\vartheta - M_{xx} R\mathrm{d}\vartheta - Q_x R\mathrm{d}\vartheta\mathrm{d}x + p_z R\mathrm{d}\vartheta\mathrm{d}x\frac{\mathrm{d}x}{2} = 0. \tag{16.35}$$

Neglecting the last term gives:

$$\frac{\mathrm{d}M_{xx}}{\mathrm{d}x} - Q_x = 0. \tag{16.36}$$

Thus, the third equilibrium condition in (16.10) is also derived again.

The current kinematic conditions are as follows:

$$\varepsilon^0_{xx} = \frac{\mathrm{d}v_0}{\mathrm{d}x}, \quad \varepsilon^0_{\vartheta\vartheta} = \frac{w_0}{R}, \quad \kappa^0_{xx} = -\frac{\mathrm{d}^2 w_0}{\mathrm{d}x^2}. \tag{16.37}$$

Also these equations are accessible to a very simple interpretation. It is obvious that the strain ε_{xx} will consist of two parts, namely on the one hand of a part ε^0_{xx} due to the displacement v_0, and on the other hand of a part resulting from the bending effect or the curvature κ^0_{xx}. While the fraction ε^0_{xx} can be expressed quite easily and

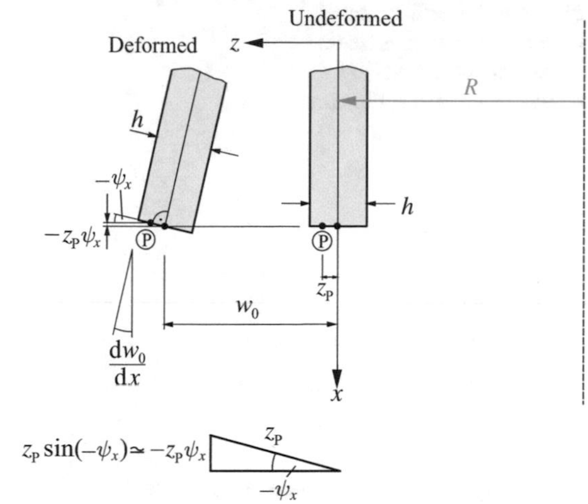

Fig. 16.5 Undeformed and deformed sectional element of the circular cylindrical shell, view in circumferential direction ϑ

without further mathematical consideration as

$$\varepsilon_{xx}^0 = \frac{\mathrm{d}v_0}{\mathrm{d}x} \tag{16.38}$$

(which corresponds to the first equation in (16.37)), the determination of the fraction due to the bending action of the shell requires a deeper discussion. For this purpose, we examine the section element of Fig. 16.5, in which a shell segment is shown in the undeformed and in the deformed state. It becomes obvious that this sectional view is very similar to that of the Kirchhoff plate. Quite analogously, therefore, $v = -z\frac{\mathrm{d}w_0}{\mathrm{d}x}$ can be inferred so that the fraction of ε_{xx} due to the deflection w_0 and the associated cross-sectional rotation is given as $-z\frac{\mathrm{d}^2 w_0}{\mathrm{d}x^2}$. Definition of the curvature κ_{xx}^0 as

$$\kappa_{xx}^0 = -\frac{\mathrm{d}^2 w_0}{\mathrm{d}x^2} \tag{16.39}$$

then leads to the third equation given in (16.37).

Regarding the ring strain $\varepsilon_{\vartheta\vartheta}^0$ the fact can be used that the deflection w_0 does not change over the circumference of the container. The resulting deformation is shown in Fig. 16.6. With the edge lengths $\mathrm{d}s_2 = R\mathrm{d}\vartheta$ and $\mathrm{d}s_2 + \Delta\mathrm{d}s_2 = (R + w_0)\mathrm{d}\vartheta$ we get $\Delta\mathrm{d}s_2 = w_0\mathrm{d}\vartheta$ such that:

$$\varepsilon_{\vartheta\vartheta}^0 = \frac{\Delta\mathrm{d}s_2}{\mathrm{d}s_2} = \frac{w_0}{R}. \tag{16.40}$$

Thus the second expression in (16.37) is also derived.

Fig. 16.6 Undeformed and deformed sectional element of the circular cylindrical container, view in x-direction

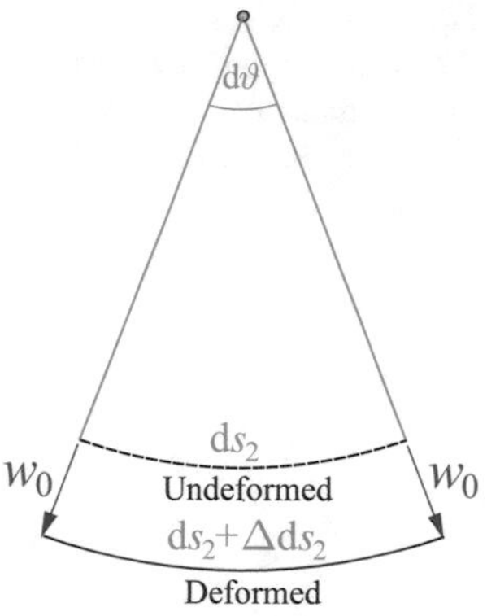

A very simple approach also consists in the energetic derivation of the basic equations from the principle of virtual displacements $\delta W_i = \delta W_a$. The virtual inner work δW_i is presently:

$$\delta W_i = \int_0^{2\pi} \int_0^H \left(N_{xx} \delta \varepsilon_{xx}^0 + N_{\vartheta\vartheta} \delta \varepsilon_{\vartheta\vartheta}^0 + M_{xx} \kappa_{xx}^0 \right) \mathrm{d}A. \quad (16.41)$$

With the surface element $\mathrm{d}A = R\mathrm{d}x\mathrm{d}\vartheta$ and the kinematic relation (16.37) it follows:

$$\delta W_i = \int_0^{2\pi} \int_0^H R \left(N_{xx} \frac{\mathrm{d}\delta v_0}{\mathrm{d}x} + N_{\vartheta\vartheta} \frac{\delta w_0}{R} - M_{xx} \frac{\mathrm{d}^2 \delta w_0}{\mathrm{d}x^2} \right) \mathrm{d}x\mathrm{d}\vartheta. \quad (16.42)$$

The virtual external work is given by the two surface loads p_x and p_z:

$$\delta W_a = \int_0^{2\pi} \int_0^H \left(p_x \delta v_0 + p_z \delta w_0 \right) R\mathrm{d}x\mathrm{d}\vartheta. \quad (16.43)$$

Equating (16.42) and (16.43) and partial integration of those terms with derivatives of the virtual displacements δv_0 and δw_0 leads to:

$$\begin{aligned} &-\int_0^{2\pi} \int_0^H R \left(\frac{\mathrm{d}N_{xx}}{\mathrm{d}x} + p_x \right) \delta v_0 \mathrm{d}x\mathrm{d}\vartheta + \int_0^{2\pi} \int_0^H \left(-R \frac{\mathrm{d}^2 M_{xx}}{\mathrm{d}x^2} + N_{\vartheta\vartheta} - R p_z \right) \delta w_0 \mathrm{d}x\mathrm{d}\vartheta \\ &+ R \int_0^{2\pi} N_{xx} \delta v_0 \mathrm{d}\vartheta \big|_0^H - R \int_0^{2\pi} M_{xx} \frac{\mathrm{d}\delta w_0}{\mathrm{d}x} \mathrm{d}\vartheta \bigg|_0^H + R \int_0^{2\pi} \frac{\mathrm{d}M_{xx}}{\mathrm{d}x} \delta w_0 \mathrm{d}\vartheta \bigg|_0^H = 0. \end{aligned} \quad (16.44)$$

It can be seen that from the bracket terms in the first line of (16.44), the equilibrium conditions (16.10) can be deduced. The remaining terms represent the corresponding boundary conditions of the given container problem and confirm those already given in Fig. 16.3.

Finally, the constitutive conditions have to be considered, which for the given circular cylindrical container are as follows:

$$\begin{aligned} N_{xx} &= A\left(\varepsilon_{xx}^0 + \nu\varepsilon_{\vartheta\vartheta}^0\right), \\ N_{\vartheta\vartheta} &= A\left(\nu\varepsilon_{xx}^0 + \varepsilon_{\vartheta\vartheta}^0\right), \\ M_{xx} &= D\kappa_{xx}^0, \\ M_{\vartheta\vartheta} &= D\nu\kappa_{xx}^0. \end{aligned} \tag{16.45}$$

16.2.2 The Container Equation

We now want to solve the given container problem by first considering in equation (16.45) the expression for $N_{\vartheta\vartheta}$ with the help of (16.37). We obtain:

$$N_{xx} = A\left(\varepsilon_{xx}^0 + \nu\varepsilon_{\vartheta\vartheta}^0\right) = A\left(\frac{\mathrm{d}v_0}{\mathrm{d}x} + \nu\frac{w_0}{R}\right). \tag{16.46}$$

Furthermore, we solve the expression (16.32) for N_{xx}:

$$N_{xx} = -\int_x p_x \mathrm{d}x + C_1. \tag{16.47}$$

Equating (16.46) and (16.47) then gives:

$$\frac{\mathrm{d}v_0}{\mathrm{d}x} + \nu\frac{w_0}{R} = \frac{1}{A}\left(-\int_x p_x \mathrm{d}x + C_1\right). \tag{16.48}$$

This is a differential equation for the two unknown displacements v_0 and w_0.

Furthermore, we consider equation (16.36). Derivation with respect to x results in:

$$\frac{\mathrm{d}^2 M_{xx}}{\mathrm{d}x^2} = \frac{\mathrm{d}Q_x}{\mathrm{d}x}. \tag{16.49}$$

Insertion into (16.34) yields:

$$R\frac{\mathrm{d}^2 M_{xx}}{\mathrm{d}x^2} - N_{\vartheta\vartheta} + Rp_z = 0. \tag{16.50}$$

Using the expression for $N_{\vartheta\vartheta}$ in (16.45) and the twofold derivative of $M_{xx} = D\kappa_{xx}^0$ in (16.45) we obtain:

$$R\frac{\mathrm{d}^2}{\mathrm{d}x^2}\left(D\frac{\mathrm{d}^2 w_0}{\mathrm{d}x^2}\right) + A\left(\nu\frac{\mathrm{d}v_0}{\mathrm{d}x} + \frac{w_0}{R}\right) - Rp_z = 0. \tag{16.51}$$

Thus, another differential equation for the two unknown displacements v_0 and w_0 has been obtained. Eliminating $\frac{\mathrm{d}v_0}{\mathrm{d}x}$ by (16.48) yields after some transformations:

$$\frac{\mathrm{d}^2}{\mathrm{d}x^2}\left(D\frac{\mathrm{d}^2 w_0}{\mathrm{d}x^2}\right) + \frac{A}{R^2}\left(1-\nu^2\right) w_0 = p_z - \frac{\nu}{R}\left(-\int_x p_x \mathrm{d}x + C_1\right). \quad (16.52)$$

This is a fourth-order differential equation for the unknown displacement w_0. If the case exists that there is no load p_x in the x-direction, then this expression simplifies again considerably, and it remains:

$$\frac{\mathrm{d}^2}{\mathrm{d}x^2}\left(D\frac{\mathrm{d}^2 w_0}{\mathrm{d}x^2}\right) + \frac{Eh}{R^2} w_0 = p_z. \quad (16.53)$$

This differential equation is also called container equation. For the special case $D =$ const. it changes into:

$$\frac{\mathrm{d}^4 w_0}{\mathrm{d}x^4} + \frac{12\left(1-\nu^2\right)}{R^2 h^2} w_0 = \frac{p_z}{D}. \quad (16.54)$$

At this point we introduce the so-called shell parameter λ, which is defined as follows:

$$\lambda^4 = \frac{3\left(1-\nu^2\right)}{R^2 h^2}, \quad (16.55)$$

so that:

$$\frac{\mathrm{d}^4 w_0}{\mathrm{d}x^4} + 4\lambda^4 w_0 = \frac{p_z}{D}. \quad (16.56)$$

16.2.3 Solutions for the Container Equation

Homogeneous solution

The solution w_0 of the inhomogeneous differential equation (16.56) is composed of a homogeneous solution $w_{0,h}$ and a particular solution $w_{0,p}$. We obtain the homogeneous solution by zeroing the right-hand side of (16.56):

$$\frac{\mathrm{d}^4 w_{0,h}}{\mathrm{d}x^4} + 4\lambda^4 w_{0,h} = 0. \quad (16.57)$$

Using the approach $w_{0,h} = e^{\eta x}$ we then obtain the following characteristic equation for the exponent η:

$$\eta^4 + 4\lambda^4 = 0, \quad (16.58)$$

which leads to the following solutions for η:

$$\eta_{1,2} = \lambda(1+i), \quad \eta_{3,4} = \lambda(1-i). \quad (16.59)$$

The homogeneous solution $w_{0,h}$ can then be written as:

$$w_{0,h} = e^{-\lambda x}\left(C_1 \cos \lambda x + C_2 \sin \lambda x\right) + e^{\lambda x}\left(C_3 \cos \lambda x + C_4 \sin \lambda x\right), \quad (16.60)$$

with the integration constants C_1, C_2, C_3, C_4. This solution, which can be decomposed into two parts, requires some interpretation. The first part describes a decaying oscillatory boundary perturbation starting from $x = 0$ in the positive x-direction (Fig. 16.7). Analogously, the second term represents an oscillatory edge perturbation decaying from the shell edge $x = H$ in the negative x-direction. The oscillation behavior, but also the decay behavior of edge perturbations is mainly controlled by the shell parameter λ, which is therefore often called decay parameter. Usually $\lambda >> 1$ is valid, so that in many cases it can be assumed that the edge perturbation will decay rapidly with increasing distance from the shell edges and will tend to zero. In order to illuminate this fact a little more closely, we determine the decay length Δx, of which we want to assume that the edge perturbation has decreased to 0.01, i.e.:

$$e^{-\lambda \Delta x} = 0.01. \quad (16.61)$$

Solving for Δx then gives:

$$\Delta x = \frac{\ln 100}{\lambda} = \frac{\ln 100}{\sqrt[4]{\frac{3(1-\nu^2)}{R^2 h^2}}} = \frac{\ln 100\sqrt{Rh}}{\sqrt[4]{3\left(1-\nu^2\right)}} \simeq 3.5\sqrt{Rh}. \quad (16.62)$$

Division by R yields:

$$\frac{\Delta x}{R} = 3.5\sqrt{\frac{h}{R}}. \quad (16.63)$$

Since $\frac{h}{R} << 1$ holds, $\frac{\Delta x}{R} << 1$ will also hold. Therefore, it can be concluded that for a sufficiently long shell with $H > R$, the two solution components of $w_{0,h}$ emanating from the ends $x = 0$ and $x = H$ practically do not affect each other at all

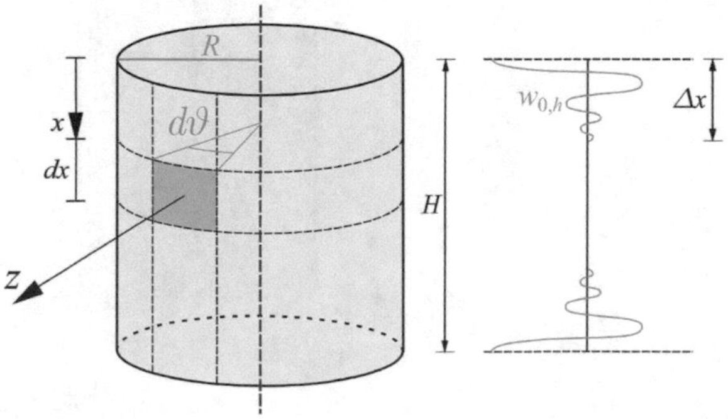

Fig. 16.7 Qualitative representation of the decay behavior of $w_{0,h}$

and therefore can be considered separately. Concerning the boundary at $x = 0$ then holds:

$$w_{0,h} = e^{-\lambda x} \left(C_1 \cos \lambda x + C_2 \sin \lambda x \right), \tag{16.64}$$

and the constants C_1 and C_2 are determined from the boundary conditions at this boundary alone. With respect to the boundary at $x = H$ we obtain:

$$w_{0,h} = e^{\lambda x} \left(C_3 \cos \lambda x + C_4 \sin \lambda x \right), \tag{16.65}$$

where the constants C_3 and C_4 are determined from the given conditions at this boundary.

If there is no edge perturbation at a shell edge, the constants in the corresponding solution fraction in $w_{0,h}$ result in zero.

As an example, consider the circular cylindrical shell of infinite length under constant boundary force flows Q_0 and boundary moment flows M_0 at the location $x = 0$ (Fig. 16.8). The boundary displacement w_0 at the location $x = 0$ is to be determined.

In this case, the container equation (16.56) is as follows

$$\frac{\mathrm{d}^4 w_0}{\mathrm{d}x^4} + 4\lambda^4 w_0 = 0, \tag{16.66}$$

so that the homogeneous solution in this case is the total solution. Thus:

$$w_0 = e^{-\lambda x} \left(C_1 \cos \lambda x + C_2 \sin \lambda x \right). \tag{16.67}$$

The following boundary conditions are given:

$$M_{xx}(x = 0) = M_0, \quad Q_x(x = 0) = Q_0. \tag{16.68}$$

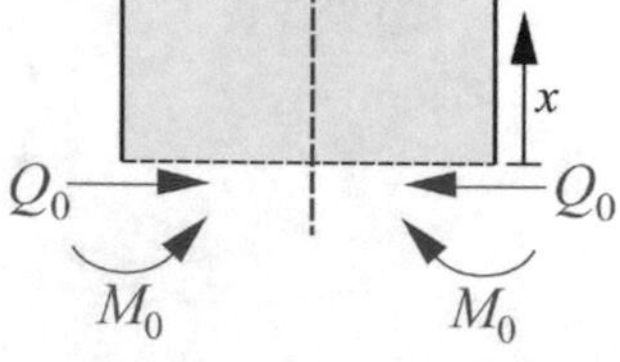

Fig. 16.8 Long circular cylindrical shell under boundary force flow Q_0 and boundary moment flow M_0

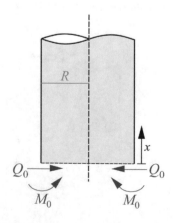

The moment flow M_{xx} is given as

$$M_{xx} = -D\frac{\mathrm{d}^2 w_0}{\mathrm{d}x^2} = -2D\lambda^2 e^{-\lambda x}\left(C_1 \sin\lambda x - C_2 \cos\lambda x\right), \tag{16.69}$$

and the transverse shear force flow Q_x can be calculated from this as

$$Q_x = \frac{\mathrm{d}M_{xx}}{\mathrm{d}x} = -2D\lambda^3 e^{-\lambda x}\left[(C_1 + C_2)\cos\lambda x + (C_2 - C_1)\sin\lambda x\right]. \tag{16.70}$$

Fitting these formulations to the boundary conditions (16.68) then yields:

$$C_1 = -\frac{1}{2D\lambda^2}\left(\frac{Q_0}{\lambda} + M_0\right), \quad C_2 = \frac{M_0}{2D\lambda^2}. \tag{16.71}$$

The displacement w_0 is then:

$$w_0 = \frac{1}{2D\lambda^2}e^{-\lambda x}\left[-\left(\frac{Q_0}{\lambda} + M_0\right)\cos\lambda x + M_0 \sin\lambda x\right]. \tag{16.72}$$

Thus, the boundary displacement $w_0(x = 0)$ can be given as:

$$w_0(x = 0) = -\frac{\frac{Q_0}{\lambda} + M_0}{2D\lambda^2}. \tag{16.73}$$

The rotation angle $\psi_x = \frac{\mathrm{d}w_0}{\mathrm{d}x}$ of the meridian tangent results as:

$$\psi_x = \frac{\mathrm{d}w_0}{\mathrm{d}x} = -\frac{1}{2D\lambda}e^{-\lambda x}\left[-\left(\frac{Q_0}{\lambda} + 2M_0\right)\cos\lambda x + \frac{Q_0}{\lambda}\sin\lambda x\right]. \tag{16.74}$$

From this, the edge rotation at the point $x = 0$ is given as:

$$\psi_x(x = 0) = \frac{\frac{Q_0}{\lambda} + 2M_0}{2D\lambda}. \tag{16.75}$$

The bending moment flow M_{xx} is:

$$M_{xx} = e^{-\lambda x}\left[\left(\frac{Q_0}{\lambda} + M_0\right)\sin\lambda x + M_0 \cos\lambda x\right]. \tag{16.76}$$

We obtain the shear force flow Q_x as the first derivative of M_{xx} with respect to x as:

$$Q_x = \lambda e^{-\lambda x}\left[\frac{Q_0}{\lambda}\cos\lambda x - \left(2M_0 + \frac{Q_0}{\lambda}\right)\sin\lambda x\right]. \tag{16.77}$$

Particular solution

The particular solution of the container equation (16.56) depends on the given load case, for which we will consider some examples below. The first problem is the circular cylinder (radius R) fixed at one end under a fluid load (Fig. 16.9, left). In

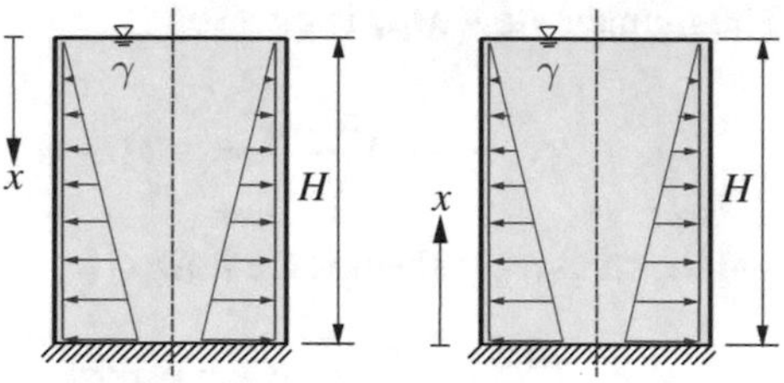

Fig. 16.9 Circular cylinder shell rigidly clamped on one side under fluid load (left), introduction of a new reference axis x (right)

this case $p_x = 0$ and $p_z = \gamma x$ holds. The container equation in this case is:

$$\frac{d^4 w_0}{dx^4} + 4\lambda^4 w_0 = \frac{\gamma x}{D}. \tag{16.78}$$

For the particular solution $w_{0,p}$ we make an approach of right-hand side type:

$$w_{0,p} = Cx, \tag{16.79}$$

with the constant C, which we can determine after substituting in (16.78) and final coefficient comparison as:

$$C = \frac{\gamma}{4\lambda^4 D}. \tag{16.80}$$

For the particular solution $w_{0,p}$ and its derivatives then holds:

$$w_{0,p} = \frac{\gamma x}{4\lambda^4 D}, \quad \frac{dw_{0,p}}{dx} = \frac{\gamma}{4\lambda^4 D}, \quad \frac{d^2 w_{0,p}}{dx^2} = 0. \tag{16.81}$$

The resulting force and moment quantities are then as follows:

$$N_{xx,p} = 0, \quad N_{\vartheta\vartheta,p} = R\gamma x,$$
$$M_{xx,p} = M_{\vartheta\vartheta,p} = 0, \quad Q_{x,p} = 0. \tag{16.82}$$

The comparison with the solution according to the membrane theory (Eq. (15.24)) shows that the particular solution determined from the container equation agrees exactly with the solution according to the membrane theory. It can be concluded that the membrane theory solution fully describes the state of the shell, except for those edge regions where edge perturbations are present, which are then described by the homogeneous solution of the container equation.

In order to treat the given example further, we introduce a new reference axis x, which runs positively upward from the boundary constraint (Fig. 16.9, right) and first determine the complete solution according to the membrane theory (marked with the index M). It results with $p_x = 0$, $p_z = \gamma(H - x)$, $N_{xx,M} = 0$:

$$\begin{aligned}
N_{\vartheta\vartheta,M} &= Rp_z = R\gamma(H-x),\\
w_{0,M} &= \frac{R}{Eh}\left(N_{\vartheta\vartheta,M} - \nu N_{xx,M}\right) = \frac{R^2\gamma}{Eh}(H-x),\\
v_{0,M} &= \frac{1}{Eh}\int_x \left(N_{xx,M} - \nu N_{\vartheta\vartheta,M}\right)\mathrm{d}x + C_2 = -\frac{\nu R\gamma x}{Eh}\left(H - \frac{x}{2}\right),\\
\psi_{x,M} &= \frac{\mathrm{d}w_0}{\mathrm{d}x} = \frac{R}{Eh}\frac{\mathrm{d}}{\mathrm{d}x}\left(N_{\vartheta\vartheta,M} - \nu N_{xx,M}\right) = -\frac{R^2\gamma}{Eh}.
\end{aligned} \tag{16.83}$$

To treat the edge perturbation problem at the clamped end of the shell, we now use the displacement solution w_0 as follows:

$$w_0 = \frac{R^2\gamma}{Eh}(H-x) + e^{-\lambda x}\left(C_1\cos\lambda x + C_2\sin\lambda x\right). \tag{16.84}$$

The first derivative with respect to x is then:

$$\frac{\mathrm{d}w_0}{\mathrm{d}x} = -\frac{R^2\gamma}{Eh} + \lambda e^{-\lambda x}\left[(C_2 - C_1)\cos\lambda x - (C_1 + C_2)\sin\lambda x\right]. \tag{16.85}$$

The boundary conditions for finding C_1 and C_2 are $w_0(x=0)=0$ and $\frac{\mathrm{d}w_0}{\mathrm{d}x}(x=0)=0$, which leads to the following expressions for C_1 and C_2:

$$C_1 = -\frac{\gamma R^2 H}{Eh}, \quad C_2 = \frac{\gamma R^2 H}{Eh}\left(\frac{1}{\lambda H} - 1\right). \tag{16.86}$$

The displacement w_0 can thus be given as:

$$w_0 = \frac{\gamma R^2}{Eh}\left[H - x + He^{-\lambda x}\left(-\cos\lambda x + \left(\frac{1}{\lambda H} - 1\right)\sin\lambda x\right)\right]. \tag{16.87}$$

The rotation $\psi_x = \frac{\mathrm{d}w_0}{\mathrm{d}x}$ of the meridian tangent follows here as:

$$\psi_x = \frac{\mathrm{d}w_0}{\mathrm{d}x} = \frac{\gamma R^2}{Eh}\left[-1 + H\lambda e^{-\lambda x}\left(\cos\lambda x + \sin\lambda x + \left(\frac{1}{\lambda H} - 1\right)(\cos\lambda x - \sin\lambda x)\right)\right]. \tag{16.88}$$

Following from this, the bending moment flow M_{xx} can be determined from the constitutive law (16.45):

$$M_{xx} = D\kappa_{xx}^0 = -D\frac{\mathrm{d}^2 w_0}{\mathrm{d}x^2} = \frac{\gamma H}{2\lambda^2}e^{-\lambda x}\left[\sin\lambda x + \left(\frac{1}{\lambda H} - 1\right)\cos\lambda x\right]. \tag{16.89}$$

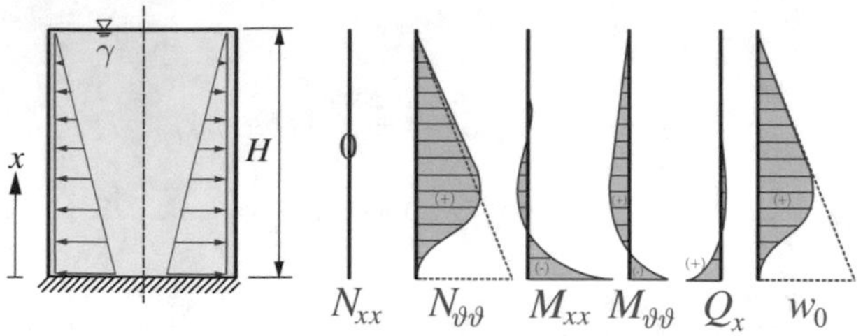

Fig. 16.10 State variables of the circular cylindrical shell rigidly clamped at one edge under fluid load

Likewise, from (16.45) follows the moment flow $M_{\vartheta\vartheta}$:

$$M_{\vartheta\vartheta} = \nu D \kappa_{xx}^0 = -\nu D \frac{d^2 w_0}{dx^2} = \frac{\nu \gamma H}{2\lambda^2} e^{-\lambda x} \left[\sin \lambda x + \left(\frac{1}{\lambda H} - 1 \right) \cos \lambda x \right]. \quad (16.90)$$

The transverse shear force flow Q_x is obtained from the third equilibrium condition in (16.10) as:

$$Q_x = \frac{dM_{xx}}{dx} = -\frac{\gamma H}{2\lambda} e^{-\lambda x} \left[\sin \lambda x - \cos \lambda x + \left(\frac{1}{\lambda H} - 1 \right) (\sin \lambda x + \cos \lambda x) \right]. \quad (16.91)$$

Finally, the force flow $N_{\vartheta\vartheta}$ can be determined from the constitutive law of the circular cylindrical shell as:

$$N_{\vartheta\vartheta} = \frac{\gamma R}{1 - \nu^2} \left[H - x + H e^{-\lambda x} \left(-\cos \lambda x + \left(\frac{1}{\lambda H} - 1 \right) \sin \lambda x \right) \right]. \quad (16.92)$$

The resulting qualitative distributions of the state variables are shown in Fig. 16.10.

16.3 The Force Method

The determination of a homogeneous solution of a given shell problem in the framework of the container theory can quickly become very involved. As an alternative, we will discuss in this section the force method on the already discussed fluid container, which turns out to be statically indeterminate due to its restraint at the lower end. To apply the force method, we make the container statically determinate, and thus apply the statically indeterminate quantities X_1 and X_2, see Fig. 16.11. This is then the statically determinate main system.

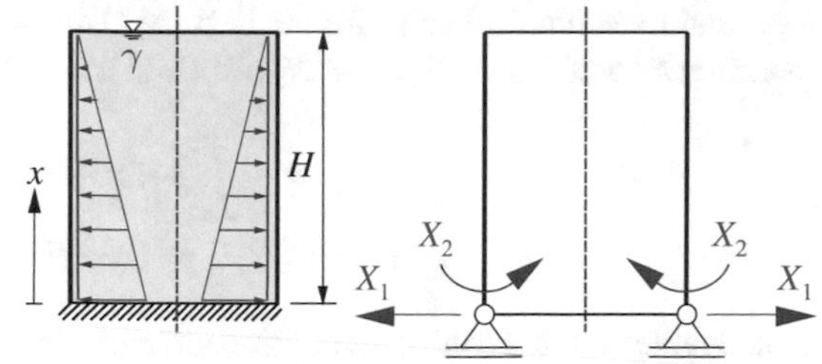

Fig. 16.11 Circular cylindrical shell rigidly clamped at one end under fluid load (left), statically determinate base system (right)

The procedure is now largely analogous to what is known as the force method from structural analysis. The displacement and rotation quantities δ_{10} and δ_{20} correspond to the boundary displacement w_0 and its first derivative $\frac{\mathrm{d}w_0}{\mathrm{d}x} = \psi_x$ at the position $x = 0$. Let the other quantities be δ_{11} and δ_{21} and $\delta_{12} = \delta_{21}$ and δ_{22} due to the statically indeterminate force and moment flows $X_1 = 1$ and $X_2 = 1$, respectively. The compatibility conditions applicable here are then (Fig. 16.12):

$$\begin{aligned} \delta_{10} + X_1\delta_{11} + X_2\delta_{12} &= 0, \\ \delta_{20} + X_1\delta_{21} + X_2\delta_{22} &= 0. \end{aligned} \tag{16.93}$$

This linear system of equations can be easily solved for the statically indeterminate quantities X_1 and X_2, and the final quantities S are obtained from the superposition of the individual cases considered:

$$S = S_0 + X_1 S_1 + X_2 S_2. \tag{16.94}$$

The same procedure can be followed for a higher number of statically indeterminate quantities.

For the application of the force method, the corresponding displacement quantities due to unit loads have to be determined for different cases and provided in a tabular overview. We will demonstrate the procedure by using the previously discussed fluid container and also draw on the results of the boundary value problem of Fig. 16.8.

The two displacements δ_{10} and δ_{20} correspond to the boundary displacement w_0 and its first derivative $\frac{\mathrm{d}w_0}{\mathrm{d}x} = \psi_x$ at the position $x = 0$ at the statically determi-

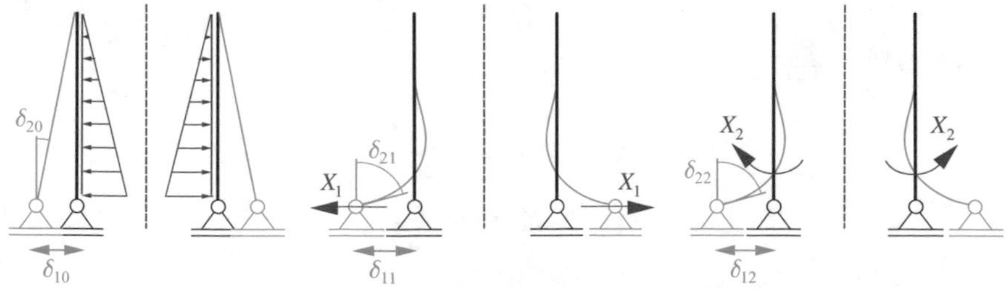

Fig. 16.12 Displacements and rotations for the statically indeterminate calculation

nate main system and can thus be derived from membrane theory. The necessary membrane force flows N_{xx} and $N_{\vartheta\vartheta}$ follow from (15.20):

$$
\begin{aligned}
N_{xx} &= -\int_x p_x \mathrm{d}x + C_1 = 0, \\
N_{\vartheta\vartheta} &= R p_z = R\gamma\,(H - x)\,.
\end{aligned}
\tag{16.95}
$$

This enables the determination of the currently required quantities:

$$
\begin{aligned}
w_0 &= \frac{R}{Eh}\left(N_{\vartheta\vartheta} - \nu N_x\right) = \frac{R^2\gamma}{Eh}\,(H - x)\,, \\
\psi_x &= \frac{\mathrm{d}w_0}{\mathrm{d}x} = \frac{R}{Eh}\frac{\mathrm{d}}{\mathrm{d}x}\left(N_{\vartheta\vartheta} - \nu N_x\right) = -\frac{R^2\gamma}{Eh}.
\end{aligned}
\tag{16.96}
$$

The corresponding boundary values of w_0 and ψ_x at $x = 0$ then correspond to the two quantities δ_{10} and δ_{20}:

$$
\begin{aligned}
\delta_{10} &= w_0(x = 0) = \frac{R^2\gamma H}{Eh}, \\
\delta_{20} &= \psi_x = -\frac{R^2\gamma}{Eh}.
\end{aligned}
\tag{16.97}
$$

We obtain the other displacement quantities directly from the solution of the boundary value problem of Fig. 16.8, where we take a unit load and a unit moment as the applied load, and also note that the unit load to be taken at present is according to Fig. 16.8 but with an opposite sign. The displacement quantities δ_{11}, δ_{21} or $\delta_{12} = \delta_{21}$ and δ_{22} due to the virtual force quantities $X_1 = 1$ or $X_2 = 1$ are then:

$$
\delta_{11} = \frac{1}{2D\lambda^3}, \quad \delta_{12} = \delta_{21} = -\frac{1}{2D\lambda^2}, \quad \delta_{22} = \frac{1}{D\lambda}.
\tag{16.98}
$$

The compatibility equations (16.93) are thus:

$$
\begin{aligned}
\frac{1}{2D\lambda^3}X_1 - \frac{1}{2D\lambda^2}X_2 &= -\frac{R^2\gamma H}{Eh}, \\
-\frac{1}{2D\lambda^2}X_1 + \frac{1}{D\lambda}X_2 &= \frac{R^2\gamma}{Eh},
\end{aligned}
\tag{16.99}
$$

or with $D\lambda^4 = \frac{Eh}{4R^2}$:

$$
\begin{aligned}
X_1 - \lambda X_2 &= -\frac{\gamma H}{2\lambda}, \\
X_1 - 2\lambda X_2 &= -\frac{\gamma}{2\lambda^2}.
\end{aligned}
\tag{16.100}
$$

Solution yields:

$$X_1 = \frac{\gamma H}{2\lambda}\left(\frac{1}{\lambda H} - 2\right), \quad X_2 = \frac{\gamma H}{2\lambda^2}\left(\frac{1}{\lambda H} - 1\right). \tag{16.101}$$

With the statically indeterminate quantities determined in this way, the state variables of the container problem treated here can be calculated from the superposition of the load stress state with the individual eigenstress states. For example, using the displacement w_0 as an example, it then follows:

$$w_0 = w_{0,0} + X_1 w_{0,1} + X_2 w_{0,2}. \tag{16.102}$$

It can be seen that the calculation according to the force method leads to the same results as the previously shown solution by means of the container equation. This is briefly demonstrated here using the deflection w_0. For the displacement $w_{0,0}$ in the framework of the membrane theory we obtain:

$$w_{0,0} = \frac{R^2\gamma}{Eh}(H - x). \tag{16.103}$$

The further necessary expressions result here as:

$$\begin{aligned} X_1 w_{0,1} &= \frac{\gamma H R^2}{Eh}\left(\frac{1}{\lambda H} - 2\right) e^{\lambda x}\cos\lambda x, \\ X_2 w_{0,2} &= \frac{\gamma H R^2}{Eh}\left(\frac{1}{\lambda H} - 1\right) e^{\lambda x}\left(-\cos\lambda x + \sin\lambda x\right), \end{aligned} \tag{16.104}$$

which leads to the following displacement function w_0:

$$w_0 = \frac{\gamma R^2}{Eh}\left[H - x + He^{-\lambda x}\left(-\cos\lambda x + \left(\frac{1}{\lambda H} - 1\right)\sin\lambda x\right)\right]. \tag{16.105}$$

Obviously, this expression agrees with (16.87).

16.4 Edge Perturbations of the Spherical Shell

Another technically relevant problem field is the container theory of the spherical shell with $r_1 = r_2 = R$ and constant wall thickness h. In the following, we consider exclusively the case of the spherical shell with boundary loading, i.e., the surface loads p_φ, p_ϑ and p_z vanish. Moreover, there is a rotationally symmetric problem, so that all derivatives $\frac{\partial}{\partial\vartheta}(\ldots)$ and the force and moment quantities $N_{\varphi\vartheta}$, Q_ϑ and $M_{\varphi\vartheta}$ vanish. The equilibrium conditions (16.8) reduce for this case as follows:

$$\frac{\mathrm{d}}{\mathrm{d}\varphi}(N_{\varphi\varphi}\sin\varphi) - N_{\vartheta\vartheta}\cos\varphi + Q_{\varphi}\sin\varphi = 0,$$
$$\frac{\mathrm{d}}{\mathrm{d}\varphi}(Q_{\varphi}\sin\varphi) - N_{\varphi\varphi}\sin\varphi - N_{\vartheta\vartheta}\sin\varphi = 0,$$
$$\frac{\mathrm{d}}{\mathrm{d}\varphi}(M_{\varphi\varphi}\sin\varphi) - M_{\vartheta\vartheta}\cos\varphi - Q_{\varphi}R\sin\varphi = 0. \tag{16.106}$$

The kinematic equations are (see (16.21)):

$$\begin{aligned}
\varepsilon^0_{\varphi\varphi} &= \frac{1}{R}\left(\frac{\mathrm{d}v_0}{\mathrm{d}\varphi} + w_0\right), \\
\varepsilon^0_{\vartheta\vartheta} &= \frac{1}{R}\left(v_0\cot\varphi + w_0\right), \\
\kappa^0_{\varphi\varphi} &= \frac{1}{R^2}\frac{\mathrm{d}}{\mathrm{d}\varphi}\left(v_0 - \frac{\mathrm{d}w_0}{\mathrm{d}\varphi}\right), \\
\kappa^0_{\vartheta\vartheta} &= \frac{\cot\varphi}{R^2}\left(v_0 - \frac{\mathrm{d}w_0}{\mathrm{d}\varphi}\right).
\end{aligned} \tag{16.107}$$

Let χ be the bending angle of the meridian tangent defined as follows:

$$\chi = \frac{1}{R}\left(v_0 - \frac{\mathrm{d}w_0}{\mathrm{d}\varphi}\right). \tag{16.108}$$

Then for the two curvatures $\kappa^0_{\varphi\varphi}$ and $\kappa^0_{\vartheta\vartheta}$ we obtain:

$$\kappa^0_{\varphi\varphi} = \frac{1}{R}\frac{\mathrm{d}\chi}{\mathrm{d}\varphi}, \quad \kappa^0_{\vartheta\vartheta} = \frac{\cot\varphi}{R}\chi. \tag{16.109}$$

The constitutive law (16.26), (16.28) for the spherical shell under rotationally symmetric load results as:

$$\begin{aligned}
N_{\varphi\varphi} &= A\left(\varepsilon^0_{\varphi\varphi} + \nu\varepsilon^0_{\vartheta\vartheta}\right), \\
N_{\vartheta\vartheta} &= A\left(\nu\varepsilon^0_{\varphi\varphi} + \varepsilon^0_{\vartheta\vartheta}\right), \\
M_{\varphi\varphi} &= D\left(\kappa^0_{\varphi\varphi} + \nu\kappa^0_{\vartheta\vartheta}\right), \\
M_{\vartheta\vartheta} &= D\left(\nu\kappa^0_{\varphi\varphi} + \kappa^0_{\vartheta\vartheta}\right),
\end{aligned} \tag{16.110}$$

where the two moment flows $M_{\varphi\varphi}$ and $M_{\vartheta\vartheta}$ can be represented as:

$$\begin{aligned}
M_{\varphi\varphi} &= \frac{D}{R}\left(\frac{\mathrm{d}\chi}{\mathrm{d}\varphi} + \nu\cot\varphi\chi\right), \\
M_{\vartheta\vartheta} &= \frac{D}{R}\left(\nu\frac{\mathrm{d}\chi}{\mathrm{d}\varphi} + \cot\varphi\chi\right).
\end{aligned} \tag{16.111}$$

For the present case of the spherical shell under rotationally symmetric load, it is useful to reduce the above 11 equations to two differential equations for the bending angle χ and the transverse shear force flow Q_φ by adequate transformation, which is shown below.

The third equilibrium condition in (16.106) is differentiated:

$$\frac{\mathrm{d}M_{\varphi\varphi}}{\mathrm{d}\varphi}\sin\varphi + M_{\varphi\varphi}\cos\varphi - M_{\vartheta\vartheta}\cos\varphi - Q_\varphi R\sin\varphi = 0. \tag{16.112}$$

Inserting the expressions (16.111) for the moment flows $M_{\varphi\varphi}$ and $M_{\vartheta\vartheta}$ gives after short calculation:

$$\frac{\mathrm{d}^2\chi}{\mathrm{d}\varphi^2} + \frac{\mathrm{d}\chi}{\mathrm{d}\varphi}\cot\varphi - \left(\nu + \cot^2\varphi\right)\chi - Q_\varphi\frac{R^2}{D} = 0. \tag{16.113}$$

This is a second-order differential equation for the bending angle χ and the transverse shear force flow Q_φ.

We further consider a spherical shell section as shown in Fig. 16.13. The equilibrium of forces in vertical direction results in:

$$Q_\varphi\cos\varphi - N_{\varphi\varphi}\sin\varphi = 0, \tag{16.114}$$

which leads to

$$N_{\varphi\varphi} = Q_\varphi\cot\varphi = 0. \tag{16.115}$$

Inserting this relation into the second equilibrium condition in (16.106), we obtain after a short calculation:

$$N_{\vartheta\vartheta} = \frac{\mathrm{d}Q_\varphi}{\mathrm{d}\varphi}. \tag{16.116}$$

From the kinematic equations (16.107), a compatibility condition can be derived as follows:

$$\left(\varepsilon^0_{\varphi\varphi} - \varepsilon^0_{\vartheta\vartheta}\right)\cot\varphi - \frac{\mathrm{d}\varepsilon^0_{\vartheta\vartheta}}{\mathrm{d}\varphi} = \frac{1}{R}\left(v_0 - \frac{\mathrm{d}w_0}{\mathrm{d}\varphi}\right) = \chi. \tag{16.117}$$

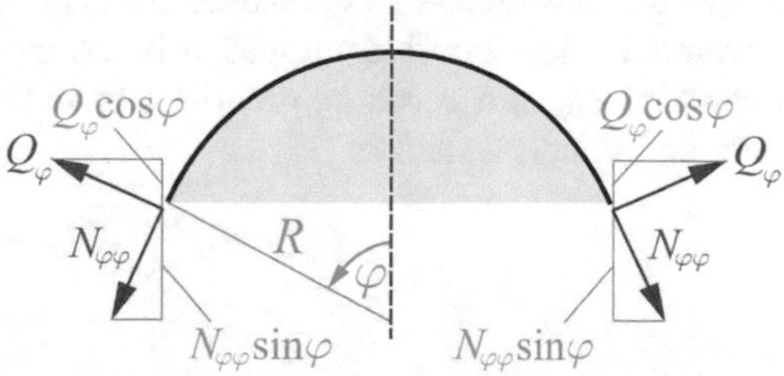

Fig. 16.13 Spherical shell section

Here the strains can be expressed by (16.110) through the force flows $N_{\varphi\varphi}$ and $N_{\vartheta\vartheta}$, and these in turn are expressed by the transverse shear force flow Q_φ with the help of (16.115) and (16.116). After some transformations we obtain:

$$\frac{\mathrm{d}^2 Q_\varphi}{\mathrm{d}\varphi^2} + \frac{\mathrm{d}Q_\varphi}{\mathrm{d}\varphi}\cot\varphi + \left(\nu - \cot^2\varphi\right) Q_\varphi + Eh\chi = 0. \tag{16.118}$$

Thus, with (16.113) and (16.118) there are two coupled differential equations for Q_φ and χ.

Introduction of the so-called Meissner operator

$$\mathbf{L} = \frac{\mathrm{d}^2}{\mathrm{d}\varphi^2} + \cot\varphi\frac{\mathrm{d}}{\mathrm{d}\varphi} - \cot^2\varphi \tag{16.119}$$

allows to represent differential equations (16.113) and (16.118) in the following form:

$$\begin{aligned} \mathbf{L}\chi - \nu\chi &= Q_\varphi \frac{R^2}{D}, \\ \mathbf{L}Q_\varphi + \nu Q_\varphi &= -Eh\chi. \end{aligned} \tag{16.120}$$

Alternatively, two decoupled differential equations for Q_φ and χ can be derived by elimination:

$$\begin{aligned} \mathbf{LL}\chi + 4\mu^4\chi &= 0, \\ \mathbf{LL}Q_\varphi + 4\mu^4 Q_\varphi &= 0. \end{aligned} \tag{16.121}$$

Herein the quantitiy μ is defined as:

$$4\mu^4 = \frac{EhR^2}{D} - \nu^2, \tag{16.122}$$

where the transverse strain number ν is generally negligible for thin shells with $\frac{R}{h} >> 1$ with respect to the first summand in (16.122). It can be easily shown that μ is consistent with the shell parameter λ according to (16.55) if the dimensionless coordinate $\xi = \frac{x}{R}$ is used instead of the coordinate x.

An exact solution of the two differential equations (16.121) can be developed in the form of hypergeometric series, which, however, generally converge poorly. Therefore, a sufficiently accurate engineering solution for not too small angles φ is presented below. For this purpose, it is assumed that the decay behavior of the state quantities Q_φ and χ can be described as in the boundary stiffness problem of the circular cylindical shell as

$$Q_\varphi \sim e^{-\mu\varphi}, \quad \chi \sim e^{-\mu\varphi}. \tag{16.123}$$

Due to $\mu >> 1$ it then follows that when the Meissner operator is applied, the second derivatives dominate over the lower derivatives. One can then derive from (16.121):

$$\frac{d^4\chi}{d\varphi^4} + 4\mu^4\chi = 0,$$
$$\frac{d^4 Q_\varphi}{d\varphi^4} + 4\mu^4 Q_\varphi = 0. \qquad (16.124)$$

These two simplified differential equations are similar in structure to the differential equation (16.56) in the framework of the container theory of the circular cylindrical shell. Accordingly, the general solution can also be found here in identical form for χ and Q_φ. For Q_φ, for example, follows using the new coordinates $\omega_1 = \varphi_1 - \varphi$ and $\omega_2 = \varphi - \varphi_2$ (see Fig. 16.14):

$$Q_\varphi = (A_1 \cos \mu\omega_1 + A_2 \sin \mu\omega_1)\, e^{-\mu\omega_1} + (A_3 \cos \mu\omega_2 + A_4 \sin \mu\omega_2)\, e^{-\mu\omega_2}. \qquad (16.125)$$

One can also represent the sum of the trigonometric functions as phase-shifted sinusoidal functions:

$$Q_\varphi = C_1 e^{-\mu\omega_1} \sin(\mu\omega_1 + \gamma_1) + C_2 e^{-\mu\omega_2} \sin(\mu\omega_2 + \gamma_2). \qquad (16.126)$$

Herein, $C_1, C_2, \gamma_1, \gamma_2$ are constants which are adapted to given boundary conditions. Once Q_φ and χ are known, the remaining force and moment quantities and the angle χ can be determined as:

$$N_{\vartheta\vartheta} = \mu\sqrt{2}\left[C_1 e^{-\mu\omega_1} \sin\left(\mu\omega_1 + \gamma_1 - \frac{\pi}{4}\right) - C_2 e^{-\mu\omega_2} \sin\left(\mu\omega_2 + \gamma_2 - \frac{\pi}{4}\right)\right],$$
$$N_{\varphi\varphi} = Q_\varphi \cot\varphi,$$
$$\chi = \frac{2\mu^2}{A\left(1-\nu^2\right)}\left[C_1 e^{-\mu\omega_1} \cos(\mu\omega_1 + \gamma_1) + C_2 e^{-\mu\omega_2} \cos(\mu\omega_2 + \gamma_2)\right],$$
$$M_{\varphi\varphi} = \frac{R}{\mu\sqrt{2}}\left[C_1 e^{-\mu\omega_1} \sin\left(\mu\omega_1 + \gamma_1 + \frac{\pi}{4}\right) - C_2 e^{-\mu\omega_2} \sin\left(\mu\omega_2 + \gamma_2 + \frac{\pi}{4}\right)\right],$$
$$M_{\vartheta\vartheta} \simeq \nu M_{\varphi\varphi}. \qquad (16.127)$$

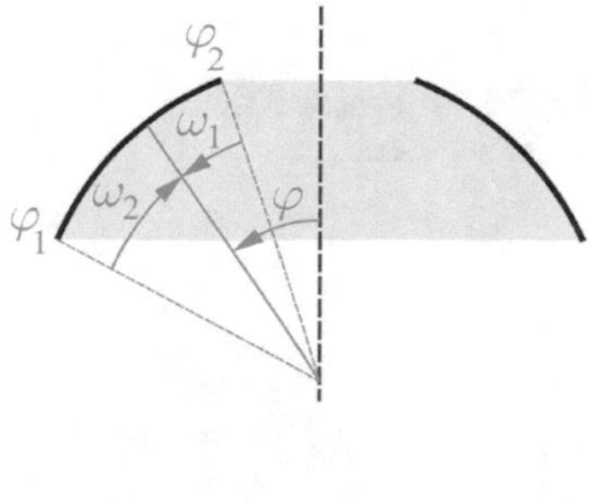

Fig. 16.14 Spherical shell with alternative coordinate system

For illustration, we consider a closed spherical shell (radius R, constant wall thickness h) under boundary loads Q_0 and M_0 (Fig. 16.15). We assume that it is sufficient to consider only the part decaying with $e^{-\mu\omega_1}$ and set the constant C_2 to zero. We will first consider the case where only the boundary moment flow M_0 occurs. At the point $\varphi = \varphi_1$ the following boundary conditions are present:

$$Q_\varphi(\varphi = \varphi_1) = 0, \quad N_{\varphi\varphi}(\varphi = \varphi_1) = 0, \quad M_{\varphi\varphi}(\varphi = \varphi_1) = -M_0. \tag{16.128}$$

From this, the two remaining constants γ_1 and C_1 can be obtained as:

$$\gamma_1 = 0, \quad C_1 = -\frac{2\mu}{R} M_0. \tag{16.129}$$

The force and moment flows of the spherical shell then follow as:

$$\begin{aligned}
Q_\varphi &= -\frac{2\mu M_0}{R} e^{-\mu\omega_1} \sin \mu\omega_1, \\
N_{\vartheta\vartheta} &= \frac{2\mu^2 M_0}{R} e^{-\mu\omega_1} (\cos \mu\omega_1 - \sin \mu\omega_1), \\
M_{\varphi\varphi} &= -M_0 e^{-\mu\omega_1} (\cos \mu\omega_1 + \sin \mu\omega_1), \\
N_{\varphi\varphi} &= Q_\varphi \cot(\varphi_1 - \omega_1), \\
M_{\vartheta\vartheta} &\simeq \nu M_{\varphi\varphi}.
\end{aligned} \tag{16.130}$$

The angular rotation $\psi_\varphi = -\chi$ results in:

$$\psi_\varphi = \frac{4\mu^3 M_0}{RA(1-\nu^2)} e^{-\mu\omega_1} \cos \mu\omega_1. \tag{16.131}$$

We also consider the case where only the boundary load Q_0 is present. The boundary conditions for $\varphi = \varphi_1$ are then:

$$Q_\varphi(\varphi = \varphi_1) = Q_0 \sin \varphi_1, \quad N_{\varphi\varphi}(\varphi = \varphi_1) = Q_0 \cos \varphi_1, \quad M_{\varphi\varphi}(\varphi = \varphi_1) = 0. \tag{16.132}$$

From this, the two constants γ_1 and C_1 can be determined as follows:

$$\gamma_1 = -\frac{\pi}{4}, \quad C_1 = -Q_0\sqrt{2} \sin \varphi_1. \tag{16.133}$$

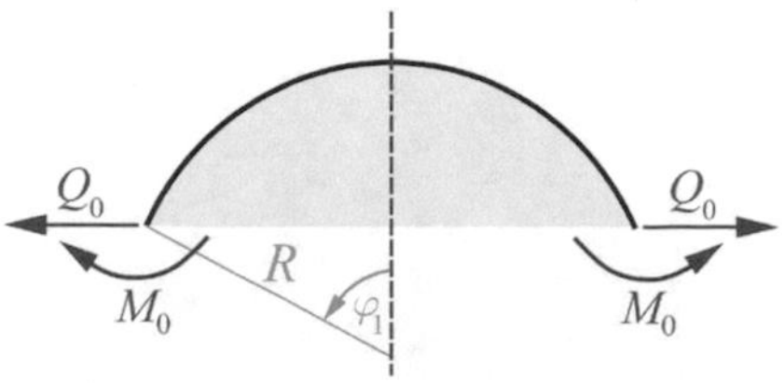

Fig. 16.15 Spherical shell under edge load

Thus, the shell force and moment flows can be determined, which, however, remains without further presentation at this point.

16.5 Edge Perturbations of Arbitrary Shells of Revolution

As in the case of spherical shells, two differential equations can be derived for shells of revolution of arbitrary shape under rotationally symmetric loading, which describe both the rotation of the meridian tangent and the transverse shear force flow Q_φ. If one uses the kinematic relations

$$\begin{aligned}
\varepsilon^0_{\varphi\varphi} &= \frac{1}{r_1}\left(\frac{\mathrm{d}v_0}{\mathrm{d}\varphi} + w_0\right),\\
\varepsilon^0_{\vartheta\vartheta} &= \frac{1}{r_2}\left(v_0 \cot\varphi + w_0\right),\\
\kappa^0_{\varphi\varphi} &= \frac{1}{r_1}\frac{\mathrm{d}}{\mathrm{d}\varphi}\left(\frac{v_0 - \frac{\mathrm{d}w_0}{\mathrm{d}\varphi}}{r_1}\right),\\
\kappa^0_{\vartheta\vartheta} &= \frac{\cot\varphi}{r_1 r_2}\left(v_0 - \frac{\mathrm{d}w_0}{\mathrm{d}\varphi}\right)
\end{aligned} \tag{16.134}$$

in the constitutive relations

$$\begin{aligned}
N_{\varphi\varphi} &= A\left(\varepsilon^0_{\varphi\varphi} + \nu\varepsilon^0_{\vartheta\vartheta}\right),\\
N_{\vartheta\vartheta} &= A\left(\nu\varepsilon^0_{\varphi\varphi} + \varepsilon^0_{\vartheta\vartheta}\right),\\
M_{\varphi\varphi} &= D\left(\kappa^0_{\varphi\varphi} + \nu\kappa^0_{\vartheta\vartheta}\right),\\
M_{\vartheta\vartheta} &= D\left(\nu\kappa^0_{\varphi\varphi} + \kappa^0_{\vartheta\vartheta}\right),
\end{aligned} \tag{16.135}$$

then one obtains:

$$\begin{aligned}
N_{\varphi\varphi} &= A\left[\left(\frac{1}{r_1}\frac{\mathrm{d}}{\mathrm{d}\varphi} + \frac{\nu\cot\varphi}{r_2}\right)v_0 + \left(\frac{1}{r_1} + \frac{\nu}{r_2}\right)w_0\right],\\
N_{\vartheta\vartheta} &= A\left[\left(\frac{\nu}{r_1}\frac{\mathrm{d}}{\mathrm{d}\varphi} + \frac{\cot\varphi}{r_2}\right)v_0 + \left(\frac{\nu}{r_1} + \frac{1}{r_2}\right)w_0\right],\\
M_{\varphi\varphi} &= \frac{D}{r_1}\left[\frac{\mathrm{d}}{\mathrm{d}\varphi}\left(\frac{v_0 - \frac{\mathrm{d}w_0}{\mathrm{d}\varphi}}{r_1}\right) + \frac{\nu\cot\varphi}{r_2}\left(v_0 - \frac{\mathrm{d}w_0}{\mathrm{d}\varphi}\right)\right],\\
M_{\vartheta\vartheta} &= \frac{D}{r_1}\left[\nu\frac{\mathrm{d}}{\mathrm{d}\varphi}\left(\frac{v_0 - \frac{\mathrm{d}w_0}{\mathrm{d}\varphi}}{r_1}\right) + \frac{\cot\varphi}{r_2}\left(v_0 - \frac{\mathrm{d}w_0}{\mathrm{d}\varphi}\right)\right].
\end{aligned} \tag{16.136}$$

With the bending angle

$$\chi = \frac{1}{r_1}\left(v_0 - \frac{\mathrm{d}w_0}{\mathrm{d}\varphi}\right) \tag{16.137}$$

it follows for the two moment flows $M_{\varphi\varphi}$ and $M_{\vartheta\vartheta}$:

$$\begin{aligned} M_{\varphi\varphi} &= D\left(\frac{1}{r_1}\frac{\mathrm{d}\chi}{\mathrm{d}\varphi} + \frac{\nu \cot\varphi}{r_2}\chi\right), \\ M_{\vartheta\vartheta} &= D\left(\frac{\nu}{r_1}\frac{\mathrm{d}\chi}{\mathrm{d}\varphi} + \frac{\cot\varphi}{r_2}\chi\right). \end{aligned} \tag{16.138}$$

In the remainder of this section, the rotation of the meridian tangent χ and the modified transverse shear force flow $U = r_2 Q_\varphi$ are treated as the unknowns to be determined. We also focus here only on the consideration of edge perturbations of arbitrary shells of revolution. The associated equilibrium conditions are:

$$\begin{aligned} \frac{\mathrm{d}}{\mathrm{d}\varphi}\left(N_{\varphi\varphi} r\right) - N_{\vartheta\vartheta}\cos\varphi r_1 + Q_\varphi r &= 0, \\ \frac{\mathrm{d}}{\mathrm{d}\varphi}\left(Q_\varphi r\right) - N_{\varphi\varphi} r - N_{\vartheta\vartheta} r_1 \sin\varphi &= 0, \\ \frac{\mathrm{d}}{\mathrm{d}\varphi}\left(M_{\varphi\varphi} r\right) - M_{\vartheta\vartheta} r_1 \cos\varphi - Q_\varphi r_1 r &= 0. \end{aligned} \tag{16.139}$$

Inserting the moment flows $M_{\varphi\varphi}$ and $M_{\vartheta\vartheta}$ according to (16.138) into the third equilibrium condition in (16.139) yields:

$$\frac{r_2}{r_1^2}\frac{\mathrm{d}^2\chi}{\mathrm{d}\varphi^2} + \left[\frac{\mathrm{d}}{\mathrm{d}\varphi}\left(\frac{r_2}{r_1}\right) + \frac{r_2}{r_1}\cot\varphi\right]\frac{1}{r_1}\frac{\mathrm{d}\chi}{\mathrm{d}\varphi} - \left[\nu + \frac{r_1}{r_2}\cot^2\varphi\right]\frac{\chi}{r_1} = \frac{U}{D}. \tag{16.140}$$

This is a differential equation for determining the two unknowns χ and U.

We derive a second equation as follows. First, we derive from a latitude circle section analogous to Fig. 16.13 the following relation between $N_{\varphi\varphi}$ and Q_φ:

$$N_{\varphi\varphi} = Q_\varphi \cot\varphi = \frac{U}{r}\cos\varphi. \tag{16.141}$$

We now also consider the first and second equilibrium conditions in (16.139). Multiplying the first equation by $\cos\varphi$ and the second by $\sin\varphi$ and summing the resulting equations, we get with $\sin^2\varphi + \cos^2\varphi = 1$ and

$$\begin{aligned} \frac{\mathrm{d}}{\mathrm{d}\varphi}\left(N_{\varphi\varphi} r\cos\varphi\right) &= \frac{\mathrm{d}}{\mathrm{d}\varphi}\left(N_{\varphi\varphi} r\right)\cos\varphi - N_{\varphi\varphi} r\sin\varphi, \\ \frac{\mathrm{d}}{\mathrm{d}\varphi}\left(Q_\varphi r\sin\varphi\right) &= \frac{\mathrm{d}}{\mathrm{d}\varphi}\left(Q_\varphi r\right)\sin\varphi + Q_\varphi r\cos\varphi \end{aligned} \tag{16.142}$$

the following relation between $N_{\vartheta\vartheta}$ and U:

$$N_{\vartheta\vartheta} = \frac{1}{r_1}\frac{\mathrm{d}}{\mathrm{d}\varphi}\left(N_{\varphi\varphi} r \cos\varphi + Q_\varphi r \sin\varphi\right) = \frac{1}{r_1}\frac{\mathrm{d}U}{\mathrm{d}\varphi}. \tag{16.143}$$

A compatibility equation between the strains $\varepsilon^0_{\varphi\varphi}$, $\varepsilon^0_{\vartheta\vartheta}$ and the meridian tangent slope χ can be established as follows:

$$\frac{\mathrm{d}}{\mathrm{d}\varphi}\left(r\varepsilon^0_{\vartheta\vartheta}\right) - r_1\varepsilon^0_{\varphi\varphi}\cos\varphi = -r_1\chi\sin\varphi. \tag{16.144}$$

Insertion of the strains

$$\begin{aligned}
\varepsilon^0_{\varphi\varphi} &= \frac{1}{Eh}\left(N_{\varphi\varphi} - \nu N_{\vartheta\vartheta}\right) = \frac{1}{Eh}\left(\frac{U}{r}\cos\varphi - \frac{\nu}{r_1}\frac{\mathrm{d}U}{\mathrm{d}\varphi}\right),\\
\varepsilon^0_{\vartheta\vartheta} &= \frac{1}{Eh}\left(N_{\vartheta\vartheta} - \nu N_{\varphi\varphi}\right) = \frac{1}{Eh}\left(\frac{1}{r_1}\frac{\mathrm{d}U}{\mathrm{d}\varphi} - \nu\frac{U}{r}\cos\varphi\right)
\end{aligned} \tag{16.145}$$

then leads to the second equation of determination for χ and U:

$$\frac{r_2}{r_1^2}\frac{\mathrm{d}^2U}{\mathrm{d}\varphi^2} + \left[\frac{\mathrm{d}}{\mathrm{d}\varphi}\left(\frac{r_2}{r_1}\right) + \frac{r_2}{r_1}\cot\varphi\right]\frac{1}{r_1}\frac{\mathrm{d}U}{\mathrm{d}\varphi} - \left[\frac{r_1}{r_2}\cot^2\varphi - \nu\right]\frac{U}{r_1} = -Eh\chi. \tag{16.146}$$

The two differential equations (16.140) and (16.146) can be transformed by introducing the Meissner operator

$$\mathbf{L} = \frac{r_2}{r_1^2}\frac{\mathrm{d}^2}{\mathrm{d}\varphi^2} + \left[\frac{\mathrm{d}}{\mathrm{d}\varphi}\left(\frac{r_2}{r_1}\right) + \frac{r_2}{r_1}\cot\varphi\right]\frac{1}{r_1}\frac{\mathrm{d}}{\mathrm{d}\varphi} - \frac{1}{r_2}\cot^2\varphi \tag{16.147}$$

to the following form:

$$\begin{aligned}
\mathbf{L}\chi - \nu\frac{\chi}{r_1} &= \frac{U}{D},\\
\mathbf{L}U + \nu\frac{U}{r_1} &= -Eh\chi.
\end{aligned} \tag{16.148}$$

Separation of variables finally results in:

$$\begin{aligned}
\mathbf{L}\mathbf{L}\chi - \nu\mathbf{L}\left(\frac{\chi}{r_1}\right) + \frac{\nu}{r_1}\left[\mathbf{L}\chi - \nu\frac{\chi}{r_1}\right] &= -\frac{12\left(1-\nu^2\right)}{h^2}\chi,\\
\mathbf{L}\mathbf{L}U + \nu\mathbf{L}\left(\frac{U}{r_1}\right) - \frac{\nu}{r_1}\left[\mathbf{L}U + \nu\frac{U}{r_1}\right] &= -\frac{12\left(1-\nu^2\right)}{h^2}U.
\end{aligned} \tag{16.149}$$

16.6 Circular Cylindrical Shell under Arbitrary Load

16.6.1 Basic Equations

In this section, we consider the problem of a circular cylindrical shell (Fig. 16.16) under an arbitrary load, which may consist of boundary loads as well as the surface loads p_x, p_ϑ and p_z (Göldner et al. 1985). The shell has the radius R and the constant wall thickness h. The trapezoidal effect is taken into account and the dimensionless coordinate $\xi = \frac{x}{R}$ is used in addition to the longitudinal coordinate x. To derive the necessary equations, we first consider the free body image of Fig. 16.17 for an infinitesimal sectional element $\mathrm{d}x\, R\mathrm{d}\vartheta$ of the circular cylindrical shell. We formulate the equilibrium of forces with respect to the x-direction and can read directly from Fig. 16.17:

$$\frac{\partial N_{xx}}{\partial \xi} + \frac{\partial N_{\vartheta x}}{\partial \vartheta} + Rp_x = 0. \tag{16.150}$$

For the equilibrium of forces with respect to the latitudinal tangent, the free body image of Fig. 16.18, top, is considered. The force balance results with $\sin \frac{\mathrm{d}\vartheta}{2} \simeq \frac{\mathrm{d}\vartheta}{2}$ and $\cos \frac{\mathrm{d}\vartheta}{2} \simeq 1$ neglecting terms of higher order and taking into account the contribution of $N_{x\vartheta}$:

$$\frac{\partial N_{\vartheta\vartheta}}{\partial \vartheta} + \frac{\partial N_{x\vartheta}}{\partial \xi} + Q_\vartheta + Rp_\vartheta = 0. \tag{16.151}$$

Analogously, it follows from Fig. 16.18, top, for the sum of forces in z-direction taking into account the contribution of Q_x:

$$\frac{\partial Q_x}{\partial \xi} + \frac{\partial Q_\vartheta}{\partial \vartheta} - N_{\vartheta\vartheta} + Rp_z = 0. \tag{16.152}$$

The sum of moments with respect to the longitudinal axis x yields:

$$\frac{\partial M_{\vartheta\vartheta}}{\partial \vartheta} + \frac{\partial M_{x\vartheta}}{\partial \xi} - Q_\vartheta R = 0. \tag{16.153}$$

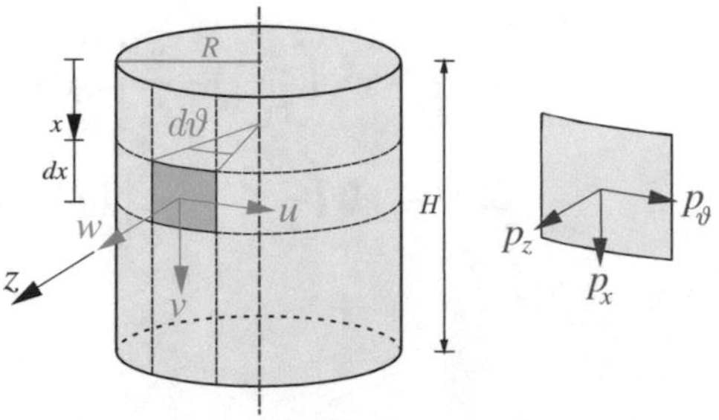

Fig. 16.16 Circular cylindrical shell (left), shell loads (right)

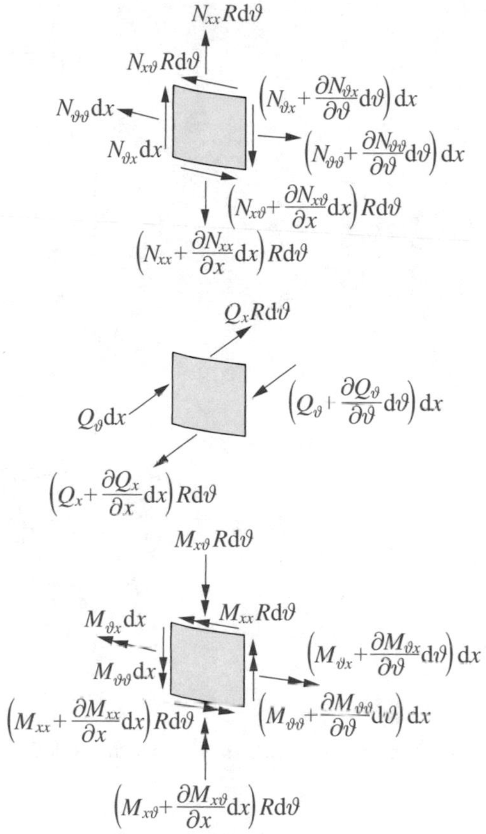

Fig. 16.17 Infinitesimal shell element with force and moment flows

For establishing the moment sum with respect to the latitude circle tangent, the moment flow $M_{\vartheta x}$ is decomposed as shown in Fig. 16.18, bottom. Considering the contributions of M_{xx} and Q_x we obtain:

$$\frac{\partial M_{xx}}{\partial \xi} + \frac{\partial M_{\vartheta x}}{\partial \vartheta} - Q_x R = 0. \tag{16.154}$$

For the moment balance with respect to the z-direction, it should be noted that in addition to the force flows $N_{x\vartheta}$ and $N_{\vartheta x}$, the moment flow $M_{\vartheta x}$ also contributes a relevant component, as can be seen directly from Fig. 16.18, bottom. It follows:

$$N_{x\vartheta} R - N_{\vartheta x} R - M_{\vartheta x} = 0. \tag{16.155}$$

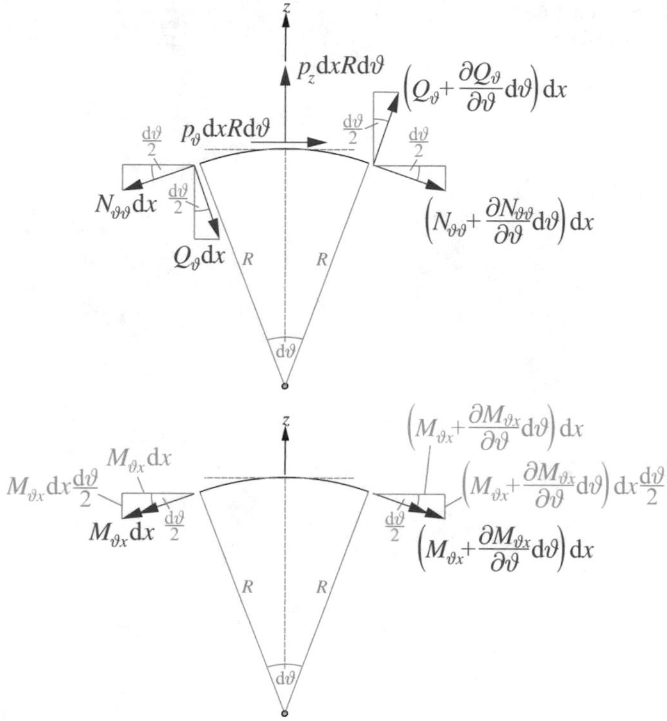

Fig. 16.18 Section of the circular cylindrical shell

Taking into account the trapezoidal effect, the force and moment flows of the circular cylindrical shell are:

$$
\begin{aligned}
N_{xx} &= \int_{-\frac{h}{2}}^{+\frac{h}{2}} \sigma_{xx}\left(1+\frac{z}{R}\right)\mathrm{d}z, \quad N_{\vartheta\vartheta} = \int_{-\frac{h}{2}}^{+\frac{h}{2}} \sigma_{\vartheta\vartheta}\mathrm{d}z, \\
N_{x\vartheta} &= \int_{-\frac{h}{2}}^{+\frac{h}{2}} \tau_{x\vartheta}\left(1+\frac{z}{R}\right)\mathrm{d}z, \quad N_{\vartheta x} = \int_{-\frac{h}{2}}^{+\frac{h}{2}} \tau_{\vartheta x}\mathrm{d}z, \\
M_{xx} &= \int_{-\frac{h}{2}}^{+\frac{h}{2}} \sigma_{xx}\left(1+\frac{z}{R}\right)z\mathrm{d}z, \quad M_{\vartheta\vartheta} = \int_{-\frac{h}{2}}^{+\frac{h}{2}} \sigma_{\vartheta\vartheta}z\mathrm{d}z, \\
M_{x\vartheta} &= \int_{-\frac{h}{2}}^{+\frac{h}{2}} \tau_{x\vartheta}\left(1+\frac{z}{R}\right)z\mathrm{d}z, \quad M_{\vartheta x} = \int_{-\frac{h}{2}}^{+\frac{h}{2}} \tau_{\vartheta x}z\mathrm{d}z. \qquad (16.156)
\end{aligned}
$$

Obviously, due to the trapezoidal effect, the force flows $N_{x\vartheta}$ and $N_{\vartheta x}$ are not identical. The same is true for the two moment flows $M_{x\vartheta}$ and $M_{\vartheta x}$.

The stress components in the shell are obtained from Hooke's law as follows:

$$\sigma_{xx} = \frac{E}{1-\nu^2}\left(\varepsilon_{xx} + \nu\varepsilon_{\vartheta\vartheta}\right),$$
$$\sigma_{\vartheta\vartheta} = \frac{E}{1-\nu^2}\left(\varepsilon_{\vartheta\vartheta} + \nu\varepsilon_{xx}\right),$$
$$\tau_{x\vartheta} = \tau_{\vartheta x} = G\gamma_{x\vartheta} = \frac{E}{2(1+\nu)}\gamma_{x\vartheta}. \tag{16.157}$$

The kinematic equations to be applied here are:

$$\varepsilon_{xx} = \frac{1}{R}\frac{\partial v}{\partial \xi},$$
$$\varepsilon_{\vartheta\vartheta} = \frac{1}{R+z}\left(\frac{\partial u}{\partial \vartheta} + w\right),$$
$$\gamma_{x\vartheta} = \frac{1}{R}\frac{\partial u}{\partial \xi} + \frac{1}{R+z}\frac{\partial v}{\partial \vartheta}. \tag{16.158}$$

The displacement field can be deduced from Fig. 16.19 and follows as:

$$u = \frac{R+z}{R}u_0 - \frac{z}{R}\frac{\partial w_0}{\partial \vartheta},$$
$$v = v_0 - \frac{z}{R}\frac{\partial w_0}{\partial \xi},$$
$$w = w_0. \tag{16.159}$$

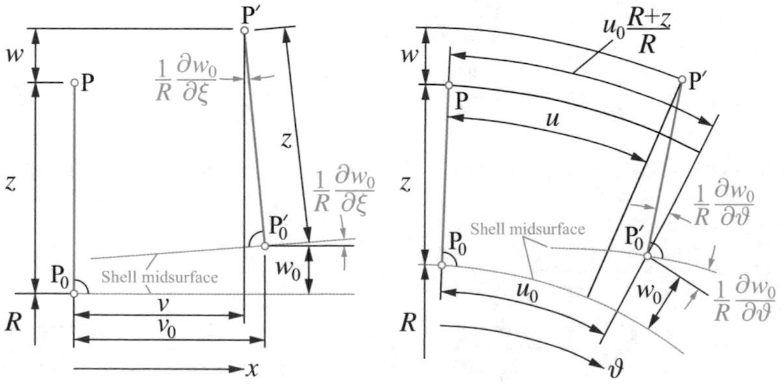

Fig. 16.19 Kinematics, section ϑ = const. (left), section x = const. (right)

The strains (16.158) result as:

$$\begin{aligned}
\varepsilon_{xx} &= \frac{1}{R}\frac{\partial v_0}{\partial \xi} - \frac{z}{R^2}\frac{\partial^2 w_0}{\partial \xi^2},\\
\varepsilon_{\vartheta\vartheta} &= \frac{1}{R}\frac{\partial u_0}{\partial \vartheta} - \frac{z}{R\,(R+z)}\frac{\partial^2 w_0}{\partial \vartheta^2} + \frac{w_0}{R+z},\\
\gamma_{x\vartheta} &= \frac{R+z}{R^2}\frac{\partial u_0}{\partial \xi} + \frac{1}{R+z}\frac{\partial v_0}{\partial \vartheta} - \frac{z}{R}\left(\frac{1}{R} + \frac{1}{R+z}\right)\frac{\partial^2 w_0}{\partial \xi \partial \vartheta}.
\end{aligned} \tag{16.160}$$

Substituting the expressions (16.160) into the stress-strain equations (16.157) and this, in turn, into the definitions (16.156), one obtains:

$$\begin{aligned}
N_{xx} &= \frac{A}{R}\left(\frac{\partial v_0}{\partial \xi} + \nu\frac{\partial u_0}{\partial \vartheta} + \nu w_0\right) - \frac{D}{R^3}\frac{\partial^2 w_0}{\partial \xi^2},\\
N_{\vartheta\vartheta} &= \frac{A}{R}\left(\frac{\partial u_0}{\partial \vartheta} + w_0 + \nu\frac{\partial v_0}{\partial \xi}\right) + \frac{D}{R^3}\left(w_0 + \frac{\partial^2 w_0}{\partial \vartheta^2}\right),\\
N_{\vartheta x} &= \frac{A}{R}\frac{1-\nu}{2}\left(\frac{\partial v_0}{\partial \vartheta} + \frac{\partial u_0}{\partial \xi}\right) + \frac{D}{R^3}\frac{1-\nu}{2}\left(\frac{\partial v_0}{\partial \vartheta} + \frac{\partial^2 w_0}{\partial \xi \partial \vartheta}\right),\\
N_{x\vartheta} &= \frac{A}{R}\frac{1-\nu}{2}\left(\frac{\partial v_0}{\partial \vartheta} + \frac{\partial u_0}{\partial \xi}\right) + \frac{D}{R^3}\frac{1-\nu}{2}\left(\frac{\partial u_0}{\partial \xi} - \frac{\partial^2 w_0}{\partial \xi \partial \vartheta}\right),\\
M_{xx} &= -\frac{D}{R^2}\left(\frac{\partial^2 w_0}{\partial \xi^2} + \nu\frac{\partial^2 w_0}{\partial \vartheta^2} - \frac{\partial v_0}{\partial \xi} - \nu\frac{\partial u_0}{\partial \vartheta}\right),\\
M_{\vartheta\vartheta} &= -\frac{D}{R^2}\left(w_0 + \frac{\partial^2 w_0}{\partial \vartheta^2} + \nu\frac{\partial^2 w_0}{\partial \xi^2}\right),\\
M_{x\vartheta} &= -\frac{D}{R^2}(1-\nu)\left(\frac{\partial^2 w_0}{\partial \xi \partial \vartheta} - \frac{\partial u_0}{\partial \xi}\right),\\
M_{\vartheta x} &= -\frac{D}{R^2}(1-\nu)\left(\frac{\partial^2 w_0}{\partial \xi \partial \vartheta} + \frac{1}{2}\frac{\partial v_0}{\partial \vartheta} - \frac{1}{2}\frac{\partial u_0}{\partial \xi}\right).
\end{aligned} \tag{16.161}$$

The transverse shear force flows can be determined from the equilibrium conditions as:

$$\begin{aligned}
RQ_x &= \frac{\partial M_{xx}}{\partial \xi} + \frac{\partial M_{\vartheta x}}{\partial \vartheta}\\
&= -\frac{D}{R^2}\left(\frac{\partial^3 w_0}{\partial \xi^3} + \frac{\partial^3 w_0}{\partial \xi \partial \vartheta^2} - \frac{\partial^2 v_0}{\partial \xi^2} + \frac{1-\nu}{2}\frac{\partial^2 v_0}{\partial \vartheta^2} - \frac{1+\nu}{2}\frac{\partial^2 u_0}{\partial \xi \partial \vartheta}\right),\\
RQ_\vartheta &= \frac{\partial M_{\vartheta\vartheta}}{\partial \vartheta} + \frac{\partial M_{x\vartheta}}{\partial \xi}\\
&= -\frac{D}{R^2}\left(\frac{\partial w_0}{\partial \vartheta} + \frac{\partial^3 w_0}{\partial \vartheta^3} + \frac{\partial^3 w_0}{\partial \xi^2 \partial \vartheta} - (1-\nu)\frac{\partial^2 u_0}{\partial \xi^2}\right).
\end{aligned} \tag{16.162}$$

Substituting these definitions into the force equilibrium conditions (16.150), (16.151) and (16.152) one obtains a coupled system of partial differential equations in the three displacements u_0, v_0, w_0:

$$\begin{aligned}
&\frac{\partial^2 v_0}{\partial \xi^2}+\frac{1-\nu}{2}\frac{\partial^2 v_0}{\partial \vartheta^2}+\frac{1+\nu}{2}\frac{\partial^2 u_0}{\partial \xi \partial \vartheta}+\nu\frac{\partial w_0}{\partial \xi}\\
&+k\left(\frac{1-\nu}{2}\frac{\partial^2 v_0}{\partial \vartheta^2}-\frac{\partial^3 w_0}{\partial \xi^3}+\frac{1-\nu}{2}\frac{\partial^3 w_0}{\partial \xi \partial \vartheta^2}\right)+\frac{p_x R^2}{A}=0,\\
&\frac{1+\nu}{2}\frac{\partial^2 v_0}{\partial \xi \partial \vartheta}+\frac{\partial^2 u_0}{\partial \vartheta^2}+\frac{1-\nu}{2}\frac{\partial^2 u_0}{\partial \xi^2}+\frac{\partial w_0}{\partial \vartheta}\\
&+k\left(\frac{3}{2}(1-\nu)\frac{\partial^2 u_0}{\partial \xi^2}-\frac{3-\nu}{2}\frac{\partial^3 w_0}{\partial \xi^2 \partial \vartheta}\right)+\frac{p_\vartheta R^2}{A}=0,\\
&\nu\frac{\partial v_0}{\partial \xi}+\frac{\partial u_0}{\partial \vartheta}+w_0+k\left(\frac{1-\nu}{2}\frac{\partial^3 v_0}{\partial \xi \partial \vartheta^2}-\frac{\partial^3 v_0}{\partial \xi^3}-\frac{3-\nu}{2}\frac{\partial^3 u_0}{\partial \xi^2 \partial \vartheta}\right.\\
&\left.+\frac{\partial^4 w_0}{\partial \xi^4}+2\frac{\partial^4 w_0}{\partial \xi^2 \partial \vartheta^2}+\frac{\partial^4 w_0}{\partial \vartheta^4}+2\frac{\partial^2 w_0}{\partial \vartheta^2}+w_0\right)-\frac{p_z R^2}{A}=0,
\end{aligned} \tag{16.163}$$

with

$$k=\frac{h^2}{12R^2}. \tag{16.164}$$

One can reduce this differential equation system for the case $p_x = p_\vartheta = 0$ to a single differential equation of a higher-order function F (Göldner et al. 1985). With the definitions

$$\begin{aligned}
u_0 &= -\frac{\partial^3 F}{\partial \vartheta^3}-(2+\nu)\frac{\partial^3 F}{\partial \xi^2 \partial \vartheta}+k\left(2\frac{\partial^5 F}{\partial \xi^2 \partial \vartheta^3}-\frac{\partial^3 F}{\partial \vartheta^3}+2\frac{\partial^5 F}{\partial \xi^4 \partial \vartheta}\right)+k^2\frac{3-\nu}{2}\frac{\partial^5 F}{\partial \xi^2 \partial \vartheta^3},\\
v_0 &= \frac{\partial^3 F}{\partial \xi \partial \vartheta^2}-\nu\frac{\partial^3 F}{\partial \xi^3}+k\left(\frac{\partial^5 F}{\partial \xi^5}-\frac{\partial^5 F}{\partial \xi \partial \vartheta^4}-3\nu\frac{\partial^3 F}{\partial \xi^3}\right)\\
&\quad+k^2\left(3\frac{\partial^5 F}{\partial \xi^5}-\frac{3}{2}(1-\nu)\frac{\partial^5 F}{\partial \xi^3 \partial \vartheta^2}\right),\\
w_0 &= \frac{\partial^4 F}{\partial \xi^4}+2\frac{\partial^4 F}{\partial \xi^2 \partial \vartheta^2}+\frac{\partial^4 F}{\partial \vartheta^4}\\
&\quad+k\left(\frac{\partial^5 F}{\partial \vartheta^5}+3\frac{\partial^4 F}{\partial \xi^4}+2(1-\nu)\frac{\partial^4 F}{\partial \xi^2 \partial \vartheta^2}\right)+3k^2\frac{1-\nu}{2}\frac{\partial^4 F}{\partial \xi^2 \partial \vartheta^2}
\end{aligned} \tag{16.165}$$

the first two conditions in (16.163) are satisfied, and from the third equation it follows:

$$\begin{aligned}
&\Delta\Delta\Delta\Delta F+2\frac{\partial^6 F}{\partial \vartheta^6}+2(4-\nu)\frac{\partial^6 F}{\partial \xi^2 \partial \vartheta^4}+6\frac{\partial^6 F}{\partial \xi^4 \partial \vartheta^2}+2\nu\frac{\partial^6 F}{\partial \xi^6}\\
&+\frac{\partial^4 F}{\partial \vartheta^4}+2(2-\nu)\frac{\partial^4 F}{\partial \xi^2 \partial \vartheta^2}+\frac{1-\nu^2}{k}\frac{\partial^4 F}{\partial \xi^4}=\frac{p_z R^2}{Ak}.
\end{aligned} \tag{16.166}$$

16.6.2 Approximation According to Donnell

A simplification of the developed basic equations developed by Donnell[1] is achieved by setting $\frac{D}{R^2} << A$ due to $k << 1$. Bending moment flows are proportional to the curvatures

$$\kappa_{xx}^0 = -\frac{\partial^2 w_0}{\partial \xi^2}, \quad \kappa_{\vartheta\vartheta}^0 = -\frac{\partial^2 w_0}{\partial \vartheta^2}, \tag{16.167}$$

and the twisting moment flow is proportional to the twist $-\frac{\partial^2 w_0}{\partial \xi \partial \vartheta}$. Thus, the trapezoidal effect is neglected, and the following simplified definitions for force and moment flows remain:

$$\begin{aligned}
N_{xx} &= \frac{A}{R}\left(\frac{\partial v_0}{\partial \xi} + \nu \frac{\partial u_0}{\partial \vartheta} + \nu w_0\right), \\
N_{\vartheta\vartheta} &= \frac{A}{R}\left(\frac{\partial u_0}{\partial \vartheta} + w_0 + \nu \frac{\partial v_0}{\partial \xi}\right), \\
N_{\vartheta x} = N_{x\vartheta} &= \frac{A}{R}\frac{1-\nu}{2}\left(\frac{\partial v_0}{\partial \vartheta} + \frac{\partial u_0}{\partial \xi}\right), \\
M_{xx} &= -\frac{D}{R^2}\left(\frac{\partial^2 w_0}{\partial \xi^2} + \nu \frac{\partial^2 w_0}{\partial \vartheta^2}\right), \\
M_{\vartheta\vartheta} &= -\frac{D}{R^2}\left(\frac{\partial^2 w_0}{\partial \vartheta^2} + \nu \frac{\partial^2 w_0}{\partial \xi^2}\right), \\
M_{x\vartheta} = M_{\vartheta x} &= -\frac{D}{R^2}(1-\nu)\frac{\partial^2 w_0}{\partial \xi \partial \vartheta}, \\
Q_x &= -\frac{D}{R^3}\left(\frac{\partial^3 w_0}{\partial \xi^3} + \frac{\partial^3 w_0}{\partial \xi \partial \vartheta^2}\right), \\
Q_\vartheta &= -\frac{D}{R^3}\left(\frac{\partial^3 w_0}{\partial \vartheta^3} + \frac{\partial^3 w_0}{\partial \xi^2 \partial \vartheta}\right).
\end{aligned} \tag{16.168}$$

The displacement differential equations then reduce to:

$$\begin{aligned}
\frac{\partial^2 v_0}{\partial \xi^2} + \frac{1-\nu}{2}\frac{\partial^2 v_0}{\partial \vartheta^2} + \frac{1+\nu}{2}\frac{\partial^2 u_0}{\partial \xi \partial \vartheta} + \nu \frac{\partial w_0}{\partial \xi} + \frac{p_x R^2}{A} &= 0, \\
\frac{1+\nu}{2}\frac{\partial^2 v_0}{\partial \xi \partial \vartheta} + \frac{\partial^2 u_0}{\partial \vartheta^2} + \frac{1-\nu}{2}\frac{\partial^2 u_0}{\partial \xi^2} + \frac{\partial w_0}{\partial \vartheta} + \frac{p_\vartheta R^2}{A} &= 0, \\
\nu \frac{\partial v_0}{\partial \xi} + \frac{\partial u_0}{\partial \vartheta} + w_0 + k\left(\frac{\partial^4 w_0}{\partial \xi^4} + 2\frac{\partial^4 w_0}{\partial \xi^2 \partial \vartheta^2} + \frac{\partial^4 w_0}{\partial \vartheta^4}\right) - \frac{p_z R^2}{A} &= 0.
\end{aligned} \tag{16.169}$$

[1] Lloyd Hamilton Donnell, 1895–1997, U.S. American engineer.

With the help of the definitions

$$u_0 = -\frac{\partial^3 F}{\partial \vartheta^3} - (2+\nu)\frac{\partial^3 F}{\partial \xi^2 \partial \vartheta},$$
$$v_0 = \frac{\partial^3 F}{\partial \xi \partial \vartheta^2} - \nu \frac{\partial^3 F}{\partial \xi^3},$$
$$w_0 = \frac{\partial^4 F}{\partial \xi^4} + 2\frac{\partial^4 F}{\partial \xi^2 \partial \vartheta^2} + \frac{\partial^4 F}{\partial \vartheta^4} = \Delta\Delta F \quad (16.170)$$

for the case of the exclusive load by p_z the reduction of the problem to a differential equation in the function F can be achieved here as well:

$$\Delta\Delta\Delta\Delta F + \frac{1-\nu^2}{k}\frac{\partial^4 F}{\partial \xi^4} = \frac{p_z R^2}{Ak}. \quad (16.171)$$

16.6.3 Solution of the Basic Equations

At this point we consider closed circular cylindrical shells of length H. The shell is loaded exclusively by boundary loads, the surface loads p_x, p_ϑ, p_z are assumed to be zero. For such a shell situation, the internal forces and moments as well as the deformations are periodic functions of ϑ with period 2π. For the displacements u_0, v_0, w_0 a Fourier series expansion is applied as follows:

$$u_0 = \sum_{m=1}^{\infty} U_m(x)\sin m\vartheta,$$
$$v_0 = \sum_{m=1}^{\infty} V_m(x)\cos m\vartheta,$$
$$w_0 = \sum_{m=1}^{\infty} W_m(x)\cos m\vartheta. \quad (16.172)$$

Substituting these expressions into the displacement differential equations (16.163), we obtain for a given value m a system of coupled ordinary differential equations in the functions U_m, V_m, W_m as follows:

$$\frac{d^2 V_m}{d\xi^2} - \frac{1-\nu}{2}m^2 V_m + \frac{1+\nu}{2}m\frac{dU_m}{d\xi} + \nu\frac{dW_m}{d\xi}$$
$$-k\left(\frac{1-\nu}{2}m^2 V_m + \frac{d^3 W_m}{d\xi^3} + \frac{1-\nu}{2}m\frac{dW_m}{d\xi}\right) = 0,$$
$$-\frac{1+\nu}{2}m\frac{dV_m}{d\xi} - m^2 U_m + \frac{1-\nu}{2}\frac{d^2 U_m}{d\xi^2} - mW_m$$

$$+k\left(\frac{3}{2}(1-\nu)\frac{d^2U_m}{d\xi^2}+\frac{3-\nu}{2}m\frac{d^2W_m}{d\xi^2}\right)=0,$$

$$\nu\frac{dV_m}{d\xi}+mU_m+W_m+k\left(-\frac{1-\nu}{2}m^2\frac{dV_m}{d\xi}-\frac{d^3V_m}{d\xi^3}-\frac{3-\nu}{2}m\frac{d^2U_m}{d\xi^2}\right.$$
$$\left.+\frac{d^4W_m}{d\xi^4}-2m^2\frac{d^2W_m}{d\xi^2}+m^4W_m-2m^2W_m+W_m\right)=0. \tag{16.173}$$

For the functions U_m, V_m, W_m, an exponential approach of the following form is used:

$$U_m=A_m e^{\lambda\xi},\quad V_m=B_m e^{\lambda\xi},\quad W_m=C_m e^{\lambda\xi}. \tag{16.174}$$

Insertion into the differential equation system (16.173) leads to a homogeneous equation system for the determination of the constants A_m, B_m, C_m. Developing the coefficient determinant and zeroing leads with $k << 1$ to the following fourth order characteristic polynomial for λ^2:

$$\lambda^8-2\left(2m^2-\nu\right)\lambda^6+\left(\frac{1-\nu^2}{k}+6m^2\left(m^2-1\right)\right)\lambda^4$$
$$-2m^2\left(2m^4-(4-\nu)m^2+(2-\nu)\right)\lambda^2+m^4\left(m^2-1\right)^2=0. \tag{16.175}$$

In the case of Donnell's theory, the following polynomial results:

$$\lambda^8-4\lambda^6m^2+\left(\frac{1-\nu^2}{k}+6m^4\right)\lambda^4-4\lambda^2m^6+m^8=0. \tag{16.176}$$

The roots of (16.175) can be given as follows:

$$\begin{aligned}\lambda_1&=-\kappa_1+i\mu_1,\quad \lambda_2=-\kappa_1-i\mu_1,\\ \lambda_3&=-\kappa_2+i\mu_2,\quad \lambda_4=-\kappa_2-i\mu_2,\\ \lambda_5&=+\kappa_1+i\mu_1,\quad \lambda_6=+\kappa_1-i\mu_1,\\ \lambda_7&=+\kappa_2+i\mu_2,\quad \lambda_8=+\kappa_2-i\mu_2.\end{aligned} \tag{16.177}$$

The general solution for U_m is given by Flügge as:

$$\begin{aligned}U_m&=e^{-\kappa_1\xi}\left(A_{1m}e^{i\mu_1\xi}+A_{2m}e^{-i\mu_1\xi}\right)+e^{-\kappa_2\xi}\left(A_{3m}e^{i\mu_2\xi}+A_{4m}e^{-i\mu_2\xi}\right)\\ &+e^{\kappa_1\xi}e^{-\kappa_1\frac{H}{R}}\left(A_{5m}e^{i\mu_1\xi}+A_{6m}e^{-i\mu_1\xi}\right)+e^{\kappa_2\xi}e^{-\kappa_2\frac{H}{R}}\left(A_{7m}e^{i\mu_2\xi}+A_{8m}e^{-i\mu_2\xi}\right).\end{aligned} \tag{16.178}$$

Analogous expressions can be given for the functions V_m and W_m.

In the case that instead of using the approaches (16.172) an approach for the function F of the form

$$F = C_m e^{\lambda\xi} \cos m\vartheta \tag{16.179}$$

is used, a polynomial for the determination of λ results after insertion into (16.166) and into (16.171), respectively, which agrees with the two polynomials (16.175) and (16.176). Therefore, no further elaboration is given at this point.

16.6.4 Boundary Conditions

The boundary conditions resulting from the approach (16.172) with (16.174) are discussed for the closed circular cylindrical shell for an edge at location x = const.

If a rigidly clamped edge is given, then the following applies:

$$u_0 = 0, \quad v_0 = 0, \quad w_0 = 0, \quad \frac{\partial w_0}{\partial \xi} = 0. \tag{16.180}$$

A simply supported boundary that can be moved in x-direction is characterized by:

$$u_0 = 0, \quad w_0 = 0, \quad N_{xx} = 0, \quad M_{xx} = 0. \tag{16.181}$$

At a free edge without any bearing, the five force and moment quantities N_{xx}, $N_{x\vartheta}$, Q_x, M_{xx}, $M_{x\vartheta}$ must vanish. However, since only four conditions are available at a shell edge, equivalent force flows are introduced analogously to the Kirchhoff plate, namely an equivalent transverse shear force flow $\bar{Q}_x$ and additionally an equivalent shear force flow $\bar{N}_{x\vartheta}$. These can be derived using Fig. 16.20. Due to the inclination of the forces resulting from the twisting moment flow $M_{x\vartheta}$, an equivalent shear force flow $\bar{N}_{x\vartheta}$ is generated in addition to the equivalent transverse shear force flow $\bar{Q}_x$ already known from the Kirchhoff plate. These two equivalent force flows result as:

$$\bar{Q}_x = Q_x + \frac{1}{R}\frac{\partial M_{x\vartheta}}{\partial \vartheta},$$
$$\bar{N}_{x\vartheta} = N_{x\vartheta} + \frac{1}{R} M_{x\vartheta}. \tag{16.182}$$

At a free shell edge thus the following holds:

$$N_{xx} = 0, \quad \bar{N}_{x\vartheta} = 0, \quad \bar{Q}_x = 0, \quad M_{xx} = 0. \tag{16.183}$$

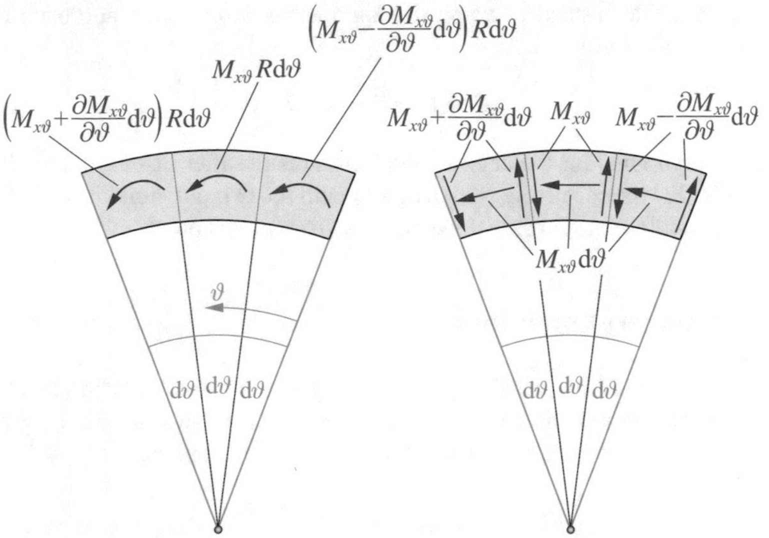

Fig. 16.20 Equivalent transverse shear force flow $\bar{Q}_x$ and equivalent shear force flow $\bar{N}_{x\vartheta}$

16.7 Laminated Shells

In this section, we consider laminated circular cylindrical shells. For reasons of clarity, the explanations are deliberately restricted to the elementary shell type of the circular cylindrical shell (Altenbach et al. 2004; Ambartsumyan 1964), but in principle arbitrary shell problems can be formulated accordingly without further problems.

A laminate is a multilayer structure composed of any number of individual layers (Chap. 13). The properties of the individual layers in terms of material, orientation of the material principal axes, and layer thicknesses can be selected arbitrarily. A schematic representation is shown in Fig. 16.21. In the following, only single layers reinforced by unidirectional fibers will be considered.

16.7.1 Basic Equations

Again, we start from the premises of the Classical Shell Theory and assume the Kirchhoff-Love hypothesis. Let u_0, v_0, w_0 be the displacements of the midsurface of the circular cylindrical shell (radius R) in the circumferential direction, in the direction of the axis of rotation and in the direction of the surface normal, which are all functions of x and ϑ. Then the kinematic equations concerning the shell

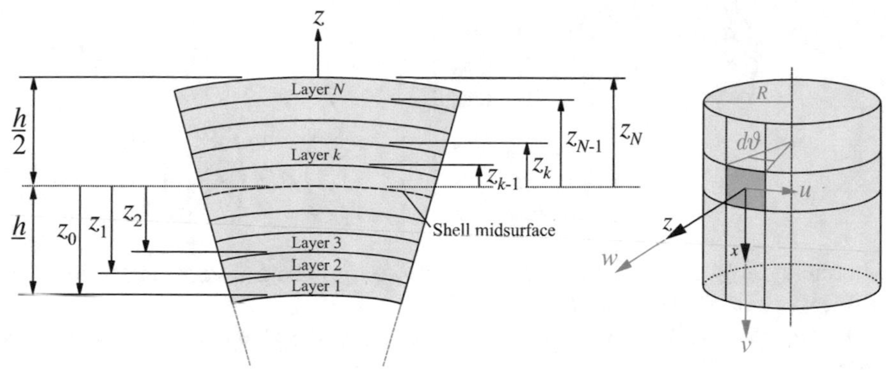

Fig. 16.21 Nomenclature (left); laminated shell (right)

midsurface can be written as follows:

$$\begin{aligned}
\varepsilon_{xx}^0 &= \frac{\partial v_0}{\partial x},\\
\varepsilon_{\vartheta\vartheta}^0 &= \frac{1}{R}\left(\frac{\partial u_0}{\partial \vartheta} + w_0\right),\\
\gamma_{x\vartheta}^0 &= \frac{1}{R}\frac{\partial v_0}{\partial \vartheta} + \frac{\partial u_0}{\partial x},\\
\kappa_{xx}^0 &= -\frac{\partial^2 w_0}{\partial x^2},\\
\kappa_{\vartheta\vartheta}^0 &= \frac{1}{R^2}\frac{\partial}{\partial \vartheta}\left(u_0 - \frac{\partial w_0}{\partial \vartheta}\right),\\
\kappa_{x\vartheta}^0 &= \frac{1}{R}\left(\frac{\partial u_0}{\partial x} - 2\frac{\partial^2 w_0}{\partial x \partial \vartheta}\right).
\end{aligned} \tag{16.184}$$

The total strains, consisting of the strains as well as the curvatures and the twist of the shell midsurface are then obtained as:

$$\begin{aligned}
\varepsilon_{xx} &= \varepsilon_{xx}^0 + z\kappa_{xx}^0,\\
\varepsilon_{\vartheta\vartheta} &= \varepsilon_{\vartheta\vartheta}^0 + z\kappa_{\vartheta\vartheta}^0,\\
\gamma_{x\vartheta} &= \gamma_{x\vartheta}^0 + z\kappa_{x\vartheta}^0.
\end{aligned} \tag{16.185}$$

In each single layer k of the laminated shell, a plane stress state is assumed with respect to the z-direction, so that the following representation holds for Hooke's law:

$$\begin{pmatrix} \sigma_{xx} \\ \sigma_{\vartheta\vartheta} \\ \tau_{x\vartheta} \end{pmatrix}_k = \begin{bmatrix} \bar{Q}_{11} & \bar{Q}_{12} & \bar{Q}_{16} \\ \bar{Q}_{12} & \bar{Q}_{22} & \bar{Q}_{26} \\ \bar{Q}_{16} & \bar{Q}_{26} & \bar{Q}_{66} \end{bmatrix}_k \begin{pmatrix} \varepsilon_{xx} \\ \varepsilon_{\vartheta\vartheta} \\ \gamma_{x\vartheta} \end{pmatrix}$$
$$= \begin{bmatrix} \bar{Q}_{11} & \bar{Q}_{12} & \bar{Q}_{16} \\ \bar{Q}_{12} & \bar{Q}_{22} & \bar{Q}_{26} \\ \bar{Q}_{16} & \bar{Q}_{26} & \bar{Q}_{66} \end{bmatrix}_k \left[\begin{pmatrix} \varepsilon^0_{xx} \\ \varepsilon^0_{\vartheta\vartheta} \\ \gamma^0_{x\vartheta} \end{pmatrix} + z \begin{pmatrix} \kappa^0_{xx} \\ \kappa^0_{\vartheta\vartheta} \\ \kappa^0_{x\vartheta} \end{pmatrix} \right], \quad (16.186)$$

wherein $z_{k-1} \leq z \leq z_k$. In this, the quantities $\bar{Q}_{ij}$ are the transformed reduced stiffnesses of the layer materials.

The constitutive law for a laminate according to the Classical Laminates Plate Theory is as follows:

$$\begin{pmatrix} N_{xx} \\ N_{\vartheta\vartheta} \\ N_{x\vartheta} \\ M_{xx} \\ M_{\vartheta\vartheta} \\ M_{x\vartheta} \end{pmatrix} = \begin{bmatrix} A_{11} & A_{12} & A_{16} & B_{11} & B_{12} & B_{16} \\ A_{12} & A_{22} & A_{26} & B_{12} & B_{22} & B_{26} \\ A_{16} & A_{26} & A_{66} & B_{16} & B_{26} & B_{66} \\ B_{11} & B_{12} & B_{16} & D_{11} & D_{12} & D_{16} \\ B_{12} & B_{22} & B_{26} & D_{12} & D_{22} & D_{26} \\ B_{16} & B_{26} & B_{66} & D_{16} & D_{26} & D_{66} \end{bmatrix} \begin{pmatrix} \varepsilon^0_{xx} \\ \varepsilon^0_{\vartheta\vartheta} \\ \gamma^0_{x\vartheta} \\ \kappa^0_{xx} \\ \kappa^0_{\vartheta\vartheta} \\ \kappa^0_{x\vartheta} \end{pmatrix}, \quad (16.187)$$

wherein here we define the force and moment flows as follows:

$$N_{xx} = \int_{-\frac{h}{2}}^{+\frac{h}{2}} \sigma_{xx} \mathrm{d}z, \quad N_{\vartheta\vartheta} = \int_{-\frac{h}{2}}^{+\frac{h}{2}} \sigma_{\vartheta\vartheta} \mathrm{d}z, \quad N_{x\vartheta} = \int_{-\frac{h}{2}}^{+\frac{h}{2}} \tau_{x\vartheta} \mathrm{d}z,$$
$$M_{xx} = \int_{-\frac{h}{2}}^{+\frac{h}{2}} \sigma_{xx} z \mathrm{d}z, \quad M_{\vartheta\vartheta} = \int_{-\frac{h}{2}}^{+\frac{h}{2}} \sigma_{\vartheta\vartheta} z \mathrm{d}z, \quad M_{x\vartheta} = \int_{-\frac{h}{2}}^{+\frac{h}{2}} \tau_{x\vartheta} z \mathrm{d}z. \quad (16.188)$$

This is equivalent to assuming Love's assumption $1 + \frac{z}{R} \simeq 1$, i.e., that the differences between the shell surfaces above and below the shell midsurface $z = 0$ are negligible. The entries of the laminate stiffness matrix are:

$$A_{ij} = \int_{-\frac{h}{2}}^{+\frac{h}{2}} Q_{ij} \mathrm{d}z, \quad B_{ij} = \int_{-\frac{h}{2}}^{+\frac{h}{2}} Q_{ij} z \mathrm{d}z, \quad D_{ij} = \int_{-\frac{h}{2}}^{+\frac{h}{2}} Q_{ij} z^2 \mathrm{d}z. \quad (16.189)$$

The condensed equilibrium conditions are after elimination of the shear force flows:

$$\frac{\partial N_{xx}}{\partial x} + \frac{1}{R}\frac{\partial N_{x\vartheta}}{\partial \vartheta} + p_x = 0,$$
$$\frac{\partial N_{x\vartheta}}{\partial x} + \frac{1}{R}\frac{\partial N_{\vartheta\vartheta}}{\partial \vartheta} + \frac{1}{R}\left(\frac{1}{R}\frac{\partial M_{\vartheta\vartheta}}{\partial \vartheta} + \frac{\partial M_{x\vartheta}}{\partial x}\right) + p_\vartheta = 0,$$
$$\frac{\partial^2 M_{xx}}{\partial x^2} + \frac{2}{R}\frac{\partial^2 M_{x\vartheta}}{\partial x \partial \vartheta} + \frac{1}{R^2}\frac{\partial^2 M_{\vartheta\vartheta}}{\partial \vartheta^2} - \frac{N_{\vartheta\vartheta}}{R} + p_z = 0. \quad (16.190)$$

Substituting the constitutive law (16.187) with the aid of the kinematic equations (16.184) yields the following system of coupled differential equations describing the static behavior of an arbitrarily layered circular cylindrical shell:

$$\begin{aligned}
&\left[\left(A_{16}+\frac{B_{16}}{R}\right)\frac{\partial^2}{\partial x^2}+\frac{1}{R}\left(A_{12}+A_{66}+\frac{B_{12}+B_{66}}{R}\right)\frac{\partial^2}{\partial x\partial\vartheta}\right.\\
&\qquad\left.+\frac{1}{R^2}\left(A_{26}+\frac{B_{26}}{R}\right)\frac{\partial^2}{\partial\vartheta^2}\right]u_0\\
&\quad+\left[A_{11}\frac{\partial^2}{\partial x^2}+\frac{2A_{16}}{R}\frac{\partial^2}{\partial x\partial\vartheta}+\frac{A_{66}}{R^2}\frac{\partial^2}{\partial\vartheta^2}\right]v_0\\
&\quad+\left[-B_{11}\frac{\partial^3}{\partial x^3}-\frac{3B_{16}}{R}\frac{\partial^3}{\partial x^2\partial\vartheta}-\frac{B_{12}+2B_{66}}{R^2}\frac{\partial^3}{\partial x\partial\vartheta^2}\right.\\
&\qquad\left.-\frac{B_{26}}{R^3}\frac{\partial^3}{\partial\vartheta^3}+\frac{A_{12}}{R}\frac{\partial}{\partial x}+\frac{A_{26}}{R^2}\frac{\partial}{\partial\vartheta}\right]w_0=-p_x,\\
&\left[\left(A_{66}+\frac{2B_{66}}{R}+\frac{D_{66}}{R^2}\right)\frac{\partial^2}{\partial x^2}+\frac{2}{R}\left(A_{26}+\frac{2B_{26}}{R}+\frac{D_{26}}{R^2}\right)\frac{\partial^2}{\partial x\partial\vartheta}\right.\\
&\quad\left.+\frac{1}{R^2}\left(A_{22}+\frac{2B_{22}}{R}+\frac{D_{22}}{R^2}\right)\frac{\partial^2}{\partial\vartheta^2}\right]u_0+\left[\left(A_{16}+\frac{B_{16}}{R}\right)\frac{\partial^2}{\partial x^2}\right.\\
&\left.+\frac{1}{R}\left(A_{12}+A_{66}+\frac{B_{12}+B_{66}}{R^2}\right)\frac{\partial^2}{\partial x\partial\vartheta}+\frac{1}{R^2}\left(A_{26}+\frac{B_{26}}{R}\right)\frac{\partial^2}{\partial\vartheta^2}\right]v_0\\
&+\left[-\left(B_{16}+\frac{D_{16}}{R}\right)\frac{\partial^3}{\partial x^3}-\frac{1}{R}\left(B_{12}+2B_{66}+\frac{D_{12}+2D_{66}}{R}\right)\frac{\partial^3}{\partial x^2\partial\vartheta}\right.\\
&\quad-\frac{3}{R^2}\left(B_{26}+\frac{D_{26}}{R}\right)\frac{\partial^3}{\partial x\partial\vartheta^2}-\frac{1}{R^3}\left(B_{22}+\frac{D_{22}}{R}\right)\frac{\partial^3}{\partial\vartheta^3}\\
&\quad\left.+\frac{1}{R}\left(A_{26}+\frac{B_{26}}{R}\right)\frac{\partial}{\partial x}+\frac{1}{R^2}\left(A_{22}+\frac{B_{22}}{R}\right)\frac{\partial}{\partial\vartheta}\right]w_0=-p_\vartheta,\\
&\left[\left(B_{16}+\frac{D_{16}}{R}\right)\frac{\partial^3}{\partial x^3}+\frac{1}{R}\left(B_{12}+2B_{66}+\frac{D_{12}+2D_{66}}{R}\right)\frac{\partial^3}{\partial x^2\partial\vartheta}\right.\\
&\quad+\frac{3}{R^2}\left(B_{26}+\frac{D_{26}}{R}\right)\frac{\partial^3}{\partial x\partial\vartheta^2}+\frac{1}{R^3}\left(B_{22}+\frac{D_{22}}{R}\right)\frac{\partial^3}{\partial\vartheta^3}\\
&\quad\left.-\frac{1}{R}\left(A_{26}+\frac{B_{26}}{R}\right)\frac{\partial}{\partial x}-\frac{1}{R^2}\left(A_{22}+\frac{B_{22}}{R}\right)\frac{\partial}{\partial\vartheta}\right]u_0\\
&\quad\left[B_{11}\frac{\partial^3}{\partial x^3}+\frac{3B_{16}}{R}\frac{\partial^3}{\partial x^2\partial\vartheta}+\frac{1}{R^2}\left(B_{12}+2B_{66}\right)\frac{\partial^3}{\partial x\partial\vartheta^2}\right.\\
&\qquad\left.+\frac{B_{26}}{R^3}\frac{\partial^3}{\partial\vartheta^3}-\frac{A_{12}}{R}\frac{\partial}{\partial x}-\frac{A_{26}}{R^2}\frac{\partial}{\partial\vartheta}\right]v_0\\
&\quad+\left[-D_{11}\frac{\partial^4}{\partial x^4}-\frac{4D_{16}}{R}\frac{\partial^4}{\partial x^3\partial\vartheta}-\frac{2}{R^2}\left(D_{12}+2D_{66}\right)\frac{\partial^4}{\partial x^2\partial\vartheta^2}\right.\\
&\qquad-\frac{4D_{26}}{R^3}\frac{\partial^4}{\partial x\partial\vartheta^3}-\frac{D_{22}}{R^4}\frac{\partial^4}{\partial\vartheta^4}+\frac{2B_{12}}{R}\frac{\partial^2}{\partial x^2}\\
&\qquad\left.+\frac{4B_{26}}{R^2}\frac{\partial^2}{\partial x\partial\vartheta}+\frac{2B_{22}}{R^3}\frac{\partial^2}{\partial\vartheta^2}-\frac{A_{22}}{R^2}\right]w_0=-p_z.
\end{aligned}\tag{16.191}$$

Depending on the specific laminate structure, the equations for various special cases can be derived from this. If the laminate is symmetric, this means that all bending-extension coupling stiffnesses B_{ij} disappear. In addition, if the laminate is a symmetric cross-ply laminate, all stiffnesses involving index pairs 16 and 26 disappear. The special case of a plane laminate can be derived from this with $R \to \infty$. For the isotropic circular cylindrical shell, the set of equations (16.191) contains the differential equations (16.30) as a special case.

16.7.2 Cross-ply Laminated Cylindrical Shells under Rotationally Symmetric Load

We now consider the case of a cross-ply laminated shell with $A_{16} = A_{26} = 0$ and $D_{16} = D_{26} = 0$ under rotationally symmetric load, where we want to restrict the loads to $p_x = 0$, $p_s = 0$ and $p_z = p_z(x)$. Thus, there is a rotationally symmetric load p_z which depends exclusively on x. Under these conditions it follows that all partial derivatives $\frac{\partial}{\partial \vartheta}$ as well as the state variables v, $N_{x\vartheta}$, $M_{x\vartheta}$ become zero. The equilibrium conditions (16.190) then reduce after elemination of the transverse shear force flow Q_x to:

$$\frac{\mathrm{d}N_{xx}}{\mathrm{d}x} = 0, \qquad \frac{\mathrm{d}^2 M_{xx}}{\mathrm{d}x^2} - \frac{N_{\vartheta\vartheta}}{R} = -p_z. \tag{16.192}$$

The corresponding kinematic equations follow from (16.184):

$$\varepsilon_{xx}^0 = \frac{\mathrm{d}v_0}{\mathrm{d}x}, \quad \varepsilon_{\vartheta\vartheta}^0 = \frac{w_0}{R}, \quad \kappa_{xx}^0 = -\frac{\mathrm{d}^2 w_0}{\mathrm{d}x^2}. \tag{16.193}$$

All other strain quantities are zero in the present case.

The constitutive law is currently simplifying to:

$$\begin{aligned}
N_{xx} &= A_{11}\varepsilon_{xx}^0 + A_{12}\varepsilon_{\vartheta\vartheta}^0 + B_{11}\kappa_{xx}^0,\\
N_{\vartheta\vartheta} &= A_{12}\varepsilon_{xx}^0 + A_{22}\varepsilon_{\vartheta\vartheta}^0 + B_{12}\kappa_{xx}^0,\\
M_{xx} &= B_{11}\varepsilon_{xx}^0 + B_{12}\varepsilon_{\vartheta\vartheta}^0 + D_{11}\kappa_{xx}^0,\\
M_{\vartheta\vartheta} &= B_{12}\varepsilon_{xx}^0 + B_{22}\varepsilon_{\vartheta\vartheta}^0 + D_{12}\kappa_{xx}^0.
\end{aligned} \tag{16.194}$$

Evaluating the equilibrium conditions (16.192) yields with the help of (16.194) and (16.193):

$$\begin{aligned}
&A_{11}\frac{\mathrm{d}^2 v_0}{\mathrm{d}x^2} + \frac{A_{12}}{R}\frac{\mathrm{d}w_0}{\mathrm{d}x} - B_{11}\frac{\mathrm{d}^3 w_0}{\mathrm{d}x^3} = 0,\\
&\left(\frac{A_{11}D_{11} - B_{11}^2}{A_{11}}\right)\frac{\mathrm{d}^4 w_0}{\mathrm{d}x^4} + \frac{2}{R}\left(\frac{A_{12}B_{11} - A_{11}B_{12}}{A_{11}}\right)\frac{\mathrm{d}^2 w_0}{\mathrm{d}x^2}\\
&+ \frac{1}{R^2}\left(\frac{A_{11}A_{22} - A_{12}^2}{A_{11}}\right) w_0 = p_z - \frac{A_{12}}{A_{11}}\frac{N_{xx}}{R}.
\end{aligned} \tag{16.195}$$

At this point, the following abbreviations are introduced:

$$D^R = \frac{A_{11}D_{11} - B_{11}^2}{A_{11}}, \quad 4\lambda^4 = \frac{1}{D^R R^2}\frac{A_{11}A_{22} - A_{12}^2}{A_{11}}. \tag{16.196}$$

Then the second equation in (16.195) takes on the following form:

$$\frac{d^4 w_0}{dx^4} + \frac{2}{D^R R}\left(\frac{A_{12}B_{11} - A_{11}B_{12}}{A_{11}}\right)\frac{d^2 w_0}{dx^2} + 4\lambda^4 w_0 = \frac{1}{D^R}\left(p_z - \frac{A_{12}}{A_{11}}\frac{N_{xx}}{R}\right). \tag{16.197}$$

In the case of a symmetrical cross-ply laminate, $B_{ij} = 0$ and $D^R = D_{11}$ are obtained, so that

$$\frac{d^4 w_0}{dx^4} + 4\lambda^4 w_0 = \frac{1}{D_{11}}\left(p_z - \frac{A_{12}}{A_{11}}\frac{N_{xx}}{R}\right), \tag{16.198}$$

where the following abbreviation was used:

$$4\lambda^4 = \frac{1}{D_{11}R^2}\frac{A_{11}A_{22} - A_{12}^2}{A_{11}}. \tag{16.199}$$

As in the isotropic case, the solution of equation (16.198) is composed of a homogeneous solution and a load-dependent particular solution: $w_0 = w_{0,h} + w_{0,p}$. For the homogeneous solution, the same applies as in the case of the isotropic circular cylindrical shell:

$$w_{0,h} = e^{-\lambda x}\left(C_1 \cos\lambda x + C_2 \sin\lambda x\right) + e^{\lambda x}\left(C_3 \cos\lambda x + C_4 \sin\lambda x\right). \tag{16.200}$$

References

Altenbach, H., Altenbach, J., Kissing, W.: Mechanics of Composite Structural Elements. Springer, Berlin (2004)

Ambartsumyan, S.A.: Theory of Anisotropic Shells. NASA Technical Translation TT F-118. Washington DC (1964)

Becker, W., Gross, D.: Mechanik elastischer Körper und Strukturen. Springer, Berlin (2002)

Eschenauer, H., Schnell, W.: Elastizitätstheorie. Bibliographisches Institut, Mannheim (1993)

Eschenauer, H., Olhoff, N., Schnell, W.: Applied Structural Mechanics. Springer, Berlin (1997)

Flügge, W.: Statik und Dynamik der Schalen, 3rd edn. Springer, Berlin (1981)

Göldner, H., Altenbach, J., Eschke, K., Garz, K.F., Sähn, S.: Lehrbuch Höhere Festigkeitslehre, vol. 1. Physik Verlag, Weinheim (1979)

Göldner, H., Altenbach, J., Eschke, K., Garz, K.F., Sähn, S.: Lehrbuch Höhere Festigkeitslehre, vol. 2. Physik Verlag, Weinheim (1985)

Ramm, E., Müller, J.: Schalen. Universität Stuttgart, Institut für Baustatik und Baudynamik, Stuttgart (1995)

Reddy, J.N.: Theory and Analysis of Elastic Plates and Shells. CRC Press, Boca Raton (2006)

Timoshenko, S., Woinowsky-Krieger, S.: Theory of Plates and Shells, 2nd edn. McGraw Hill, New York (1964)

Index

A
ABD matrix, 455
Airy stress function, 106, 168, 422, 429
  Polar coordinates, 223
Angle-ply laminate, 461
Anisotropy, 23
  Full, 29
  Orthogonal, 32

B
Beam-type disks, 123
Bending angles, 361
Bending theory, 523
  Circular cylindrical shell under arbitrary load, 556
  Constitutive equations, 531
  Donnell shell theory, 562
  Equilibrium conditions, 523, 556
  Kinematic equations, 527
Bending-extension coupling, 457
Bending-twisting coupling, 457
Boundary conditions, 28
Boundary perturbations
  Decay behaviour, 134, 232
Boundary value problems, 27
Boundary values, 28
Buckling, 391
Buckling load, 396
Buckling mode, 395, 396, 398

C
Cartesian coordinates, 4, 40
Cauchy's theorem, 8
Circular cylindrical shell, 499, 516, 526, 530, 532
  Arbitrary load, 556
Circular plates, 340
  Asymmetric bending, 353
  Solutions, 341
  Strain energy, 356
Circular ring plate, 350
Classical Laminated Plate Theory, 446
  Angle-ply laminates, 461
  Assumptions, 447
  Boundary conditions, 467
  Condensed plate equilibrium, 465
  Constitutive law, 455
  Coupling effects, 455
  Cross-ply laminates, 460
  Displacement differential equations, 465
  Displacement field, 450
  Equilibrium conditions, 463
  Kinematics, 448
  Laminate code, 446
  Navier solution, 468
  Quasi-isotropic laminate, 462
  Stiffnesses, 454
  Strain field, 450
  Stress field, 452
  Symmetric laminates, 460
Classical Shell Theory, 483
Compatibility condition, 21, 47, 165
  Disk, 103, 220, 223
  Plane strain state, 46
  Polar coordinates, 56
Complementary strain energy
  Euler-Bernoulli beam, 69
Compliance matrix, 24, 36
Compliance tensor, 24
Compliances, 24, 35, 36, 55
  Isotropic, 37

C. Mittelstedt, *Theory of Plates and Shells*,
https://doi.org/10.1007/978-3-662-66805-4

Condensed plate equilibrium, 271, 301, 305, 465
Plate buckling, 394
Configuration
Momentary configuration, 14
Reference configuration, 14
Conical shell, 507, 516
Constitutive equations, 3, 47, 52, 54
Polar coordinates, isotropic, 56
Polar coordinates, orthotropic, 56
Constitutive law, 22
Container equation, 537
Container theory, 533
Laminated shell, 570
Shell parameter, 538
Coordinates
Eulerian, 15
Lagrange, 14
Coupling effects, 29, 455
Anisotropy, 24
Coupling stiffnesses, 454
Cross-ply laminate, 460
Cylindrical coordinates, 40

D

Deformation
Eulerian, 15
Lagrangian, 15
Deformations, 14
Direction cosine, 9
Disk, 99
Airy stress function, 223
Anisotropic, 219
Arbitrary boundary, 114
Beam-type, 123
Beam-type, anisotropic, 228
Boundary conditions, 108, 173
Cartesian coordinates, 99
Circular holes, 192
Compatibility condition, 103, 165, 220, 223
Displacement method, 103, 166
Displacements, 107
Elementary cases, 120, 178, 226
Equilibrium conditions, 103, 164, 220, 223
Finite elements, 214
Force flows, 100, 164
Force method, 104, 167
Hooke's law, 165
Isotropic, 99, 163
Kinematic equations, 102, 164, 220, 223
Layered, 243
Non-rotationally symmetric circular disk, 186
Orthotropic, 54
Polar coordinates, 163
Principle of the minimum of the total elastic potential, 112
Principle of virtual displacements, 110, 114
Ritz method, 201
Rotationally symmetric, 165, 179
St. Venant's principle, 132
Strain energy, 109, 170
Wedge-shaped disk, 190
Disk equation, 106
Anisotropic, 220
Polar coordinates, 168
Solutions, 117, 168
Solutions, anisotropic, 221
Disk stiffness, 102, 165
Displacement gradient, 16
Displacement vector, 14
Displacements, 3, 4, 14, 42
Donnell shell theory, 562

E

Effective width
Beams, 151
Load introduction, 157
Elastic foundation, 307
Elasticity, 22
Elasticity tensor, 23, 24
Symmetry properties, 26
Elasticity theory, 3
Energy, 63
Internal, 62
Internal, rod, 65
Potential, 63
Engineering constants, 35, 44, 52, 53
Equilibrium conditions, 3, 12, 47, 164
Disk, 103, 220, 223
First-Order Shear Deformation Theory, 365
Geometrically nonlinear analysis, 421
Local, 13, 27, 40
Plate, 337
Polar coordinates, 55
Shell, 494, 523, 556
Third-Order Shear Deformation Theory, 384

Equivalent force flows
Shell, 565
Equivalent transverse shear forces, 272

F
Fiber-reinforced plastic, 29, 32–34
Field equations, 28
Finite elements
Disk, 214
First-Order Shear Deformation Theory, 360
Assumptions, 360
Bending angles, 361
Boundary conditions, 366
Constitutive law, 361, 363
Displacement field, 361
Equilibrium conditions, 365
Geometrically nonlinear analysis, 440
Kinematics, 360
Lévy-type solutions, 372
Navier solution, 371
Plate strips, 368
Principle of the minimum of the total elastic potential, 373
Ritz method, 372
Shear correction factor, 363, 364
Strain energy, 368
Strain field, 361
Stress field, 361
Transverse shear stiffnesses, 362
Transverse shear strains, 361
Transverse shear stresses, 362
Force method, 104, 544
Fourier integral, 145
Fourier series, 128, 130, 135, 140, 142, 144, 283, 371, 385, 468
Full anisotropy, 29

G
Galerkin method, 93, 329
Geometrically nonlinear analysis, 433
Garland curve, 397
Gaussian curvature, 481
General principle of work and energy of elastostatics, 72
Generalized Hooke's law, 23
Geometrically nonlinear analysis, 415
Boundary terms, 424
Constitutive law, 421
Equilibrium conditions, 421
External potential, 427
First-Order Shear Deformation Theory, 440
Galerkin method, 433
Mixed formulation, 430
Plate bending, 432
Ritz method, 434
Strain energy, 426
Green-Dirichlet minimum principle, 84
Green-Lagrangian strain tensor, 17

H
Half-plane
Isotropic, 134, 139
Non-periodic load, 145
Periodic boundary load, 139
Higher-order plate theories, 359
Hooke's law, 3, 22, 28
Generalized, 35, 43
Inverted, 24
Hypothesis of plane cross-sections, 254, 448

I
Infinitesimal strain tensor, 19
Invariant, 11
Material properties, 31
Stress tensor, 11
Sum of normal stresses, 51
Isotropy, 23, 34, 35, 37
Transversal, 34

K
Kinematic equations, 3, 20, 27, 42, 45, 47, 56, 164
Cylindrical coordinates, 43
Disk, 102, 220, 223
Plate, 255
Shell, 513, 527, 559
Kirchhoff plate theory
Approximation methods, 313
Arbitrary boundary, 303
Assumptions, 254
Boundary conditions, 272, 275, 301, 339
Cartesian coordinates, 253
Circular plates, 340
Condensed plate equilibrium, 271, 301, 305
Constitutive law, 260, 336
Curvatures, 257
Displacement field, 257, 335
Effective stiffnesses, 265

Elastic foundation, 307
Elementary solutions, 278
Equilibrium conditions, 337
Equivalent transverse shear force, 272, 336
Force and moment flows, 260
Galerkin method, 329
Kinematics, 255
Lévy-type solutions, 291
Navier solution, 283
Plate equation, 272, 338
Plate equilibrium, 271
Polar coordinates, 333
Principle of the minimum of the total elastic potential, 298
Principle of virtual displacements, 301
Ritz method, 313
Special load cases, 288
Stiffnesses, 261
Strain field, 257, 335
Stress field, 258
Transformation equations, 264
Twist, 257
Kirchhoff-Love approximation, 486, 492

L
Lévy-type solutions
First-Order Shear Deformation Theory, 372
Kirchhoff plate theory, 291
Laminate, 268, 445
Shell, 566
Laminate code, 446
Laminated shell, 566
Laplace operator, 335
Polar coordinates, 168
Law of conservation of energy, 63
Line load, 5
Linear elasticity, 22

M
Material constants, 24, 30, 32, 35
Material law, 22
Material symmetries, 29, 30
Membrane, 310, 431
Membrane stiffnesses, 454
Membrane theory, 491
Assumptions, 491
Circular cylindrical shell, 499, 516
Conical shell, 507, 516
Constitutive equations, 513
Displacements, 514
Equilibrium conditions, 494
Force flows, 498
Kinematic equations, 513
Kinematics, 508
Solutions for shells of revolution, 499
Spherical shell, 502, 515
Modulus of elasticity, 23, 35, 37
Monoclinic material, 30
Monotropy, 30

N
Navier solution, 468
Normal hypothesis, 255, 360, 448
Normal strain, 19, 23
Normal stress, 5, 6, 23
Principal, 50

O
Off-axis layer, 456
Off-axis system, 38, 51
On-axis system, 38, 51
Orthotropy, 32, 39, 54
Cylindrical, 43

P
Perpetuum Mobile, 63
Plane strain state, 44, 54
Plane stress state, 46, 52
Polar coordinates, 55
Plate, 253
First-Order Shear Deformation Theory, 360
Geometrically nonlinear analysis, 415
Higher-order theories, 359
Normal hypothesis, 360
Third-Order Shear Deformation Theory, 376
Plate buckling, 391
Biaxial load, 398
Buckling half-waves, 396
Buckling load, 396
Buckling mode, 396, 398
Energy methods, 401
Garland curve, 397
Navier solution, 395
Rayleigh quotient, 403
Ritz method, 408
Total elastic potential, 403
Plate equation, 272, 340
Polar coordinates, 335, 338

Solution, 339
Plate stiffnesses, 454
Plate strip, 280
First-Order Shear Deformation Theory, 368
Point load, 5
Poisson's ratio, 35, 37
Polar coordinates, 55
Potential
External, 84
Internal, 83
Potential energy, 63
Principal axes, 11
Material, 29, 38, 51, 53
Stresses, 50
Principal direction, 29, 34
Normal stresses, 50
Principal normal stresses, 50
Principal shear stresses, 11, 51
Principal stresses, 10–12
Principle of the minimum of the total elastic potential, 84, 90, 298
Disk, 112, 173
Euler-Bernoulli beam, 87
First-Order Shear Deformation Theory, 373
Rod, 84
Principle of the stationary value of the total elastic potential, 83, 84
Principle of virtual displacements, 74, 84
Continuum, 79
Disk, 110, 114, 172
Euler-Bernoulli beam, 82
First-Order Shear Deformation Theory, 366
Geometrically nonlinear analysis, 416
Kirchhoff plate theory, 301
Membrane Theory, 518
Third-Order Shear Deformation Theory, 382
Principle of virtual total potential, 84
Principle of work and energy, 62

Q
Quasi-isotropic laminate, 462

R
Radial strain, 42
Rayleigh quotient, 403
Reduced compliances, 55
Reduced Stiffnesses, 53
Transformed, 54
Ritz method, 90
Approach functions, 317
Disk, 201
Displacement based, 201
Euler-Bernoulli beam, 91
First-Order Shear Deformation Theory, 372
Force-based, 211
Geometrically nonlinear analysis, 434
Kirchhoff plate theory, 313
Plate buckling, 408
Ritz constants, 90
Ritz equations, 91
Stiffness matrix, 93
Third-Order Shear Deformation Theory, 387

S
Shear correction factor, 69, 363, 364, 376
Shear coupling, 32, 227, 456
Shear modulus, 23, 35, 37
Shear number, 233
Shear strain, 19, 20, 23, 42
Shear strain energy, 364
Shear stress, 5, 6, 23
Extremal, 50
Shell parameter, 538
Shells, 477
Bending stiffness, 531
Bending theory, 523
Boundary conditions, 532, 565
Classical Shell Theory, 483
Constitutive equations, 513, 531
Container equation, 537
Container theory, 533
Displacements, 487, 514
Donnell shell theory, 562
Edge perturbations, 539, 547, 553
Equilibrium conditions, 494, 523, 556
Equivalent shear force flows, 565
Extensional stiffness, 513
Force and moment flows, 484, 558
Force flows, 498
Force method, 544
Kinematic equations, 513, 527, 559
Kinematics, 508
Kirchhoff-Love approximation, 486, 492
Laminate, 566
Membrane stiffness, 531
Membrane theory, 491

Principle of virtual displacements, 518
Shells of revolution, 477, 478
Strains, 487
Stresses, 484, 486
Trapezoidal effect, 485, 558
Shells of revolution, 478
Single-layer theory, 263
Spherical shell, 502, 515, 525, 529, 547
St. Venant's principle, 132
Stiffness matrix, 29
Anisotropic, 24
Stiffnesses, 35, 36
Anisotropy, 24
Isotropic, 37
Reduced, 53
Strain energy, 25, 62
Circular plates, 356
Complementary, continuum, 71
Complementary, rod, 66
Continuum, 71
Disk, 109, 170
Euler-Bernoulli beam, 68
First-Order Shear Deformation Theory, 368
Geometrically nonlinear analysis, 426
Kirchhoff plate theory, 299
Rod, 65
Third-Order Shear Deformation Theory, 387
Virtual, continuum, 77
Strain energy density, 26
Complementary, continuum, 71
Complementary, rod, 66
Continuum, 71
Euler-Bernoulli beam, 68
Euler-Bernoulli beam, complementary, 69
Incremental, 64
Rod, 64, 65
Virtual, 76
Virtual, continuum, 77
Strain state
Plane, 54
Strain tensor
Green-Lagrange, 17, 19
Infinitesimal, 19, 22
von Kármán, 18, 416
Strain transformation, 38, 53
Strains, 3, 19, 42
Local, 14
Stress, 3, 5
Index, 5
Invariants, 11
Stress state, 5
Plane, 46, 52, 55
Stress tensor, 7
Stress transformation, 9, 38, 53
Stress vector, 5, 9
Substitute elastic modulus, 45
Substitute Poisson's ratio, 45
Superposition principle, 28
Surface load, 4
Symmetry plane, 30

T

Tangential strain, 42
Tetrahedron
Equilibrium, 7
Third-Order Shear Deformation Theory, 376
Displacement field, 378
Equilibrium conditions, 384
Kinematic equations, 378
Kinematics, 376
Navier solution, 385
Principle of virtual displacements, 382
Ritz method, 387
Strain energy, 387
Total elastic potential, 84
Transformation matrix, 39
Strains, 20
Stresses, 9
Transformation rules, 8
Stresses, 49
Compliances, 39
Reduced stiffnesses, 54
Stiffnesses, 39, 40
Strains, 53
Stresses, 53, 56
Transformed reduced stiffnesses, 54
Transversal isotropy, 34
Transverse strain, 23
Trapezoidal effect, 485, 558

V

Variation, 74
First variation, 77
Variational operator, 77
Virtual displacements, 74
Virtual work, 75
Continuum, 77
External, 75, 84

External, beam, 75
External, continuum, 79
Internal, 75, 83
Internal, beam, 76
Internal, continuum, 77, 79
Volume force, 4
Von-Kármán strain tensor, 18

W

Work, 25
External, 61
External, solid body, 62
Increment, 60, 64, 65
Internal, 61

Printed in the United States
by Baker & Taylor Publisher Services